HANDBOOK OF SOLID STATE DIFFUSION: VOLUME 1

HANDBOOK OF SOLID STATE DIFFUSION: VOLUME 1

Diffusion Fundamentals and Techniques

Edited by

ALOKE PAUL

SERGIY DIVINSKI

ELSEVIER

elsevier.com

Elsevier
Radarweg 29, PO Box 211, 1000 AE Amsterdam, Netherlands
The Boulevard, Langford Lane, Kidlington, Oxford OX5 1GB, United Kingdom
50 Hampshire Street, 5th Floor, Cambridge, MA 02139, United States

Notices

Knowledge and best practice in this field are constantly changing. As new research and experience broaden our understanding, changes in research methods, professional practices, or medical treatment may become necessary.

Practitioners and researchers must always rely on their own experience and knowledge in evaluating and using any information, methods, compounds, or experiments described herein. In using such information or methods they should be mindful of their own safety and the safety of others, including parties for whom they have a professional responsibility.

To the fullest extent of the law, neither the Publisher nor the authors, contributors, or editors, assume any liability for any injury and/or damage to persons or property as a matter of products liability, negligence or otherwise, or from any use or operation of any methods, products, instructions, or ideas contained in the material herein.

Library of Congress Cataloging-in-Publication Data
A catalog record for this book is available from the Library of Congress

British Library Cataloguing-in-Publication Data
A catalogue record for this book is available from the British Library

ISBN: 978-0-12-804287-8

Publisher: Matthew Deans
Acquisition Editor: Christina Gifford
Editorial Project Manager: Jennifer Pierce
Production Project Manager: Julie-Ann Stansfield
Designer: Mark Rogers

Typeset by VTeX

CONTENTS

BIOGRAPHIES

Editors and contributors:

Professor Aloke Paul heads a research group working on various aspects of diffusion in solids in the Department of Materials Engineering, Indian Institute of Science, Bangalore, India. Major research areas include developing new phenomenological models, materials in electronic packaging, bond coat in jet engine applications, the growth of A15 intermetallic superconductors, etc. He teaches a postgraduate level course on Diffusion in Solids. He has guided several PhD and ME students and co-authored around 100 articles in various international journals. During his PhD studies under the guidance of Prof. Frans J.J. van Loo and Dr. Alexander Kodentsov (Eindhoven University of Technology), he was part of one of the most important discoveries of recent times on previously unknown phenomena related to the Kirkendall effect. After joining the Indian Institute of Science, his group developed new methods for estimation of the diffusion coefficients, for example, a physicochemical approach that relates microstructural evolution with the rate of diffusing components and a pseudo-binary method in multicomponent diffusion. These methods are included in the course curriculum of many universities and also in the books written on this topic. Professor Paul has also coauthored a textbook titled Thermodynamics, Diffusion and the Kirkendall effect in Solids. Currently, he is an editorial board member of Journal of Electronic Materials, published by Springer-TMS, and of Scientific Reports, published by Nature Publishing Group.

Dr. Sergiy Divinski is a Privat-Dozent at the Institute of Materials Physics, University of Münster, Germany, where he leads the radiotracer laboratory, which represents one of the most reputed and internationally recognized diffusion schools. His research is focused on diffusion and especially grain boundary diffusion in metals and alloys. He teaches graduate and postgraduate courses on Diffusion in Solids, Numerical methods in Material Science, and different aspects of Materials Science. He has coauthored more than 150 articles in various international journals, several book chapters in the field of Diffusion in Solids, and a textbook titled Thermodynamics, Diffusion and the Kirkendall effect in Solids.

Contributors:

Dr. Tomi Laurila received his DSc degree (with honors) in 2001 and an adjunct professorship in 2006, both from Helsinki University of Technology (HUT). Currently he acts as Associate Professor in the field of microsystem technology in Aalto University. His research is focused on the study of interfacial reactions between dissimilar materials used in microsystems, biocompatibility issues related to different types of (bio)materials, properties and synthesis of hybrid carbon nanomaterials, and electrochemical measurements of different biomolecules under in vitro and in vivo conditions. He has contributed to two textbooks dealing with interfacial compatibility issues and thermodynamics of solid state diffusion as well as published over 80 scientific papers and several book chapters.

Le Zhou is currently a postdoctoral researcher at University of Central Florida, where he received his PhD degree in Materials Science and Engineering in 2016. His research focuses on diffusion phenomena and application of diffusion couples as a combinatorial tool to discover new materials and investigate composition-dependent material properties. He also has expertise in physical/powder metallurgy and electron microscopy. He has worked on different materials, including high temperature alloys and coating, shape memory alloys and lightweight materials. He has published and coauthored more than 15 journal articles and conference proceedings and serves as a reviewer for many international journals.

Dr. Yongho Sohn is a UCF Pegasus Professor of Materials Science and Engineering, and Associate Director for Materials Characterization Facility (MCF) at University of Central Florida. His research and teaching focus on microstructure analysis and control for various applications including gas turbine engines for power generation, both aircraft and utilities, advanced nuclear fuels with low enrichment, and superlight-weight metal matrix composites for structural applications. He has published 8 book chapters, over 140 journal papers, and 60 proceedings papers. He gave over 400 presentations, including 97 invited lectures at conferences around the globe. He is a Fellow of ASM International (FASM), recipient of NSF CAREER Award (2003–08), Outstanding Materials Engineer Award from Purdue University (2016), UCF's 2012 and 2006 research incentive awards, UCF's 2007 and 2013 teaching incentive award. He is an associate editor for Journal of Phase Equilibria and Diffusion and a member of editorial board for Metallurgical and Materials Transactions.

Professor Mysore A. Dayananda has been a member of the faculty of Materials Engineering at Purdue University in West Lafayette, Indiana, USA, for nearly 5 decades. He has held Visiting Professorships at Münster University in Germany (1980) and at the Indian Institute of Science (1992) in Bangalore, India. He has mentored about 50 graduate students for MS and PhD degrees, published more than 120 technical papers and edited four books. His expertise is in the broad area of multicomponent, multiphase diffusion in metallic systems and intermetallic compounds, including aluminides, silicides, stainless steels, high temperature alloys, coatings, nuclear fuels, and claddings. His main contributions are in the areas of intrinsic and interdiffusion, vacancy wind effects, diffusion structures, diffusion paths, interface stability, microstructural evolution, and thermotransport in multicomponent systems. He developed new approaches involving concepts of integrated and average effective interdiffusion coefficients, and characteristic diffusion depths in the analysis of multicomponent diffusion. He has also developed novel transfer matrix methods for the determination of interdiffusion coefficients and assessment of diffusional constraints in multicomponent diffusion couples. He discovered the phenomenon of zero-flux planes and flux reversals, and was honored at a special TMS symposium held in 2006 for his various contributions in multicomponent diffusion. He is a Fellow of ASM International, a member of The Metallurgical Society, Sigma Xi, ASEE and the Advisory Board of Diffusion and Defect Data journal.

Dr. Teh Yu Tan is Professor of Materials Science in the Department of Mechanical Engineering and Materials Science, Duke University, in Durham NC, USA. His main research interest is in the area of defects and diffusion in semiconductors, and in gettering of metallic impurities in silicon for improving device fabrication yield in both integrated circuits and silicon solar cells. He has pioneered the intrinsic gettering method nowadays widely used in the integrated circuit industry for improving fabrication yield.

Dr. Lijun Zhang currently holds a full professor chair in State Key Laboratory of Powder Metallurgy, Central South University, P.R. China. His research interests include thermodynamics and diffusion kinetics in alloys using a combination of experimental and CALPHAD techniques, as well as development of new phase-field models and phase-field simulation of microstructure in alloys using software and house-made codes. He gives courses on Fundamentals for Materials Science, Materials Science for Powder Metallurgy, Thermodynamics, Diffusion and Materials Design to undergraduates and postgraduates. He has coauthored more than 130 papers in refereed journals such as Acta Materialia, Scripta Materialia, and so on, as well as 7 publications in books (i.e., Landolt–Börnstein) and conference proceedings. Currently, he is an editorial board member of Journal of Mining and Metallurgy, Section B: Metallurgy, and serves as a reviewer for more 20 international journals. He is also a committee member of Computational Materials Science Branch of Chinese Materials Research Society.

Dr. Qing Chen is a Senior R&D Scientist and Director of Database Development and Innovation Research at Thermo-Calc Software AB, Stockholm, Sweden. He received his PhD degree from the Department of Materials Science and Engineering at the Central South University, China in 1995. His postdoctoral research was undertaken at KTH (The Royal Institute of Technology, Sweden) and the Ohio State University. Dr. Qing Chen is a leading expert on modeling phase diagrams, phase transformations and microstructural evolution. His main research interests include developing thermodynamic/kinetic models and databases, simulating phase transformations and microstructure evolution with both sharp and diffusive interface (phase field) approaches, and linking first principles calculation with CALPHAD to model thermodynamic and thermophysical properties using physical parameters. Dr. Qing Chen has about 60 publications in peer-reviewed journals such as Acta Materialia, Scripta Materialia, Journal of Phase Equilibria and Diffusion, and CALPHAD. He has won the Best CALPHAD Paper Award twice and been frequently invited as a reviewer for many well-known scientific journals. Dr. Qing Chen is one of the principal developers of Thermo-Calc, its programming interface TQ, and most recently the precipitation simulation software TC-PRISMA.

Dr. Abhik Choudhury has been an assistant professor in the Department of Materials Engineering at the Indian Institute of Science, Bangalore since Nov. 2013. He has graduated from the Indian Institute of Technology with a dual degree in 2008 from the Department of Metallurgical and Materials Engineering, with a major in Materials Engineering and a minor in Theoretical Computer Science. Thereafter he received his PhD from the University of Karlsruhe (Germany) under the supervision of Prof. Dr. Britta Nestler for the thesis "Quantitative phase-field modeling of multi-component diffusion coupled phase transformations", which was completed in 2012. Dr. Choudhury continued as a post-doctoral researcher with Prof. Dr. Mathis Plapp at the Ecole Polytechnique Paris, where he worked on developing a phase-field model for understanding morphological instabilities originating during electrochemical reactions, and subsequently joined the Indian Institute of Science as a faculty member in 2013. His research interests include studies on pattern formation during phase transformations such as solidification, electrochemical reactions, and stress-mediated evolution of microstructures.

Franz Dieter Fischer is Professor Emeritus of Mechanics at the Institute of Mechanics, Montanuniversität Leoben, Austria. He has been working on mechanics and thermodynamics of materials for more than 30 years. He cooperates with several groups such as those at the Institute of Physics of Materials of the Academy of Sciences, Czech Republic, the Max-Planck Institute für Eisenforschung, Germany, and the Max-Planck Institute for Colloids and Interfaces/Biomaterials, Germany. He is author or coauthor of more than 650 papers in international journals, book chapters and in conference proceedings. He was awarded the Erwin Schrödinger Prize of Austrian Academy of Sciences and the Humboldt Research Prize of the Alexander von Humboldt Foundation, Germany.

Dr. Jiří Svoboda is a senior researcher in the Institute of Physics of Materials of the Academy of Sciences of the Czech Republic. He deals with development of thermodynamic micromechanical models for evolution of solid-state systems at high temperatures involving diffusion and interface migration. A majority of his models utilize the Thermodynamic Extremal Principle formulated in discrete characteristic variables for which the principle provides evolution equations. He is author or coauthor of about 150 papers in peer-reviewed journals.

Professor Graeme E. Murch is the Professor of Materials Science at the University of Newcastle in NSW, Australia, where he is also the Director of the Center for Mass and Thermal Transport in Engineering Materials. His research areas cover theory and computer simulations of mass and thermal diffusion in solids and liquids. Computer simulation methods include Monte Carlo, molecular dynamics, first principles and finite element analysis. In these areas he has published over 400 journal papers, 15 book chapters, one textbook, and 33 edited books. He is the Editor of the journals: Diffusion Foundations and Solid State Phenomena. For his past services, he was recently appointed to the positions of honorary editor of the journals Materials Science Forum and Defect and Diffusion Forum.

Professor Irina Belova is the Professor of Computational Engineering at the University of Newcastle in NSW, Australia, where she is also the Joint Director of the Center for Mass and Thermal Transport in Engineering Materials. Her research areas cover all aspects of theory along with computer simulation of mass diffusion in solids. Computer simulation methods cover Monte Carlo, molecular dynamics and finite element analysis. In these areas she has published over 270 journal papers, 10 book chapters, and 9 edited books. She is one of the Editors of Diffusion Foundations and an Associate Editor of the Journal of Phase Equilibria and Diffusion.

PREFACE

Diffusion in solids is one of the most important topics in Materials Science and Solid State Physics. Understandably, numerous books were written in last 100 years by different authors. After going through the books available, one can easily understand that the topics covered in a particular book are naturally biased towards the research fields and the knowledge of the authors. There are no books available which cover many important aspects with equal importance from basic fundamentals and materials applications, experimental methods, and simulations to diffusion in ordered compounds, thin films, and nanomaterials. There are few books available which introduce the methods, estimation steps, and a particular analysis in various material systems in such a way that students and researchers can learn these by reading from a single book. Various types of diffusion coefficients and their physical significance may not be easy to follow from the existing literature.

Therefore, this book was planned such that the important aspects mentioned above were covered in one place with equal importance. As the title suggests, the topics are introduced in a very concise manner with a straightforward guidance for estimation of the data and their analysis. Chapters in Volume 1 mainly include the fundamental methods, whereas studies in various materials systems are covered in Volume 2. It is true that most of the chapters are mainly written on metallic systems, since, as a first step, we had to make this effort possible. In the next edition, many new topics and material systems will be included.

ACKNOWLEDGMENT

We are grateful to all the authors of this book for making it successful. Along with scientific contributions, their continuous and unconditional support to clear all the hurdles we have faced is greatly acknowledged. We do not find proper (or enough) words to thank all of them.

We thank our mentors, Prof. Frans J.J. van Loo and Dr. Alexander Kodentsov, Eindhoven University of technology (of Aloke Paul), Prof. Leonid N. Larikov, National Academy of Science, Kiev, Ukraine and Prof. Christian Herzig, University of Münster, Germany (of Sergiy V. Divinski) for teaching us this fascinating subject. We were very lucky to work with such pleasant and knowledgeable personalities. We still cherish those golden periods! Even today our thinking and analysis of the scientific problems are greatly influenced by them.

We thank our students, collaborators, and colleagues for their help on many aspects throughout our professional careers.

Professional career and achievements are meaningless without an excellent personal life. In that sense, we are fortunate for all the support and joyful moments that were given to us by our families:

Bhavna, wife, Pihu, daughter of Aloke Paul;

Sveta, wife, Veronika and Alisa, daughters of Sergiy Divinski.

CHAPTER 1

Defects, Driving Forces and Definitions of Diffusion Coefficients in Solids

Aloke Paul*, Tomi Laurila†, Sergiy Divinski‡
*Indian Institute of Science, Bangalore, India
†Aalto University, Espoo, Finland
‡University of Münster, Münster, Germany

Diffusion of components plays an important role in many physical and mechanical properties. In many applications, durability of a product is controlled by the diffusion rate of the components. On the other hand, many products are prepared utilizing this phenomenon with a controlled microstructure. These aspects are discussed extensively in other chapters. The aim of this chapter is to familiarize the reader with different types of defects assisting transport of components and correspondingly defined diffusion coefficients. Further, diffusion coefficients are also classified based on the absence or presence of different types of driving forces.

1.1 DEFECTS IN CRYSTALLINE SOLID

1.1.1 Zero-Dimensional Point Defects

In a perfect crystalline solid, all the atoms or molecules arrange themselves ideally. However, imperfections, i.e., crystal defects, are always present. Thermodynamically we can classify them as *equilibrium* and/or *non-equilibrium* defects. Addition of certain defects up to an equilibrium concentration decreases the free energy, making the material thermodynamically more stable. These are in general point (or zero-dimensional) defects such as thermal vacancies, self-interstitials, impurities, and antisites (antistructure atoms). Equilibrium concentrations depend on type of bonding and structure of a material. On the other hand, when addition of defects increases the free energy, they are called non-equilibrium defects. That means that, in the presence of a sufficient thermal energy, a system will try to decrease the concentration of these defects. Irradiation- or deformation-induced point defects, dislocations, grain boundaries, interfaces are examples of such defects. A further classification, especially suitable for semiconductors, refers to defects as *intrinsic* (present in a pure material at equilibrium) or *extrinsic* (e.g., induced by doping or alloying, or by presence of an extended defect-like grain boundary or surface) in nature.

Handbook of Solid State Diffusion, Volume 1
DOI: 10.1016/B978-0-12-804287-8.00001-4

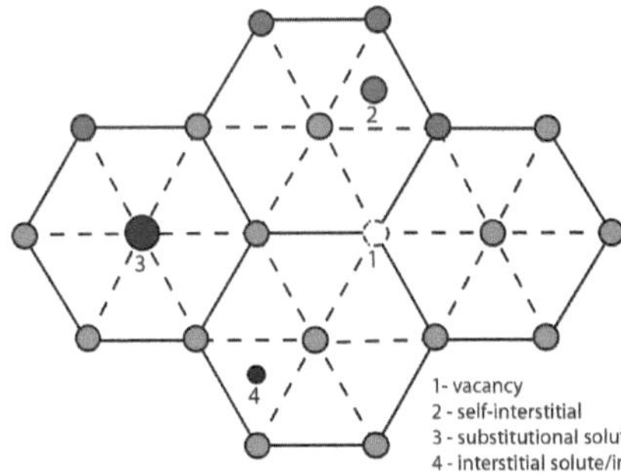

Figure 1.1 Different types of point defect such as vacancy, self-interstitial, substitutional solute/impurity, and interstitial solute/impurity shown for the basal plane of a model HCP lattice.

At a constant pressure and temperature, the change in free energy ΔG (kJ/mol) is the result of the change in enthalpy ΔH (kJ/mol) and entropy ΔS (kJ/mol s) since [1]

$$\Delta G = \Delta H - T\Delta S \tag{1.1}$$

where T is the temperature in Kelvin (K). For thermodynamical equilibrium defects, up to a certain concentration, the increase in entropy due to their production dominates favorably over the corresponding increase in enthalpy such that the material becomes more stable by decreasing the free energy. However, in the case of non-equilibrium defects, the increase in enthalpy always dominates over the increase in entropy, which results in an increase of the free energy violating the material stability. Therefore, the thermodynamically stable defects are present naturally, whereas unwanted thermodynamically unstable defects are present because of unavoidable (or sometimes enforced) situations!

Defects are generally classified according to their dimensionality. Point defects are the zero-dimensional defects. There are different point defects such as vacancies, self-interstitials, solutes/impurities, or anti-sites. As shown in Fig. 1.1, when components (atoms or ions) are missing from regular lattice points, these are called vacancies. When these host atoms are located at the interstitial position instead of regular lattice positions, these are called self-interstitials (note that these are different from impurity atoms present at interstitial sites, which is explained next). Since in a typical metallic lattice, the size of a void at interstitial position is small to accommodate a host atom displaced from a regular lattice point, the corresponding formation energy is high and the concentration of self-interstitials in pure metal or substitutional alloy is negligibly small. Therefore, this kind of defect is relatively uncommon in metals unless, e.g., the material is subjected to irradiation. Vacancies, on the other hand, are dominant equilibrium defects in a metal, as explained in Section 1.1.1.1. At above 0 K, vacancy-free material is impossible physically. In a covalent solid (e.g., semiconductors) the situation is more involved.

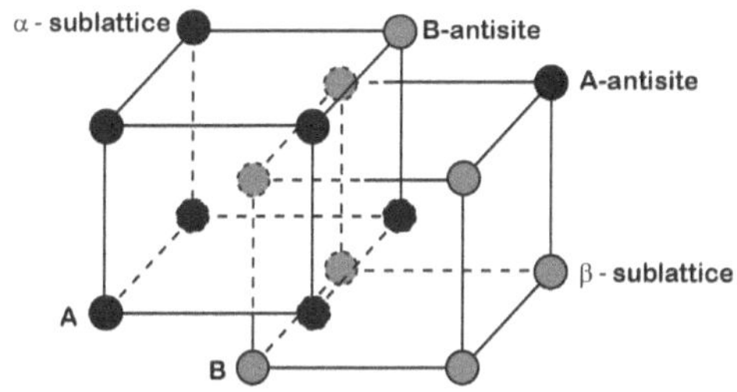

Figure 1.2 Antisite defects are shown for an ordered compound with two interpenetrating sublattices for A (dark colored spheres) and B (light colored spheres) atoms.

Both self-interstitials (e.g., in Si) or vacancies (e.g., in Ge) can be dominant intrinsic defects at equilibrium. Furthermore, a charge state of these defects, i.e., of vacancies and self-interstitials, has to be taken into account, too.

A further class of point defects is represented by foreign atoms. These could be found isolated inside the lattice or under certain conditions agglomerated or segregated to higher-dimensional defects, like dislocations or grain boundaries. When these foreign atoms are added intentionally, they are typically called solutes. These are termed as impurities if present although unwanted. When foreign atoms (typically of comparable atomic size or larger than that of the host atoms) occupy regular lattice sites, these are called substitutional solute/impurities. When the foreign atoms (typically much smaller in size than the host atoms) occupy the interstitial sites, these are called interstitial solute/impurities. Solutes and impurities are always present up to a certain concentration since they are thermodynamically stable, as explained in Section 1.1.1.2. Equilibrium concentration of interstitials mainly depends on the corresponding void size and the size of the foreign atom.

As shown in Fig. 1.2, another type of thermodynamical equilibrium defects, namely, antisites (or antistructure atoms), are present in the compounds. In a compound being prone to ordering, different sublattices may be defined for different components. However, because of (configurational) entropy gain, certain concentrations of components occupy a wrong sublattice, i.e., the sublattice for another component. Suppose that α-sublattice is designated for A, whereas β-sublattice is designated for B. When A occupies β-sublattice and B occupies α-sublattice, these defects are called A-antisite (A_B) and B-antisite (B_A). When these are present even at the stoichiometric composition because of thermodynamic reasons, such defects are called thermal defects. When they are present to accommodate a composition deviation from the stoichiometric one, they are called structural (constitutional) defects. Note, however, that this distinction is illustrative and useful for didactic purposes only. In reality, the antisites cannot be separated into thermal and structural, i.e., they do not carry any special "mark" and are just defects, but the corresponding concentrations can be determined and treated separately. Further description is given in Section 1.1.1.3.

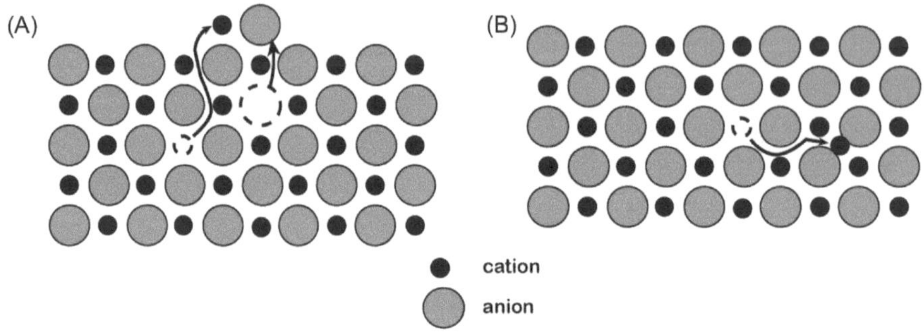

Figure 1.3 (A) Schottky and (B) Frenkel defects in MeXe compound.

Depending on the source, point defects are classified into two categories: intrinsic and extrinsic. When defects are created by the displacement of own atoms in an impurity-free crystal, they are called intrinsic defects. Therefore, vacancies, self-interstitial and antisites are the intrinsic defects. When defects are created by adding foreign components, they are called extrinsic defects. Solutes and impurities are classified as the extrinsic defects. In ionic compounds, as is explained latter, vacancies and interstitials can also be created extrinsically, i.e., by adding alloying components.

In ionic compounds, unlike in metallic systems, defects are present in ionic form. Defect chemistry is complex because of the need for charge neutrality along with the mass conservation and the fixed ratio of sites of anions and cations depending on the stoichiometric composition. Along with point defects, electronic defects (such as holes and excited electrons) are also created. However, in this chapter only point defects will be discussed, which influence the diffusion of components. Intrinsic thermal defects can be created in different ways. These are explained using a hypothetical compound, MeXe, as an example, where Me is the cation and Xe is the anion forming component. When equal numbers of anion and cation vacancies are created, it is called *Schottky* defect as shown in Fig. 1.3A. Ions leave their positions and travel to the surface leaving behind vacant sites. Therefore, such a defect creates volume expansion, leading to a decrease in crystal density. When equally many cations and anions transport from the surface to interstitial positions, such defects are known as *anti-Schottky* defects. They increase the density of the material. When movement of cations from a regular site to interstitial positions creates vacant sites and interstitial cations, it is called a Frenkel defect as shown in Fig. 1.3B. When movement of anions creates the same type of defect, namely, vacant anion sites and anion interstitials, it is called an anti-Frenkel defect. As a matter of fact, all types of defect are possible in a particular compound; however, one type typically dominates over the others based on favorable enthalpy of formation. Most common are the Schottky and Frenkel defects.

Defects are also created extrinsically or by doping, i.e., adding a foreign compound, and hence the need for charge neutrality. For example, when $CaCl_2$ is added into NaCl,

2 Cl^{-1} occupy the anion sublattices. However, since only one Ca^{+2} is there for $2Cl^{-1}$ in $CaCl_2$, unlike $2Na^{+1}$ for $2Cl^{-1}$ in NaCl, a vacancy is created on the cation sublattice. One has to take into account the fact that alloying changes the position of the Fermi level that may affect the electron and hole concentrations, too.

Ionic compounds are also (like intermetallics in metallic systems) not always present with a perfect stoichiometric composition that leads to formation of structural (or chemical) defects. Some of the compounds, although expressed with stoichiometric ratio, do not even exist as perfect compounds. For example, FeO is a cation-deficient oxide and this is promoted by vacant cations. In general, the deviation from a stoichiometry is possible in four different ways. Among MeXe compounds, cation-dominant compounds can exist because of excess cation ($Me_{1+\delta}Xe$) or shortage of anion ($MeXe_{1-\delta}$). Similarly, anion dominant compounds are possible because of shortage of cation ($Me_{1-\delta}Xe$) or excess of anion ($MeXe_{1+\delta}$). There is a difference between defect types in intermetallics and ionic compounds. In most of the intermetallic compounds, deviations from the stoichiometric composition are favored due to presence of antisite defects. However, in ionic compounds, antisites are not created by an exchange of positions between cations and anions because of size and charge differences and also related energy costs. Rather antisites could be created when more than one cation is present, for example, in ABO_3 compounds by exchanging positions between two cation-forming components, A and B.

Most of the point defects discussed above are present at a thermodynamic equilibrium. On the other hand, higher-dimensional defects are thermodynamically unstable and are produced because of unavoidable circumstances. Dislocations are line defects (one-dimensional), which are introduced mainly because of mechanical work or plastic strain induced during phase transformation in a material. Grain boundaries, interfaces (for example, solid/liquid and solid/vapor), and interphase boundaries are two-dimensional defects. When two crystals grow simultaneously, grain boundaries are created because of a mismatch in crystal orientation between them. Solid–liquid and solid–vapors interfaces are natural interfaces because of simultaneous presence of two different forms of phases of a particular material. Interphase boundaries are created when two different phases, materials or compounds, are forced to stay together. Cracks and voids are three-dimensional defects present in a material. Cracks are produced because of mechanical reasons, whereas voids could be a result of both mechanical action and supersaturation of vacancies. Given a favorable condition, concentrations of all these inequilibrium defects (one- to three-dimensional) try to decrease to yield a defect-free stable material. Note that grain boundaries may represent defects at a local equilibrium and that stability will be forced by external conditions.

The concepts behind the formation of these defects are described in detail in the next few sections.

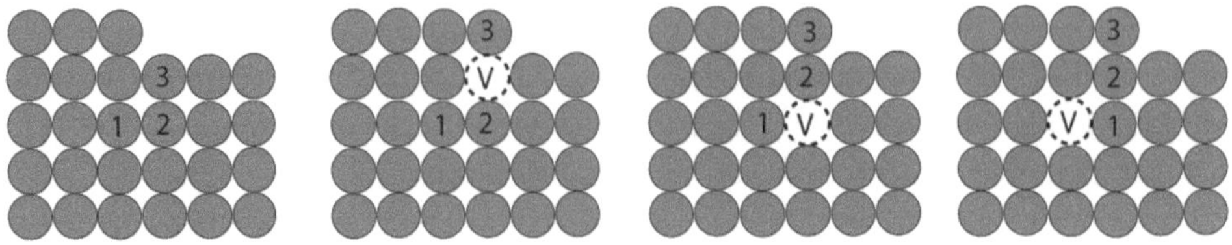

Figure 1.4 Creation of a vacancy by the transport of atom to surface (the Schottky mechanism).

1.1.1.1 Thermal Vacancies in Metallic Systems

Stability of a system at a constant pressure is defined by the value of Gibbs free energy, G (kJ/mol) expressed as $G = H - TS$, where H (kJ/mol) is the enthalpy, S (kJ/mol K) is the entropy, and T is the temperature in Kelvin. A phase is stable when it has minimum free energy, which means that there is not enough free energy to do a particular work or make a phase transformation possible. Any change/transformation in the system is possible when there is a possibility for the system to transform to a state with lower free energy. This means that a transformation is possible when $\Delta G < 0$, i.e., $(\Delta H - T\Delta S) < 0$.

Thermal vacancies are created when atoms diffuse from the interior of a crystal to the free surface, leaving behind vacant lattice sites by the Schottky mechanism as explained in Fig. 1.4. The broken bonds due to the creation of vacancies lead to an increase in enthalpy, i.e., positive change in enthalpy (ΔH_v). This is unfavorable for the system. However, at the same time, there is an increased contribution of entropy both because of irregular vibration of atoms surrounding the vacancies and entropy of mixing (configurational entropy), i.e., different possibilities of mixing arrangements of components and vacancies. This favorably decreases the free energy of the system. Therefore, the creation of the vacancies is influenced by the change in unfavorable enthalpy and favorable entropy. This can be further understood from the thermodynamic description of the vacancy creation. Note that the case of non-interacting defects is considered here. Since the equilibrium concentration of vacancies is typically small, the enthalpy change ($\Delta H \approx X_v \Delta H_v$) due to broken bonds and entropy change ($\Delta S \approx X_v \Delta S_v$) due to irregular vibrations could be considered to vary linearly with the vacancy concentration. Here ΔH_v and ΔS_v are the changes in enthalpy and entropy, respectively, for one mole of vacancies, and X_v is the mole fraction of vacancies present. Entropy contribution due to change in vibration pattern is much smaller compared to the enthalpy change due to broken bonds. However, introduced vacancies also give rise to considerable change of configurational entropy (ΔS_{Config}) because of increased possibilities of arrangements between components and vacancies. This is expressed (per mole of components) as

$$\Delta S_{Config} = k \ln W = k \ln \frac{(N_o + n_v)!}{N_o! n_v!} \tag{1.2}$$

where k is the Boltzmann constant and W is the number of arrangements possible between the filled sites for one mole of atoms, N_o, and vacant sites, n_v. N_o is the Avogadro number, and the total number of sites is $N_o + n_v$. Note that, according to the Schottky mechanism, creating of a vacant site (i.e., a vacancy) is accompanied by formation of a further lattice position (at the surface). Applying Stirling's approximation ($\ln N! \approx N \ln N - N$), we can write

$$\Delta S_{Config} = k\left[(N_o + n_v)\ln(N_o + n_v) - (N_o + n_v) - N_o \ln N_o + N_o - n_v \ln n_v + n_v\right],$$

which gives

$$\Delta S_{Config} = kN_o\left[\ln\frac{N_o + n_v}{N_o} + \frac{n_v}{N_o}\ln\frac{N_o + n_v}{n_v}\right] = R\left[\ln(1 + X_v) + x_v \ln\frac{1 + X_v}{X_v}\right]$$

and finally

$$\Delta S_{Config} = -R\left[X_v \ln X_v - (1 + X_v)\ln(1 + X_v)\right], \tag{1.3}$$

where R is the gas constant ($R = kN_o$) and $X_v = n_v/N_o$ is the mole fraction of vacancies. Therefore, the overall free energy change due to introduction of vacancies, following Eq. (1.1), can be rewritten as

$$\Delta G = X_v(\Delta H_v - T\Delta S_v) + TR\left[X_v \ln X_v - (1 + X_v)\ln(1 + X_v)\right], \tag{1.4a}$$

or equivalently,

$$\Delta G = X_v \Delta G_v + RT\left[X_v \ln X_v - (1 + X_v)\ln(1 + X_v)\right], \tag{1.4b}$$

since $\Delta G_v = \Delta H_v - T\Delta S_v$. In the relations above, and as shown in Fig. 1.5A, the contribution from ΔH_v is always higher than that of $-T\Delta S_v$, which makes ΔG_v positive, i.e., unfavorable for vacancy formation. However, for a smaller concentration of vacancies, the second part, i.e., the configurational entropy part, dominates over the ΔG_v, making the overall change of the free energy of system, ΔG, negative. Note here that X_v is smaller than 1, making $\ln X_v$ negative, whereas $(1 + X_v) \approx 1$. Therefore $RT[X_v \ln X_v - (1 + X_v)\ln(1 + X_v)] \approx RTX_v \ln X_v$ is negative. The gradient of the configurational entropy is infinite for a very minute content of vacancy. This can be understood from $\lim_{X_v \to 0} \frac{d\Delta S_{Config}}{dX_v}$, computed using Eq. (1.3). As shown in Fig. 1.5A, initially, the free energy drops with the addition of a small amount of vacancies due to a sharp change in configuration entropy. However, with a further increase in the vacancy concentration, the free energy increases because of change in enthalpy. The equilibrium concentration of vacancies is related to the maximum change (decrease) of the free energy (ΔG_v^*) from a defect-free material. Therefore, the equilibrium concentration is derived from

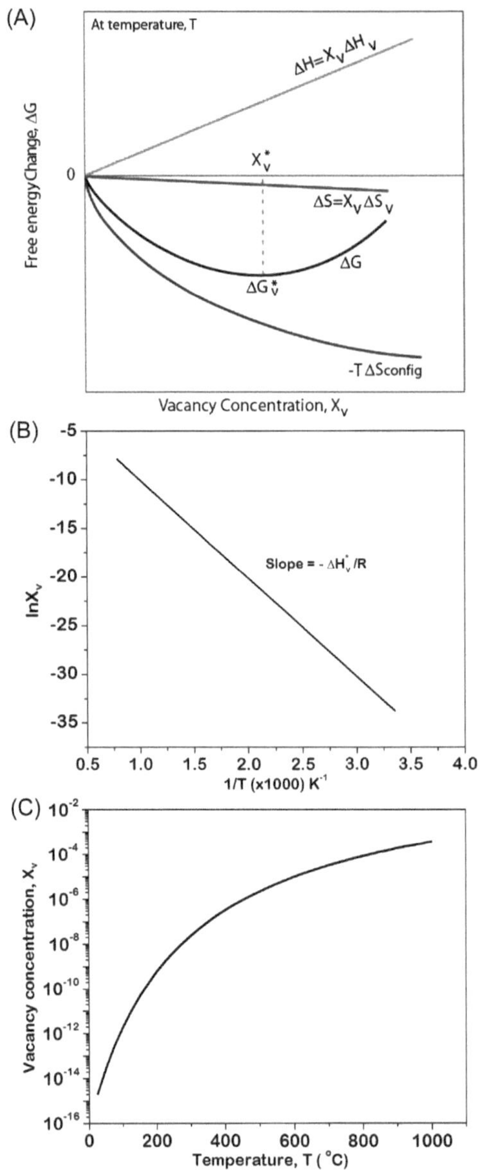

Figure 1.5 (A) Enthalpy, entropy, and fee energy change in a pure component because of addition of vacancies. (B) Arrhenius plot for estimation of the activation enthalpy for vacancy. (C) Variation of equilibrium concentration vacancies in Cu with temperature.

$\frac{d(\Delta G)}{dX_v} = 0$ using Eq. (1.4) and utilizing the fact that $(1 + X_v) \approx 1$ for a negligibly small vacancy concentration. This leads to

$$X_v^* = \exp\left(-\frac{\Delta G_v^*}{RT}\right), \tag{1.5a}$$

or equivalently,

$$X_v^* = \exp\left(-\frac{\Delta H_v^* - T\Delta S_v^*}{RT}\right) = X_v^o \exp\left(-\frac{\Delta H_v^*}{RT}\right), \tag{1.5b}$$

where $X_v^o = \exp(\frac{\Delta S_v^*}{R})$ is the pre-exponential factor and ΔH_v^* is the activation enthalpy. Note again that ΔH_v^* is the change in enthalpy due to broken bonds and ΔS_v^* is the entropy change due to irregular vibration of the atoms for the equilibrium concentration X_v^*. Configurational entropy is not included in the formation energy.

An alternative, Frenkel mechanism of vacancy formation in a solid is shown in Fig. 1.3B. The formation of a Frenkel defect includes simultaneous formation of a vacancy and of a corresponding self-interstitial atom, but not necessarily in nearest-neighbor positions. It should be noted here that in metallic systems Frenkel defect is not present due to the need of very high activation energy to create self-interstitial. This is rather common in ionic systems as explained later. Therefore, the determination of the contribution of this mechanism here is rather illustrative from a didactic point of view and also useful for the discussion on point defects in ionic solids. The main steps are similar as presented above in Eqs. (1.1)–(1.5), but with a few important modifications. Again, we are considering the limit of non-interacting defects and their negligible concentrations. The enthalpy change is now $\Delta H \approx X_F \Delta H_F$ (X_F is here the concentration of the Frenkel defects, represented by vacancy–self-interstitial pairs, and ΔH_F is the corresponding enthalpy of formation; both values are in per mole); and the entropy change is similarly equal to $\Delta S \approx X_F \Delta S_F$, again because of a modification of the vibration spectra of atoms due to defect formation. The creation of Frenkel defects gives rise to a considerable change of configurational entropy, ΔS_{Config}, and it is now (per mole)

$$\Delta S_{Config} = k \ln W = k \ln \frac{(N_o)!}{(N_o - n_F)!n_F!} \frac{(AN_o)!}{(AN_o - n_F)!n_F!} \tag{1.6}$$

where W is the number of arrangements of vacancies and self-interstitials. In fact, the vacancies are arranged between N_o sites, with $N_o - n_F$ being filled, and the self-interstitials are arranged between AN_o sites, with $AN_o - n_F$ of them being filled. Here A is a constant, which is the number of interstitial positions per number of atoms. As explained later (in Section 1.1.1.2), it depends on the crystal structure and type of interstitial site such as octahedral and tetrahedral. For example, if we consider the octahedral voids, in a BCC crystal $A = 3$, whereas in an FCC crystal $A = 1$. Again, applying Stirling's approximation, after a few simple steps one may show that the configurational entropy is

$$\Delta S_{Config} = -R\left[2X_F \ln X_F + (A - X_F)\ln(A - X_F) + (1 - X_F)\ln(1 - X_F)\right], \quad (1.7)$$

where $X_F = n_F/N_o$ is the mole fraction of Frenkel defects. Therefore, the overall free energy change due to introduction of Frenkel defects following Eq. (1.1) can be rewritten as

$$\Delta G = X_F \Delta G_F + RT\left[2X_F \ln X_F + (A - X_F)\ln(A - X_F) + (1 - X_F)\ln(1 - X_F)\right], \quad (1.8)$$

where $\Delta G_F = \Delta H_F - T\Delta S_F$. Again, the contribution from ΔH_F is always greater than that from $-T\Delta S_F$, which makes ΔG_F positive, i.e., unfavorable for vacancy formation via the Frenkel mechanism. However, for a smaller concentration of Frenkel defects (if present), the second part, i.e., the configurational entropy part, could dominate over the ΔG_F, making the overall change of the free energy of system, ΔG, negative. Again and this is extremely important, the gradient of the configurational entropy is infinite for a very minute addition of Frenkel defects, since $\lim_{X_F \to 0} \frac{d\Delta S_{Config}}{dX_F}$ computed using Eq. (1.7) is infinite. However, with a further increase of the concentration of Frenkel defects, the enthalpy term starts to dominate and the free energy increases. The equilibrium concentration of Frenkel defects is related to the maximum change (decrease) of the free energy (ΔG_F^*) from a defect-free material. Therefore, the equilibrium concentration is derived from $\frac{d(\Delta G)}{dX_F} = 0$ using Eq. (1.8) and utilizing the fact that $(1 - X_F) \approx 1$ and $(A - X_F) \approx A$ for a negligibly small defect concentration. This leads to

$$X_F^* = (A)^{1/2} \exp\left(-\frac{\Delta G_F^*}{2RT}\right), \quad (1.9a)$$

or equivalently,

$$X_F^* = (A)^{1/2} \exp\left(-\frac{\Delta H_F^* - T\Delta S_F^*}{2RT}\right) = X_F^o \exp\left(-\frac{\Delta H_F^*}{2RT}\right) \quad (1.9b)$$

where $X_F^o = (A)^{1/2} \exp(\frac{\Delta S_F^*}{2R})$ is the pre-exponential factor and ΔH_F^* is the activation enthalpy. Note again that ΔH_F^* is the change in enthalpy due to broken and rearranged bonds and ΔS_F^* is the change in entropy from irregular vibration of the atoms for the equilibrium concentration X_F^*. Configurational entropy is not included in the formation energy.

The most important difference between the expressions for concentration of Schottky and Frenkel defects is the appearance of power 1/2 (i.e., 1/2 inside the exponential term) in Eq. (1.9b). This is a direct consequence of the fact that two defects, i.e., a vacancy and a self-interstitial atom, are formed as soon as a Frenkel defect is formed. The factor $(A)^{1/2}$ results from availability of interstitial positions per atom in a given lattice.

Typically, in a metallic system, ΔH_F^* is significantly larger than ΔH_v^* and the Schottky mechanism dominates over the Frenkel one with a consequence that practically no self-interstitials are present at thermodynamic equilibrium. However, both types of defect are very common in ionic systems as discussed later, since void sizes of interstitial sites are much bigger.

There are different techniques available for the determination of vacancy concentrations in a solid, such as by the measurement of thermal expansion or resistivity, and the most accurately by the measurement of positron annihilation. After measuring the equilibrium concentration of vacancies at a particular temperature, one could estimate the formation free energy (ΔG_v^*) directly from Eq. (1.5a). However, in experiments, instead of measuring the absolute vacancy concentration at a particular temperature, we actually measure the relative change in the vacancy concentration because of a change in temperature. Therefore, we actually measure the activation enthalpy following Eq. (1.5b) from the plot of $\ln X_v$ *vs.* $1/T$, as shown in Fig. 1.5B, in which the slope is equal to $-H_v^*/R$. The entropy term can be calculated from the pre-exponential factor, but it could also be avoided since it is generally too small and may incur high error in calculation. The variation of equilibrium vacancy concentration with temperature in pure Cu is shown in Fig. 1.5C. It can be seen that even at 1000°C, which is close to the melting point of Cu, the mole fraction of vacancy is only 3.7×10^{-4}. This means that one vacancy is present for ~2700 atoms. This further means that, for example, only one vacancy is present in a cube of ~14 atoms on each side. This low vacancy concentration is nevertheless enough for a significant rate of substitutional diffusion as discussed in Section 1.3.

1.1.1.2 Solutes in Metallic Systems

Solutes, i.e., extrinsic defects, could be substitutional or interstitial in nature. Substitutional solutes occupy the lattice positions and are typically randomly distributed. Different conditions by which the solute concentration is limited in a solvent are governed by empirical Hume–Rothery rules:

- **(i)** (Size) When the atomic radii of solute and solvent differ by less than 15%, solid solution is favorable. With a higher than 15% difference, the solubility limit is very small.
- **(ii)** (Structure) For a completely solid solution, the crystal structures of both components should be the same. A difference in the crystal structures leads to a limited solubility.
- **(iii)** (Valency) If valency is the same for both the solvent and solute, then a complete miscibility is possible. Solubility is higher when the solvent has a higher valence.
- **(iv)** (Electronegativity) Similar electronegativity favors a higher solubility. The higher the difference (i.e., the greater the chance of chemical reactivity), the greater the possibility of new phase formation.

It should be recollected here that unwanted solutes, i.e., impurities, are always present because of an increase in stability of solid solution by a decrease of the free energy mainly due to a gain in the configurational entropy.

In the case of interstitial impurities, such as C, H, N, and O, size factor plays a crucial role. In this case, the relative sizes of the interstitial atoms and interstitial voids (i.e., the maximum size of a sphere which can be positioned at an interstitial position and which does not overlap with neighboring matrix atoms) become very important. Enthalpy increases because of strain created on the host atoms surrounding the impurity at interstitial positions. Entropy increases because of change in vibration pattern of the atoms surrounding the impurity and an increase in the configurational entropy. Therefore, the overall discussion for the estimation of the equilibrium concentration is similar to that of the vacancy formation with an added complication of the crystal structure. In general, most of the impurities (other than H) occupy the octahedral voids [1]. To explain this further, octahedral and tetrahedral voids in BCC (Body Centered Cubic) and FCC (Face Centered Cubic) crystals are shown in Fig. 1.6. In general, interstitial atoms (other than H) are bigger than the void size available and strain the crystal to increase the enthalpy of the system, which is unfavorable. In a BCC crystal, the ratio of the radius of interstitial void, r_i, to the atomic radius, r_a, is 0.29 and 0.155 for the tetrahedral and octahedral voids, respectively. However, the open (i.e., atom-free) space in this structure is asymmetric. When an interstitial atom occupies a tetrahedral void, it strains all four atoms surrounding it. When it occupies an octahedral void, it mainly strains two atoms, numbered as 1 and 2 in Fig. 1.6A, which are at a distance of $a/2$ from the octahedral position. The other four atoms, numbered as 3, 4, 5 and 6, are located at a further distance, $a/\sqrt{2}$, and not strained to the same extent. Therefore, in a BCC crystal, although the octahedral void is smaller than the tetrahedral void, which is actually the lowest radius in an asymmetrical void, interstitials prefer to occupy the octahedral voids. In an FCC crystal, the r_i-to-r_a ratio is 0.225 and 0.414 for the tetrahedral and octahedral voids, respectively. Therefore, the interstitial atoms occupy the octahedral voids. To summarize, because of different reasons most of the interstitial atoms in both types of crystal occupy the octahedral voids.

The discussion of the change in enthalpy and entropy due to interstitial (or self-interstitial) atom addition is similar to that of the vacancy formation. When an interstitial is added, the enthalpy (ΔH_I) increases because of the strain created, and the entropy (ΔS_I) increases because of change in thermal vibration around the interstitial. Moreover, a modification is required for the configurational entropy. As shown in Fig. 1.6, in a BCC unit cell, there are 3 octahedral voids present for each host atom. Therefore, considering the Avogadro number, there are $3N_o$ voids present, for example, for n_I (corresponding to mole fraction X_I) interstitial atoms to reside in one mole of material, i.e.,

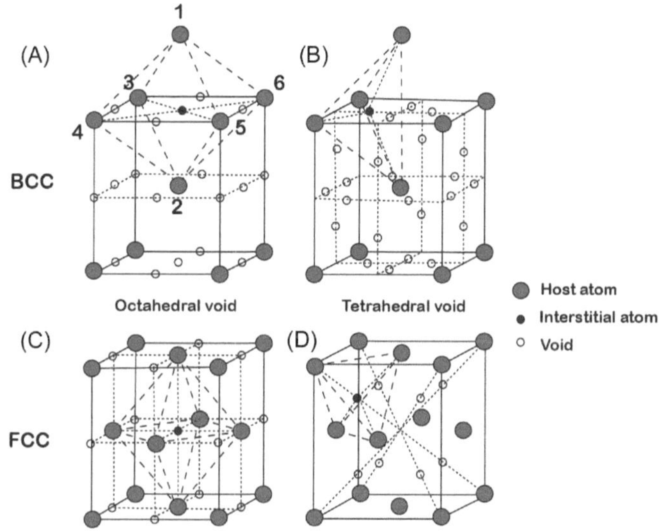

Figure 1.6 (A) Octahedral and (B) tetrahedral positions in BCC crystal. (C) Octahedral and (D) tetrahedral positions in FCC crystal.

among N_o atoms. Therefore, the configurational entropy is expressed as

$$\Delta S_{Config} = k \ln W = k \ln \frac{(3N_o)!}{(3N_o - n_I)!n_I!}. \tag{1.10}$$

Furthermore, the equilibrium concentration can be derived following a similar line of analysis as for vacancy formation presented in [1]

$$X_I = 3\exp\left(-\frac{\Delta G_I^*}{RT}\right) = 3\exp\left(-\frac{\Delta H_I^* - T\Delta S_I^*}{RT}\right) = X_I^o \exp\left(-\frac{\Delta H_I^*}{RT}\right) \tag{1.11}$$

where $X_I^o = 3\exp(\frac{\Delta S_I^*}{R})$ is the pre-exponential factor, ΔG_I^* is the activation energy, and ΔH_I^* is the activation enthalpy.

In an FCC crystal, one octahedral void is present for each host atom, and the equilibrium concentration is derived similarly as follows:

$$X_I = \exp\left(-\frac{\Delta G_I^*}{RT}\right) = \exp\left(-\frac{\Delta H_I^* - T\Delta S_I^*}{RT}\right) = X_I^o \exp\left(-\frac{\Delta H_I^*}{RT}\right) \tag{1.12}$$

where $X_I^o = \exp(\frac{\Delta S_I^*}{R})$ is the pre-exponential factor. Therefore, the pre-exponential factor depends on the ratio of voids and host atoms in a particular crystal structure. In a phase diagram, for a small concentration of impurities (<1 at.%), the relations above can correctly predict the phase boundary compositions for Me(x) solid solution, in which Me is the host metal and x is the impurity atom [2].

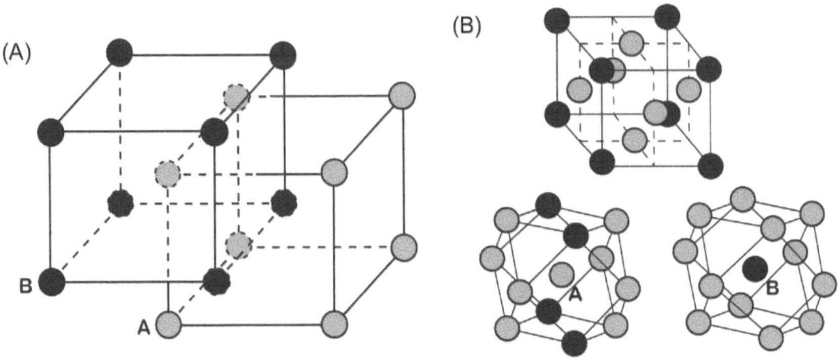

Figure 1.7 Nearest neighbors are shown for different components in (A) AB compound with L10 and (B) A_3B compound with $L1_2$ structures.

For example, C concentration in BCC α-Fe (ferrite) is expressed as [2]

$$X_I = 11.2\exp\left(-\frac{77.3\ \text{kJ/mol}}{RT}\right), \tag{1.13a}$$

$$\textit{Weight percent} = 240\exp\left(-\frac{77.3\ \text{kJ/mol}}{RT}\right). \tag{1.13b}$$

It gives 0.022 weight percent of C at 727°C according to the Fe–C phase diagram. This is equivalent to a mole fraction of $\sim 10^{-3}$. This means that the equilibrium concentration of impurity atoms could be as low as 1 atom in a cube of 10 atoms on each side. The activation enthalpy can be determined following a similar method as that employed for vacancies, i.e., by measuring the impurity concentration at different temperatures.

1.1.1.3 Point Defects in Intermetallic Compounds

Different components are designated to occupy different sublattices in an ordered or intermediate phase, also known as intermetallic compound. For example, in an AB compound (A:B 50:50) with the B2 structure (for example, β-NiAl, CoAl, CoGa, NiGa, AuZn, NiZn, or AgMg), A and B occupy body corner and body center positions, respectively, in a derivative BCC lattice, as shown in Fig. 1.7A. Therefore, A and B are surrounded by unlike atoms.

In an A_3B (A:B 75:25) compound with the $L1_2$ structure (for example, Ni_3Al, Ni_3Ga, or Zr_3Al), illustrated in Fig. 1.7B, B atoms occupy the body corner positions, whereas A atoms occupy the face center positions of the FCC derivative structure. In this phase, A is surrounded by 8 A atoms and 4 B atoms, and B is surrounded by 12 A atoms. However, these descriptions are *ideally* true only in a perfect stoichiometric compound at 0 K. At an elevated temperature, even at stoichiometric composition, two

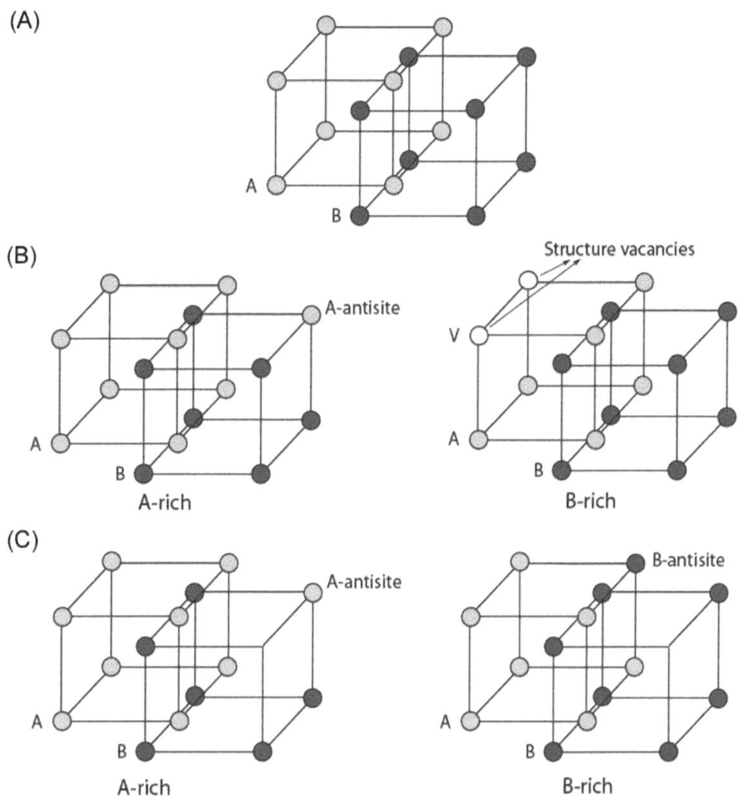

Figure 1.8 Structural (constitutional) defects present in B2 structure are shown. (A) Defect free structure, (B) A-rich side with antisite defects and B-rich side with structure vacancies, (C) both A-rich and B-rich sides with antisite defects.

types of intrinsic defect are created: vacancies and antisites (or antistructures). Vacancies, as already explained, are the vacant lattice sites and are found on both sublattices but not necessarily with same concentrations. These are denoted as V_A and V_B for vacancies on sublattices of A and B. When one particular atom occupies the wrong sublattice, i.e., the sublattice of the other component, it is called an antisite. When A occupies the sublattice designated for B, it is called antisite A and denoted as A_B. Similarly, the antisite of B is denoted as B_A. Different concentration of defects might be found on different sublattices depending on the composition.

Along with thermal defects, most of the ordered phases have structural (constitutional) defects. As a matter of fact, there are no compounds which would only be present with exactly stoichiometric composition. These have a very small to a very large composition range depending on the system. Deviation from the stoichiometric composition is possible because of structural defects. These could again be vacancies and/or antisites on different sublattices. Few B2 compounds with formula AB, such as NiAl, CoAl, NiGa, or CoGa, are of special, so-called triple defect type, in which two different types of structural defect are found depending on A- or B-rich sides. In the A-rich side, the antisite defects compensate the deviation from the stoichiometry, as explained in Fig. 1.8B, left. In this figure, A and B are designated to occupy the α- and β-sublattices, respectively. With deviation towards the A-rich side, excess of A occupies the β-sublattice to produce A antisites (A_B). The higher the deviation, the higher the concentration of A_B. On the other hand, in the B-rich side *structure vacancies* are present. In this case, excessive B atoms do not go to the α-sublattice. Instead, the B atoms remain on the β-sublattice, and two vacancies are formed on α-sublattice (V_A) for each excessive B atom; see Fig. 1.8B, right. The thermal disorder in this compound corresponds to formation of a triple defect, i.e., of two A vacancies and one A antisite (A_B). It is important that the thermal disorder via formation of triple defects does not mean that these defects are bounded entities and reside at nearest neighboring positions. Instead, the triple defect disorder refers to the total concentration of elementary defects which corresponds to two A vacancies per each A antisite. The formation of the triple defect resembles, in fact, the Schottky mechanism in ionic AB compounds where, due to the charge neutrality condition, only pairs of A and V vacancies have to be formed. In an ordered AB compound, vacancy formation corresponds to creation of a pair of A and B vacancies and the corresponding atoms are placed at the surface, adding a new structure unit, AB, which does not affect the composition. As a matter of fact, in such a compound a B vacancy is energetically unfavorable (or even mechanically unstable) against hoping of a nearest neighbor A atom to it, which creates an A antisite (the A atom on the B sublattice) and leaves an A vacancy. As a result, a triple defect, $2V_A + A_B$, is created instead of the vacancy pair, $V_A + V_B$.

However, compounds with triple defects are not very common, and in many B2 compounds (AgMg, NiZn, CuZn, AuZn) the deviations from the stoichiometric composition is compensated by creation of antisite defects, as shown in Fig. 1.5C.

Therefore, by definition, structural or constitutional defects are those defects which accommodate the deviation from the stoichiometric composition and remain in thermal equilibrium even at 0 K in its *maximum disordered* state. The difference between the concentration of defects at an elevated temperature and 0 K gives the concentration of thermal defects. However, the material does not spot defects as thermal or constitutional, and defect concentration variation on different sublattices are presented overall. A few examples of these for the NiAl system are given in Fig. 1.9 at $T = 0.70T_m$, where T_m is the melting point of the compound. Defect concentrations are computed at 1338 K for NiAl (the B2 structure) and at 1169 K for Ni_3Al (the $L1_2$ structure) [1].

The calculation of point defects in ordered phases is explained below. We will use the chemical-reaction approach and give numerical estimates for the example of NiAl with the B2 structure. The vibrational energy and entropy contributions are neglected.

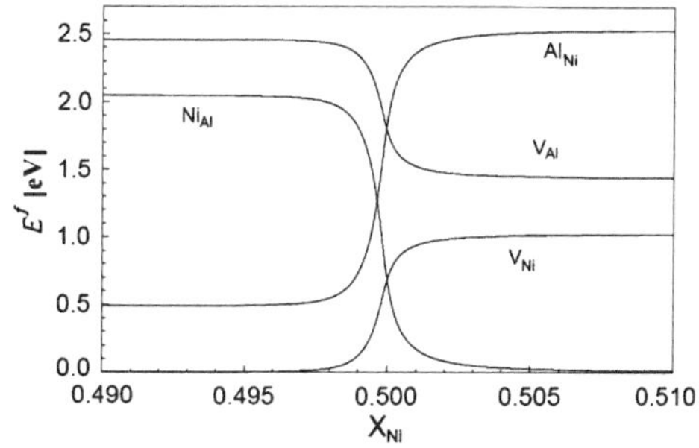

Figure 1.9 Effective formation energies E^f of the point defects as a function of composition X_{Ni} for the B2 compound NiAl.

In the case of an intermetallic compound, the defect formation energies have to be calculated in a different way as outlined above, e.g., for pure metals. Let us consider a vacancy. In a pure metal, the energy difference between a perfect crystalline block (N atoms) and that containing a vacancy (and thus $N-1$ atoms) plus a cohesive energy for 1 atom (in this way the number of atoms is conserved) gives the energy cost of vacancy formation, i.e., the vacancy formation energy. In an intermetallic compound, for any given defect, the energy difference between a perfect crystalline block and that containing the defect under consideration gives a value which may be called as a "raw" formation energy of the defect. However, the creation of a defect in the crystalline block may change the ratio of A and B atoms, i.e., affect the composition, or change the number of atoms in the block (it is *not* the case if we just swap A and B atoms between two sublattices). Since introduction of a single defect generally violates the composition of the compound (e.g., creation of an A vacancy by removal of the A atom from a crystalline block affects the ratio of A and B atoms in the compound), these "raw" values by themselves cannot represent the thermodynamical quantities. The effective formation energies, which correspond to the Arrhenius approximations of the temperature dependence of the defect concentrations, can be used with this aim.

The equilibrium defect concentrations in an intermetallic compound can be calculated using either canonical or grand canonical ensemble formalism, and both methods will give the same result, so the choice of the approach is a matter of convenience. Alternatively, the point defects can be treated as individual species and different point-defect reactions can be considered as chemical reactions in a multi-component lattice gas. This approach is sketched here and it gives the same results as the previous two, but, maybe, in a more elegant way.

The point-defect concentrations are assumed to be small, so that the correlations in the defect distributions are neglected. Formally, each type P of point defects is described by a chemical potential

$$\mu_P = \varepsilon_P + kT \ln Y_P. \qquad (1.14)$$

Here ε_P is the "raw" formation energy of a single defect, Y_P is the occupation probability of the defect P per unit site in its own sublattice. Thus, the defect concentrations (in mole fractions) X_P expressed as the site fractions of the number of defects are related with Y_P by the obvious relation for the B2 structures: $2X_P = Y_P$.

The four unknown equilibrium point-defect concentrations $\{Y_P\}$ in AB compound (two types of vacancy and two types of antisite) can be determined from the kinetic equilibrium conditions with respect to three point-defect reactions, supplemented by the conservation law for a fixed alloy composition. The latter can be written in the form

$$\xi(1 - 2Y_{V_A} - 2Y_{V_B}) = \frac{1}{4}(Y_{V_B} - Y_{V_A}) + \frac{1}{2}(Y_{A_B} - Y_{B_A}) \tag{1.15}$$

where

$$\xi = X_A - \frac{1}{2} \tag{1.16}$$

is the deviation of the alloy composition (mole fraction of A, X_A) from the ideal stoichiometric composition ($X_A = 0.5$). Neglecting the terms like ξ^2 in Eq. (1.15) for small deviations from the stoichiometric composition and small concentrations of the point defects, the above equation can be presented as

$$\xi = \frac{1}{4}(Y_{V_B} - Y_{V_A}) + \frac{1}{2}(Y_{A_B} - Y_{B_A}). \tag{1.17}$$

The choice of three point-defect reaction is a matter of convenience and depends on whether we deal with a triple-defect or anti-site disorder compound. For an AB alloy, a convenient set of reactions is:

$$2V_A + A_B + AB \leftrightarrow 0, \tag{1.18}$$

$$A_B + B_A \leftrightarrow 0, \tag{1.19}$$

$$2V_B + AB \leftrightarrow A_B. \tag{1.20}$$

Eq. (1.18) represents the triple-defect equilibrium in the system. It simply shows that a triple defect (two A vacancies and an A antisite) can be canceled by adding a structural unit AB to the system. Similarly, Eq. (1.19) represents an antisite equilibrium and shows that a pair of antistructure atoms can be created or canceled. Finally, Eq. (1.20) shows that, if we initially have two B vacancies and add a structural unit AB, then B will cancel one of the vacancies, while A will turn the other vacancy in an A_B-antistructure atom.

The dynamic equilibrium in the above defect reactions, Eqs. (1.18)–(1.20), can be reached by the equality of corresponding chemical potentials. The structural unit, AB,

Table 1.1 The "raw" formation energies of single point defects in NiAl [1]. V is the vacancy, and the subscript denotes the sublattice. ε_0 is the cohesive energy of the alloy

	εV_{Ni}	εV_{Al}	εNi_{Al}	εAl_{Ni}	ε_0
"Raw" Formation Energy, eV/atom	5.978	5.471	−0.920	3.457	−4.494

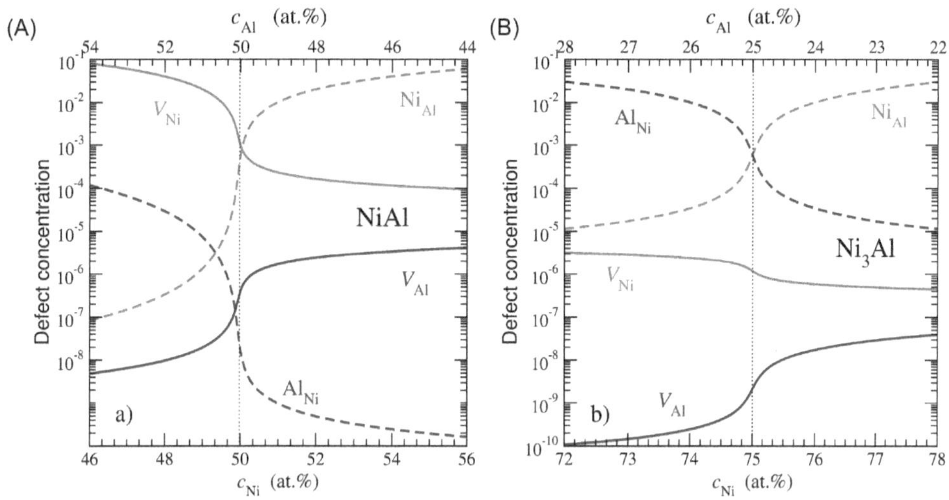

Figure 1.10 Defect concentration variations at $T = 0.75$ Tm in (A) NiAl and (B) Ni_3Al.

will be presented by the double value of the cohesive energy per atom, $2\varepsilon_0$. This will give

$$\varepsilon_{td} + kT \ln Y_{V_A}^2 Y_{A_B} = 0, \tag{1.21}$$

$$\varepsilon_{A_B} + \varepsilon_{B_A} + kT \ln Y_{A_B} Y_{B_A} = 0, \tag{1.22}$$

$$2\varepsilon_{V_B} - \varepsilon_{A_B} + 2\varepsilon_0 + kT \ln\left(\frac{Y_{V_B}^2}{Y_{A_B}}\right) = 0. \tag{1.23}$$

Here $\varepsilon_{td} = 2\varepsilon_0 + 2\varepsilon_{V_A} + \varepsilon_{A_B}$ is the so-called triple-defect energy.

Eqs. (1.15) and (1.21)–(1.23) could be solved numerically for the four defect concentrations depending on the composition of the alloy. Then, the effective activation energies can be estimated by linear fitting in logarithm concentration against inverse temperature coordinates within the given temperature interval. The analysis shows that the temperature dependence of the point defect concentrations can be adequately treated with single formation energy. The results for the NiAl compound with the B2 structure with the numerical values from Table 1.1 are given in Fig. 1.10.

Besides, Eqs. (1.15) and (1.21)–(1.23) can be solved analytically in an approximate manner by using the fact that in a triple defect compound AB, Y_{V_A} and Y_{A_B} are much

Table 1.2 Analytical expressions and numerical values for NiAl of the effective formation energies Q_P^f of the point defects P, $P = V_{\mathrm{Ni}}$, V_{Al}, $\mathrm{Ni_{Al}}$, and $\mathrm{Al_{Ni}}$ (in eV/atom) depending on composition $\xi = X_{\mathrm{Ni}} - 1/2$

	$\xi < 0$	$\xi = 0$	$\xi > 0$
$Q^f_{V_{\mathrm{Ni}}}$	0	$\varepsilon_{td}/3$	$\varepsilon_{td}/2$
	0	0.683	1.024
$Q^f_{V_{\mathrm{Al}}}$	$2\varepsilon_0 + \varepsilon_{V_{\mathrm{Al}}} + \varepsilon_{V_{\mathrm{Ni}}}$	$(4\varepsilon_0 + 3\varepsilon_{V_{\mathrm{Al}}} + \varepsilon_{V_{\mathrm{Ni}}} - \varepsilon_{\mathrm{Ni_{Al}}})/3$	$\varepsilon_0 + \varepsilon_{V_{\mathrm{Al}}} - \frac{1}{2}\varepsilon_{\mathrm{Ni_{Al}}}$
	2.461	1.778	1.437
$Q^f_{\mathrm{Ni_{Al}}}$	ε_{td}	$\varepsilon_{td}/3$	0
	2.048	0.683	0
$Q^f_{\mathrm{Al_{Ni}}}$	$\varepsilon_{\mathrm{Al_{Ni}}} - 2\varepsilon_0 - 2\varepsilon_{V_{\mathrm{Ni}}}$	$(-2\varepsilon_0 - 2\varepsilon_{V_{\mathrm{Ni}}} + 3\varepsilon_{\mathrm{Al_{Ni}}} + 2\varepsilon_{\mathrm{Ni_{Al}}})/3$	$\varepsilon_{\mathrm{Al_{Ni}}} + \varepsilon_{\mathrm{Ni_{Al}}}$
	0.489	1.854	2.537

greater than the other two concentrations. Then, Eq. (1.15) can be rewritten in the form

$$\xi = \frac{1}{2}Y_{A_B} - \frac{1}{4}Y_{V_A}. \tag{1.24}$$

From Eq. (1.24) we have $Y_{V_A} = 2Y_{A_B}$ for the stoichiometric composition, $\xi = 0$. For the A-rich compositions, $\xi > 0$, the A antistructure atoms are the main defects, $Y_{V_A} << Y_{A_B}$, and Eq. (1.20) transforms into $Y_{A_B} = 2\xi$. On the other hand, for the B-rich compositions, $\xi < 0$, the A vacancies are mainly formed, $Y_{V_A} >> Y_{A_B}$, and Eq. (1.24) will be read as $Y_{V_A} = -4\xi$. These relations allow us to analytically solve the system of Eqs. (1.21)–(1.23) and to find the concentrations of all four point defects under the given approximations. The analytical expressions for the effective formation energies are given in Table 1.2. The comparison with Fig. 1.10 suggests that these results correspond well to the numerical data at the stoichiometric composition and far from the stoichiometry.

As shown in Fig. 1.9, the calculations suggest that the Ni vacancies and the Ni anti-structure atoms are the main defects in NiAl and the concentration of Al vacancies is by a few orders of magnitude lower than the concentration of Ni vacancies in accordance with experimental observations.

Similar analytic solution can be used for other compounds with different lattices. One obviously has to modify the mass-balance Eq. (1.15) in a proper way and probably defect reactions, Eqs. (1.21)–(1.23). The results for NiAl (the B2 structure) and $\mathrm{Ni_3Al}$ (the $\mathrm{L1_2}$ structure) are presented in Figs. 1.9A and 1.9B. These figures demonstrate a few important features of defect behavior. It is obvious that $\mathrm{Ni_3Al}$ belongs to the antistructure-defect type of intermetallic compounds, since antistructure atoms are predominantly generated to accommodate the deviation from the stoichiometry. In contrast, NiAl reveals a triple-defect-type of point defect disorder and constitutional Ni vacancies exist in NiAl on the Al-rich side. Moreover, the Ni vacancy concentration is very large also on the Ni-rich side, e.g., $X_{V_A} \sim 10^{-4}$ at $T = 0.7T_m$. In Ni_3Al the vacancies are mainly concentrated on the transition-metal sublattice, too, and their concentration reaches about 10^{-6} to 10^{-5} at $T = 0.7T_m$. These are also the typical vacancy concentrations in close-packed pure metals at the same reduced temperature. The vacancy concentration on the Al sublattice is remarkably smaller.

Concentration of these intrinsic defects could be changed drastically when alloying elements (solutes) are added to the intermetallic compound. For example, when Pt was added to β-NiAl, it substitutionally replaced Ni, yet decreased defect formation energy for both Ni and Al [3], leading to a higher concentration of defects and diffusion rates of components [4,5]. In general, most of the alloying addition replaces a particular component in a compound; however, there are few instances in which it replaces both components in different sublattices. Further, as explained below, the interstitial addition of pure components, e.g., atoms such as H, C, N, and O, also occupies the interstitial positions in the intermetallic compounds. This is also true, at least partially, for Fe and Ni in Ti_3Al, for which a dissociative mechanism was observed.

1.1.1.4 Point Defects in Ionic Solids

Like in metallic systems, as already explained, defects can be produced intrinsically and extrinsically in ionic compounds. However, the need for charge neutrality makes the defect chemistry more cumbersome. Before learning about the defects, it is necessary to familiarize with the Kröger–Vink notations [6]. Instead of being denoted by actual charges, for the sake of convenience, the effective (or relative) charges of the defects are expressed with respect to a perfect crystal as the reference state. Therefore, cations and anions in an ionic compound MeXe at their regular sites are denoted as Me_{Me} and Xe_{Xe}. Following Kröger–Vink notations, $'$ and $\cdot$ are used for single negative and positive signs to emphasize a difference between -1 and $+1$ used normally to describe a charge of ionic crystal. A negatively charged electron is denoted by e' and a positively charged hole is denoted by $h^{\cdot}$. Me^{+1} and Xe^{-1} in an ionic crystal are written as Me^0 and Xe^0 in the reference state. Therefore, when vacancies are created at these sites, they are denoted by V'_{Me} and $V^{\cdot}_{Xe}$. Similarly, cation and anion interstitials are denoted by $Me^{\cdot}_i$ and Xe'_i (these are self-interstitials if the terminology for metals is used). Furthermore, when an anion is removed, it is not present anymore to hold the electron. Therefore, the electron is removed and the vacant anion site becomes positively charged. When an anion is moved to an interstitial, it is presented with an electron and becomes negatively charged. The opposite is true for a cation vacancy and interstitial.

(i) Intrinsic Defects in Ionic Compounds

Description of the defects in ionic systems depends on the type of the compound, i.e., whether it is stoichiometric and non-stoichiometric.

Defects in Stoichiometric Compounds

There are mainly four types of defect present in a stoichiometric compound. In a compound MeXe with monovalent cations and anions (arguments are similar for higher-valent components), the reactions for different types of defect (keeping the total charge of the system as zero) can be written as

(Schottky defects)	$Null \rightleftharpoons V'_{Me} + V^{\cdot}_{Xe}$	(1.25a)
(Anti-Schottky defects)	$Me_{Me}Xe_{Xe} \rightleftharpoons Me^{\cdot}_i + Xe'_i$	(1.25b)
(Frenkel defects)	$Me_{Me} \rightleftharpoons Me^{\cdot}_i + V'_{Me}$	(1.25c)
(Anti-Frenkel defects)	$Xe_{Xe} \rightleftharpoons Xe'_i + V^{\cdot}_{Xe}$	(1.25d)

Note that the condition of charge neutrality holds in all these equations as it should. As already mentioned, theoretically all types of defect could be present in a particular material, however, one type typically dominates based on the low enthalpy of formation of that type of defect in a particular system compared to the others. In general, Frenkel and Schottky defects are more common. Anti-Schottky defects are most difficult to produce. Formation of two interstitials is accompanied with creation of larger strains in the crystal. Similarly, creation of anti-Frenkel defects by producing large anion interstitial leads to large strain.

Equilibrium concentration of defects can be calculated following similar reasoning as explained for vacancies and interstitials in metallic systems. There will be an increase in enthalpy due to strain created by interstitial atoms or broken bonds by the creation of vacancies. Increase in entropy should also be noticed because of change in vibration patterns of the components next to the defects. Additionally, there will be a contribution from configurational entropy. In the case of Schottky defects, this is expressed as

$$\Delta S_{Config} = k \ln w = k \ln\left[\frac{(N_o + n_{V'_{Me}})!}{(N_o)! n_{V'_{Me}}!}\right]\left[\frac{(N_o + n_{V^{\cdot}_{Xe}})!}{(N_o)! n_{V^{\cdot}_{Xe}}!}\right]. \tag{1.26}$$

The difference in Eq. (1.22) compared to Eq. (1.2) is that now two types of vacancies are present. Since in a compound with monovalent cation and anion there is equal number of vacancies on both sublattices, $n_v^{Me} = n_v^{Xe} = n_v$, the configurational entropy can be written as

$$\Delta S_{Config} = k \ln w = 2k \ln\left[\frac{(N_{o+n_v})!}{(N_o)! n_v!}\right]. \tag{1.27}$$

Again following similar explanation as for vacancies in a metallic system, the mole fraction of vacancies can be derived as

$$X_{V'_{Me}} = X_{V^{\cdot}_{Xe}} = \exp\left(-\frac{\Delta G_S^*}{2RT}\right) = \exp\left(-\frac{\Delta H_S^* - T\Delta S_S^*}{2RT}\right) = X_S^o \exp\left(-\frac{\Delta H_S^*}{2RT}\right), \quad (1.28)$$

where $X_S^o = \exp(\frac{\Delta S_S^*}{2R})$ and the subscript "S" refers to the Schottky defects.

The defect equilibrium in ionic systems can also be understood based on chemical reaction kinetics. For example, we consider a reaction equation

$$mM + nN \rightleftharpoons pP + qQ. \quad (1.29)$$

The free energy for this case can be written as

$$\Delta G = \Delta G^o + RT \ln \frac{a_P^p . a_Q^q}{a_M^m . a_N^n} = \Delta G^o + RT \ln K, \quad (1.30)$$

where ΔG^o is the free energy change at the standard state and a is the activity, m, n, p and q are the numbers of moles, and M, N, P and Q are the components. At equilibrium concentrations, we have $\Delta G = 0$. Therefore, we can write

$$K = \exp\left(-\frac{\Delta G^o}{RT}\right). \quad (1.31)$$

Considering ideal solution, activity could be taken equal to the concentration in mole fraction of the defects, i.e., $a_i = X_i = [i]$, where i is the component considered. In the reaction equation, concentrations are generally expressed by a bracketed term. Therefore, the equilibrium constant for Schottky defects (Eq. (1.25a)) can be written as

$$K_S = X_{V'_{Me}} . X_{V^{\cdot}_{Xe}} = \frac{[V'_{Me}]}{[Me_{Me}]} . \frac{[V^{\cdot}_{Xe}]}{[Xe_{Xe}]} = \exp\left(-\frac{\Delta G_S^*}{RT}\right).$$

Since concentrations of defects are generally very small, we can consider $Me_{Me} = 1$ and $Xe_{Xe} = 1$.

$$[V'_{Me}] = [V^{\cdot}_{Xe}] = K_S^{1/2} = \exp\left(-\frac{\Delta H_S^* - T\Delta S_S^*}{2RT}\right) = X_S^o \exp\left(-\frac{\Delta H_S^*}{2RT}\right). \quad (1.32)$$

Following similar arguments (see also Eqs. (1.6)–(1.9)), for Frenkel defects, we can write:

$$X_{Me^{\cdot}_i} = X_{Xe'_i} = \exp\left(-\frac{\Delta G_F^*}{2RT}\right) = \exp\left(-\frac{\Delta H_F^* - T\Delta S_F^*}{2RT}\right) = X_F^o \exp\left(-\frac{\Delta H_F^*}{2RT}\right), \quad (1.33)$$

$$K_F = X_{Me^{\cdot}_i} . X_{Xe'_i} = \frac{[Me^{\cdot}_i]}{[Me_{Me}]} . \frac{[Xe'_i]}{[Xe_{Xe}]} = \exp\left(-\frac{\Delta G_F^*}{RT}\right),$$

$$[Me_i^{\cdot}] = [Xe_i^{\prime}] = K_F^{1/2} = X_F^o \exp\left(-\frac{\Delta H_F^*}{2RT}\right) \tag{1.34}$$

where F stands for Frenkel defects.

Defects in Non-stoichiometric Compounds

The defect concentrations in a stoichiometric compound do not depend on the partial pressures. If there is a composition deviation from the stoichiometry, the partial pressure should be considered. As already explained, the composition deviation in non-stoichiometric compounds can be found in four different ways. Considering MeXe compounds, cation-dominant compounds exist because of excess cation ($Me_{1+\delta}Xe$) or shortage of anion ($MeXe_{1-\delta}$). Similarly, anion-dominant compounds exist because of shortage of cation ($Me_{1-\delta}Xe$) or excess of anion ($MeXe_{1+\delta}$).

For clarification, let us consider the example of a metal-oxide (MeO), in which non-stoichiometry is very common. Cation-dominant oxide is possible due to interstitial cations or anion vacancies. Similarly, anion-dominant oxides could exist due to presence of vacant cations or interstitial anions.

A defect reaction equation for a cation-dominant compound with interstitial cations ($Me_{1+\delta}O$) can be written as

$$Me_{Me} + O_o \rightleftharpoons Me_i^{\cdot\cdot} + \frac{1}{2}O_2 \uparrow + 2e^{\prime}. \tag{1.35}$$

Therefore, the reaction constant can be written as

$$K \doteq \frac{[Me_i^{\cdot\cdot}].p_{O_2}^{1/2}.n_e^2}{[Me_{Me}].[O_o]}. \tag{1.36}$$

Enforcing charge neutrality, since two electrons are created for every fully ionized cation with double valency, i.e., the number of electrons is double that of interstitial cations, we can write

$$n_e = 2[Me_i^{\cdot\cdot}]. \tag{1.37}$$

Substituting Eq. (1.37) into Eq. (1.36) and considering small concentration of defects, leading to $[Me_{Me}] = [O_o] \approx 1$, we get

$$K = \frac{n_e}{2}.p_{O_2}^{1/2}.n_e^2 \tag{1.38a}$$

and from here

$$n_e = (2K)^{1/3} p_{O_2}^{-1/6}. \tag{1.38b}$$

Therefore, unlike for stoichiometric compounds, the defect concentrations in non-stoichiometric compounds do depend on the oxygen partial pressure as shown in Fig. 1.11.

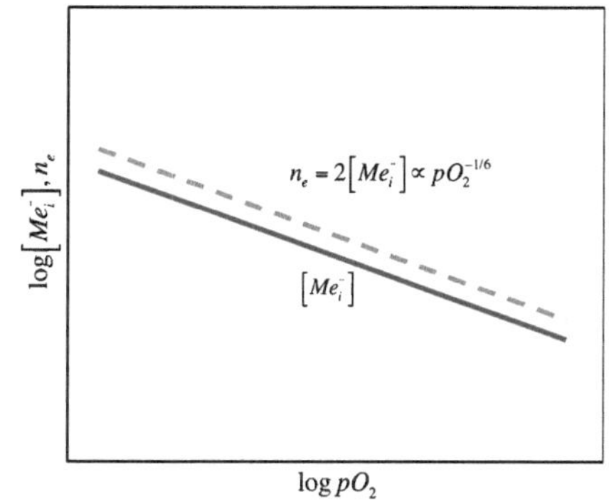

Figure 1.11 Variation of defect concentration with oxygen partial pressure.

Cation-dominant oxide ($MeO_{1-\delta}$) can also be produced by oxygen vacancies. Defect reaction can be written as

$$O_O \rightleftharpoons V_O^{\cdot\cdot} + \frac{1}{2}O_2 \uparrow + 2e'. \tag{1.39}$$

The reaction constant can be taken as

$$K = [V_O^{\cdot\cdot}].p_{O_2}^{1/2}.n_e^2. \tag{1.40}$$

Keeping charge neutrality, we can write

$$n_e = 2[V_O^{\cdot\cdot}]. \tag{1.41}$$

Substituting Eq. (1.41) into Eq. (1.40), we have

$$n_e = (2K)^{1/3} p_{O_2}^{-1/6} \tag{1.42a}$$

and

$$[V_O^{\cdot\cdot}] = \frac{n_e}{2}. \tag{1.42b}$$

Therefore, in both considered cation dominant oxides, concentration of defects decreases with the increase in oxygen partial pressure.

Anion dominant oxides could be of two types, cation deficient ($Me_{1-\delta}O$) or anion excessive ($MeO_{1+\delta}$). The first example, ($Me_{1-\delta}O$), is produced by cation vacancies following the reaction equation

$$\frac{1}{2}O_2 \rightleftharpoons O_O + V_M'' + 2h^{\cdot}. \tag{1.43}$$

Equilibrium constant can be written as

$$K = \frac{[O_o][V''_M].n_h^2}{p_{O_2}^{1/2}}. \tag{1.44}$$

Since $[O_o] \approx 1$ and $n_h = 2[V''_M]$, we can write

$$n_h = (2K)^{1/3} p_{O_2}^{1/6}, \tag{1.45a}$$

$$[V''_M] = \frac{n_h}{2}. \tag{1.45b}$$

For excess oxygen as interstitials ($MeO_{1+\delta}$), the reaction equation can be written as

$$\frac{1}{2}O_2 \rightleftharpoons Me_{Me} + O''_i + 2h^{\cdot}. \tag{1.46}$$

Therefore, the equilibrium constant is

$$K = \frac{[Me_{Me}][O''_i].n_h^2}{p_{O_2}^{1/2}}. \tag{1.47}$$

Since $[Me_{Me}] \approx 1$ and $n_h = 2[O''_i]$, we can write

$$n_h = (2K)^{1/3} p_{O_2}^{1/6}, \tag{1.48a}$$

$$[O''_i] = \frac{n_h}{2}. \tag{1.48b}$$

Therefore, in an oxygen-rich (or metal-deficient) non-stoichiometric oxide, the defect concentration increases with an increase in the oxygen partial pressure.

(ii) Extrinsic Defects in Ionic Compounds

In ionic solids, the addition of solutes extrinsically also leads to the formation of charged defects. Defects again can be substitutional and interstitial in nature. For example, when CaO is added to ZrO_2, there are two possibilities to create defects:

$$\mathrm{CaO} \xrightarrow{\mathrm{ZrO_2}} \mathrm{Ca}''_{\mathrm{Zr}} + \mathrm{O_O} + V^{\cdot\cdot}_{\mathrm{O}},$$

$$2\mathrm{CaO} \xrightarrow{\mathrm{ZrO_2}} \mathrm{Ca}''_{\mathrm{Zr}} + \mathrm{Ca}^{\cdot\cdot}_i + 2\mathrm{O_O}.$$

Following the first reaction, substitutional anion and oxygen vacancy are created, whereas in the second reaction, substitutional anion and interstitial cation are created.

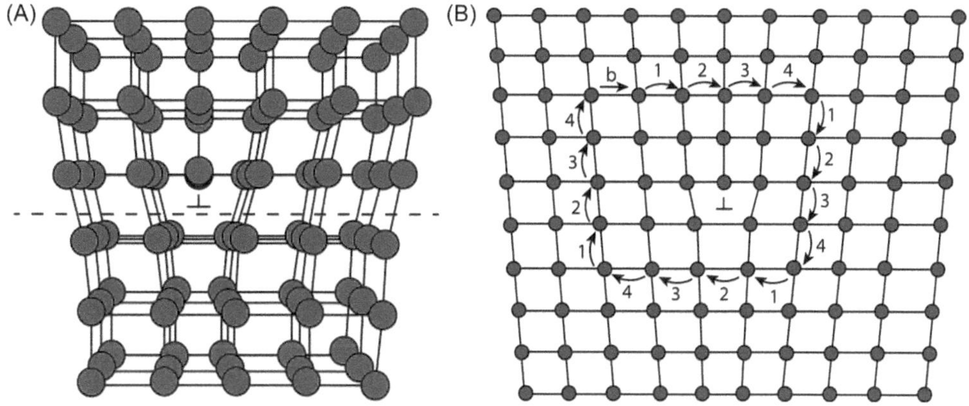

Figure 1.12 (A) Edge dislocation and (B) Burger vector, b.

In general, in a particular system, only one type of defect can be realized. Lattice parameter measurement and density calculation indicate that oxygen vacancies are created due to addition of CaO into ZrO_2 in this substitutional solid solution, following the reaction equations of the first type at 1600°C. At 1800°C, however, there is a change in defect type from oxygen vacancies to cation interstitials with the increase in CaO content beyond a certain concentration [7].

When Y_2O_3 is added to ZrO_2, there are two possibilities:

$$Y_2O_3 \xrightarrow{ZrO_2} 2Y'_{Zr} + 3O_o + V_O^{\cdot\cdot},$$

$$2Y_2O_3 \xrightarrow{ZrO_2} 3Y'_{Zr} + 6O_o + Y_i^{\cdot}.$$

X-ray studies and density calculation indicate that mobile oxygen vacancies are created in this system.

1.1.2 One-Dimensional Line Defects

Dislocations are line defects and generally of two types, edge and screw. The edge dislocations can be visualized by the presence of an extra half-plane, as shown in Fig. 1.12A. Creation of an edge dislocation is illustrated in Fig. 1.13. If force is applied such that atoms in the upper portion of a crystal slip relative to the lower portion, it creates an extra plane. This plane is called the glide (or slip) plane. The line along the edge of the extra plane of atoms is called the dislocation line.

A Burger circuit is drawn to find the Burger vector or slip vector, as shown in Fig. 1.12B. When equally many atoms are counted on each edge around a dislocation, failure to close the loop indicates presence of a dislocation and "b" is the Burger vector.

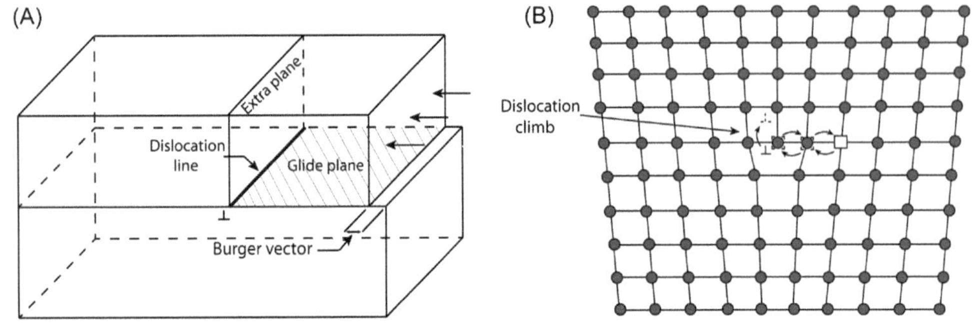

Figure 1.13 (A) Dislocation glide and (B) dislocation climb.

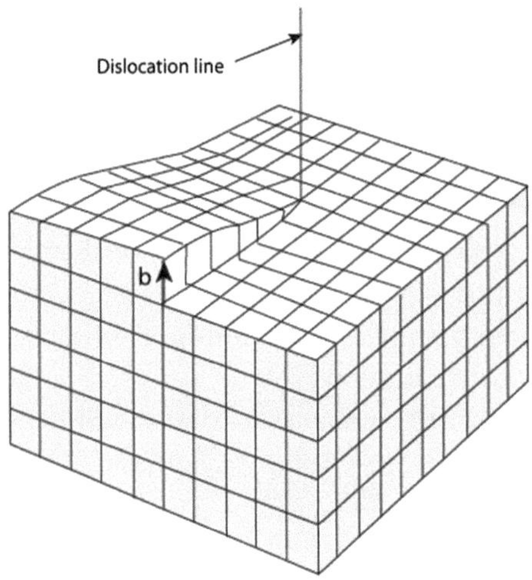

Figure 1.14 Screw dislocation.

In the case of edge dislocation, the dislocation line is perpendicular to the Burger vector. When the dislocation moves along the direction of the Burger vector, it is called dislocation glide. This can be visualized in Fig. 1.13A, which moves towards left with applied force. When dislocation moves up or down, this is called dislocation climb. In fact, the climbing dislocation acts as a vacancy source or sink, depending on the direction of the dislocation climb, as explained in Fig. 1.13B.

Another type of dislocation is known as a screw dislocation, which is shown in Fig. 1.14. In this dislocation, unlike the edge dislocation, the Burger vector is parallel to the dislocation line.

In reality, dislocations are, in general, of a mixed type instead being pure edge or screw. This is illustrated by a schematic diagram, as shown in Fig. 1.15. At one end of the plane above which the locations of the atoms are dislocated, the Burger vector b is

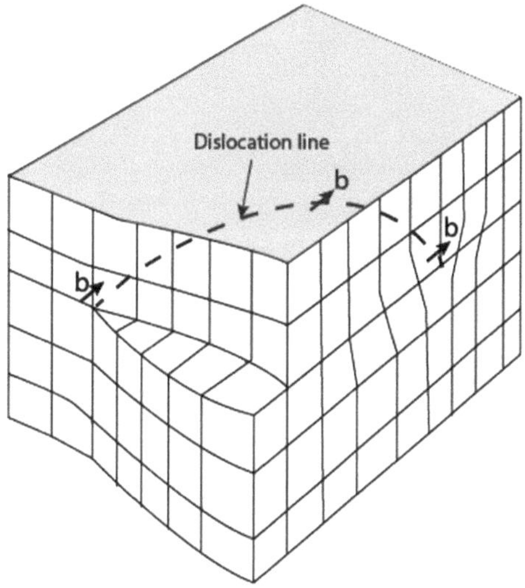

Figure 1.15 Mixed dislocation.

parallel to the dislocation line, and therefore here we have a pure screw dislocation. At the other end, the Burger vector b is perpendicular to the dislocation line, and therefore it is a pure edge dislocation. However, in the middle, it is of a mixed type, with both screw and edge components.

1.1.3 Two-Dimensional Planar Defects

There are mainly two types of two-dimensional defect present in a structure: surfaces and interfaces. A solid body free surface is surrounded by a gas phase or vacuum. There can be different interfaces present. For example, an interface can be created when two different phases such as solid–liquid and two different solids are placed next to each other. This is called an interphase interface (or boundary). An interface can also be created in the same solid phase when two crystals with different orientations are placed together. This is called a grain boundary. These are shown in Fig. 1.16, in which there are two interfaces between different phases such as Nb/Nb_3Sn $Nb_3Sn/Cu(Sn)$ solid solution. At the same time, different grains can be seen inside Nb_3Sn [8].

Interphase boundaries can be of three different types depending on the extent of mismatch of the neighboring lattices: coherent, semi-coherent, and incoherent. In coherent boundary, lattice planes across the interface are continuous, as shown in Fig. 1.17A. In practical examples, the lattice parameters of two different phases are never exactly the same, and therefore are elastically strained when maintaining continuity. This type of interface is found between Ni(Al) solid solution and Ni_3Al intermetallic phases in a Ni-based superalloy. In the absence of lattice plane continuity, incoherent interfaces

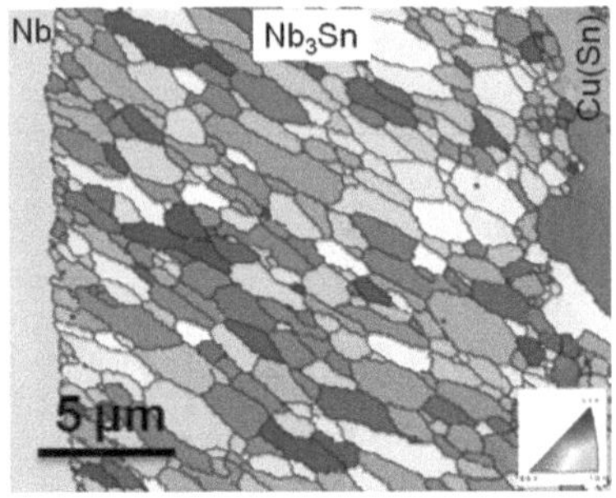

Figure 1.16 Microstructure of an Nb/Cu(Sn) diffusion couple and Nb_3Sn as the product phase.

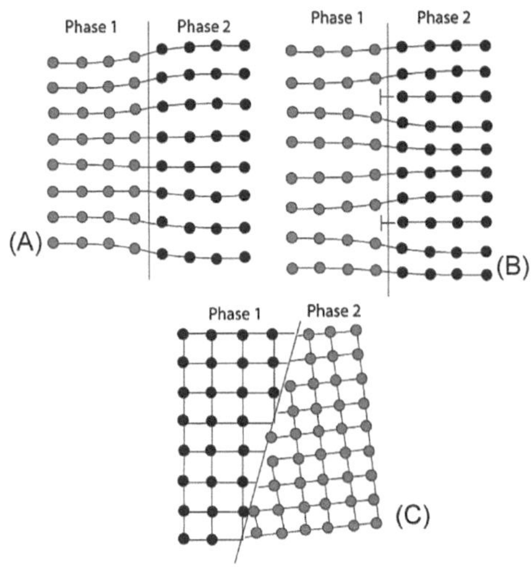

Figure 1.17 (A) Coherent, (B) incoherent, and (C) semi-coherent interfaces.

are found, as shown in Fig. 1.17B. In general, the interface between impurities and the matrix are incoherent. Many systems, however, do not have completely coherent or incoherent interfaces. In the example shown in Fig. 1.17C, both continuous and discontinuous lattice planes are found. Therefore, during substitutional diffusion, creation or annihilation of vacancies can be found in semi-incoherent and incoherent interfaces, but not in coherent interfaces by lattice rearrangements following a similar approach, as already discussed for dislocation climb (Fig. 1.13B).

A grain boundary represents an interface between two crystals. Therefore each grain boundary can be characterized by the following *macroscopic* and *microscopic* parameters (see Fig. 1.18): the matrix of rotation between the two crystalline coordinate systems, $\{x_1, x_2, x_3\}$, e.g., the axis $\vec{\omega}$ and angle θ of rotation (3 parameters); inclination of the GB plane with respect to a chosen direction, $\vec{n}$ (2 parameters), and the matrix of atomic

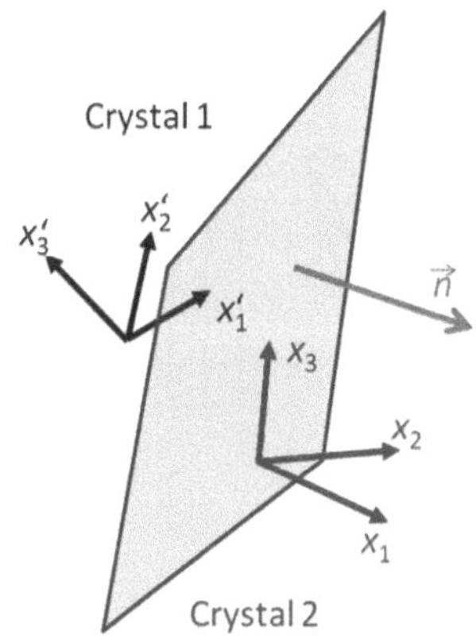

Figure 1.18 Orientations of two neighboring crystals, 1 & 2, and the normal to the grain boundary.

translation $\vec{t} = (t_x, t_y, t_z)$ of one crystal with respect to another. The set of parameters $(\vec{\omega}, \theta, \vec{n})$ can be chosen on demand by setting the crystals' misorientations and properly choosing the required position of the interface. The parameters $\vec{t}$ cannot be influenced, and thus they are called microscopic.

Depending on the given set of the parameters $(\vec{\omega}, \theta, \vec{n})$, one distinguishes:

- Low-angle grain boundary with $\theta \leq 15°$;
- High-angle GBs with $\theta > 15°$;
- Special (low sigma) GBs with special values of $(\vec{\omega}, \theta, \vec{n})$ so that the value of the inverse density of equivalent (coinciding) positions of the two lattices is small, e.g., $\Sigma 5$, $\Sigma 7$, etc. The values of $\Sigma 1$ and $\Sigma 3$ correspond to the low-angle and twin boundaries, respectively;
- Pure tilt GBs with $\vec{\omega} \perp \vec{n}$, i.e., when the misorientation axis lies in the GB plane;
- Pure twist GBs with $\vec{\omega} \parallel \vec{n}$, i.e., when the misorientation axis is perpendicular to the GB plane;
- GBs of mixed type;
- Symmetric and asymmetric GBs.

Some of these grain boundaries are specified below. The tilt and twist boundaries are shown in Fig. 1.19, and θ is the misorientation angle. A low-angle tilt boundary can be defined by a vertical array of edge dislocations as shown in Fig. 1.20. Similarly, a low-angle twist boundary can be visualized by an arrangement of screw dislocations.

High-angle grain boundaries could be understood by the mismatch of the orientations of two different crystals. This cannot be described by the arrays of dislocations. Note here that interfaces, unlike point defects discussed above, are non-equilibrium defects. Therefore, the total area of interfaces decreases with the input of thermal energy, if they are not mechanically or externally stabilized.

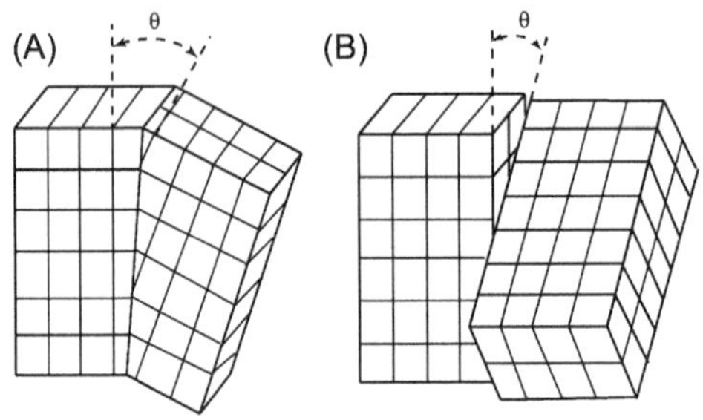

Figure 1.19 (A) Tilt and (B) twist boundaries.

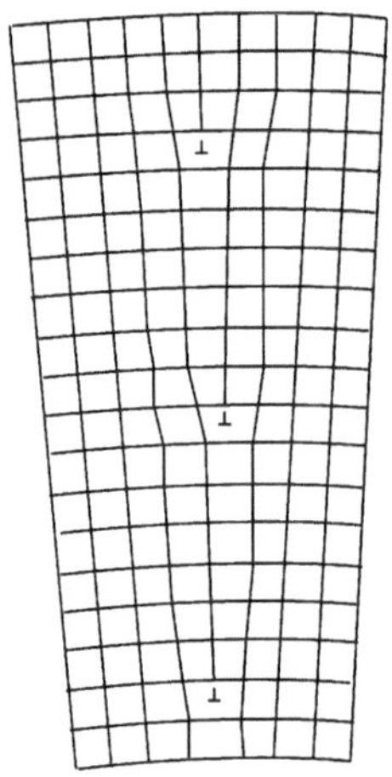

Figure 1.20 Low angle grain boundary.

1.1.4 Three-Dimensional Volume Defects

Volume defects are aggregates of foreign components or vacancies. These can be defined differently depending on whether these are created by a natural process, added intentionally, or unwanted but difficult to avoid. For example, precipitates are produced by solid state reactions because of supersaturation of components in a given condition. Sometimes these are produced intentionally for a beneficial role (precipitation hardening), but sometimes these are produced during service because of excessive use of alloying elements (for example, in Ni-based superalloys) and deleterious to the structure [5]. Sometimes foreign particles are added to enhance the physicomechanical properties; they are called dispersants. When foreign particles are present but unwanted, they are called inclusions, which are deleterious to the material properties. Cracks are often created mechanically; voids are created because of entrapment of air/gas during solidification or supersaturation of vacancies during physical process. These are also undesirable and may lead to failure by degrading the mechanical properties of the structure.

1.2 DRIVING FORCES FOR DIFFUSION

Diffusion may occur in the absence or presence of driving forces. In the absence of any driving forces, components will diffuse randomly, driven by the maximization of the entropy of the system. This is realized, for example, during diffusion of small quantity of radioisotope A* in A. Diffusion between them is just one or two neutron masses. Diffusion occurs even if any entropic driving force is absent, for example, diffusion of component A in a block of the same material. In the latter case, we are not able to follow this process with marked atoms, since a gradient of marked atoms induces an entropic driving force. Still, random atomic jumps could be registered by nuclear resonance methods like perturbed angular correlation; see, e.g., [9]. These methods give information about the atomic jump frequencies, but not the long range diffusion coefficient [10].

In the presence of a driving force, diffusion rate becomes driving force direction-dependent. Generally, many different driving forces can be present in a system. For example, chemical (or thermodynamic) driving force is present in most of material applications because of difference in activity (or chemical potential) of components at different locations of the system. Electromigration or electrotransport is a phenomenon in which an electric field acts as an external driving force. When the driving force is the thermal gradient, the process is called thermomigration or thermotransport. A stress gradient can also affect the diffusion of components. All these are discussed in detail below in different sections.

1.2.1 Thermodynamic Driving Force

Thermodynamic driving force occurs because of a composition variation, which is typically present in most material systems. This can be easily understood from the free energy diagram of a simple binary isomorphous system, as shown in Fig. 1.21. Fig. 1.21A shows a phase diagram of a model A–B binary system. A schematic free energy diagram, i.e., molar Gibbs free energy, G vs. X_B at one particular temperature (marked by T in Figs. 1.21A) in solid state is shown in Fig. 1.21B. At a given composition, X_B^*, the chemical potentials of A, i.e., μ_A^*, and of B, i.e., μ_B^*, can be estimated by taking a slope on the free energy curve and then extending it to $X_B = 0$ and 1, respectively. Subsequently, the activities of the components, a_i, can be determined from the relation $\mu_i^* - \mu_i^o = RT \ln a_i$. Here μ_i^o is the chemical potential of the component in the standard state, i.e., under atmospheric pressure, at 25°C, and with its most stable crystal structure. Now suppose, as shown in Fig. 1.21C, that we construct a diffusion couple out of two alloys with compositions X_B^M and X_B^P. Amounts of these alloys are chosen such that the overall composition of these two blocks sums up to be X_B^N. The schematic representation of the diffusion couple is shown in Fig. 1.21D. It can be seen in the Fig. 1.21C that when alloys M and P are kept together without mixing at temperature T, the free energy of the overall system is $G_N' = X_B^M G_M + X_B^P G_P$. This would

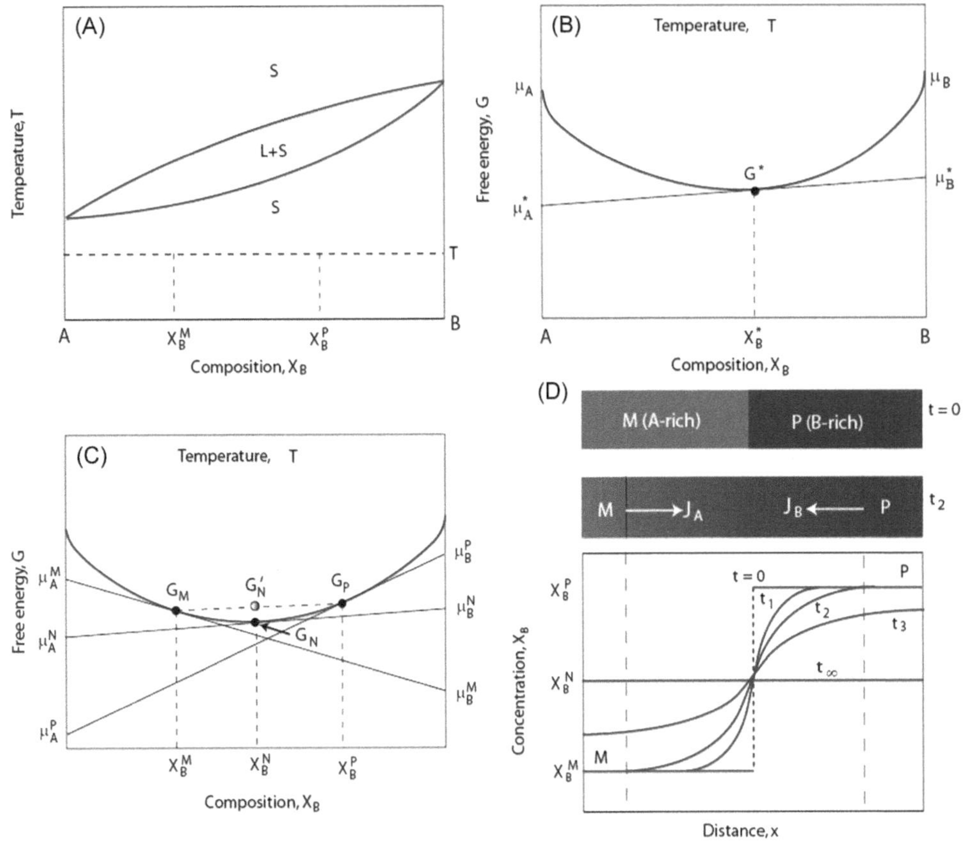

Figure 1.21 Low-angle tilt boundary: (A) an imaginary A–B isomorphous phase diagram, (B) G vs. X_B at temperature T, (C) chemical potentials of components at different compositions, and (D) composition profiles of a diffusion couple at temperature T and annealing time of $t = 0 \rightarrow t_1 \rightarrow t_2 \rightarrow t_\infty$.

correspond to a purely mechanical mixture of M and P without any intermixing. Since the equilibrium free energy of the alloy N, i.e., G_N is lower than G'_N, the system will try to decrease its free energy. This is possible if the alloys M and P are mixed to produce an alloy of composition N. Since the blocks are held together in the solid state, mixing is possible only by solid-state diffusion of the components A and B. From Fig. 1.21C, it can be seen by considering the chemical potentials of A in different alloys that $\mu_A^M > \mu_A^P$. One has μ_A^N in the final equilibrium alloy, which is lower than μ_A^M but higher than μ_A^P. Therefore, components should diffuse in such a way that the chemical potential of A decreases in alloy M and increases in alloy P. The chemical potential of A can be decreased by increasing the mole fraction of B or by decreasing the mole fraction of A. Therefore, alloy M should loose and alloy P should gain component A to increase the chemical potential A. Therefore, component A will diffuse from A-rich M towards A-lean P. Based on the chemical potential of B, diffusion direction of this component is just the opposite, i.e., from B-rich P to B-lean M. This will continue until a single composition alloy N is produced after infinite annealing time. Infinite time, which is the time required to produce a homogeneous alloy, depends on the diffusion rates of the components at a particular temperature and the amount of material in the couple, and is therefore more or less symbolic. What is important is that, due to the chemical driving force, components will always diffuse down their chemical potential gradients. From the phenomenological point of view, diffusing flux of component, i, expressed as J_i (mol/m^2 s) is proportional to the chemical potential gradient

$$J \propto -\frac{d\mu_i}{dx}. \tag{1.49}$$

The negative sign stems from the fact that components diffuse down the chemical potential gradient. If, for the sake of discussion, we consider that chemical potential varies linearly between M and P, then we can write

$$J_B = L_B \frac{\mu_B^M - \mu_B^P}{\Delta x} = -L_B \frac{\mu_B^P - \mu_B^M}{\Delta x} = -L_B \frac{d\mu_B}{dx},$$

$$J_B \propto -\frac{d\mu_B}{dx}, \tag{1.50a}$$

$$J_A = L_A \frac{\mu_A^P - \mu_A^M}{\Delta x} = -L_A \frac{\mu_A^M - \mu_A^P}{\Delta x} = -L_A \frac{d\mu_A}{dx},$$

$$J_A \propto -\frac{d\mu_A}{dx}. \tag{1.50b}$$

Note that $\mu_B^M < \mu_B^P$ and $\mu_A^P < \mu_A^M$. L_A and L_B are the phenomenological constants, and Δx is the length of the diffusion couple considered.

With respect to the Fick's first law of diffusion, the same can be written as

$$J_B = D_B \frac{C_B^M - C_B^P}{\Delta x} = -D_B \frac{C_B^P - C_B^M}{\Delta x} = -D_B \frac{dC_B}{dx},$$

$$J_B \propto -\frac{dC_B}{dx}, \tag{1.51a}$$

$$J_A = D_A \frac{C_A^P - C_A^M}{\Delta x} = -D_A \frac{C_A^M - C_A^P}{\Delta x} = -D_A \frac{dC_A}{dx},$$

$$J_A \propto -\frac{dC_A}{dx}, \tag{1.51b}$$

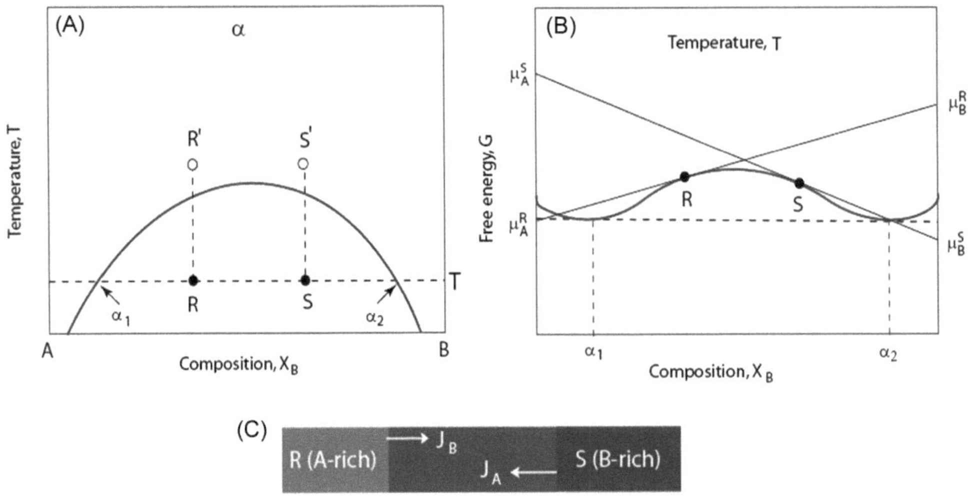

Figure 1.22 (A) An imaginary A–B phase diagram with a miscibility gap, (B) chemical potentials of components in different alloys, and (C) direction of diffusing components.

where D_B and D_A are the diffusion coefficients of components B and A. Since $C_A^M > C_A^P$ and component A diffuses from M to P, it is a normal downhill diffusion with respect to concentration gradient of the component. The same is true for the component B since $C_B^P > C_B^M$ and it diffuses from P to M.

There can be situations where a particular component diffuses against its own concentration gradient. This is called an uphill diffusion. It should be noted that, although components diffuse against their concentration gradient, they always diffuse down their chemical potential gradient, as shown in Fig. 1.22. Let us consider a binary A–B system with a miscibility gap in the lower temperature range, as shown in Fig. 1.22A. If we consider the temperature of interest T, an imaginary free energy diagram across the composition range can be drawn as shown in Fig. 1.22B. Any alloys inside the miscibility gap, for example, R or S, if held at temperature T, will dissociate to produce thermodynamically stable compositions α_1 and α_2 by decreasing the free energy of the system. Suppose that we quench the alloys R' and S' to R and S, i.e., from the miscible to immiscible part of the phase diagram, and then construct a diffusion couple. Interdiffusion of components will then take place between these two blocks, inducing compositional changes in the two alloys as follows. Chemical potential of components A and B can be determined by drawing a tangent on the free energy curve at the composition of interest as shown in Fig. 1.22B. It can be seen that the chemical potential of A is lower in the A-rich alloy R and higher in the A-lean alloy S. Similarly, chemical potential of B is higher in B-lean alloy R and lower in B-rich alloy S. Therefore, A will diffuse from the A-lean S to A-rich R, whereas B will diffuse from the B-lean R to B-rich S. Thus components diffuse against their concentration gradients, and this is called an uphill diffusion, although the direction of diffusing component is always down the chemical potential gradient (Fig. 1.22):

$$J_B = L_B \frac{\mu_B^S - \mu_B^R}{\Delta x} = -L_B \frac{\mu_B^R - \mu_B^S}{\Delta x} = -L_B \frac{d\mu_B}{dx},$$

$$J_B \propto -\frac{d\mu_B}{dx}, \tag{1.52a}$$

$$J_A = L_A \frac{\mu_A^R - \mu_A^S}{\Delta x} = -L_A \frac{\mu_A^S - \mu_A^R}{\Delta x} = -L_A \frac{d\mu_A}{dx},$$

$$J_A \propto -\frac{d\mu_A}{dx}. \tag{1.52b}$$

One important point should be stressed from the relations between the flux and the concentration gradient, namely,

$$J_B = D_B \frac{C_B^S - C_B^R}{\Delta x} = D_B \frac{dC_B}{dx},$$

$$J_B \propto \frac{dC_B}{dx}, \tag{1.53a}$$

$$J_A = D_A \frac{C_A^R - C_A^S}{\Delta x} = D_A \frac{dC_A}{dx},$$

$$J_A \propto \frac{dC_A}{dx}. \tag{1.53b}$$

In the above equations the negative sign is missing since the diffusion occurs against the concentration gradient. In general, the relations for estimating the diffusion parameters are derived following the equation, which considers diffusion of components down the concentration gradient, i.e., $J_i = -D\frac{dC_i}{dx}$. Therefore, in the presence of an uphill diffusion, the diffusion coefficient will have a negative value.

To continue our discussion as shown in Fig. 1.23A, let us consider an A–B binary system, in which, for example, at temperature T, A-rich solid solution phase α, B-rich solid solution phase γ, and one intermediate phase β are present. Now the question is: what do we expect to grow in an interdiffusion zone of a diffusion couple of A and B? To get an answer, first we need to draw the free energy diagram and the chemical potential variation at temperature of interest T, as shown in Figs. 1.23B and 1.23C. It has three single phase regions, α, β, and γ, and two phase regions, $\alpha + \beta$ and $\beta + \gamma$. Note that the variation of μ_B and the variation of μ_A are just the opposite. It can be

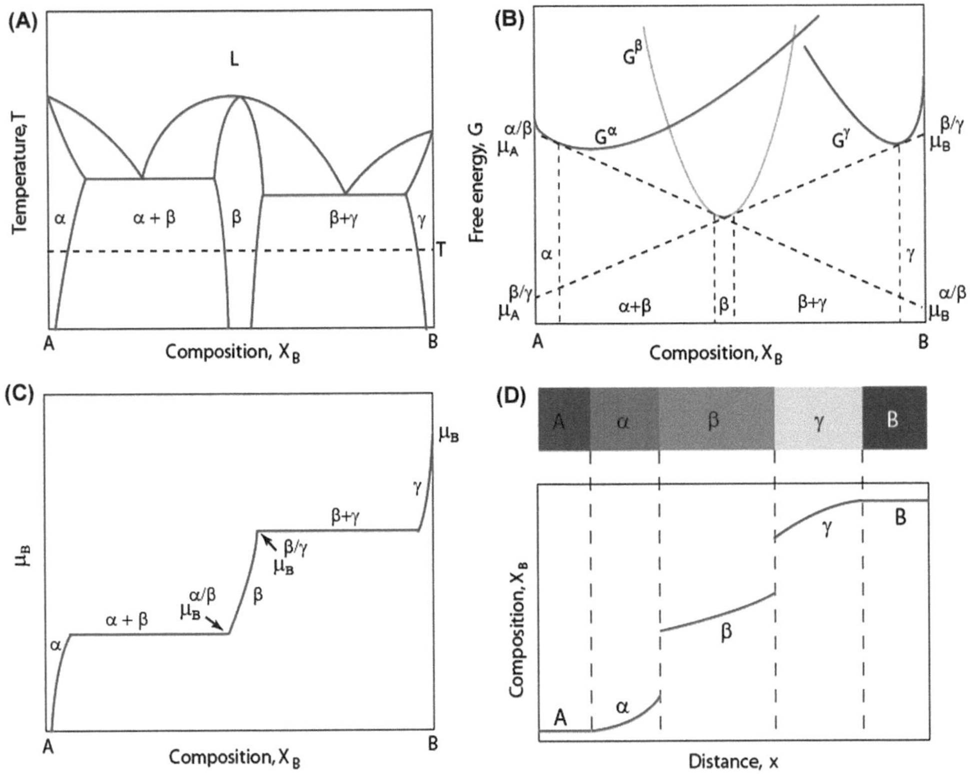

Figure 1.23 (A) An imaginary A–B phase diagram, (B) chemical potentials from a free energy diagram, (C) variation of μ_B with X_B, and (D) diffusion couple of A and B.

seen that inside the two phase region, the chemical potentials of the components are the same and constant so they are in equilibrium. This situation is referred to as a local equilibrium condition. Under this condition, the equilibrium (i.e., equality of chemical potentials of components) is established only at the interface. This type of equilibrium is frequently met in practical applications. On the other hand, chemical potentials vary continuously inside the one phase regions. This can be further extended to understand the evolution of the phases in an interdiffusion zone between pure A and B. Following similar reasoning, as explained for a completely solid solution in Fig. 1.21, α and γ solid solution phases will develop a composition profile up to their compositions at the phase boundaries. However, a phase mixture in a binary system cannot grow in an interdiffusion zone. For example, if $\alpha + \beta$ phase mixture exists in the interdiffusion zone then components cannot diffuse through this layer since the chemical potential for a particular component is the same at different locations inside the phase mixture region and there is no driving force for diffusion. On the other hand, the diffusion couple A–B is not in equilibrium till it reaches the final composition with minimum free energy, and therefore the system will not allow forming a two phase mixture which otherwise will cease the diffusion of components. This is true for any binary system. In a ternary system, because of an extra degree of freedom, a two phase mixture, but not a three phase mixture, can exist inside the reaction zone [1]. In an interdiffusion zone, in fact, chemical potential varies continuously without any jump, but the composition profile can experience a jump because of missing phase mixture, as shown in Fig. 1.23D. The chemical potential diagram also indicates the driving force for the whole diffusion couple or between different interfaces in the reaction zone. Note that chemical potential of B varies from an insignificantly low value at the α-phase to μ_B at the β-phase. The opposite is true for μ_A. These differences in the chemical potentials at two ends of the diffusion couple of A and B are the overall driving forces for components A and B. Note that in Eqs. (1.50), the values of the driving forces for diffusion of A and B (concentration gradients in Eq. (1.51)) depend on the changes in the value of $\frac{d\mu_i}{dx}$ at a particular position as a function time. It should be noted here that the chemical potentials of the components are related by $\sum_{i=0}^{n} X_i d\mu_i = 0$ according to the Gibbs–Duhem equation [1]. We can also obtain the driving forces for the diffusion of components in a particular phase considering the chemical potentials at different interfaces. For example, the difference of $\mu_B^{\beta/\gamma}$ and $\mu_B^{\alpha/\beta}$ is the driving force for B in the β-phase. Similarly, the difference of $\mu_A^{\alpha/\beta}$ and $\mu_A^{\beta/\gamma}$ is the driving force for A in the same phase. Note also that before the growth of β-phase can take place, there must be diffusion of A into γ and of B into α so that the interfacial compositions corresponding to values of $\mu_A^{\alpha/\beta}(\mu_B^{\alpha/\beta})$ and $\mu_B^{\gamma/\beta}(\mu_A^{\gamma/\beta})$ have been reached.

In order to emphasize the importance of chemical potential difference over the composition difference in solid state diffusion, two further situations are considered. Suppose that we form a couple from two alloys, P and Q, as shown in Fig. 1.24. Both alloys have a phase mixture of α and β but with different volume fractions, which can be determined by the Lever rule. Despite the fact that these two alloys have different compositions, the chemical potential of components in the two alloys are the same. Therefore, there will be no interdiffusion between these two alloys. Next suppose that we couple alloys R and S. R is a phase mixture of α and β, and S is a phase mixture of β and γ. Since there is a chemical potential difference between these two alloys, interdiffusion should occur. Since a phase mixture cannot grow in the interdiffusion zone in this binary system, only the β-phase will grow. Note also the continuous change in chemical potential over the interdiffusion zone although there is a composition jump at the α/β- and β/γ-interfaces.

One important aspect should be remembered here. Thermodynamics indicates only the possibility of the growth of different phases in an interdiffusion zone. The thickness of the phases, however, depends on the diffusion kinetics (although it can be influenced by the thermodynamic driving forces). For example, as shown in Fig. 1.25, when V and Si are coupled at 1200°C for 16 h out of four intermetallic compounds in the phase

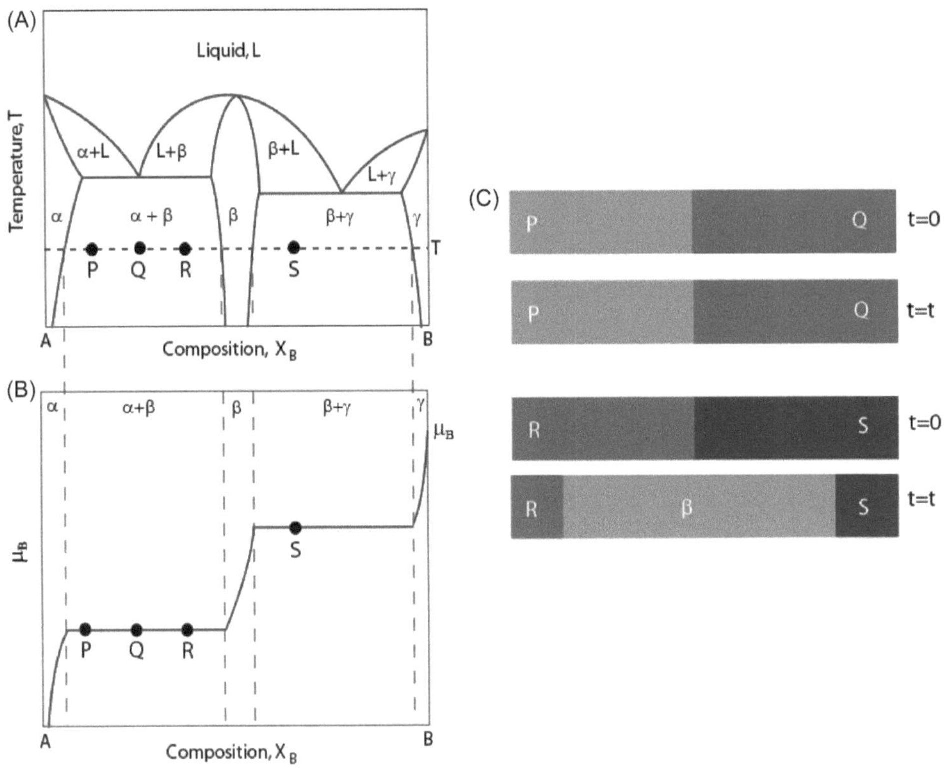

Figure 1.24 (A) An imaginary A–B phase diagram, (B) chemical potential variation, and (C) two different examples of diffusion couples.

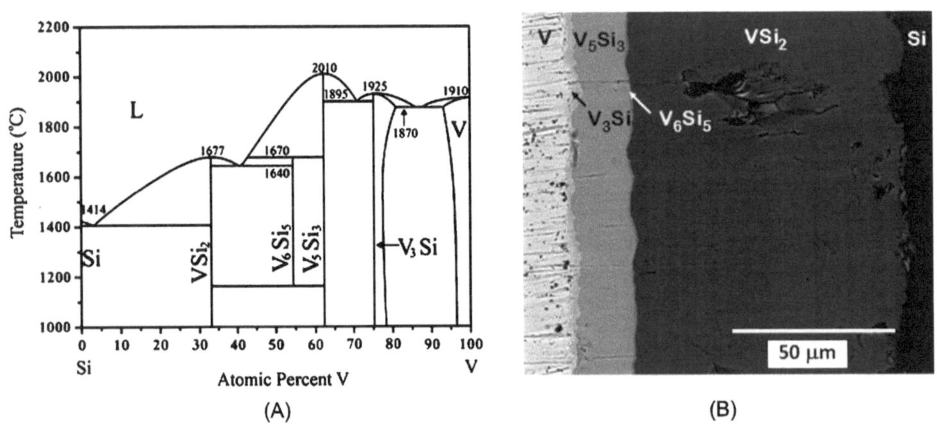

Figure 1.25 (A) V–Si phase diagram [11] and (B) V/Si diffusion couple annealed at 1200°C for 16 h [12].

diagram [11], only VSi_2 and V_5Si_3 are found with significant thickness [12]. On the other hand, V_3Si and V_6Si_5 are found with much lower thickness. V(Si) solid solution phase must also be present as a very thin layer, but it is difficult to resolve with a scanning electron microscope. This indicates that diffusion rates of the components are much lower in V_3Si and V_6Si_5 compared to VSi_2 and V_5Si_3. Regarding this, it should be noted that defect chemistry, which heavily influences the transport of components in ordered intermetallic compounds, may play a much bigger role compared to the thermodynamic driving forces on growth kinetics of the phases [5]. There are also examples where thermodynamic driving forces play a dominating role [13,14]. Another example can be seen with Ta/Si couple annealed at 1250°C for 9 h, as shown in Fig. 1.26 [15]. Although four intermetallics should be found in the interdiffusion zone [16], only two phases, $TaSi_2$ and Ta_5Si_3, can be detected in the interdiffusion zone. By removing Si from the Ta/Si couple, and annealing further another phase, Ta_2Si grows with detectable thickness in the interdiffusion zone, again indicating a much lower growth rate of this phase compared to other detectable phases found in the Ta/Si couple. As explained in Chapter 3 of Volume 1, in an incremental diffusion couple, i.e., when couples are not made out of pure components, the growth kinetics of the phases is higher. Many other examples of such behavior can be found in the literature [1,17,18]. It should also be noted here that we have till now ignored any nucleation issues in our discussions. As shown above, by utilizing the chemical potentials as driving forces, one can rationalize the evolution of the phases in the interdiffusion zone. However, this is true only if the phases do not have difficulty to nucleate and therefore grow simultaneously from the start of annealing of the diffusion couple. In bulk diffusion couples, this is more or less true, except for one or two examples [1]. However, in thin film reaction couples, nucleation issues are not so uncommon because of the presence of very high stress at the interface, hindering nucleation. Sometimes sequential, instead of simultaneous growth of the phases may be observed [19].

1.2.2 Other External Driving Forces

As discussed in the previous section, the thermodynamic driving force results from the chemical potential differences at different locations inside an inhomogeneous material system. There are also frequent examples of the external driving forces influencing the diffusion of components. One of the most commonly encountered is the electrotransport or electromigration. In a pure material, in the absence of any external driving forces, components diffuse because of thermal vibration. However, it has to go through a barrier (activation barrier) created by the surrounding components. In the middle of the jump distance, from one saddle point to another, it needs maximum energy. At this position, the diffusing component is said to be at the activated state. Components at the activated state are heavily influenced by the flow of electrons. Therefore, the force created by the flow of electrons, known as electron wind, influences the transport rate

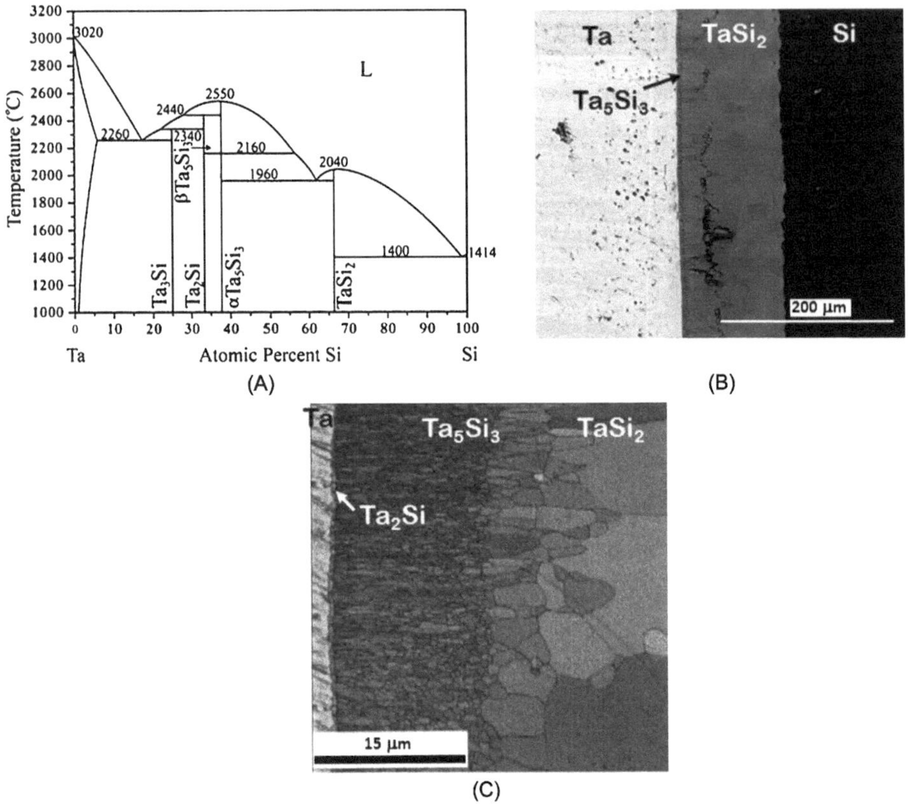

Figure 1.26 (A) Ta/Si phase diagram [16], (B) Ta/Si diffusion couple annealed at 1250°C for 9 h, and (C) Ta/$TaSi_2$ diffusion couple annealed at 1350°C for 9 h [15].

of components. There is also another component, called the direct field effect, that will influence the movement of the atoms under electromigration conditions, but it is typically of minor importance in comparison to electron wind. Electrotransport becomes important when current density exceeds the range of 10^4–10^6 A/cm^2 depending on the material. In a material, when both chemical and electrical potential gradients are present, the flux with respect to the phenomenological constants in Eq. (1.49) is modified to be

$$J_i = -L_i \frac{d\mu_i}{dx} - L_e \frac{d\varphi}{dx}, \tag{1.54}$$

where L_e is the phenomenological constant for transport due to electron flow and φ is the electric potential.

In a single component metallic film used in integrated circuits, electrotransport occurs by vacancy mediated diffusion. Therefore, voids can grow at one location because

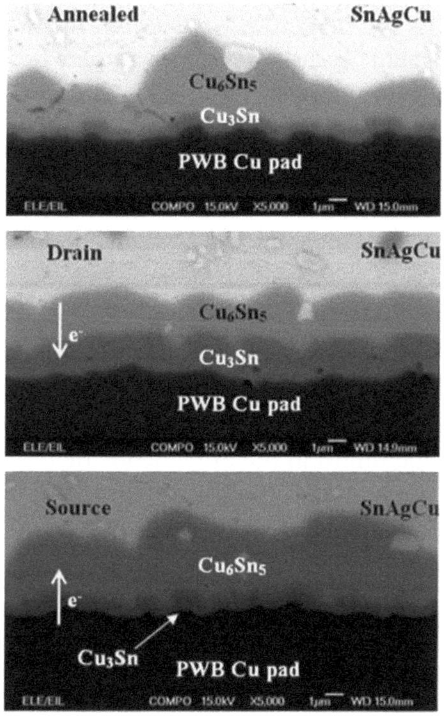

Figure 1.27 Growth of phases in Cu/Sn system depending on the direction of flow of electrons [19].

of loss of components by vacancy supersaturation and protrusions at another location because of accumulation of material. Coalescence of voids may lead to discontinuity of the electron flow. Likewise extrusions may lead to short-circuit failures. Electron flow affects the intrinsic fluxes of diffusing components differently in an interdiffusion zone depending on the direction of the flow of electrons and diffusion of components. If the directions are the same then electron flow will enhance, whereas, if they are opposite, it will hinder the diffusion rate of components. The effect of electric current on diffusion controlled growth of the phases in a diffusion couple of Cu/SnAgCu-solder can be seen in Fig. 1.27 [20]. Because of chemical potential gradient, Cu diffuses towards solder, whereas Sn diffuses towards Cu. In the absence of any current (annealed at 110°C for 750 h), the thickness of Cu_3Sn, (Δx_{Cu_3Sn}) is less than that of Cu_6Sn_5, $(\Delta x_{Cu_6Sn_5})$ leading to $\frac{\Delta x_{Cu_6Sn_5}}{\Delta x_{Cu_3Sn}} \approx 3.8$. When electron flow is in the direction of diffusion of Sn, the thickness of both layers becomes comparable resulting in $\frac{\Delta x_{Cu_6Sn_5}}{\Delta x_{Cu_3Sn}} \approx 1.1$. When the electron flow is opposite (in the direction of flow of Cu), the overall thickness of the interdiffusion zone increases a bit compared to other two conditions with the thickness ratio of $\frac{\Delta x_{Cu_6Sn_5}}{\Delta x_{Cu_3Sn}} \approx 4.2$. The above short discussion has touched only the changes in the intrinsic fluxes of the diffusing components. However, the growth process by reaction and dissociation of components (as in this case) is complex and can be understood only by utilizing the physicochemical models [20], which are introduced in Chapters 3 and 4 of Volume 2 of this handbook.

In an ionic material, the origin of electrotransport is different than in metallic systems. In a metallic system, as discussed above, current is carried because of the electron flow. On the other hand, in an ionic solid, current arises because of flow of ions, i.e., diffusion of ionic components. Therefore, conductivity of these materials depends on the diffusion of components mediated by the defects. It should be noted that the concentration of these defects depends heavily on the presence of impurities and their relative valences with respect to host atoms, and so forth.

In thermotransport or thermomigration, a temperature gradient leads to demixing of a homogeneous alloy and diffusion of components. In a material when both chemical and thermal gradients are present, the flux with respect to the phenomenological constants in Eq. (1.49) is modified to be

$$J_i = -L_i \frac{d\mu_i}{dx} - L_T \frac{dT}{dx}, \tag{1.55}$$

where L_T is the phenomenological constant for transport due to temperature gradient. In an interstitial alloy, the interstitial atom could jump either up or down the temperature gradient depending on the type of interstitial and host components. In a vacancy mediated substitutional alloy, diffusion process is influenced additionally because of difference in vacancy concentration at different temperatures.

Diffusion is also influenced in the presence of stress applied externally or present internally because of presence of dislocations or cracks under load. In thin films, interfacial stress is generated during deposition and greatly influences the nucleation and diffusion.

1.3 DEFINITIONS OF DIFFERENT TYPES OF DIFFUSIONS

In a given material system, as discussed above, diffusion of components can take place in the absence or presence of driving forces. The atoms will diffuse in their elemental or ionic form through the lattice, along dislocations, via grain boundaries or free surface. Depending on various conditions, different definitions of diffusion coefficients are introduced to recognize the type of diffusion. In this section, we will introduce the definitions and concepts behind different terms starting from the very basics, i.e., self-diffusion.

Self-diffusion is the diffusion of components by itself just because of thermal vibration and in the absence of any other internal or external driving forces. It means that

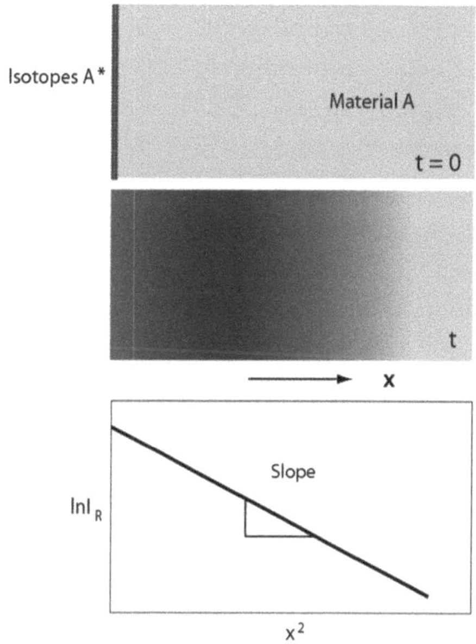

Figure 1.28 Schematic representation of the measurement of the tracer diffusion coefficient.

components jump just because of presence of thermal activation at a given temperature. Self-diffusion coefficients (D_S) should be reasonably high at elevated temperatures. However, they are not easy to detect or follow. For example, the diffusion rate of C in γ-Fe is on the order of 10^{-10} m^2/s at 1100°C. With respect to jump frequency (Γ in s^{-1}) and jump length (Δx in m), we can write $D_S \propto \Gamma \Delta x^2 = g\Gamma \Delta x^2$, where g is the geometrical factor that depends on the crystal structure [1]. If we ignore g and consider typical jump length on the order of 10^{-10} m (1 Å), then C changes position with the rate of 10^{10} times per second! We don't have any technique that can follow such an astronomical number of jumps in just one second. Therefore, we do not have any technique to measure the self-diffusion coefficient directly by measuring the number of jumps in a given time in an arbitrary case. Another, more fundamental, problem is that we have no means to identify the diffusing atoms as they are all alike. One existing possibility, for example, is the usage of perturbed angular correlation (PAC) technique which is sensitive to single jumps of specific nucleus between lattice positions with different fine crystalline fields. For example, the jump frequency of Cd isotopes in $LnIn_3$ could be measured [9], without measurement of long-range diffusion, examining if these jumps are successive or if they represent reverse jumps between two positions!

To circumvent this problem with respect to long-range diffusion, the concept of the tracer diffusion coefficient is established. From definition, one can understand that special components are used to trace and measure the diffusion coefficients of other components. In general, these are radioisotopes, which could be detected even in very small quantities because of emission of different radioactive rays. This is explained with the help of a schematic diagram in Fig. 1.28. For example, radioisotopes A* are deposited on a block of component A to determine the self-diffusion coefficient of A by measuring its tracer diffusion coefficient, D^*. As it will be discussed next, these two values are not necessarily the same and depend on whether we consider substitutional diffusion or interstitial impurity diffusion with very low concentration of interstitial impurities. When the impurity diffusion coefficient is measured by using a radiotracer of impurity components (for example, B^* in material A) by substitutional or interstitial diffusion, it is called an impurity diffusion coefficient, D^*_{imp}. After deposition, the block is annealed at one particular temperature for a desired period of time. Then sectioning is done from the top of the block and the side at which radioisotopes were deposited, and concentrations of the radioisotopes at different depths are measured. This way one can measure the relative intensity of the radioisotopes at different depths with respect to the intensity before annealing. The thin film solution of the Fick's second law can be written as [1]

$$I_R(x) = \frac{1}{\sqrt{\pi D^*_A t}} \exp\left(-\frac{x^2}{4D^*_A t}\right), \tag{1.56a}$$

$$\ln I_R = \ln\left(\frac{1}{\sqrt{\pi D^*_A t}}\right) - \frac{x^2}{4D^*_A t}, \tag{1.56b}$$

where I_R is the relative intensity, x is the position parameter, and t is the annealing time. After measuring the relative intensities at different locations in the diffusion zone, D^*_A can be measured from the slope $(-\frac{1}{4D^*_A t})$ of $\ln I_R$ vs. x^2 plot using known annealing time.

To understand the exact difference between the self- and the tracer diffusion coefficients, let us first discuss the vacancy mediated substitutional diffusion. Fig. 1.29 shows a schematic diagram of substitutional diffusion of a component, say, A, by the vacancy mechanism. The atomic arrangement is for (111) plane in an FCC crystal. In substitutional diffusion, a component jumps to a vacancy lattice site available in the neighboring position. This can also be seen as exchange of places between the component and vacancy. Concentrations of radioisotopes used and vacancies present in the material are very low and not necessarily always placed next to each other. Suppose that at a certain point radioisotope A* and vacancy V are placed next to each other and they exchange their positions. Note here that all other components surrounding the vacancy have practically equal probabilities to exchange positions since the effect of mass difference of radiotracer and other components is very small [1]. Then the question is what would be the next jump position for the radioisotope. The next most probable jump, of course, is the reverse jump. This means that two successive jumps cancel the

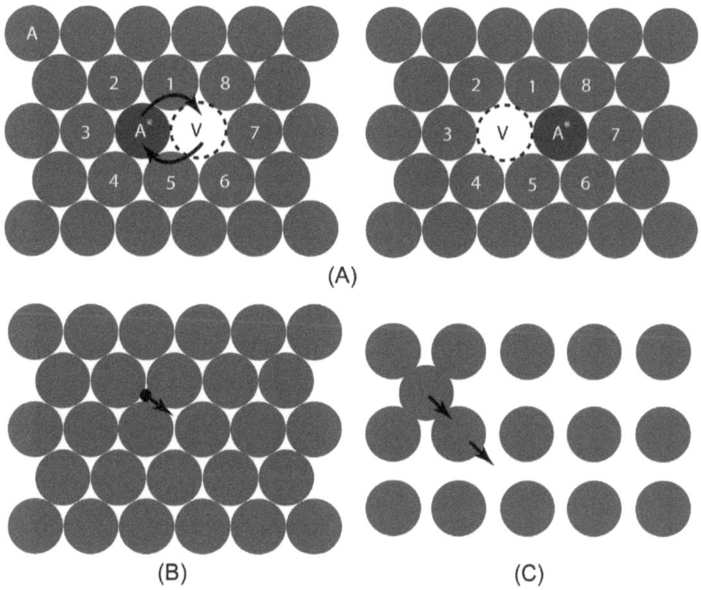

Figure 1.29 (A) Vacancy mediated substitutional diffusion, (B) interstitial diffusion, and (C) interstitialcy diffusion.

diffusion or transport of the radioisotope. However, the experimental results do indicate that radioisotopes make successful jumps and go forward, as shown in the schematic diagram of Fig. 1.28. Therefore, there should be other jump options. The next most probable jump for the radioisotope is to positions 1 and 5 under the condition that the vacancy has already exchanged places and reached those positions. Jumping to positions 6 and 8 has lower probability. Jumping to position 7 has the lowest probability because a vacancy needs to make a few jumps before it arrives to this position and makes the exchange between the radioisotope and vacancy possible. Note here that the jump to positions 6, 7, and 8 will help the radioisotope to go forward. Therefore, the jumps of the radioisotopes in substitutional diffusion are correlated and depend on jumps of vacancies. On the other hand, vacancy is free to exchange position with any of the host atoms or radioisotopes. Therefore, a single vacancy contributes to the diffusion of many components. This means that vacancy or the host components will go through a random jump. Thus, the diffusion of vacancies can be defined as self-diffusion. It must also be clear now that the tracer diffusion coefficient must be less than the self-diffusion coefficient. The two are related by [1]

$$D^* = fD_s, \tag{1.57a}$$

$$f = \lim_{n\to\infty}\left(1 + \frac{2}{n}\overline{\sum_{j=1}^{n-1}\sum_{i=1}^{n-j}\langle\cos\theta_{i,i+j}\rangle}\right), \tag{1.57b}$$

where f is the correlation factor. $n \to \infty$ means that infinitely many jumps are considered for derivation of this relation, i and j are the different steps of the jumps. Readers are referred to [1] for more details. The estimation of the correlation factor is not very easy to visualize. However, the subsequent simple discussion may give better understanding even though it is a very rough estimation.

Here we would like to draw attention to the fact that a combination of two basically different measurements, of tracer diffusion (e.g., with radiotracer technique) and that of jump frequencies (applying PAC method), will provide access to this fundamental quantity, the correlation factor. So far, no such measurements were undertaken.

Following the same reasoning as above, the probability that the tracer atom first exchanges position with vacancy is $\frac{1}{Z}$, where Z is the coordination number. The probability of the opposite jump immediately is again $\frac{1}{Z}$. Therefore, a rough estimate of the probability of successful jump, i.e., the correlation factor considering the two successive jumps not resulting in diffusion of components, can be written as

$$f = 1 - \frac{2}{Z}. \tag{1.58}$$

Probabilities are added since these two successive jumps are not related to each other and are separate events. This is a very rough estimate since there are many other possibilities for the tracer atom to come back to the same position after making a few jumps. However, that number is much smaller because of random jumps of the components. Table 1.3 gives the values of the rough estimates and actual values of the correlation factors for different crystal structures. It can be noticed that the correlation factor increases with the increase in the coordination number since there are more possibilities to avoid two successive opposite jumps. At the same time the exact value of correlation factor increases not so strongly with Z as it would follow from the rough estimation, since there are increasingly many possibilities for the component to come back to the same location following different random routes, if the coordination factor is larger. Note that in the simple approximation, Eq. (1.58), we considered all jumps which did not result in an immediate exchange of the vacancy with the chosen atom, as *escaping* jumps. In reality, if a vacancy performs many random jumps around a chosen atom and the total escaping frequency [10] is smaller since the same vacancy may return after several jumps around the chosen atom.

From the discussion above it should be clear that the self- and tracer diffusion coefficients are not the same for substitutional diffusion. However, they are the same in the case of interstitial impurity diffusion of components such as H, C, and N in a host material with a small quantity of these impurities, as shown by black dots in Fig. 1.29B.

Table 1.3 Roughly estimated $(1-\frac{2}{Z})$ and actual values, f, of the correlation factor for different crystal structures

Crystal structure	Z	$1-\frac{2}{Z}$	f
Simple cubic	6	0.6667	0.65549
Body centered cubic	8	0.75	0.72149
Face centered cubic	12	0.8333	0.78145

This can be understood from a simple discussion based on typical concentration of these defects in a dilute alloy. For example, in α-Fe, the equilibrium concentration of C at 727°C is around 10^{-3}, i.e., one C for every 1000 Fe. This means that in a cube of 10 Fe atoms on each side only one C is present on average. Therefore, most of the interstitial voids are free for C to transport and experience random jumps. This is similar to the jumps of vacancies in substitutional diffusion. Therefore, the tracer and self-diffusion coefficients are equal, with the correlation factor being 1 for the interstitial impurity diffusion. As shown in Fig. 1.29C, there are other less common diffusion mechanisms in which host atoms can diffuse via the interstitial positions, which is called interstitialcy diffusion mechanism.

Until now, we discussed the diffusion of components in the absence of any driving forces. In most of the practical examples, as discussed above, diffusion occurs under the influence of thermodynamic driving forces. The diffusion coefficient of the component moving under the driving force is called the intrinsic diffusion coefficient and expressed as D_i for component i. These are related to the tracer diffusion coefficients in a binary A–B system as [21]:

$$D_A = \frac{v_m}{\overline{v}_B} D_A^* \emptyset (1 + W_A), \tag{1.59a}$$

$$D_B = \frac{v_m}{\overline{v}_A} D_B^* \emptyset (1 - W_B), \tag{1.59b}$$

$$W_i = \frac{2X_i(D_A^* - D_B^*)}{M_o(X_A D_A^* + X_B D_B^*)}, \tag{1.59c}$$

where W_i is the vacancy wind effect which can be neglected in many instances by taking it to be zero. v_m is the molar volume, and $\overline{v}_i$ is the partial molar volume of component i. At a particular mole fraction, these are related by $v_m = X_A\overline{v}_A + X_B\overline{v}_B$. M_o is the structure factor which depends on the crystal structure [21]. $\emptyset = \frac{d \ln a_A}{d \ln N_A} = \frac{d \ln a_B}{d \ln N_B}$ is the thermodynamic factor accounting for the influence of the thermodynamic driving force. This is the same for both components in a binary system, according to the Gibbs–Duhem equation, if vacancy equilibrium is maintained. The relations are much more complex in a ternary system, as can be read in [1].

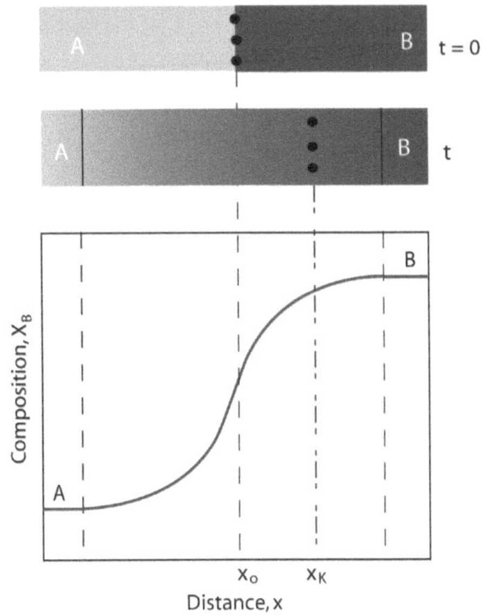

Figure 1.30 Composition profile in a diffusion couple prepared with inert markers indicating the location of the Kirkendall plane (x_K).

As already explained, the tracer diffusion coefficient can be measured directly from the radiotracer technique, as explained in Chapter 2 of Volume 1 in more detail. If thermodynamic driving forces are known, then one can indirectly estimate the intrinsic diffusion coefficients. On the other hand, the diffusion couple technique is used to estimate the intrinsic diffusion coefficients directly, as explained in detail in Chapter 3 of Volume 1. As shown in Fig. 1.30, two blocks with different compositions are coupled at an elevated temperature of interest for desired annealing time to facilitate diffusion of components. From the composition profile developed in the interdiffusion zone, one can directly measure the interdiffusion coefficient, $\tilde{D}$, at any composition, because of mixing of components, which is a kind of average of intrinsic diffusion coefficients expressed as

$$\tilde{D} = C_B \overline{v}_B D_A + C_A \overline{v}_A D_B, \tag{1.60}$$

where $C_i = \frac{X_i}{V_m}$ is the concentration of component i. The experimental methods to estimate the interdiffusion coefficients in binary and multicomponent systems are explained in Chapter 3 and Chapter 4 of Volume 1. After plotting $\tilde{D}$ vs. X_B and then extending it to $X_B = 0$ and 1, one can indirectly estimate the impurity diffusion coefficient of B in A ($D^B_{imp,A}$) and that of A in B ($D^A_{imp,B}$). To estimate the intrinsic diffusion coefficients, a Kirkendall marker experiment should be designed, in which inert particles are used at the initial bonding interface to examine the relative movements of these markers depending on the relative mobilities of the components. Then the intrinsic diffusion coefficients can be measured at Kirkendall marker plane, x_K, only, since this is the only plane detected by using inert markers, which moves from the annealing time $t=0$ from the initial contact or Matano plane, x_o.

Transport of components, i.e., self- or intrinsic diffusion, can happen via different modes such as lattice, dislocations, grain boundaries, and free surfaces. Diffusion via lattice (D_l) is facilitated by the presence of point defects, for example, vacancies in substitutional alloys, and by the presence of impurities in the interstitial diffusion. In intermetallics, diffusion is assisted additionally by the presence of antisites. In a polycrystalline material, grain boundaries or dislocations are always present, and when transport of components is measured, for example, through the grain boundaries, it is called the grain boundary diffusion coefficient, D_{gb}. In a polycrystalline material, diffusion is typically always found from both – lattice diffusion through the inside of the grain and grain boundary diffusion through the grain boundaries.

Accurate analysis of diffusion enhancement caused by short-circuits, i.e., by grain boundaries or dislocations, is not a simple task, and the situation crucially depends on the corresponding kinetic regime; for details see Chapter 2 in Volume 1. For example, if diffusion is considered at relatively high temperatures, when atoms have enough time to explore lattice between short-circuits (formally, $\sqrt{D_l t} > d$ where d is the average distance between short-circuits and t is the diffusion time), diffusion proceeds in the A regime, and we actually measure an apparent diffusion coefficient, D_{app}. In the case of grain boundary diffusion, the apparent (effective) diffusion coefficient will be expressed as

$$D_{app} = (1-\delta)D_l + \delta D_{\text{gb}} \approx D_l + \delta D_{\text{gb}}, \tag{1.61}$$

since the volume fraction of grain boundary, δ, is in general much lower than the volume fraction lattice, $\delta << 1$. It should be noted that in metallic materials the grain boundary diffusion coefficient is typically larger than the lattice diffusion coefficient because of the more open structure of the grain boundaries. However, the overall diffusion process might be lattice or grain boundary diffusion, controlled depending on the temperature at which the experiments are conducted and the relative concentration of the defects present, as shown in Fig. 1.31 (here and below, the analysis is done for the A regime case). At relatively high temperatures, the concentration of point defects is high enough, and the diffusion will be lattice diffusion controlled. At a lower temperature, concentration of point defects decreases exponentially, yielding a grain boundary diffusion controlled process. Therefore, the transition from lattice and grain boundary diffusion depends on the volume fraction of the grain boundaries, as shown in the figure considering two different volume fractions, $\delta_2 > \delta_1$. In a nanocrystalline material, due to a very high volume fraction, diffusion process is always controlled by the grain bound-

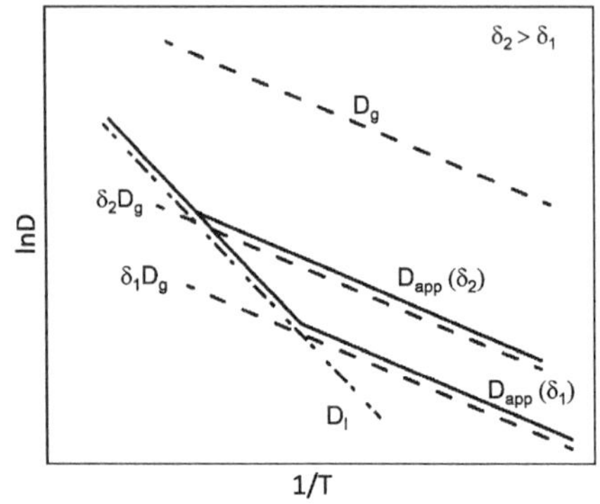

Figure 1.31 Change in diffusion controlled process from lattice to grain boundary diffusion depending on the volume fraction of grain boundaries.

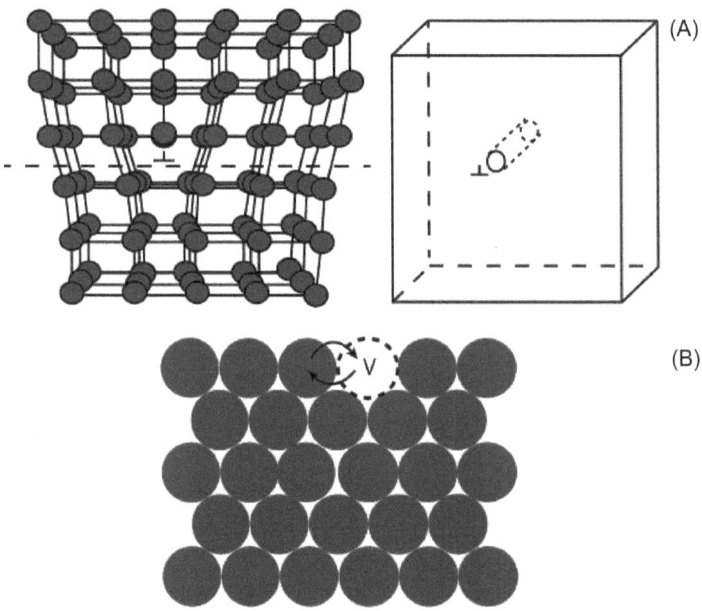

Figure 1.32 (A) Pipe diffusion through an edge dislocation and (B) surface diffusion.

ary diffusion. This discussion is not that straightforward, and actual in-depth analysis and discussion can be read in Chapter 2 of Volume 1.

As shown in Fig. 1.32A, when diffusion occurs via the edge dislocation, it is called pipe diffusion, since it appears as the components are diffusing via a pipe. When diffusion occurs via a free surface, surface of a crack or a void it is called surface diffusion, which is illustrated in Fig. 1.32B. Surface diffusion is the easiest because constraints from neighboring components are induced only from one side.

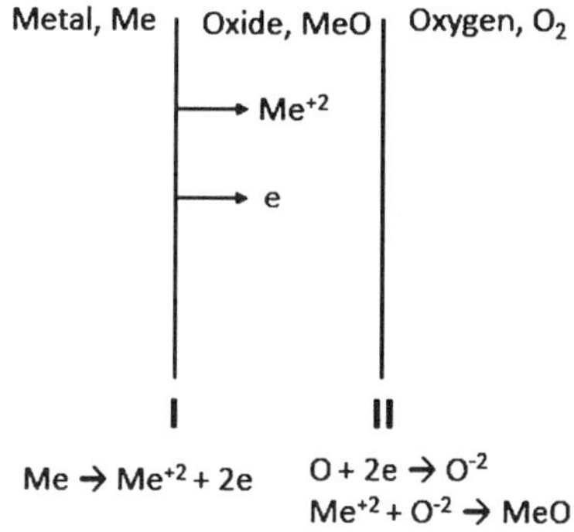

Figure 1.33 Coupled diffusion in an ionic solid.

Almost all the discussions above are true also for oxide or ionic solids. However, additional complexity comes from the requirement of charge neutrality, leading to transport of electronic defects, i.e., electrons or holes along with components, making it coupled diffusion. This is explained for an oxide layer growth on a metal as shown in Fig. 1.33. The metal component is first dissociated to metal ion (Me^{+2}) and electrons (e) at the Me/MeO interface. Then, these two diffuse through the oxide layer separately to reach the MeO/O_2 interface. At this interface, O_2 converts to O^{-2} by accepting the dissociated electrons from the metal component and produces an oxide (MeO) by reaction with Me^{+2}. In many oxides, the diffusion of oxygen is slow compared to that of metal. However, the diffusion of this anion can be explained in a similar fashion. It is apparent that the diffusion rates of electronic defects (electrons or holes) are much greater than the diffusion rate of the ionic components. Therefore, the growth of the oxide layer is limited by the diffusion of the slower component.

REFERENCES

[1] A. Paul, T. Laurila, V. Vuorinen, S. Divinski, Thermodynamics, Diffusion and the Kirkendall Effect in Solids, Springer, Heidelberg, 2014.

[2] R. Abbaschian, L. Abbaschian, R.E. Reed-Hill, Physical Metallurgy Principles, 4th edition, Wadsworth Publishing Co Inc, UK, 2008.

[3] K.A. Marino, Emily A. Carter, Ni and Al diffusion in Ni-rich NiAl and the effect of Pt additions, Intermetallics 18 (2010) 1470–1479.

[4] P. Kiruthika, A. Paul, A pseudo-binary interdiffusion study in the β-Ni(Pt)Al phase, Philosophical Magazine Letters 95 (2015) 138–144.

[5] P. Kiruthika, S.K. Makineni, C. Srivastava, K. Chattopadhyay, A. Paul, Growth mechanism of the interdiffusion zone between platinum modified bond coats and single crystal superalloys, Acta Materialia 105 (2016) 438–448.

[6] W.D. Kingery, H.K. Bowen, Donald R. Uhlmann, Introduction to Ceramics, 2nd edition, Wiley-Interscience, USA, 1976.

[7] A. Diness, R. Roy, Experimental confirmation of major change of defect type with temperature and composition in ionic solids, Solid State Communications 3 (1965) 123–125.
[8] Sangeeta Santra, PhD thesis, Indian Institute of Science, Bangalore, India, 2016.
[9] G.S. Collins, A. Favrot, L. Kang, E.R. Nieuwenhuis, D. Solodovnikov, J. Wang, M.O. Zacate, Hyperfine Interactions 159 (2005) 1–8.
[10] H. Mehrer, Diffusion in Solids, Springer, Heidelberg, Germany, 2007.
[11] J.F. Smith, The Si–V (silicon–vanadium) system: addendum, Bulletin of Alloy Phase Diagrams 6 (1985) 266–271.
[12] S. Prasad, A. Paul, Reactive diffusion between vanadium and silicon, Journal of Phase Equilibria and Diffusion 32 (2011) 212–218.
[13] T. Laurila, V. Vuorinen, A.K. Kumar, A. Paul, Diffusion and growth mechanism of Nb_3Sn superconductor grown by bronze technique, Applied Physics Letters 96 (2010) 2319101-3.
[14] S. Santra, A. Paul, Effect of Ga content in Cu(Ga) on the growth of V_3Ga following bronze technique, Intermetallics 70 (2016) 1–6.
[15] S. Roy, A. Paul, Growth mechanism of tantalum silicides by interdiffusion, Philosophical Magazine 92 (2012) 4215–4229.
[16] M.E. Schlesinger, The Si–Ta (silicon–tantalum) system, Journal of Phase Equilibria 15 (1994) 90–95.
[17] A. Paul, C. Ghosh, W.J. Boettinger, Diffusion parameters and growth mechanism of phases in the Cu–Sn system, Metallurgical and Materials Transactions A 42A (2011) 952–963.
[18] S. Prasad, A. Paul, Growth mechanism of phases by interdiffusion and diffusion of species in the Nb–Si system, Acta Materialia 59 (2011) 1577–1585.
[19] K.N. Tu, G. Ottaviani, U. Gösele, H. Föll, Intermetallic compound formation in thin-film and in bulk samples of the Ni–Si binary system, Journal of Applied Physics 54 (1983) 758–763.
[20] T. Laurila, J. Karppinen, V. Vuorinen, A. Paul, M. Paulasto-Kröckel, Effect of constant and cyclic current stressing on the evolution of intermetallic compound layers, Journal of Electronic Materials 40 (2011) 1517–1526.
[21] J.R. Manning, Diffusion and the Kirkendall shift in binary alloys, Acta Metallurgica 21 (1973) 817–826.

CHAPTER 2

Tracer Diffusion and Understanding the Atomic Mechanisms of Diffusion

Sergiy Divinski
University of Münster, Münster, Germany

2.1 INTRODUCTION

The experimental methods of diffusion measurements can conventionally be separated into two large groups, namely *direct* and *indirect* methods:

- The **direct methods** utilize theoretical descriptions of diffusion based on continuum or random walk approaches. In this case we are following the motion of atoms against the gradient of chemical potential (often simply the gradient of concentration) and determine the rate of change of the concentration gradient.
- The **indirect methods** are related to different physical phenomena in solid and liquid states, which are affected by thermal motion of atoms. Generally, thermally activated jumps of atoms do not require nonzero gradients of chemical potential, still it is often convenient to follow different relaxation processes (such as mechanical, magnetic, dielectric relaxation, nuclear magnetic relaxation – NMR, etc.), which invoke jumps of constituent atoms to flatten the induced deviations from equilibrium. The Mössbauer spectroscopy and quasi-elastic neutron scatter are two further microscopic methods often used to quantify the diffusion behavior of atoms.

The typical ranges of the diffusion coefficients which are accessible by different techniques are illustrated in Fig. 2.1. The following methods are most commonly used for diffusion investigations in solids and liquids:

- Tracer methods with sectioning, applying mechanical grinding/chemical etching or ion-beam sputtering for depth profiling;
- Secondary Ion Mass Spectrometry (SIMS);
- Auger electron spectroscopy (AES);
- Electron microprobe analysis (EMPA);
- Neutron reflectivity (NR);
- Pulsed field gradient NMR (pulsed-NMR);
- Field gradient NMR (FG-NMR);
- Rutherford back-scattering (RBS);
- Spreading resistance profiling (SRP) for semiconductors;
- Impedance spectroscopy in *dc* or *ac* modes for ionic conductors;
- Mechanical spectroscopy using after effect, internal friction (IF) or Gorsky effect;

Handbook of Solid State Diffusion, Volume 1
DOI: 10.1016/B978-0-12-804287-8.00002-6

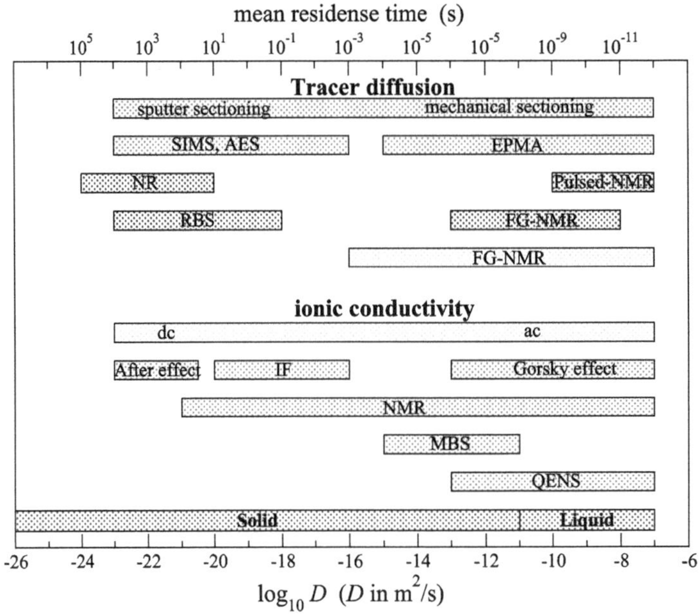

Figure 2.1 Typical ranges of the diffusion coefficients D measured by selected techniques. For the designations see text. The corresponding typical mean residence times τ for diffusing atoms, i.e., the times between successive diffusion jumps, are given on the upper abscissa axis.

- Nuclear magnetic relaxation (NMR) applying line-shape spectroscopy, spin lattice relaxation spectroscopy, or spin alignment experiments;
- Mössbauer spectroscopy (MBS);
- Quasi-elastic neutron scattering (QENS).

The typical ranges which characterize the diffusion coefficients of atoms in solid and liquid states are shown at the bottom in Fig. 2.1. In simple cases the diffusion coefficient can be represented by the following Einstein–Smoluchowski expression:

$$D = f\frac{\lambda^2}{6\tau} \tag{2.1}$$

where f is the correlation factor (equal, e.g., just to 1 for interstitial diffusion, and being less than unity for vacancy-mediated diffusion), λ is the length of diffusion jumps, and τ is the mean residence time or mean time between successive diffusion jumps. Using Eq. (2.1) and considering the typical lengths of diffusion jumps of the order of several angstroms, the corresponding ranges of mean residence time, which correspond to the given diffusion coefficients, are shown in Fig. 2.1, ordinate axis, too.

In this chapter the tracer methods are presented. For a detailed analysis of other methods, the reader is referred to famous diffusion books by Philibert [1] or Mehrer [2].

2.2 (RADIO-)TRACER METHOD

Self-diffusion is one of the most basic processes in solids. Thermal activation is responsible for atomic jumps in neighboring vacant positions, i.e., vacancies, in crystalline materials (for substitutional diffusion mechanism) or direct jumps between different interstitial positions (interstitial diffusion mechanism). Tracer methods are used to follow directly the jumps of marked atoms (e.g., of radioactive atoms of the given kind) in a solid. For this the tracer atoms have to be different from the main atoms available in the solid, e.g., to represent a different isotope for self-diffusion studies. A necessary condition is a sufficiently low amount, typically below 1%, of the given isotope in the natural isotope mixture for the given chemical element. In such a case the diffusion jumps of, e.g., ^{64}Ni isotopes in pure Ni can be followed using a mass-spectroscopic technique (as SIMS) if a source of highly enriched ^{64}Ni is deposited on sample's surface, since the amount of ^{64}Ni in natural Ni is about 0.3%.

Alternatively, the radioactive ^{63}Ni isotope can be used in combination with spectroscopic techniques for detection of radioactive decays in order to investigate self-diffusion in pure Ni. Note that not all elements have suitable stable isotopes for the self-diffusion measurements, e.g., pure Al, Cu or Ti. Correspondingly, not all elements have suitable radioactive isotopes, e.g., Al.

Although radioactive isotopes exist for all chemical elements, only relatively long-lived radioisotopes can be used for reliable experiments. In most common cases isotopes with half-times of more than several days are used, like ^{59}Fe with the half-time of 49 days or ^{63}Ni with the half-time of 100 years. Only a limited number of studies were published on diffusion measurements applying the relatively short-lived radioisotopes, like ^{32}Si with the half-time of about 2 hours or ^{64}Cu with the 12 hour half-time.

For some important elements, there is no suitable radioactive isotope, e.g., for Al, see the corresponding discussion in Chapter 10, in view of extremely high cost of producing the long-lived ^{26}Al isotope with the half-time of 10^5 years and its extremely low specific activity. The ^{29}Al isotope with the half-time of 6.6 min may potentially be applied for an investigation of very fast diffusion processes, e.g., diffusion in metallic materials at about their melting point. Still, no such experiments were performed so far.

Typical tracer diffusion experiments consists of the following steps, Fig. 2.2:

- sample preparation;
- tracer deposition;
- diffusion annealing;
- serial sectioning;
- concentration measurements;
- final analysis of a penetration profile.

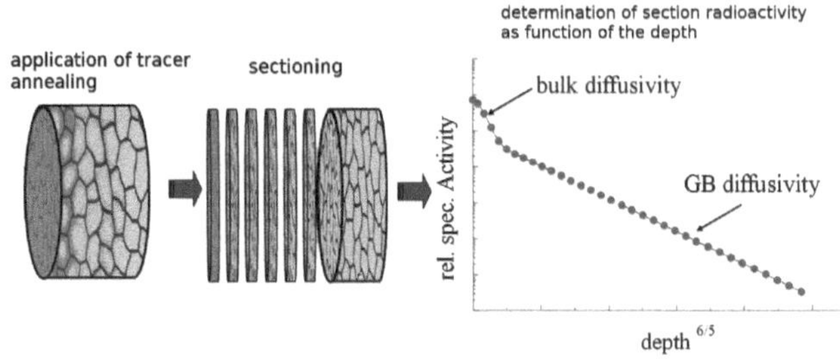

Figure 2.2 A scheme of a typical radiotracer diffusion experiment. As a result of the measurements, a concentration profile is determined, i.e., the depth dependence of the relative specific radioactivity (which is proportional to the layer tracer concentration).

2.2.1 Sample Preparation

Typically, cylindrical (disc-shaped) specimens of diameter from 5 to 11 mm with the thickness of 1 to 3 mm are prepared from the initial material.

One face of specimens is polished by standard metallographical procedures to a mirror-like quality. The samples have to be sealed in quartz tubes or put in high-vacuum chamber and then subjected to pre-diffusion annealing treatments first at a reasonably high temperature (to remove mechanical stresses induced by polishing) and secondly at the temperature and for at least the time of the intended diffusion annealing. The latter step guarantees quasi-equilibrium conditions, including grain boundary segregation of all impurity elements available in the material during the subsequent diffusion experiment.

These last steps are sometimes omitted when a recovery of treatment-induced state is investigated, as in the case of severely deformed samples, for example.

Thermal treatment of alloys has to be done in special containers from the same material to prevent preferential evaporation of one of the components and a change of the composition.

2.2.2 Tracer Deposition

After pre-annealing, the radiotracer in question or their mixture is deposited on the prepared surface, Fig. 2.2, upper-left panel. There are several procedures for tracer deposition:

- drop-and-dry
- electrolytic deposition
- evaporation in a high-vacuum chamber
- sputtering
- implantation

The first technique is the simplest and most often used. The tracer is dropped in the form of a highly dilute acid solution on the prepared surface of each specimen. After drying the sample is ready for diffusion annealing treatment.

Several examples for different tracer applications are described below.

Ag Tracer. The radiotracer ^{110m}Ag (half-life of 252 days) is typically used in a form of nitrate solution. Most reliable results can be obtained with tracers with a well-defined specific activity. In this case, a silver chip should be exposed to neutron irradiation, and the isotope ^{110m}Ag can be produced from the isotope ^{109}Ag according to the nuclear reaction

$$^{109}_{47}\mathrm{Ag}(n,\gamma) \rightarrow {}^{110m}_{47}\mathrm{Ag}. \tag{2.2}$$

The activated Ag chip then has to be dissolved in 30% nitric acid and, before tracer application onto the specimen surface, the solution has to be further dissolved in distilled water (at least with the ratio of 1:100). In this case one can achieve a required initial activity of the tracer solution (typically of about 10 kBq) without affecting the surface by the acid solution.

Similarly, several other isotopes, like ^{207}Bi (half-life of 38 years) and ^{44}Ti (half-life of 40 years), which are also available as nitrate solutions, or ^{64}Cu (half-life of 12.7 hours) and ^{59}Fe (half-life of 45 days), which are typically dissolved in HCl, can be used.

Nb Tracer. It is a quite involved task to reliably measure the diffusion rate of, e.g., the ^{95}Nb radioisotope. This radioisotope (766 keV γ-radiation; half-life of 35 days) is typically available only in a radiochemical equilibrium with its mother nuclide ^{95}Zr (757 keV γ-radiation; half-life of 65.5 days). As a result, the concentration of ^{95}Nb in a given section of a specimen is determined by several factors:

1. direct diffusion of ^{95}Nb atoms;
2. diffusion of ^{95}Zr isotope and its decay rate into the ^{95}Nb isotope;
3. the decay of ^{95}Nb itself.

Generally, all these factors have to be taken into account to determine the ^{95}Nb diffusivity. Moreover, the energies of the corresponding γ-lines for the ^{95}Zr and ^{95}Nb isotopes differ by only 9 keV, which imposes relatively strong requirements on the energy resolution of the counting system.

The mother ^{95}Zr radioisotope is typically produced by neutron irradiation of a high purity zirconium chip. The activated Zr chip is dissolved in a mixture of 50 vol.% H_2O, 47 vol.% HNO_3, and 3 vol.% HF and then further diluted with double-distilled water. It is important that the tracer material be stored for a month in order to build up an appropriate amount of the daughter ^{95}Nb radioisotope. Then, the solution containing the mixture of the two tracers is dropped on the polished face of the specimen and dried.

In some experiments, only the ^{95}Nb radioisotope may be detected after diffusion annealing, and the absence of the Zr radioisotope is then explained by its preferred evaporation. This circumstance can facilitate remarkably the determination of the Nb diffusivity, since no corrections of the continuous decay of its mother ^{95}Zr radionuclide had to be done. However, in general it is necessary to correct the activities because of the continuous decay of both tracers.

2.2.3 Diffusion Annealing Treatments

After application of the tracer solution, the samples are subjected to annealing treatment. Most reliable results will be obtained when the samples are placed in containers wrapped with refractory-metal foil (e.g., from Ta) in the case of intermetallic compounds. In this case preferential evaporation of some alloy constituent can be prohibited.

Diffusion annealing treatments have to be done at selected temperatures for given times. Typical diffusion times are in the range of 10^2 to 10^6 s. In the case of short annealing times, a heating correction has to be done. Such a correction is typically not necessary if the diffusion times are longer than 10^4 s.

2.2.4 Sectioning

After given diffusion annealing treatments, samples have to be first reduced in diameter to remove the effect of lateral-surface diffusion in the case of subsequent mechanical sectioning. This step is not necessary in the case of a SIMS analysis or the use of ion-beam sputtering method. The penetration profiles are commonly determined by various modifications of the sectioning technique. Typically, diffusion experiments are intended to be performed in as wide as possible temperature ranges. This requires different techniques for measuring the diffusion profiles with relatively large (more than 100 μm) as well as very small (less than 1 μm) penetration depths.

Several such techniques are explained below.

2.2.4.1 Mechanical Grinding

For determination of the diffusion penetration profiles extending over more than 10 μm, the radiotracer serial sectioning technique is applied by measuring the average tracer concentration in thin sections parallel to the diffusion source as a function of the penetration depth, Fig. 2.2. In the case of hard and brittle materials, e.g., of intermetallics or severely deformed materials, sectioning is performed using a precision parallel-grinding device with special abrasive mylar foils. The thickness of each section is calculated from the mass reduction, measured on a microbalance, the diameter, and the density of the specimens. Examples of the penetration profiles determined by grinding for Ti diffusion in TiAl [3,4] are given in Fig. 2.3.

By this technique, the penetration profiles over tens or hundreds of microns can reliably be processed. By applying mechanical grinding, a minimum thickness of a section is typically larger than about 0.5 μm. This value is determined by a minimum value of

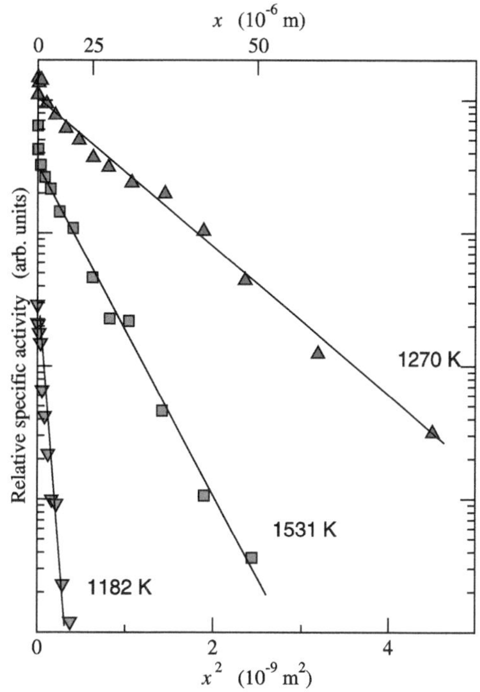

Figure 2.3 Examples of concentration profiles measured by parallel grinding for the case of Ti diffusion in TiAl [3] at several selected temperatures. The profiles follow Gaussian solutions (solid lines).

the section mass which can be reliably weighed, about 0.3 μg, by the sample geometry (i.e., by the sample diameter), and by the density of the material. If smaller sections are required, one can use larger samples (with a larger diameter). The diffusion coefficients which can be determined by this technique can be estimated as follows:

minimum section thickness	≈ 0.5 μm
minimum 10 sections	≈ 5 μm
	$\sqrt{Dt} \geq 5 \times 10^{-6}$ m
diffusion time	$\leq 10^7$ s
low limit for D	$D \geq 2.5 \times 10^{-18}$ m^2/s

This is a "hard" limit for the value of the diffusion coefficient which can be measured by the radiotracer technique and serial sectioning by mechanical grinding. Diffusion annealing treatments for 3 months (about 10^7 s) are seldom used, typical diffusion times are 10^3–10^6 s. It is also hard to measure very fast diffusion processes which are characterized by the diffusion coefficients $\geq 10^{-8}$ m^2/s. Diffusion time of 100 s would already correspond to the diffusion length of $\geq 10^{-3}$ m and mm-thick sections would have to

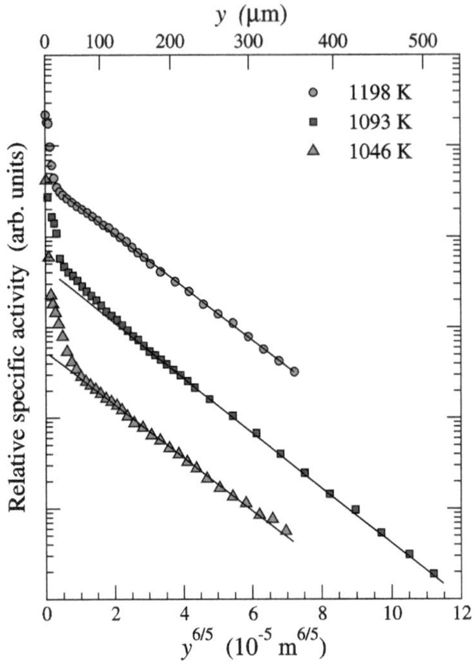

Figure 2.4 Examples of concentration profiles measured by microtome sectioning for the case of Bi grain boundary diffusion in pure Cu [5]. The profiles are fitted by the corresponding solution of the grain boundary diffusion problem (solid lines).

be removed. Correspondingly, the sample has to be very large to minimize the effects of radial and surface diffusion.

2.2.4.2 Sectioning by Microtome

The microtome is used in the case of relatively ductile materials, e.g., copper and copper-based alloys. Two to ten micrometer sections can be cut. The sections are to be weighed on a microbalance to determine the penetration depth exactly.

An example of the penetration profiles measured by microtome for the case of Bi diffusion in Cu is presented in Fig. 2.4.

The penetration profiles over hundreds of microns can reliably be processed. Let us estimate the lowest diffusion coefficient which can be measured by this technique:

minimum section thickness	$\approx 2\ \mu m$
minimum 10 sections	$\approx 20\ \mu m$
	$\sqrt{Dt} \geq 2 \times 10^{-5}$ m
diffusion time	$\leq 10^7$ s
low limit for D	$D \geq 4 \times 10^{-17}\ m^2/s$

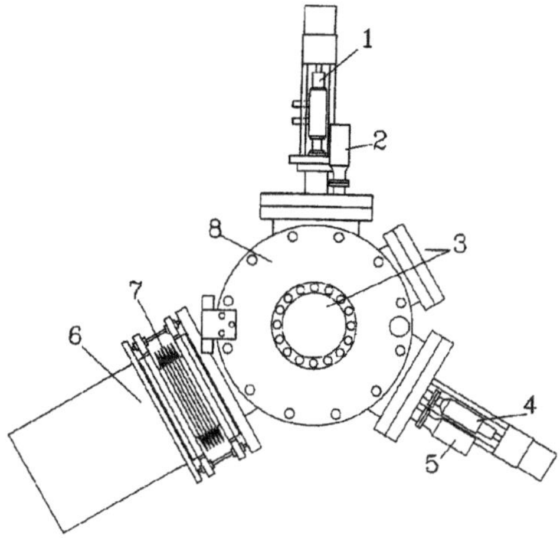

Figure 2.5 Frontal section of the sputter device: 1 is the sample holder which is constantly rotated and cooled down by water; 2 and 5 are the electrical contacts; 3 is the window; 4 is the set-up with camera-like device to collect the sputtered layers with radioisotopes; 6 is the ion source. From Wenwer et al. [6].

2.2.4.3 Ion-Beam Sputtering

If the measured penetration depths extend over less than several micrometers, the ion-beam sputtering technique can be applied for parallel sectioning. A scheme of the sputtering device developed by Wenwer et al. [6] at the Institute of Materials Physics, University of Münster, Germany is shown in Fig. 2.5.

Sputtering is performed in a high-vacuum chamber using Ar^+ ions with the energy of 1.1 keV and a beam current of 20 mA. A special mask is applied to sputter off a central region of each sample with a diameter from about 3 to 13 mm. The specimen holder is rotated to achieve homogeneous sputtering.

The sputtered material is collected by a camera-like device on special plastic foil. The foil is spooled stepwise in the given time intervals. The sputtering rate is kept constant during sectioning. The corresponding beam current is recorded, and its value is proved to be constant within ±5%.

The sputtering time is converted to the respective penetration depth, assuming a constant sputtering rate. The total penetration depth is determined by measuring the crater size and the weight loss of the specimen with the known density. Examples of the determined penetration profiles for the case of Ti diffusion in TiAl alloys [3] are shown in Fig. 2.6.

2.2.4.4 Effect of Sectioning Technique on the Measured Profile

Each method for determination of concentration profiles introduces certain modifications of the real distribution, which are related to the “grinding-in” or “sputtering-in”

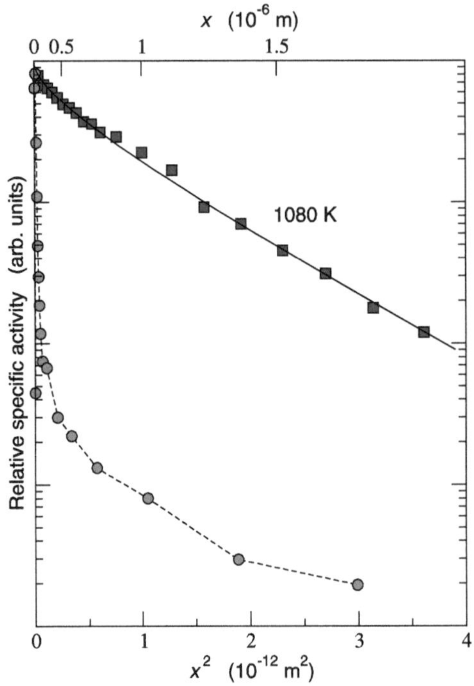

Figure 2.6 Concentration profile determined for Ti diffusion in TiAl [3] using the ion-beam sputter device (squares). Solid line represents fit according to the error-function solution of the diffusion problem. The circles correspond to the "zero-diffusion" profile.

effects (microtome usage was found to negligibly modify the real tracer distribution). Such effects are most important for SIMS and ion-beam sputtering due to intensive forward-mixing.

These effects have to be determined in each particular case. A convenient approach is the determination of the penetration profile which is measured without any diffusion annealing treatment. In Fig. 2.6 such a profile is presented (circles) and compared with the profile measured after annealing at 1080 K (squares). The "zero-diffusion profile" is characterized by a sum of a Gaussian-like component and a low-activity long tail. The real profile should then be determined by deconvolution of the measured tracer distribution and the "zero-diffusion profile." Since convolution of two Gaussian functions is also Gaussian, one immediately recognizes that ion-beam sputtering induces certain broadening of the real Gaussian distribution in addition to a characteristic low-activity tail (note that this low-activity tail is formally similar to the contribution of the grain boundary diffusion and it has to be carefully analyzed). The most fortunate case is when Gaussian broadening due to the "zero-diffusion profile" is marginal; a particular case is shown in Fig. 2.6.

In general numerical deconvolution has to be performed. A simplified approach is to fit the "zero-diffusion profile" by a sum of suitable functions, generally a Gaussian and simple exponential functions. Then the deconvolution can be done analytically.

Let the real tracer distribution be a function $r(x)$ and the "device function" (or "zero-diffusion profile") be $G(x)$. The latter can often be represented as a sum of a narrow Gaussian function and several, say m, exponentially decaying functions $g(x)$. Approximating the narrow Gaussian by a delta-function, we have

$$G(x) = (1-\eta)\cdot\delta(x) + \eta\cdot g(x) \tag{2.3}$$

where η is a weighting factor, and $g(x)$ is by definition

$$g(x) = \sum_{j=1}^{m} C_j \exp(-b_j x). \tag{2.4}$$

This representation of $g(x)$ corresponds to the assumption that some fraction (which may depend on the actual depth) of the radioisotope from a given section is redeposited (or grinded-in) to deeper layers at each step. As a result, the measured penetration profile, $e(x)$, is a convolution of the device function G and the true activity profile r,

$$e(x) = \int_0^x r(y)\cdot G(x-y)dy. \tag{2.5}$$

Eq. (2.5) reflects the fact that the activity in a deeper section is affected by the previously measured sections. Then we have

$$e(x) = (1-\eta)\cdot r(x) + \eta\int_0^x r(x)\cdot g(x-y)dy. \tag{2.6}$$

In a simple case, the real tracer distribution $r(x)$ is as a superposition of n Gaussians,

$$r(x) = \sum_{i=1}^{n} A_i \exp(-a_i x^2) \tag{2.7}$$

where A_i and a_i are constants. By definition, the parameters a_i are related to the diffusion coefficients which have to be determined. Combining the above-written equations, we arrive at the following expression:

$$e(x) = (1-\eta)\sum_{i=1}^{n} A_i \exp(-a_i x^2) + \eta\sum_{i=1}^{n}\sum_{j=1}^{m} A_i C_j \frac{\exp(b_j^2/4a_i)}{2\sqrt{a_i/\pi}} \times \left[\operatorname{erf}\left(\frac{b_j}{2\sqrt{a_i}}\right) + \operatorname{erf}\left(\sqrt{a_i}\cdot x - \frac{b_j}{2\sqrt{a_i}}\right)\right]. \tag{2.8}$$

As an important conclusion one immediately sees that the bulk diffusion contribution, $r(x)$, is almost not disturbed but superimposed by a low-intensity contribution. Thus, if the broadening due to grinding-in effects is small (delta-function), reliable bulk diffusion coefficients will be determined.

The second term on the right-hand side of Eq. (2.8) describes the distortion of the diffusion coefficient determined from the experimentally measured profile caused by forward mixing. The term in brackets vanishes for $x \to 0$ and converges to a constant value of $1 + \mathrm{erf}(b_j/2\sqrt{a_i})$ as $x \to \infty$. As a result, forward mixing may formally transform a Gaussian-like profile into $\ln \bar{C}$ vs. x dependence at large penetration depths, which corresponds to the above-mentioned "ghost" GB diffusion contribution. Such behavior implies that significant correction, if any, will be needed only for the fast diffusion section of activity profile measured at large depths.

The minimal diffusion coefficient which can be measured by the ion-beam sputtering method is:

minimum section thickness	≈ 2 nm
minimum 10 sections	≈ 20 nm
	$\sqrt{Dt} \geq 2 \times 10^{-8}$ m
diffusion time	$\leq 10^7$ s
low limit for D	$D \geq 4 \times 10^{-23}$ m^2/s

2.2.4.5 SIMS Profiling

In addition to the radiotracer experiments, in some fortunate cases self-diffusion can be analyzed by applying the secondary ion mass spectrometry (SIMS). This approach is illustrated for the case of Ni self-diffusion in Ni aluminides. A highly-enriched stable isotope ^{64}Ni is used as a tracer. From the natural abundance of only 0.3 at.%, a dynamic range of about two orders of magnitude in concentration along a diffusion profile can be expected – after background correction – which is sufficient for reliable diffusion profile detection.

A thin layer of less than 20 nm of the highly enriched (>98%) ^{64}Ni has to be evaporated under vacuum conditions (10^{-4} Pa) onto the polished surface of a specimen. After the diffusion annealing treatment, the ^{64}Ni penetration profiles can be determined by a suitable SIMS equipment using, e.g., oxygen ions O_2^+, accelerated through 5.5 kV, as primary ions. The primary ion current is typically 1 μA, and the ion probe of 75 μm in diameter is scanned over an area of 250 × 250 μm^2. For the analysis, an aperture lens system extracts only positively charged secondary ions, Al^+ and Ni^+, that came from the central area of the crater of about 60 μm in diameter. Diffusion annealing times have to be adjusted to generate profiles of several μm in depth.

The time of the analysis is recalculated to the crater depth by assuming a constant sputter rate by the primary beam. The latter is determined from the total time of the

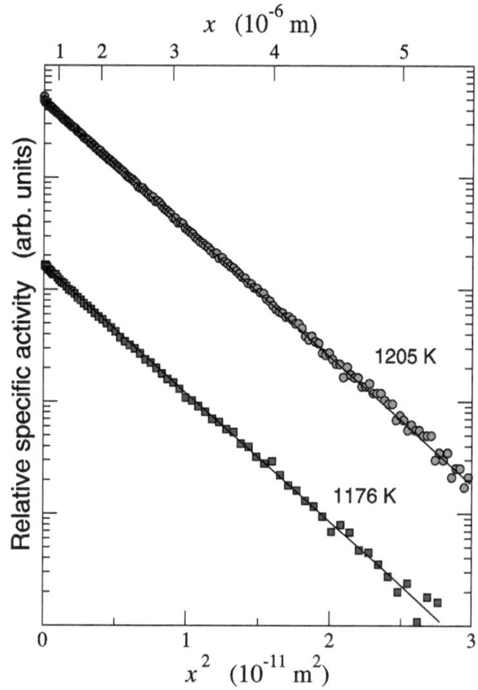

Figure 2.7 Examples of concentration profiles determined for Ni self-diffusion in NiAl [7] using SIMS. Solid lines represent fits according to the Gaussian solution of the diffusion problem.

analysis and the final crater depth, which is measured by a surface profilometer by averaging over at least five individual measurements. The calibration of the obtained profiles with the measured crater depths results in quantitative penetration profiles for the Ni self-diffusion in NiAl, see the concentration profiles in Fig. 2.7.

The natural background of ^{64}Ni in the samples has to be detected very precisely to provide an accurate background correction of the measured relative ^{64}Ni-concentration along the diffusion profile. At least two SIMS profiles should be measured on each sample. The discrepancy between the diffusion coefficients thus obtained is usually below 10%.

A similar approach can be applied for investigation of solute diffusion in intermetallics, too. In this case natural elements are used.

A drawback of SIMS with respect to, e.g., ion-beam sputtering corresponds to the fact that secondary ions are followed by conventional SIMS only and the ionization probability of sputtered atoms depends crucially on the surface preparation. Ion-beam sputtering records a large fraction of radioactive isotopes removed from the sample.

Recently, secondary neutral mass spectrometry (SNMS) was elaborated which, as the name states, allows the detection of neutral atoms sputtered from a sample [8].

Such instrument works with noble gas plasma and the bombarding ion current has an extremely high lateral homogeneity. The low bombarding energies (in the order of 100 eV) and the homogeneous plasma profile result in an outstanding depth resolution (<2 nm) [8] and typical detection limit of the SNMS is about 10 ppm.

Since the minimum section thicknesses are similar to those for ion-beam sputtering, the same estimate holds for the minimum diffusion coefficients which could be measured:

minimum section thickness	≈ 2 nm
minimum 10 sections	≈ 20 nm
	$\sqrt{Dt} \geq 2 \times 10^{-8}$ m
diffusion time	$\leq 10^7$ s
low limit for D	$D \geq 4 \times 10^{-23}$ m^2/s

2.2.5 Activity Measurements

The activity, i.e., concentration of the tracer isotopes, in each section can be detected with high efficiency by liquid scintillation counting, using a scintillation analyzer combined with an automatic sample changer in the case of β-decays, i.e., for the ^{63}Ni, ^{207}Bi, ^{64}Cu, ^{110m}Ag, ^{59}Fe isotopes.

The activity of γ-radiation can be detected with high efficiency using a well-type intrinsic Ge γ-detector with energy resolution, which is especially important in the case of niobium diffusion to clearly distinguish between the decays of the ^{95}Nb isotope with 766 keV energy from the decay of the ^{95}Zr nuclei with the nearby energy of 757 keV.

Further details and specific application can be found, e.g., in [7,9,10].

An alternative method, which was originally proposed first, is the so-called Grusin's or residual activity measurement [11]. It is based on repeated measurements of the residual activity of a sample after each sectioning step. It is an appropriate technique for radioactive elements which are either very soft β-emitters (like ^{63}Ni, only β decays with the energy of 68 keV) or very hard γ-emitters (e.g., ^{65}Zn and ^{22}Na which emit γ-quanta with the energies of 1115 keV and 1275 keV, respectively).

In the first case the β-quanta from deep layers are effectively adsorbed by the sample and the radioactivity of a thin sub-surface layer is selectively measured (e.g., the absorption length of β-quanta of ^{63}Ni is about 4 μm in metal). This method was recently applied even for grain boundary diffusion measurements [12].

In the opposite case the γ-quanta are practically not absorbed by the sample and the total radioactivity of the sample is successively measured. Since a layer of the material is progressively removed, the total radioactivity drops and the relative radioactivity of the removed layer can be easily determined.

2.2.6 Profile Processing/Analysis

The determined penetration profiles have to be analyzed to provide the diffusion coefficients. To this end, the relative specific activity ($A^*(x) = A(x)/m(x)$ as determined by the radiotracer experiments, where A is the relative activity of the section in counts per second – background subtracted – and m is the section mass, and x is the coordinate of the section center) or concentration ($C(x)$) profiles are compared with Gaussian (instantaneous source),

$$A^*(x) \propto C(x) = \frac{M}{\sqrt{\pi D t}} \exp\left(-\frac{x^2}{4Dt}\right), \tag{2.9}$$

or error-function solutions (constant source),

$$A^*(x) \propto C(x) = c_0 \operatorname{erfc}\left(\frac{x}{2\sqrt{Dt}}\right) = c_0 \left[1 - \operatorname{erf}\left(\frac{x}{2\sqrt{Dt}}\right)\right]. \tag{2.10}$$

Here D is the diffusion coefficient to be determined, t is the diffusion time, M is the amount of applied tracer material, c_0 is the initial tracer concentration on the open surface, which is kept constant during annealing for the case of constant surface conditions.

Typically, we are using tiny amounts of the tracer material and the chemical composition of the sample is not affected by diffusion of the tracer. Therefore, a chemically homogeneous system is investigated and a tracer diffusion coefficient is determined. This approach is commonly used for investigation of self-diffusion in pure elements and compounds.

One can differentiate the following diffusion coefficients:

A^* in A	tracer self-diffusion in A	$D_A^{A^*}$,
A^* in A_xB_{1-x}	tracer self-diffusion in A_xB_{1-x}	$D_{AB}^{A^*}$,
B^* in A_xB_{1-x}	tracer self-diffusion in A_xB_{1-x}	$D_{AB}^{B^*}$.

2.2.6.1 *Requirements and Benefits of the Thin Layer Conditions*

The following conditions have to be guaranteed in order to apply the thin layer solution to the given bulk diffusion problem:

1. The layer thickness of the applied tracer material, Δ, has to be small with respect to the mean diffusion length:

$$\Delta \ll \sqrt{Dt};$$

2. The total sample thickness, H, and the sample radius, R, have to be considerably larger than the mean diffusion distance:

$$H \gg \sqrt{Dt} \quad \text{and} \quad R \gg \sqrt{Dt};$$

3. The section thickness, Δ_x, must be sufficiently smaller than the mean diffusion length:

$$\Delta_x \ll \sqrt{Dt};$$

4. The thickness of radially removed layer before sectioning, Δ_R, – in the case of mechanical grinding or microtome sectioning – has to be larger than the mean diffusion length:

$$\Delta_R \ll \sqrt{Dt}.$$

The latter condition guarantees the absence of lateral diffusion fluxes, which would affect the tracer concentration profile, and ensures an effective 1D geometry of the diffusion problem.

A very important advantage of the thin layer solution is that the exact thickness of the initial tracer layer and absolute concentrations of tracer atoms in individual sections need not be determined. Taking a logarithm of both sides in Eq. (2.9), we arrive at

$$\ln A^*(x) \propto \ln C(x) = \ln\left(\frac{M}{\sqrt{\pi Dt}}\right) - \frac{x^2}{4Dt}, \tag{2.11}$$

where the unknown diffusion coefficient, D, can be directly determined measuring the **slope** of the concentration profile in the coordinates of logarithm of concentration vs. the penetration depth squared,

$$D = \frac{1}{4t}\left(-\frac{\partial \ln A^*}{\partial x^2}\right)^{-1} = \frac{1}{4t}\left(-\frac{\partial \ln C}{\partial x^2}\right)^{-1} \tag{2.12}$$

for the given diffusion time.

The penetration profiles can deviate from the predicted solutions. There are several reasons of such behavior, for example,

- Existence of short-circuit diffusion paths in the sample. Dislocations, grain boundaries and triple junctions are examples of such short-circuits since they provide paths for enhanced diffusion in metallic materials. In a fortunate case, i.e., the so-called B type regime of grain boundary or dislocation diffusion,[1] a distinct branch on the concentration profile can appear which can be distinguished and, if necessary, processed separately. In this case, the volume diffusion coefficients will be determined with sufficient reliability.

  In the case of regime A of the short-circuit diffusion, the diffusion fluxes along, e.g., grain boundaries and in the grain bulk are overlapped and an effective diffusion coefficient will be determined.

[1] For the analysis of grain boundary diffusion, see, e.g., [14]. The diffusion regimes are treated here according to the classification of Harrison [15].

The usage of single crystals offers a possibility to reliably measure the volume diffusion coefficients.

- Final thickness of the initial tracer source on the sample surface. In this case the concentration profile can be fitted by the following solution [13]:

$$C(X, t) = \frac{1}{2} C_0 \left[\operatorname{erf}\left(\frac{h - x}{2\sqrt{D_v t}} \right) + \operatorname{erf}\left(\frac{h + x}{2\sqrt{D_v t}} \right) \right] \tag{2.13}$$

where h is the thickness of the initial tracer layer.

- Evaporation of the tracer material, which induces a deviation of the concentration profiles from the Gaussian solution in near-surface region. Moreover, if the measurements in these conditions will be performed for different (increasing) diffusion times, the concentrations will successively drop, keeping the overall shape of the profile almost unchanged.

2.3 SOLUTE (IMPURITY) DIFFUSION

Diffusion of an element B in pure A or in a homogeneous alloy A_xC_{1-x} corresponds to solute (or impurity) diffusion, if the concentration of the solute element B is sufficiently small. The data on solute diffusion rates are required for all practical applications, which include heat treatments and/or exposition to atmospheres.

The corresponding diffusion coefficients can be determined with the same tracer methods as described above. It is important that the composition of the samples are not affected by the tracer diffusion. We can introduce the following diffusion coefficients:

B^* in A	(tracer-) solute diffusion coefficient of B in A	$D_{B^*}^{A}$,
B^* in A_xC_{1-x}	(tracer-) solute diffusion of B in A_xC_{1-x}	$D_{B^*}^{AC}$.

2.4 EXPERIMENTAL DETERMINATION OF THE DIFFUSION MECHANISM

In this chapter we are concentrating on diffusion in crystalline bulk. Diffusion in a solid is promoted by point defects, see detailed discussion in Chapter 1. In Fig. 1.1, possible point defects in a pure metal are schematically presented.

With respect to self-diffusion in pure metals, two main **diffusion mechanisms** can be differentiated:

- vacancy mechanism;
- self-interstitial mechanism.

In densely-packed metals self-diffusion occurs typically via a thermally-activated vacancy mechanism, Fig. 1.29(A). The self-interstitials, Fig. 1.29(B), are practically absent in metals under normal conditions, though they can be created under irradiation as a result of atomic collisions. Point defects in compounds are more complex, see Chapters 1 and 10.

For a given diffusion mechanism, the self-diffusion coefficient D is

$$D = g\nu_0 f a^2 \exp\left\{\frac{\Delta S}{R}\right\}\exp\left\{-\frac{\Delta H}{RT}\right\}, \tag{2.14}$$

or in short

$$D = D_0 \exp\left\{-\frac{\Delta G}{RT}\right\}. \tag{2.15}$$

Here g is the geometric factor, ν_0 the attempt frequency at a site position, f the tracer correlation factor, and $\Delta G = -T\Delta S + \Delta H$ the difference in the Gibbs energy of the system with jumping atom at site and saddle-point positions. D_0 is then the corresponding pre-exponential factor. If ΔG is a constant (a temperature-independent quantity), the plot of $\ln D$ vs. $1/T$ should be linear, and many experimentally measured diffusion coefficients do follow this dependence, see, e.g., the examples in Chapter 10.

Occasionally, several diffusion mechanisms may contribute to the atomic transport, and in this case the resulting diffusion coefficient is

$$D = D_I + D_{II} + \cdots = D_0^I \exp\left\{-\frac{\Delta G_I}{RT}\right\} + D_0^{II}\exp\left\{-\frac{\Delta G_{II}}{RT}\right\} + \cdots. \tag{2.16}$$

If the diffusion data are analyzed in a limited temperature interval, an effective activation energy can be established according to

$$D = D_0^{eff} \exp\left\{-\frac{\Delta G^{eff}}{RT}\right\} \tag{2.17}$$

where the effective activation energy is

$$\Delta G^{eff} = \frac{d\ln D}{d\left(-\frac{1}{RT}\right)} = \frac{1}{D}\frac{dD}{d\left(-\frac{1}{RT}\right)} = \Delta G_I \frac{D_I}{D} + \Delta G_{II}\frac{D_{II}}{D} + \cdots. \tag{2.18}$$

Do we have *experimental* methods to establish the diffusion mechanism in a solid?

There are generally two main approaches to shed light on the underlying diffusion mechanism by investigation of **pressure dependence** (i.e., by measuring the **activation volume**) of diffusion and via quantification of the **isotope effect** of diffusion.

2.4.1 Activation Volume of Diffusion

Generally, the difference of the Gibbs energy of the system with a tracer at a site position and at a saddle-point configuration, which determines the diffusion rate, is written as

$$\Delta G = -T\Delta S + \Delta U + p\Delta V, \tag{2.19}$$

see Chapter 1. Rewriting Eq. (2.15) we arrive at

$$\Delta G = -RT \ln \frac{D}{D_0}. \tag{2.20}$$

Using a thermodynamic relation

$$\frac{\partial \Delta G}{\partial p} = \Delta V, \tag{2.21}$$

the activation volume can be determined from Eq. (2.20) as

$$\Delta V = \frac{\partial \left(-RT \ln \frac{D}{D_0} \right)}{\partial p} = -RT \left. \frac{\partial \ln D}{\partial p} \right|_T + RT \left. \frac{\partial \ln \left(g \cdot f \cdot \nu_0 \cdot a^2 \right)}{\partial p} \right|_T . \tag{2.22}$$

Taking the geometric factor g as unity (this holds for cubic lattices) and neglecting the deviation of the correlation factor f from unity, the second term in brackets is approximately the product of the isothermal compressibility, K_T, and the Grüneisen constant, γ_G, namely

$$\left. \frac{\partial \ln \left(\nu_0 \cdot a^2 \right)}{\partial p} \right|_T \approx K_T \cdot \gamma_G. \tag{2.23}$$

Using typical values of these constants, we arrive at

$$\left. \frac{\partial \ln \left(\nu_0 \cdot a^2 \right)}{\partial p} \right|_T \approx (0.01 \div 0.03)\,\Omega \tag{2.24}$$

where Ω is the atomic volume. Thus we conclude that with sufficient accuracy the activation volume of diffusion can be determined from the pressure dependence of the diffusion coefficient as

$$\Delta V \cong \frac{\partial \left(-RT \ln \frac{D}{D_0} \right)}{\partial p}. \tag{2.25}$$

In Table 2.1 we list typical values of the activation volumes measured for self-diffusion and impurity diffusion in selected solids.

The first group of data represent the measured activation volumes for self-diffusion in FCC metals and the values about 1 Ω, i.e., about one atomic volume, were determined. These values are considered to be proofs for a vacancy-mediated diffusion mechanism in simple metals.

The activation volume for a vacancy-mediated mechanism, ΔV, can be split in two contributions,

$$\Delta V = \delta V_f + \Delta V_m \tag{2.26}$$

Table 2.1 Experimental data for the diffusion activation volume measured for self- and impurity diffusion in some elements

Element	ΔV (Ω)	Reference
Cu	$+(0.93 \div 1.09)$	[16]
Ag	$+(0.66 \div 0.88)$	[17]
Na	$+(0.40 \div 0.75)$	[18]
N in α-Fe	$+0.05$	[19]
Ge in Si	$-(0.28 \div 0.68)$	[20]

corresponding to the activation volumes of vacancy formation, ΔV_f, and migration, ΔV_m. It is a current wisdom that the vacancy formation volume, ΔV_f, largely determines the value of ΔV and $\Delta V - m \approx (0.2 \div 0.3)\,\Omega$.

The activation volume for N diffusion in α-Fe is very small but positive, Table 2.1, which clearly indicates the interstitial diffusion mechanism in this case. The negative values of the activation volume of Ge diffusion in Si indicate a complex diffusion mechanism with probable contribution of both interstitial and vacancy mechanisms of diffusion of the substitutional solute Ge.

2.4.2 Isotope Effect

A basic information about the diffusion mechanism can be gain if the corresponding correlation factor f is known. Its value can be estimated (to what extent we will see below) if the diffusion coefficients of two isotopes of the same element, say A_1 and A_2, are measured. Examples can be the isotopes ^{110m}Ag and ^{105}Ag or ^{57}Co and ^{61}Co. The corresponding diffusion coefficients, e.g., for the vacancy-mediated self-diffusion, are:

$$D_1 = g \cdot f_1 \cdot a^2 \cdot c_v \cdot \omega_1 \quad \text{and} \quad D_2 = g \cdot f_2 \cdot a^2 \cdot c_v \cdot \omega_2. \tag{2.27}$$

It is obvious that the geometric factors g, the lattice parameter a, and the vacancy concentration c_v are the same for the two isotopes. The jump frequency ω_i of a given isotope, $i = 1, 2$, is simply

$$\omega_i = \nu_0^i \cdot \exp\left\{-\frac{\Delta G_i}{RT}\right\} \tag{2.28}$$

where ν_0^i is the corresponding attempt frequency (the frequency of atom vibrations) and the ΔG_i is the energy barrier. It is generally assumed that the barriers are the same for the two isotopes, since their are determined by inter-atomic interactions, which depends on the kind of the chemical element but not on its mass,

$$\Delta G_1 = \Delta G_2 = \Delta G. \tag{2.29}$$

According to the Einstein model of atomic vibrations, a crystal is represented by a set of independent atomic oscillators, whose vibration frequency is inversely proportional to the square root of the atom mass, $\nu_0^i \propto m_i^{-0.5}$. Under this approximation, the ratio of jump frequencies is inversely proportional the square root of the inverse ratio of the isotope mass,

$$\frac{\omega_1}{\omega_2} = \frac{\nu_0^1}{\nu_0^2} = \sqrt{\frac{m_2}{m_1}}. \tag{2.30}$$

In order to make a next step, we have to analyze why the correlation factor depends on the isotope mass. The reason is that careful analysis of the correlations for a vacancy mechanism results in the following form of the correlation factor f [21,22]:

$$f = \frac{2H}{\omega + 2H}. \tag{2.31}$$

Here H is the *escape frequency*, i.e., the frequency of events when a particular vacancy escapes the given tracer atom. It is important that the escape frequency does not depend on the frequency of tracer–vacancy exchanges, ω. Then Eq. (2.31) can be written as

$$f_1 = \frac{2H}{\omega_1 + 2H} \text{ and } f_2 = \frac{2H}{\omega_2 + 2H}. \tag{2.32}$$

Using Eqs. (2.32), the correlation factor f_2 can be written via the value of f_1. First

$$2H = \frac{f_1\omega_1}{1 - f_1} \tag{2.33}$$

and then

$$f_2 = f_1 \frac{\omega_1}{\omega_2(1 - f_1) + f_1\omega_1}. \tag{2.34}$$

Correspondingly the ratio of the diffusion coefficients reads as

$$\frac{D_1}{D_2} = 1 + f_1\left(\frac{\omega_1}{\omega_2} - 1\right), \tag{2.35}$$

or

$$\frac{D_1}{D_2} - 1 = f_1\left(\frac{\omega_1}{\omega_2} - 1\right). \tag{2.36}$$

In view of Eq. (2.30), Eq. (2.36) transforms to

$$\frac{\frac{D_1}{D_2} - 1}{\frac{\omega_1}{\omega_2} - 1} = f_1, \tag{2.37}$$

or

$$\frac{\frac{D_1}{D_2} - 1}{\sqrt{\frac{m_2}{m_1}} - 1} = f_1. \tag{2.38}$$

We conclude that within the Einstein approximation the measurements of the isotope effect, i.e., the ratio of the diffusion coefficients of two isotopes of the same chemical species, give direct access to the value of the correlation factor f, and this fact allows us to draw a certain conclusion about the diffusion mechanism.

If the multibody effects at the saddle point are taken into account [23,24], Eq. (2.30) is modified to

$$\frac{\omega_1}{\omega_2} = \Delta K \sqrt{\frac{m_2}{m_1}}, \tag{2.39}$$

where ΔK defines the so-called *kinetic energy transfer factor* as a fraction of the kinetic energy, which is lost by the hopping atom at the saddle-point configuration to neighboring atoms [24]. If the whole energy is lost, $\Delta K = 0$; and $\Delta K = 1$, if no kinetic energy is transferred. In view of Eq. (2.39), Eq. (2.38) can be rewritten in its final expression:

$$\frac{\frac{D_1}{D_2} - 1}{\sqrt{\frac{m_2}{m_1}} - 1} = f_1 \Delta K \equiv E, \tag{2.40}$$

where E is the *isotope effect* or *isotope effect parameter*. In principle, there are two isotope effect parameters, the first entering the above expression with f_1 and the other entering the same expression for f_2 (if one determines the inverse ratio of the diffusion coefficients, D_2/D_1). Nevertheless, the difference between f_1 and f_2 is generally considered to be small, and one neglects it, i.e.,

$$f_1 \approx f_2 \approx f_0 \tag{2.41}$$

where f_0 is the (geometric or lattice) correlation factor for self-diffusion on the given lattice (if self-diffusion is considered). Thus, Eq. (2.40) determines the product of the lattice correlation factor and ΔK. Since both $f < 1$ and $\Delta K < 1$, we conclude that $E < 1$. In an ideal case, when the motion of the jumping atom is completely decoupled from that of neighboring atoms, $\Delta K = 1$ and $E = f_0$.

In Table 2.2 experimentally measured isotope effect parameters are shown. Assuming a mono-vacancy diffusion mechanism, e.g., $f_0 = 0.7815$ for FCC lattice, the estimates of the kinetic transfer factor, ΔK, are given.

As a general trend one can recognize that the isotope effect decreases as temperature increases. We can explain this fact by an increased amount of the kinetic energy which is transferred from the jumping atom to its neighboring atoms at the saddle-point due to anharmonic effects. An alternative explanation would be based on an influence of

Table 2.2 The experimental data on the isotope parameter E for self-diffusion of different elements. The energy transfer factor ΔK is determined assuming mono-vacancy diffusion mechanism for substitutional diffusion, $\Delta K = E/f_0$, where f_0 is the correlation factor for the given lattice

Element	T (K)	E	f_0	ΔK	Reference
Au	1040	0.71	0.7815	0.91	[25]
	1127	$0.65 \div 0.73$		$0.83 \div 0.93$	
	1327	$0.62 \div 0.71$		$0.79 \div 0.91$	
Na	423	0.19	0.727	0.26	[26]
Fe	1282	0.68	0.727	0.94	[27]
	1688	0.57		0.78	
	1918	0.37		0.51	
Cu	1220	0.74	0.7815	0.95	[28]
	1282	0.63		0.81	
	1351	0.58		0.74	
Co in α-Ti	985	0.08	1	0.08	[29]

the second diffusion mechanism at higher temperatures which would be characterized by a lower value of the correlation factor, and the divacancy mechanism was assumed to give such a contribution [2]. Recent finite temperature DFT-based calculations have proven the dominance of anharmonic effects for the deviation from Arrhenius-type temperature dependence of self-diffusion in FCC lattices at high temperatures and a relatively small (negligible) contribution of the divacancy mechanism [30].

REFERENCES

[1] J. Philibert, Atom Movements – Diffusion and Mass Transport in Solids, Les Editions de Physique, Les Ulis, 1991.

[2] H. Mehrer, Diffusion in Solid Matter, Springer, Heidelberg, 2007.

[3] S.V. Divinski, F. Hisker, Chr. Klinkenberg, Chr. Herzig, Niobium and titanium diffusion in the high niobium-containing Ti–54Al–10Nb alloy, Intermetallics 14 (2006) 792.

[4] S.V. Divinski, C. Klinkenberg, C. Herzig, Tracer diffusion of niobium and titanium in binary and ternary titanium aluminides, J. Phase Equilibria Diffusion 26 (2005) 452–457.

[5] V. Divinski, M. Lohmann, Chr. Herzig, Grain boundary diffusion and segregation of Bi in Cu: radiotracer measurements in B and C diffusion regimes, Acta Mater. 52 (2004) 3973.

[6] F. Wenwer, A. Gude, G. Rummel, M. Eggersmann, T. Zumkley, N. Stolwijk, H. Mehrer, A universal ion-beam-sputtering device for diffusion studies, Measurement Science Technology 7 (1996) 632–640.

[7] St. Frank, S.V. Divinski, U. Södervall, Chr. Herzig, Acta Mater. 49 (2001) 1399.

[8] A. Lakatos, A. Csik, G.A. Langer, G. Erdelyi, G.L. Katona, L. Daroczi, K. Vad, J. Toth, D.L. Beke, Investigations of failure mechanisms at Ta and TaO diffusion barriers by secondary neutral mass spectrometry, Vacuum 84 (2010) 130–133.

[9] S.V. Divinski, F. Hisker, W. Löser, U. Södervall, Chr. Herzig, Ni radiotracer diffusion in B2 ordered NiFeAl alloys, Intermetallics 14 (2006) 308–314.

[10] S.V. Divinski, Y.-S. Kang, W. Löser, U. Södervall, Chr Herzig, Ni and Fe tracer diffusion in $Ni_{40}Fe_{10}Al_{50}$ ternary alloy, Intermetallics 12 (2004) 511.

[11] P.L. Gruzin, Doklady Akademii Nauk SSSR 86 (1952) 289.
[12] V. Rothová, A simple measurement of the Ni-63 diffusion profiles by residual activity method using a liquid scintillation counter, Nuclear Instruments Methods in Physics Research A 729 (2013) 702–706.
[13] J. Crank, The Mathematics of Diffusion, Clarendon Press, Oxford, 1975.
[14] A. Paul, T. Laurila, V. Vuorinen, S.V. Divinski, Thermodynamics, Diffusion and the Kirkendall Effect in Solids, Springer Int. Publ. Switzerland, 2014.
[15] L.G. Harrison, Influence of dislocations on diffusion kinetics in solids with particular reference to alkali halides, Trans. Faraday Soc. 57 (1961) 1191–1197.
[16] M. Beyeler, Y. Adda, J. Phys. (Paris) 29 (1968) 345.
[17] G. Rein, H. Mehrer, Philos. Mag. A 45 (1982) 767.
[18] J.N. Mundy, Phys. Rev. B 3 (1971) 2431.
[19] A.J. Bosman, P.E. Brommer, G.W. Rathenau, Physica 23 (1957) 1001.
[20] E. Nygren, M.J. Aziz, D. Turnbull, J.F. Hays, J.M. Poate, D.C. Jacobson, R. Hull, Mat. Res. Soc. Symp. Proc. 86 (1988) 77.
[21] A.B. Lidiard, Philos. Mag. 46 (1955) 1218.
[22] A.B. Lidiard, Philos. Mag. 5 (1960) 1171.
[23] J.G. Mullen, Phys. Rev. 121 (1961) 1649.
[24] A.D. Le Claire, Philos. Mag. 14 (1966) 1271.
[25] Chr. Herzig, H. Eckseler, W. Bussmann, D. Cardis, J. Nucl. Mater. 69/70 (1978) 61.
[26] J.N. Mundy, W.D. McFall, Phys. Rev. B 7 (1973) 4363.
[27] M.G. Coleman, C.A. Wert, R.F. Peart, Phys. Rev. 175 (1968) 788.
[28] S. Fujikawa, S. Ushino, Unpublished, 1990.
[29] H. Nakajima, S. Ishioka, M. Koiwa, Phil. Mag. A 52 (1985) 743.
[30] A. Glensk, B. Grabowski, T. Hickel, J. Neugebauer, Physical Review X 4 (2014) 011018.

CHAPTER 3

Estimation of Diffusion Coefficients in Binary and Pseudo-Binary Bulk Diffusion Couples

Aloke Paul
Indian Institute of Science, Bangalore, India

Since many topics are to be introduced in a single chapter, only very important aspects for the estimation of the diffusion parameters will be discussed. For a more detailed analysis, the readers are referred to the book coauthored by the present author [1]. This and the next chapter (Chapter 4) cover almost all the issues in binary and multicomponent systems. However, readers should be aware that different authors express equations for flux and concentrations differently, which are the main differences between these two chapters. Otherwise, the next chapter can be considered as a continuation of this chapter from binary and pseudo-binary to multicomponent systems.

Although the tracer diffusion method is explained in the previous chapter (Chapter 2), a very basic concept is also introduced in this chapter for establishing relations with the data estimated by the diffusion couple technique. Similarly, a very basic concept of the multicomponent systems is also introduced to explain the pseudo-binary approach.

3.1 FICK'S LAWS OF DIFFUSION

Adolf Fick [2] proposed that the flux in a one-dimensional system is proportional to the concentration gradient, namely

$$J \propto -\frac{\partial C}{\partial x} = -D\frac{\partial C}{\partial x}, \tag{3.1}$$

where J (mol/m^2 s) is the flux, D (m^2/s) is the proportionality constant, i.e., the diffusion coefficient, C (mole/m^3) is the concentration, and x (m) is the position parameter. The negative sign emphasizes that diffusion occurs down the concentration gradient, which is not necessarily true in every system as discussed in Chapter 1 of Volume 1. Diffusion of the component is governed by the chemical potential gradient. Components always diffuse down the chemical potential gradient, and if this leads to the diffusion of components up the concentration gradient, it is called uphill diffusion. The relation

Handbook of Solid State Diffusion, Volume 1
DOI: 10.1016/B978-0-12-804287-8.00003-8

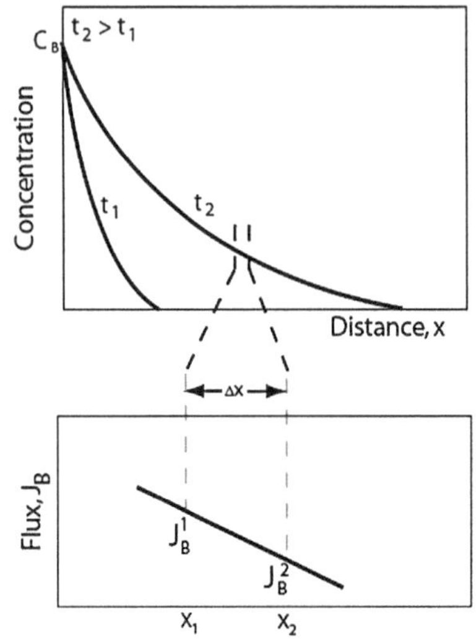

Figure 3.1 Non-steady-state diffusion profile and explanation of the derivation of Fick's second law.

expressed in Eq. (3.1) is known as Fick's first law. This is used to derive the relations for the estimation of the diffusion coefficients as explained in the following sections. In a normal downhill diffusion, the estimated diffusion coefficient, therefore, gives a positive value; however, it gives a negative value in the case of an uphill diffusion.

Although Fick's first law is used to derive the relations for the estimation of the diffusion coefficients, it is alone not sufficient to explain the time-dependent, i.e., non-steady state diffusion profile. It can be used directly only if there is no change in concentration profile with time, which is almost impossible to find in most of the practical examples. To circumvent this problem, Fick's second law is derived with the help of Fick's first law and mass conservation. Consider a situation in which the concentration of B on the surface of a material A is kept constant with a continuous supply. With increasing time, the diffusion profile of component B will increase, as shown in Fig. 3.1. It means that at one particular location composition changes with time and therefore creates a non-steady state situation. We consider a very thin slab with thickness $\Delta x (= x_2 - x_1)$ in which the flux coming in is J_B^1 and that going out is J_B^2 such that $J_B^1 > J_B^2$. Therefore, the total amount of B coming into the thin slab with a unit cross-section in a short time δt is $J_B^1 \delta t$. Similarly, the total amount of material going out through the plane x_2 is $J_B^2 \delta t$. Therefore, the increase in concentration of B can be written as

$$\delta C_B = \frac{(J_B^1 - J_B^2)\delta t}{\Delta x}. \tag{3.2}$$

In a very thin slab, the variation of flux can be considered linear such that

$$\frac{\partial J_B}{\partial x} = \frac{J_B^2 - J_B^1}{\Delta x} = -\frac{J_B^1 - J_B^2}{\Delta x}. \tag{3.3}$$

Substituting Eq. (3.3) into Eq. (3.2), we get

$$\frac{\partial C_B}{\partial t} = -\frac{\partial J_B}{\partial x}. \tag{3.4}$$

Replacing Fick's first law (expressed in Eq. (3.1)) in Eq. (3.4), we get

$$\frac{\partial C_B}{\partial t} = \frac{\partial}{\partial x}\left(D_B \frac{\partial C_B}{\partial x}\right). \tag{3.5a}$$

If the diffusion coefficient is constant throughout the concentration profile, Eq. (3.5a) can be written as

$$\frac{\partial C_B}{\partial t} = D_B \frac{\partial^2 C_B}{\partial x^2}. \tag{3.5b}$$

Eq. (3.5) is known as Fick's second law, which can be used to describe a non-steady state diffusion profile.

3.2 SOLUTIONS OF FICK'S SECOND LAW CONSIDERING CONSTANT DIFFUSION COEFFICIENTS

In general, diffusion coefficients vary with composition because of the change in thermodynamic driving forces and defect concentrations. However, considering composition-dependent diffusion coefficients is not easy, as discussed in the sections starting from Section 3.3. On the other hand, there are systems in which diffusion coefficients do not change significantly with composition, and a solution to Fick's second law considering a constant diffusion coefficient could be handy. In this section, three such solutions and their limitations are explained in detail.

3.2.1 Solution for a Thin-Film Condition

Let us first consider a system in which a very thin layer of B is sandwiched between two blocks of A, as shown in Fig. 3.2. We are considering a very small amount of B which can diffuse and mix as an impurity in A instead of producing a compound or any new alloy. From the experimental point of view, component B exponentially decays in composition and can be expressed as [3]

$$C_B(x) = \frac{C_0}{t^{1/2}} \exp\left(-\frac{x^2}{4D_B t}\right) \tag{3.6}$$

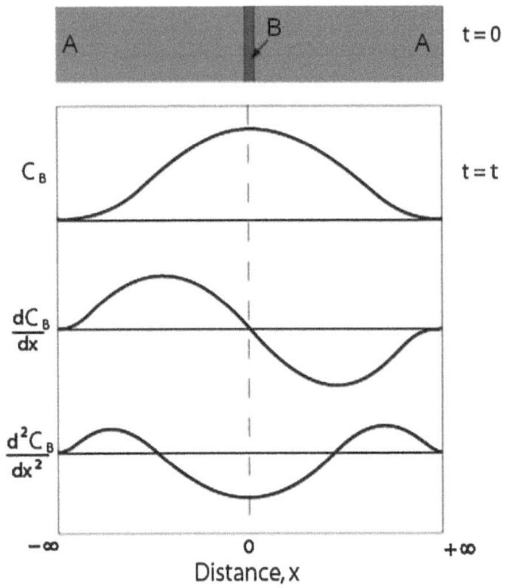

Figure 3.2 Diffusion profile of component B when sandwiched between two blocks of component A with respect to C_B, $\frac{dC_B}{dx}$ and $\frac{d^2C_B}{dx^2}$, with x after annealing for time t.

where $C_B(x)$ is the concentration of material B at position x, C_0 is the constant that depends on the amount of material B, D_B is the diffusion coefficient of component B, and t is annealing time. One can establish the correctness of the solution as expressed in Eq. (3.6) by differentiating with respect to t and x and then replacing obtained derivatives in Eq. (3.5). Since we are considering a constant diffusion coefficient, the distribution profile of B is symmetric with respect to position $x = 0$, as shown in Fig. 3.2. Therefore, the positive side is an exact reflection of the negative side. The constant C_0 can be replaced with the amount of material B, i.e., M_B. M_B can be expressed as

$$M_B = \int_{-\infty}^{+\infty} C_B dx. \tag{3.7}$$

Substituting Eq. (3.6) into Eq. (3.7), we get

$$M_B = \int_{-\infty}^{+\infty} \frac{C_0}{t^{1/2}} \exp\left(-\frac{x^2}{4D_B t}\right) dx. \tag{3.8}$$

Let us assume $\lambda = \frac{x}{2\sqrt{D_B t}}$ such that $dx = (2\sqrt{D_B t})d\lambda$. Since C_0 is constant, Eq. (3.8) can be written as

$$M_B = 2C_0\sqrt{D_B} \int_{-\infty}^{+\infty} \exp(-\lambda^2)\, d\lambda = 2C_0\sqrt{\pi D_B}, \tag{3.9}$$

since $\int_{-\infty}^{+\infty} \exp(-\lambda^2) d\lambda = \sqrt{\pi}$.

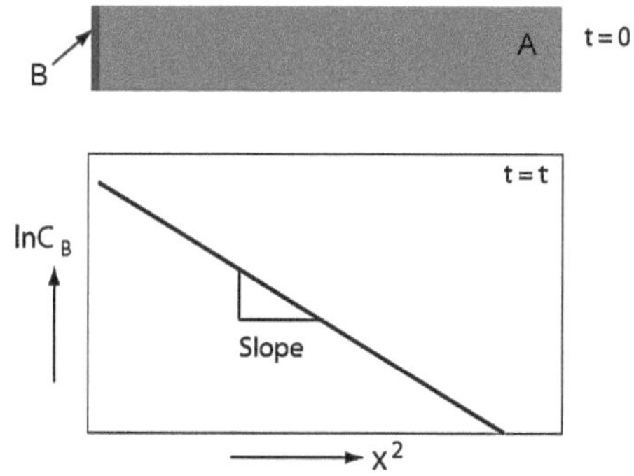

Figure 3.3 Estimation of the diffusion data following the method of thin film solution.

Substituting Eq. (3.9) into Eq. (3.6), we get

$$C_B(x) = \frac{M_B}{2\sqrt{\pi D_B t}} \exp\left(-\frac{x^2}{4D_B t}\right). \tag{3.10}$$

As shown in Fig. 3.2, following Fick's first law of diffusion, the graph of $\frac{dC_B}{dx}$ vs. x reflects the flux at different planes after annealing time t. The $\frac{d^2C}{dx^2}$ vs. x profile indicates the region where the system is losing the diffusing component, i.e., $(\frac{\partial^2 C_B}{\partial x^2}) < 0$, and the region where it is gaining, i.e., $(\frac{\partial^2 C_B}{\partial x^2}) > 0$. If component B is not sandwiched between two blocks of A but deposited on the surface of one block, as shown in Fig. 3.3, Eq. (3.10) can be modified to

$$C_B = \frac{M_B}{\sqrt{\pi D_B t}} \exp\left(-\frac{x^2}{4D_B t}\right). \tag{3.11}$$

The only difference between Eqs. (3.10) and (3.11) is the factor 2, which is removed from the denominator in Eq. (3.10) since component B diffuses towards one side only. It can be further written as

$$\ln C_B = \ln \frac{M_B}{\sqrt{\pi D_B t}} - \frac{x^2}{4D_B t}. \tag{3.12}$$

If $\ln C_B$ is plotted with respect to x^2, one can estimate the diffusion coefficient from the slope $-\frac{1}{4D_B t}$ utilizing the known annealing time t.

As already mentioned in the beginning, the amount of component to be diffused is so small that it can mix in the substrate at the impurity level without forming a new alloy or compound. The main use of this technique is the measurement of the self-diffusion coefficient or impurity diffusion coefficient by radiotracer technique. As explained in Chapter 1 of Volume 1, components diffuse even in the absence of any

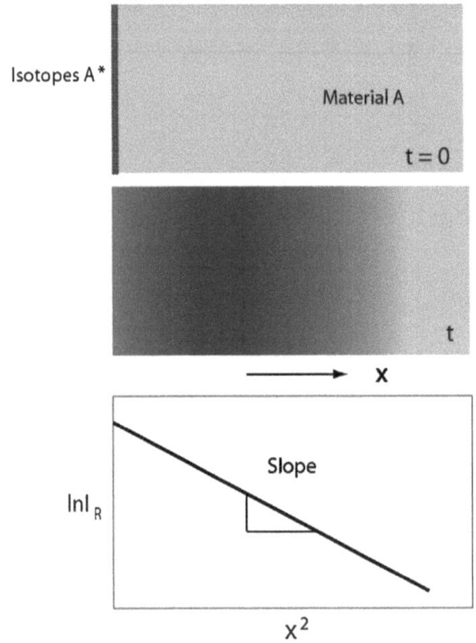

Figure 3.4 Schematic description of the estimation of the tracer diffusion coefficient utilizing the thin film solution.

driving forces just because of thermal energy. The radiotracer technique is used to measure this self-diffusion coefficient, for example, diffusion of component A in A or impurity diffusion coefficient, for example, diffusion of B in A. To clarify further, suppose that component A has enough thermal energy so that diffusion of atoms is happening at a reasonable rate. However, because of very small size, we cannot follow the diffusion. Therefore, radioisotope A* is deposited on A and annealed at the desired temperature, as explained in Fig. 3.4. The difference between these two types of atom is just one or two neutron masses. Different atoms diffuse different lengths because of a random walk [1]. After experimenting for a certain time, the specimen is sliced at different known distances along the x direction. Then the relative concentration of radioisotopes is measured at these distances by measuring intensities of the emitted rays. With respect to relative intensities, Eq. (3.11) can be expressed as

$$I_R(x) = \frac{1}{\sqrt{\pi D_A^* t}} \exp\left(-\frac{x^2}{4D_A^* t}\right), \tag{3.13a}$$

or equivalently,

$$\ln I_R = \ln\left(\frac{1}{\sqrt{\pi D_A^* t}}\right) - \frac{x^2}{4D_A^* t}, \tag{3.13b}$$

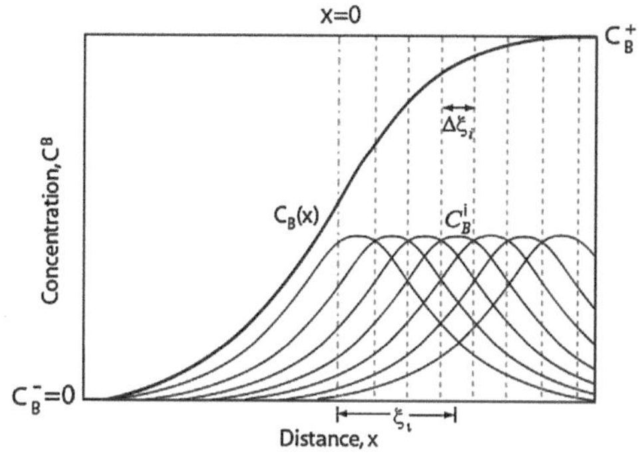

Figure 3.5 The derivation of the error function analysis.

where I_R is the specific or relative intensity measured, D_A^* is the tracer diffusion coefficient of component A. From the slope $-\frac{1}{4D_A^*t}$ in the $\ln I_R$ vs. x^2 plot, as shown in Fig. 3.4, D_A^* can be calculated using the known annealing time. As already discussed in Chapter 1 of Volume 1, self- and tracer diffusion coefficients are related by [1]

$$D_A^* = fD_S^A \tag{3.14}$$

where f is the correlation factor and D_S^A is the self-diffusion coefficient of A. In a similar experimental method when the diffusion rate of a radioisotope of B is studied, it is called impurity diffusion coefficient of B in A, D_{imp,B^*}^A. It should be noted here that there is no problem in considering a constant value of D for this thin film solution of Fick's second law since the diffusion coefficient is measured in a fixed composition. The radiotracer method is discussed in detail in Chapter 2 of Volume 1.

3.2.2 Error Function Analysis in a Semi-infinite Diffusion Couple

A semi-infinite diffusion couple is the couple of two dissimilar materials at an elevated temperature for a certain annealing time such that the ends of the materials are not affected by diffusion of components. In general, the lengths of the blocks are considered to be semi-infinite when the diffusion length is $> 10\sqrt{Dt}$. In a broader sense, even the presence of a very thin unaffected layer at the ends of the couple is sufficient for the analysis. The treatment is similar when a piece of material is exposed to the gaseous medium, for example, as in the case of steel hardening by carburization, similar to the profile shown in Fig. 3.1. With increasing annealing time, the penetration depth increases such that the concentration changes at a particular position. For our derivation, we consider a diffusion couple of pure component A ($C_B^- = 0$) and an alloy of A–B binary system ($C_B = C_B^+$), as shown in Fig. 3.5. The initial boundary conditions are [4]

$$C_B^- = 0 \text{ if } x < 0 \text{ at time } t = 0,$$
$$C_B = C_B^+ \text{ if } x > 0 \text{ at time } t = 0.$$

Note here that following the sign convention in the diffusion community, a minus sign is used in the superscript for the material placed on the left-hand side, and a positive sign is used in the superscript for the material placed on the right-hand side of the diffusion couple. It does not indicate which has a larger value.

The solution in this case can be developed using the solution already described for the thin-film condition in the previous section. We assume that the block with the concentration C_B^+ is made of n very thin slices of thickness $\Delta\xi$ and diffusion out of any component from any of the blocks is not affected by the other component. In slice $\Delta\xi_i$, the total amount of B can be written as $M_B = C_B^+ \Delta\xi_i$. Therefore, the diffusion profile from this slice can be written as [4]

$$C_B^i(x) = \frac{C_B^+ \Delta\xi_i}{2\sqrt{\pi D_B t}} \exp\left[-\frac{(x-\xi_i)^2}{4D_B t}\right]. \tag{3.15}$$

Note here that x in Eq. (3.11) is replaced by $(x - \xi_i)$, since the slice of interest for the present case is located at a distance ξ_i from the initial contact plane, $x = 0$. Similarly, we can write diffusion of components from all the slices, and the overall concentration at any position x after superposition can be written as

$$C_B(x) = \frac{C_B^+}{2\sqrt{\pi D_B t}} \sum_{i=1}^{n} \exp\left[-\frac{(x-\xi_i)^2}{4D_B t}\right] \Delta\xi_i. \tag{3.16}$$

The concentration in the integral form can be written as

$$C_B(x) = \frac{C_B^+}{2\sqrt{\pi D_B t}} \int_0^{\infty} \exp\left[-\frac{(x-\xi)^2}{4D_B t}\right] d\xi. \tag{3.17}$$

Let us assume that

$$\eta = \frac{x-\xi}{2\sqrt{D_B t}}. \tag{3.18}$$

Differentiating we get

$$d\eta = -\frac{d\xi}{2\sqrt{D_B t}}. \tag{3.19}$$

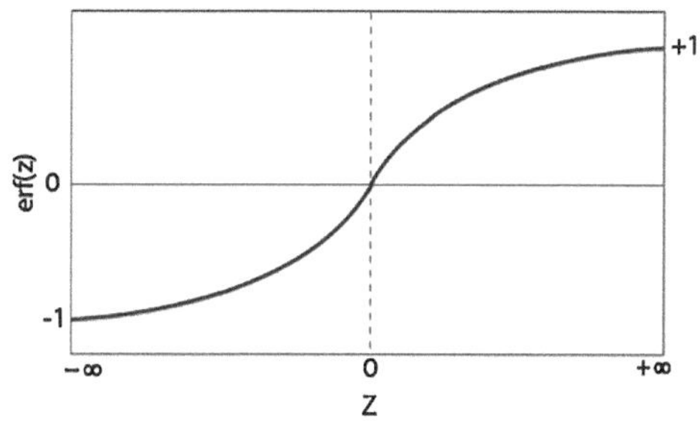

Figure 3.6 The variation of erf(z) with z.

When $\xi = 0$, we have $\eta = \frac{x}{2\sqrt{D_B t}}$, and for $\xi = \infty$, we obtain $\eta = -\infty$. Therefore, Eq. (3.17) can be written as

$$C_B(x) = -\frac{C_B^+}{\sqrt{\pi}} \int_{x/(2\sqrt{D_B t})}^{-\infty} \exp(-\eta^2) d\eta = \frac{C_B^+}{\sqrt{\pi}} \int_{-\infty}^{x/(2\sqrt{D_B t})} \exp(-\eta^2) d\eta. \quad (3.20)$$

The integral in Eq. (3.20) is not straightforward to compute. Therefore, it is written in terms of the error function. The error function of z, i.e., the value of the integral from 0 to z is expressed as

$$\text{erf(z)} = \frac{2}{\sqrt{\pi}} \int_0^z \exp(-\eta^2) d\eta. \quad (3.21)$$

It can be expanded as

$$\text{erf}(z) = \frac{2}{\sqrt{\pi}} \sum_{n-0}^{\infty} \frac{(-1)^n z^{2n+1}}{n!(2n+1)} = \frac{2}{\sqrt{\pi}} \left(z - \frac{z^3}{1! \times 3} + \frac{z^5}{2! \times 5} - \frac{z^7}{3! \times 7} \cdots \right). \quad (3.22)$$

The values of erf(z) are listed in Table 3.1 and plotted in Fig. 3.6 for both positive and negative values of z. It should be noted that erf($-z$) = − erf(z).

Eq. (3.20) can be rewritten as

$$C_B(x) = \frac{C_B^+}{\sqrt{\pi}} \left[\int_{-\infty}^{0} \exp(-\eta^2) d\eta + \int_0^{x/(2\sqrt{D_B t})} \exp(-\eta^2) d\eta \right],$$

which yields

$$C_B(x) = \frac{C_B^+}{\sqrt{\pi}} \left[-\int_0^{-\infty} \exp(-\eta^2) d\eta + \int_0^{x/(2\sqrt{D_B t})} \exp(-\eta^2) d\eta \right]$$

and so

$$C_B(x) = \frac{C_B^+}{2} \left[-\text{erf}(-\infty) + \text{erf}\left(\frac{x}{2\sqrt{D_B t}} \right) \right]. \quad (3.23)$$

Table 3.1 Values of erf(z) for different z

z	0	1	2	3	4	5	6	7	8	9
0.0	0.00000	0.01128	0.02256	0.03384	0.04511	0.05637	0.06762	0.07886	0.09008	0.10128
0.1	0.11246	0.12362	0.13476	0.14587	0.15695	0.16800	0.17901	0.18999	0.20094	0.21184
0.2	0.22270	0.23352	0.24430	0.25502	0.26570	0.27633	0.28690	0.29742	0.30788	0.31828
0.3	0.32863	0.33891	0.34913	0.35928	0.36936	0.37938	0.38933	0.39921	0.40901	0.41874
0.4	0.42839	0.43797	0.44747	0.45689	0.46623	0.47548	0.48466	0.49375	0.50275	0.51167
0.5	0.52050	0.52924	0.53790	0.54646	0.55494	0.56332	0.57162	0.57982	0.58792	0.59594
0.6	0.60386	0.61168	0.61941	0.62705	0.63459	0.64203	0.64938	0.65663	0.66378	0.67084
0.7	0.67780	0.68467	0.69143	0.69810	0.70468	0.71116	0.71754	0.72382	0.73001	0.73610
0.8	0.74210	0.74800	0.75381	0.75952	0.76514	0.77067	0.77610	0.78144	0.78669	0.79184
0.9	0.79691	0.80188	0.80677	0.81156	0.81627	0.82089	0.82542	0.82987	0.83423	0.83851
1.0	0.84270	0.84681	0.85084	0.85478	0.85865	0.86244	0.86614	0.86977	0.87333	0.87680
1.1	0.88021	0.88353	0.88679	0.88997	0.89308	0.89612	0.89910	0.90200	0.90484	0.90761
1.2	0.91031	0.91296	0.91553	0.91805	0.92051	0.92290	0.92524	0.92751	0.92973	0.93190
1.3	0.93401	0.93606	0.93807	0.94002	0.94191	0.94376	0.94556	0.94731	0.94902	0.95067
1.4	0.95229	0.95385	0.95538	0.95686	0.95830	0.95970	0.96105	0.96237	0.96365	0.96490
1.5	0.96611	0.96728	0.96841	0.96952	0.97059	0.97162	0.97263	0.97360	0.97455	0.97546
1.6	0.97635	0.97721	0.97804	0.97884	0.97962	0.98038	0.98110	0.98181	0.98249	0.98315
1.7	0.98379	0.98441	0.98500	0.98558	0.98613	0.98667	0.98719	0.98769	0.98817	0.98864
1.8	0.98909	0.98952	0.98994	0.99035	0.99074	0.99111	0.99147	0.99182	0.99216	0.99248
1.9	0.99279	0.99309	0.99338	0.99366	0.99392	0.99418	0.99443	0.99466	0.99489	0.99511
2.0	0.99532	0.99552	0.99572	0.99591	0.99609	0.99626	0.99642	0.99658	0.99673	0.99688
2.1	0.99702	0.99715	0.99728	0.99741	0.99753	0.99764	0.99775	0.99785	0.99795	0.99805
2.2	0.99814	0.99822	0.99831	0.99839	0.99846	0.99854	0.99861	0.99867	0.99874	0.99880
2.3	0.99886	0.99891	0.99897	0.99902	0.99906	0.99911	0.99915	0.99920	0.99924	0.99928
2.4	0.99931	0.99935	0.99938	0.99941	0.99944	0.99947	0.99950	0.99952	0.99955	0.99957
2.5	0.99959	0.99961	0.99963	0.99965	0.99967	0.99969	0.99971	0.99972	0.99974	0.99975
2.6	0.99976	0.99978	0.99979	0.99980	0.99981	0.99982	0.99983	0.99984	0.99985	0.99986
2.7	0.99987	0.99987	0.99988	0.99989	0.99989	0.99990	0.99991	0.99991	0.99992	0.99992
2.8	0.99992	0.99993	0.99993	0.99994	0.99994	0.99994	0.99995	0.99995	0.99995	0.99996
2.9	0.99996	0.99996	0.99996	0.99997	0.99997	0.99997	0.99997	0.99997	0.99997	0.99998
3.0	0.99998	0.99998	0.99998	0.99998	0.99998	0.99998	0.99998	0.99999	0.99999	0.99999
3.1	0.99999	0.99999	0.99999	0.99999	0.99999	0.99999	0.99999	0.99999	0.99999	0.99999
3.2	0.99999	0.99999	0.99999	1.00000	1.00000	1.00000	1.00000	1.00000	1.00000	1.00000

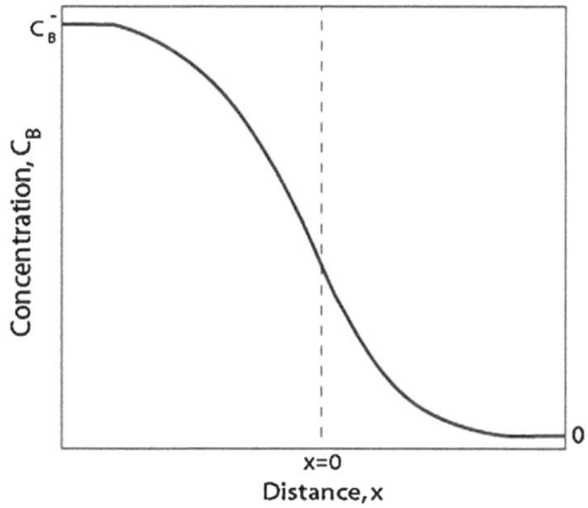

Figure 3.7 An imaginary composition profile of a diffusion couple prepared with $C_B = C_B^-$ and $C_B = 0$.

Since $\text{erf}(-\infty) = -1$, Eq. (3.23) can be written as

$$C_B(x) = \frac{C_B^+}{2}\left[1 + \text{erf}\left(\frac{x}{2\sqrt{D_B t}}\right)\right], \tag{3.24a}$$

or equivalently,

$$\frac{C_B(x)}{C_B^+} = \frac{1}{2}\left[1 + \text{erf}\left(\frac{x}{2\sqrt{D_B t}}\right)\right]. \tag{3.24b}$$

The position x is the distance from the initial contact plane ($x = 0$) of the diffusion couple. A positive and negative sign should be considered depending on the relative position of interest with respect to the initial contact plane. It can be immediately noticed that for $x = 0$, $C_B(x) = \frac{C_B^+}{2}[1 + 0] = \frac{C_B^+}{2}$. Therefore, the location of the initial contact plane is in the middle of the concentration profile, since our analysis is performed assuming a constant diffusion coefficient at all concentrations in the diffusion couple.

Readers might get confused sometimes because of a negative instead of a positive sign used in Eq. (3.24) as expressed in Eq. (3.25). This is written when the concentration profile is just the opposite so that it is the diffusion couple of C_B^- and $C_B^+ = 0$, as shown in Fig. 3.7:

$$\frac{C_B(x)}{C_B^-} = \frac{1}{2}\left[1 - \text{erf}\left(\frac{x}{2\sqrt{D_B t}}\right)\right] = \frac{1}{2}\text{erfc}\left(\frac{x}{2\sqrt{D_B t}}\right), \tag{3.25}$$

where $\text{erfc}(z)$ is the error function complement.

Practically one can consider any of the equations keeping in mind that a positive sign is used for the part in which concentration increases from the initial contact plane, and a negative sign should be used for the part in which concentration decreases from the initial contact plane. In the examples above, the concentration of one of the end-

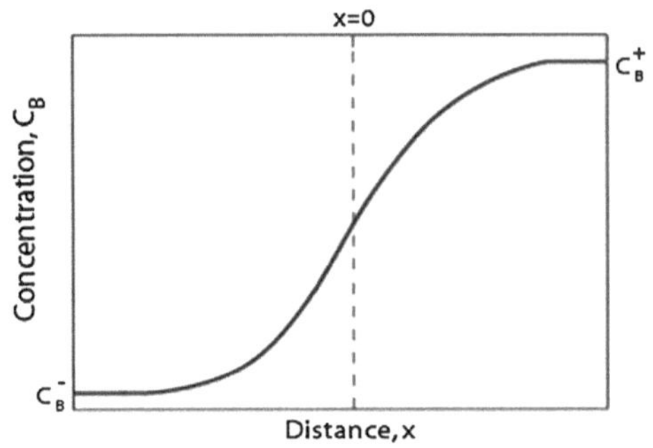

Figure 3.8 An imaginary composition profile in a diffusion couple of C_B^- and C_B^+ such that $C_B^+ > C_B^-$.

members is $C_B = 0$. If we consider the diffusion couple of C_B^- and C_B^+, as shown in Fig. 3.8, Eq. (3.24b) can be rewritten by normalizing the concentration profile with respect to the concentrations of the end members of the diffusion couple as

$$\frac{C_B(x) - C_B^-}{C_B^+ - C_B^-} = \frac{1}{2}\left[1 + \operatorname{erf}\left(\frac{x}{2\sqrt{D_B t}}\right)\right] \tag{3.26a}$$

and

$$C_B(x) = \frac{C_B^+ + C_B^-}{2} + \frac{C_B^+ - C_B^-}{2}\operatorname{erf}\left(\frac{x}{2\sqrt{D_B t}}\right). \tag{3.26b}$$

Therefore, the location of the initial contact plane can be found from

$$C_B(x=0) = \frac{C_B^- + C_B^+}{2}. \tag{3.26c}$$

The error function analysis is commonly used to calculate the diffusion depth of a carburized steel. In this process, a piece of steel is kept in a gas mixture of CH_4 and CO at the carburization temperature of interest. The carbon concentration at the surface can be kept constant depending on the ratio of the gasses supplied. With the increase in annealing time, the concentration profile of the carbon varies, as shown in Fig. 3.9. We assume that the initial concentration of carbon in the steel is C_0 and the carbon concentration at the surface is kept as C_S. Therefore, Eq. (3.26) can be rewritten to suit this situation as

$$\frac{C(x) - C_o}{C_S - C_o} = 1 - \operatorname{erf}\left(\frac{x}{2\sqrt{Dt}}\right) \tag{3.27a}$$

and

$$C(x) = C_S - (C_S - C_0)\operatorname{erf}\left(\frac{x}{2\sqrt{Dt}}\right). \tag{3.27b}$$

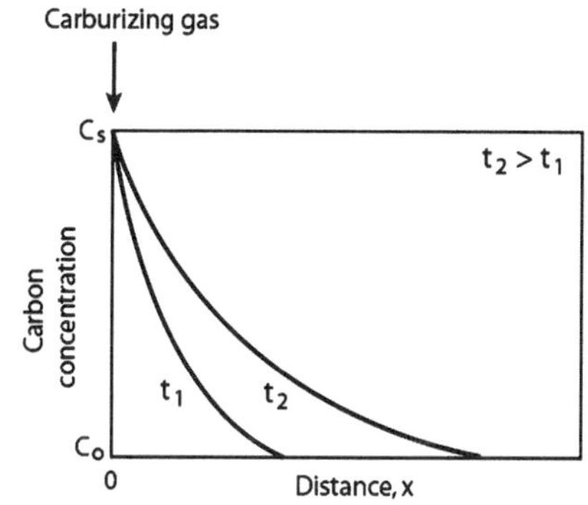

Figure 3.9 Imaginary composition profile variations with increasing annealing time during carburization.

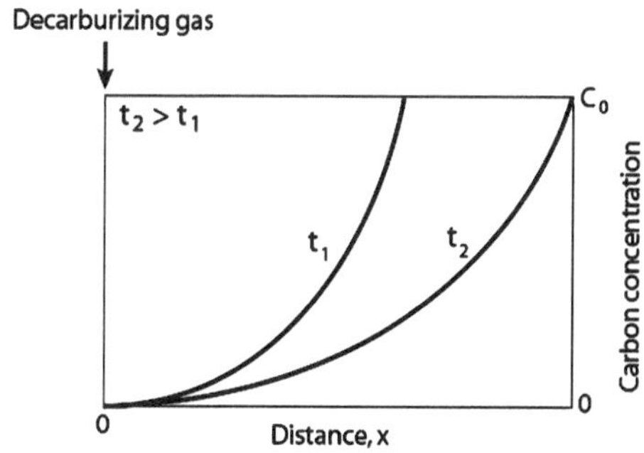

Figure 3.10 Imaginary composition profiles for different annealing times during decarburization.

Notice two differences when comparing Eqs. (3.26a) and (3.27a). A negative sign is used before the error function in Eq. (3.27a) since we are considering the situation in which concentration decreases from the surface. Secondly, the factor 1/2 is missing in Eq. (3.27a) since here we have only one side of the profile from $x = 0$ compared to the two sides in the previous example. When the material is used with almost zero carbon concentration, i.e., $C_o = 0$, Eq. (3.27) reduces to

$$C(x) = C_S\left[1 - \text{erf}\left(\frac{x}{2\sqrt{Dt}}\right)\right] = C_S\text{erfc}\left(\frac{x}{2\sqrt{Dt}}\right). \tag{3.28}$$

For decarburization treatment, as shown in Fig. 3.10, of steel with concentration C_0, the latter function can be written as

$$C(x) = C_0\,\text{erf}\left(\frac{x}{2\sqrt{Dt}}\right). \tag{3.29}$$

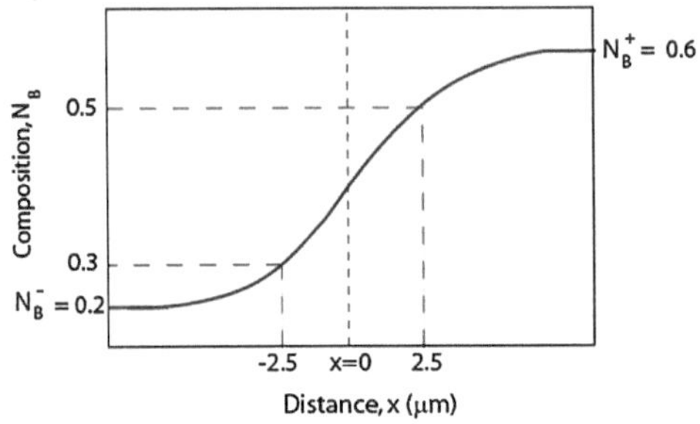

Figure 3.11 Explanation of the estimation of diffusion coefficient following the error function analysis.

Suppose the diffusion coefficient of carbon at the annealing temperature T_1 is D_1, and at T_2 it is D_2. If we wish to relate the time difference required to achieve a particular concentration at these two different temperatures, we can write Eq. (3.27a) as

$$\frac{x_1^2}{D_1 t_1} = \frac{x_2^2}{D_2 t_2}. \tag{3.30}$$

Estimation of Different Parameters Utilizing the Error Function Analysis

Suppose that we would like to estimate the diffusion coefficient in a diffusion couple made of two alloys in the binary A–B system. The composition (in atomic or mole fraction) of the left-hand side of the couple is $N_B^- = 0.2$ and the right-hand side of the couple is $N_B^+ = 0.6$ as shown in Fig. 3.11. Since we can use the error function analysis in a system with constant diffusion coefficient, we consider that this system has indeed constant diffusivity at all compositions. Further, we consider that the lattice parameter, and therefore the molar volume, does not vary with the composition. Therefore, Eq. (3.26) can be rewritten to suit this condition as

$$\frac{N_B(x) - N_B^-}{N_B^+ - N_B^-} = \frac{1}{2}\left[1 + \text{erf}\left(\frac{x}{2\sqrt{D_B t}}\right)\right] \quad \text{for } x > 0, \tag{3.31a}$$

$$\frac{N_B(x) - N_B^-}{N_B^+ - N_B^-} = \frac{1}{2}\left[1 - \text{erf}\left(\frac{x}{2\sqrt{D_B t}}\right)\right] \quad \text{for } x < 0. \tag{3.31b}$$

Note that $C_B = \frac{N_B}{v_m} = \frac{X_B}{v_m}$,

$$\text{since } N_B = \frac{n_B}{n_A + n_B} = \frac{n_B/N_o}{n_A/N_o + n_B/N_o} = \frac{x_B}{x_A + x_B} = X_B, \tag{3.32}$$

where composition of B, N_B, is in atomic fraction, n_A and n_B are the total numbers of atoms of component A and B in the system, respectively, N_o is the Avogadro number

(6.022×10^{23} per mole), x_A and x_B are the total numbers of moles of A and B, and X_B is the mole fraction of B. Suppose that a composition profile, as shown in Fig. 3.11, is developed after annealing for 25 hours at the temperature of interest. The location of the initial bonding plane, i.e., $x=0$ at which $\text{erf}(0)=0$ should have the composition of $N_B(x=0)=\frac{N_B^+-N_B^-}{2}+N_B^-=\frac{N_B^++N_B^-}{2}=\frac{0.6+0.2}{2}=0.4$.

Since it has constant diffusivity, if composition at the location of 2.5 μm is $N_B=0.5$, the composition at the location of −2.5 μm should be 0.3. Further, the same diffusivity should be estimated at any location. To test the correctness of the statement, let us first estimate the diffusivity at 2.5 μm at which the composition is 0.5. We have $N_B^+=0.6$, $N_B^-=0.2$, $x=2.5\times10^{-6}$ m, and $t=25\times60\times60$ s. Therefore, Eq. (3.31a) can be rewritten as

$$\frac{0.5-0.2}{0.6-0.2}=\frac{1}{2}\left[1+\text{erf}\left(\frac{2.5\times10^{-6}}{2\sqrt{D_B\times25\times60\times60}}\right)\right],\quad \text{where}$$
$$\text{erf}\left(\frac{2.5\times10^{-6}}{2\sqrt{D\times90000}}\right)=0.5. \tag{3.33a}$$

In Table 3.1, the value of z exactly giving $\text{erf}(z)=0.5$ is not listed. We have the values $\text{erf}(z)=0.49375$ for $z=0.47$ and $\text{erf}(z)=0.50275$ for $z=0.48$. Since the values are very close to each other, considering a linear interpolation, the value of z for $\text{erf}(z)=0.5$ can be estimated as

$$\frac{z-0.47}{0.48-0.47}=\frac{0.5-0.49375}{0.50275-0.49375}\qquad z=0.477.$$

Therefore, we can write

$$\text{erf}(0.477)\approx0.5. \tag{3.33b}$$

Comparing Eqs. (3.33a) and (3.33b), we get

$$\frac{2.5\times10^{-6}}{2\sqrt{90000\times D}}=0.477,$$

which yields

$$D=7.6\times10^{-17}\ \text{m}^2/\text{s}.$$

Now we consider $N_B=0.3$ at the location of −2.5 μm. Then, Eq. (3.31b) can be written as

$$\frac{0.3-0.2}{0.6-0.2}=\frac{1}{2}\left[1-\text{erf}\left(\frac{2.5\times10^{-6}}{2\sqrt{90000\times D}}\right)\right],$$

which gives

$$\operatorname{erf}\left(\frac{2.5 \times 10^{-6}}{2\sqrt{D \times 90000}}\right) = 0.5.$$

Therefore, it has the same relation as expressed in Eq. (3.33a) and will lead to the same value of the estimated diffusion coefficient.

Now let us do another type of useful calculation. Sometimes we know the time taken for a particular diffusion profile to develop at a certain temperature. If we are interested in knowing the annealing time required at another temperature at which the diffusion coefficient is known, then we can directly use Eq. (3.30) since concentration values are the same in both cases. Suppose that we are interested in annealing at a higher temperature where the diffusion coefficient is larger than the already estimated value of $D = 1.05 \times 10^{-16}$ m^2/s. We are interested in estimating how much time it will take to achieve $N_B = 0.5$ at the position $x = 2.5$ μm. Therefore, Eq. (3.30) can be rewritten as $t_2 = \frac{D_1 t_1}{D_2}$ where $D_1 = 7.6 \times 10^{-17}$ m^2/s, $t_1 = 25$ hours and $D_2 = 1.05 \times 10^{-16}$ m^2/s. Thus we estimate that the time required is 18.1 hours.

3.2.3 Solution for Homogenization (Separation of Variables)

We now turn our discussion to the situation where the material system reaches complete homogenization after relatively long annealing times. In a particular system, the concentration varies with respect to two variables such as time t and position x. Therefore, we can write [4]

$$C(x, t) = X(x)T(t). \tag{3.34}$$

Partial differentiation leads to

$$\frac{dC}{dt} = X\frac{dT}{dt}, \tag{3.35a}$$
$$\frac{d^2C}{dx^2} = T\frac{d^2X}{dx^2}. \tag{3.35b}$$

Replacing Eq. (3.35) in Fick's second law as expressed in Eq. (3.5b), we can write

$$X\frac{dT}{dt} = DT\frac{d^2X}{dx^2},$$
$$\frac{1}{DT}\frac{dT}{dt} = \frac{1}{X}\frac{d^2X}{dx^2}. \tag{3.36}$$

Therefore, the equation is separated into two parts with respect to t and x. Both sides of the equation should have an equal value. For the sake of convenience and physical significance, which will be understood in the end of derivation, we consider it as $-\lambda^2$.

Therefore, we can write

$$\frac{1}{DT}\frac{dT}{dt} = -\lambda^2,$$
$$\frac{dT}{T} = -\lambda^2 Ddt. \tag{3.37}$$

Integration yields

$$\int_{T_o}^{T}\frac{dT}{T} = -\lambda^2 D\int_{o}^{t} dt,$$
$$\ln\frac{T}{T_o} = -\lambda^2 Dt,$$
$$T(t) = T_o\exp\left(-\lambda^2 Dt\right), \tag{3.38}$$

where T_o is the value at time $t = 0$. The reason for considering a negative sign for the constant value in Eq. (3.37) can be understood here. It indicates that T decreases exponentially with the increasing annealing time. Similarly,

$$\frac{1}{X}\frac{d^2X}{dx^2} = -\lambda^2,$$
$$\frac{d^2X}{dx^2} + \lambda^2 X = 0.$$

Integration gives

$$X(x) = \left(A'\sin\lambda x + B'\cos\lambda x\right), \tag{3.39}$$

where A' and B' are the constants.

Substituting Eqs. (3.38) and (3.39) into Eq. (3.34), we get

$$C(x,t) = \left(A'\sin\lambda x + B'\cos\lambda x\right)T_o\exp\left(-\lambda^2 Dt\right) = \left(A\sin\lambda x + B\cos\lambda x\right)\exp\left(-\lambda^2 Dt\right)$$

where $A = A'T_o$ and $B = B'T_o$. It should be noted that this solution is applicable for any values of λ. When there are different values of λ, the solution should be the sum of all the values and can be expressed as

$$C(x,t) = \sum_{m=1}^{\infty}\left(A_m\sin\lambda_m x + B_m\cos\lambda_m x\right)\exp\left(-\lambda_m^2 Dt\right), \tag{3.40}$$

where A_m, B_m and λ_m can be determined in a particular system of interest. We further consider a thin sheet of thickness h with a constant concentration C_o. Suppose, the concentration at the surfaces of the sheet is kept at zero such that the boundary conditions can be written as

$$C = C_o \quad \text{for } 0 < x < h \text{ at } t = 0,$$
$$C = 0 \quad \text{for } x = 0 \text{ and } x = h \text{ at } t > 0.$$

The second boundary condition, i.e., the zero concentration at $x = 0$ at any time, can be met only if $B_m = 0$. Moreover, C being zero at $x = h$ is possible only when $\lambda_m = m\pi/h$, where m is a positive integer. Replacing these values in Eq. (3.40), we get

$$C(x,t) = \sum_{m=1}^{\infty}\left(A_m \sin\frac{m\pi}{h}x\right)\exp\left(-\frac{m^2\pi^2}{h^2}Dt\right). \tag{3.41}$$

Now from the first boundary condition for $t = 0$, we can write

$$C_o = \sum_{m=1}^{\infty}\left(A_m \sin\frac{m\pi}{h}x\right). \tag{3.42}$$

Next step is to find the value of A_m. Multiplying both sides by $\sin\frac{n\pi}{h}xdx$ and integrating from 0 to h, we get

$$\int_o^h C_o \sin\left(\frac{n\pi}{h}x\right)dx = \sum_{m=1}^{\infty}\left(A_m \int_o^h \sin\frac{n\pi}{h}x \sin\frac{m\pi}{h}xdx\right). \tag{3.43}$$

All the integrals on the right-hand side are zero except for $n = m$ at which the value of the integral is $h/2$. Therefore, we can write

$$A_m = \frac{2}{h}\int_o^h C_o \sin\left(\frac{m\pi}{h}x\right)dx. \tag{3.44}$$

The value of integration is 0 for all even values of m and is $4C_o/m\pi$ for all odd values of m. To use this relation for every integer, the summation over the integer values can be changed such that it gives odd values for every integer. This can be written as

$$A_m = A_j = \frac{4C_o}{(2j+1)\pi} \quad j = 0, 1, 2\ldots$$

Eq. (3.41) can therefore be written as

$$C(x,t) = \frac{4C_o}{\pi}\sum_{j=0}^{\infty}\frac{1}{2j+1}\sin\frac{(2j+1)\pi x}{h}\exp\left(-\frac{(2j+1)^2\pi^2}{h^2}Dt\right). \tag{3.45}$$

This solution considers the summation over all integers j from zero to infinity. It should be noted from the negative sign inside the exponent that the value for every successive j

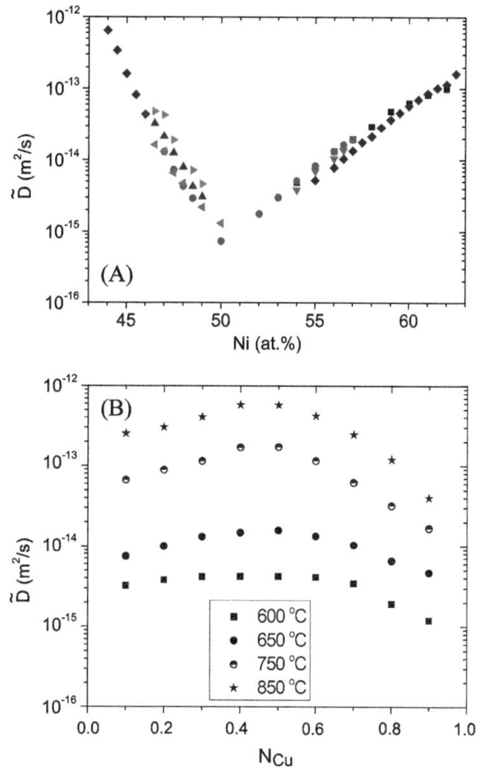

Figure 3.12 Variation of the diffusion coefficient with the composition in (A) β-NiAl at 1000°C [5] and (B) Au–Cu system [6].

decreases exponentially. It should, therefore, be possible to represent the concentration with the first few values of j.

3.2.4 Limitations of Analysis Considering Constant Diffusion Coefficient

In many systems, diffusion coefficients might vary by one to several orders of magnitude with composition, as shown in Fig. 3.12 [5,6]. Therefore, the error function analysis assuming a constant diffusion coefficient cannot be used to estimate the interdiffusion coefficients. It is not possible to solve Fick's second law of diffusion considering the variation in the diffusion coefficient with composition since a particular composition changes location with time in the interdiffusion zone. To clarify, considering the variation of interdiffusion coefficient with composition, Fick's second law can be written as

$$\frac{\partial C}{\partial t} = \frac{\partial}{\partial x}\left(\tilde{D}\frac{\partial C}{\partial x}\right) = \frac{\partial \tilde{D}}{\partial x}\frac{\partial C}{\partial x} + \tilde{D}\frac{\partial^2 C}{\partial x^2}. \tag{3.46}$$

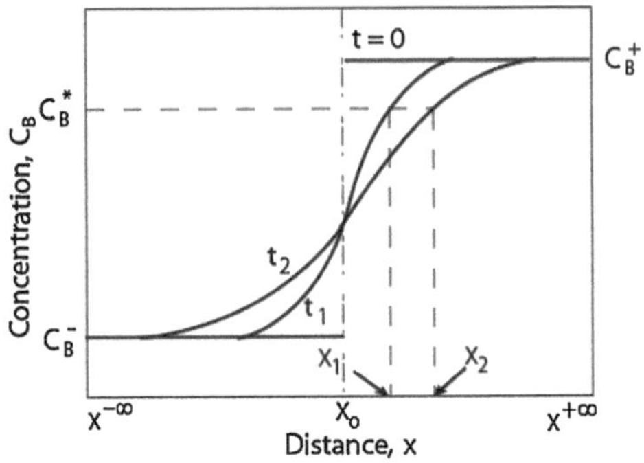

Figure 3.13 Explanation of Boltzmann parameter.

The term $\partial\tilde{D}/\partial x$ makes the equation inhomogeneous, and solving it in closed form is not possible.

However, instead of finding a solution for $C(x, t)$ as done in Section 3.2.2, it is possible to find the solution for $\tilde{D} = \tilde{D}(C)$, that is, the variation of the interdiffusion coefficient with concentration following the Matano–Boltzmann treatment.

3.3 MATANO–BOLTZMANN ANALYSIS FOR THE ESTIMATION THE VARIABLE INTERDIFFUSION COEFFICIENTS

Fick's second law was solved by Matano [7] with the help of Boltzmann parameter for the estimation of the composition-dependent interdiffusion coefficients. Although this procedure has its own serious limitation, especially in the systems with a non-ideal variation of the molar volume, it is important to go through this analysis, since modified treatments for the estimation of the diffusion coefficients are developed based on it. Occasionally, this method can be even followed, when molar volume variation from the ideal condition is not significant, as discussed in Section 3.7.

3.3.1 Derivation of the Relation Given by Matano

We should first understand the significance of Boltzmann parameter before proceeding to the derivation of Matano relation. Suppose that two alloys with initial compositions of C_B^- and C_B^+ are diffusion coupled and annealed at the desired temperature for a particular time. Time should be short enough such that ends of the diffusion couple are unaffected by interdiffusion process as shown in Fig. 3.13. Therefore, the boundary conditions are

$$\begin{aligned} C_B &= C_B^- \quad \text{for } x < 0 \text{ at } t = 0, \\ C_B &= C_B^+ \quad \text{for } x > 0 \text{ at } t = 0. \end{aligned} \tag{3.47}$$

As previously mentioned "−" and "+" represent the left- and right-hand ends of the unaffected parts of the diffusion couple. They do not reflect the relative values of the concentrations.

We introduce Boltzmann parameter

$$\lambda = \lambda(C_B) = (x - x_o)/t^{\frac{1}{2}} = x/t^{\frac{1}{2}}. \tag{3.48}$$

The location of the initial contact plane (bonding interface) is $x_o = 0$. This is often called Matano plane, and x is the location of a concentration with respect to Matano plane. The relation expressed in Eq. (3.48) indicates that every concentration, for example, C_B^* moves from Matano plane such that Boltzmann parameter for this concentration $\lambda^* = \lambda(C_B^*) = x^*/t^{\frac{1}{2}}$ has a fixed value. This is clarified in Fig. 3.13. Suppose after annealing time t_1, concentration C_B^* is located at x_1. With increasing annealing time t_2, suppose the same concentration is shifted to x_2. According to Boltzmann parameter, when parabolic diffusion-controlled process prevails, the plane corresponding to C_B^* moves such that

$$\lambda^* = \lambda\left(C_B^*\right) = x_1/t_1^{\frac{1}{2}} = x_2/t_2^{\frac{1}{2}} = \text{const.}$$

Therefore, every concentration in the interdiffusion zone has a fixed value of λ. Of course, this value for a particular concentration in a system will be different depending on the concentrations of the end members in different diffusion couples. The diffusion-controlled process can be verified by plotting λ vs. $x/t^{\frac{1}{2}}$ after conducting the experiments for different annealing times. The variation of λ, estimated for different annealing times, with respect to $x/t^{\frac{1}{2}}$ will be the same.

Utilizing Boltzmann parameter, Matano derived the expression for the estimate of the interdiffusion coefficient as described below. From Eq. (3.48), we can write

$$\begin{aligned} \frac{\partial C_B}{\partial t} &= \frac{\partial C_B}{\partial \lambda}\frac{\partial \lambda}{\partial t} = -\frac{1}{2}\frac{x}{t^{\frac{3}{2}}}\frac{\partial C_B}{\partial \lambda} \\ \frac{\partial C_B}{\partial x} &= \frac{\partial C_B}{\partial \lambda}\frac{\partial \lambda}{\partial x} = \frac{1}{t^{\frac{1}{2}}}\frac{\partial C_B}{\partial \lambda} \end{aligned} \tag{3.49}$$

Substituting the relations of Eq. (3.49) into Fick's second law as expressed in Eq. (3.5a) gives

$$\begin{aligned} &\frac{\partial C_B}{\partial t} = \frac{\partial}{\partial x}\left(\tilde{D}\frac{\partial C}{\partial x}\right) \\ &-\frac{x}{2t^{\frac{3}{2}}}\frac{\partial C_B}{\partial \lambda} = \frac{\partial}{\partial x}\left(\frac{\tilde{D}}{t^{\frac{1}{2}}}\frac{\partial C_B}{\partial \lambda}\right) = \frac{\partial}{t^{\frac{1}{2}}\partial x}\left(\tilde{D}\frac{\partial C_B}{\partial \lambda}\right). \end{aligned} \tag{3.50}$$

Further, for a fixed annealing time, we can write $\partial\lambda = \partial x/t^{\frac{1}{2}}$. Therefore, Eq. (3.50) can be written as

$$-\frac{x}{2t^{\frac{3}{2}}}\frac{\partial C_B}{\partial\lambda} = \frac{1}{t}\frac{\partial}{\partial\lambda}\left(\tilde{D}\frac{\partial C_B}{\partial\lambda}\right).$$

Utilizing Eq. (3.48), with respect to the variable λ we obtain

$$-\frac{\lambda}{2}\frac{dC_B}{d\lambda} = \frac{d}{d\lambda}\left(\tilde{D}\frac{dC_B}{d\lambda}\right), \tag{3.51}$$

$$-\frac{1}{2}\lambda dC_B = d\left(\tilde{D}\frac{dC_B}{d\lambda}\right). \tag{3.52}$$

The initial boundary conditions expressed in Eq. (3.47) at time $t = 0$ can be rewritten as

$$\begin{aligned} C &= C_B^- \quad \text{at } \lambda = -\infty \\ C &= C_B^+ \quad \text{at } \lambda = +\infty \end{aligned} \tag{3.53}$$

After integrating Eq. (3.52) from the initial unaffected end member concentration, C_B^-, to the concentration of interest, C_B^*, to measure the interdiffusion coefficient, $\tilde{D}$, we obtain

$$-\frac{1}{2}\int_{C_B^-}^{C_B^*}\lambda dC_B = \tilde{D}\frac{dC_B}{d\lambda}\bigg|_{C_B^-}^{C_B^*}. \tag{3.54}$$

The data is collected after a fixed annealing time so that t is constant. After replacing the Boltzmann parameter, we get

$$-\frac{1}{2}\int_{C_B^-}^{C_B^*} xdC_B = \tilde{D}t\frac{dC_B}{dx}\bigg|_{C_B^-}^{C_B^*} = \tilde{D}t\left(\frac{dC_B}{dx}\right)_{C_B^*} - \tilde{D}t\left(\frac{dC_B}{dx}\right)_{C_B^-} = \tilde{D}t\left(\frac{dC_B}{dx}\right)_{C_B^*}. \tag{3.55}$$

Note that $(\frac{dC_B}{dx})_{C_B^-}$ is zero at the unaffected part of the diffusion couple, where it has a fixed concentration of C_B^-, as shown in the Fig. 3.14. Therefore, Eq. (3.55) can be written as

$$\tilde{D}(C_B^*) = -\frac{1}{2t}\left(\frac{dx}{dC_B}\right)_{C_B^*}\int_{C_B^-}^{C_B^*} xdC_B. \tag{3.56}$$

Interdiffusion coefficient at the concentration of interest, for example, C_B^*, can be determined utilizing Eq. (3.56). It is easier to integrate with respect to x instead of C_B, which can be written after integration by parts as

$$\tilde{D}(C_B^*) = -\frac{1}{2t}\left(\frac{dx}{dC_B}\right)_{C_B^*}\left[x^*(C_B^* - C_B^-) - \int_{x^{-\infty}}^{x^*}(C_B - C_B^-)dx\right]. \tag{3.57a}$$

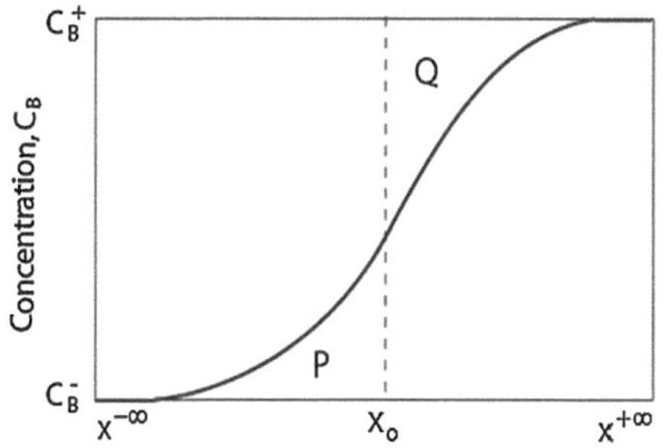

Figure 3.14 Finding the initial contact plane (Matano plane).

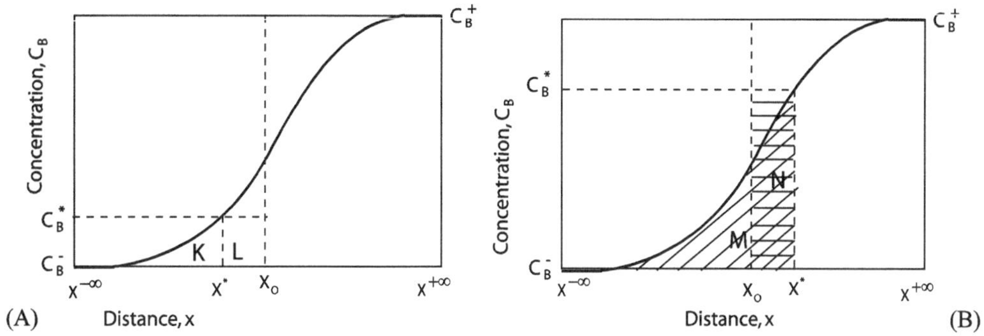

Figure 3.15 Different steps for estimation of the interdiffusion coefficient following Matano–Boltzmann analysis.

Therefore, the interdiffusion flux following Fick's first law of diffusion can be written as

$$\tilde{J}_B^* = \frac{1}{2t}\left[x^*\left(C_B^* - C_B^-\right) - \int_{x^{-\infty}}^{x^*}\left(C_B - C_B^-\right)dx\right]. \qquad (3.57b)$$

Note that $x^* = x^* - x_o$, since the location parameter is measured with respect to x_o. $x^{-\infty}$ is the location on the left at the unaffected end member of the diffusion couple. The estimation procedure is explained in Fig. 3.14. The location of the initial contact plane, i.e., Matano plane, can be determined by equalizing the areas P and Q. Once this location is set as zero, one can estimate the variation of interdiffusion coefficients across the affected interdiffusion zone. If we are interested in estimating this parameter at the negative position with respect to the initial contact plane, as shown in Fig. 3.15A, the areas K and L give the values inside the brackets in Eq. (3.57a), which can be rewritten as

$$\tilde{D}\left(C_B^*\right) = \frac{1}{2t}\left(\frac{dx}{dC_B}\right)_{C_B^*}[L + K]. \qquad (3.58)$$

Note that the interdiffusion coefficient will have a positive value, since the values of K and L are positive and negative, respectively, in this particular case for a concentration of interest, which is located before the initial contact plane. For the other side of the initial contact plane, as shown in Fig. 3.15B, Eq. (3.57) can be rewritten as

$$\tilde{D}(C_B^*) = \frac{1}{2t}\left(\frac{dx}{dC_B}\right)_{C_B^*}[M - N]. \tag{3.59}$$

Area M is shown by diagonal dashed lines; and area N is shown by straight lines. Since the value of M will always be larger than that of N, in this case also the interdiffusion coefficient will always be positive. Note here that in certain cases (uphill diffusion), the diffusion coefficient will have a negative value, but it will develop a different type of concentration profile. The concentration profile at a certain part will go up instead of going down monotonously [8].

For the analysis, the first important step is to locate the initial contact plane. From the mass balance condition, one can write

$$\int_{C_B^-}^{C_B^+} x dC_B = 0,$$

$$\int_{C_B^-}^{C_B^o} x dC_B + \int_{C_B^o}^{C_B^+} x dC_B = 0, \tag{3.60a}$$

$$\int_{x^{-\infty}}^{x_o} (C_B - C_B^-) dx = \int_{x_o}^{x^{+\infty}} (C_B^+ - C_B) dx. \tag{3.60b}$$

Note that the other term of integration by parts on both the sides is equal to zero since $x_o = 0.x^{+\infty}$ is the location at the unaffected end member of the diffusion couple on the right. Following Eq. (3.60), the location of x_o is found by equalizing the areas, as explained in Fig. 3.14.

If we assume a constant molar volume, Eq. (3.57a) reduces to

$$\tilde{D}(N_B^*) = -\frac{1}{2t}\left(\frac{dx}{dN_B}\right)_{N_B^*}\left[x^*(N_B^* - N_B^-) - \int_{x^{-\infty}}^{x^*} (N_B - N_B^-) dx\right]. \tag{3.61}$$

3.3.2 Limitation of Matano–Boltzmann Analysis

Matano–Boltzmann analysis is important in the sense that one can estimate the variation of the interdiffusion coefficient with a composition which cannot be done by the error function analysis. However, this has a serious limitation. This is related to the difficulties in locating the initial contact plane, i.e., Matano plane. This cannot be located exactly in most of the practical examples, and therefore it generates errors in estimates of the interdiffusion coefficients.

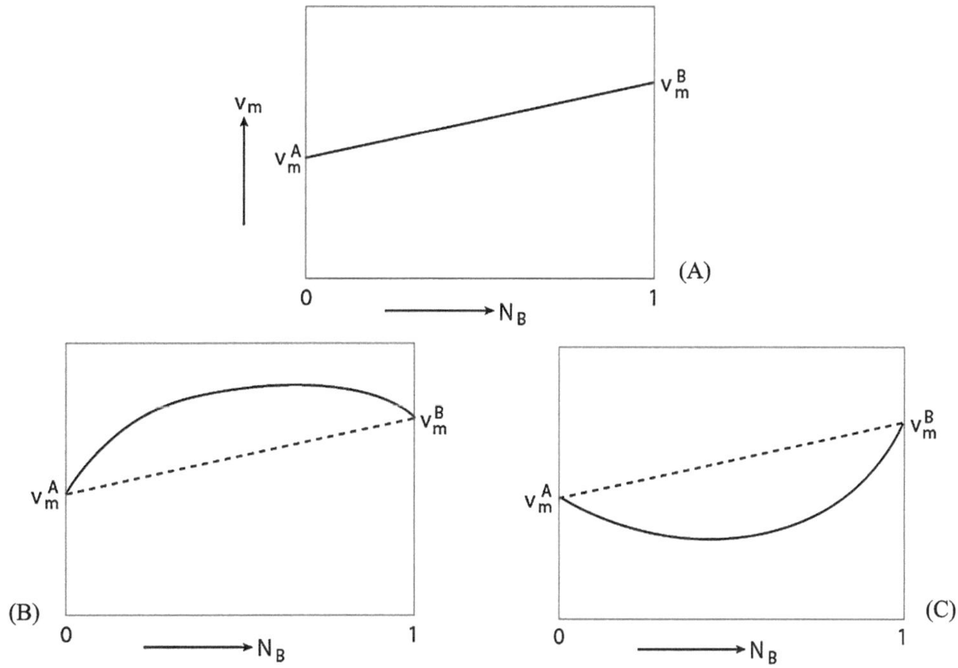

Figure 3.16 Molar volume variation with composition: (A) ideal, (B) positive deviation, and (C) negative deviation.

We have shown previously that the location of the initial contact plane can be located from the measured concentration profile C_B vs. x, utilizing the relation $\int_{x^{-\infty}}^{x_o}(C_B - C_B^-)dx = \int_{x_o}^{x^{+\infty}}(C_B^+ - C_B)dx$. This is done by equalizing the areas of P and Q in Fig. 3.14 since, from the concept of mass balance, the amount of a component lost from one end member is gained by the other. Similarly, we should be able to locate the same plane at the same position with respect to the concentration profile of A, i.e., C_A vs. x, utilizing the relation $\int_{x^{-\infty}}^{x_o}(C_A^- - C_A)dx = \int_{x_o}^{x^{+\infty}}(C_A - C_A^+)dx$. Note that this plane is unique, and therefore ideally the analysis should not give two different values when determined from the composition profiles of different components. However, this is true only when molar volume variation is constant or follows Vegard's law. Considering a constant molar volume over a wide composition range is not practical, but there could be some systems in which molar volume varies almost according to Vegard's law, as shown in Fig. 3.16A. In this case, the location of the initial contact plane determined by any profile of the components will be more or less similar if the deviation is not significant, and there will be hardly any change in total volume of the diffusion couple. However, most of the systems deviate positively or negatively from Vegard's law. This deviation might be significant in the presence of intermetallic compounds in the system. When the molar volume deviates positively, the interdiffusion zone grows with the compositions which have a larger molar volume than expected from Vegard's law, i.e., greater values compared to the value estimated by linear variation (dotted line) between v_m^A and v_m^B. Therefore, there will be an overall increase in the total volume of the diffusion couple, as shown in Fig. 3.16B, and two different values of the initial contact plane will be estimated when determined from C_A vs. x and C_B vs. x profiles. The opposite is true in the case of negative deviation, as shown in Fig. 3.16C, and there will be shrinkage in the diffusion couple. In this case also, two different values of the initial contact plane will be estimated. In fact, both determined locations will be wrong, and the actual location is not easy to determine. Therefore, it will lead to incorrect estimated values of the interdiffusion coefficients when determined following Matano–Boltzmann analysis. Even more, measured values with respect to the composition profile of A and B will give different values, leading to confusion. This will be shown mathematically in Section 3.6.

Therefore, there has been intense search for a relation in which there is no need to determine the initial contact plane. Ballufi [9] was the first to derive a relation for the estimation of the interdiffusion coefficient in a system with non-ideal molar volume variation. Sauer and Freise [10] generalized the relation developed by Matano for a similar condition. Later Wagner [11] developed the same relation as did Sauer and Freise. However, the steps made by Wagner are much easier to understand for the reader. Immediately after that den Broeder [12] proposed a graphical interpretation to directly modify Matano–Boltzmann analysis. In this chapter, the Den Broeder approach is explained first, since it is just a continuation of Matano–Boltzmann analysis. Then, Wagner's method is explained, which also lays the foundation for the estimation of a different type of diffusion parameter (integrated interdiffusion coefficient) used especially for the line compounds, in which one cannot estimate the concentration gradient.

It is important to note the term used in the literature related to the Matano plane, which is not strictly followed in this chapter. The plane, which is determined based on the concentration profiles, is called Matano plane. This may or may not coincide with the initial contact plane or the original interface depending on the variation of the molar volume. Therefore, it is stated that, for an ideal variation of the molar volume, Matano plane coincides with the initial contact plane whereas, when the molar volume variation is not ideal, Matano planes determined from the concentration profiles are different from the actual initial contact plane. In a broader sense, this is an attempt to determine the location of the initial contact plane. Therefore, since we have considered an ideal variation of the molar volume, we are able to locate the initial contact plane correctly, whereas we are unable to do it in the case of non-ideal variation of the molar volume.

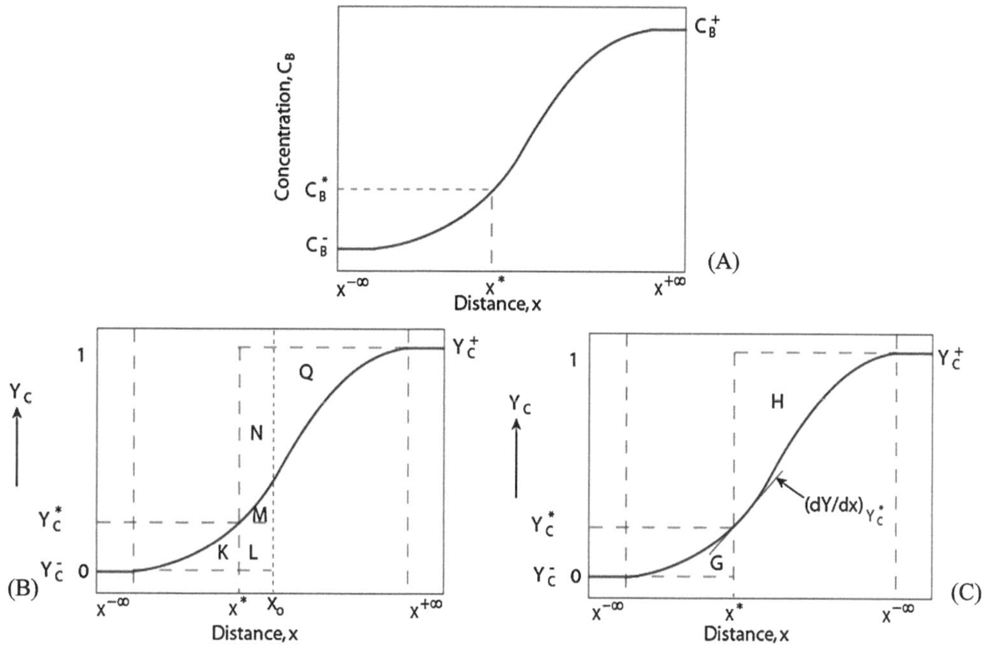

Figure 3.17 Estimation steps of the interdiffusion coefficients following den Broeder approach.

3.4 DEN BROEDER APPROACH TO DETERMINE THE INTERDIFFUSION COEFFICIENT

Den Broeder [12] developed an easy to follow graphical approach for the estimation of interdiffusion coefficients utilizing Matano–Boltzmann analysis. Let Y_C be a concentration normalizing variable expressed as

$$Y_C = \frac{C_B - C_B^-}{C_B^+ - C_B^-}. \tag{3.62}$$

Then we can write

$$dY_C = \frac{dC_B}{C_B^+ - C_B^-}. \tag{3.63}$$

Substituting Eqs. (3.62) and (3.63) into Eq. (3.57), we get

$$\tilde{D}(Y_C^*) = -\frac{1}{2t}\left(\frac{dx}{dY_C}\right)_{Y_C^*}\left[x^* Y_c^* - \int_{x^{-\infty}}^{x^*} Y_C dx\right]. \tag{3.64}$$

Therefore, first we need to plot the concentration profile with respect to Y_C vs. x, as shown in Fig. 3.17B. Since the concentration profile is normalized with respect to the unaffected end member concentrations, we have $Y_C^- = 0$ and $Y_C^+ = 1$. Eq. (3.64) can be expressed with respect to the areas, as shown in Fig. 3.17B, by

$$\tilde{D}(Y_C^*) = \frac{1}{2t}\left(\frac{\partial x}{\partial Y_C}\right)_{Y_C^*}[L + K]. \tag{3.65}$$

The minus sign is no longer there in Eq. (3.65), since L has a negative value and K has a positive value. Den Broeder proposed the following steps to develop a relation, in which we do not have to locate the initial contact plane for the estimation of the interdiffusion coefficients. As already discussed, every diffusion couple has a particular initial contact plane, which might be difficult to find. However, with respect to this position, we can write

$$K + L + M = Q. \tag{3.66}$$

This is similar to the fact that $P = Q$ in Fig. 3.14B, where $K + L + M = P$.

After adding N to both the sides, we have

$$K + L + M + N = Q + N,$$

or equivalently,

$$L + M + N = Q + N - K. \tag{3.67}$$

We can now write

$$L + M + N = (x^* - x_o)Y_C^+ = x^*. \tag{3.68a}$$

Note that $Y_C^+ = 1$ and so

$$Q + N = \int_{x^*}^{x^{+\infty}} (1 - Y_C)dx, \tag{3.68b}$$

$$K = \int_{x^{-\infty}}^{x^*} Y_C dx. \tag{3.68c}$$

Substituting Eqs. (3.68a)–(3.68c) into (3.67), we have

$$x^* = \int_{x^*}^{x^{+\infty}} (1 - Y_C)dx - \int_{x^{-\infty}}^{x^*} Y_C dx. \tag{3.69}$$

Therefore, we obtain

$$L = x^* Y_C^* = Y_C^* \int_{x^*}^{x^{+\infty}} (1 - Y_C)dx - Y_C^* \int_{x^{-\infty}}^{x^*} Y_C dx, \tag{3.70}$$

which gives

$$L+K=Y_C^*\int_{x^*}^{x^{+\infty}}(1-Y_C)dx-Y_C^*\int_{x^{-\infty}}^{x^*}Y_C dx+\int_{x^{-\infty}}^{x^*}Y_C dx$$
$$=(1-Y_C^*)\int_{x^{-\infty}}^{x^*}Y_C dx+Y_C^*\int_{x^*}^{x^{+\infty}}(1-Y_C)dx. \quad (3.71)$$

Replacing Eq. (3.71) in Eq. (3.65) gives

$$\tilde{D}(Y_C^*)=\frac{1}{2t}\left(\frac{dx}{dY_C}\right)_{Y_C^*}\left[(1-Y_C^*)\int_{x^{-\infty}}^{x^*}Y_C dx+Y_C^*\int_{x^*}^{x^{+\infty}}(1-Y_C)dx\right]. \quad (3.72a)$$

Therefore, the interdiffusion flux following Fick's first law of diffusion can be written as

$$\tilde{J}_B^*=-\frac{C_B^+-C_B^-}{2t}\left[(1-Y_C^*)\int_{x^{-\infty}}^{x^*}Y_C dx+Y_C^*\int_{x^*}^{x^{+\infty}}(1-Y_C)dx\right]. \quad (3.72b)$$

As shown in Fig. 3.17C, Eq. (3.72a) can be rewritten as

$$\tilde{D}(Y_C^*)=\frac{1}{2t}\left(\frac{dx}{dY_C}\right)_{Y_C^*}[(1-Y_C^*)G+Y_C^*H]. \quad (3.73)$$

Note that in Eqs. (3.72) and (3.73) there is no need to locate the initial contact plane. After the diffusion couple experiment, we measure the composition profile, i.e., N_B vs. x using composition measuring equipment. N_B is the atomic fraction of component B. Following the concentration profile, C_B vs. x is plotted after dividing atomic fraction by molar volume, that is, $C_B=\frac{X_B}{v_m}=\frac{N_B}{v_m}$ (see Eq. (3.32)). The concentration profile is then converted into the plot of concentration normalizing variable Y_C vs. x. Next, the interdiffusion coefficient is determined from the estimated gradient at the concentration of interest and measured areas G and H, as explained in Fig. 3.17C. In a similar fashion, if required, one can estimate the interdiffusion flux utilizing Eq. (3.72b), in which there is no need to find the initial contact plane.

3.5 WAGNER'S APPROACH FOR THE CALCULATION OF THE INTERDIFFUSION COEFFICIENT [11]

A few standard thermodynamic relations are expressed as [1]

$$dC_B=\frac{\bar{v}_A}{v_m^2}dN_B; \qquad dC_A=\frac{\bar{v}_B}{v_m^2}dN_A, \quad (3.74a)$$

$$N_A+N_B=1, \qquad dN_A+dN_B=0, \quad (3.74b)$$

$$N_B\bar{v}_B + N_A\bar{v}_A = v_m, \tag{3.74c}$$

$$\bar{v}_A C_A + \bar{v}_B C_B = \bar{v}_A \frac{N_A}{v_m} + \bar{v}_B \frac{N_B}{v_m} = 1, \tag{3.74d}$$

$$\bar{v}_A dC_A + \bar{v}_B dC_B = 0. \tag{3.74e}$$

Following Fick's first law and the utilizing the relations above, we can write

$$\tilde{J}_B = -\tilde{D}\frac{\partial C_B}{\partial x} = -\tilde{D}\frac{\bar{v}_A}{v_m^2}\frac{\partial N_B}{\partial x}, \tag{3.75a}$$

$$\tilde{J}_A = -\tilde{D}\frac{\partial C_A}{\partial x} = -\tilde{D}\frac{\bar{v}_B}{v_m^2}\frac{\partial N_A}{\partial x} = \tilde{D}\frac{\bar{v}_B}{v_m^2}\frac{\partial N_B}{\partial x} = \frac{\bar{v}_B}{\bar{v}_A}\tilde{D}\frac{\bar{v}_A}{v_m^2}\frac{\partial N_B}{\partial x} = -\frac{\bar{v}_B}{\bar{v}_A}\tilde{J}_A, \tag{3.75b}$$

$$\bar{v}_A\tilde{J}_A + \bar{v}_B\tilde{J}_B = 0. \tag{3.76}$$

Note that the interdiffusion coefficient is a material constant, and it has a fixed value at particular concentration irrespective of the components profile used for the estimation as explained in Section 3.17. On the other hand, the interdiffusion flux and the concentration gradient depend on the component we are considering for the estimation.

Utilizing Eq. (3.74c) in (3.75a), we have

$$\tilde{J}_B\frac{(N_B\bar{v}_B + N_A\bar{v}_A)}{v_m} = -\tilde{D}\frac{\bar{v}_A}{v_m^2}\frac{\partial N_B}{\partial x},$$

$$\tilde{D} = -\frac{v_m(N_B\bar{v}_B\tilde{J}_B + N_A\bar{v}_A\tilde{J}_B)}{\bar{v}_A(\partial N_B/\partial x)}.$$

Using Eq. (3.76), we get

$$\tilde{D} = -\frac{v_m(-N_B\bar{v}_A\tilde{J}_A + N_A\bar{v}_A\tilde{J}_B)}{\bar{v}_A(\partial N_B/dx)}$$

which yields

$$\tilde{D} = \frac{v_m(N_B\tilde{J}_A - N_A\tilde{J}_B)}{\partial N_B/\partial x}. \tag{3.77}$$

Fick's second law with respect to the interdiffusion flux, as expressed in Eq. (3.4), can be rewritten as

$$\frac{\partial}{\partial t}\left(\frac{N_B}{v_m}\right) = \frac{\partial C_B}{\partial t} = -\frac{\partial \tilde{J}_B}{\partial x}, \tag{3.78a}$$

$$\frac{\partial}{\partial t}\left(\frac{N_A}{v_m}\right) = \frac{\partial C_A}{\partial t} = -\frac{\partial \tilde{J}_A}{\partial x}. \tag{3.78b}$$

A composition-normalized variable is expressed as

$$Y_B = \frac{N_B - N_B^-}{N_B^+ - N_B^-}, \tag{3.79}$$

which can be rearranged as

$$N_B = N_B^+ Y_B + N_B^-(1 - Y_B). \tag{3.80a}$$

Utilizing (3.74b), we can replace $N_B = 1 - N_A$ as follows:

$$\begin{aligned}
1 - N_A &= N_B^+ Y_B + N_B^-(1 - Y_B),\\
1 - N_A &= N_B^+ Y_B + N_B^-(1 - Y_B) + Y_B - Y_B,\\
N_A &= 1 - N_B^+ Y_B - N_B^-(1 - Y_B) - Y_B + Y_B,\\
N_A &= \left(Y_B - N_B^+ Y_B\right) + \left[1 - Y_B - N_B^-(1 - Y_B)\right],\\
N_A &= \left(1 - N_B^+\right) Y_B + \left(1 - N_B^-\right)(1 - Y_B).
\end{aligned} \tag{3.80b}$$

Substituting Eqs. (3.80a) and (3.80b) into Eqs. (3.78a) and (3.78b), respectively, we can write

$$N_B^+ \frac{\partial}{\partial t}\left(\frac{Y_B}{v_m}\right) + N_B^- \frac{\partial}{\partial t}\left(\frac{1 - Y_B}{v_m}\right) = -\frac{\partial \tilde{J}_B}{\partial x}, \tag{3.81a}$$

$$(1 - N_B^+)\frac{\partial}{\partial t}\left(\frac{Y_B}{v_m}\right) + (1 - N_B^-)\frac{\partial}{\partial t}\left(\frac{1 - Y_B}{v_m}\right) = -\frac{\partial \tilde{J}_A}{\partial x}. \tag{3.81b}$$

Differentiating Boltzmann parameter in Eq. (3.48), we obtain

$$\begin{aligned}
d\lambda &= -\frac{1}{2}\frac{x}{t^{3/2}}dt = -\frac{1}{2}\frac{\lambda}{t}dt,\\
dt &= -2t\frac{d\lambda}{\lambda}.
\end{aligned} \tag{3.82}$$

Replacing Eq. (3.82) in (3.81), we get

$$\frac{\lambda}{2t}\left[N_B^+ \frac{d}{d\lambda}\left(\frac{Y_B}{v_m}\right) + N_B^- \frac{d}{d\lambda}\left(\frac{1 - Y_B}{v_m}\right)\right] = \frac{\partial \tilde{J}_B}{\partial x}, \tag{3.83a}$$

$$\frac{\lambda}{2t}\left[(1 - N_B^+)\frac{d}{d\lambda}\left(\frac{Y_B}{v_m}\right) + (1 - N_B^-)\frac{d}{d\lambda}\left(\frac{1 - Y_B}{v_m}\right)\right] = \frac{\partial \tilde{J}_A}{\partial x}. \tag{3.83b}$$

Our aim now is to write the equations with respect to $\frac{Y_B}{v_m}$ and $\frac{1-Y_B}{v_m}$. Therefore, we multiply Eq. (3.83a) by $(1 - N_B^-)$ and Eq. (3.83b) by N_B^-. Then after subtracting the corresponding sides, we have

$$-\frac{\lambda}{2t}(N_B^+ - N_B^-)\frac{d}{d\lambda}\left(\frac{Y_B}{v_m}\right) = N_B^-\frac{\partial \tilde{J}_A}{\partial x} - (1 - N_B^-)\frac{\partial \tilde{J}_B}{\partial x}. \tag{3.84a}$$

Following similar steps, we multiply Eq. (3.83a) by $(1 - N_B^+)$ and Eq. (3.83b) by N_B^+. After subtracting the corresponding sides, we get

$$\frac{\lambda}{2t}(N_B^+ - N_B^-)\frac{d}{d\lambda}\left(\frac{1 - Y_B}{v_m}\right) = N_B^+\frac{\partial \tilde{J}_A}{\partial x} - (1 - N_B^+)\frac{\partial \tilde{J}_B}{\partial x}. \tag{3.84b}$$

Differentiating Boltzmann parameter (Eq. (3.48)) yields

$$d\lambda = \frac{dx}{t^{\frac{1}{2}}}.$$

After multiplying by $d\lambda$ the left-hand side and by $dx/t^{1/2}$ the right hand side of Eq. (3.84a), we get

$$-\frac{\lambda}{2t}(N_B^+ - N_B^-)d\left(\frac{Y_B}{v_m}\right) = \frac{1}{t^{1/2}}[N_B^- d\tilde{J}_A - (1 - N_B^-)d\tilde{J}_B].$$

Using the same Boltzmann parameter, we can write for $x = x^{-\infty}$, $\lambda = \lambda^{-\infty}$ and for $x = x^*$, $\lambda = \lambda^*$.

After integration from $\lambda = \lambda^{-\infty}$ to a particular position of interest $\lambda = \lambda^*$ (corresponding to the mole fraction N_B^*) for a fixed annealing time t of the diffusion couple, we have

$$-\frac{1}{2t}(N_B^+ - N_B^-)\int_{\lambda^{-\infty}}^{\lambda^*}\lambda d\left(\frac{Y_B}{v_m}\right) = \frac{1}{t^{1/2}}\left[N_B^-\int_o^{\tilde{J}_A^*}d\tilde{J}_A - (1 - N_B^-)\int_o^{\tilde{J}_B^*}d\tilde{J}_B\right],$$

since the interdiffusion flux at the unaffected part of the diffusion couple, $x = x^{-\infty}$ or $\lambda = \lambda^{-\infty}$, is zero.

After integration by parts of the left-hand side, we have

$$-\frac{1}{2t}(N_B^+ - N_B^-)\left[\frac{\lambda^* Y_B^*}{v_m^*} - \int_{\lambda^{-\infty}}^{\lambda^*}\frac{Y_B}{v_m}d\lambda\right] = \frac{1}{t^{\frac{1}{2}}}[N_B^-\tilde{J}_A^* - (1 - N_B^-)\tilde{J}_B^*],$$

$$\frac{1}{2t}(N_B^+ - N_B^-)\left[-\frac{\lambda^* Y_B^*}{v_m^*} + \int_{\lambda^{-\infty}}^{\lambda^*}\frac{Y_B}{v_m}d\lambda\right] = \frac{1}{t^{\frac{1}{2}}}[N_B^-\tilde{J}_A^* - (1 - N_B^-)\tilde{J}_B^*]. \tag{3.85a}$$

Similarly, Eq. (3.84b) is multiplied by $d\lambda$ on the left and by $dx/t^{1/2}$ on the right. After integration from $\lambda = \lambda^*$ to $\lambda = \lambda^{+\infty}$ for a fixed annealing time of the diffusion couple and after integration by parts of the left-hand side of the equation, we obtain

$$\frac{1}{2t}\left(N_B^+ - N_B^-\right)\left[\frac{\lambda^*(1-Y_B^*)}{v_m^*} + \int_{\lambda^*}^{\lambda+\infty} \frac{(1-Y_B)}{v_m} d\lambda\right] = \frac{1}{t^{\frac{1}{2}}}\left[N_B^+ \tilde{J}_A^* - (1-N_B^+)\tilde{J}_B^*\right].$$

A positive sign appears after integration of parts because of $(1 - Y_B)$:

$$\frac{1}{2t}\left(N_B^+ - N_B^-\right)\left[-\frac{\lambda^*(1-Y_B^*)}{v_m^*} - \int_{\lambda^*}^{\lambda+\infty} \frac{(1-Y_B)}{v_m} d\lambda\right]$$
$$= \frac{1}{t^{\frac{1}{2}}}\left[-N_B^+ \tilde{J}_A^* + (1-N_B^+)\tilde{J}_B^*\right]. \quad (3.85b)$$

Note that $\tilde{J}_A^*$ and $\tilde{J}_B^*$ are the fluxes at the particular concentration of interest that corresponds to $\lambda = \lambda^*$. Eq. (3.85a) is multiplied by $(1 - Y_B^*)$, and Eq. (3.85b) is multiplied by Y_B^*. Then after subtracting corresponding sides, we get

$$\frac{1}{2t}\left(N_B^+ - N_B^-\right)\left[(1-Y_B^*)\int_{\lambda-\infty}^{\lambda^*} \frac{Y_B}{v_m} d\lambda + Y_B^* \int_{\lambda^*}^{\lambda+\infty} \frac{(1-Y_B)}{v_m} d\lambda\right]$$
$$= \frac{1}{t^{\frac{1}{2}}}\left[N_B^* \tilde{J}_A^* - (1-N_B^*)\tilde{J}_B^*\right]. \quad (3.86)$$

The right-hand side of the relation is derived after replacing $Y_B^* = \frac{N_B^* - N_B^-}{N_B^+ - N_B^-}$ and $1 - Y_B^* = \frac{N_B^+ - N_B^*}{N_B^+ - N_B^-}$.

Eq. (3.86) can be rewritten using the relation $d\lambda = \frac{dx}{t^{\frac{1}{2}}}$ as

$$\frac{1}{2t}\left(N_B^+ - N_B^-\right)\left[(1-Y_B^*)\int_{x-\infty}^{x^*} \frac{Y_B}{v_m} dx + Y_B^* \int_{x^*}^{x+\infty} \frac{(1-Y_B)}{v_m} dx\right]$$
$$= \left[N_B^* \tilde{J}_A^* - (1-N_B^*)\tilde{J}_B^*\right]. \quad (3.87)$$

At the composition of interest for a particular annealing time in a particular diffusion couple, Eq. (3.77) can be written as

$$\tilde{D}(N_B^*) = \frac{v_m^*\left(N_B^* \tilde{J}_A^* - N_A^* \tilde{J}_B^*\right)}{(dN_B/dx)_{x^*}}. \quad (3.88)$$

From Eqs. (3.87) and (3.88), we get

$$\tilde{D}(N_B^*) = \frac{(N_B^+ - N_B^-)v_m^*}{2t(dN_B/dx)_{x^*}}\left[(1-Y_B^*)\int_{x-\infty}^{x^*} \frac{Y_B}{v_m} dx + Y_B^* \int_{x^*}^{x-\infty} \frac{(1-Y_B)}{v_m} dx\right]. \quad (3.89)$$

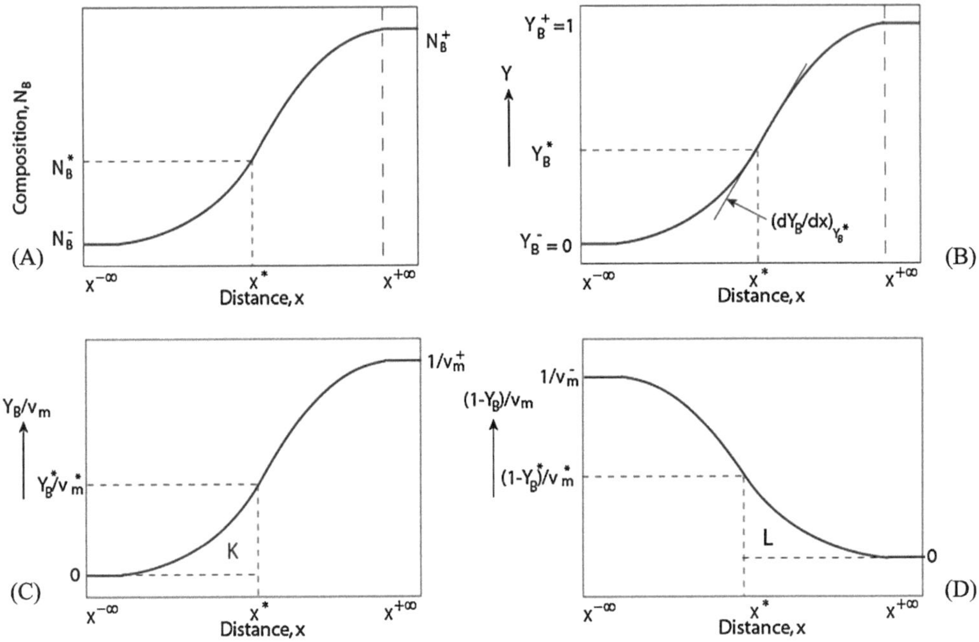

Figure 3.18 Estimation steps of the interdiffusion coefficients following Wagner's approach.

Eq. (3.79) gives

$$dY_B = \frac{dN_B}{N_B^+ - N_B^-}. \tag{3.90}$$

Replacing Eq. (3.90) in Eq. (3.89), we can write the final equation for the estimation of the interdiffusion coefficient as

$$\tilde{D}(Y_B^*) = \frac{v_m^*}{2t(dY_B/dx)_{x^*}}\left[(1-Y_B^*)\int_{x^{-\infty}}^{x^*}\frac{Y_B}{v_m}dx + Y_B^*\int_{x^*}^{x^{+\infty}}\frac{(1-Y_B)}{v_m}dx\right]. \tag{3.91a}$$

Therefore, the interdiffusion flux following Fick's first law of diffusion (see Eq. (3.75a)) can be written as

$$\tilde{J}_B^* = -\frac{\bar{v}_A^*}{v_m^*}\frac{(N_B^+ - N_B^-)}{2t}\left[(1-Y_B^*)\int_{x^{-\infty}}^{x^*}\frac{Y_B}{v_m}dx + Y_B^*\int_{x^*}^{x^{+\infty}}\frac{(1-Y_B)}{v_m}dx\right]. \tag{3.91b}$$

The estimation steps for the interdiffusion coefficient followed by Wagner's method are explained in Fig. 3.18. As a first step, the composition profile is measured experimentally, which might be similar to the profile shown in Fig. 3.18A, especially in a solid solution. Following the composition, the profile is normalized utilizing Eq. (3.79) and plotted as Y_B vs. x, shown in Fig. 3.18B. From the known values of molar volumes estimated from the lattice parameter variation in the system of interest, $\frac{Y_B}{v_m}$ vs. x and $\frac{1-Y_B}{v_m}$ vs. x are plotted, as shown in Figs. 3.18C and 3.18D, respectively. Suppose, we are interested in estimating the interdiffusion coefficient at the composition, N_B^*. Then we first estimate the gradient $\frac{dY}{dx}$ at Y^*. Next the areas $K = \int_{x^{-\infty}}^{x^*} \frac{Y_B}{v_m} dx$ and $L = \int_{x^*}^{x^{+\infty}} \frac{(1-Y_B)}{v_m} dx$ are determined to estimate the interdiffusion coefficients utilizing Eq. (3.91a) from the known annealing time, t, of the diffusion couple. In a similar fashion, if interested, one can estimate the interdiffusion flux following Eq. (3.91b), in which there is no need to locate the initial contact plane.

3.6 DEVIATION FROM IDEAL MOLAR VOLUME AND ERROR IN LOCATING THE INITIAL CONTACT PLANE (OR MATANO PLANE)

We have explained (without providing any mathematical proof) in Section 3.3.2 that finding the location of the initial contact plane correctly in a diffusion couple is almost impossible if there is deviation from ideal molar volume. Two different locations are found when we analyze employing C_A vs. x and C_B vs. x profiles. In fact, neither is correct and the actual location is somewhere between these two values [1]. Moreover, as shown below, the difference between these two locations is equal to the change in total volume of the diffusion couple.

We consider a diffusion couple of two alloys N_B^- (left-hand side of the couple) and N_B^+ (right-hand side of the couple). In this couple, we consider that $N_B^+ > N_B^-$, which further means that N_B^- is the B-lean alloy and N_B^+ is the B-rich alloy. Therefore, with respect to N_A, the composition profile is just the opposite, meaning that $N_A^+ < N_A^-$. Previously, we explained that the initial contact plane can be located with respect to the concentration profiles C_B vs. x or C_A vs. x. Similarly, it can be estimated with respect to the plots of composition normalized variables Y_B/v_m vs. x or Y_A/v_m vs. x, where $Y_B = \frac{N_B - N_B^-}{N_B^+ - N_B^-}$ and $Y_A = \frac{N_A - N_A^+}{N_A^- - N_A^+}$.

We can write

$$Y_A = \frac{N_A - N_A^+}{N_A^- - N_A^+} = \frac{1 - N_A^+ - 1 + N_A}{1 - N_A^+ - 1 + N_A^-} = \frac{N_B^+ - N_B}{N_B^+ - N_B^-} = 1 - \frac{N_B - N_B^-}{N_B^+ - N_B^-} = 1 - Y_B. \quad (3.92)$$

Therefore, the initial contact plane can be located using Y_B/v_m vs. x or $(1 - Y_B)/v_m$ vs. x plots, as shown in Fig. 3.19, by equating the areas M and N. Let us consider that the location of the initial contact plane is found at x_o^I from the Y_B/v_m vs. x plot and at x_o^{II} from the $(1 - Y_B)/v_m$ vs. x plot. Mathematically, M = N in these two plots can be written as

$$\int_{x^{-\infty}}^{x_o^I} \frac{Y_B}{v_m} dx = \int_{x_o^I}^{x^{+\infty}} \left(\frac{1}{v_m^+} - \frac{Y_B}{v_m} \right) dx, \quad (3.93a)$$

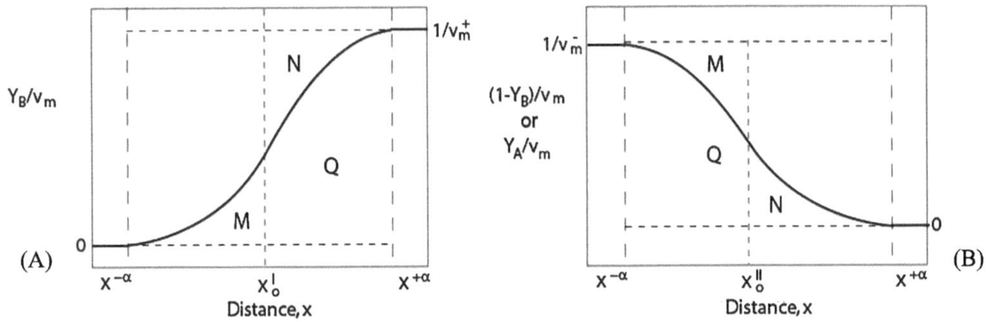

Figure 3.19 Finding the location of the initial contact plane.

$$\int_{x^{-\infty}}^{x_o^{II}} \left(\frac{1}{v_m^-} - \frac{(1-Y_B)}{v_m} \right) dx = \int_{x_o^{II}}^{x^{+\infty}} \frac{(1-Y_B)}{v_m} dx, \tag{3.93b}$$

where v_m^- is the molar volume of the composition N_B^- (and N_A^-) and v_m^+ is the molar volume of the composition N_B^+ (and N_A^+).

Adding Q on both sides, we can write $M+Q=N+Q$. On the basis of this equation, we can write Eq. (3.93a) as

$$\int_{x^{-\infty}}^{x_o^{I}} \frac{Y_B}{v_m} dx + \int_{x_o^{I}}^{x^{+\infty}} \frac{Y_B}{v_m} dx = \int_{x_o^{I}}^{x^{+\infty}} \left(\frac{1}{v_m^+} - \frac{Y_B}{v_m} \right) dx + \int_{x_o^{I}}^{x^{+\infty}} \frac{Y_B}{v_m} dx,$$

$$\int_{x^{-\infty}}^{x^{+\infty}} \frac{Y_B}{v_m} dx = \frac{1}{v_m^+} \int_{x_o^{I}}^{x^{+\infty}} dx,$$

$$\frac{x^{+\infty} - x_o^I}{v_m^+} = \int_{x^{-\infty}}^{x^{+\infty}} \frac{Y_B}{v_m} dx,$$

$$x^{+\infty} - x_o^I = v_m^+ \int_{x^{-\infty}}^{x^{+\infty}} \frac{Y_B}{v_m} dx. \tag{3.94a}$$

Similarly, from Eq. (3.93b) we obtain

$$\int_{x_o^{-\infty}}^{x_o^{II}} \frac{(1-Y_B)}{v_m} dx + \int_{x^{-\infty}}^{x_o^{II}} \left(\frac{1}{v_m^-} - \frac{(1-Y_B)}{v_m} \right) dx = \int_{x_o^{-\infty}}^{x_o^{II}} \frac{(1-Y_B)}{v_m} dx + \int_{x_o^{II}}^{x^{+\infty}} \frac{(1-Y_B)}{v_m} dx,$$

$$\frac{1}{v_m^-} \int_{x^{-\infty}}^{x_o^{II}} dx = \int_{x^{-\infty}}^{x^{+\infty}} \frac{(1-Y_B)}{v_m} dx,$$

$$\frac{x_o^{II} - x^{-\infty}}{v_m^-} = \int_{x^{-\infty}}^{x^{+\infty}} \frac{(1-Y_B)}{v_m} dx,$$

$$x_o^{II} - x^{-\infty} = v_m^- \int_{x^{-\infty}}^{x^{+\infty}} \frac{(1-Y_B)}{v_m} dx. \tag{3.94b}$$

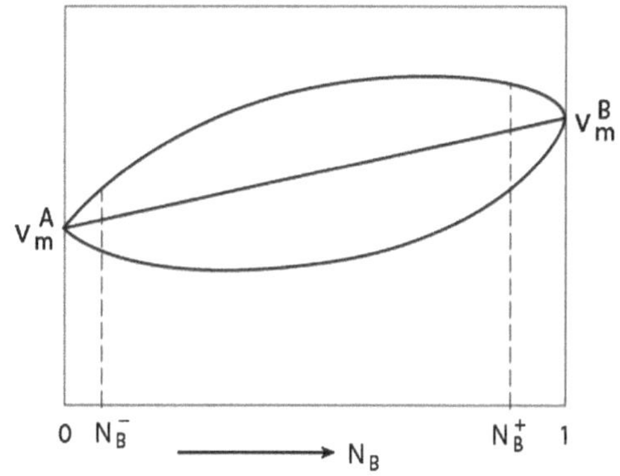

Figure 3.20 Molar volume variations.

Adding Eqs. (3.94a) and (3.94b) gives

$$x_o^{II} - x^{-\infty} + x^{+\infty} - x_o^I = v_m^- \int_{x^{-\infty}}^{x^{+\infty}} \frac{(1 - Y_B)}{v_m} dx + v_m^+ \int_{x^{-\infty}}^{x^{+\infty}} \frac{Y_B}{v_m} dx,$$

$$x_o^{II} - x^{-\infty} + x^{+\infty} - x_o^I = \int_{x^{-\infty}}^{x^{+\infty}} \frac{v_m^- + (v_m^+ - v_m^-) Y_B}{v_m} dx. \tag{3.95}$$

Now consider Fig. 3.20, in which all three types of imaginary molar volume variation are shown. In the case of ideal variation, molar volume v_m^- is at N_B^- and v_m^+ is at N_B^+. This variation with respect to the composition normalized variable can be written as v_m^- for $Y_B = 0$ and v_m^+ for $Y_B = 1$. Therefore, this linear variation can be expressed as

$$v_m = v_m^- + \left(v_m^+ - v_m^-\right) Y_B. \tag{3.96}$$

The negative ($-\Delta v_m$) or positive ($+\Delta v_m$) deviation of the molar volume at a particular composition can be imposed as

$$v_m = v_m^- + \left(v_m^+ - v_m^-\right) Y_B \pm \Delta v_m,$$

$$v_m \mp \Delta v_m = v_m^- + \left(v_m^+ - v_m^-\right) Y_B. \tag{3.97}$$

Substituting Eq. (3.97) into (3.95) yields

$$x_o^{II} - x^{-\infty} + x^{+\infty} - x_o^I = \int_{x^{-\infty}}^{x^{+\infty}} \frac{v_m \mp \Delta v_m}{v_m} dx,$$

$$x_o^{II} - x^{-\infty} + x^{+\infty} - x_o^I = x^{+\infty} - x^{-\infty} \mp \int_{x^{-\infty}}^{x^{+\infty}} \frac{\Delta v_m}{v_m} dx,$$

$$x_o^I - x_o^{II} = \pm \int_{x^{-\infty}}^{x^{+\infty}} \frac{\Delta v_m}{v_m} dx. \tag{3.98}$$

Since Δv_m is the deviation from the ideal molar volume at a particular composition located at a particular position after the diffusion annealing, $\int_{x-\infty}^{x+\infty} \frac{\Delta v_m}{v_m} dx$ gives the total volume change. This further indicates that the difference between the locations of the initial contact planes determined by different profiles is exactly equal to the change in total volume. In the case of ideal variation of the molar volume ($\Delta v_m = 0$), the location of the initial contact plane can be found at the same place from any of the profiles, and there is no confusion in the use of Matano–Boltzmann parameter for the estimation of the interdiffusion coefficients. However, in the case of positive ($x_o^I > x_o^{II}$) and negative ($x_o^I < x_o^{II}$) deviation of the molar volume, it is safe to use the methods proposed by Den Broeder or Sauer–Freise/Wagner. The influence of molar volume can further be understood in the next section.

3.7 COMPARISON OF THE INTERDIFFUSION COEFFICIENTS ESTIMATED BY DIFFERENT METHODS

For comparison, we consider three different approaches: Matano–Boltzmann (MB), den Broeder (dB), and Wagner (WG) [13]. We consider an imaginary composition profile, as shown in Fig. 3.21A. This type of composition profile is developed in a diffusion couple prepared in a system having complete solid solution. We consider two different molar volume variations: ideal (v_m^{id}) and non-ideal with positive variation (v_m^{nid}), as shown in Fig. 3.21B. Fig. 3.21C indicates that the maximum deviation of non-ideal molar volume variation is less than 15% from the ideal molar volume variation. As already mentioned in the previous sections, initial contact plane positions can be found from C_B vs. x (let's say x_o^I) or C_A vs. x (let's say x_o^{II}) plots. They can be equivalently found by using Y_B/v_m vs. x (for x_o^I) or $Y_A/v_m = (1 - Y_B)/v_m$ vs. x (for x_o^{II}) plots. In the case of an ideal variation of the molar volume, this position is located at 175 μm. It is found at the same location from both profiles by equalizing the shaded areas, as shown in Fig. 3.22. In the case of non-ideal variation, the locations are found at $x_o^I = 183$ μm and $x_o^{II} = 170$ μm. As already discussed in the case of non-ideal variation, both are incorrect, and the actual location must be somewhere in the middle of the two, which cannot be identified. Also note that in the case of non-ideal variation the location of the initial contact plane is not the same as that estimated for the ideal case.

Now compare the values of interdiffusion coefficients estimated by different methods for ideal and non-ideal molar volume variations, as shown in Fig. 3.23. It can be seen in Fig. 3.23A that $\tilde{D}$ values are very different when estimated using two different initial contact planes. The difference increases as we move towards B-rich composition mainly because of a sharp change in the data estimated from x_o^{II}. This indicates that the actual location of the initial contact plane (which cannot be determined) must be closer to x_o^I. When we compare the data estimated by dB and WG methods, the variation of molar volume has a stronger influence on dB method. However, an interesting fact can

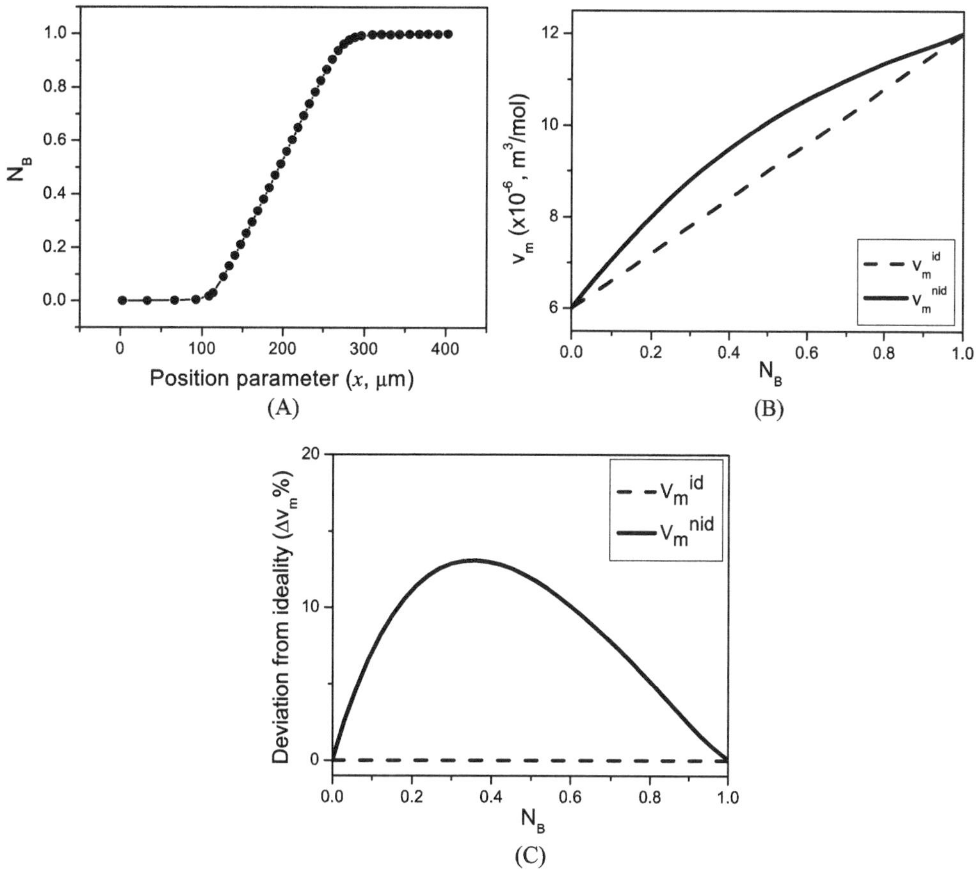

Figure 3.21 (A) A diffusion profile, (B) a molar volume variation, and (C) positive deviation from the ideal molar volume [13].

be noticed here that molar volume deviation is larger in the range of $N_B = 0.3$–0.4, as shown in Fig. 3.21C. Interestingly, the peak of $\tilde{D}$ falls in this composition range when measured following the dB method. On the other hand, $\tilde{D}$ is not influenced that significantly when measured following the WG method, and stays more or less the same for both the ideal and non-ideal variations of the molar volume.

Now we compare the data determined by different approaches for ideal and non-ideal variations of the molar volume separately, as shown in Fig. 3.24. It can be seen in Fig. 3.24A that there are no big differences in the data estimated following different approaches for the ideal variation of the molar volume. However, there are differences, especially when the data are estimated following MB method using x_o^{II}. Since all other data are close to each other, this indicates that x_o^I does not generate a very large error; however, this statement is true only for this particular example. When only MB method

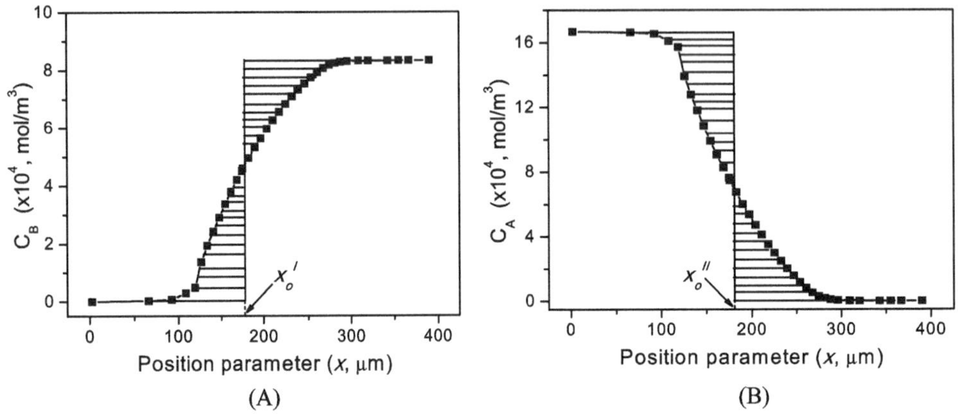

Figure 3.22 Finding the initial contact plane from (A) C_B vs. x and (B) C_A vs. x plot for ideal variation of the molar volume [13].

is followed, it is not easy to understand which location of the initial contact plane will generate a smaller error. In certain cases, in fact, both x_o^I and x_o^{II} will generate a large error. Therefore, it is safe to use dB or WG methods.

3.8 THE CONCEPT OF THE INTEGRATED INTERDIFFUSION COEFFICIENT FOR THE PHASES WITH NARROW HOMOGENEITY RANGE

One of the important steps to estimate the interdiffusion coefficients following any method is the estimation of the concentration gradient, dC_i/dx (in Matano–Boltzmann and den Broeder methods) or gradient of the composition normalized variable, dY_i/dx (in Wagner method). Here, i represents the component considered for the estimation of the interdiffusion coefficient. However, in many systems, line compounds or the compounds with very narrow homogeneity range grow in an interdiffusion zone. These compounds might have a reasonable thickness; however, because of a very small range of composition or concentration, it is almost impossible to measure the concentration or composition normalized gradient. To circumvent this problem, Wagner [11] introduced a concept of a new diffusion parameter called integrated interdiffusion coefficient. It is defined as the interdiffusion coefficient integrated over the unknown composition range of the phase of interest and expressed as

$$\tilde{D}_{\text{int}}^{\beta} = \int_{N_B^{\beta 1}}^{N_B^{\beta 2}} \tilde{D} dN_B. \tag{3.99a}$$

Then using Fick's first law of diffusion and assuming a constant molar volume ($\tilde{J}_B = -\tilde{D}\frac{dC_B}{dx} = -\tilde{D}\frac{1}{V_m}\frac{dN_B}{dx}$), which is reasonable to consider in a phase with narrow homo-

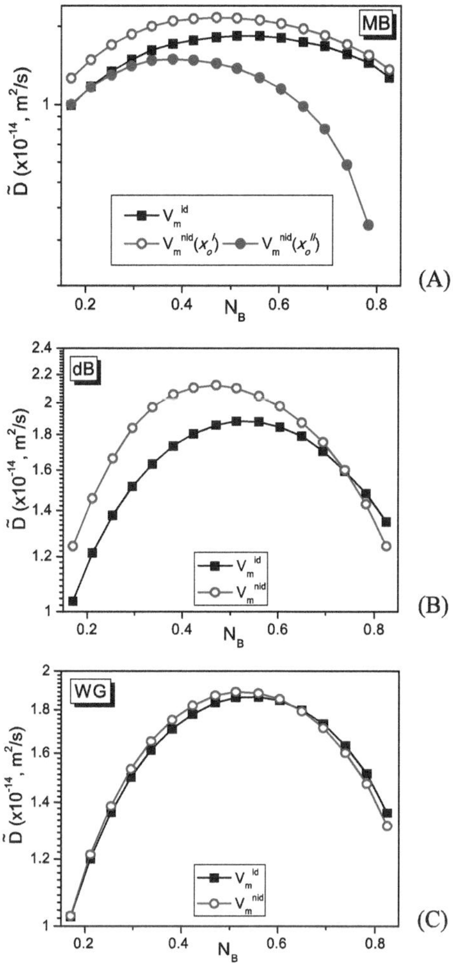

Figure 3.23 Interdiffusion coefficients determined following different approaches: (A) Matano–Boltzmann (MB), (B) den Broeder (dB), and (C) Wagner (WG) [13].

geneity range, we can express it with respect to the interdiffusion flux as

$$\tilde{D}_{\text{int}}^{\beta} = -v_m^{\beta} \int_{x^{\beta_1}}^{x^{\beta_2}} \tilde{J}_B dx \tag{3.99b}$$

where β is the phase of interest having an unknown composition range of $\Delta N_B^{\beta} = N_B^{\beta_2} - N_B^{\beta_1}$, and $\Delta x_{\beta} = x_{\beta_2} - x_{\beta_1}$ is the thickness of the phase layer of interest in an interdiffusion zone. Sometimes we may assume that the interdiffusion coefficient does not vary significantly over the small composition range of the phase from an average

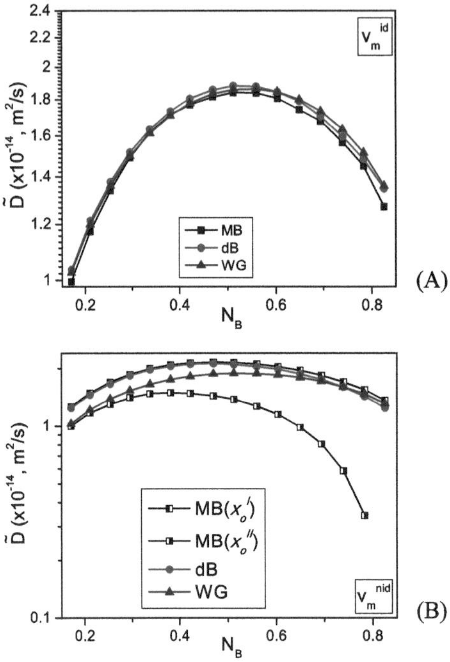

Figure 3.24 Variation of interdiffusion coefficients with composition measured by different approaches for (A) ideal and (B) non-ideal variation of the molar volume [13].

value of $\tilde{D}_{av}$ such that

$$\tilde{D}^{\beta}_{\text{int}} = \tilde{D}\left(N_B^{\beta_2} - N_B^{\beta_1}\right) = \tilde{D}_{av}\Delta N_B^{\beta}. \tag{3.100}$$

In some systems, intermetallic compounds might have a small but known composition range (ΔN_B^{β}) such that it is still difficult to estimate the composition gradient. However, one can estimate $\tilde{D}_{av}$ utilizing the estimated value of $\tilde{D}^{\beta}_{\text{int}}$ as

$$\tilde{D}_{av} = \frac{\tilde{D}^{\beta}_{\text{int}}}{\Delta N_B^{\beta}}. \tag{3.101}$$

Wagner [11] derived the relations for the estimation of $\tilde{D}^{\beta}_{\text{int}}$ from a composition profile directly. Consider an imaginary phase diagram, as shown in Fig. 3.25A. Suppose that we prepare a diffusion couple by joining blocks of solid solutions α and β. Based on the thermodynamic arguments presented in Chapter 1 of Volume 1, a possible composition profile of the interdiffusion zone is shown in Fig. 3.25B. It has three parts, α and γ-phases with wide composition range, which are grown from the end members of the same phase and a β-phase with narrow homogeneity range in the middle. Therefore,

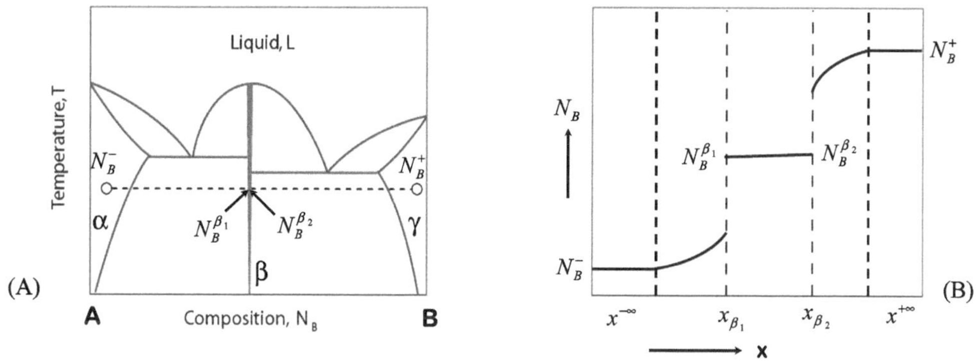

Figure 3.25 (A) An imaginary phase diagram that has a phase, β, with narrow homogeneity range and (B) evolution of the diffusion profile in this system.

Wagner's relation in Eq. (3.91a) can be rewritten as

$$\tilde{D}(Y_B^\beta) = \frac{v_m^\beta}{2t}\left(\frac{dx}{dY_B}\right)_\beta \left[(1 - Y_B^\beta)\int_{x^{-\infty}}^{x_{\beta_1}} \frac{Y_B}{v_m}dx + (1 - Y_B^\beta)\int_{x_{\beta_1}}^{x_{\beta_2}} \frac{Y_B}{v_m}dx + Y_B^\beta \int_{x_{\beta_2}}^{x^{+\infty}} \frac{1 - Y_B}{v_m}dx\right]. \quad (3.102)$$

The phase of interest, β, for the estimation of the diffusion data has an almost fixed composition, N_B^β, with a fixed molar volume v_m^β. Therefore the composition normalized variable Y_B^β also has a fixed value, and we can write

$$\tilde{D}(Y_B^\beta) = \frac{v_m^\beta}{2t}\left(\frac{dx}{dY_B}\right)_\beta \left[(1 - Y_B^\beta)\int_{x^{-\infty}}^{x_{\beta_1}} \frac{Y_B}{v_m}dx + \frac{(1 - Y_B^\beta)Y_B^\beta}{v_m^\beta}\int_{x_{\beta_1}}^{x_{\beta_2}} dx + Y_B^\beta \int_{x_{\beta_2}}^{x^{+\infty}} \frac{1 - Y_B}{v_m}dx\right], \quad (3.103)$$

$$\tilde{D}(Y_B^\beta) = \frac{v_m^\beta}{2t}\left(\frac{dx}{dY_B}\right)_\beta \left[(1 - Y_B^\beta)\int_{x^{-\infty}}^{x_{\beta_1}} \frac{Y_B}{v_m}dx + \frac{(1 - Y_B^\beta)Y_B^\beta}{v_m^\beta}\Delta x_\beta + Y_B^\beta \int_{x_{\beta 2}}^{x^{+\infty}} \frac{1 - Y_B}{v_m}dx\right]. \quad (3.104)$$

The composition normalized gradient can be rewritten with respect to the thickness of the β phase, $\Delta x_\beta = x_{\beta_2} - x_{\beta_1}$, as

$$\left(\frac{dx}{dY_B}\right)_\beta = (N_B^+ - N_B^-)\left(\frac{dx}{dN_B}\right)_\beta = (N_B^+ - N_B^-)\frac{\Delta x_\beta}{\Delta N_\beta}. \quad (3.105)$$

From Eqs. (3.104) and (3.105), we can write

$$\tilde{D}_{int}^{\beta} = \tilde{D}\Delta N_{\beta} = \frac{v_m^{\beta}}{2t}\Delta x_{\beta}(N_B^{+} - N_B^{-})\left[(1 - Y_B^{\beta})\int_{x^{-\infty}}^{x_{\beta 1}} \frac{Y_B}{v_m}dx + \frac{(1 - Y_B^{\beta})Y_B^{\beta}}{v_m^{\beta}}\Delta x_{\beta} + Y_B^{\beta}\int_{x_{\beta 2}}^{x^{+\infty}} \frac{1 - Y_B}{v_m}dx\right].$$

The composition normalized variable is expressed as $Y_B = \frac{N_B - N_B^{-}}{N_B^{+} - N_B^{-}}$ so that $Y_B^{\beta} = \frac{N_B^{\beta} - N_B^{-}}{N_B^{+} - N_B^{-}}$, and we can write

$$\tilde{D}_{int}^{\beta} = \frac{(\Delta x_{\beta})^2}{2t}\frac{(N_B^{\beta} - N_B^{-})(N_B^{+} - N_B^{\beta})}{N_B^{+} - N_B^{-}} + \frac{v_m^{\beta}}{2t}(N_B^{+} - N_B^{-})\Delta x_{\beta}\left[(1 - Y_B^{\beta})\int_{x^{-\infty}}^{x_{\beta 1}} \frac{Y_B}{v_m}dx + Y_B^{\beta}\int_{x_{\beta_2}}^{x^{+\infty}} \frac{1 - Y_B}{v_m}dx\right],$$

$$\tilde{D}_{int}^{\beta} = \frac{(N_B^{\beta} - N_B^{-})(N_B^{+} - N_B^{\beta})}{N_B^{+} - N_B^{-}}\frac{(\Delta x_{\beta})^2}{2t} + \frac{\Delta x_{\beta}}{2t}\left[\frac{N_B^{+} - N_B^{\beta}}{N_B^{+} - N_B^{-}} \times \int_{x^{-\infty}}^{x^{\beta 1}} \frac{v_m^{\beta}}{v_m}(N_B - N_B^{-})dx + \frac{N_B^{\beta} - N_B^{-}}{N_B^{+} - N_B^{-}} \times \int_{x^{\beta 2}}^{x^{+\infty}} \frac{v_m^{\beta}}{v_m}(N_B^{+} - N_B)dx\right]. \quad (3.106a)$$

Therefore, the first term corresponds to the line compound for which we are interested in estimating the integrated interdiffusion coefficient. The first term inside the brackets corresponds to the diffusion profile on the left-hand side, and the second term corresponds to the diffusion profile on the right-hand side of the phase of interest. There are many systems in which the interdiffusion zone consists of several line compounds with negligible diffusion profile at the ends of the diffusion couples. In this situation, Eq. (3.106a) can be modified to

$$\tilde{D}_{int}^{\beta} = \frac{(N_B^{\beta} - N_B^{-})(N_B^{+} - N_B^{\beta})}{N_B^{+} - N_B^{-}}\frac{(\Delta x_{\beta})^2}{2t} + \frac{\Delta x_{\beta}}{2t} \times \left[\frac{(N_B^{+} - N_B^{\beta})\sum_{i=2}^{\beta-1}\frac{v_m^{\beta}}{v_m^{i}}(N_B^{i} - N_B^{-})\Delta x_i + (N_B^{\beta} - N_B^{-})\sum_{i=\beta+1}^{n-1}\frac{v_m^{\beta}}{v_m^{i}}(N_B^{+} - N_B^{i})\Delta x_i}{N_B^{+} - N_B^{-}}\right]. \quad (3.106b)$$

Numbers 1 and n correspond to the end members, and numbers from 2 to $(n - 1)$ in the middle correspond to the phases grown in the interdiffusion zone. To understand Eq. (3.106b), consider a diffusion couple in which three phases, α, β and γ, are grown in the interdiffusion zone, as shown in Fig. 3.26. If we are interested in estimating the

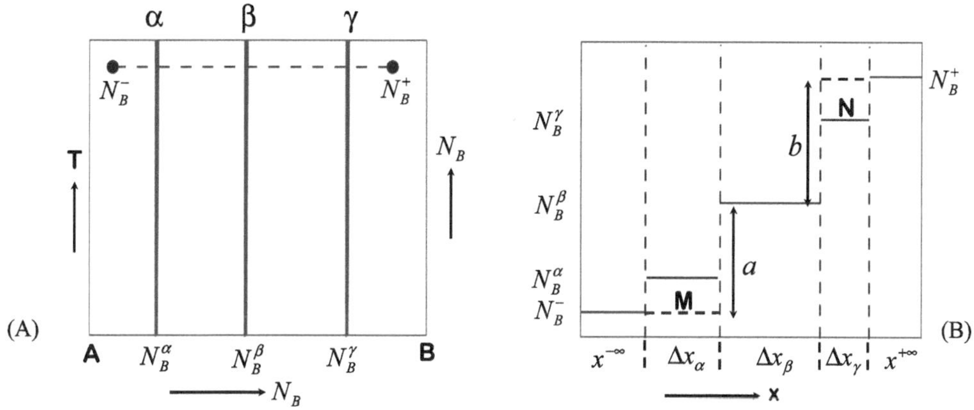

Figure 3.26 (A) An imaginary phase diagram that has three line compounds and (B) evolution of the diffusion profile in this system along with an explanation of estimation of the integrated diffusion coefficient.

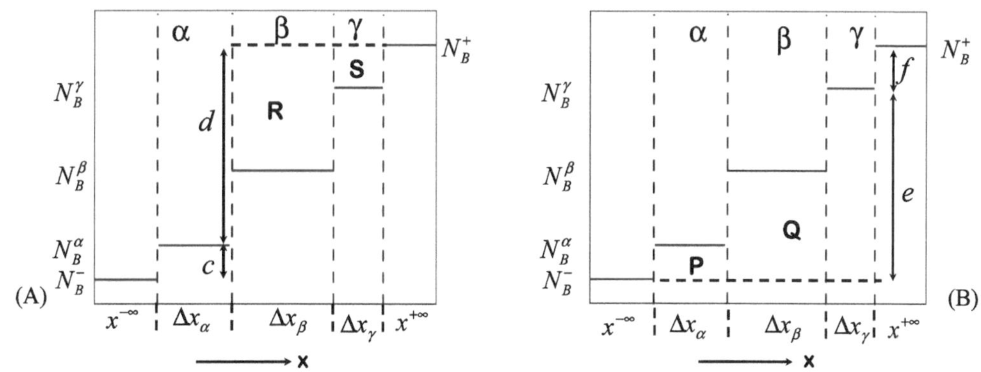

Figure 3.27 Explanation of the integrated diffusion coefficients in other phases in a profile that developed in an imaginary phase diagram, as shown in Fig. 3.26A.

integrated interdiffusion coefficient of phase β, $\tilde{D}^{\beta}_{int}$, Eq. (3.106b) modifies to

$$\tilde{D}^{\beta}_{int} = \frac{a \times b}{a+b}\frac{(\Delta x_{\beta})^2}{2t} + \frac{\Delta x_{\beta}}{2t}\left[\frac{b(\frac{v^{\beta}_m}{v^{\alpha}_m}M) + a(\frac{v^{\beta}_m}{v^{\gamma}_m}N)}{a+b}\right] \tag{3.107a}$$

where $a = N^{\beta}_B - N^{-}_B$, $b = N^{+}_B - N^{\beta}_B$, $M = \Delta x_{\alpha} \times (N^{\alpha}_B - N^{-}_B)$, and $N = \Delta x_{\gamma} \times (N^{+}_B - N^{\gamma}_B)$.

Similarly, as shown in Fig. 3.27A, the integrated interdiffusion coefficients for the α phase can be estimated as

$$\tilde{D}^{\alpha}_{int} = \frac{c \times d}{c+d}\frac{(\Delta x_{\alpha})^2}{2t} + \frac{\Delta x_{\alpha}}{2t}\left[\frac{d(0) + c(\frac{v^{\alpha}_m}{v^{\beta}_m}R + \frac{v^{\alpha}_m}{v^{\gamma}_m}S)}{c+d}\right] \tag{3.107b}$$

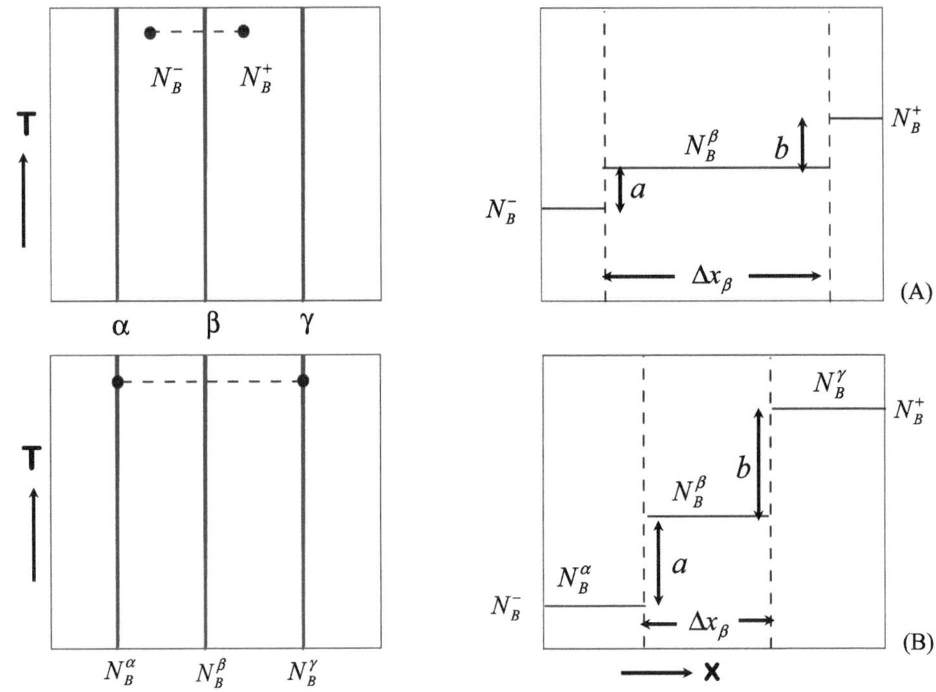

Figure 3.28 Explanation of the integrated diffusion coefficient in a diffusion couple with a single product phase when grows from (A) end members with phase mixtures and (B) pure phases.

where $c = N_B^\alpha - N_B^-$, $d = N_B^+ - N_B^\alpha$, $R = \Delta x_\beta \times (N_B^+ - N_B^\beta)$, and $S = \Delta x_\gamma \times (N_B^+ - N_B^\gamma)$. The first term inside the brackets is zero since no other phases are grown on the left-hand side of the phase of interest. As shown in Fig. 3.27B, the integrated interdiffusion coefficient of the γ phase can be estimated as

$$\tilde{D}_{\text{int}}^\gamma = \frac{e \times f}{e+f}\frac{(\Delta x_\gamma)^2}{2t} + \frac{\Delta x_\gamma}{2t}\left[\frac{f\left(\frac{v_m^\gamma}{v_m^\alpha}P + \frac{v_m^\gamma}{v_m^\beta}Q\right) + e(0)}{e+f}\right] \tag{3.107c}$$

where $e = N_B^\gamma - N_B^-$, $f = N_B^+ - N_B^\gamma$, $P = \Delta x_\alpha \times (N_B^\alpha - N_B^-)$ and $Q = \Delta x_\beta \times (N_B^\beta - N_B^-)$. The second term inside the brackets is zero since no other phases are grown on the right-hand side of the phase of interest.

Following similar arguments, as shown in Fig. 3.28, when only one line compound grows in an interdiffusion zone without developing any significant diffusion profile in the end member alloys, Eq. (3.106b) is reduced to

$$\tilde{D}_{\text{int}}^\beta = \frac{a \times b}{a+b}\frac{(\Delta x_\beta)^2}{2t} = \frac{a \times b}{a+b}k_p^\beta \tag{3.107d}$$

where k_p^β is the parabolic growth constant of the β phase. This is the next topic for discussion. One important point to be noted here is that integrated interdiffusion coefficient of a particular phase at one particular temperature is constant. Now, if we compare Figs. 3.28A and 3.28B, the end member compositions in the first diffusion couple (Fig. 3.28A) are closer to each other, making the value of $\frac{a \times b}{a+b}$ smaller than that of the second example (Fig. 3.28B). Therefore, the layer thickness in the first example should be larger compared to the second example for the same annealing temperature and time. The reason could be understood better following the physicochemical approach, as explained in Chapter 4, Volume 2. In the first example, the end members are mixtures of the $\alpha + \beta$ and $\beta + \gamma$ phases. Therefore, along with the formation of the product phase β from α and γ in two end members, some part of already existing β will join the product phase. On the other hand, in the second example, the β phase is not present in the end members and it has to grow by dissociation and reaction from α and γ phases.

One important point should be noted here that instead of following the steps provided by Wagner, as explained above, for the estimation of the integrated interdiffusion coefficient, one can measure it utilizing the relation given in Eq. (3.99b) directly. The interdiffusion flux required in this equation can be estimated following a suitable method, namely Matano–Boltzmann (Eq. (3.57b)), den Broeder (Eq. (3.72b)) or Wagner (Eq. (3.91b)), after considering a constant molar volume in the phase of interest.

3.9 PARABOLIC GROWTH CONSTANT

In an interdiffusion zone with wide composition range, one can check the veracity of the data by conducting diffusion couple experiments for different annealing times such that Boltzmann parameter fits the following equation: $\lambda = \lambda(C_B) = x_1/t_1^{1/2} = x_2/t_2^{1/2}$, where $x = x - x_o$ and $x_o = 0$ is the initial contact plane, or Matano plane. It means that every composition/concentration in a diffusion couple with particular end members will have a particular value of λ irrespective of different annealing times. This further means that a particular composition will be located at a certain position after the certain annealing time such that the Boltzmann parameter for that composition/concentration is fixed. However, the same procedure cannot be followed in the case of line compound due to very small composition range of the phase. This can be verified with respect to parabolic growth constant, indicating the parabolic nature, that is, the diffusion controlled growth of the phase. Consider initially that a single product phase with narrow homogeneity range grows in the interdiffusion zone. If it is grown by a diffusion controlled process, the rate of increase of the layer thickness will be inversely proportional to its own thickness since the amount of the diffusing flux decreases with the increase in layer thickness.

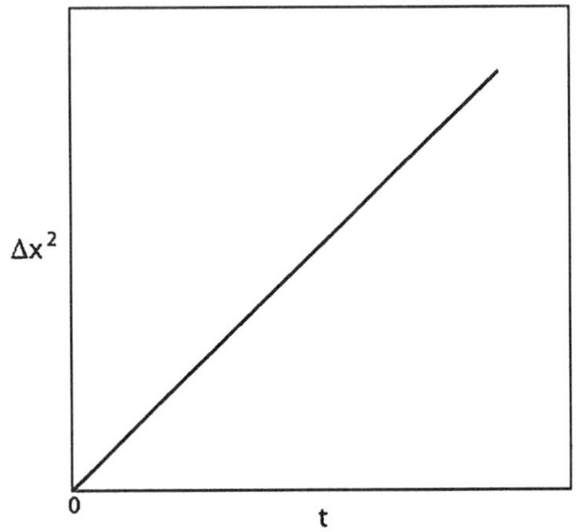

Figure 3.29 Time-dependent growth of a product phase plotted with respect to Δx^2 vs. t.

$$\frac{dx}{dt} \propto \frac{1}{x}, \tag{3.108a}$$

$$\frac{dx}{dt} = k\frac{1}{x}, \tag{3.108b}$$

where k is a proportionality constant, x is the layer thickness, and t is the annealing time. Eq. (3.108b) can be integrated from annealing time zero for the zero layer thickness to annealing time t for the layer thickness of Δx. Therefore, we have

$$\Delta x^2 = 2k_p t. \tag{3.109}$$

Here k_p is the parabolic growth constant. Therefore, the diffusion-controlled growth of a phase layer can be verified by conducting time-dependent experiments and by plotting Δx^2 vs. t, as shown in Fig. 3.29. The slope will be linear with a single value of gradient strictly under the condition that there is no change in the diffusion mechanism controlling the growth of the phase layer. Then the parabolic growth constant can be estimated utilizing Eq. (3.109). It is a matter of choice to plot Δx vs. t or Δx^2 vs. t; however, Pieraggi [14] discussed this issue extensively, specializing on the right choice for the growth of the oxide layer. One can also read the manuscript written by van Loo [15] summarizing this problem.

As already mentioned, under ideal conditions, there will be no change in gradient of the Δx^2 vs. t plot, and the extension of the slope after fitting the experimental data points to $t = 0$ will intersect at $\Delta x^2 = 0$. However, very frequently, different situations are found in different systems. It should be noted that both lattice and grain boundary diffusion occur in polycrystalline material. Depending on the material system and temperature of experiment, either lattice or grain boundary diffusion might play

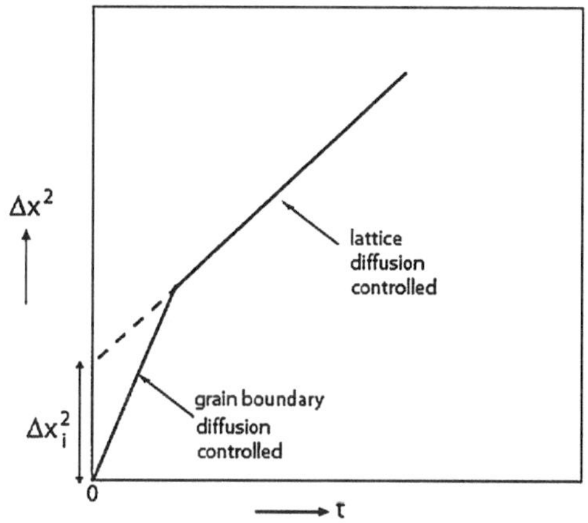

Figure 3.30 Time-dependent growth of a product phase plotted with respect to Δx^2 vs. t, which is affected by the grain boundary diffusion controlled growth before lattice diffusion controlled growth.

a dominating role. For example, at a relatively higher temperature in the presence of high concentration of vacancies in a material with big grains, lattice diffusion plays a dominating role. It means that the total flux through the lattice is greater than the total flux through grain boundaries simply because of relatively small grain boundary area, as explained in Chapter 1 of Volume 1. On the other hand, at a lower temperature, due to the lack of vacancies even with small grain boundary area, the contribution from this could be larger than that from lattice diffusion, making it grain boundary controlled growth. However, even if the growth of the phase layer is mainly controlled by the lattice diffusion, at the very initial stage the situation can be different. Due to the presence of very small grains, the grain boundary area could be sufficiently high to have greater flux through the grain boundaries compared to the lattice. With the continuation of annealing, grains also grow in size, and therefore lattice diffusion starts playing a domination role. In this situation, the transition from grain boundary to lattice diffusion controlled process could be clear from the Δx^2 vs. t plot, as shown in Fig. 3.30. However, it will be difficult to detect if this transition happens at a very small annealing time with a small thickness of the phase layer. However, one can understand this change in diffusion mechanism, when the data are extrapolated to $t = 0$, which will intersect at a positive Δx_i. Therefore, the parabolic growth constant is calculated from

$$\Delta x^2 - \Delta x_i^2 = 2k_p t. \tag{3.110}$$

It further means that k_p can be estimated directly from the gradient of the line drawn for the higher range of annealing times.

Many times a different situation is observed. The product phase might not start growing immediately after the start of annealing of the diffusion couple, as shown in

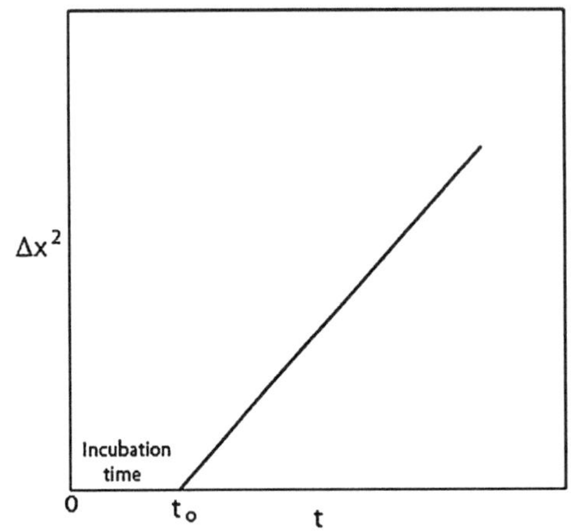

Figure 3.31 The time dependent growth of a product phase plotted with respect to Δx^2 vs. t which has incubation period at the early stage of growth.

Fig. 3.31. Sometimes a very thin oxide layer that is grown on the ground and a polished end member may hinder the diffusion of components for a certain incubation time (t_o) before breaking to resume a diffusion-controlled growth process [15,16]. Sometimes the broken pieces are found back inside the product phase [16]. In this situation, the parabolic growth constant is estimated from

$$\Delta x^2 = 2k_p(t - t_o). \tag{3.111}$$

Sometimes [15,17] impurities like carbon or phosphorous are segregated on the surface in very small concentrations and hinder the initial stage of the growth of the phase layer, leading to reaction-controlled growth. Therefore, at the initial stage, the layer grows linearly with time before diffusion-controlled parabolic time-dependent growth, as shown in Fig. 3.32. In a bulk diffusion couple, the growth of the product phase is not largely affected by the stress. However, in thin-film couples, the stress developed during the deposition process could be high enough to hinder the diffusion-controlled growth. High interfacial energy leads to difficulties in nucleation of the phase, making it a reaction-controlled growth. Sometimes even non-equilibrium phases are also found to grow.

The activation energy for the growth is determined based on temperature-dependent experiments following the Arrhenius equation

$$k_p = k_p^o \exp\left(-\frac{Q_g}{RT}\right). \tag{3.112}$$

Here k_p^o is the pre-exponential factor, R is the gas constant, and Q_g is the activation energy for the parabolic growth. The activation energy can be determined from the plot of

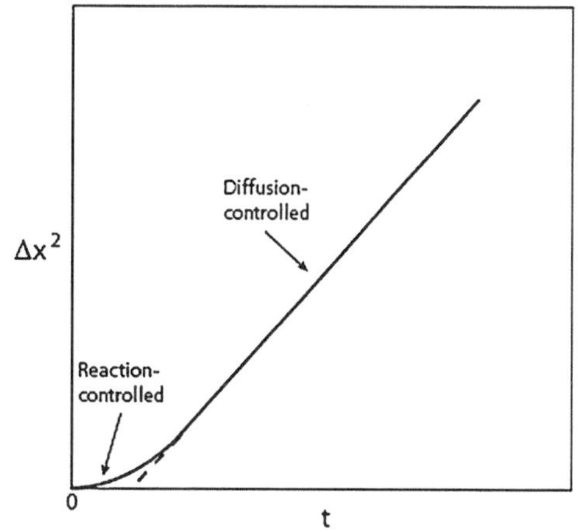

Figure 3.32 The time dependent growth of a product phase plotted with respect to Δx^2 vs. t which goes through a reaction controlled process at the early stage of growth.

$\ln k_p$ vs. $1/T$, where the slope is $-Q_g/R$. A couple of important points should be noted here. If all individual phases grow parabolically with time, the complete interdiffusion zone with many product phases will also grow parabolically with time. The only importance of estimating the parabolic growth constant is to validate the diffusion-controlled growth before estimating the diffusion coefficients. Parabolic growth constant is not a material constant and depends on the end member compositions and also on the number of phases grown in the interdiffusion zone since a phase layer grows with different thickness in a single layer interdiffusion zone compared to the growth in a multiphase interdiffusion zone along with other phases. Even the activation energies for the growth will be different in these two cases. On the other hand, diffusion coefficient is a material constant and it should have the same value for a line compound or a fixed composition/concentration, irrespective of the end member compositions at a fixed annealing temperature. Therefore, one should be careful when drawing any conclusions about the diffusion mechanism based on estimated parabolic growth constant and the activation energy determined from it, especially if it is estimated from a multiphase interdiffusion zone. However, one can still discuss diffusion mechanism based on the activation energy estimated from the parabolic growth constants when only a single phase layer grows in an interdiffusion zone, since the value of this activation energy is the same as the integrated diffusion coefficient, which can happen due to Eqs. (3.107d) and (3.112), leading to

$$\tilde{D}^{\beta}_{\text{int}} = \frac{a \times b}{a+b} k^{\beta}_{p} = \frac{a \times b}{a+b} k^{o}_{p} \exp\left(-\frac{Q_g}{RT}\right) = \tilde{D}^{0}_{\text{int}} \exp\left(-\frac{Q}{RT}\right), \tag{3.113}$$

where $\tilde{D}^{0}_{\text{int}}$ is the pre-exponential factor for integrated interdiffusion coefficient and Q is the activation energy for interdiffusion. It should be noted that, when considering activation energy estimated from the interdiffusion or integrated diffusion coefficients, it is not always easy to discuss the atomic mechanism of diffusion, unless the absolute intrinsic diffusivities (diffusion rates of the components) or their ratio is known as discussed in Section 3.20.

3.10 ESTIMATION OF THE INTRINSIC DIFFUSION COEFFICIENTS OF COMPONENTS

Till now we have explained different approaches to estimate the diffusion parameters. For example, following the thin film solution (see Section 3.2.1), as also explained in Chapter 1 of Volume 1, one can estimate the tracer diffusion coefficients and impurity diffusion coefficients using radioisotopes. This is a measuring technique of the self-diffusion coefficient, i.e., the diffusion coefficient of a component in the absence of thermodynamic driving forces. Readers can go through Chapter 2 of Volume 1 to learn the experimental technique for the estimation of diffusion parameters utilizing radioisotopes. On the other hand, following the error function analysis, Matano–Boltzmann, den Broeder or Wagner methods, we measure the interdiffusion coefficient of the components. Interdiffusion coefficient is a kind of average of diffusion coefficients of the components under the influence of a thermodynamic driving force. These are called the intrinsic diffusion coefficients of the components. Therefore, the main difference between the tracer and intrinsic diffusion coefficients of a component is the influence of thermodynamic factor. Tracer diffusion coefficients, which are directly related to the self-diffusion coefficients by a correlation factor, are important for the understanding of the basic diffusion mechanism of a component and the influence of defects present in the structure. A comparison of the tracer and intrinsic diffusion coefficients explains the role of thermodynamic force on diffusion. The intrinsic diffusion coefficient is an important parameter to understand the diffusion process in inhomogeneous material systems in applications. On the other hand, since interdiffusion coefficient is a kind of average of the intrinsic diffusion coefficients, it is not necessarily straightforward to draw a conclusion on the atomic mechanism of diffusion, as discussed in Section 3.20. However, it is very important to understand the diffusion (composition) profile that develops in an interdiffusion zone between two dissimilar materials. In this section, the estimation procedure of the intrinsic diffusion coefficients is introduced. In the next section, the indirect method to estimate the tracer diffusion coefficients from a diffusion couple is explained.

For a long time, it was a common belief in the scientific community that diffusion rates of all the components are the same in a particular material. Kirkendall along with his student Smigelkas proved that different components diffuse with different rates based on an experiment utilizing the inert markers indicating the relative mobilities of the species [18]. This is known as Kirkendall effect in the scientific community,

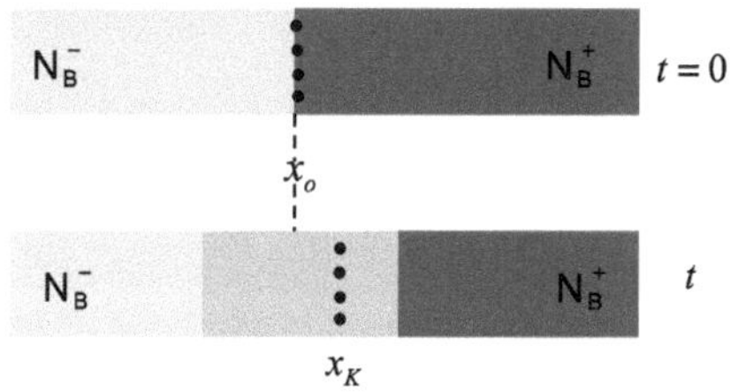

Figure 3.33 Movement of the inert Kirkendall markers from the initial contact plane to Kirkendall plane with diffusion annealing.

which helped Huntington and Seitz establish the vacancy diffusion mechanism in a substitutional alloy. Readers interested in the historical perspective and dramatic development related to these inventions are referred to the textbook [1]. Immediately after Kirkendall's finding, Darken published an analysis relating interdiffusion with intrinsic diffusion coefficients. Darken [19] developed the relations without considering the molar volume effect, which was later modified to relate the practical systems in which the molar volume is an important parameter to consider. Before explaining the estimation procedure for the intrinsic diffusion of components, the relations between different diffusion coefficients are introduced.

An imaginary diffusion couple of components A and B is shown in Fig. 3.33, where end members are prepared with N_B^- (A-rich) and N_B^+ (B-rich) alloys. Inert particles, which do not take part in the diffusion process but are replaced depending on the relative mobilities of the components, are placed at the bonding interface. Once the interdiffusion process starts, the markers get trapped at one particular composition, depending on the end member compositions, and always remain with the same composition, although the location of this composition or the marker plane moves depending on the annealing time. In the present example, the marker plane (x_K) is moved towards the right-hand side from the initial position, x_o, indicating that the intrinsic diffusion coefficient of component B (D_B) is greater than the intrinsic diffusion coefficient of component A (D_A). A larger amount of B is moved from the right to the left compared to the movement of A from the left to the right, leading to the shift of the inert markers towards the right with respect to the initial contact plane. The intrinsic fluxes and the intrinsic diffusion coefficients of the components can be related to Fick's first law as

$$J_A = -D_A\left(\frac{\partial C_A}{\partial x}\right)_K, \tag{3.114a}$$

$$J_B = -D_B\left(\frac{\partial C_B}{\partial x}\right)_K. \tag{3.114b}$$

Darken [19] established the relation between the interdiffusion ($\tilde{J}_A$ and $\tilde{J}_B$) and intrinsic fluxes (J_A and J_B) based on a logical argument involving the velocity of Kirkendall marker plane v_K as

$$\tilde{J}_A = J_A + v_K C_A, \tag{3.115a}$$
$$\tilde{J}_B = J_B + v_K C_B. \tag{3.115b}$$

As we have derived before (Eq. (3.76)),

$$\bar{v}_A \tilde{J}_A + \bar{v}_B \tilde{J}_B = \tilde{J}_A^{vol} + \tilde{J}_B^{vol} = 0. \tag{3.116}$$

$\tilde{J}_A^{vol}$ and $\tilde{J}_B^{vol}$ are defined to be the volume flux (= partial molar volume of the component, $\bar{v}_i \times$ molar flux) when estimated with respect to components A and B. From Eq. (3.15), we arrive at

$$J_A^{vol} = \bar{v}_A \tilde{J}_A = \bar{v}_A J_A + v_K \bar{v}_A C_A, \tag{3.117a}$$
$$J_B^{vol} = \bar{v}_B \tilde{J}_B = \bar{v}_B J_B + v_K \bar{v}_B C_B. \tag{3.117b}$$

Substituting Eq. (3.117) into (3.116) and then utilizing Eq. (3.114), we get

$$-\bar{v}_A D_A \left(\frac{\partial C_A}{\partial x}\right)_K + \bar{v}_A C_A v_K - \bar{v}_B D_B \left(\frac{\partial C_B}{\partial x}\right)_K + \bar{v}_B C_B v_K = 0,$$
$$(\bar{v}_A C_A + \bar{v}_B C_B) v_K = \bar{v}_A D_A \left(\frac{\partial C_A}{\partial x}\right)_K + \bar{v}_B D_B \left(\frac{\partial C_B}{\partial x}\right)_K. \tag{3.118}$$

Utilizing the standard thermodynamic relations expressed in Eqs. (3.74d) and (3.74e) yields

$$v_K = -(\bar{v}_B J_B + \bar{v}_A J_A) = \bar{v}_B (D_B - D_A) \left(\frac{\partial C_B}{\partial x}\right)_K. \tag{3.119}$$

If one is interested in estimating it with respect to the composition profile of A, the same relation can be derived as

$$v_K = \bar{v}_A (D_A - D_B) \left(\frac{\partial C_A}{\partial x}\right)_K. \tag{3.120}$$

From the standard thermodynamic relation expressed in Eq. (3.74a), Eq. (3.119) can be written as

$$v_K = \frac{\bar{v}_A \bar{v}_B}{v_m^2} (D_B - D_A) \left(\frac{\partial N_B}{\partial x}\right)_K. \tag{3.121}$$

Considering the constant molar volume (as assumed by Darken [19]) such that $\bar{v}_A = \bar{v}_B = v_m$, we get

$$v_K = (D_B - D_A) \left(\frac{\partial N_B}{\partial x}\right)_K. \tag{3.122}$$

Kirkendall marker plane velocity can also be determined based on the movement from the initial contact plane where these were kept before annealing and into the interdiffusion zone, that is, from the initial contact plane, x_o, to the new position of the markers after annealing for a certain time, that is, the Kirkendall marker plane, x_K. Then Boltzmann parameter expressed in Eq. (3.48) becomes

$$\lambda_K = \lambda(C_K) = \frac{x_K - x_o}{t^{\frac{1}{2}}} = \frac{x_K}{t^{\frac{1}{2}}},$$

and so

$$v_K = \frac{dx_K}{dt} = \frac{d(\lambda_K t^{\frac{1}{2}})}{dt} = \frac{\lambda_K}{2t^{\frac{1}{2}}} = \frac{x_K - x_o}{2t} = \frac{x_K}{2t}. \tag{3.123}$$

Eq. (3.115) can be rewritten with respect to Fick's first law as

$$-\tilde{D}\frac{\partial C_B}{\partial x} = -D_B\frac{\partial C_B}{\partial x} + C_B v_K.$$

Utilizing the relation in Eq. (3.120), we have

$$-\tilde{D}\left(\frac{\partial C_B}{\partial x}\right)_K = -D_B\left(\frac{\partial C_B}{\partial x}\right)_K + C_B\bar{v}_B(D_B - D_A)\left(\frac{\partial C_B}{\partial x}\right)_K,$$

$$-\tilde{D}\left(\frac{\partial C_B}{\partial x}\right)_K = -(1 - C_B\bar{v}_B)D_B\left(\frac{\partial C_B}{\partial x}\right)_K - C_B\bar{v}_B D_A\left(\frac{\partial C_B}{\partial x}\right)_K.$$

Utilizing the standard thermodynamic relation in Eq. (3.74d), we have

$$-\tilde{D}\left(\frac{\partial C_B}{\partial x}\right)_K = -(C_A\bar{v}_A D_B + C_B\bar{v}_B D_A)\left(\frac{\partial C_B}{\partial x}\right)_K,$$

$$\tilde{D} = \bar{v}_A C_A D_B + \bar{v}_B C_B D_A. \tag{3.124}$$

Considering Darken's assumption of constant molar volume yields

$$\tilde{D} = N_A D_B + N_B D_A. \tag{3.125}$$

Eq. (3.125) is known as Darken's relation.

One should note an important point that the interdiffusion coefficients can be estimated at any compositions in the interdiffusion zone. However, the intrinsic diffusion coefficients can only be estimated at the composition of the Kirkendall marker plane, which indicates the relative mobilities of the components and is affected from the very beginning of the diffusion annealing. There are two main methods to estimate the intrinsic diffusion coefficients. Heumann [20] first developed an estimation method in which one needs to locate the initial contact plane and therefore allow error in the determination of the diffusion parameters in practical examples where molar volume deviates from ideal [13]. Much later, van Loo [21] developed a method in which we do not need to locate the initial contact plane. More recently, the present author derived the same relation following Wagner's line of treatment for the interdiffusion coefficient [22]. Huemann's derivation and that developed by the current author are explained in the next two subsections. Following both methods, one can estimate the intrinsic diffusion coefficients at the marker plane only. To circumvent this problem, a multifoil technique is developed to estimate these parameters at many compositions in the interdiffusion zone. However, the experimental method is a bit more tedious, which is explained in the following subsection.

3.10.1 Heumann's Method for Estimation of the Intrinsic Diffusion Coefficients [20]

Inert markers are trapped at one particular composition at the very beginning and move along with it as annealing continues. Since the diffusion coefficients are material constants at a particular composition at a particular temperature, the intrinsic flux at Kirkendall marker plane can be integrated over the annealing time t as follows:

$$\Phi_B = -\int_0^t J_B^K \, dt = D_B \int_0^t \left(\frac{dC_B}{dx}\right)_K dt. \tag{3.126}$$

We have taken a negative sign for the diffusion of component B since this component diffuses from the right to the left (opposite to the increasing position parameter), as shown in Fig. 3.34. "K" corresponds to Kirkendall marker plane at which the intrinsic diffusion coefficients are estimated. We can write

$$\left(\frac{dC_B}{dx}\right)_K = \left(\frac{dC_B}{d\lambda}\right)_K \frac{d\lambda}{dx}, \tag{3.127a}$$

where $\lambda = \lambda(C) = \frac{x-x_0}{t^{1/2}} = \frac{x}{t^{1/2}}$ is Boltzmann parameter. And so we get

$$\left(\frac{dC_B}{dx}\right)_K = \left(\frac{dC_B}{d\lambda}\right)_K \frac{1}{t^{1/2}}. \tag{3.127b}$$

Therefore, Eq. (3.126) can be written as

$$\Phi_B = D_B \left(\frac{dC_B}{d\lambda}\right)_K \int_0^t \frac{dt}{t^{1/2}} = 2D_B t^{1/2} \left(\frac{dC_B}{d\lambda}\right)_K. \tag{3.128}$$

Using Eq. (3.127b) in Eq. (3.128) and utilizing the standard thermodynamic relation expressed in Eq. (3.74a), we get

$$\Phi_B = 2D_B t \left(\frac{dC_B}{dx}\right)_K = 2D_B t \frac{\bar{v}_A}{v_m^2} \left(\frac{dN_B}{dx}\right)_K. \tag{3.129a}$$

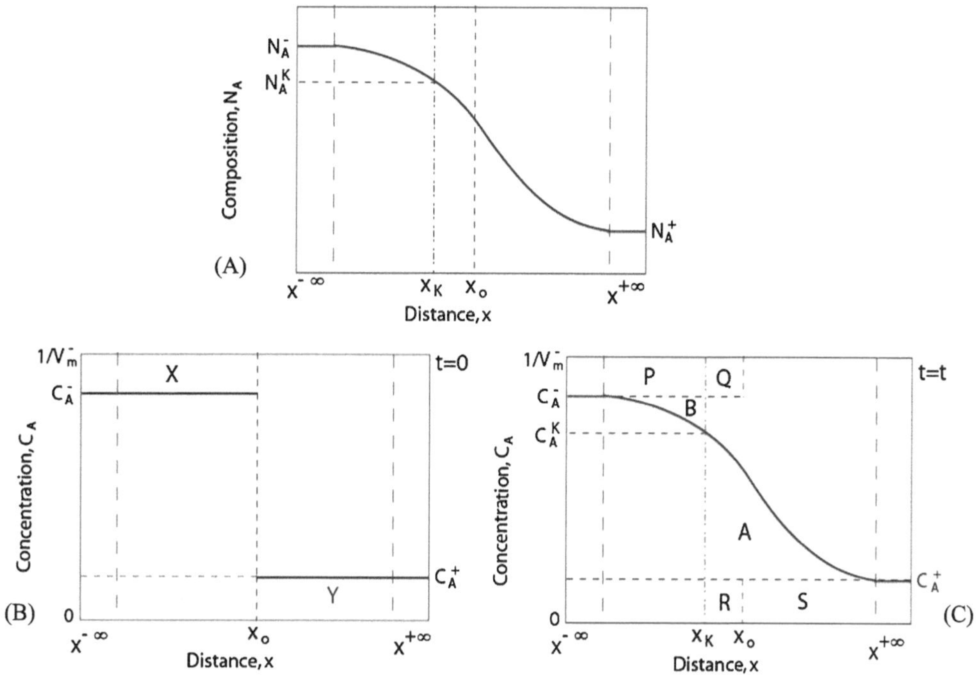

Figure 3.34 Estimation steps of the ratio of the intrinsic diffusion coefficients following Heumann's method.

Similarly, since A diffuses from the left to the right, we derive

$$\Phi_A = \int_0^t J_A^K dt = -2D_A t\left(\frac{dC_A}{dx}\right)_K = -2D_A t\frac{\bar{v}_B}{v_m^2}\left(\frac{dN_A}{dx}\right)_K = 2D_A t\frac{\bar{v}_B}{v_m^2}\left(\frac{dN_B}{dx}\right)_K. \quad (3.129b)$$

Therefore, the ratio of the intrinsic diffusion coefficients can be written as

$$\frac{D_B}{D_A} = \frac{\bar{v}_B}{\bar{v}_A}\frac{\Phi_B}{\Phi_A}. \quad (3.130)$$

The cumulative fluxes Φ crossing Kirkendall marker plane can be estimated graphically by comparing the concentration or composition profile (for fixed molar volume) developed after annealing time $t = 0$ and $t = t$, as shown in Fig. 3.34. On the left-hand side of Kirkendall marker plane x_K, the composition profile is developed because of diffusion of B from the other side of the end member (N_B^+). Therefore, we can state the following:

Before the start of the annealing at time $t = 0$, the area in the concentration profile, as shown in Fig. 3.34B, is X. After annealing for time t, the area in the concentration profile following the new position of the marker plane x_K, as shown in Fig. 3.34C, is

B + P. Therefore, the difference between these two areas can be expressed as

$$\Phi_B = B + P - X = B + P - (P + Q) = B - Q = B + x_K\left(\frac{1}{v_m^-} - C_A^-\right).$$

Note that a negative sign is used for x_K with respect to $x_o = 0$ in the present example.

Similarly, the concentration profile in the other (right-hand) side of Kirkendall marker plane is developed by the diffusion of component A. Therefore, we can state the following:

Before annealing at time $t = 0$, as shown in Fig. 3.34B, the area in the concentration profile is Y. After annealing for time t, as shown in Fig. 3.34C, the area is A + R + S. Therefore, the difference between these two areas can be expressed as

$$\Phi_A = A + R + S - Y = A + R = A - x_K C_A^+.$$

Note that here S = Y.

Therefore, we can write

$$\frac{D_B}{D_A} = \frac{\bar{v}_B}{\bar{v}_A}\frac{B + x_K(\frac{1}{v_m^-} - C_A^-)}{A - x_K C_A^+} = \frac{\bar{v}_B}{\bar{v}_A}\frac{B + x_K(\frac{1}{v_m^-} - \frac{N_A^-}{v_m^-})}{A - x_K\frac{N_A^+}{v_m^+}}. \tag{3.131a}$$

If we assume a constant molar volume ($v_m = \bar{v}_A = \bar{v}_B$), following similar arguments in Figs. 3.34B and 3.34C, we can write

$$\frac{D_B}{D_A} = \frac{B + x_K(1 - N_A^-)}{A - x_K N_A^+}. \tag{3.131b}$$

From the calculated value of the ratio of the intrinsic diffusion coefficients, one can estimate the absolute values utilizing the relation expressed in Eqs. (3.124) or (3.125).

Similar to the calculation of the interdiffusion coefficients following Matano–Boltzmann analysis, we need to locate the initial contact plane following Heumann's method. Since it is impossible to locate this plane correctly in practical examples where the molar volume deviates from ideal, it draws a certain amount of error in the calculated data [13], as explained in Section 3.10.3.

3.10.2 van Loo's Relations for the Intrinsic Diffusion Coefficients [21] Developed by Paul [22] Following the Line of Treatment Proposed by Wagner

Eqs. (3.85a) and (3.85b) can be rewritten at Kirkendall marker plane as

$$\frac{1}{2t}\left(N_B^+ - N_B^-\right)\left[-\frac{\lambda^K Y_B^K}{v_m^K} + \int_{\lambda-\infty}^{\lambda^K}\frac{Y_B}{v_m}d\lambda\right] = \frac{1}{t^{\frac{1}{2}}}\left[N_B^- \tilde{J}_A^K - (1 - N_B^-)\tilde{J}_B^K\right], \tag{3.132a}$$

$$\frac{1}{2t}\left(N_B^+ - N_B^-\right)\left[-\frac{\lambda^K(1-Y_B^K)}{v_m^K} - \int_{\lambda^K}^{\lambda^{+\infty}} \frac{(1-Y_B)}{v_m} d\lambda\right] = \frac{1}{t^{\frac{1}{2}}}\left[-N_B^+ \tilde{J}_A^K + (1-N_B^+)\tilde{J}_B^K\right]. \tag{3.132b}$$

The aim is now to rewrite the equations with respect to $\tilde{J}_B^K$ and derive the relation for D_B. Eq. (3.132a) is multiplied by N_B^+, and Eq. (3.132b) multiplied by N_B^-, then both are added, and after rearranging we obtain

$$\frac{1}{2t}\left(N_B^+ - N_B^-\right)\left[-\lambda^K\left\{\frac{Y_B^K}{v_m^K}N_B^+ + \frac{(1-Y_B^K)}{v_m^K}N_B^-\right\} + \left\{N_B^+\int_{\lambda^{-\infty}}^{\lambda^K}\frac{Y_B}{v_m}d\lambda - N_B^-\int_{\lambda^K}^{\lambda^{+\infty}}\frac{(1-Y_B)}{v_m}d\lambda\right\}\right]$$
$$= \frac{1}{t^{\frac{1}{2}}}\left[-N_B^+(1-N_B^-)\tilde{J}_B^K + N_B^-(1-N_B^+)\tilde{J}_B^K\right].$$

Replacing composition normalized variables at Kirkendall marker plane, $Y_B^K = \frac{N_B^K - N_B^-}{N_B^+ - N_B^-}$ and $1 - Y_B^K = \frac{N_B^+ - N_B^K}{N_B^+ - N_B^-}$, we get

$$\frac{1}{2t}\left(N_B^+ - N_B^-\right)\left[-\lambda^K\frac{N_B^K}{v_m^K} + \left\{N_B^+\int_{\lambda^{-\infty}}^{\lambda^K}\frac{Y_B}{v_m}d\lambda - N_B^-\int_{\lambda^K}^{\lambda^{+\infty}}\frac{(1-Y_B)}{v_m}d\lambda\right\}\right]$$
$$= \frac{1}{t^{\frac{1}{2}}}\left[-(N_B^+ - N_B^-)\tilde{J}_B^K\right],$$
$$\tilde{J}_B^K = \frac{t^{\frac{1}{2}}}{2t}\left[\lambda^K\frac{N_B^K}{v_m^K} - N_B^+\int_{\lambda^{-\infty}}^{\lambda^K}\frac{Y_B}{v_m}d\lambda + N_B^-\int_{\lambda^K}^{\lambda^{+\infty}}\frac{(1-Y_B)}{v_m}d\lambda\right],$$
$$\tilde{J}_B^K = \frac{t^{\frac{1}{2}}}{2t}\lambda^K\frac{N_B^K}{v_m^K} - \frac{t^{\frac{1}{2}}}{2t}\left[N_B^+\int_{\lambda^{-\infty}}^{\lambda^K}\frac{Y_B}{v_m}d\lambda - N_B^-\int_{\lambda^K}^{\lambda^{+\infty}}\frac{(1-Y_B)}{v_m}d\lambda\right].$$

Replacing Boltzmann parameter $\lambda^K = \frac{x_K}{t^{1/2}}$, utilizing the relation of Kirkendall marker velocity, $v_K = \frac{x_K - x_o}{2t} = \frac{x_K}{2t}$ (Eq. (3.123)), and concentration $C_B^K = \frac{N_B^K}{v_m^K}$, we get

$$\tilde{J}_B^K = v_K C_B^K - \frac{t^{1/2}}{2t}\left[N_B^+\int_{\lambda^{-\infty}}^{\lambda^K}\frac{Y_B}{v_m}d\lambda - N_B^-\int_{\lambda^K}^{\lambda^{+\infty}}\frac{(1-Y_B)}{v_m}d\lambda\right].$$

Exploiting Boltzmann parameter, we obtain $d\lambda = \frac{dx}{t^{\frac{1}{2}}}$. Therefore, we get

$$\tilde{J}_B^K = v_K C_B^K - \frac{1}{2t}\left[N_B^+\int_{x^{-\infty}}^{x_K}\frac{Y_B}{v_m}dx - N_B^-\int_{x_K}^{x^{+\infty}}\frac{(1-Y_B)}{v_m}dx\right].$$

Comparing with Eq. (3.115b), we can write the expression for the intrinsic flux of B as

$$J_B^K = -\frac{1}{2t}\left[N_B^+ \int_{x^{-\infty}}^{x_K} \frac{Y_B}{v_m} dx - N_B^- \int_{x_K}^{x^{+\infty}} \frac{(1-Y_B)}{v_m} dx\right]. \tag{3.133a}$$

Utilizing Fick's first law, we have

$$D_B = \frac{1}{2t}\left(\frac{\partial x}{\partial C_B}\right)_K \left[N_B^+ \int_{x^{-\infty}}^{x_K} \frac{Y_B}{v_m} dx - N_B^- \int_{x_K}^{x^{+\infty}} \frac{(1-Y_B)}{v_m} dx\right]. \tag{3.133b}$$

Therefore the intrinsic flux and the intrinsic diffusion coefficients of component B at Kirkendall marker plane can be directly estimated from Eqs. (3.133a) and (3.133b).

Utilizing the standard thermodynamic relation given in Eq. (3.74a), we can rewrite it as

$$D_B = \frac{1}{2t}\frac{v_m^2}{\bar{v}_A}\left(\frac{\partial x}{\partial N_B}\right)_K \left[N_B^+ \int_{x^{-\infty}}^{x_K} \frac{Y_B}{v_m} dx - N_B^- \int_{x_K}^{x^{+\infty}} \frac{(1-Y_B)}{v_m} dx\right]. \tag{3.134}$$

Similarly, to derive the relation for D_A, we first need to derive the relation for $\tilde{J}_A$. This can be achieved by multiplying Eq. (3.132a) by $(1-N_B^+)$ and Eq. (3.132b) by $(1-N_B^-)$ and then following similar steps, yielding

$$J_A = -\frac{1}{2t}\left[N_A^+ \int_{x^{-\infty}}^{x_K} \frac{Y_B}{v_m} dx - N_A^- \int_{x_K}^{x^{+\infty}} \frac{(1-Y_B)}{v_m} dx\right], \tag{3.135a}$$

$$D_A = \frac{1}{2t}\left(\frac{\partial x}{\partial C_A}\right)_K \left[N_A^+ \int_{x^{-\infty}}^{x_K} \frac{Y_B}{v_m} dx - N_A^- \int_{x_K}^{x^{+\infty}} \frac{(1-Y_B)}{v_m} dx\right]. \tag{3.135b}$$

Therefore the intrinsic flux and the intrinsic diffusion coefficients of component A at Kirkendall marker plane can be directly estimated from Eqs. (3.135a) and (3.135b).

Again utilizing the standard thermodynamic relation in Eq. (3.74a), we obtain

$$D_A = \frac{1}{2t}\frac{v_m^2}{\bar{v}_B}\left(\frac{\partial x}{\partial N_A}\right)_K \left[N_A^+ \int_{x^{-\infty}}^{x_K} \frac{Y_B}{v_m} dx - N_A^- \int_{x_K}^{x^{+\infty}} \frac{(1-Y_B)}{v_m} dx\right]. \tag{3.136}$$

Utilizing Eqs. (3.134) and (3.135) and Fig. 3.35, we can write

$$\frac{D_B}{D_A} = \frac{\bar{v}_B}{\bar{v}_A}\left[\frac{N_B^+ \int_{x^{-\infty}}^{x_K} \frac{Y_B}{v_m} dx - N_B^- \int_{x_K}^{x^{+\infty}} \frac{(1-Y_B)}{v_m} dx}{-N_A^+ \int_{x^{-\infty}}^{x_K} \frac{Y_B}{v_m} dx + N_A^- \int_{x_K}^{x^{+\infty}} \frac{(1-Y_B)}{v_m} dx}\right] = \frac{\bar{v}_B}{\bar{v}_A}\left[\frac{N_B^+\varphi - N_B^-\psi}{-N_A^+\varphi + N_A^-\psi}\right], \tag{3.137}$$

where ϕ and ψ are the areas shown in Fig. 3.35. The estimation procedure is straightforward compared to the Heumann's method with an added benefit that there is no need to locate the initial contact plane and therefore the error in the calculation must be smaller [13].

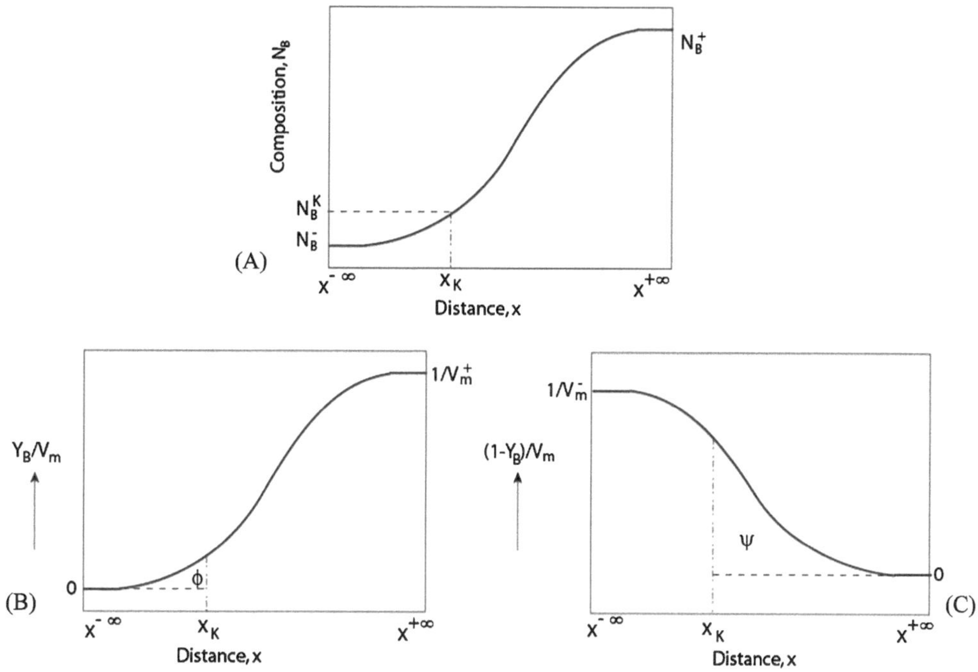

Figure 3.35 Estimation steps of ratio of the intrinsic diffusion coefficients following the method developed by extending Wagner's approach, which is equivalent to the approach followed by van Loo [21].

3.10.3 Comparison of the Data Estimated by Different Methods

For this discussion, we consider the growth of the intermetallic compound V_3Ga in a diffusion couple of V/V (0.4Ga) annealed at 1000°C for 25 hours. The interdiffusion zone is shown in Fig. 3.36A, and the composition profile is shown in Fig. 3.36B [13]. In general, variation of the lattice parameter data is not available for many intermetallic compounds; however; these details are available for V_3Ga. When variation of the lattice parameter data is not available, a constant molar volume over the whole composition range is considered, which is estimated based on the data available at the stoichiometric composition ($V_{0.75}Ga_{0.25}$). Both types of data are shown in Fig. 3.36C. It is interesting to note that the difference is very small between these two variations. Therefore, we shall examine how this small difference influences the estimated data.

The total layer thickness of the product phase is 54 μm. Kirkendall marker plane is found to be at 26.1 at.%Ga and 15 μm from the V/V_3Ga interface. Considering constant molar volume, the initial contact plane is found to be at the distance of 18 μm from the V/V_3Ga interface. When the actual molar volume variation is considered, the plane was located at 17.8 μm and 18.5 μm from the concentration profile of V and Ga, respectively. As expected, the difference between two estimated initial contact planes is found to be very small.

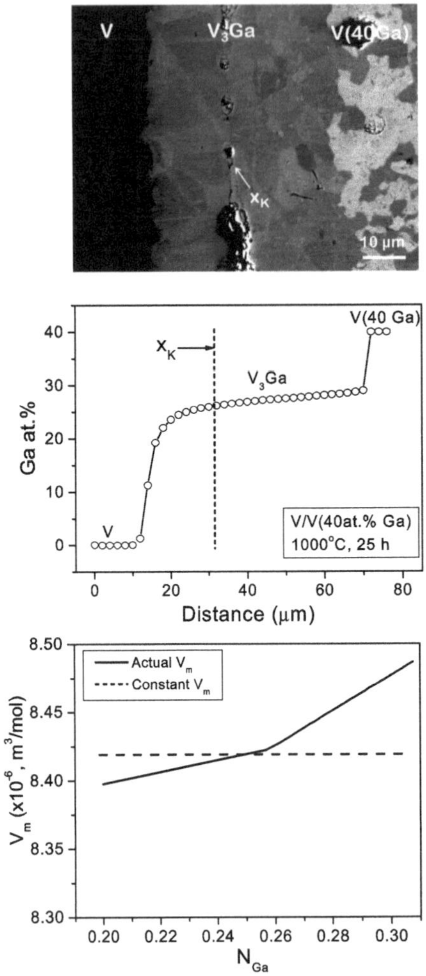

Figure 3.36 (A) Growth of V_3Ga in a diffusion couple of V/V(40Ga) annealed at 1000°C for 25 hrs, (B) composition profile, and (C) different molar volume variations considered for estimation of the diffusion coefficients [13].

The interdiffusion coefficient following Matano–Boltzmann method using constant molar volume at Kirkendall marker plane (26.1 at.%Ga) is estimated as 2.12×10^{-14} m^2/s. Using the actual volume variation, the same coefficient is estimated as 2.19×10^{-14} m^2/s and 2.28×10^{-14} m^2/s using two different initial contact planes. As expected, the difference between these estimated values is small. Following Wagner's method, we do not need to locate the initial contact plane. These values at the same Kirkendall marker plane composition are estimated as 2.11×10^{-14} m^2/s and 2.06×10^{-14} m^2/s for the constant and actual variation of the molar volume, respectively. As expected (see Section 3.7), Wagner's method gives an even smaller difference. This is an important point to make here.

Next, we check the influence on the intrinsic diffusion coefficients following different methods. The ratio of the partial molar volumes of the components, $\frac{\bar{v}_{Ga}}{\bar{v}_V}$ for constant molar volume, is 1, whereas it is 1.16 at Kirkendall marker plane. Note that the partial molar volumes are estimated by extending the gradient at the molar volume of interest to 0 and 100 at.%Ga for $\bar{v}_V$ and $\bar{v}_{Ga}$, respectively. Following Heumann's method, the values are estimated as $\frac{D_{Ga}}{D_V} = 0.56$ for constant molar volume. The same ratio is estimated as 0.206 and 0.212 for two different initial contact planes for the actual variation of the molar volume. Although the difference is less for two different initial contact planes (due to the small difference between these two values), however there is a significant difference when a constant molar volume is considered. Following the relation developed by van Loo (the same relation developed by Paul following Wagner's treatment), we estimate the ratio as $\frac{D_{Ga}}{D_V} = 0.84$ for constant molar volume and 0.82 for the actual variation of the molar volume. Therefore, this method gives a much smaller error with an added advantage that we do not need to locate the initial contact plane. In a system with a greater deviation from the ideal molar volume, the data estimated following Heumann's method will be extremely large. Therefore, it is safe to use this approach compared to Heumann's method.

3.11 IDENTIFYING THE LOCATION OF KIRKENDALL MARKER PLANE

It must be clear from the description above that we can estimate the intrinsic diffusion coefficients at Kirkendall marker plane only. This plane moves from the very beginning of the diffusion annealing process depending on the relative mobilities of the components. Therefore, detecting the location of this plane correctly is very important. The most straightforward approach is, of course, to use inert particles at the bonding interface before the start of the diffusion annealing and then find them in the interdiffusion zone after completion of the experiment.

Depending on the material systems of interest, inert markers are selected. Kirkendall himself [18] and many other researchers in that era used wires as an inert marker. However, currently, powders are employed more frequently mainly due to ease of use. In low-temperature material systems, i.e., when the temperature of diffusion annealing is relatively low, W and Mo wires or powders are used as inert markers since they even do not react with many metallic systems in that temperature range. However, to be on the safe side, refractory metal oxides are suggested for use. ThO_2 is found to be one of the best choices simply because particles of this oxide are easily found in a scanning electron image with white contrast in the interdiffusion zone, as shown in Fig. 3.37A [23]. However, it is not easy to obtain this oxide, since the use of Th-based materials is strictly monitored in most countries. TiO_2 and Y_2O_3 are used more commonly

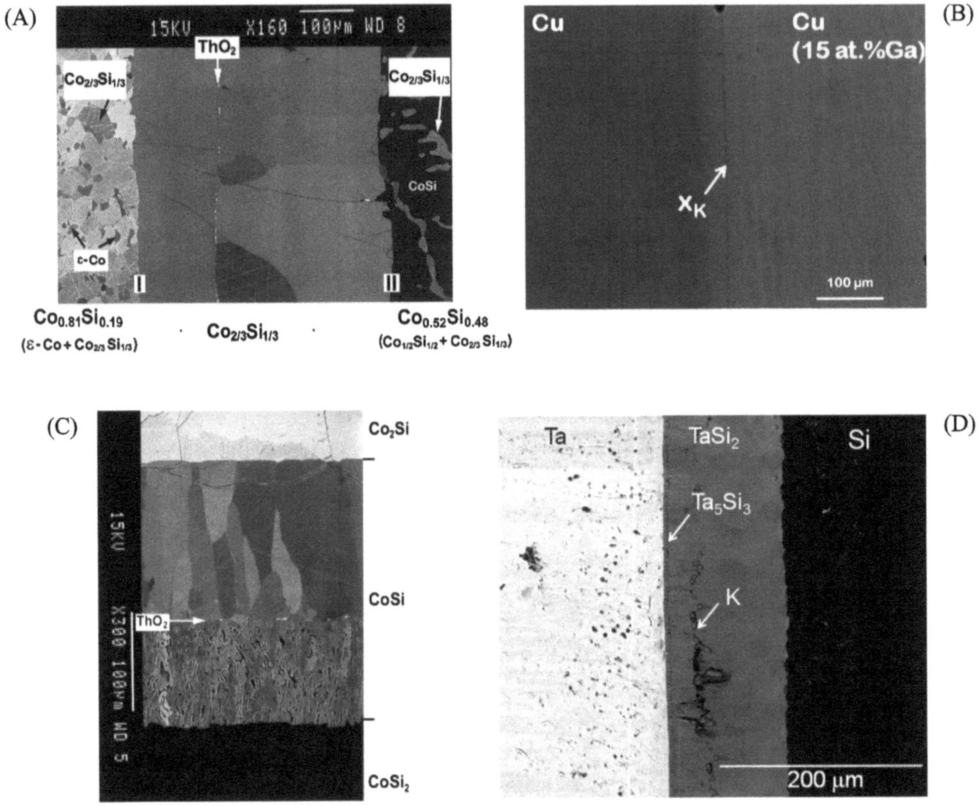

Figure 3.37 Different ways to locate Kirkendall marker plane are explained based on the experimental results in various systems: (A) Co–Si [23], (B) Cu–Ga [24], Co–Si [26], and Ta–Si [29].

and are easily available. In oxide systems, Au and Pt particles are used since these are inert in such systems. Materials used as ideal markers should not react with the diffusion component and the size should be small enough such that they do not hinder the flux of components. Therefore, there should be enough space in-between particles, and clustering of many particles at one place should be avoided as much as possible. If very big particles are used, they may hinder the flux of components. Sometimes a gap might be created between the end members of a diffusion couple prepared with very hard materials, for example, refractory metal–silicon systems. As a result, there will no bond between the end members, and the experiment will fail. A too high load for making a contact is not advisable since this may break (when a brittle material is used) or creep the diffusion couple. On the other hand, too small particles might be dragged by the grain boundaries into the interdiffusion zone. We have witnessed this phenomenon especially when the particle size had been less than 0.1 μm. We found that the particle sizes in the range of 0.5–5 μm are more or less suitable for most systems; however, the range of 1–2 μm is preferable. These particles are dispersed in acetone and a small amount is dropped on the surface of a metallographically prepared end member. Once the acetone is evaporated, particle distribution can be checked under an optical microscope. Ideal random distribution with roughly 20–30 μm gap between different particles is preferred. Clustering of particles cannot be avoided but could be minimized simply by shaking the end member lightly after dropping the dispersed particle. This might not be achieved in one attempt and can be repeated a few times after cleaning the surface with acetone till it is acceptable. Marker plane position can be located in the interdiffusion zone in a cross-sectioned and metallographically prepared diffusion couple by detecting X-ray peak of an element originated from the marker in an SEM (scanning electron microscope) attached with EDS (energy dispersive spectrometer). Sometimes debris introduced during metallographic preparation could also act as markers, which are found at Kirkendall marker plane. To prepare a successful diffusion couple, the flatness of the end members is given more importance than the almost scratch free surface. Flatness is lost especially in the edges of relatively soft materials when metallographic steps are followed for too long to get a completely scratch-free material. Scratches rather help in detecting the location of Kirkendall marker plane in many systems. They can be seen as a negative surface with a line void at the interface through which materials cannot diffuse. Therefore, these voids, if not healed during diffusion annealing, also move along with Kirkendall marker plane like inert particles and can be found as lines of pores. As shown in Fig. 3.37B [24], these were found along with TiO_2 particles at Kirkendall marker plane. Sometimes markers are lost during the metallographic preparation, leaving behind the line of pores. We faced this problem when ThO_2 particles were used in the Au/Sn system. Then, the use of W marker was successful [22,25], which helped establish the presence of bifurcation of Kirkendall marker plane in this system.

In many systems, even in the bulk diffusion couples, it is not easy to use inert particles. For example, both materials in the metal–silicon system are very hard and do not make diffusion bond because of the even small gap created by the presence of even small inert particles. Also the line of pores created by the scratches may not be found in all diffusion couples. However, one can easily locate this plane from the evolution of the microstructure in an interdiffusion zone, as shown in Fig. 3.37C [26]. A duplex morphology is found in an interdiffusion zone when the marker plane is present in the middle of a phase layer. Otherwise, it will have a uniform morphology with continuous grains over the whole thickness of the phase layer, if the marker plane is present at one of the interfaces. This statement is true in most systems except for a few in which grains are repeatedly nucleated during interdiffusion. This is explained in much detail in Chapter 4 of Volume 2 based on a physicochemical approach developed by the present author [27,28]. As shown in Fig. 3.37A, sometimes the location of this plane is evident from both inert markers and the microstructure. Other times, this is evident by the presence of both line of pores and microstructure, as shown in Fig. 3.37D [29].

One should be careful when locating the marker plane by the presence of a line of pores. The microstructure might have different types of voids created by various reasons such as grinding artifacts or the formation of Kirkendall voids when vacancies are not absorbed because of unavailability of enough sites acting as sinks. Interested readers can go through [1] for further details. The presence of Kirkendall voids is very prominent on one side of Kirkendall marker plane in the CoSi phase, as shown in Fig. 3.37C. Locating the marker plane just with the help of microstructure is very handy since in many systems, especially in thin films, we cannot use any markers with a suitable size. Therefore, the location of this plane can be easily found by revealing the microstructure after mapping of grain by EBSD (electron back-scattered electron diffraction) or in a TEM (transmission electron microscope).

3.12 MULTIFOIL TECHNIQUE TO ESTIMATE THE INTRINSIC DIFFUSION COEFFICIENTS FOR MANY COMPOSITIONS FROM A SINGLE DIFFUSION COUPLE

As mentioned in the previous section, we can estimate the intrinsic diffusion coefficients at Kirkendall marker plane. Only one marker plane is found in most of the diffusion couples except for a few rare cases when more than one marker plane is found [1,5,27, 28]. Therefore, if we are interested in estimating these parameters for many compositions in a composition range, we need to prepare many diffusion couples with different end members such that Kirkendall markers are trapped at different compositions in different couples. This was done extensively in the β-NiAl phase [5]. Naturally, this is a time-consuming process since the success rate of diffusion couple experiments could be very low in many systems. To circumvent this problem, a multifoil technique is established such that these parameters can be estimated for many compositions from a single diffusion couple. Since very thin foils are used in this method to produce the diffusion couple, it can only be followed using the materials which can be produced as thin foils and are not very brittle. Therefore, this method is mostly restricted to the systems in which metallic foils are used as the end members for the growth of complete solid solution and/or intermetallic compound in the interdiffusion zone. This is described with the help of experimental results in the Ni–Pt system [30].

Before designing the experiment, one needs to decide on the number and thickness of the foils required to make a diffusion couple. Since the preferable marker size is around 1–2 μm, the thickness of foils in the range of 25–50 μm should be suitable for the multifoil experiment. Thinner foils of a relatively soft metal are not easy to handle. Based on our experience, 40–50 μm foils are suitable. However, even 25 μm foils are manageable if these are not too soft, which were used for the experiment in the Ni–Pt system. The number of foils to be used depends on the total thickness of the interdiffusion zone. It is important to note that a few foils present at the both ends of diffusion couple should be unaffected by the diffusing components to fulfill the boundary conditions of the equations used for the estimation of the diffusion coefficients. Before stacking the foils, Y_2O_3 particles were used as inert markers at the interface of all the foils of Ni and Pt. A back scattered electron image of the multifoil experiment in the Pt/Ni system is shown in Fig. 3.38A. The velocity, v, of different marker planes is determined by the relation developed by different researchers [31–33]:

$$v = \frac{1}{2t}\left(y - x_o\frac{dy}{dx_o}\right), \tag{3.138}$$

where x_o is the initial position of marker plane before interdiffusion and x is the new position after the interdiffusion. Therefore $y = (x - x_0)$ is the displacement because of interdiffusion. Note that the markers located at $x_0 = 0$ move to establish Kirkendall marker plane with a velocity of $v = y/2t$ according to Eq. (3.138). The composition profile of the interdiffusion zone is shown in Fig. 3.38B. The displacement of different marker planes is directly measured from the backscattered electron image, as shown in Fig. 3.38A, and then y vs. x plot is shown in Fig. 3.38C. Note that the marker planes outside the interdiffusion zone also show some displacement which might be due to deformation of the foils. To bring this effect into account, a base line is drawn connecting the unreacted ends of the diffusion couple, as shown in Fig. 3.38D [30]. This indicates that we are assuming an equal and constant movement for all the foils. Therefore the actual displacement is the difference between the displacement measured directly from the micrograph and the base line values. Note here that Kirkendall marker might not have the maximum displacement, as it is found in this system. Then the displacement is plotted with respect to y vs. $x_0(= x - y)$ to estimate the values of dy/dx_o. Utilizing these values, one can estimate the values of v for different marker planes, as shown in Fig. 3.38E. Now we need to derive the equation relating v with the intrinsic diffusion coefficients.

We can use the standard thermodynamic relation in Eq. (3.74d) [1] to get

$$\frac{D_B}{D_A} = \frac{\bar{v}_B D_B}{\bar{v}_B D_A},$$
$$\frac{D_B}{D_A} = \frac{\bar{v}_B(\bar{v}_B C_B + \bar{v}_A C_A)D_B}{\bar{v}_B(\bar{v}_B C_B + \bar{v}_A C_A)D_A},$$
$$\frac{D_B}{D_A} = \frac{\bar{v}_B(\bar{v}_A C_A D_B + \bar{v}_B C_B D_B)}{\bar{v}_B \bar{v}_A C_A D_A + \bar{v}_B \bar{v}_B C_B D_A},$$
$$\frac{D_B}{D_A} = \frac{\bar{v}_B[(\bar{v}_A C_A D_B + \bar{v}_B C_B D_A) + \bar{v}_B C_B(D_B - D_A)]}{\bar{v}_B(\bar{v}_A C_A D_B + \bar{v}_B C_B D_A) - \bar{v}_A \bar{v}_B C_A(D_B - D_A)}.$$

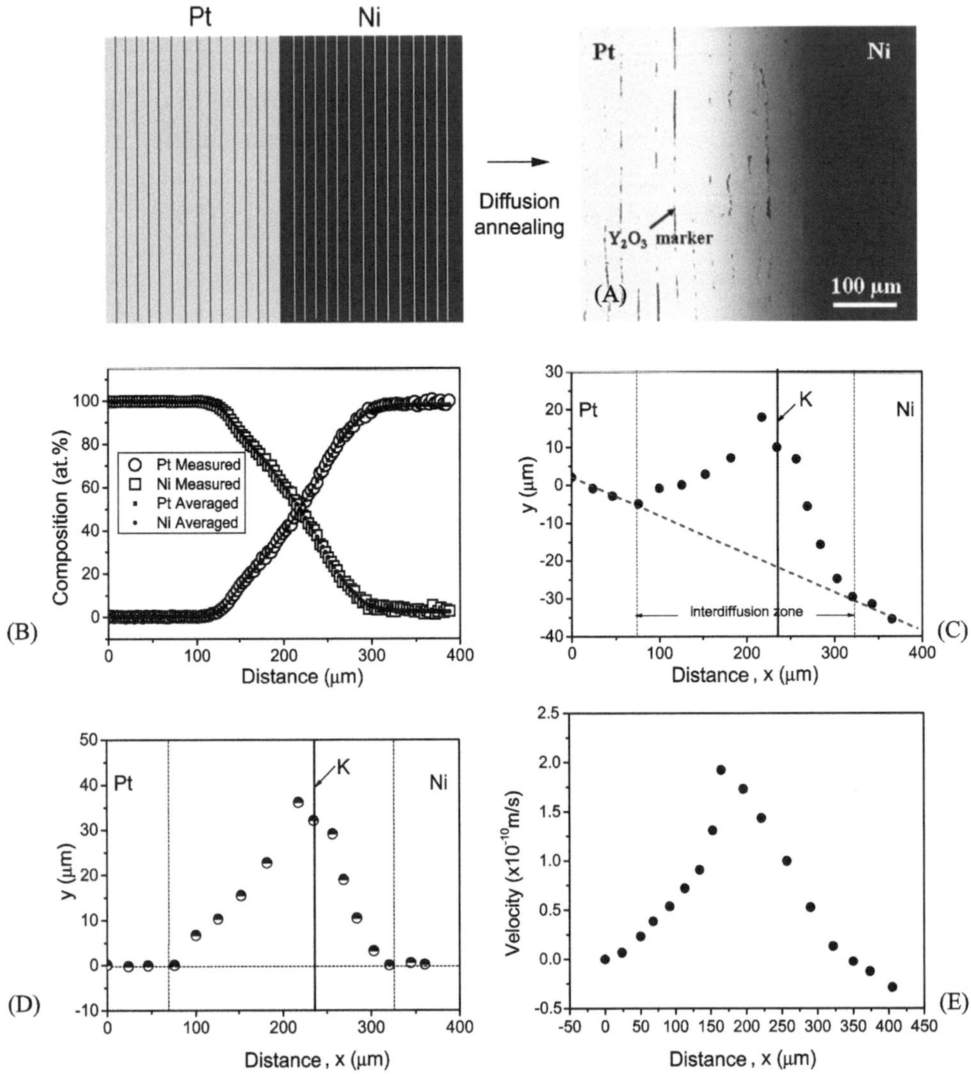

Figure 3.38 Explanation of estimation of the diffusion coefficients following the multifoil experiment in the Ni–Pt system [30].

Utilizing the relation between the interdiffusion and intrinsic diffusion coefficients ($\tilde{D} = \bar{v}_A C_A D_B + \bar{v}_B C_B D_A$) expressed in Eq. (3.124) yields

$$\frac{D_B}{D_A} = \frac{\bar{v}_B[\tilde{D} + \bar{v}_B C_B(D_B - D_A)]}{\bar{v}_B\tilde{D} - \bar{v}_A\bar{v}_B C_A(D_B - D_A)},$$

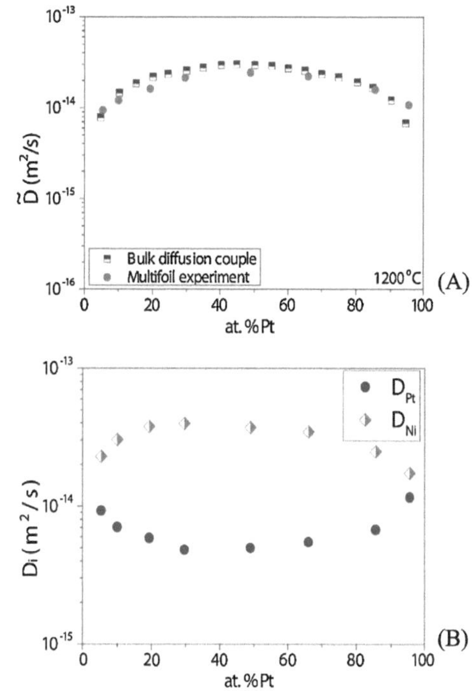

Figure 3.39 The estimated diffusion coefficients in the Ni–Pt system following the multifoil diffusion couple [30].

$$\frac{D_B}{D_A} = \frac{\tilde{D} + \bar{v}_B C_B (D_B - D_A)}{\tilde{D} - \frac{\bar{v}_A \bar{v}_B}{\bar{v}_B} C_A (D_B - D_A)},$$

$$\frac{D_B}{D_A} = \frac{\tilde{D}\left(\frac{\partial C_B}{\partial x}\right) + \bar{v}_B (D_B - D_A)\left(\frac{\partial C_B}{\partial x}\right) C_B}{\tilde{D}\left(\frac{\partial C_B}{\partial x}\right) - \bar{v}_B (D_B - D_A)\left(\frac{\partial C_B}{\partial x}\right) \frac{\bar{v}_A}{\bar{v}_B} C_A}.$$

Utilizing Eq. (3.119), in which velocity is related to the intrinsic diffusion coefficients as $v = \bar{v}_B (D_B - D_A)\left(\frac{\partial C_B}{\partial x}\right)$, gives

$$\frac{D_B}{D_A} = \frac{\tilde{D}(\partial C_B/dx) + v C_B}{\tilde{D}(\partial C_B/dx) - v \frac{\bar{v}_A}{\bar{v}_B} C_A},$$

$$\frac{D_B}{D_A} = \frac{1 + \frac{v}{\tilde{D}(\partial C_B/dx)} C_B}{1 - \frac{v}{\tilde{D}(\partial C_B/dx)} C_A \frac{\bar{v}_A}{\bar{v}_B}},$$

$$\frac{D_B}{D_A} = \frac{1 + f C_B}{1 - f C_A (\bar{v}_A/\bar{v}_B)} \quad \text{where } f = \frac{v}{\tilde{D}(\partial C_B/dx)}. \tag{3.139}$$

Therefore, the interdiffusion coefficients and the concentration gradients are estimated at different marker planes, and then the ratio of the intrinsic diffusion coefficients is estimated using the relation in Eq. (3.139). Next the intrinsic diffusion coefficients are estimated using Eq. (3.124). To cross-check if the multifoil diffusion couple has produced reliable data, the interdiffusion data estimated from a bulk diffusion couple is compared with the data estimated from the multifoil diffusion couple. It can be seen in Fig. 3.39A that the data points in both couples are comparable and are within the range of experimental error. The estimated intrinsic diffusion coefficients are shown in Fig. 3.39B.

3.13 ESTIMATION OF THE TRACER DIFFUSION COEFFICIENTS INDIRECTLY FROM DIFFUSION COUPLE EXPERIMENTS

As explained in Chapter 2 of Volume 1, the tracer diffusion coefficients are conventionally measured using the radioisotopes. Knowledge of these parameters is important for understanding the fundamental atomic mechanism of diffusion in the absence of thermodynamic driving forces. Therefore, these indicate about the defects present in the structure. To understand the difference between different types of diffusion parameter, readers are suggested to read Chapter 1 of Volume 1. However, experiments with radioisotopes are not straightforward, and very few laboratories are presently using them for diffusion studies. On the other hand, the diffusion couple technique has emerged as one of the reliable techniques to estimate the tracer diffusion coefficients indirectly based on the relations developed by Darken [19] and with the addition of a correction factor due to Manning [34,35]. These are explained in the next two subsections.

3.13.1 Darken's Formulism's Relating the Intrinsic and Interdiffusion Coefficients With the Tracer Diffusion Coefficients

For Darken's analysis, we need to understand the thermodynamical aspect of the interdiffusion process. As already explained in Chapter 1 of Volume 1, the chemical potential gradient explains the direction of the component and not necessarily the concentration gradient in every system. In an n-component system, an equilibrium condition is defined with respect to temperature, T, pressure, P, and the chemical potentials of the components, $\mu_1, \mu_2 \ldots \mu_n$. Therefore, the flux of component i can be expressed based on the thermodynamical point of view as

$$J_i = -L_{i1}\frac{d\mu_1}{dx} - L_{i2}\frac{d\mu_2}{dx} \ldots - L_{in}\frac{d\mu_n}{dx}. \tag{3.140}$$

As explained in Chapter 1 of Volume 1, additional terms are added for the diffusion under the influence of electric potential and stress. At constant temperature and pressure in a binary system, when vacancy (V) diffusion mechanism is operative and in the absence of any other driving forces, Eq. (3.140) is modified to

$$J_A = -L_{AA}\frac{d\mu_A}{dx} - L_{AB}\frac{d\mu_B}{dx} - L_{AV}\frac{d\mu_v}{dx}, \tag{3.141a}$$

$$J_B = -L_{BA}\frac{d\mu_A}{dx} - L_{BB}\frac{d\mu_B}{dx} - L_{BV}\frac{d\mu_v}{dx}, \tag{3.141b}$$

$$J_v = -L_{VA}\frac{d\mu_A}{dx} - L_{VB}\frac{d\mu_B}{dx} - L_{VV}\frac{d\mu_v}{dx}, \tag{3.141c}$$

where L_{ij} is the phenomenological constant of component i that depends on the chemical potential gradient of component j, i.e., $\frac{d\mu_j}{dx}$. Vacancies are also treated as a component in this analysis. Darken analysis is based on two important assumptions: (i) concentration of vacancies remains in equilibrium over the whole interdiffusion zone at every concentration during diffusion annealing, and (ii) all the off-diagonal phenomenological constants are zero. Therefore, we can write: $\mu_v \approx 0$, $\frac{d\mu_v}{dx} \approx 0$, $L_{AB} = 0$, $L_{BA} = 0$, $L_{AV} = 0$, $L_{BV} = 0$, $L_{VA} = 0$, and $L_{VB} = 0$. Therefore, Eq. (3.141) reduces to

$$J_A = -L_{AA}\frac{d\mu_A}{dx}, \tag{3.142a}$$

$$J_B = -L_{BB}\frac{d\mu_B}{dx}. \tag{3.142b}$$

We consider that the force $F = -\frac{d\mu_B}{dx}$ is coming from the chemical potential gradient so that component B move with velocity v_B. The mobility of B is $M_B = v_B/F$ such that the flux of component B can be expressed as

$$J_B = C_B v_B = M_B F C_B, \tag{3.143}$$

$$J_B = -M_B C_B \frac{d\mu_B}{dx}. \tag{3.144}$$

Comparing with the relations in Eq. (3.142), the phenomenological constant for component B can be expressed as $L_B = M_B C_B$. Comparing with Fick's first law, we get

$$D_B = M_B C_B \frac{d\mu_B}{dC_B} = M_B \frac{N_B}{v_m}\frac{d\mu_B}{dC_B}. \tag{3.145}$$

Utilizing the standard thermodynamic relation given by Eq. (3.74a), we have

$$D_B = \frac{v_m}{\bar{v}_A} M_B N_B \frac{d\mu_B}{dN_B} = \frac{v_m}{\bar{v}_A} M_B \frac{d\mu_B}{d\ln N_B}. \tag{3.146}$$

From the equation relating chemical potential with activity, a, and activity coefficient, γ, we know [1] that

$$\mu_B = \mu_B^o + RT\ln a_B = \mu_B^o + RT(\ln N_B + \ln \gamma_B). \tag{3.147}$$

Comparing Eqs. (3.146) and (3.147), we get

$$D_B = \frac{v_m}{\bar{v}_A} M_B RT \frac{d\ln a_B}{d\ln N_B} = \frac{v_m}{\bar{v}_A} M_B RT\left(1 + \frac{d\ln \gamma_B}{d\ln N_B}\right). \tag{3.148}$$

As explained in Chapters 1 and 2 of Volume 1, there is no role of thermodynamic driving force in the measurement of the tracer diffusion coefficients and it is measured in an alloy with a fixed composition. Therefore, for the tracer diffusion coefficient, Eq. (3.148) reduces to

$$D_B^* = M_B^* RT \tag{3.149}$$

This is known as Nernst–Einstein relation.

Replacing Eq. (3.149) in Eq. (3.148), we have

$$D_B = D_B^* \frac{v_m}{\bar{v}_A} \frac{d\ln a_B}{d\ln N_B} = D_B^* \frac{v_m}{\bar{v}_A}\left(1 + \frac{d\ln \gamma_B}{d\ln N_B}\right). \tag{3.150a}$$

Following similar derivation, the relation between the intrinsic and tracer diffusion coefficients of component A can be expressed as

$$D_A = D_A^* \frac{v_m}{\bar{v}_B} \frac{d\ln a_A}{d\ln N_A} = D_A^* \frac{v_m}{\bar{v}_B}\left(1 + \frac{d\ln \gamma_A}{d\ln N_A}\right). \tag{3.150b}$$

According to Gibbs–Duhem relation, $\frac{d\ln a_A}{d\ln N_A} = \frac{d\ln a_B}{d\ln N_B}$ [1]. Utilizing this relation and replacing Eq. (3.150) in Eq. (3.124), we get

$$\tilde{D} = (N_A D_B^* + N_B D_A^*)\left(\frac{d\ln a_B}{d\ln N_B}\right) = (N_A D_B^* + N_B D_A^*)\Theta \tag{3.151}$$

where $\Theta = (\frac{d\ln a_B}{d\ln N_B})$ is the thermodynamic factor. The ratios of intrinsic and tracer diffusion coefficients are related by

$$\frac{D_B}{D_A} = \frac{\bar{v}_B}{\bar{v}_A}\frac{D_B^*}{D_A^*}. \tag{3.152}$$

Therefore utilizing the relation in Eq. (3.137), the ratio of the tracer diffusion coefficients can be directly estimated at Kirkendall marker plane as

$$\frac{D_B^*}{D_A^*} = \left[\frac{N_B^+ \int_{x^{-\infty}}^{x_K} \frac{Y_B}{v_m} dx - N_B^- \int_{x_K}^{x^{+\infty}} \frac{(1-Y_B)}{v_m} dx}{-N_A^+ \int_{x^{-\infty}}^{x_K} \frac{Y_B}{v_m} dx + N_A^- \int_{x_K}^{x^{+\infty}} \frac{(1-Y_B)}{v_m} dx}\right] = \left[\frac{N_B^+ \varphi - N_B^- \psi}{-N_A^+ \varphi + N_A^- \psi}\right]. \tag{3.153}$$

Similarly, the velocity of Kirkendall marker plane as in Eq. (3.119) can be rewritten as

$$v_K = \bar{v}_B\left(D_B^* \frac{v_m}{\bar{v}_A} - D_A^* \frac{v_m}{\bar{v}_B}\right)\Theta\left(\frac{\partial C_B}{\partial x}\right)_K. \tag{3.154}$$

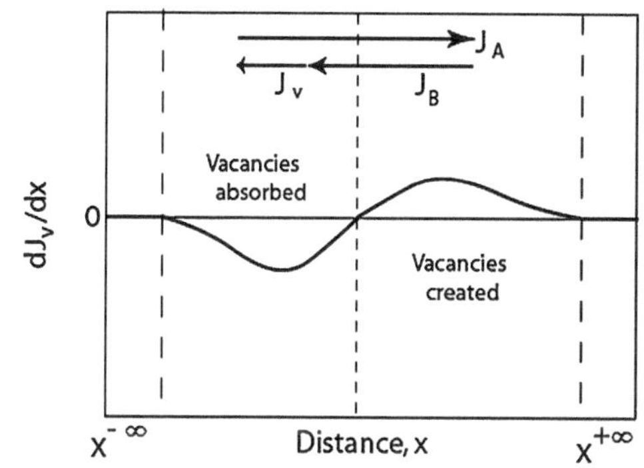

Figure 3.40 The regions in which in a diffusion couple vacancies are absorbed and created.

3.13.2 Manning's Correction and the Concept of the Vacancy-Wind Effect

In a diffusion process controlled by vacancy mechanism, the net flow of component in one direction is compensated by the flux of vacancies in the opposite direction. For example, as shown in Fig. 3.40, we have $J_A > J_B$. Therefore, the net molar flux is $(J_A + J_B)$ in the direction of diffusion direction of A. Note that it expresses the net flux correctly, since intrinsic fluxes diffuse in opposite direction and have opposite sign. Therefore, the flow of vacancies is opposite to the net flux of components in substitutional diffusion. This controls the shift of Kirkendall marker plane from the initial contact plane $x_K = (x_K - x_o)$. With respect to the volume flux (molar flux×partial molar volume) considering unit cross-section, the marker shift can be written as

$$v_K = v_m J_V = -(\bar{v}_B J_B + \bar{v}_A J_A). \tag{3.155}$$

Therefore, $\frac{dJ_V}{dx}$ in Fig. 3.40 indicates the creation of vacancies at one side and absorption at another side of the diffusion couple. Manning proposed, based on a mathematical treatment, that the flow of vacancies creates a vacancy-wind effect [34,35]. Because of this effect, extra vacancies will change place along with the net flow depending on the intrinsic fluxes of the components. The transport of extra vacancies will increase the diffusion rate of the component which has higher diffusion rate and will decrease the diffusion rate of the component which has lower diffusion rate. Therefore, correction factors are added to the relations developed by Darken, and they are known as Darken–Manning equations.

Manning considered a random alloy model and assumed no supersaturation or binding of vacancies to add the correction factor to Eq. (3.150) as follows:

$$D_A = \frac{v_m}{\bar{v}_B} D_A^* \Theta (1 + W_A), \tag{3.156a}$$

$$D_B = \frac{v_m}{\bar{v}_A} D_B^* \Theta (1 - W_B), \tag{3.156b}$$

where $W_i = \frac{2N_i(D_A^* - D_B^*)}{M_o(N_A D_A^* + N_B D_B^*)}$ is the vacancy-wind effect, and constant M_o depends on the crystal structure. Utilizing Eq. (3.156) in Eq. (3.124) and after a few simple steps, we arrive at

$$\tilde{D} = (N_A D_B^* + N_B D_A^*)\Theta W_{AB}, \qquad (3.157)$$
$$W_{AB} = 1 + \frac{2N_A N_B (D_A^* - D_B^*)^2}{M_o(N_A D_B^* + N_B D_A^*)(N_A D_A^* + N_B D_B^*)}.$$

Kirkendall marker velocity is modified to

$$v_K = \bar{v}_B\left(\frac{v_m}{\bar{v}_A} D_B^* - \frac{v_m}{\bar{v}_B} D_A^*\right)\alpha\Theta\frac{\partial C_B}{\partial x} = \frac{1}{v_m}(\bar{v}_B D_B^* - \bar{v}_A D_A^*)\alpha\Theta\frac{\partial N_B}{\partial x} \qquad (3.158)$$

where $\alpha = 1/f$ and f is the correlation factor relevant for tracer diffusion as discussed in Chapter 1 of Volume 1. The values of M_o are 7.15, 5.33, 3.77, and 2, and those of α are 1.280, 1.375, 1.531, and 2, for a face-centered cubic, body-centered cubic, simple cubic, and diamond structure, respectively.

3.14 INTRINSIC AND TRACER DIFFUSION COEFFICIENTS IN A PHASE WITH NARROW HOMOGENEITY RANGE

As discussed in Section 3.8, the interdiffusion coefficient cannot be estimated in a phase with narrow homogeneity range since we cannot estimate the concentration or composition gradient. To circumvent this problem, the concept of the integrated interdiffusion coefficient is introduced. Due to the same reason, we cannot estimate the absolute values of the intrinsic diffusion coefficients at Kirkendall marker plane. One can instead estimate the ratio of intrinsic diffusion coefficients following the relation in Eq. (3.137). Since partial molar volumes cannot be estimated in most phases with narrow homogeneity range, we actually measure the ratio of the tracer diffusion coefficients following Eq. (3.153); however, after neglecting the contribution from the vacancy-wind effect as in Eq. (3.156). Vacancy wind effect cannot be easily estimated in an intermetallic compound because of the unknown value of M_o in Eq. (3.156). One can estimate it following a theoretical approach, which is not necessarily very straightforward [1].

The estimation procedure can be understood with the help of Fig. 3.41. In Eq. (3.153), we need to estimate the integrals $\varphi = \int_{x^{-\infty}}^{x_K} \frac{Y_B}{v_m} dx$ and $\psi = \int_{x_K}^{x^{+\infty}} \frac{(1-Y_B)}{v_m} dx$, which are related to the areas in the composition profile on two different sides of Kirkendall marker plane x_K. As shown in Fig. 3.41A, sometimes a single line compound with more or less fixed composition is present between two solid solutions with variable compositions. Therefore, the areas on either side of the marker plane can be expressed

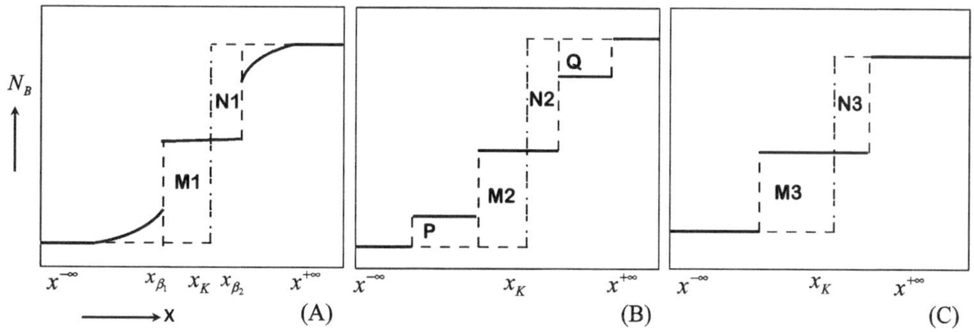

Figure 3.41 Estimation of the ratio of intrinsic and tracer diffusion coefficients in a line compound.

as [1]

$$\phi = \left(\int_{x^{-\infty}}^{x_{\beta_1}} \frac{(N_B - N_B^-)}{v_m} dx + \frac{(N_B^\beta - N_B^-)}{v_m^\beta} \Delta x_{\beta_1} \right) = \left(\int_{x^{-\infty}}^{x_{\beta_1}} \frac{(N_B - N_B^-)}{v_m} dx + \frac{M1}{v_m^\beta} \right), \tag{3.159a}$$

$$\psi = \left(\int_{x_{\beta_2}}^{x^{+\infty}} \frac{(N_B^+ - N_B)}{v_m} dx + \frac{(N_B^+ - N_B^\beta)}{v_m^\beta} \Delta x_{\beta_2} \right) = \left(\int_{x_{\beta_2}}^{x^{+\infty}} \frac{(N_B^+ - N_B)}{v_m} dx + \frac{N1}{v_m^\beta} \right). \tag{3.159b}$$

While converting Y_B to N_B, the term $(N_B^+ - N_B^-)$ is not included since it is present both in the numerator and denominator. In many systems, only the line compounds grow in the interdiffusion zone, as shown in Fig. 3.41B. Therefore, we can write

$$\varphi = \frac{(N_B^\alpha - N_B^-)}{v_m^\alpha} \Delta x_\alpha + \frac{(N_B^\beta - N_B^-)}{v_m^\beta} \Delta x_{\beta_1} = \frac{P}{v_m^\alpha} + \frac{M2}{v_m^\beta}, \tag{3.160a}$$

$$\psi = \frac{(N_B^+ - N_B^\gamma)}{v_m^\gamma} \Delta x_\gamma + \frac{(N_B^+ - N_B^\beta)}{v_m^\beta} \Delta x_{\beta_2} = \frac{Q}{v_m^\gamma} + \frac{N2}{v_m^\beta}. \tag{3.160b}$$

In an incremental diffusion couple, when only one phase grows in the interdiffusion zone, as shown in Fig. 3.41C, we can write

$$\phi = \frac{(N_B^\beta - N_B^-)}{v_m^\beta} \Delta x_{\beta_1} = \frac{M3}{v_m^\beta}, \tag{3.161a}$$

$$\psi = \frac{(N_B^+ - N_B^\beta)}{v_m^\beta} \Delta x_{\beta_2} = \frac{N3}{v_m^\beta}. \tag{3.161b}$$

Although we cannot estimate the absolute values of the intrinsic diffusion coefficients directly from a diffusion couple experiment, one can still estimate the absolute

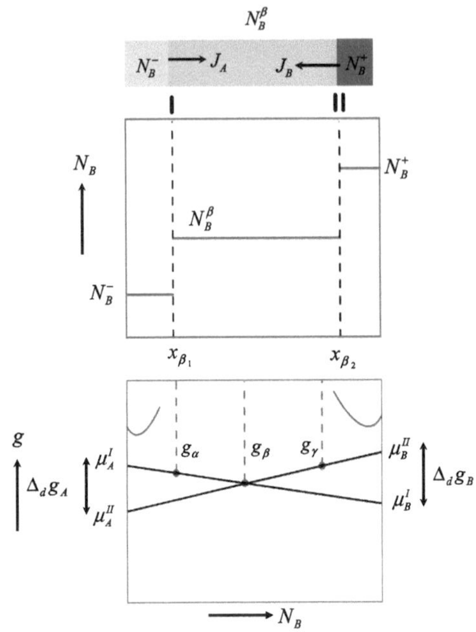

Figure 3.42 Estimation of the absolute values of the tracer diffusion coefficients in line compound.

value of the tracer diffusion coefficients. Fig. 3.42 shows an imaginary diffusion couple of two line compounds α and γ, in which β phase is grown in the interdiffusion zone. Since α is rich in component A and γ is rich in component B, A will dissociate from α to produce the product phase β. Diffused component A reacts with γ at the interface II to produce the product phase. Similarly, B dissociates from the B-rich γ to produce the product phase. The dissociated B diffuses and reacts with the other end member at the interface I to grow the product phase. This process is explained in detail in Chapter 4 of Volume 2. The reason for the direction of the diffusion of components can be realized from the values of chemical potential of components in a different phase, as shown in Fig. 3.42. As explained in Chapter 1 of Volume 1, since $\mu_A^I > \mu_A^{II}$, component A diffuses from interface I to II. The opposite is true for component B. Replacing Eq. (3.157) in Eq. (3.99a) and neglecting the vacancy-wind effect (*i.e.* same as Eq. (3.151)), the relation between the integrated diffusion coefficient of β and the tracer diffusion coefficients can be expressed as

$$\tilde{D}_{\text{int}} = \int_{N_B^{\beta 1}}^{N_B^{\beta 2}} (N_A^\beta D_B^* + N_B^\beta D_A^*)\left(\frac{d\ln a_B}{d\ln N_B}\right) dN_B,$$

$$\tilde{D}_{\text{int}} = \int_{a_B^I}^{a_B^{II}} (N_A^\beta D_B^* + N_B^\beta D_A^*) N_B^\beta \, d\ln a_B. \tag{3.162}$$

We can consider the tracer diffusion coefficient values as constant, since we are considering here a line compound with a very small composition range. Therefore, we can write

$$\tilde{D}_{\text{int}} = \int_{a_B^I}^{a_B^{II}} (N_A^\beta D_B^* + N_B^\beta D_A^*) N_B^\beta d\ln a_B = (N_A^\beta D_B^* + N_B^\beta D_A^*) N_B^\beta [\ln a_B^{II} - \ln a_B^I]. \quad (3.163)$$

The chemical potentials and activities are related by [1]:

$$\mu_B^I = \mu_B^o + RT\ln a_B^I, \quad (3.164a)$$
$$\mu_B^{II} = \mu_B^o + RT\ln a_B^{II}, \quad (3.164b)$$

where μ_B^o is the chemical potential at the standard state. Then, based on the discussion in Chapter 1 of Volume 1, the driving force for diffusion of component B can be written as

$$\Delta_d g_B = \mu_B^I - \mu_B^{II} = -RT(\ln a_B^{II} - \ln a_B^I). \quad (3.165)$$

Employing Eqs. (3.163) and (3.165), we can write

$$\tilde{D}_{\text{int}}^\beta = -(N_A^\beta D_B^* + N_B^\beta D_A^*)\frac{N_B^\beta \Delta_d g_B}{RT}. \quad (3.166a)$$

Since $N_B^\beta \Delta_d g_B = N_A^\beta \Delta_d g_A$ [1], this relation with respect to the driving force for diffusion of component A can be written as

$$\tilde{D}_{\text{int}}^\beta = -(N_A^\beta D_B^* + N_B^\beta D_A^*)\frac{N_A^\beta \Delta_d g_A}{RT}. \quad (3.166b)$$

If a single line compound, β, exists between two pure components in a binary A–B system, Eq. (3.166) is modified to

$$\tilde{D}_{\text{int}}^\beta = -(N_A^\beta D_B^* + N_B^\beta D_A^*)\frac{\Delta_f g_\beta}{RT}, \quad (3.167)$$

where $\Delta_f g_\beta$ is the formation energy for the β phase. Therefore, one can estimate the absolute values of the tracer diffusion coefficients by estimating the ratio of the tracer diffusion coefficients at the marker plane using Eq. (3.153) and integrated diffusion coefficient using Eq. (3.106).

In the above example, the driving force is calculated in a couple, in which a single-phase layer grows in an interdiffusion zone. Note that the integrated coefficient or the ratio of the tracer diffusion coefficients can be estimated for a phase in multiphase diffusion zone or in a phase that is grown from the end members prepared with a phase mixture. These parameters are material constants and should give the same values

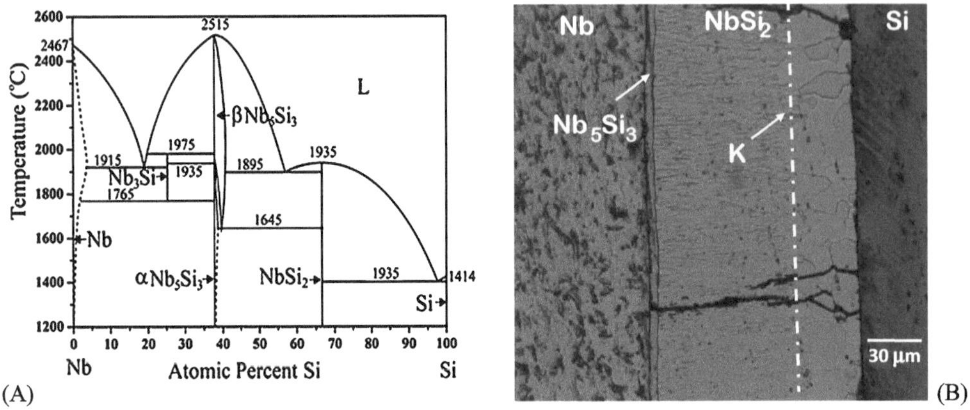

Figure 3.43 (A) Nb–Si phase diagram [37] and (B) growth of the phases in the interdiffusion zone of a Nb/Si diffusion couple annealed at 1250°C for 24 hrs [36].

within the range of experimental error when calculated from diffusion couples with different end members. In other words, the thickness of the phase layers and the location of Kirkendall marker plane are found differently in different diffusion couples such that these parameters remain the same. However, the driving force for diffusion should be estimated considering the growth of a single product phase between two phases surrounding it in the respective phase diagram. Moreover, these phases should be considered in their pure form. Sometimes a phase that is present in a phase diagram may not grow with significant growth rate to detect it in an interdiffusion zone. This can be ignored for the estimation of the diffusion coefficients since it has a negligible thickness in the composition profile. However, it should be considered for the estimation of the driving force. For example, suppose that $A/\alpha/\beta/B$ is the phase sequence in a binary A–B phase diagram. When A and B are diffusion coupled, α may not grow with a detectable thickness because of very low diffusion coefficient and its presence can be ignored while estimating the diffusion parameters in the β phase. However, for the estimation of the driving force, we need to think as if the phase β is grown between α and B.

To explain the estimation procedure, let us consider two examples. First, we look at the growth of phases in Nb/Si couple which was annealed at 1250°C for 24 hours [36]. As shown in the phase diagram in Fig. 3.43A [37], two intermetallic compounds Nb_5Si_3 and $NbSi_2$ exist at the temperature of interest. Therefore, these are grown in an interdiffusion zone, as shown in Fig. 3.43B. Composition analysis indicates that Nb(Si) and Si(Nb) solid solutions do not grow with significant thickness. Kirkendall marker plane is detected from the morphological evolution, as indicated by "K". The thickness of the phases is estimated as 4.7 μm for Nb_5Si_3 and 110.2 μm for $NbSi_2$. Kirkendall marker plane divides the $NbSi_2$ phase into two sublayers of thicknesses 76.2 and 34 μm,

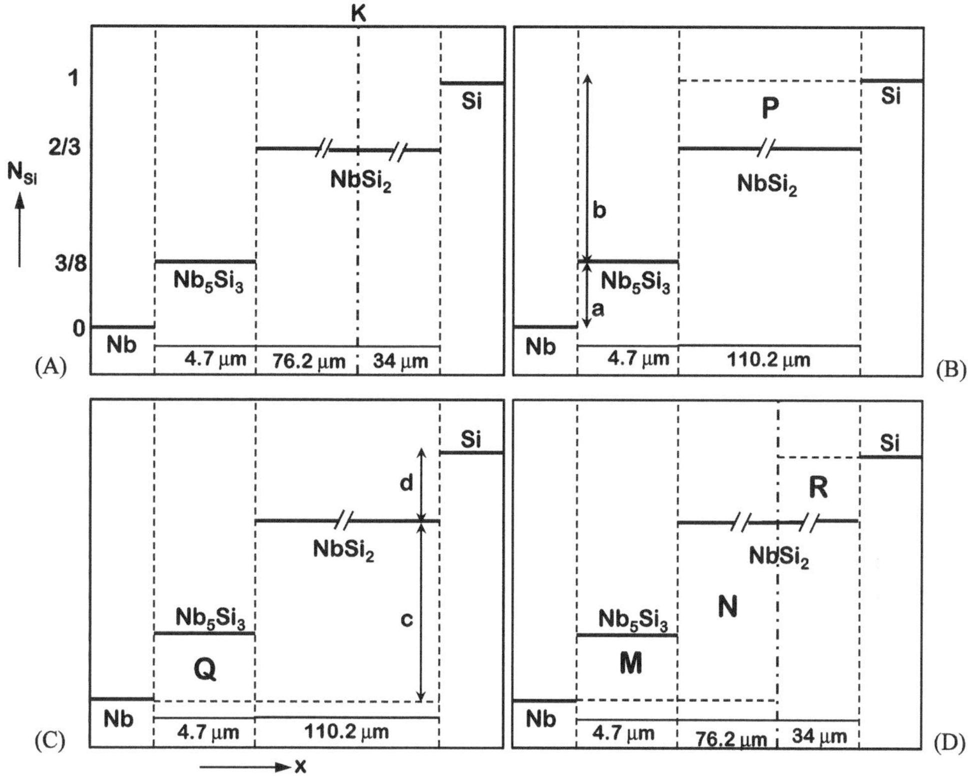

Figure 3.44 Estimation steps of the diffusion coefficients in the Nb/Si diffusion couple as shown in Fig. 3.43.

respectively. From the values of lattice parameter data available in the literature, the molar volume values are estimated as $v_m^{Nb_5Si_3} = 9.6 \times 10^{-6}$ and $v_m^{NbSi_2} = 8.7 \times 10^{-6}$ m^3/mol. The calculation is explained with the help of Fig. 3.44.

The integrated diffusion coefficient of the phases can be estimated following the explanation in Section 3.8 as follows [1]:

$$\tilde{D}_{\text{int}}^{Nb_5Si_3} = \frac{a \times b}{a+b}\frac{\Delta x^2_{Nb_5Si_3}}{2t} + \frac{\Delta x_{Nb_5Si_3}}{2t}\left[\frac{b(0) + a(\frac{v_m^{Nb_5Si_3}}{v_m^{NbSi_2}}P)}{a+b}\right] = 4.4 \times 10^{-16} \text{ m}^2/\text{s}$$

since $a = \frac{3}{8} - 0 = \frac{3}{8}$, $b = 1 - \frac{3}{8} = \frac{5}{8}$, $\Delta x_{Nb_5Si_3} = 4.7 \times 10^{-6}$ m, $2t = 2 \times 24$ hours $=$ 172800 s, $P = (1 - \frac{2}{3}) \times 110.2 \times 10^{-6}$ m,

$$\tilde{D}_{\text{int}}^{NbSi_2} = \frac{c \times d}{c+d}\frac{\Delta x^2_{NbSi_2}}{2t} + \frac{\Delta x_{NbSi_2}}{2t}\left[\frac{d(\frac{v_m^{NbSi_2}}{v_m^{Nb_5Si_3}}Q) + c(0)}{c+d}\right] = 1.6 \times 10^{-14} \text{ m}^2/\text{s},$$

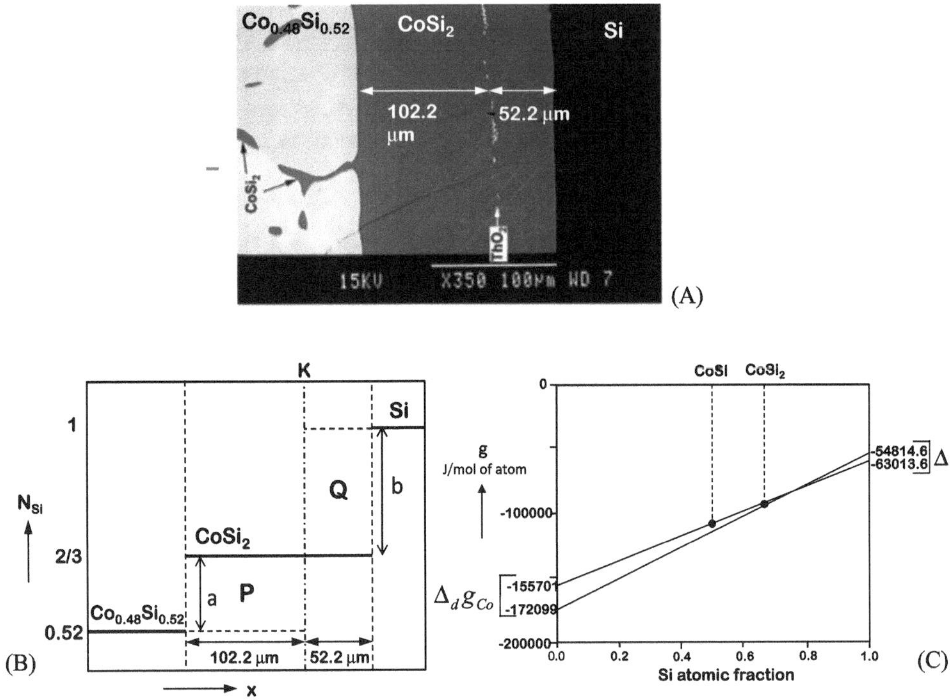

Figure 3.45 (a) $Co_{0.48}Si_{0.52}$/Si diffusion couple annealed at 1186°C for 100 hrs, (B) the composition profile, and (C) estimation of the driving forces for diffusion of components [38].

since $c = \frac{2}{3} - 0 = \frac{2}{3}$, $d = 1 - \frac{2}{3} = \frac{1}{3}$, $\Delta x_{NbSi_2} = 110.2 \times 10^{-6}$ m, $2t = 2 \times 24$ hours $=$ 172800 s, $Q = (1 - \frac{3}{8}) \times 4.7 \times 10^{-6}$ m. Similarly, the ratio of tracer diffusion coefficients can be estimated as explained in this section, namely

$$\frac{D^*_{Si}}{D^*_{Nb}} = \left[\frac{N^+_{Si}\phi - N^-_{Si}\psi}{-N^+_{Nb}\phi + N^-_{Nb}\psi}\right] = 4.6$$

since $\phi = \frac{M}{v_m^{Nb_5Si_3}} + \frac{N}{v_m^{NbSi_2}}$, $\psi = \frac{R}{v_m^{NbSi_2}}$, $M = (\frac{3}{8} - 0) \times 4.7 \times 10^{-6}$ m, $N = (\frac{2}{3} - 0) \times 76.2 \times 10^{-6}$ m, $R = (1 - \frac{2}{3}) \times 34 \times 10^{-6}$ m, $N^-_{Si} = 0$, $N^+_{Si} = 1$, $N^-_{Nb} = 1$, $N^+_{Nb} = 0$.

Let us now estimate the absolute values of tracer diffusion coefficients in the Co–Si system [26,38]. The incremental diffusion couple between the alloy $Co_{0.48}Si_{0.52}$ and pure Si, which was annealed at 1186°C for 100 hours is shown in Fig. 3.45A. $Co_{0.48}Si_{0.52}$ alloy has mainly CoSi with a small quantity of $CoSi_2$. If required the volume fractions can be determined by the lever rule. As expected, based on the arguments given in Chapter 1 of Volume 1, only one phase $CoSi_2$ is grown in the interdiffusion zone. The average total thickness of the phase layer is 154.4 µm, which is divided into parts (102.2 µm and 52.2 µm) by Kirkendall marker plane detected by the presence of

Table 3.2 Thermodynamic parameters for the estimation of the free energies of different phases in the Co–Si system [26]

$C_p = A + BT - \frac{C}{T^2}$

	CoSi	$CoSi_2$	Si
A	49.16	70.86	23.698
$B \times 10^{-3}$	12.1	18.66	3.305
$C \times 10^5$	7.54	9.93	4.351
Δh_{298}	43.2	64.2	18.81
S_{298}	−94600	−98700	0

ThO_2 particles. Therefore the diffusion parameters can be estimated with the help of Fig. 3.45B as

$$\tilde{D}_{\text{int}}^{CoSi_2} = \frac{ab}{a+b}\frac{\Delta x_{CoSi_2}^2}{2t} = \frac{(2/3 - 0.52)(1 - 2/3)}{1 - 0.52}\frac{(154.4 \times 10^{-6})^2}{2 \times 100 \times 60 \times 60} = 3.3 \times 10^{-15}\ \text{m}^2/\text{s},$$

$$\frac{\nu_{Co}}{\nu_{Si}}\frac{D_{Si}}{D_{Co}} = \frac{D_{Si}^*}{D_{Co}^*} = \left[\frac{N_{Si}^+ P - N_{Si}^- Q}{-N_{Co}^+ P + N_{Co}^- Q}\right] = 0.7,$$

$P = (2/3 - 0.52) \times 102.2 \times 10^{-6}$, $Q = (1 - 2/3) \times 52.2 \times 10^{-6}$, $N_{Si}^- = 0.52$, $N_{Si}^+ = 1$, $N_{Co}^- = 0.48$, $N_{Co}^+ = 0$.

As already discussed before, the driving forces for diffusion of components should be estimated as if $CoSi_2$ has grown between CoSi and Si. Free energies, g (J/mol of atom), at the temperature of interest can be estimated from the calculated values of enthalpy, h (J/mol of atom), and entropy, s (J/mole of atom.K), using the relation $g = h - Ts$ [1], where the temperature, T, is in Kelvin. These are estimated from the values of $C_p = A + BT - \frac{C}{T^2}$. The values of A, B, and C are given in Table 3.2 [38]. From the thermodynamic relations, we can write

$$h_T = \Delta h_{298} + \int_{298}^{T} C_P dT, \tag{3.168a}$$

$$s_T = s_{298} + \int_{298}^{T} \frac{C_P}{T} dT. \tag{3.168b}$$

Note that the values given in the table are given for one mole of compound CoSi and $CoSi_2$. The free energies are estimated as $g = -218714.9$ J/mole of CoSi, $g = -281727.8$ J/mol of $CoSi_2$, and $g(Si) = -54814.57$ J/mole of atom Si. For the estimation of driving forces, we need these values for one mole of components. Therefore, the free energy value of CoSi should be divided by 2, and the free energy values of $CoSi_2$ should be divided by 3. Therefore, we have $g(\text{CoSi}) = -109357.45$ J/mole of atom, and $g(\text{CoSi}_2) = -93909.27$ J/mole of atom. The free energy calculated for Si remains unchanged. Note that frequently the basic parameters for the estimation of these free energies are given per mole of the atom and there is no need to complete the last step. These values are added in the plot in Fig. 3.45C, and the estimated driving forces for components are calculated as

$$\Delta_d g_{Si} = -8199 \text{ J/mol of Si},$$
$$\Delta_d g_{Co} = -16398 \text{ J/mol of Co}.$$

Further, note that $N_{Si}\Delta_d g_{Si} = N_{Co}\Delta_d g_{Co}$ considering $N_{Si} = 2/3$ and $N_{Co} = 1/3$ in the product phase $CoSi_2$. Therefore, we can write [1]

$$\tilde{D}_{int} = 3.3 \times 10^{-15} \text{ m}^2/\text{s} = -\left(N_{Co}D^*_{Si} + N_{Si}D^*_{Co}\right)\frac{N_{Si}\Delta_d g_{Si}}{RT},$$
$$\frac{D^*_{Si}}{D^*_{Co}} = 0.7.$$

These lead to

$$D^*_{Co} = 8.25 \times 10^{-15} \text{ m}^2/\text{s}$$
$$D^*_{Si} = 5.77 \times 10^{-15} \text{ m}^2/\text{s}.$$

Writing simple reaction equations at the interfaces of a diffusion couple made of pure phases of CoSi and Si in which $CoSi_2$ grows at the interface, we can differently show that $N_{Si}\Delta_d g_{Si} = N_{Co}\Delta_d g_{Co}$. To visualize, imagine $Co_{0.48}Si_{0.52}$ as pure CoSi. Following the physicochemical approach explained in Chapter 4 of Volume 2, the reaction dissociation of the components in this diffusion couple can be described. Therefore Co will dissociate from interface I and diffuse to interface II. Both steps will produce $CoSi_2$ at two different interfaces. Si will diffuse from interface II to I and react with CoSi to produce $CoSi_2$ at this interface only. With respect to equations, we can write [1] the following:

Interface I: (CoSi/CoSi₂)

$$2\text{CoSi} = [\text{Co}]_\text{d} + \text{CoSi}_2, \tag{3.169a}$$
$$\text{CoSi} + [\text{Si}]_\text{d} = \text{CoSi}_2; \tag{3.169b}$$

Interface II (CoSi₂/Si)

$$2\text{Si} + [\text{Co}]_\text{d} = \text{CoSi}_2. \tag{3.169c}$$

The free energy change per mole of Si atom, using Eq. (3.169b), can be written as

$$\Delta_d g_{Si} = g(CoSi_2) - g(CoSi) - g(Si). \tag{3.170a}$$

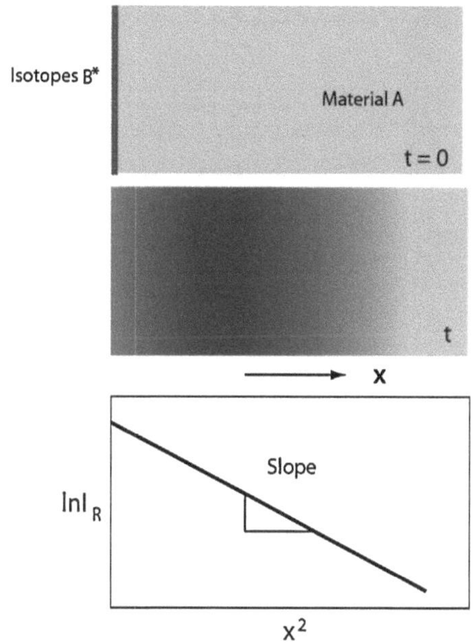

Figure 3.46 Schematic representation of the impurity diffusion coefficient by radioisotope method.

Considering dissociation of Co at the interface I following Eq. (3.169a) and reaction of the same at the interface II following Eq. (3.169c), leading to the formation of the product phase, we can add these two phenomena and write

$$2CoSi + 2Si = 2CoSi_2.$$

Therefore, the free energy change per mole of Co atom can be written as

$$\Delta_d g_{Co} = 2g(CoSi_2) - 2g(CoSi) - 2g(Si). \tag{3.170b}$$

By comparing Eqs. (3.170a) and (3.170b), we have

$$\begin{aligned} \Delta_d g_{Co}^{CoSi_2} &= 2\Delta_d g_{Si}^{CoSi_2}, \\ \frac{1}{3}\Delta_d g_{Co}^{CoSi_2} &= \frac{2}{3}\Delta_d g_{Si}^{CoSi_2}, \\ N_{Co}^{CoSi_2}\Delta_d g_{Co}^{CoSi_2} &= N_{Si}^{CoSi_2}\Delta_d g_{Si}^{CoSi_2}. \end{aligned}$$

3.15 ESTIMATION OF THE IMPURITY DIFFUSION COEFFICIENTS

Impurity diffusion coefficient, for example, diffusion coefficient of B in A (expressed as $D_{B(A)}$ or D_{imp}), can be directly measured by the radiotracer technique as also explained

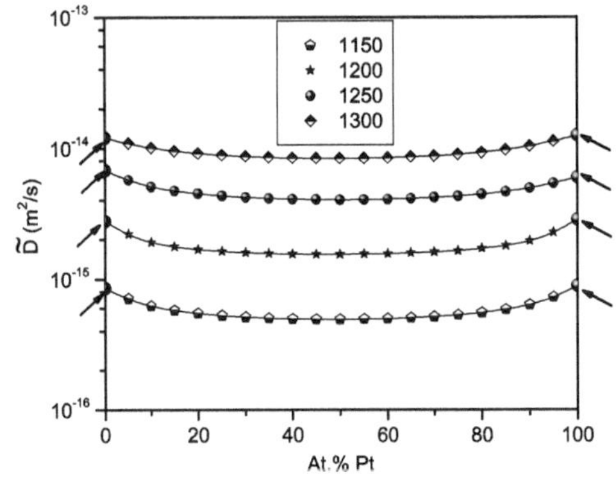

Figure 3.47 Estimation of the impurity diffusion coefficient by diffusion couple technique in Pd–Pt system [39].

in Chapter 1, Volume 1. Radioisotope B* is deposited in a block of A and annealed at the temperature of interest for diffusion, as shown in Fig. 3.46. As described in Section 3.2.1, for the thin film solution and the measured intensities of radiotracer B*, we can follow Eq. (3.13b) [1] and write a similar equation:

$$\ln I_R = \ln\left(\frac{1}{\sqrt{\pi D^*_{B(A)} t}}\right) - \frac{x^2}{4D^*_{B(A)} t}. \tag{3.171}$$

Then the impurity diffusion coefficient can be determined from the slope. Note that "$*$" is used as a superscript since the measurement is conducted with the help of radioisotope.

This diffusion parameter can also be determined by the diffusion couple technique. Very often it is measured from the estimated composition-dependent interdiffusion coefficients. From Eqs. (3.124) and (3.151), we know that

$$\tilde{D} = \bar{v}_A C_A D_B + \bar{v}_B C_B D_A,$$
$$\tilde{D} = \left(N_A D^*_B + N_B D^*_A\right)\left(\frac{d\ln a_B}{d\ln N_B}\right).$$

Since $N_A \approx 1$, $\bar{v}_A = v^A_m$, and the thermodynamic parameter can be neglected, we can write $\tilde{D}_{N_A\approx 1} = D_{B(A)}$. Since it is measured indirectly with the help of interdiffusion coefficients and without the use of radioisotopes, "$*$" in the superscript is not used. An example of such a measurement in the Pd–Pt system is shown in Fig. 3.47 [39]. In this figure, interdiffusion coefficients are measured following Wagner's method in the range of 5–95 at.% Pt. Then they are extended to 0 at.% Pt for $D_{Pt(Pd)}$ and to 100 at.% Pt for $D_{Pd(Pt)}$. These values are indicated by arrows in the figures. In general, this is a reliable method if the composition profile in an interdiffusion zone does not change very sharply. Otherwise, this method can induce significant error mainly due to the problem of measuring the concentration gradient. To circumvent this problem, Hall [40] developed an estimation procedure from the composition profile measured at the end of the diffusion couple, which was then modified by Crank [3,41]. This is expressed as

$$D_{imp} = \frac{1}{h^2} + \frac{k\sqrt{\pi}}{h^2}(1 + \text{erf}u)\exp(u^2), \tag{3.172a}$$

$$\frac{C}{C^{\infty}} = \frac{1}{2}(1 + \text{erf}u), \tag{3.172b}$$

$$u = h\lambda + k, \tag{3.172c}$$

where λ is Boltzmann parameter, C^{∞} is the initial concentration, and C is the concentration of the initial part of the interdiffusion zone. Values of u should be estimated first from the concentration profile utilizing Eq. (3.172b) and then they are plotted in the graph of u versus λ; on a semi-probability plot h is the slope and k is the intercept of the straight line obtained at the very low concentration range. More recently, Ahmed et al. [42] extended Hall method for estimation of the interdiffusion coefficients over the whole composition range and compared the results with estimated data obtained by other standard methods such as Matano–Boltzmann and Sauer–Freise (an equivalent of Wagner's relation). They found that the extended method can only be applied in a composition range with constant diffusion coefficients similar to the error function analysis. When the interdiffusion coefficient varies with composition, other established methods are more reliable.

3.16 A PSEUDO-BINARY APPROACH IN MULTICOMPONENT DIFFUSION

Although the aim of this chapter is mainly to discuss the binary system, it is necessary to introduce a few equations of the ternary systems to understand the advantages and limitations of the pseudo-binary approach developed by the group of the present author [43–45].

Complications in estimating the diffusion coefficients (following the diffusion couple technique) in a multicomponent system increase rapidly with the increase in the number of components. In a binary system, as already explained, the variation of the interdiffusion coefficients can be estimated over the whole composition range from a single diffusion couple. In a multicomponent system, the extended Fick's law of diffusion is expressed as [1]

$$\tilde{J}_i = -\sum_{j=1}^{n-1} \tilde{D}_{ij}^{n} \frac{\partial C_j}{\partial x}, \tag{3.173a}$$

$$\sum_{i=1}^{n} \bar{v}_i \tilde{J}_i = \bar{v}_1 \tilde{J}_1 + \bar{v}_2 \tilde{J}_2 + \cdots + \bar{v}_n \tilde{J}_n = 0, \tag{3.173b}$$

where $\tilde{J}_i$ is the interdiffusion flux of component i in an n-component system. Therefore $(n-1)^2$ interdiffusion coefficients are needed to relate with the interdiffusion fluxes of $(n-1)$ independent components. The nth component is considered as the dependent variable. In a three-component system, the latter system of equations reduces to

$$\tilde{J}_1 = -\tilde{D}_{11}^3 \frac{\partial C_1}{\partial x} - \tilde{D}_{12}^3 \frac{\partial C_2}{\partial x}, \tag{3.174a}$$

$$\tilde{J}_2 = -\tilde{D}_{22}^3 \frac{\partial C_2}{\partial x} - \tilde{D}_{21}^3 \frac{\partial C_1}{\partial x}, \tag{3.174b}$$

$$\bar{v}_1 \tilde{J}_1 + \bar{v}_2 \tilde{J}_2 + \bar{v}_3 \tilde{J}_3 = 0. \tag{3.174c}$$

According to the last equation, the interdiffusion flux of component 3 can be determined from the interdiffusion fluxes of components 1 and 2 by utilizing Eqs. (3.174a) and (3.174b) simply because the atomic fractions are related by $N_1 + N_2 + N_3 = 1$. Third component is considered as the dependent variable in this case. The details of considering a particular component as the dependent variable are discussed by Kirkaldy and Young [46]. It can be realized that in a multicomponent system, the interdiffusion fluxes are related to the interdiffusion coefficients in a much more complicated manner compared to the binary system. For example, $\tilde{D}_{11}^3$ and $\tilde{D}_{22}^3$ are the main (or direct) interdiffusion coefficients. The contribution to the interdiffusion flux due to these interdiffusion coefficients depends on the concentration gradient of the same components, as shown in the above equations. $\tilde{D}_{12}^3$ and $\tilde{D}_{21}^3$ are the cross- (or indirect) interdiffusion coefficients. The contribution to the interdiffusion flux due to these interdiffusion coefficients depends on the concentration gradient of the other components.

Interdiffusion fluxes can be estimated from similar equations as explained for binary systems, for example, Eq. (3.57b) when following Matano–Boltzmann method, Eq. (3.72b) when following den Broeder approach, and Eq. (3.91b) when following Wagner's approach. However, to estimate the four interdiffusion coefficients, just two equations (viz. Eqs. (3.174a) and (3.174b)) are not enough. Therefore, two diffusion couples are required to estimate the four interdiffusion coefficients. For example, it can be done at the composition of the intersection of the two diffusion couples (DC1 and DC2), as shown on Gibbs triangle of Co–Ni–Pt system in Fig. 3.48A [8]. As an example, composition profile of DC1 is shown in Fig. 3.48B. In a quarternary system, three diffusion couples should intersect at one composition for the estimation of nine interdiffusion coefficients. However, it is impossible to predict the combination of three diffusion couples such that these intersect at one common composition [1].

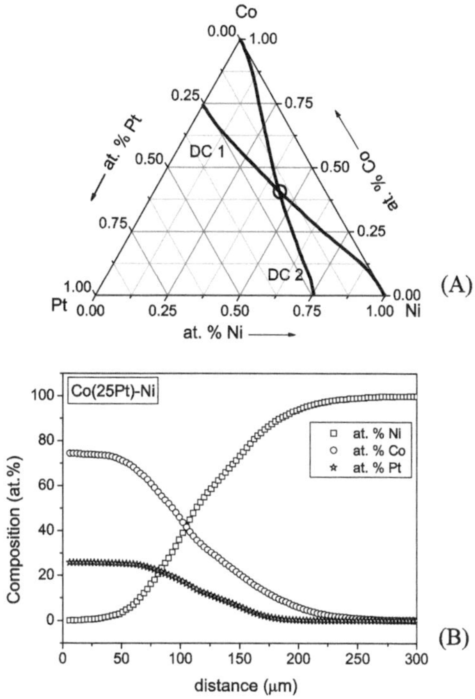

Figure 3.48 Conventional method for estimating the diffusion coefficients in the Ni–Co–Pt ternary system [8].

Similarly, intrinsic fluxes and intrinsic diffusion coefficients are related by [1] (see the next chapter on multicomponent diffusion):

$$J_1 = -D_{11}^3 \frac{\partial C_1}{\partial x} - D_{12}^3 \frac{\partial C_2}{\partial x}, \tag{3.175a}$$

$$J_2 = -D_{21}^3 \frac{\partial C_1}{\partial x} - D_{22}^3 \frac{\partial C_2}{\partial x}, \tag{3.175b}$$

$$J_3 = -D_{31}^3 \frac{\partial C_1}{\partial x} - D_{32}^3 \frac{\partial C_2}{\partial x}. \tag{3.175c}$$

The interdiffusion and intrinsic fluxes are related by [1]:

$$\tilde{J}_1 = (1 - N_1)J_1 - N_1(J_2 + J_3) = (N_2 + N_3)J_1 - N_1(J_2 + J_3), \tag{3.176a}$$

$$\tilde{J}_2 = (1 - N_2)J_2 - N_2(J_1 + J_3) = (N_1 + N_3)J_2 - N_2(J_1 + J_3), \tag{3.176b}$$

$$\tilde{J}_3 = (1 - N_3)J_3 - N_3(J_1 + J_2) = (N_1 + N_2)J_3 - N_3(J_1 + J_2). \tag{3.176c}$$

Therefore, one can determine the intrinsic fluxes from the estimated interdiffusion fluxes utilizing Eq. (3.176). Then, the intrinsic diffusion coefficients can be estimated from Eq. (3.75). However, again there are six intrinsic diffusion coefficients related to three intrinsic fluxes. Hence, we need two diffusion couples to intersect at one common composition to estimate these coefficients. However, it is extremely complicated since Kirkendall markers in both diffusion couples should also be found at the composition of intersection [1]! It is almost impossible to predict for which combinations of end member alloys in two different diffusion couples this can be achieved since the diffusion profiles are in general serpentine in nature on Gibbs triangle and go through different composition paths, as shown in Fig. 3.48A. Therefore, there is almost no chance of finding the markers at the same plane, unless by incident [1].

To circumvent these problems, different approaches were developed by different groups over many decades to study ternary or multicomponent systems (mostly without much success). Only the very few important approaches are mentioned here. Krishtal et al. considered constant diffusion coefficients and developed a method of estimating diffusion coefficients in a multicomponent system [47]. Thompson and Morral [48] also developed a square-root diffusivity approach, assuming constant diffusion coefficients over a composition range. Therefore, these methods can be used in a reasonably small composition range of a particular system in which composition-dependent diffusion coefficients do not vary significantly. These methods were followed in a few systems to estimate the diffusion parameters with small composition range of 5 at.% [49,50]. It was shown by Stalker et al. [48] that interdiffusion flux could be estimated back for such a small composition range with minimal error. Dayananda and Sohn [51] developed a method for the estimation of average effective interdiffusion coefficients. It has an advantage that one can estimate the diffusion parameters from a single diffusion couple, which can be utilized to calculate back the concentration profile of the same diffusion couple from which these are estimated [52]. However, it has one major disadvantage that the values are averaged over a wide composition range. Moreover, these parameters are not material constants. They depend on the particular diffusion couple and the composition range considered for the estimation. Although they can be used to calculate back the interdiffusion fluxes of the same diffusion couple from which these were estimated, one cannot get an idea on the composition-dependent diffusion coefficients and therefore the atomic mechanism of diffusion.

More recently, a pseudo-binary approach was proposed by the group of the present author [43–45]. As usual, this approach also has many advantages along with certain limitations, which are explained in detail in this section. It should be noted that the term "pseudo-binary approach" was used previously by different groups for the estimation of diffusion coefficients in a ternary system [53,54], but the experimental methods followed by authors did not support the use of this term for their analysis. In a true sense, the diffusion couple in ternary or multicomponent systems should be made in such a way that only two components are active in developing the diffusion profile. Composition profiles of other components should remain the same throughout the diffusion couple starting from the end members to fulfill the mathematical constraints for the estimation of the diffusion coefficients following this approach. This is demonstrated with the help of the experimental results in two ternary systems, Cu(Sn,Ga) solid solution [44] and β-Ni(Pt)Al intermetallic compound [45], since the estimation strategy should be followed differently in these two different types of phases. The method explained in these two ternary systems can be extended to multicomponent systems following similar steps.

3.16.1 Application of the Pseudo-Binary Approach in Cu(Sn,Ga) Solid Solution [44]

Fig. 3.49 shows the pseudo-binary diffusion profile in the Cu(Sn,Ga) system in which Cu content is fixed at 92 at.% throughout the diffusion couple. The composition of Ga and Sn varies from 0 to 8 at.%, and only these two develop the diffusion profile fulfilling the condition of a pseudo-binary diffusion couple. In Eqs. (3.174), if we consider the composition profile and the interdiffusion flux of component 2 as zero, then the conditions become similar to the binary system. However, component 2 still exists in the system such that $N_A + N_B + N_C = 1$. Therefore, the relations for the interdiffusion and the intrinsic diffusion coefficients developed for the binary systems should be modified to suit this condition in a pseudo-binary diffusion couple. In a random solid solution of a ternary A–B–C system, suppose that the composition of the third component remains fixed throughout the diffusion couple and the diffusion coefficients are estimated with respect to the composition profiles of components 1 or 2. Therefore, the modified compositions of these two diffusive components can be written as

$$M_A = N_A + \frac{1}{2}N_C, \tag{3.177a}$$

$$M_B = N_B + \frac{1}{2}N_C, \tag{3.177b}$$

such that

$$M_A + M_B = 1. \tag{3.177c}$$

The modified concentration normalized variable, similar to that in Eq. (3.62) used in den Broeder relation, should be written as

$$Y_{M_{CB}} = \frac{M_{C_B} - M_{C_B}^-}{M_{C_B}^+ - M_{C_B}^-}, \tag{3.178}$$

where the modified concentrations estimated with respect to component B are expressed as

$$M_{C_B} = \frac{M_B}{v_m}, \qquad M_{C_B}^- = \frac{M_B^-}{v_m^-}, \qquad M_{C_B}^+ = \frac{M_B^+}{v_m^+}. \tag{3.179}$$

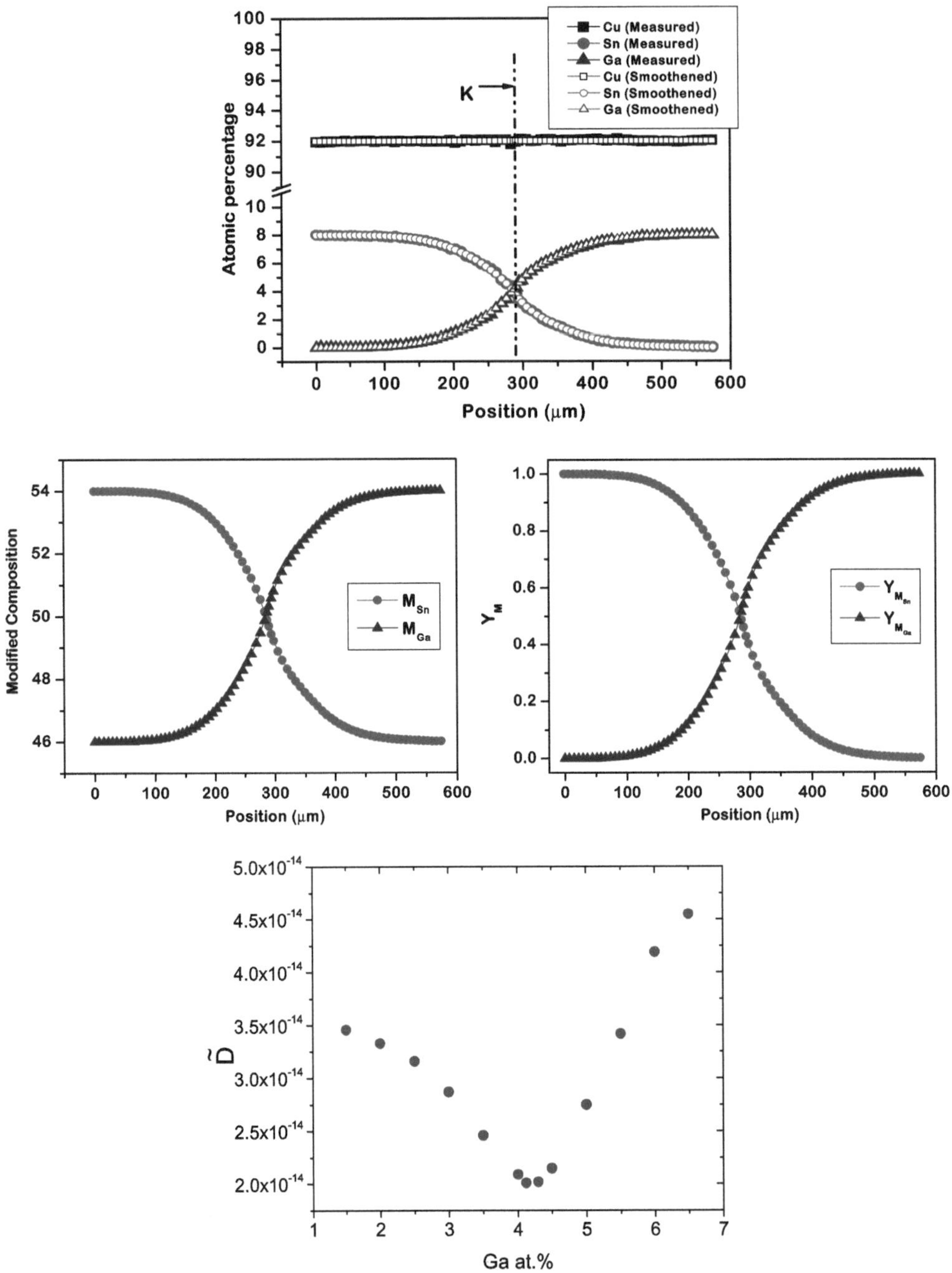

Figure 3.49 Different steps of estimating the diffusion data following a pseudo-binary approach with the experimental result in the Cu–Ga–Sn system [44].

Therefore, the interdiffusion coefficients following den Broeder approach can be estimated after modifying the relation given in Eq. (3.72a) as

$$\tilde{D}(Y^*_{MC_B}) = \frac{1}{2t}\left(\frac{dx}{dY_{MC_B}}\right)_{Y^*_{MC_B}}\left[\left(1 - Y^*_{MC_B}\right)\int_{x^{-\infty}}^{x^*} Y_{MC_B}\,dx + Y^*_{MC_B}\int_{x^*}^{x^{+\infty}}(1 - Y_{MC_B})\,dx\right]. \quad (3.180)$$

If someone is interested in following Wagner's approach, the modified composition normalized variable similar to that in Eq. (3.79) can be written as

$$Y_{M_B} = \frac{M_B - M_B^-}{M_B^+ - M_B^-}. \quad (3.181)$$

Therefore, Wagner's relation given in Eq. (3.91a) becomes

$$\tilde{D}(Y^*_{M_B}) = \frac{v^*_M}{2t(\frac{dY^*_{M_B}}{dx})}\left[\left(1 - Y^*_{M_B}\right)\int_{x^{-\infty}}^{x^*}\frac{Y_{M_B}}{v_M}dx + Y^*_{M_B}\int_{x^*}^{x^{+\infty}}\frac{(1 - Y_{M_B})}{v_M}dx\right]. \quad (3.182)$$

Similarly, the relation between the ratio of intrinsic diffusion coefficients given in Eq. (3.137) and the ratio of the tracer diffusion coefficients given in Eq. (3.153) can be modified as

$$\frac{D_B}{D_A} = \frac{\bar{v}_B}{\bar{v}_A}\frac{[M_B^+\int_{x^{-\infty}}^{x_K}\frac{Y_{M_B}}{v_M}dx - M_B^-\int_{x_K}^{x^{+\infty}}\frac{1-Y_{M_B}}{v_M}dx]}{[-M_A^+\int_{x^{-\infty}}^{x_K}\frac{Y_{M_B}}{v_M}dx + M_A^-\int_{x_K}^{x^{+\infty}}\frac{1-Y_{M_B}}{v_M}dx]}, \quad (3.183a)$$

$$\frac{D^*_B}{D^*_A} = \frac{[M_B^+\int_{x^{-\infty}}^{x_K}\frac{Y_{M_B}}{v_M}dx - M_B^-\int_{x_K}^{x^{+\infty}}\frac{1-Y_{M_B}}{v_M}dx]}{[-M_A^+\int_{x^{-\infty}}^{x_K}\frac{Y_{M_B}}{v_M}dx + M_B^-\int_{x_K}^{x^{+\infty}}\frac{1-Y_{M_B}}{v_M}dx]}. \quad (3.183b)$$

Therefore, the interdiffusion and intrinsic diffusion coefficients measured by pseudo-binary approach in the ternary system is modified from Eq. (3.124) to

$$\tilde{D} = M_{C_A}\bar{v}_A D_B + M_{C_B}\bar{v}_B D_A. \quad (3.184)$$

It should be noted that molar volumes and partial molar volumes are estimated from the variation of molar volumes with respect to actual composition and not the modified composition.

The estimation steps in Cu(Ga,Sn) solid solution are shown in Fig. 3.49 [44]. The variations of lattice parameter data with compositions are not available in this ternary system. The estimated molar volume of Cu(8at.%Ga) is 7.49×10^{-6} and that of Cu(8at.%Sn) is 7.84×10^{-6} m^3/mol [44]. Since the difference in molar volume of these two end members is very small, we can assume a constant molar volume variation, leading to $v_M \approx \bar{v}_{Ga} \approx \bar{v}_{Sn}$ at every composition. Under such a condition, we have shown that the data estimated by any of the methods (Matano–Boltzmann, den Broeder/Wagner) will give more or less the same value. Figs. 3.49B and 3.49C show the modified composition and modified composition normalized variable useful for the data estimated by Wagner's method. For a constant molar volume, both den Broeder (Eq. (3.180)) and Wagner's (Eq. (3.182)) methods reduce to

$$\tilde{D}(Y_{M_B}^*) = \frac{1}{2t(\frac{dY_{M_B}^*}{dx})}\left[(1 - Y_{M_B}^*)\int_{x^{-\infty}}^{x^*} Y_{M_B}dx + Y_{M_B}^*\int_{x^*}^{x^{+\infty}} (1 - Y_{M_B})dx\right]. \quad (3.185)$$

The estimated values are shown in Fig. 3.48D. The location of Kirkendall marker plane is found at 4.2 at.%Ga. The ratio of diffusivity is estimated by Eq. (3.183) as

$$\frac{D_{Ga}}{D_{Sn}} = \frac{D_{Ga}^*}{D_{Sn}^*} = 1.3 \pm 0.3$$

(since a constant molar volume is considered). One should note that, like in a binary system, exactly the same values will be estimated, irrespective of the use of M_A or M_B vs. x profiles as explained in Section 3.17.

3.16.2 Application of the Pseudo-Binary Approach in β-Ni(Pt)Al Intermetallic Compound [45]

This phase is present in the turbine blades of a jet engine and provides oxidation protection to the Ni-based superalloy. Pt-modified bond coats are used for many decades to enhance the stability and oxidation resistance of the material system at the service temperature in a harsh environment. However, the exact role of Pt on the diffusion rates of Ni and Al is still not clear; they play an important role in the formation of the bond coat and also in the evolution of the interdiffusion zone along with many refractory-rich brittle precipitates between the bond coat and the superalloy. The difficulties of diffusion studies following the conventional method in this ternary intermetallic compound can be understood from the imaginary diffusion profiles drawn on the phase diagram in Fig. 3.50A [55]. As already explained, two diffusion couples should intersect at a composition to allow for estimation of the diffusion coefficients. Diffusion profiles on Gibbs triangle follow a serpentine line. Therefore, it is difficult to predict the exact composition of intersection a priori. Moreover, to study the role of Pt, one would be interested in comparing the data estimated at different compositions, keeping the ratio of $(N_{Ni} + N_{Pt}) : N_{Al}$ the same, since Pt replace Ni on the same sublattice. With this aim, even if the compositions of the end members are changed systematically, there is still little chance that these diffusion couples will intersect at a composition, maintaining the

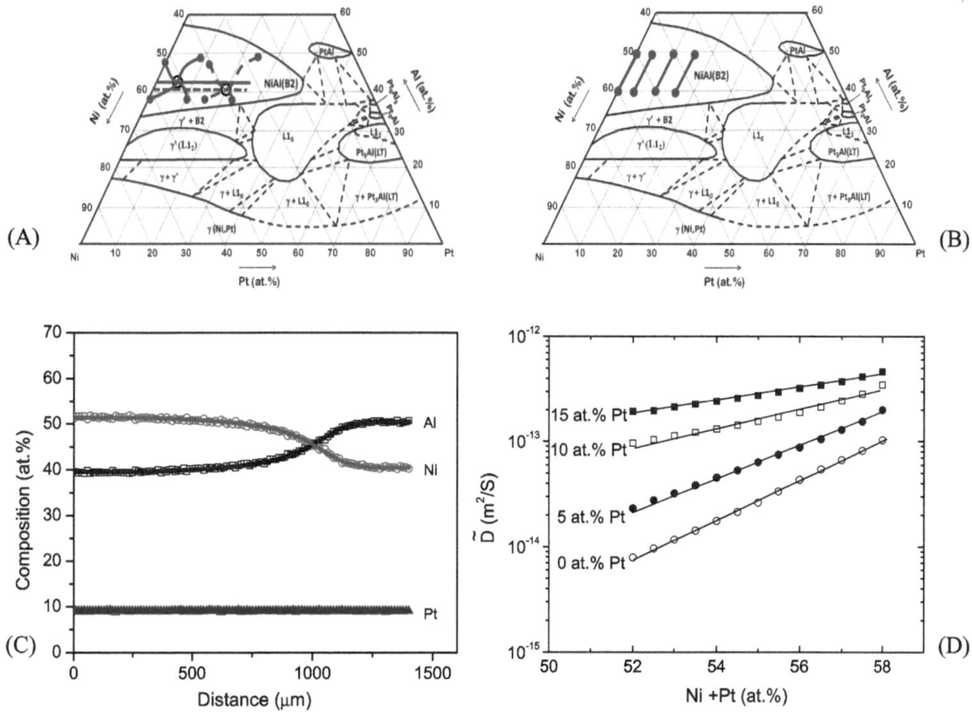

Figure 3.50 Estimation of the diffusion data in the Ni–Pt–Al system following the pseudo-binary approach [45].

ratio of $(N_{Ni} + N_{Pt}) : N_{Al}$, as shown in Fig. 3.50A. Therefore, we end up estimating the data which cannot be compared to develop an understanding, unless a fairly large number of experiments are conducted [8]. The development of the serpentine nature of the diffusion profiles can be read in the articles written by van Loo et al. [56,57] and Thompson and Morral [58,59]. Therefore, a pseudo-binary approach is rather suitable, as explained in Fig. 3.50B. Different diffusion couples are shown, in which Pt content was kept constant at 0, 5, 10, and 15 at.%Pt in both end members of the diffusion couples. In such cases, there is a chance to develop a constant composition profile of Pt such that only Ni and Al develop the diffusion profile, as shown in Fig. 3.50C. These are controlled by the variation of the activities of the components [45].

The condition of a pseudo-binary approach is maintained in the diffusion couples in the β-Ni(Pt)Al, as shown in Fig. 3.50C. The experimental approach in an intermetallic compound is similar to the already discussed approach for solid solution, however, a different step should be considered for analysis. Although this phase has a very wide composition range because of the presence of structural defects, components cannot occupy the lattice sites randomly like in the Cu(Ga,Sn) solid solution. Different sublattices are designated for two different components Ni and Al in a binary system. When the third component Pt is added to the alloy, it replaces Ni and occupies the same sublattice. Therefore, instead of adding equal parts of Pt to both Ni and Al for the estimation of the modified composition profile, Pt should be added to Ni only. Therefore, the modified composition profiles are estimated by

$$M_{Al} = N_{Al}, \tag{3.186a}$$

$$M_{Ni} = N_{Ni} + N_{Pt}, \tag{3.186b}$$

$$Y_{M_{Al}} = \frac{M_{Al} - M_{Al}^-}{M_{Al}^+ - M_{Al}^-}, \tag{3.186c}$$

$$Y_{M_{Ni}} = Y_{(N_{Ni}+N_{Pt})} = \frac{M_{Ni} - M_{Ni}^-}{M_{Ni}^+ - M_{Ni}^-}. \tag{3.186d}$$

Now, similar steps can be followed to estimate the interdiffusion coefficients, using the composition profile of any of the components, as explained in the previous Section 3.16.1 in the Cu(Ga,Sn) solid solution. The estimated interdiffusion coefficients for different Pt content are shown in Fig. 3.50D. If correct steps are followed, one should be able to estimate the same values using any of the profiles of M_{Al} or M_{Ni} vs. x, which is discussed again in Section 3.17. In a binary system, in the absence of Pt, it is already established that the interdiffusion coefficient increases with the increase in Ni content because of increase in Ni antisite defects [1]. When Pt is added, a similar trend is found, but the interdiffusion coefficients increase overall over the whole composition range used in the experiment, indicating an increase in the concentration of defects because of Pt addition [45,60]. In fact, by following the pseudo-binary approach, for the first time, a link is established between the fundamental diffusion studies and the evolution of phase in a multicomponent system used in an actual application [60]. Further, the role of Pt is explained based on the change in driving forces and a possible change in defect concentrations [45]. This is further explained in Section 3.19.

3.16.3 Comparison of the Pseudo-Binary Approach With Other Existing Methods

Although the author of this chapter has just recently introduced the estimation steps following the pseudo-binary approach; the term pseudo-binary was used previously by others. As per the author's knowledge, Guillemot et al. [53] used this term, for the first time, while estimating the diffusion coefficient in the Ti–$Mo_{65}Ta_{35}$ diffusion couple, in which all the components develop diffusion profiles. For the estimation of the diffusion coefficients, they added the profiles of Mo and Ta to consider it as Ti–(Mo,Ta) diffusion couple. This is very similar to the situation of diffusion profile in the Ni–(0.75Co0.25Pt) diffusion couple, as shown in Fig. 3.48B. In this case, even if we add the profiles of Co and Pt, we cannot consider the system as a pseudo-binary diffusion couple. In a true sense, except for two components, all other components should have zero interdiffusion flux, as discussed at the beginning of Section 3.16.1 to fulfill the mathematical conditions. Fukaya et al. [54] followed similar steps in Ni–0.23Al/Ni–0.23Al–0.02X (X = V, Ti, Nb) diffusion couples, even though none of the components had constant compositions profiles throughout the diffusion couple to fulfill the mathematical conditions of the pseudo-binary approach. Sometimes a component may still develop a diffusion profile even if a fixed composition is used in both end members of a diffusion couple, and hence we cannot use the pseudo-binary approach. It can still be used without much error if the deviation of the composition profile from the line connecting the fixed compositions of the end members is not very high. However, one should be very careful in averaging the composition profile, which will otherwise lead to error in calculation, as explained in the next section.

Now the readers should be aware of the advantages and disadvantages of the pseudo-binary approach compared to other approaches. For this, along with the pseudo-binary approach, we consider two other approaches, the conventional method, in which two diffusion couples are required to intersect at one composition, and the estimation of the average effective interdiffusion coefficients developed by Dayananda and Sohn [51] as explained in the next chapter written by them and also in the book chapter written by the present author [1].

Interdiffusion coefficients are important for relating them directly with the interdiffusion fluxes, and therefore for understanding the evolution of the diffusion profile in a certain diffusion couple. In this aspect, the conventional method is the most suitable, since these values are estimated without any assumptions (other than considering a constant molar volume because of unknown lattice parameters). Moreover, one can understand the effect of different components on the interdiffusion flux following Eq. (3.174) by comparing the direct (depending on the concentration gradient of the same component) and cross- (depending on the concentration gradient of the other component) interdiffusion coefficients. However, one cannot estimate the systematic variation of the interdiffusion coefficients from a limited number of diffusion couple experiments since we do not have much control on the composition of the intersection from different diffusion couples. Moreover, the error of estimation is also large since two diffusion couples are required to estimate the data at one composition. This is confirmed by estimating the data from a different combination of couples using the composition at which three diffusion couples intersect [8]. Further, it is almost impossible to estimate the intrinsic diffusion coefficients because of stringent requirements, and one cannot draw any conclusion on the atomic mechanism of diffusion as discussed in Section 3.20.

Because of the strict requirement on the two diffusion couples for the estimation of the interdiffusion coefficients at one composition, Dayananda and Sohn [51] developed an innovative method by which both main and cross- interdiffusion coefficients can be estimated from one single diffusion couple; however, only the average values over a large composition range can be obtained. Therefore, one cannot get an idea about the variation of the interdiffusion coefficients with composition. These values depend on the composition range considered. Therefore, by changing the composition range of the diffusion couple, one will find different values, since the composition ranges considered are different. Moreover, again like in the conventional method, there is no possibility of estimating the intrinsic diffusion coefficients. From the estimated average effective interdiffusion coefficients, one can calculate back the interdiffusion flux and therefore the composition profile of the same diffusion couple from which these were estimated; however, one cannot calculate back the profile for another diffusion couple from which these parameters are not estimated. To take it forward, Cermak and Rothova [61] justified that, by considering a very small composition range, the estimated values will be almost the same to the values determined following the conventional method from two diffusion couples at the composition of the intersection. However, Cheng et al. [62] argued that this is not true, and therefore this topic is still open for further analysis and discussion.

In the search for a new analysis, the group of the present author established the pseudo-binary approach for the estimation of the diffusion coefficients in a multicomponent system. The advantage of following this approach can be understood immediately, as explained in Section 3.16.2. As already explained based on the data in β-Ni(Pt)Al phase, one can fairly easily generate composition-dependent interdiffusion coefficients very systematically (Fig. 3.50D), which is not possible following the other methods in a system with more than two components. Following the conventional method, diffusion data cannot be estimated in a system with more than three components. However, by this method one can estimate the data in any multicomponent system (i.e., with any number of components) as long as it is possible to produce the diffusion couple with only two components developing the diffusion profile and the composition of other components remaining fixed in the interdiffusion zone. Moreover, one can even estimate the intrinsic diffusion coefficients, which is almost impossible even in a ternary system, because of this stringent requirement. Diffusion data estimated by this method is useful to understand the role of defects and atomic mechanism of diffusion. One should be aware of the disadvantages of this approach also. To fulfill the mathematical conditions, only two components should develop diffusion profiles, while the composition of other components is kept the same. Sometimes a component might still develop a diffusion profile, even if the composition is kept the same in both end members. This was found in $Ni_{0.5}Al_{0.5}/Fe_{0.5}Al_{0.5}$ diffusion couple, in which Al also developed a minor diffusion profile near the middle of the composition range, i.e., at $(Ni_{0.25}Fe_{0.25})Al_{0.5}$, which depends on the variation of activity of the components in the composition range of interest. Then, a successful pseudo-binary diffusion couple was produced, avoiding the composition range at which Al developed a diffusion profile. Moreover, one cannot

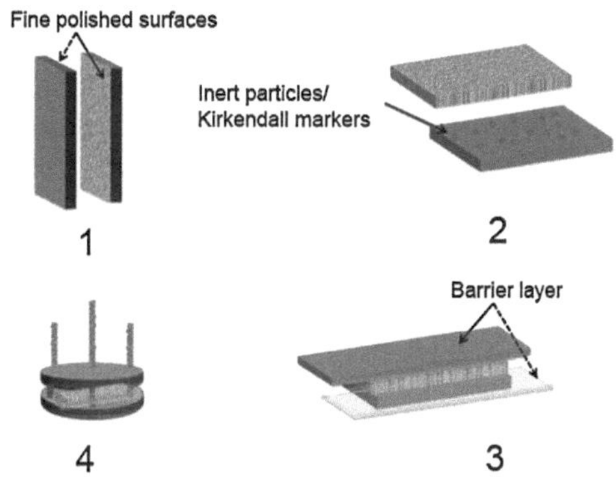

Figure 3.51 Different steps of making a diffusion couple.

estimate the cross-interdiffusion coefficients following this method, which is possible by the conventional method and is important for understanding the effect of one component on the other. Irrespective of these difficulties, for the first time, the research group of the present author developed a link between diffusion data estimated in binary and ternary (pseudo-binary) systems, for a multicomponent material of practical importance incorporating the discussion on thermodynamics and defects assisting the diffusion process [60].

3.17 IMPORTANT STEPS FOR ESTIMATION OF THE DIFFUSION PARAMETERS

In the previous sections, estimation procedure for different types of diffusion coefficients was explained. In this section, our aim is to explain the correct or practical steps one should follow to estimate the data with minimal error.

(i) *Preparation of a bulk diffusion couple*

Different steps for making a diffusion couple are shown in Fig. 3.51. Different groups are comfortable in using different thickness and cross-sectional area of the samples. Based on our experience, we have found that a cross-section, equivalent to 7–8 mm in diameter, and a thickness of 1–3 mm are suitable for metallographic sample preparation manually, i.e., without mounting in a resin and without using an automated grinding and polishing machine. We prefer to do it manually so that there is no damage introduced while removing from the resin before making a diffusion couple. Samples can be square, rectangular, or round, but it is advisable to use similar dimensions for both end members of the diffusion couple for better bonding. Both flatness (of both sides) and fine grinding/polishing (of one side) of the end members are very important for good bonding. We have frequently experienced that excessive grinding to get a very smooth surface leads to rounded edges and improper joining, especially in the edges of the diffusion couple. If good bonding in a large area is not developed, the end members might get separated during handling of the samples. On the other hand, uneven or non-parallel surface with many scratches will lead to no bonding or disturbance at the interface. Therefore, it is very important to keep a balance of both. As a first step, both surfaces of both the end members are ground with rough emery paper to ensure flatness. Then, only one of the surfaces is fine ground and polished to produce a smooth surface which will produce a bonding with the smooth surface of another end member. The presence of a few scratches is fine; they sometimes have a beneficial role in detecting Kirkendall marker plane, as explained Section 3.11.

The procedure above is suitable in most metallic systems which are ductile. When the materials, such as β-Ni(Pt)Al, are very brittle, these will break during manual metallographic preparation. Therefore, after cutting a relatively thick piece of 2–3 mm (such that it does not break during cutting) by slow-speed diamond saw, it is advisable to mount the sample in a (transparent) resin. The thickness of the resin should not be much larger than the thickness of the material. Then, both sides should be ground with rough grinding paper or wheel in an automatic grinding machine to expose the material from both sides and to achieve flat surfaces. Next, one of the sides should be ground and polished further to make a smooth bonding surface. To remove the sample from the resin, the mounted block can be placed in an oven for a small time, above the temperature at which the resin softens such that the sample comes out easily. One should avoid removing it using mechanical force since the prepared material might be damaged. If the formation of an oxide layer is suspected, then a light grinding step with very fine paper or wheel could follow. This procedure is not suitable for materials with very low melting temperature.

Sometimes materials such as Si and Ge wafers can be acquired with flat and smooth surfaces and there is no need to follow any sample preparation steps as discussed above. Once both end members of the diffusion couple are prepared metallographically and cleaned ultrasonically in acetone, inert particles are spread over one of the finely polished end members. TiO_2, HfO_2, Y_2O_3, ZrO_2, W, or Mo powders of 1–5 μm are commonly used for this purpose. Powders are preferred over wires since they are easy to handle. Powders are dispersed in acetone and then dropped on the desired surface with a dropper. Distribution of the particles can be checked under an optical microscope once acetone is evaporated, leaving behind the particles. The very ideal spread is not easy to get without any agglomeration; however, one can try the procedure a few times after cleaning the sample in acetone to achieve a desirable or acceptable distribution. 20–30 micrometer distance between the particles is desirable for the successful bonding. Then, the smooth surface of another end member is placed on the end member with inert particles, and then two diffusion barrier foils are placed on opposite sides. Al_2O_3 thin pieces or thin foils of Mo or W can be used as a diffusion barrier layer depending on the materials used for making a diffusion couple. This stack is then placed in a fixture such that a small load or pressure can be applied for an efficient contact. Fixture can be made of stainless steel, molybdenum or tungsten, again depending on the annealing temperature. There is no particular rule for applicable load or pressure. For example, in a metal–silicon system, tightening the screw by hand is enough to get a good contact if annealed at a reasonably high temperature. Pressure should be enough to bring the smooth surfaces tightly together. If both end members are very hard (for example, refractory metal–silicon diffusion couples), these may not create a good contact when inert particles are used as Kirkendall marker. In such a situation, it is advisable not to use inert markers, and the location of Kirkendall plane can be detected from morphological evolution or from the presence of a line of pores as explained in Section 3.11. Next, the fixture is placed in a furnace equipped with high vacuum system or under the flow of inert gas. If the furnace is not equipped with controlled atmosphere, the diffusion couple can be sealed in a glass or quartz capsule under vacuum or after filling it with Ar or any other inert gas depending on the annealing temperature such that capsules are not broken or destroyed at the annealing temperature. After annealing of the sample at desired temperature for a certain time in a calibrated furnace, it is quenched if the interface is not very weak, like in metal–silicon systems. Otherwise, slow but controlled cooling in vacuum or air is desirable. Then, the sample is removed from the fixture, cross-sectioned by slow speed diamond saw, and metallographically prepared for examination under optical or scanning electron microscope. Annealing time at certain temperature should be such that the ends of the diffusion couple are not affected by the diffusion of components. Otherwise, it cannot be used to estimate the diffusion coefficients since it will not fulfill the mathematical boundary conditions of a semi-infinite diffusion couple. Theoretically, even one unaffected atomic layer at the ends of the couple is enough to consider it as a semi-infinite couple; however, it is common practice to select the thickness of the material and annealing conditions such that half of the thickness is not affected.

For multifoil Kirkendall marker experiment, the foils acquired should be smooth enough such that there is no need for metallographic preparation before making the diffusion couple. The thickness of the foils should not be very low depending on the strength of the material. Based on our experience, one can even handle the thickness of 25 μm of many metals such as Ni, Co, Pt, Fe, etc. A diffusion couple experiment with bulk pieces should be conducted first to decide on the thickness and number of foils required for the multifoil experiment. Foils with 8–10 mm diameter or equivalent cross-section are easier to handle during stacking. After cutting into the desired size, these should be pressed lightly with a flat material for a flat surface, which otherwise will create gaps between the foils. After spreading the inert particles as Kirkendall markers, one could stack these foils between the same metals with higher thickness and cross-section. Once this stack is placed in a fixture, enough pressure should be applied to ensure no gap between the foils and good contact. To verify the results, one should first compare the interdiffusion coefficients measured in bulk and multifoil experiments, as shown in Fig. 3.39A.

(ii) *Smoothing and averaging the composition profiles*

The bulk diffusion couple is annealed in such a way that the diffusion length is enough to measure the composition profile in an electron probe microanalyzer (EPMA) with the help of wavelength dispersive spectrometer (WDS). This is one of the desirable techniques if pure components are used as standard along with a right combination of voltage and current depending on the components present in the couple. With this technique, it is not difficult to bring down the error of measurement within 0.5 at.%. When WDS is not available, very frequently, the composition profiles are measured in a scanning electron microscope (SEM) equipped with energy dispersive spectrometer (EDS). In this technique, the measurement error is larger, in the range of 1–3 at.%, depending on the sample preparation, use of standards, components which are measured, and the right choice of area of the energy peaks produced by the detector.

The next very important step is to smoothen the composition profile. This is a very important step, especially if the composition profile is measured by EDS and may not be easy to catch the actual trend, especially when the composition profile changes very sharply. There is no particular or single equation which should always be applied for smoothing the data, since composition profile may be very different in different systems. A few examples are shown in Fig. 3.52, which were measured in different systems such as Cu–Ga [24], Au–Cu [6], and Co–Nb [63]. In some cases, it is wise to consider different parts of the diffusion couple separately for smoothing and then join them together to achieve a desirable smoothened composition profile. Sometimes there is no other option but to follow this step, especially when different phases have different types of composition profiles, as it is found in the Co–Nb system.

One should be warned that even if smoothened composition profile looks almost perfect, there can still be a problem locally at many places, which could be understood only if different parts are enlarged and cross-checked. One typical situation is explained in Fig. 3.53. At one place, a few measured points have a larger error (open circles) than others (filled circles). When the smoothened data profile is considered, based on the polynomial fit of the experimentally measured composition data points, the fitted line may follow the dotted line because of bad points instead of the solid line. To understand the effect of considering the dotted line for the estimation of the interdiffusion coefficients, we need to understand the influence of different parts of the equations used for such estimation. If we consider den Broeder's approach (Eq. (3.72a)), it has three parts $\frac{dx}{dY_C}$, $\int_{x^{-\infty}}^{x^*} Y_C dx$, and $\int_{x^*}^{x^{+\infty}} (1 - Y_C)dx$ to be estimated separately to get the interdiffusion coefficients. It can be immediately understood that the gradient of concentration normalized variable $\frac{dY_C}{dx}$ will give a large error in estimation since it depends on the data surrounding the point at which it is calculated. On the other hand, many data points over a long distance are used to estimate the values of $\int_{x^{-\infty}}^{x^*} Y_C dx$ and $\int_{x^{-\infty}}^{x^*} (1 - Y_C)dx$.

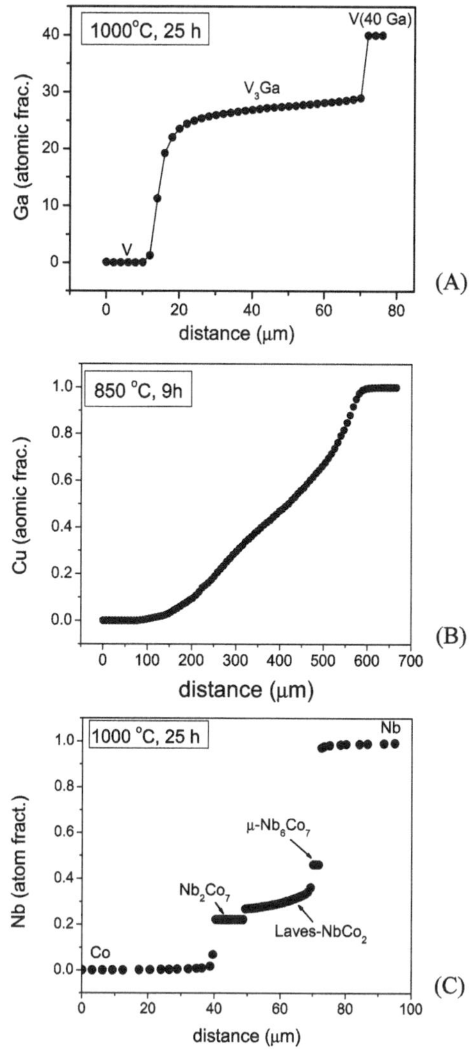

Figure 3.52 Diffusion profiles developed in (A) V–Ga [13], (B) Au–Cu [6], and (C) Co–Nb [63] systems.

Therefore, these will not induce much error because of error in few data points. Therefore, the estimated interdiffusion coefficient will give an error in the calculation if the measured data are not smoothened correctly mainly due to the problem of estimation of the gradient of concentration normalized variable. This can be avoided sometimes by smoothing the estimated values of $\frac{dY_C}{dx}$ in a $\frac{dY_C}{dx}$ vs. N_B plot. Similarly, if Wagner's relation (Eq. (3.91a)) is used, the error will originate mainly in the estimate of the composition normalized variable gradient, $\frac{dY_B}{dx}$, which should be smoothened following the steps

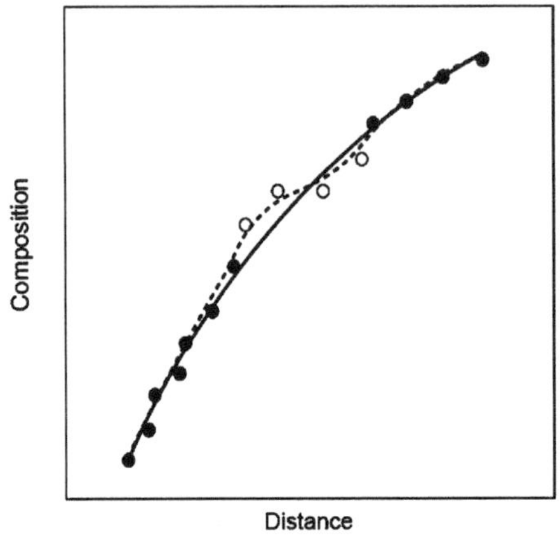

Figure 3.53 Averaging of a composition profile.

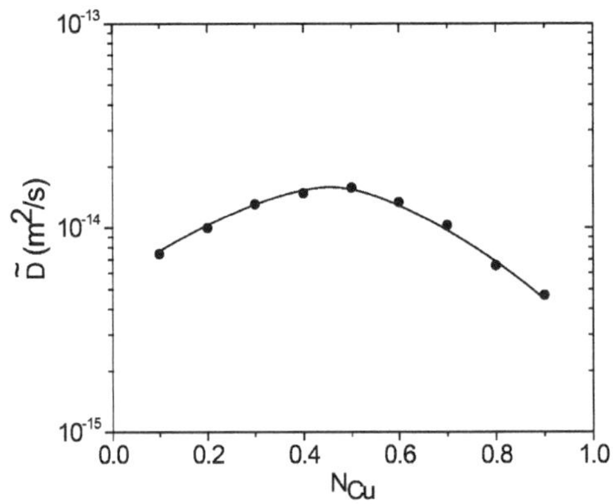

Figure 3.54 Interdiffusion coefficients measured in the Au–Cu system at 650°C [6].

discussed above. Fig. 3.54 shows the interdiffusion coefficients estimated in the Au–Cu system at 650°C [6]. It can be seen that when proper measures are taken during the estimation steps, the interdiffusion coefficients do not deviate much from the average line. When the data are scattered without indicating the trend of variation of the interdiffusion coefficient with composition, one should repeat the calculation following proper steps. To ensure even smaller estimation error, one can smooth the data of $\int_{x^{-\infty}}^{x^*} Y_C dx$ and $\int_{x^{-\infty}}^{x^*}(1 - Y_C)dx$ when den Broeder approach is followed. $\int_{x^{-\infty}}^{x^*} \frac{Y_B}{v_m} dx$ vs. N_B and $\int_{x^{-\infty}}^{x^*}\left(\frac{1-Y_B}{v_m}\right)dx$ vs. N_B should be plotted when Wagner's approach is used. On the other hand, these steps are not required, if composition profiles are smoothened properly.

An important fact should be noted here, when temperature and composition dependent interdiffusion coefficients are measured. Sometimes it is noticed that different types

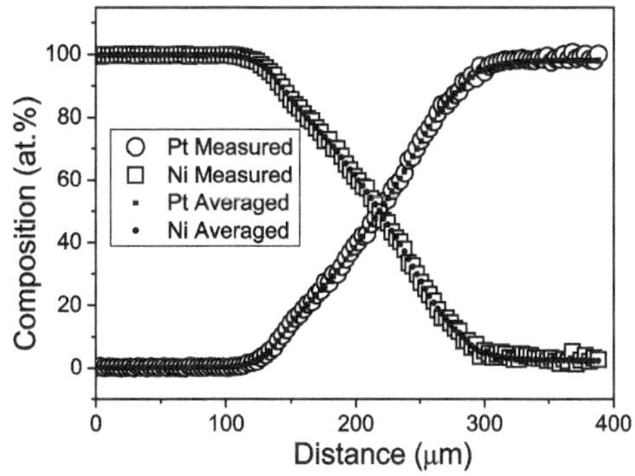

Figure 3.55 Composition profile in the Ni–Pt diffusion couple [30].

of composition-dependent variations are reported at different temperatures. This might be a problem due to faulty experiments, wrong measurement of composition profiles or subsequent incorrect steps during estimation of the data, unless there is a certain phase, structure or magnetic transformation with the change in temperature. In absence of such transformations, the type of variation of the interdiffusion coefficients with the composition at different temperatures is expected to be similar, as shown in Fig. 3.12B [6]. One can cross-check the data at different temperatures even by plotting diffusion coefficients at one composition following the Arrhenius equation, $\tilde{D} = \tilde{D}_o \exp(-\frac{Q}{RT})$, leading to a straight line fit without much deviation or error. Here $\tilde{D}_o$ is the pre-exponential factor, Q is the activation energy, and T is the temperature in Kelvin. Then, the variation of the activation energies calculated at different compositions will have a smooth variation with composition if the estimation procedure of diffusion coefficients at all compositions and temperatures is followed properly from correct diffusion couples. This is, of course, true only when the phase does not go through any phase or magnetic transformation since otherwise the diffusion mechanism will change, leading to different activation energy.

A very important fact should be noted while averaging the measured composition profiles. This is explained with the composition profile measured in the Ni–Pt diffusion couple, as shown in Fig. 3.55. Ideally, at every location in a binary diffusion couple should fulfill $N_{Ni} + N_{Pt} = 1$ when measured in atomic fraction (or $N_{Ni} + N_{Pt} = 100$ when measured in atomic percentage). However, when these two profiles are averaged and smoothened independently, this may lead to $N_{Ni} + N_{Pt} \neq 1$. In such a situation, $\bar{v}_A \tilde{J}_A + \bar{v}_B \tilde{J}_B = 0$ as expressed in Eq. (3.76) will not hold. As a result, this will lead to different interdiffusion coefficients when estimated from composition profile of different components because of error in averaging and smoothing of profile. To prove this in more detail, following Fick's first law of diffusion, Eq. (3.76) can be expressed as

$$-\tilde{D}(A)\frac{\bar{v}_A dC_A}{dx} - \tilde{D}(B)\frac{\bar{v}_B dC_B}{dx} = 0 \tag{3.187a}$$

where $\tilde{D}(A)$ and $\tilde{D}(B)$ are the interdiffusion coefficients measured from composition profiles of component A and B, respectively.

From the standard thermodynamic relation in Eq. (3.74e), we have

$$\bar{v}_A dC_A = -\bar{v}_B dC_B.$$

Therefore, we can write

$$\tilde{D}(A) = \tilde{D}(B). \tag{3.187b}$$

This indicates that we should have the same value of interdiffusion coefficient at one particular composition, irrespective of the choice of the composition profile of component A or B for the estimation of the interdiffusion coefficient. This will be true only when correct estimation steps are followed. Therefore, instead of averaging both profiles independently, one should average and smooth one composition profile only (for example, N_B) and then calculate the other composition profile (for example, N_A) from $N_A = 1 - N_B$.

This is true for the pseudo-binary approach and is a very important step to consider. Tsai et al. [64,65] has reported two different values of interdiffusion coefficient when the interdiffusion coefficients had been measured from the composition profiles of Cr and Mn in a pseudo-binary diffusion couple prepared in an alloy of NiCoFeMnCr. In this diffusion couple, the composition of Ni, Co, and Fe were kept constant in both end members, also leading to the same and constant compositions in the interdiffusion zone. Therefore, as a correct step, if the composition profiles of Mn, Ni, Co, and Fe are averaged and smoothened, the composition profile of Cr should be calculated from $N_{Cr} = 1 - (N_{Mn} + N_{Ni} + N_{Co} + N_{Fe})$ instead of being averaged and smoothened separately. As a better approach, which is explained in Ref. [64], both the profiles of Cr and Mn should be optimized and smoothened together such that the total is equal to 1. Then only the values of interdiffusion coefficients will be the same when measured from Cr or Mn composition profiles. It is possible that this step was not followed, and two different values were found when interdiffusion coefficients were measured independently from the composition profiles of Cr and Mn.

(iii) *Estimation of molar volume and partial molar volumes*

Estimation of the molar volume is an important prerequisite for the estimation of the diffusion coefficients. This is estimated from the equation

$$v_m = \frac{v_{cell}}{n_a} N_o \tag{3.188}$$

Table 3.3 Details used in Pearson symbols and formulas of molar volumes for different crystal structures

(A) Family of the crystal

a	triclinic (anorthic)
m	monoclinic
o	orthorhombic
t	tetragonal
h	hexagonal
c	cubic

(B) Type of the lattice

P	Primitive
S	One side face centered
I	Body centered (innenzentriert)
R	Rhombohedral centered
F	Center of all faces

(C) Volume of the unit cells:

Cubic ($a=b=c$; $\alpha=\beta=\gamma=90°$)	a^3
Tetragonal ($a=b\neq c$; $\alpha=\beta=\gamma=90°$)	a^2c
Hexagonal ($a=b\neq c$; $\alpha=\beta=90°$, $\gamma=120°$) (Note that one third of the hexagon is considered)	$(\sqrt{3}/2)a^2c$
Orthorhombic ($a\neq b\neq c$; $\alpha=\beta=\gamma=90°$)	abc
Monoclinic ($a\neq b\neq c$; $\alpha=\gamma=90°\neq\beta$)	$abc\sin\beta$
Rhombohedral ($a=b=c$; $\alpha=\beta=\gamma$ but $\neq 90°$)	$a^3\sqrt{1-3\cos^2\alpha+2\cos^3\alpha}$
Triclinic ($a\neq b\neq c$; $\alpha\neq\beta\neq\gamma\neq 90°$)	$a^3\sqrt{1-\cos^2\alpha-\cos^2\beta-\cos^2\gamma+2\cos\alpha\cos\beta\cos\gamma}$

where v_{cell} is the volume of the unit cell, n_a is the number of atoms in the unit cell, and N_o is Avogadro number (6.022×10^{23}/mol). For this, along with lattice parameter data, it is advisable to know Pearson symbol, which indicates the type of unit cell and number of atoms present in that unit cell. For example, Pearson symbol of intermetallic compound Nb_3Sn is cP8 (cubic, primitive with 8 atoms in the unit cell). The descriptions of different letters in Pearson symbols are given in Tables 3.3A and 3.3B. The lattice parameter at the stoichiometric composition is $a=0.5169$ nm. Since it is a cubic structure, the molar volume of this phase is calculated as

$$v_m=\frac{v_{cell}}{n_a}N_o=\frac{a^3}{n_a}N_o=\frac{(0.5169)^3\times 10^{-27}}{8}\times 6.022\times 10^{23}=10.4\times 10^{-6}\ \mathrm{m^3/mol}.$$

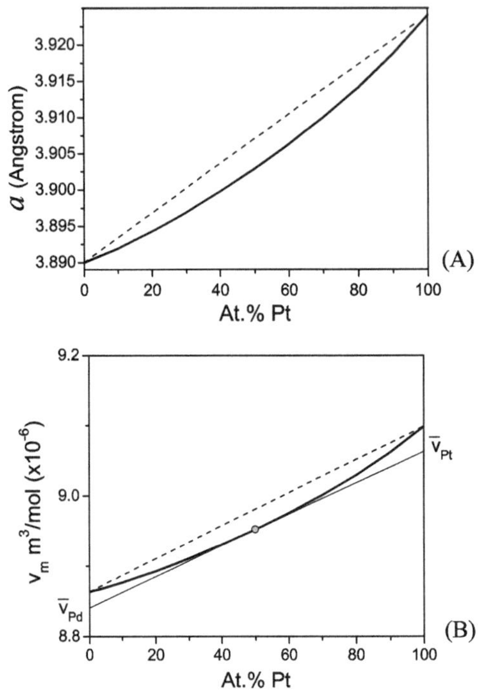

Figure 3.56 (A) Lattice parameter and (B) molar volume variation in the Pd–Pt system [39].

Equations for the estimation of volumes for different types of unit cells are listed in Table 3.3C.

In most phases, the number of atoms is equal to the number of lattice sites in the unit cell. However, in very few compounds, due to the presence of very high concentration of vacancies, one needs to carefully estimate the actual number of atoms. For example, β-NiAl phase [5] has a very wide homogeneity range both in Al- and Ni-rich sides. In the Ni-rich side, Ni antisite defects occupy the Al sublattice, and there is no need for special care during calculation. However, in the Al-rich side, a large concentration of structural vacancies is present. For example, at 45 at.%Al, which is 5 at.% from the stoichiometric composition, it has 10% structural vacancies. Pearson symbol of this phase at the stoichiometric composition is cP2. Therefore, the number of atoms in a unit cell at 45 at.%Al should be $n_a = 1.8$, which is less by 10% from the number of atoms at the stoichiometric composition. This practice is not important in the phases with only thermal vacancies since the concentration of thermal vacancies is very small.

Partial molar volumes can be estimated when molar volumes at a few more points (at least) around the composition of interest are known, as shown in the Pd–Pt solid solution in Fig. 3.56 [39]. The actual variation of molar volume is calculated, which deviates from the ideal (dotted line) based on Vegard's law. The partial molar volumes

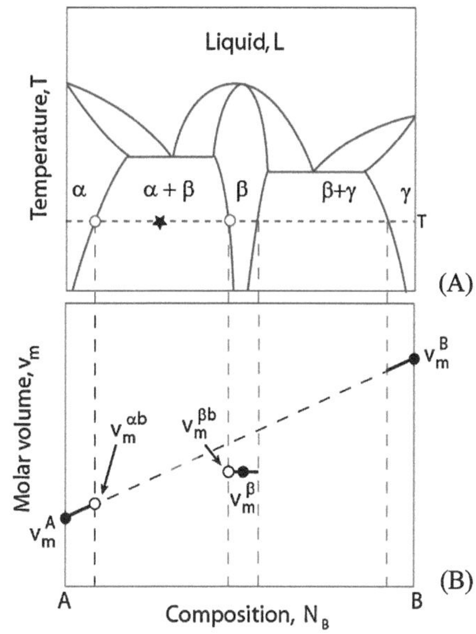

Figure 3.57 Calculation of the average molar volume.

of Pd and Pt are estimated by taking a tangent at the composition of interest and then extending it to $N_{Pt} = 0$ and $N_{Pt} = 1$, respectively. Very often the composition-dependent lattice parameters are not available, and a practical step can be followed in such a situation as shown in Fig. 3.57. Suppose that at the temperature of interest, there are three phases, α and γ solid solutions and β intermetallic compound. We have lattice parameter data for only pure A, B, and stoichiometric composition of the β phase. Therefore, we can estimate the molar volumes only at three compositions as shown by filled circles, yet we also need data at other compositions inside all the phases for the estimation of the diffusion coefficients. In such a situation, one can follow Vegard's law for α and γ solid solutions as shown by solid lines based on the equation

$$v_m = N_A v_m^A + N_B v_m^B \tag{3.189a}$$

where v_m is the molar volume of the alloy at the composition of interest, v_m^A is the molar volume of pure A, and v_m^B is the molar volume of pure B. Therefore, the partial molar volumes in this solid solutions are $\bar{v}_A = v_m^A$ and $\bar{v}_B = v_m^B$. Intermetallic compounds always have a very different crystal structure than the pure components and also its molar volume may deviate positively or negatively (as considered in the present case) from the dotted line drawn following Vegard's law. Therefore, we cannot extend the molar volume estimated at the stoichiometric composition following Vegard's law to other compositions. It is a common practice to consider it to be constant over the whole composition range inside this phase such that $\bar{v}_A = \bar{v}_B = v_m^\beta$.

Very frequently a phase mixture is used as one of the end-members, for example, the composition as shown by a start on the phase diagram. One can estimate the fraction of phases' α (f_α) and β (f_β) present in the alloy by lever rule from the phase boundary compositions of the phases at that temperature. If the molar volumes of the phase boundary compositions are $v_m^{\alpha b}$ and $v_m^{\beta b}$, the average molar volume of the alloy can be estimated as

$$v_m = f_\alpha v_m^{\alpha b} + f_\beta v_m^{\beta b}. \tag{3.189b}$$

3.18 ANALYSIS OF DIFFUSION DATA FOR UNDERSTANDING THE ROLE OF THERMODYNAMIC DRIVING FORCE AND DEFECTS

(i) *Understanding the role of thermodynamic driving forces and defects on variation of interdiffusion coefficients with composition*

From Eqs. (3.156), the intrinsic and tracer diffusion coefficients are related by

$$D_A = \frac{v_m}{\bar{v}_B} D_A^* \Theta (1 + W_A),$$
$$D_B = \frac{v_m}{\bar{v}_A} D_B^* \Theta (1 - W_B).$$

Intrinsic diffusion coefficients are related to the interdiffusion coefficient (Eq. (3.124)) by

$$\tilde{D} = \bar{v}_A C_A D_B + \bar{v}_B C_B D_A.$$

Tracer diffusion coefficients indicate the diffusion rate of components in the absence of any thermodynamic driving forces, and they depend only on the defects present in the structure. The intrinsic diffusion coefficients show additionally the influence of thermodynamic driving force $\Theta = \frac{d \ln a_A}{d \ln N_A} = \frac{d \ln a_B}{d \ln N_B}$ when compared with the tracer diffusion coefficients. Subsequently, the interdiffusion coefficients, and therefore the evolution of the diffusion profile, depend on the intrinsic diffusion coefficients. There can be systems in which both defects and thermodynamic driving force affect composition-dependent diffusion coefficients significantly. However, many times one of the factors dominates over another, and it is possible to understand this from the estimated diffusion coefficients. This is explained based on the estimated diffusion parameters in the Ni–Pt and Ni–Al systems.

Experimental data measured in the Ni–Pt system at 1200°C are shown in Fig. 3.58 [30]. The average grain size of Ni and Pt was in millimeter range, and since the experiments were conducted at a relatively high homologous temperature, transport of components was controlled by lattice diffusion. Estimated data indicates that Ni has a

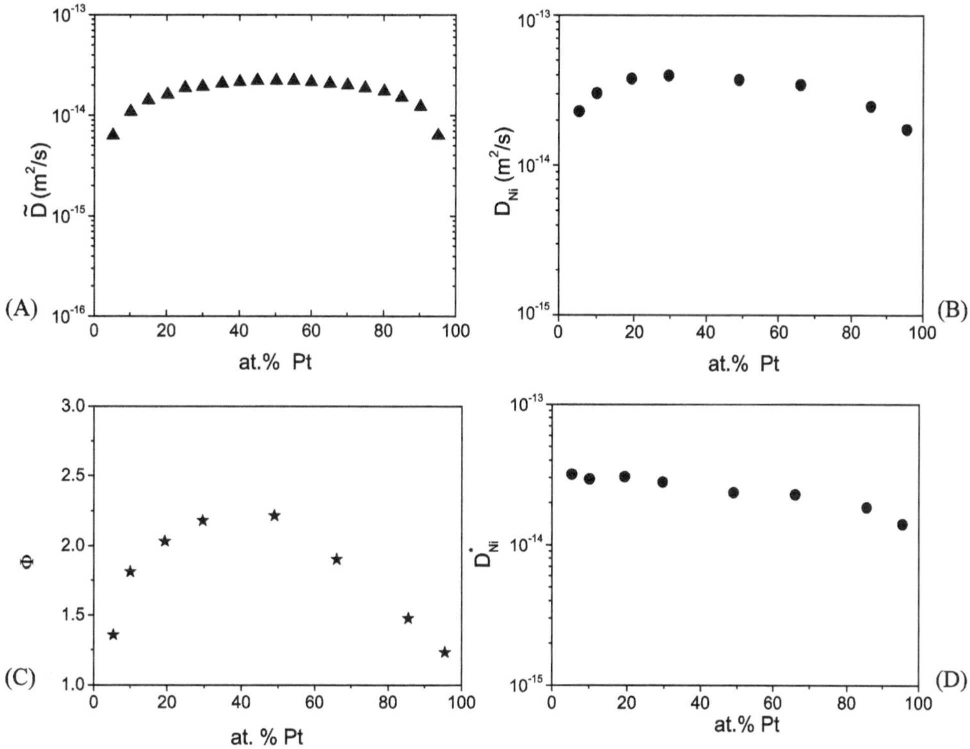

Figure 3.58 (A) Estimated interdiffusion coefficients, (B) intrinsic diffusion coefficients, (C) thermodynamic parameters, and (D) tracer diffusion coefficients at 1200°C [30].

higher diffusion rate compared to Pt at all compositions. Since interdiffusion process is controlled by the component with a higher diffusion rate, the diffusion rate of Ni is important to consider for discussion in this system. It can be seen that D^*_{Ni} does not vary significantly with composition, which indicates that the concentration of defects (vacancies in a solid solution) does not vary significantly with composition. On the other hand, the trend of variation of intrinsic diffusion coefficients and the thermodynamic driving force are similar in nature, which indicates the important role of the thermodynamic driving force on the diffusing components in this system. Therefore, the variation of the interdiffusion coefficients also had a similar trend.

To take the discussion one step further, let us examine the diffusion data [5], thermodynamic driving force, and defects present in β-NiAl phase in the Ni–Al binary system. It can be seen in Fig. 3.59A [66] that this phase has a very wide homogeneity range on both Al-rich and Ni-rich sides of the stoichiometric compositions. This is achieved because of the presence of a very high concentration of defects, as shown in Fig. 3.59B, which was calculated by Divinski [1]. In the Al-rich side, triple defects

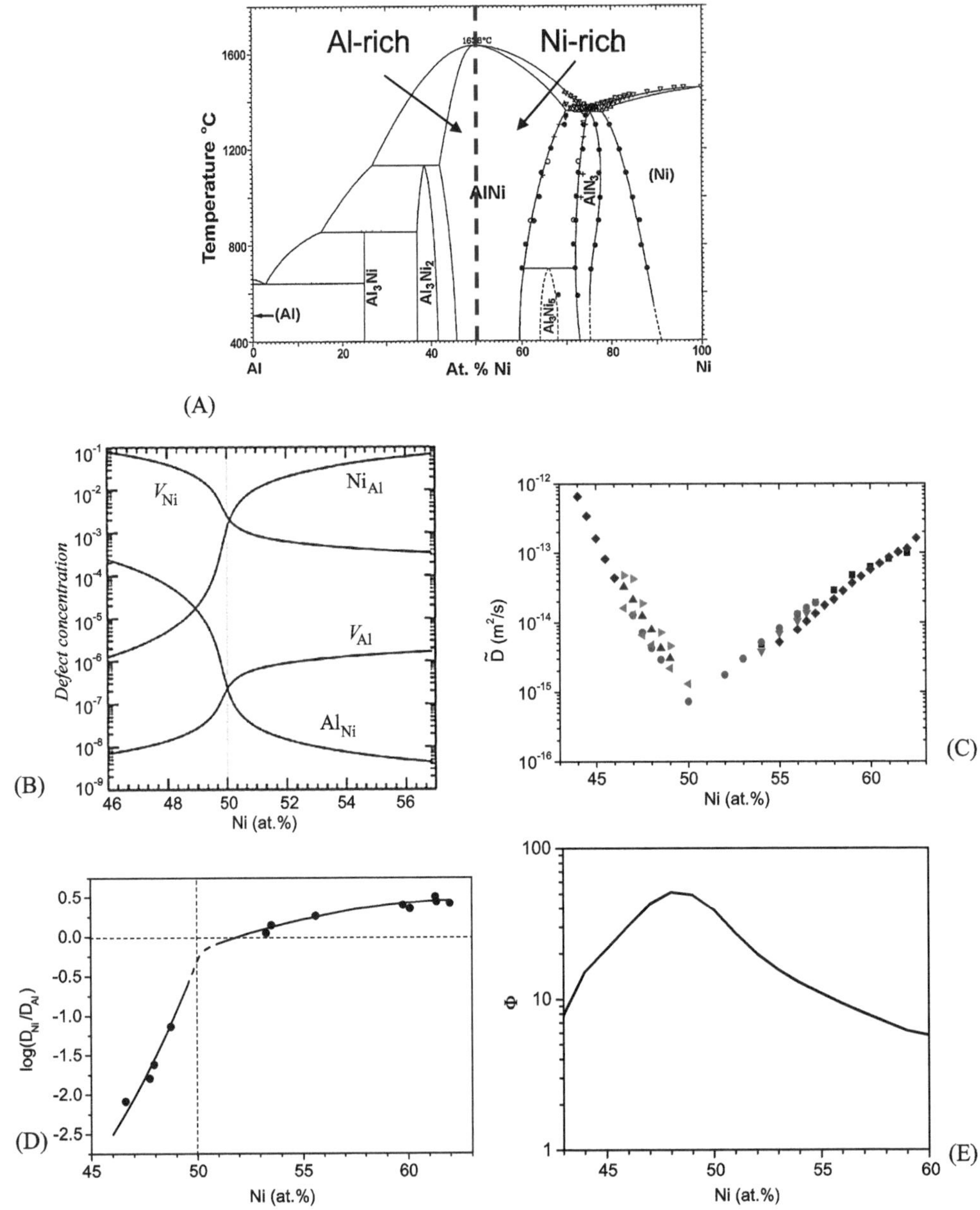

Figure 3.59 (A) Ni–Al phase diagram [66], (B) variation of defect concentrations [60], (C) interdiffusion coefficients [5], (D) ratio of the intrinsic diffusion coefficients [5], and (E) thermodynamic parameter in the β-NiAl phase [60].

dominate, which are constituted by two vacancies on Ni sublattice (V_{Ni}) for one Al antisite on Ni sublattice (Al_{Ni}). With increasing deviation, the concentration of these defects increases. On the other hand, deviation on the Ni-rich side is achieved due to the presence of Ni antisites (Ni_{Al}). Since defects assist the diffusion of components, interdiffusion coefficients increase with the increase in deviation from the stoichiometry with a minimum value close to the stoichiometric composition as shown in Fig. 3.59C [5]. These values increase faster in the Al-rich side compared to the Ni-rich side because of the presence of extra structural vacancies (V_{Ni}). Note that the system does not differentiate between thermal and structural vacancies. However, it is a known fact that structural defects have much higher concentration particularly in this phase, especially at the non-stoichiometric compositions. The ratio of diffusivities is measured over the whole composition range from multiple diffusion couples, as shown in Fig. 3.59D [5]. In the Al-rich side, Al has a much higher diffusion rate compared to Ni, whereas on the Ni-rich side Ni has a higher diffusion rate compared to Al (Fig. 3.59D). It can be seen that the trend of variation of the interdiffusion coefficient is similar to the dominant defects such as V_{Ni} and Al_{Ni} in the Al-rich side, and Ni_{Al} and V_{Ni} on the Ni-rich side. On the other hand, the thermodynamic driving force (Fig. 3.59E), which is equal for both components, has just opposite trend. Therefore, in this system, defects dominate the thermodynamic driving force when considering the variation of diffusion coefficients at a particular temperature.

Therefore, there are examples in which thermodynamic driving force plays a dominating role over defects. In other systems, defects play a dominating role over the thermodynamic driving force. However, one should be very careful in analyzing the role of these factors, since in many systems both might have a significant role and it is not easy to differentiate the exact contributions. In the examples above, we considered the systems in which lattice diffusion plays a dominating role. However, the roles of both defects and thermodynamic driving force are significant even when a product phase is grown by grain boundary diffusion. The only difference is that instead of considering point defects, which are relevant for the lattice diffusion, one should examine the grain size and hence the grain boundary area which is important for grain boundary diffusion [67].

3.19 PREDICTING THE DEFECTS PRESENT BASED ON THE ESTIMATED DIFFUSION COEFFICIENTS IN INTERMETALLIC COMPOUNDS

Estimating the concentration of point defects (experimentally or theoretically) in an intermetallic compound is not an easy task, since different types of defects such as antisites and vacancies are present. These are present in different concentration on different sublattices. An example is already shown for the case of β-NiAl phase (Fig. 3.59B). Also, these are almost impossible to determine following the experimental techniques. They are not even easy to estimate by theoretical methods, unless the bond energies

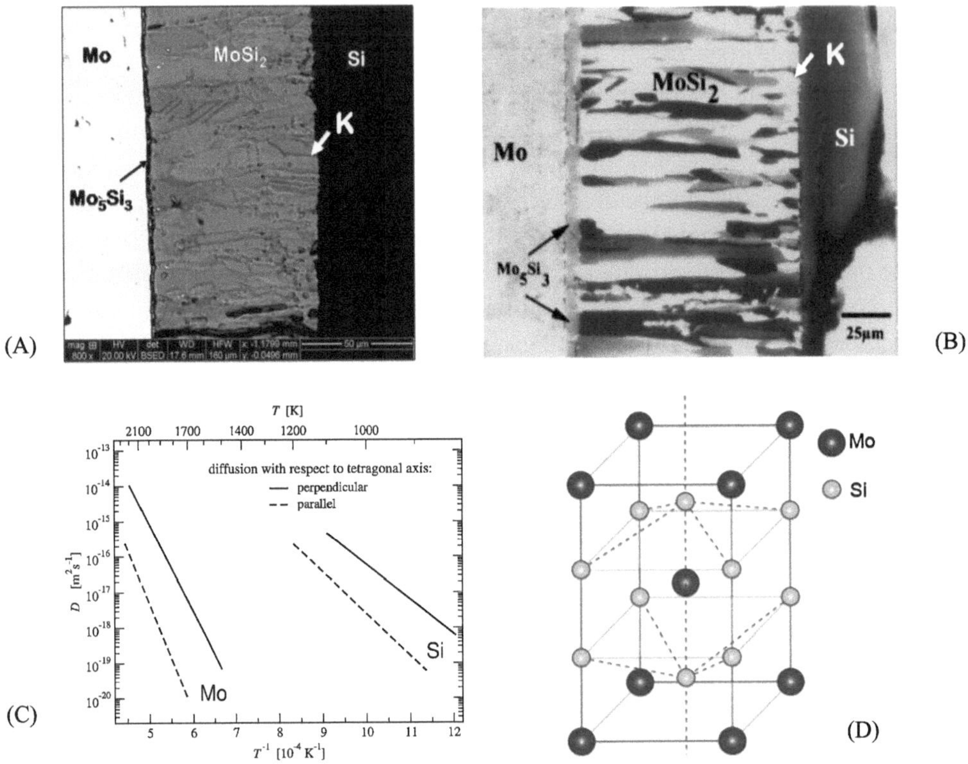

Figure 3.60 (A)–(B) Interdiffusion zone in Mo–Si systems [68,69], (C) tracer diffusion coefficients [70], and (D) the crystal structure of the $MoSi_2$ phase.

are available or calculated. Therefore, these concentrations are available only in very few systems. It is a general practice that the variation of the estimated diffusion coefficients is explained with the known variation of the defect concentrations. However, the group of the present author has demonstrated that the estimated diffusion data can indicate the presence of the type and the relative (not the absolute) concentration of defects. Knowledge of defects is important for understanding the many physical and mechanical properties of the materials.

To facilitate this discussion, let us consider the experimental results in the refractory metal–silicon systems. We mainly consider the disilicide phase in various systems. In the beginning, to establish the veracity of the arguments, let us first examine the experimental results in the Mo–Si diffusion couple. As can be seen in Figs. 3.60A and 3.60B [68,69] based on morphological evolution, the location of Kirkendall marker plane is found to be at the Si/$MoSi_2$ interface. This indicates that the $MoSi_2$ phase grows because of diffusion of Si from the Mo_5Si_3/$MoSi_2$ interface. Mo has negligible diffusion rate through this phase. The growth mechanism can be better understood following the physicochemical approach [27,28] developed by the present author as explained in Chapter 4 of Volume 2. Indication of the diffusion rates of Mo and Si can be confirmed from the tracer diffusion coefficients measured in this phase, as shown in Fig. 3.60C [70]. It can be seen that Si has a few orders of magnitude higher diffusion rate compared to Mo. This phase possesses the tetragonal crystal structure (tI6, $C11_b$) in which the refractory metal component Mo is surrounded by 10 Si, whereas Si is surrounded by 5 Si and 5 Mo. Therefore, following sublattice diffusion mechanism [1], as discussed in Chapter 10 of Volume 1, Si can diffuse via its own sublattice on the condition that vacancies (V_{Si}) are present on the same sublattice. On the other hand, Mo cannot diffuse even if vacancies are present on the same sublattice, since Mo is surrounded by only Si atoms. Mo antisites on Si sublattice (Mo_{Si}) are only allowed to diffuse by making a jump to the next neighbor position since they only can occupy the sublattice of Si. Therefore, we can conclude that the concentration of Mo_{Si} must be negligible in this phase, since Mo diffusion rate is negligible. A study on defects in this phase indicates that vacancies are mainly present on the Si sublattice, only supporting this discussion [71].

Now consider systems in which tracer diffusion coefficients are not known or there is no knowledge of defects present. However, we can draw a conclusion about defects based on the estimated integrated interdiffusion coefficients and the ratio of diffusivities by the diffusion couple technique, as shown in Fig. 3.61. Data are estimated in group IVB and VB (in the periodic table) refractory metal–silicon systems [36,72–76]. From group IVB, Ti, Zr, and Hf with increasing atomic number are considered to couple with Si. In all the systems, disilicide phase grows with enough thickness for the estimation of the diffusion coefficients. The estimated integrated diffusion coefficients are plotted with respect to the inverse of homologous temperature, i.e., T_m/T, where T_m is the melting point of the phase and T is the temperature at which these data are measured. It can be seen that with increasing atomic number, the integrated diffusion coefficients increase. Further, the marker plane was always found at the Si/MSi_2 (M stands for refractory metal Ti, Zr, or Hf) interface. This indicates that the disilicide phase is grown by the diffusion of Si only, and the diffusion rate of the refractory metal is negligible. Therefore the ratio of diffusivities (D^*_{Si}/D^*_M) of components is estimated as infinite, by utilizing Eq. (3.153). It should be noted that the ratio is not actually infinite, but one component has a much higher diffusion rate compared to the other. In such a case, the diffusion coefficient of the slower species cannot be measured by the diffusion couple technique and Eq. (3.153) gives a value of infinity. As already mentioned, in a line compound we actually measure the ratio of the tracer diffusion coefficients (neglecting the vacancy-wind effect) since the partial molar volumes are not known. At the temperature range of interest, as shown in Fig. 3.62A, $TiSi_2$ has a C54 structure in which Ti is surrounded by 4 Ti and 10 Si. Similarly, Si is surrounded by 5 Ti and 9 Si. As shown in Fig. 3.62B, $ZrSi_2$ and $HfSi_2$ have C49 structure. In this system, the refractory metal component (M = Zr and Hf) is surrounded by 10 Si (4 Si-I and 6 Si-II) and 6

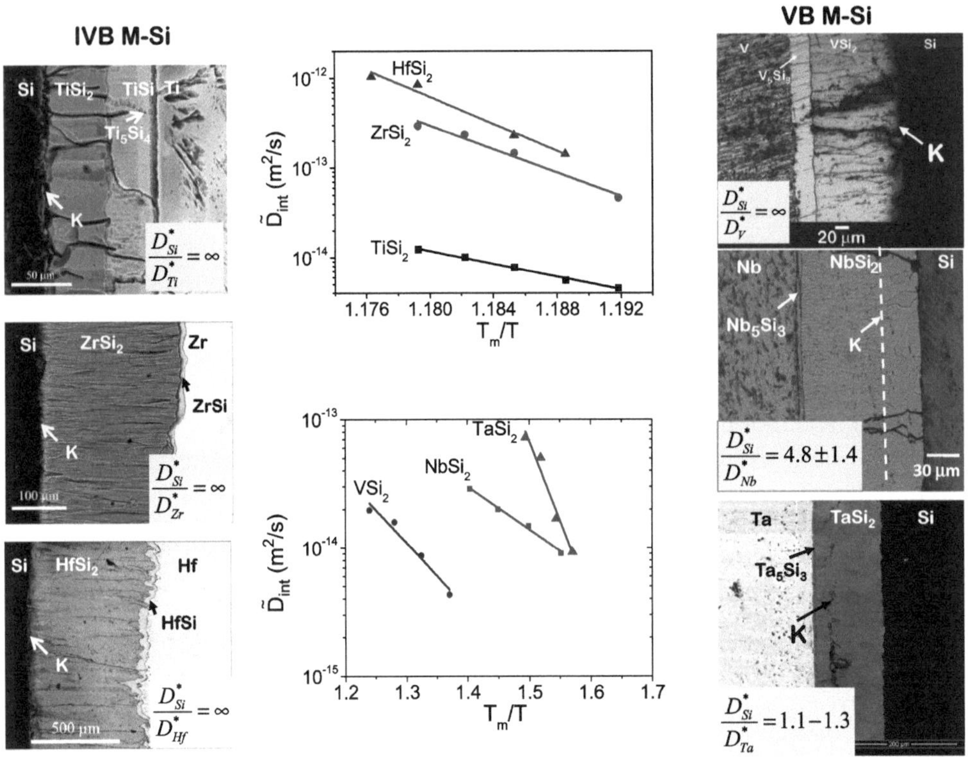

Figure 3.61 Interdiffusion zone in different refractory metal–silicon systems [36,72–76].

M. Si has two different sublattices. Si-I is surrounded by 12 Si (8 Si-I and 4 Si-II), and Si-II is surrounded by 10 Si (6 Si-I and 4 Si-II). There is one fact common in all the disilicides, $TiSi_2$, $ZrSi_2$ and $HfSi_2$, that is, both refractory metal component and Si are surrounded by each other. Therefore, if vacancies are present on all the sublattices, both M and Si can diffuse by the sublattice diffusion mechanism [1], by making the next nearest neighbor jump, although this might happen with different rates depending on vacancy migration and formation energies. However, following our estimated diffusion coefficients, we know that Si has a much higher diffusion rate compared to the metal component in all the three disilicide phases. Therefore, it indicates that only Si vacancies are present with higher concentration, and vacancies on the metal component sublattice must be very low. Further, we have the data on integrated diffusion coefficients, which increase with the atomic number of the metal component from the lowest value for $TiSi_2$ to the highest value for $HfSi_2$. Since the interdiffusion process is controlled by the diffusion of Si, there must be an increase in vacancy concentration on the Si sublattice with the increase in an atomic number of the metal component, which depends on the bond energies between different atoms.

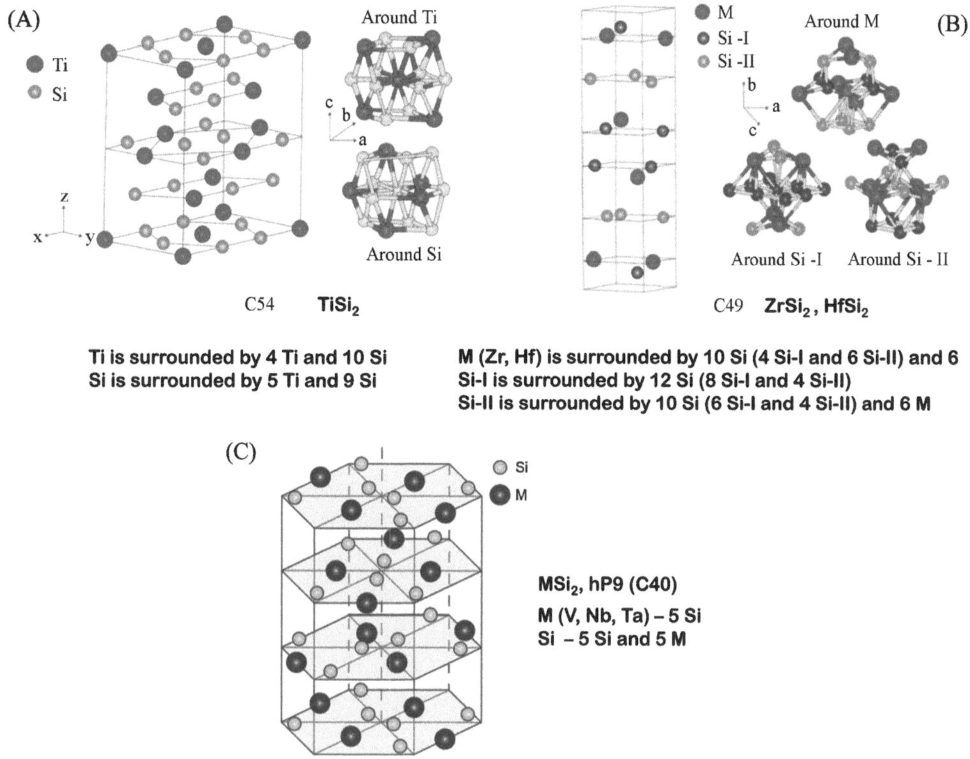

Figure 3.62 Crystal structures of different disilicide phases.

Now we consider the diffusion data estimated in the group VB (in the periodic table) refractory metal (V, Nb, and Ta)–silicon systems, as shown in Fig. 3.61. It can be seen again that the integrated diffusion coefficients increase with the increase in atomic number when plotted with respect to the inverse of homologous temperature. However, there is a difference in the location of Kirkendall marker plane, and therefore the ratio of diffusivities is estimated. Location of the marker plane indicates that the VSi_2 phase grows by the diffusion of Si only ($D^*_{Si}/D^*_V = \infty$). With the increase in atomic number of the refractory metal component, Kirkendall marker plane is found inside the $NbSi_2$ phase, indicating also the diffusion of the metal component Nb ($D^*_{Si}/D^*_{Nb} = 4.8 \pm 1.4$). With further increase in the atomic number of the refractory component, Kirkendall marker plane is found even further inside the $TaSi_2$ phase, indicating even higher relative diffusivity of the refractory component ($D^*_{Si}/D^*_V = 1.1 - 1.3$). As shown in Fig. 3.62C, all three phases have the same crystal structure hP9. In this structure the metal component is surrounded by 5 Si, whereas Si is surrounded by 5 Si and 5 M (V, Nb, Ta) atoms. Therefore, Si can diffuse easily through its own sublattice under the condition that vacancies are present on this sublattice. On the other hand, only

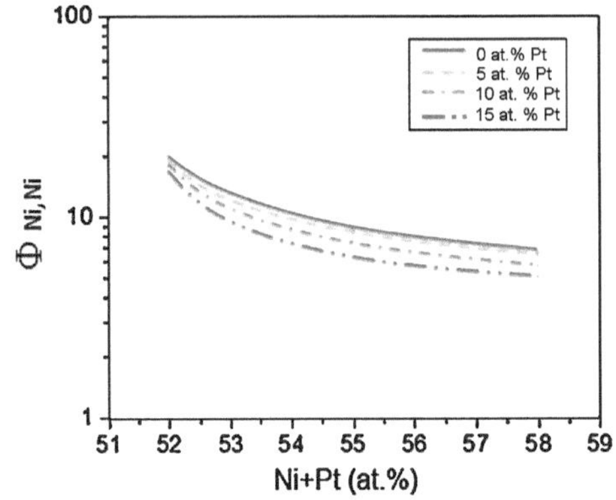

Figure 3.63 Thermodynamic parameters for Ni in β-Ni(Pt)Al phase [60].

the antisite metal components are allowed to diffuse, since they can only occupy the Si sublattice. The relative increase in diffusion rate of the metal component indicates the increase in metal antisite defects with the increasing atomic number. Moreover, since the integrated diffusion coefficient increases with the increase in atomic number, there must be an overall increase in defect concentrations.

Let us now continue the discussion based on the diffusion data measured in the β-Ni(Pt)Al phase following the pseudo-binary approach, as shown in Fig. 3.50D [60]. It can be seen that the interdiffusion coefficients increase at a particular (Ni+Pt):Al composition with the increase in Pt content, indicating the increase in the concentration of defects because of Pt addition. Since Ni has a higher diffusion rate compared to Al on the Ni-rich side, the role of thermodynamic driving force can be understood from the thermodynamic factor estimated for Ni, i.e., $\Phi_{Ni,Ni} = \frac{d \ln a_{Ni}}{d \ln N_{Ni}}$, as shown in Fig. 3.63. It can be seen that thermodynamic factors decrease with the increase in Pt content. Therefore, defects must have a dominating role in the diffusion of components, like in the binary β-NiAl phase, as shown in Fig. 3.59. A theoretical calculation indicates that the defect formation and migration energies decrease due to added Pt, explaining the increase in interdiffusion coefficients. Additionally, one can draw an additional conclusion from the estimated interdiffusion data, as shown in Fig. 3.50D. Near the stoichiometric composition, interdiffusion coefficients increase with a higher rate with the addition of Pt compared to the (Ni+Pt)-rich side of the phase. We already know from the defect and diffusion data in the binary system that the diffusion rate on the Ni-rich side is higher compared to the stoichiometric composition because of presence of antisite defects. Therefore, we can conclude that a certain percentage of Pt atoms absorb Ni antisites. The higher the concentration of antisites, the higher the replacement of Ni antisites by Pt. Therefore, the addition of Pt is less effective on the (Ni+Pt)-rich or Al-lean side for the increase in diffusion coefficients compared to the stoichiometric composition.

3.20 PHYSICAL SIGNIFICANCE OF THE ESTIMATED DIFFUSION COEFFICIENTS

Till now different diffusion coefficients and their estimation procedures were described in various sections. However, readers should be aware of the physical significance and use of these parameters for the discussion of a particular topic.

Tracer diffusion coefficient is the most basic diffusion coefficient, which helps understand the atomic mechanism when considering the crystal structure and defects present in a material, in the absence of any driving force for diffusion (as explained in Chapters 2 and 10 of this volume). Please note that the self-diffusion coefficient (D_S^i) and the tracer diffusion coefficient (D_i^*) are related by $D_i^* = fD_S^i$, as explained in Chapter 1 [1]. The experimental method for the estimation of the tracer diffusion coefficients is straightforward under the condition that the laboratory is equipped with safety measures for handling the radioisotopes. Activation enthalpy is measured from the temperature-dependent experimental data utilizing Arrhenius equation

$$D_i^* = D_{o,i}^* \exp\left(-\frac{Q_i^*}{RT}\right) \tag{3.190a}$$

where $D_{o,i}^*$ is the pre-exponential factor and Q_i^* is the activation enthalpy of the tracer diffusion coefficient. Then, the theoretically calculated value based on simulation is compared with the experimentally determined value for developing a clear understanding of the atomic mechanism of diffusion.

However, knowledge of the tracer diffusion coefficient is not sufficient for understanding the diffusion process in a practical system with inhomogeneous composition in which the diffusion rates of components are also driven by the chemical potential gradient. Therefore, knowledge of intrinsic diffusion coefficients is important for understanding the role of thermodynamics in addition to the defects present in a material. The intrinsic and tracer diffusion coefficients are related by Eq. (3.156). Similar to the tracer diffusion coefficient, one could be interested in determining the activation enthalpy for intrinsic diffusion coefficients from the temperature-dependent experiments utilizing Arrhenius equation:

$$D_i = D_{o,i} \exp\left(-\frac{Q_i}{RT}\right) \tag{3.190b}$$

where $D_{o,i}$ (m^2/s) is the pre-exponential factor and Q_i (kJ/mol) is the activation enthalpy for the intrinsic diffusion coefficients. Unlike the activation enthalpy for tracer diffusion coefficients (from radioisotopes technique), the activation enthalpy for the intrinsic diffusion coefficients (from diffusion couple technique) is not easy to determine, especially in solid solutions or a phase with a wide homogeneity range. As already explained, such data can be only measured at the composition of Kirkendall marker plane in a diffusion couple. The main problem comes from the fact that with the change in experimental temperature, the rates of the intrinsic diffusion coefficients will change differently based on their own pre-exponential factors and activation enthalpies. Therefore, the location of Kirkendall marker plane will be found at different composition even if the end members of the diffusion couple are the same. These diffusion coefficients are material constants and their values are different at different compositions in a phase depending on the defects present and the thermodynamic driving forces at the compositions of interest. Therefore, the intrinsic diffusion coefficients measured at different compositions at different temperatures cannot be used to determine the activation enthalpy, unless the marker plane is found at the same composition in the temperature range of interest.

Otherwise, this can be done only if a series of experiments are conducted first at each temperature with different compositions of the diffusion couple end-members such that Kirkendall markers are found at different compositions to measure the variation of intrinsic diffusion coefficients with composition. Once the composition-dependent intrinsic diffusion coefficients are estimated at different temperatures, the activation enthalpies of the intrinsic diffusion coefficients can be determined by extracting and plotting the data with respect to Arrhenius equation at different compositions independently. Similarly, the multifoil experiments can be conducted at different temperatures with the same purpose. The composition dependent intrinsic diffusion coefficients were determined in β-Ni(Pt)Al [5] (from several diffusion couples with different end-member compositions) and Ni–Pt, Co–Pt [30] (multifoil technique) but only at a single temperature. The same procedure should be followed at different temperatures for estimation of the activation enthalpy of intrinsic diffusion coefficients, which is, of course, a tiresome process.

This difficulty is automatically avoided in the case of line compounds or the phases with a very large homogeneity range. Therefore, the diffusion coefficients are practically measured at one average composition from a single diffusion couple. Therefore, one should be able to estimate the activation enthalpies of diffusion of components from the data estimated at different temperatures. However, as already discussed in Section 3.13.1, we do not know the partial molar volumes of the components in a line compound, and therefore we actually measure the ratios of the tracer diffusion coefficients by Eq. (3.153) or the absolute values of the tracer diffusion coefficients by Eq. (3.166) [38,77]. This is rather an indirect method of estimating the tracer diffusion coefficients, but not the intrinsic diffusion coefficients. Nonetheless, we can use the data to establish the atomic mechanism of diffusion.

Because of the difficulties explained above, (ratio of) the intrinsic diffusion coefficients are mostly determined at one particular temperature. Nevertheless, these are very useful to understand the atomic mechanism of diffusion, as already explained in Sections 3.18 and 3.19 based on the experimental results in different systems. One should be aware of the error in the calculation of the data by the diffusion couple technique. The calculation error is very large if the ratio of diffusivities falls outside the range 0.1–1. Beyond this range, a large error in estimation of the diffusion rate of the slower component is introduced. If the difference between the diffusion rates of components is a few orders of magnitude, then, eventually, we cannot measure the diffusion rate of the slower component by this technique. This is the reason that the ratio of diffusivities is estimated as infinite in the disilicide phases of a few refractory metal (Ti, Zr, Hf)–silicon systems (Fig. 3.61). In such a case, the tracer diffusion coefficient of all the components can be measured by the radioisotope method.

Unlike the intrinsic diffusion coefficients, the interdiffusion coefficients can be estimated at all compositions in the interdiffusion zone of a diffusion couple in a binary system. Therefore, one can easily estimate the activation enthalpies of the interdiffusion coefficients by conducting the experiments at different temperatures. This is the reason that these data are easily available in many systems. However, we need to understand the usefulness and limitations of the interdiffusion coefficients and the activation enthalpies estimated utilizing such data at different temperatures. We already know that the interdiffusion coefficients are averages of intrinsic diffusion coefficients expressed by the relation given in Eq. (3.124) or (3.125) (for constant molar volume). Even if we consider the simplified relation (Eq. (3.125)), we can write it with respect to Arrhenius equation in a binary A–B system as (Eq. (3.151))

$$\begin{aligned}
\tilde{D} &= N_B D_A + N_A D_B = \left(N_A D_B^* + N_B D_A^*\right)\left(\frac{d\ln a_B}{d\ln N_B}\right),\\
\tilde{D}_o \exp\left(-\frac{Q}{RT}\right) &= D_{o,A}\exp\left(-\frac{Q_A}{RT}\right) + D_{o,B}\exp\left(-\frac{Q_B}{RT}\right)\\
&= \left[D_{o,A}^*\exp\left(-\frac{Q_A^*}{RT}\right) + D_{o,B}^*\exp\left(-\frac{Q_B^*}{RT}\right)\right]\left(\frac{d\ln a_B}{d\ln N_B}\right),
\end{aligned} \tag{3.191}$$

where $\tilde{D}_o$ is the pre-exponential factor and Q is the activation enthalpy for interdiffusion coefficient. From Eq. (3.191) it must be clear that these parameters determined from the interdiffusion coefficients are related to similar parameters for intrinsic or tracer diffusion coefficients in a very complicated manner. Therefore, the values of $\tilde{D}_o$ and Q determined from the interdiffusion coefficients are vague and not useful to understand the atomic mechanism of diffusion without knowledge of the intrinsic or tracer diffusion coefficients. The interdiffusion coefficients can still be used to understand the atomic mechanism of diffusion when one component is known to have a higher diffusion rate compared to the other, as explained in Sections 3.18 and 3.19.

It is also true that the interdiffusion coefficients are material constants since the intrinsic and tracer diffusion coefficients are material constants. Therefore, the pre-exponential factor and the activation enthalpy determined from the interdiffusion coefficients are useful to extract the data at a temperature of interest under the condition that the material does not go through a phase or magnetic transition. Then, these can be used to estimate the composition profile in a particular diffusion couple. This discussion is the same for the integrated diffusion coefficient determined for the line compounds. Sometimes, when we estimate an average interdiffusion coefficient because of certain difficulties (following Eq. (3.101)), we actually introduce much higher vagueness in the estimated data. In such a situation, one should be very careful to draw any conclusion on the growth mechanism of phases or atomic mechanism of diffusion.

Since the concept of the pseudo-binary approach is introduced in this chapter, the discussion of the same topic in multicomponent system is continued here. As already mentioned, the next chapter deals with diffusion in multicomponent systems. With the increase in the number of components just by one, from a binary A–B to a ternary A–B–C system, we introduce much higher vagueness with respect to the physical significance of the interdiffusion coefficient. For example, the two main ($\tilde{D}^C_{AA}$ and $\tilde{D}^C_{BB}$) and the two cross- ($\tilde{D}^C_{AB}$ and $\tilde{D}^C_{BA}$) interdiffusion coefficients estimated at one composition of cross-section following the conventional method (as explained in the beginning of Section 3.16) are related by the six intrinsic diffusion coefficients (D^k_{ij}), as given below [1]:

$$\tilde{D}^C_{AA} = D^C_{AA} - N_A\left(D^C_{AA} + D^C_{BA} + D^C_{CA}\right), \tag{3.192a}$$

$$\tilde{D}^C_{AB} = D^C_{AB} - N_A\left(D^C_{AB} + D^C_{BB} + D^C_{CB}\right), \tag{3.192b}$$

$$\tilde{D}^C_{BA} = D^C_{BA} - N_B\left(D^C_{AA} + D^C_{BA} + D^C_{CA}\right), \tag{3.192c}$$

$$\tilde{D}^C_{BB} = D^C_{BB} - N_B\left(D^C_{AB} + D^C_{BB} + D^C_{CB}\right). \tag{3.192d}$$

Therefore, these are related in an even more complicated way in a ternary system compared to the binary system, and it is difficult to draw any reasonable conclusion on the atomic mechanism of diffusion. Therefore, when an average value of the interdiffusion coefficient, which is defined as the average effective interdiffusion coefficients [1], as explained in Chapter 4 in Volume 1, over a composition range is estimated, an even higher level of vagueness is introduced. In fact, these average diffusion coefficients are not material constants and depend on the composition range considered for the estimation. An added problem comes from the fact that the intrinsic diffusion coefficients are almost impossible to estimate in a ternary system. In a system with a higher number of components, we cannot estimate any diffusion coefficients in a diffusion couple following the conventional method. Even if this were possible, they would not carry much of physical significance.

To summarize, the interdiffusion data are important for understanding the evolution of the diffusion profile in inhomogeneous materials. However, such data are not useful to understand the atomic mechanism of diffusion, unless we know the values (or the ratio) of the intrinsic or tracer diffusion coefficients. The interdiffusion and intrinsic diffusion coefficients are related by a simple equation only in a binary system. With the increase in the number of components, relations become very complicated. The intrinsic diffusion coefficients cannot be measured by a diffusion couple technique in a system with more than two components following the conventional method. On the other hand, one can straightforwardly measure the tracer diffusion coefficients following the radioisotope method (as explained in the previous chapter) in a material with any number of components, which can help understand the atomic mechanism of diffusion. However, without the intrinsic diffusion coefficients, the phenomenological diffusion process in an inhomogeneous material cannot be understood. Since we cannot estimate these parameters, the diffusion processes in multicomponent systems in various applications are always discussed qualitatively. In that sense, the interdiffusion and the intrinsic diffusion coefficients measured by the pseudo-binary approach in a ternary or multicomponent system are easier to correlate. Following this method in a ternary β-Ni(Pt)Al phase, we were able to correlate the data estimated in the binary β-NiAl phase with the help of defects and the thermodynamic driving forces. This led to a fair understanding of the diffusion process and evolution of the microstructure in an interdiffusion zone between multicomponent single crystal superalloy and β-Ni(Pt)Al bond coat [60].

It must be clear now that we are not yet equipped with suitable methods for quantitative diffusion analysis in complicated inhomogeneous material systems used in various applications. However, instead of being obsessed with the estimation of the diffusion coefficients, one can develop a fair understanding on the evolution of microstructure in an interdiffusion zone by simply comparing the composition profiles and thicknesses [60,67] or the estimated interdiffusion fluxes of different components [78]. It should be noted here that the estimation of the interdiffusion flux is fairly straightforward using the relations expressed in Eqs. (3.57b), (3.72b) or (3.91b). These are established for binary systems; however, the same equations can be used in multicomponent systems, as it was recently shown by Sekerka et al. [79]. In the same manuscript, the authors also have argued on a very important topic (which was being questioned sometimes) that the interdiffusion fluxes can be estimated using the equations introduced in this chapter even when there is a discontinuity (or jump) in a composition profile in a multiphase interdiffusion zone.

REFERENCES

[1] A. Paul, T. Laurila, V. Vuorinen, S.V. Divinski, Thermodynamics, Diffusion and the Kirkendall Effect in Solids, Springer, Heidelberg, Germany, 2014.

[2] A. Fick, Annalen der Physik 170 (1855) 59.
[3] J. Crank, The Mathematics of Diffusion, Oxford University Press, New York, USA, 1956.
[4] P. Shewmon, Diffusion in Solids, 2nd edition, Wiley, Hoboken, USA, 1989.
[5] A. Paul, A.A. Kodentsov, F.J.J. van Loo, Acta Materialia 52 (2004) 4041.
[6] R. Ravi, A. Paul, Journal of Materials Science: Materials in Electronics 23 (2012) 2152.
[7] C. Matano, Japanese Journal of Physics 8 (1933) 109.
[8] V.D. Divya, U. Ramamurty, A. Paul, Philosophical Magazine 93 (2013) 2190.
[9] R.W. Ballufi, Acta Metallurgica 8 (1960) 871.
[10] F. Sauer, V. Freise, Zeitschrift für Elektrochemie 66 (1962) 353.
[11] C. Wagner, Acta Metallurgica 17 (1969) 99.
[12] F.J.A. den Broeder, Scripta Metallurgica 3 (1969) 321.
[13] S. Santra, A. Paul, Metallurgical and Materials Transactions 46 (2015) 3887.
[14] B. Pieraggi, Oxidation of Metals 27 (1987) 177.
[15] F.J.J. van Loo, Progress in Solid State Chemistry 20 (1990) 47.
[16] F.J.J. van Loo, G.D. Rieck, Acta Metallurgica 21 (1973) 61.
[17] G.M. Becht, F.J.J. van Loo, R. Metselaar, Reactivity of Solids 6 (1988) 45.
[18] A.D. Smigelkas, E.O. Kirkendall, Transaction of AMIE 171 (1947) 130.
[19] L.S. Darken, Transaction of Metallurgical Society AIME 175 (1948) 184.
[20] T. Heumann, Zeitschrift für Physikalische Chemie 201 (1952) 168.
[21] F.J.J. van Loo, Acta Metallurgica 18 (1970) 1107.
[22] A. Paul, The Kirkendall Effect in Solid State Diffusion, PhD Thesis, Technische Universiteit Eindhoven, Eindhoven, The Netherlands, 2004.
[23] A. Paul, A.A. Kodentsov, F.J.J. van Loo, Intermetallics 14 (2006) 1428.
[24] S. Santra, H. Dong, T. Laurila, A. Paul, Proceedings of the Royal Society A 470 (2014) 20130464.
[25] A. Paul, A.A. Kodentsov, F.J.J. van Loo, Zeitschrift für Metallkunde 95 (2004) 913–920.
[26] M.J.H. Dal van, D.G.G.M. Huibers, A.A. Kodentsov, F.J.J. van Loo, Intermetallics 9 (2001) 409.
[27] C. Ghosh, A. Paul, Acta Materialia 55 (2007) 1927.
[28] C. Ghosh, A. Paul, Acta Materialia 57 (2009) 493.
[29] S. Roy, A. Paul, Philosophical Magazine 92 (2012) 4215.
[30] V.D. Divya, U. Ramamurty, A. Paul, Journal of Materials Research 26 (2011) 238431.
[31] J. Levasseur, J. Philibert, Journal of Physica Status Solidi B 21 (1967) K1.
[32] J.F. Cornet, Journal of Physics and Chemistry of Solids 35 (1974) 1247.
[33] F.J.J. van Loo, G.F. Bastin, G.D. Rieck, Science of Sintering 11 (1979) 9.
[34] J.R. Manning, Physical Review 124 (1961) 470.
[35] J.R. Manning, Acta Metallurgica 15 (1967) 817.
[36] S. Prasad, A. Paul, Acta Materialia 59 (2011) 1577.
[37] M.E. Schlesinger, H. Okamoto, A.B. Gokhale, R. Abbaschian, Journal of Phase Equilibria and Diffusion 14 (1993) 502.
[38] M.J.H. Dal van, A.A. Kodentsov, F.J.J. van Loo, Intermetallics 9 (2001) 451.
[39] V.A. Baheti, R. Ravi, A. Paul, Journal of Materials Science: Materials in Electronics 24 (2013) 2833.
[40] L. Hall, Journal of Chemical Physics 21 (1953) 87.
[41] N. Sarafinos, Journal of Materials Science 21 (1986) 2283.
[42] T. Ahmed, I.V. Belova, A.V. Evteev, E.V. Levchenko, G.E. Murch, Journal of Phase Equilibria and Diffusion 36 (2015) 366.
[43] A. Paul, Philosophical Magazine 93 (2013) 2297.
[44] S. Santra, A. Paul, Scripta Materialia 103 (2015) 18.
[45] P. Kiruthika, A. Paul, Philosophical Magazine Letters 95 (2015) 138.
[46] J.S. Kirkaldy, D.J. Young, Diffusion in the Condensed State, The Institute of Metals, London, 1987.
[47] S. Krishtal, A.P. Mokrov, A.V. Akimov, P.N. Zaharov, Fiz Metal Metalloved 35 (1973) 1234.

[48] M.K. Stalker, J.E. Morral, A.D. Romig, Metallurgical Transactions A 23 (1992) 3245.
[49] Y. Minamino, Y. Koizumi, T. Tsuji, Y. Yamada, T. Takahashi, Materials Transactions 44 (2003) 63.
[50] T. Takahashi, K. Hisayuki, T. Yamane, Y. Minanino, T. Hino, Materials Transactions 44 (2003) 2252.
[51] M.A. Dayananda, Y.H. Sohn, Scripta Materialia 35 (1996) 683.
[52] M.A. Dayananda, Y.H. Sohn, Metallurgical and Materials Transactions A 30 (1999) 535.
[53] F. Guillemot, J. Debuigne, D. Ansel, Metallurgical and Materials Transactions A 31 (2000) 3198.
[54] H. Fukaya, Md. Moniruzzaman, Y. Murata, M. Morinaga, T. Koyama, W. Hashimoto, K. Tanaka, H. Inui, Defect and Diffusion Forum 297–301 (2010) 384.
[55] B. Grushko, D. Kapush, Journal of Alloys and Compounds 594 (2014) 127.
[56] F.J.J. van Loo, G.F. Bastin, J.W.G.A. Vrolijk, Metallurgical Transactions A 18 (1987) 801.
[57] K.J. Rönka, A.A. Kodentsov, P.J.J. van Loon, J.K. Kivilahti, F.J.J. van Loo, Metallurgical and Materials Transactions A 27 (1996) 2229.
[58] M.S. Thompson, J.E. Morral, Acta Metallurgica 34 (1986) 339.
[59] M.S. Thompson, J.E. Morral, Acta Metallurgica 34 (1986) 2201.
[60] P. Kiruthika, S.K. Makineni, C. Srivastava, K. Chattopadhyay, A. Paul, Acta Materialia 105 (2016) 438.
[61] J. Cermak, V. Rothova, Acta Materialia 51 (2003) 4411.
[62] K. Cheng, W. Chen, D. Liu, L. Zhang, Y. Du, Scripta Materialia 76 (2014) 5.
[63] S.S.K. Balam, A. Paul, Journal of Materials Science 46 (2011) 889.
[64] A. Paul, Comments on "Sluggish diffusion in Co–Cr–Fe–Mn–Ni high-entropy alloys", accepted for publication, Scripta Materialia (2017).
[65] K.Y. Tsai, M.H. Tsai, J.W. Yeh, Acta Materialia 61 (2003) 4887.
[66] H. Okamoto, Journal of Phase Equilibria 14 (1993) 257.
[67] S. Santra, S.K. Makineni, S. Suwas, K. Chattopadhyay, A. Paul, Materials & Design 110 (2016) 404.
[68] S. Prasad, A. Paul, Intermetallics 19 (2011) 1191–1200.
[69] P.C. Tortorici, M.A. Dayananda, Metallurgical and Materials Transactions A 30A (1999) 545.
[70] M. Salamon, A. Strohm, T. Voss, P. Laitinen, I. Riihimäki, S. Divinski, W. Frank, J. Räisänen, H. Mehrer, Philosophical Magazine 84 (2004) 737.
[71] X.Y. Zhang, W. Sprengel, K. Blaurock, A.A. Rempel, K.J. Reichle, K. Reimann, H. Inui, H.E. Schaefer, Physical Review B 66 (2002) 144105.
[72] S. Roy, S. Divinski, A. Paul, Philosophical Magazine 94 (2014) 683.
[73] S. Roy, A. Paul, Materials Chemistry and Physics 143 (2014) 1309.
[74] S. Prasad, A. Paul, Journal of Phase Equilibria and Diffusion 32 (2011) 212.
[75] S. Roy, A. Paul, Philosophical Magazine 92 (2012) 4215.
[76] S. Roy, S. Prasad, S. Divinski, A. Paul, Philosophical Magazine 94 (2014) 1508.
[77] A.K. Kumar, T. Laurila, V. Vuorinen, A. Paul, Scripta Materialia 60 (2009) 377.
[78] M.A. Dayananda, G.E. Murch (Eds.), Diffusion in Solids; Recent Developments, Metallurgical Society of AMIE, Warrendale, PA, 1985.
[79] R.F. Sekerka, G.B. McFadden, W.J. Boettinger, Journal of Phase Equilibria and Diffusion 37 (2016) 640–650.

CHAPTER 4

Diffusion in Multicomponent Alloys

L. Zhou*, M.A. Dayananda†, Y.H. Sohn*
*University of Central Florida, Orlando, FL, USA
†Purdue University, West Lafayette, IN, USA

4.1 INTRINSIC DIFFUSION IN MULTICOMPONENT ALLOYS

Starting from the basic Onsager's formalism and assuming that only chemical potential gradients are present without external forces, the phenomenological equations for diffusion in a multicomponent alloy containing N components can be expressed as [1]:

$$
\begin{aligned}
J_i &= -\sum_{j=1}^{n} L_{ij}\left(\frac{\partial \mu_j}{\partial x}\right)_t \qquad (i=1,2,\ldots,n)\\
J_1 &= -L_{11}\frac{\partial \mu_1}{\partial x} - L_{12}\frac{\partial \mu_2}{\partial x} - \cdots - L_{1N}\frac{\partial \mu_N}{\partial x}\\
J_2 &= -L_{21}\frac{\partial \mu_1}{\partial x} - L_{22}\frac{\partial \mu_2}{\partial x} - \cdots - L_{2N}\frac{\partial \mu_N}{\partial x}\\
&\vdots \\
J_N &= -L_{N1}\frac{\partial \mu_1}{\partial x} - L_{N2}\frac{\partial \mu_2}{\partial x} - \cdots - L_{NN}\frac{\partial \mu_N}{\partial x}
\end{aligned} \tag{4.1}
$$

where J_i is the intrinsic diffusion flux based on a lattice-fixed frame of reference, L_{ij} is the Onsager's phenomenological coefficients and μ_j is the chemical potential of component j, and the partial derivative with respect to distance x along the diffusion direction is taken at a fixed time t.

For a ternary alloy, the phenomenological equations can be written as:

$$
\begin{aligned}
J_1 &= -L_{11}\frac{\partial \mu_1}{\partial x} - L_{12}\frac{\partial \mu_2}{\partial x} - L_{13}\frac{\partial \mu_3}{\partial x}\\
J_2 &= -L_{21}\frac{\partial \mu_1}{\partial x} - L_{22}\frac{\partial \mu_2}{\partial x} - L_{23}\frac{\partial \mu_3}{\partial x}\\
J_3 &= -L_{31}\frac{\partial \mu_1}{\partial x} - L_{32}\frac{\partial \mu_2}{\partial x} - L_{33}\frac{\partial \mu_3}{\partial x}
\end{aligned} \tag{4.2}
$$

Typically, the vacancies are assumed to maintain their local equilibrium concentrations so that the chemical potential of vacancies remains constant, and thus $\partial \mu_V/\partial x = 0$. Therefore the vacancy flux is not considered in Eqs. (4.1) and (4.2).

Handbook of Solid State Diffusion, Volume 1
DOI: 10.1016/B978-0-12-804287-8.00004-X

Under the assumption of local equilibrium and applying the Gibbs–Duhem relation,

$$C_1\frac{\partial \mu_1}{\partial x}+C_2\frac{\partial \mu_2}{\partial x}+C_3\frac{\partial \mu_3}{\partial x}=0, \tag{4.3}$$

where C_i is the mole fraction of component i, Eq. (4.2) can be rewritten for a ternary alloy as:

$$\begin{aligned}
J_1&=-\left(L_{11}-L_{13}\frac{C_1}{C_3}\right)\frac{\partial \mu_1}{\partial x}-\left(L_{12}-L_{13}\frac{C_2}{C_3}\right)\frac{\partial \mu_2}{\partial x}\\
J_2&=-\left(L_{21}-L_{23}\frac{C_1}{C_3}\right)\frac{\partial \mu_1}{\partial x}-\left(L_{22}-L_{23}\frac{C_2}{C_3}\right)\frac{\partial \mu_2}{\partial x}\\
J_3&=-\left(L_{31}-L_{33}\frac{C_1}{C_3}\right)\frac{\partial \mu_1}{\partial x}-\left(L_{32}-L_{33}\frac{C_2}{C_3}\right)\frac{\partial \mu_2}{\partial x}
\end{aligned} \tag{4.4}$$

In addition, only two of the three chemical potentials can vary independently in a ternary system, i.e.,

$$\begin{aligned}
\frac{\partial \mu_1}{\partial x}&=\frac{\partial \mu_1}{\partial C_1}\frac{\partial C_1}{\partial x}+\frac{\partial \mu_1}{\partial C_2}\frac{\partial C_2}{\partial x}\\
\frac{\partial \mu_2}{\partial x}&=\frac{\partial \mu_2}{\partial C_1}\frac{\partial C_1}{\partial x}+\frac{\partial \mu_2}{\partial C_2}\frac{\partial C_2}{\partial x}
\end{aligned} \tag{4.5}$$

Substituting Eq. (4.5) into Eq. (4.4), the following equations can be derived:

$$\begin{aligned}
J_1=&-\left[\left(L_{11}-L_{13}\frac{C_1}{C_3}\right)\frac{\partial \mu_1}{\partial C_1}+\left(L_{12}-L_{13}\frac{C_2}{C_3}\right)\frac{\partial \mu_2}{\partial C_1}\right]\frac{\partial C_1}{\partial x}\\
&-\left[\left(L_{11}-L_{13}\frac{C_1}{C_3}\right)\frac{\partial \mu_1}{\partial C_2}+\left(L_{12}-L_{13}\frac{C_2}{C_3}\right)\frac{\partial \mu_2}{\partial C_2}\right]\frac{\partial C_2}{\partial x}\\
J_2=&-\left[\left(L_{21}-L_{23}\frac{C_1}{C_3}\right)\frac{\partial \mu_1}{\partial C_1}+\left(L_{22}-L_{23}\frac{C_2}{C_3}\right)\frac{\partial \mu_2}{\partial C_1}\right]\frac{\partial C_1}{\partial x}\\
&-\left[\left(L_{21}-L_{23}\frac{C_1}{C_3}\right)\frac{\partial \mu_1}{\partial C_2}+\left(L_{22}-L_{23}\frac{C_2}{C_3}\right)\frac{\partial \mu_2}{\partial C_2}\right]\frac{\partial C_2}{\partial x}\\
J_3=&-\left[\left(L_{31}-L_{33}\frac{C_1}{C_3}\right)\frac{\partial \mu_1}{\partial C_1}+\left(L_{32}-L_{33}\frac{C_2}{C_3}\right)\frac{\partial \mu_2}{\partial C_1}\right]\frac{\partial C_1}{\partial x}\\
&-\left[\left(L_{31}-L_{33}\frac{C_1}{C_3}\right)\frac{\partial \mu_1}{\partial C_2}+\left(L_{32}-L_{33}\frac{C_2}{C_3}\right)\frac{\partial \mu_2}{\partial C_2}\right]\frac{\partial C_2}{\partial x}
\end{aligned} \tag{4.6}$$

The intrinsic diffusion coefficients, D_{ij}, in a ternary system can now be defined from Eq. (4.6):

$$\begin{aligned}
D_{11}^3 &= \left(L_{11} - L_{13}\frac{C_1}{C_3}\right)\frac{\partial \mu_1}{\partial C_1} + \left(L_{12} - L_{13}\frac{C_2}{C_3}\right)\frac{\partial \mu_2}{\partial C_1}\\
D_{12}^3 &= \left(L_{11} - L_{13}\frac{C_1}{C_3}\right)\frac{\partial \mu_1}{\partial C_2} + \left(L_{12} - L_{13}\frac{C_2}{C_3}\right)\frac{\partial \mu_2}{\partial C_2}\\
D_{21}^3 &= \left(L_{21} - L_{23}\frac{C_1}{C_3}\right)\frac{\partial \mu_1}{\partial C_1} + \left(L_{22} - L_{23}\frac{C_2}{C_3}\right)\frac{\partial \mu_2}{\partial C_1}\\
D_{22}^3 &= \left(L_{21} - L_{23}\frac{C_1}{C_3}\right)\frac{\partial \mu_1}{\partial C_2} + \left(L_{22} - L_{23}\frac{C_2}{C_3}\right)\frac{\partial \mu_2}{\partial C_2}\\
D_{31}^3 &= \left(L_{31} - L_{33}\frac{C_1}{C_3}\right)\frac{\partial \mu_1}{\partial C_1} + \left(L_{32} - L_{33}\frac{C_2}{C_3}\right)\frac{\partial \mu_2}{\partial C_1}\\
D_{32}^3 &= \left(L_{31} - L_{33}\frac{C_1}{C_3}\right)\frac{\partial \mu_1}{\partial C_2} + \left(L_{32} - L_{33}\frac{C_2}{C_3}\right)\frac{\partial \mu_2}{\partial C_2}
\end{aligned} \tag{4.7}$$

and a simplified intrinsic diffusion flux, J_i, can be expressed as:

$$\begin{aligned}
J_1 &= -D_{11}^3\frac{\partial C_1}{\partial x} - D_{12}^3\frac{\partial C_2}{\partial x}\\
J_2 &= -D_{21}^3\frac{\partial C_1}{\partial x} - D_{22}^3\frac{\partial C_2}{\partial x}\\
J_3 &= -D_{31}^3\frac{\partial C_1}{\partial x} - D_{32}^3\frac{\partial C_2}{\partial x}
\end{aligned} \tag{4.8}$$

Experimentally, the determination of the six intrinsic diffusion coefficients in a ternary system requires the use of inert particles as markers to indicate the position of the Kirkendall plane. The intrinsic flux integrated with respect to time at the marker plane is then identical to the cumulative intrinsic flux, A_i, that passes the marker plane and is expressed as [2]:

$$\begin{aligned}
A_i &= \int_0^t J_i^{x_m}\, dt = -\int_0^t \left[\left(D_{i1}^3\frac{\partial C_1}{\partial x} + D_{i2}^3\frac{\partial C_2}{\partial x}\right)_t\right]_{x_m} dt\\
&= -2t\left(D_{i1}^3\frac{\partial C_1}{\partial x} + D_{i2}^3\frac{\partial C_2}{\partial x}\right)_{x_m} \qquad (i = 1, 2, 3)
\end{aligned} \tag{4.9}$$

According to the Heumann's method [2], the value of A_i can be determined from the concentration profiles, which is similar to that in a binary system. Eq. (4.9) can be used for the determination of the six ternary intrinsic diffusion coefficients at the marker plane. To solve for the six intrinsic diffusion coefficients, a total of six equations are needed. This requires two sets of diffusion couples with markers to be experimentally prepared. More importantly, the composition at these two marker planes needs to be the same. Following this method, the ternary intrinsic diffusion coefficients at a few compositions have been estimated, for example, in Fe–Ni–Al alloys [3]. However, this requirement is extremely difficult to be fulfilled because the composition at the marker plane after diffusion cannot be properly predicted for ternary alloys. Therefore, the ternary intrinsic diffusion coefficients reported so far are limited to only a few studies.

4.2 ATOMIC MOBILITY AND VACANCY WIND EFFECT IN MULTICOMPONENT ALLOYS

For a multicomponent system, the intrinsic flux of component i, defined in terms of diffusion coefficients, can be written in terms of atomic mobility [4], β_i, by

$$J_i = -C_i\beta_i\frac{\partial \mu_i}{\partial x} \tag{4.10}$$

where C_i refers to the mole fraction, and $\frac{\partial \mu_i}{\partial x}$ is the chemical potential gradient.

Manning [5] has shown that the intrinsic flux terms can be written to include contributions from the vacancy wind effect if the diffusion occurs through vacancy mechanisms. The fundamental expression of the intrinsic flux of the component i, relative to the local lattice plane, is then given by [5,6]

$$J_i = -C_i\beta_i\frac{\partial \mu_i}{\partial x} + J_{iv} \tag{4.11}$$

where the J_{iv} term is the additional flux due to the vacancy wind effect, and is related to the vacancy flux, J_v, by [7]

$$J_{iv} = -\alpha C_i\beta_i J_v \tag{4.12}$$

where α is the vacancy wind parameter that depends on composition and crystal structure.

Furthermore, Eq. (4.12) can be expressed in terms of the net matter flow, $\sum_{j=1}^{n} J_j$, and then one can get the following equation by combining Eqs. (4.11) and (4.12):

$$J_i = -C_i\beta_i\frac{\partial \mu_i}{\partial x} + \alpha C_i\beta_i\sum_{j=1}^{n} J_j. \tag{4.13}$$

Dividing Eq. (4.13) by β_i and then summing over all species i in the system results in the following equation for α [6]:

$$\alpha = \frac{\sum_{i=1}^{n} J_i/\beta_i}{\sum_{i=1}^{n} J_i}. \tag{4.14}$$

By considering the cumulative intrinsic diffusion flux, A_i, and integrating Eq. (4.13), the following expression is derived:

$$A_i = \int_0^t J_i dt = -2tC_i\beta_i \frac{\partial \mu_i}{\partial x} + \alpha\beta_i C_i \sum_{j=1}^{n} A_j \bigg|_{x_m}. \tag{4.15}$$

And similarly, α can be expressed in terms of A_i as [6]:

$$\alpha = \frac{\sum_{i=1}^{n} A_i/\beta_i}{\sum_{i=1}^{n} A_j}. \tag{4.16}$$

Eq. (4.16) can be used to determine the mobilities and the vacancy wind parameter in a multicomponent system. Experimentally, similar to the determination of intrinsic diffusion coefficients, two sets of diffusion couples with the respective marker plane containing the same composition are needed. The mobilities and vacancy wind parameter at the marker plane can be obtained by solving Eq. (4.16). The mobilities and vacancy wind parameter in the Ag–Zn–Cd [8] and U–Pu–Zr [9] alloy systems were determined following this method, and documented that the vacancy wind contribution to the intrinsic diffusion can be significant.

4.3 INTERDIFFUSION IN MULTICOMPONENT ALLOYS

Similar to the intrinsic diffusion flux, the interdiffusion flux $\tilde{J}_i$ of a component i in a multicomponent alloy, according to Onsager's formalism [10,11] of Fick's law, can be written as

$$\tilde{J}_i = -\sum_{j=1}^{n-1} \tilde{D}_{ij}^n \frac{\partial C_i}{\partial x} \tag{4.17}$$

where $\tilde{D}_{ij}^n$ is the interdiffusion coefficients, C_i is the mole fraction of component i, and x represents the position. Here, the unit for the interdiffusion flux is mole (atomic) fraction•m/s. For a closed system, the mole fraction of different components remains constant and is related by $C_1 + C_2 + C_3 + \cdots + C_n = 1$. Similarly, the interdiffusion fluxes are related by $\sum_{i=1}^{n} \tilde{J}_i = 0$. Therefore, for an n-component alloy, the number of independent components is only $(n - 1)$. It is then assumed that component n is dependent and thus only $(n - 1)^2$ interdiffusion coefficients are required to express the $(n - 1)$ interdiffusion fluxes. Typically, but not necessarily, the solvent species is chosen as the dependent variable.

For a ternary alloy, the interdiffusion fluxes can be reduced to the following equations according to Eq. (4.17):

$$\tilde{J}_1 = -\tilde{D}_{11}^3 \frac{\partial C_1}{\partial x} - \tilde{D}_{12}^3 \frac{\partial C_2}{\partial x}$$
$$\tilde{J}_2 = -\tilde{D}_{21}^3 \frac{\partial C_1}{\partial x} - \tilde{D}_{22}^3 \frac{\partial C_2}{\partial x} \quad (4.18)$$

The interdiffusion flux $\tilde{J}_3$ for the dependent component 3 is related by:

$$\tilde{J}_1 + \tilde{J}_2 + \tilde{J}_3 = 0 \quad (4.19)$$

As seen from Eq. (4.18), a total of four independent interdiffusion coefficients, $\tilde{D}_{11}^3$, $\tilde{D}_{12}^3$, $\tilde{D}_{21}^3$, and $\tilde{D}_{22}^3$, are required to describe and analyze the interdiffusion behavior in a ternary system. $\tilde{D}_{11}^3$ and $\tilde{D}_{22}^3$ are defined as the main (or diagonal) interdiffusion coefficients, which represent the influence of the concentration gradient of component 1 (or 2) on the interdiffusion of component 1 (or 2). Coefficients $\tilde{D}_{12}^3$ and $\tilde{D}_{21}^3$ are defined as the cross (or off-diagonal) interdiffusion coefficients, which represent the influence of the concentration gradient of component 2 (or 1) on the interdiffusion of the component 1 (or 2).

Similar to the calculation of interdiffusion coefficients in a binary alloy, the Boltzmann–Matano [12,13] method can be employed to calculate the four interdiffusion coefficients in a ternary alloy. From the concentration profiles, the interdiffusion fluxes for each component can be determined without knowing the interdiffusion coefficients by utilizing the following equation [14]:

$$\tilde{J}_i(x) = \frac{1}{2t} \int_{C_i^{-\infty} or C_i^{+\infty}}^{C_i^x} (x - x_0) dC_i \quad (4.20)$$

where $C_i^{-\infty}$ and $C_i^{+\infty}$ refer to the composition of component i at the terminal ends of the diffusion couple, and x_0 is the position of Matano plane. The location of Matano plane can be determined by integrating over the concentration profile as shown in the following equation [15]:

$$\int_{C_i^{-\infty}}^{C_i^0} x dC_i + \int_{C_i^0}^{C_i^{+\infty}} x dC_i = 0 \quad (4.21)$$

where C_i^0 is the mole fraction of component i at Matano plane. The following simplified equation can be used for direct determination of Matano plane assuming the starting position is zero (assuming $x^{-\infty} = 0$):

$$x_0 = \frac{\int_{C_i^{-\infty}}^{C_i^{+\infty}} x dC_i}{C_i^{+\infty} - C_i^{-\infty}}. \quad (4.22)$$

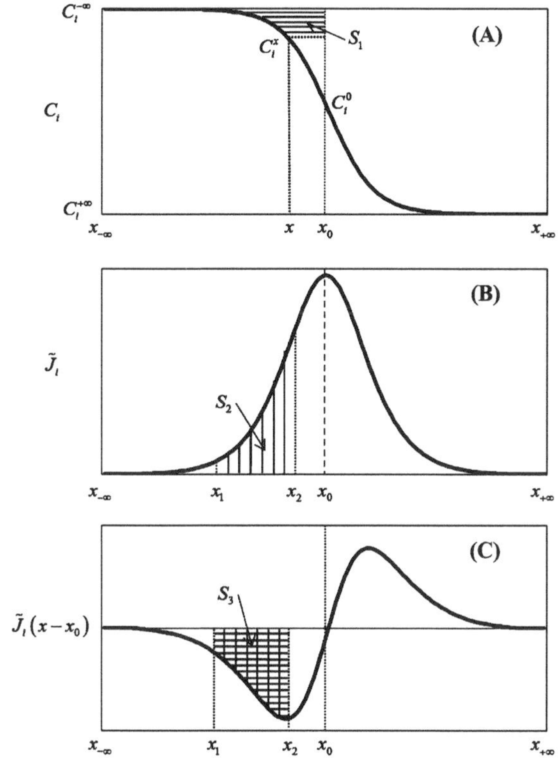

Figure 4.1 Schematic (A) concentration profiles, (B) interdiffusion fluxes, and (C) $\tilde{J}_i(x - x_0)$ for hypothetical solid–solid diffusion couples.

Fig. 4.1(A) presents a schematic concentration profile for one of the components in a multicomponent diffusion couple. The location of Matano plane, x_0, and the composition at Matano plane, C_i^0, are marked as well. An interdiffusion flux profile can be established by integrating the concentration profile following Eq. (4.20), as shown in Fig. 4.1(B). The interdiffusion flux $\tilde{J}_i$ at any position x corresponds to the area S_1 in Fig. 4.1(A).

It needs to be pointed out that the calculation of interdiffusion fluxes using Eq. (4.20) assumes that the molar volume remains constant or the variation of molar volume across the interdiffusion zone is negligible. This assumption may not be satisfied for some systems. However, in most ternary or higher order multicomponent systems, the molar volume at different compositions is not accessible.

In any case where the molar volume has to be considered, the interdiffusion flux, $\tilde{J}_i^M(x^*)$, of each component i can be determined through the relation [16–18]:

$$\tilde{J}_i^M(x^*) = -\frac{(C_i^{+\infty} - C_i^{-\infty})}{2t}\left[(1 - Y_i^*)\int_{x^{-\infty}}^{x^*} \frac{Y_i}{V_m}dx + Y_i^*\int_{x^*}^{x^{+\infty}} \frac{(1 - Y_i)}{V_m}dx\right] \quad (4.23)$$

where V_m is the molar volume at position of x^*, and Y_i is the relative concentration variable which is defined as $Y_i = \frac{C_i - C_i^-}{C_i^+ - C_i^-}$. Here, the unit for the interdiffusion flux is mol/m^2•s. Eq. (4.18) is then needed to adjust to the following:

$$\begin{aligned} \tilde{J}_1^M &= -\tilde{D}_{11}^3 \frac{1}{V_m} \frac{\partial C_1}{\partial x} - \tilde{D}_{12}^3 \frac{1}{V_m} \frac{\partial C_2}{\partial x} \\ \tilde{J}_2^M &= -\tilde{D}_{21}^3 \frac{1}{V_m} \frac{\partial C_1}{\partial x} - \tilde{D}_{22}^3 \frac{1}{V_m} \frac{\partial C_2}{\partial x} \end{aligned} \tag{4.24}$$

Both Eqs. (4.20) and (4.23) can be employed to analyze the interdiffusion fluxes in a multicomponent alloy in cases where molar volume change is negligible. There are some differences to be considered when using these equations. For Eq. (4.20), the calculation of Matano plane is inevitable, which may introduce some uncertainties if concentration profiles of different components are used for the location of Matano plane. For example, in the diffusion couple of Ti vs. $Ti_{87.4}Al_{7.6}Mo_5$ which has an interdiffusion zone approximately 1500 μm in length, the difference between the calculated Matano plane using Al and Mo concentration is approximately 50 μm [19]. The variation of the molar volume across the interdiffusion zone can result in a difference in Matano plane. Furthermore, the experimental error of obtaining the concentration profile and the improper smoothening of the concentration profiles can also contribute to the difference in Matano plane. Obviously, the advantage of using Eq. (4.23) is that it circumvents the calculation of Matano plane. However, the disadvantage of Eq. (4.23) is that it cannot be applied to a component which has a same terminal end composition. This kind of interdiffusion experiment is designed in multicomponent alloys quite often and is referred to as "Darken" type diffusion couples. In this case, the relative concentration variable Y_i may become indefinite [20].

Kirkaldy extended the Boltzmann–Matano method for a multicomponent alloy as shown in the following equation [21]:

$$\begin{aligned} \frac{1}{2t} \int_{C_1^{-\infty}}^{C_1^x} (x - x_0) dC_1 &= -\tilde{D}_{11}^3 \frac{\partial C_1}{\partial x} - \tilde{D}_{12}^3 \frac{\partial C_2}{\partial x} \\ \frac{1}{2t} \int_{C_2^{-\infty}}^{C_2^x} (x - x_0) dC_2 &= -\tilde{D}_{21}^3 \frac{\partial C_1}{\partial x} - \tilde{D}_{22}^3 \frac{\partial C_2}{\partial x} \end{aligned} \tag{4.25}$$

In order to obtain the four interdiffusion coefficients, four independent equations involving the interdiffusion coefficients are needed. Therefore, it is not possible to solve Eq. (4.25) from one single diffusion couple. Two diffusion couples are required to calculate the four interdiffusion coefficients. Moreover, since the interdiffusion coefficients are composition-dependent, it is also indispensable that the two diffusion couples intersect at a common composition point. Based on the four interdiffusion flux equations

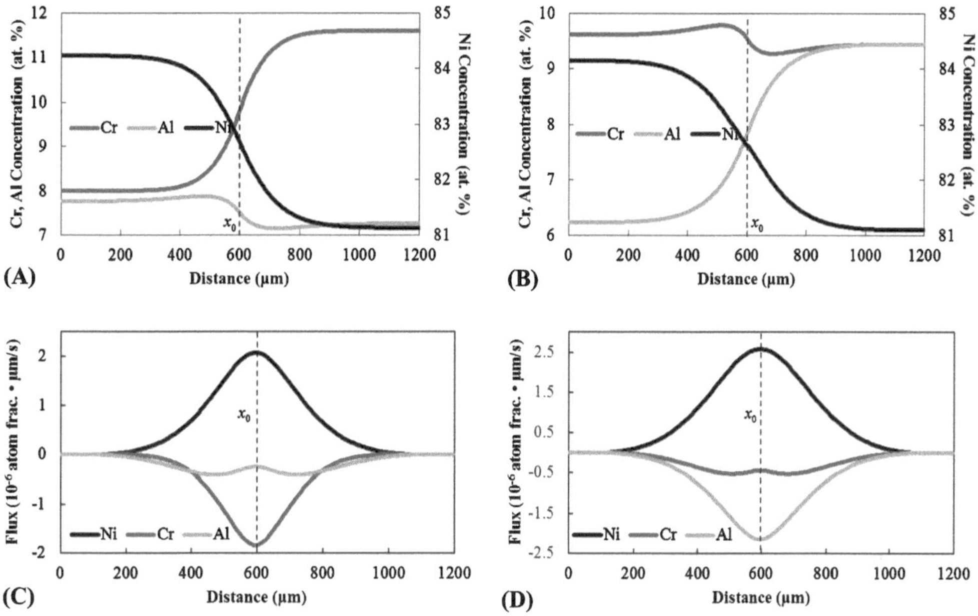

Figure 4.2 Concentration profiles (A)–(B) and interdiffusion flux profiles (C)–(D) for diffusion couples of $Ni_{84.2}Cr_{8.0}Al_{7.8}$ vs. $Ni_{81.1}Cr_{11.6}Al_{7.3}$ and $Ni_{84.2}Cr_{9.6}Al_{6.2}$ vs. $Ni_{81.0}Cr_{9.5}Al_{9.5}$, respectively [23–25].

from the two intersecting diffusion couples, the interdiffusion coefficients can be determined [22] as

$$\begin{aligned}
\tilde{D}_{11}^3 &= \frac{-(\frac{1}{2t}\int_{C_1^{-\infty}}^{C_1^x}(x-x_0)dC_1)_A - \tilde{D}_{12}^3(\frac{\partial C_2}{\partial x})_A}{(\frac{\partial C_1}{\partial x})_A} \\
\tilde{D}_{12}^3 &= \frac{(\frac{1}{2t}\int_{C_1^{-\infty}}^{C_1^x}(x-x_0)dC_1)_B(\frac{\partial C_1}{\partial x})_A - (\frac{1}{2t}\int_{C_1^{-\infty}}^{C_1^x}(x-x_0)dC_1)_A(\frac{\partial C_1}{\partial x})_B}{(\frac{\partial C_2}{\partial x})_A(\frac{\partial C_1}{\partial x})_B - (\frac{\partial C_2}{\partial x})_B(\frac{\partial C_1}{\partial x})_A} \\
\tilde{D}_{21}^3 &= \frac{-(\frac{1}{2t}\int_{C_2^{-\infty}}^{C_2^x}(x-x_0)dC_2)_A - \tilde{D}_{22}^3(\frac{\partial C_2}{\partial x})_A}{(\frac{\partial C_1}{\partial x})_A} \\
\tilde{D}_{22}^3 &= \frac{(\frac{1}{2t}\int_{C_2^{-\infty}}^{C_2^x}(x-x_0)dC_2)_B(\frac{\partial C_1}{\partial x})_A - (\frac{1}{2t}\int_{C_2^{-\infty}}^{C_2^x}(x-x_0)dC_2)_A(\frac{\partial C_1}{\partial x})_B}{(\frac{\partial C_2}{\partial x})_A(\frac{\partial C_1}{\partial x})_B - (\frac{\partial C_2}{\partial x})_B(\frac{\partial C_1}{\partial x})_A}
\end{aligned} \tag{4.26}$$

Two sets of diffusion couples, $Ni_{84.2}Cr_{8.0}Al_{7.8}$ vs. $Ni_{81.1}Cr_{11.6}Al_{7.3}$ and $Ni_{84.2}Cr_{9.6}Al_{6.2}$ vs. $Ni_{81.0}Cr_{9.5}Al_{9.5}$, were prepared in the γ phase region of the Ni–Cr–Al alloys to determine the ternary interdiffusion coefficients [23–25]. The concentration profiles are presented in Fig. 4.2, (A) and (B), respectively. The interdiffusion fluxes are calculated according to Eq. (4.20) and are presented in Fig. 4.2, (C) and (D). The Matano plane

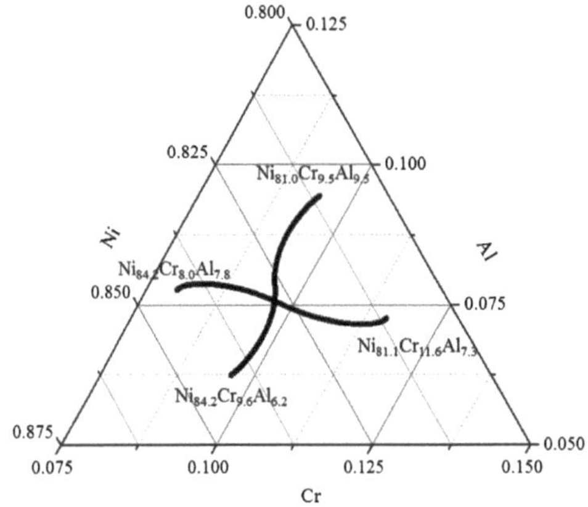

Figure 4.3 Diffusion paths for diffusion couples of $Ni_{84.2}Cr_{8.0}Al_{7.8}$ vs. $Ni_{81.1}Cr_{11.6}Al_{7.3}$ and $Ni_{84.2}Cr_{9.6}Al_{6.2}$ vs. $Ni_{81.0}Cr_{9.5}Al_{9.5}$ [23–25].

Table 4.1 Ternary interdiffusion coefficients at the intersecting composition of $Ni_{82.8}Cr_{9.7}Al_{7.5}$ between diffusion couples of $Ni_{84.2}Cr_{8.0}Al_{7.8}$ vs. $Ni_{81.1}Cr_{11.6}Al_{7.3}$ and $Ni_{84.2}Cr_{9.6}Al_{6.2}$ vs. $Ni_{81.0}Cr_{9.5}Al_{9.5}$

Composition at. %	Ternary interdiffusion coefficients (10^{-15} m^2/s)			
$Ni_{82.8}Cr_{9.7}Al_{7.5}$	$\tilde{D}^{Ni}_{CrCr}$ 11.5	$\tilde{D}^{Ni}_{CrAl}$ 7.4	$\tilde{D}^{Ni}_{AlCr}$ 8.1	$\tilde{D}^{Ni}_{AlAl}$ 23.7
	$\tilde{D}^{Al}_{NiNi}$ 31.1	$\tilde{D}^{Al}_{NiCr}$ 11.5	$\tilde{D}^{Al}_{CrNi}$ −7.4	$\tilde{D}^{Al}_{CrCr}$ 4.1
	$\tilde{D}^{Cr}_{AlAl}$ 15.6	$\tilde{D}^{Cr}_{AlNi}$ −8.1	$\tilde{D}^{Cr}_{NiAl}$ −11.5	$\tilde{D}^{Cr}_{NiNi}$ 19.6

is marked as x_0 in Fig. 4.2 based on Eq. (4.22). These two diffusion couples were designed so that their diffusion paths intersect at a common composition of approximately $Ni_{82.8}Cr_{9.7}Al_{7.5}$, as illustrated in Fig. 4.3. Therefore the use of Eqs. (4.25) and (4.26) allowed one to calculate the ternary interdiffusion coefficients at this intersecting composition reported in Table 4.1.

The signs of the cross ternary interdiffusion coefficients can be directly determined by examining the concentration profiles of the component with similar terminal compositions. For example, in Fig. 4.2, (A) and (C), Al interdiffuses up its own concentration gradient against the Ni concentration gradient, but interdiffuses down the Cr concentration. This implies that the cross ternary interdiffusion coefficients $\tilde{D}^{Cr}_{AlNi}$ are negative whereas $\tilde{D}^{Ni}_{AlCr}$ is positive. Similarly, in Fig. 4.2, (B) and (D), Cr diffuses up its own concentration gradient against the Ni concentration gradient, but down the Al concentration. This implies that the cross ternary interdiffusion coefficients $\tilde{D}^{Al}_{CrNi}$ are negative whereas $\tilde{D}^{Ni}_{CrAl}$ is positive. This is confirmed from the ternary interdiffusion coefficients reported in Table 4.1.

In the analysis, the solvent Ni has been selected as the dependent variable. The solute atoms can also be chosen as the dependent variable. In some cases this is necessary in order to understand the diffusional interactions between different atoms. The additional sets of ternary interdiffusion coefficients can be transformed from the $\tilde{D}_{ij}^3$ on the basis of the following equations [3,26]:

$$\begin{aligned}
\tilde{D}_{11}^2 &= \tilde{D}_{11}^3 - \tilde{D}_{12}^3 \\
\tilde{D}_{13}^2 &= -\tilde{D}_{12}^3 \\
\tilde{D}_{31}^2 &= \tilde{D}_{22}^3 + \tilde{D}_{12}^3 - \tilde{D}_{11}^3 - \tilde{D}_{21}^3 \\
\tilde{D}_{33}^2 &= \tilde{D}_{22}^3 + \tilde{D}_{12}^3 \\
\tilde{D}_{22}^1 &= \tilde{D}_{22}^3 - \tilde{D}_{21}^3 \\
\tilde{D}_{23}^1 &= -\tilde{D}_{21}^3 \\
\tilde{D}_{32}^1 &= \tilde{D}_{11}^3 + \tilde{D}_{21}^3 - \tilde{D}_{22}^3 - \tilde{D}_{12}^3 \\
\tilde{D}_{33}^1 &= \tilde{D}_{11}^3 + \tilde{D}_{21}^3
\end{aligned} \tag{4.27}$$

Table 4.1 also lists the ternary interdiffusion coefficients with Cr or Al as the dependent component, which can be calculated through equations in (4.27).

Since the ternary interdiffusion coefficients are material constants, it is expected that ternary interdiffusion coefficients are dependent on the alloy composition. By setting up different diffusion couples that have multiple intersecting compositions, the compositional dependent ternary interdiffusion coefficients can be obtained. For example, in Ni–Co–Pt ternary alloys, multiple diffusion couples were prepared and intersect at various compositions. The ternary interdiffusion coefficients, $\tilde{D}_{CoCo}^{Ni}$, $\tilde{D}_{PtPt}^{Ni}$, $\tilde{D}_{CoPt}^{Ni}$, and $\tilde{D}_{PtCo}^{Ni}$, were extracted and plotted in a ternary triangle, where the trend can be readily revealed [27].

However, it is apparent that in order to obtain composition-dependent ternary interdiffusion coefficients, many diffusion couples have to be carefully designed and prepared. Particularly, these diffusion couples need to intersect at common compositions, which can be hard to predict before the experiments.

An important criterion for validating the calculated ternary interdiffusion coefficients, based on thermodynamic requirements, is listed below [1]:

$$\begin{aligned}
&\tilde{D}_{11}^3 + \tilde{D}_{22}^3 > 0 \\
&\left(\tilde{D}_{11}^3 + \tilde{D}_{22}^3\right)^2 \geq 4\left(\tilde{D}_{11}^3\tilde{D}_{22}^3 - \tilde{D}_{12}^3\tilde{D}_{21}^3\right) \\
&\left(\tilde{D}_{11}^3\tilde{D}_{22}^3 - \tilde{D}_{12}^3\tilde{D}_{21}^3\right) \geq 0
\end{aligned} \tag{4.28}$$

This criterion points out that the sum of $\tilde{D}_{11}^3 + \tilde{D}_{22}^3$ must be positive, but the main interdiffusion coefficients can be negative, although it is rarely observed. In addition, the cross interdiffusion coefficients can be asymmetric and have opposite signs. The ternary interdiffusion coefficients listed in Table 4.1 satisfies all of the conditions listed in Eq. (4.28).

The interdiffusion flux of a component is related to the intrinsic flux according to the following equation [28]:

$$\tilde{J}_i = J_i - C_i \sum_{k=1}^{n} J_k. \tag{4.29}$$

Furthermore, the interdiffusion coefficients can be calculated if the intrinsic diffusion coefficients are known [3]:

$$\tilde{D}_{ij}^{n} = D_{ij}^{n} - C_i \sum_{k=1}^{n} D_{kj}^{n}. \tag{4.30}$$

Determination of interdiffusion coefficients in a ternary alloy is not straightforward experimentally. It becomes more complicated for a quaternary alloy, where a total of nine independent interdiffusion coefficients are needed according to Eq. (4.17). It requires that at least three independent diffusion couples are experimentally prepared, and these couples must intersect at a common composition point. The difficulty of determining interdiffusion coefficients has always driven research to develop new approaches, both theoretically and experimentally.

4.4 ZERO FLUX PLANE (ZFP)

Dayananda and Kim first discovered the zero flux plane (ZFP) in Cu–Ni–Zn diffusion couples [29]. The ZFP phenomenon was also identified later in diffusion couples assembled in many other ternary and quaternary systems [3,14,30–32]. ZFP refers to the location where the interdiffusion flux of a component becomes zero, and the interdiffusion flux on each side of the ZFP is opposite in sign. The development of the ZFP indicates that there is an interdiffusion flux reversal at the ZFP for the respective component [29].

The ZFPs are illustrated in the diffusion couple $Cu_{30.1}Ni_{44.7}Zn_{25.2}$ vs. $Cu_{46.0}Ni_{17.7}Zn_{36.3}$ as shown in Fig. 4.4. For the Zn component, a negative interdiffusion flux is observed throughout the interdiffusion zone, which indicates that Zn is diffusing from the right to the left terminal end. For the Ni component, the interdiffusion flux is negative from the left terminal end but goes toward positive before returning to zero. Therefore, one ZFP plane is identified for Ni component. For the Cu component, the interdiffusion flux starts with a positive value, decreases towards a negative value, then curves back to positive again before returning to zero. Therefore in this particular diffusion couple, for the Cu component, two ZFPs are observed.

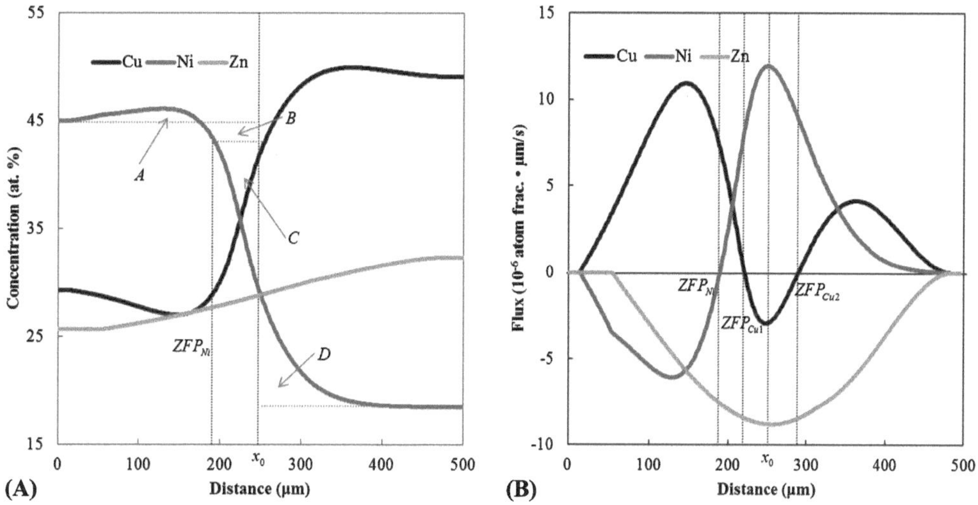

Figure 4.4 (A) Concentration profiles and (B) interdiffusion flux profiles for the diffusion couple of $Cu_{30.1}Ni_{44.7}Zn_{25.2}$ vs. $Cu_{46.0}Ni_{17.7}Zn_{36.3}$ annealed at 775°C for 48 hours [29].

According to the definition, the location of the ZFP can be calculated from Eq. (4.20) by setting the interdiffusion fluxes $\tilde{J}_i$ equal to zero:

$$\int_{C_{\pm\infty}}^{C_{ZFP}} (x - x_0)\,dC_i = 0 \tag{4.31}$$

In Fig. 4.4, the locations of Matano plane and ZFP are identified by Eqs. (4.21) and (4.31), respectively. Eq. (4.31) can be better understood from the concentration profile of Ni shown in Fig. 4.4(A). Integrating from the left terminal end, Eq. (4.31) is equivalent to $Area\,A + Area\,B = 0$. Similarly, $Area\,C + Area\,D = 0$ integrated from the right terminal end.

Direct information on the diffusional interactions among components can be extracted at the ZFP in a ternary alloy. The ratio between the main and cross interdiffusion coefficients can be obtained from the concentration gradient by setting the interdiffusion flux equal to zero from Eq. (4.18):

$$\begin{aligned} \tilde{J}_i &= -\tilde{D}_{i1}^3 \frac{\partial C_1}{\partial x} - \tilde{D}_{i2}^3 \frac{\partial C_2}{\partial x} = 0 \\ \frac{\tilde{D}_{i2}^3}{\tilde{D}_{i1}^3} &= -\left.\frac{\partial C_1}{\partial C_2}\right|_{ZFP} \end{aligned} \tag{4.32}$$

Additional information for the ternary interdiffusion coefficients can also be directly obtained through ideas similar to those seen in Eq. (4.32). At locations where the concentration gradient of a component goes to zero, one of the main or cross ternary interdiffusion coefficients can be calculated through the following equation:

$$\begin{aligned} \tilde{J}_1 &= -D^3_{11} \left.\frac{\partial C_1}{\partial x}\right|_{\frac{\partial C_2}{\partial x}=0} \\ \tilde{J}_2 &= -D^3_{21} \left.\frac{\partial C_1}{\partial x}\right|_{\frac{\partial C_2}{\partial x}=0} \end{aligned} \tag{4.33}$$

Two intersecting diffusion couples are not required in this case to obtain the ternary interdiffusion coefficients. The interdiffusion flux at a location where the concentration gradient is zero corresponds to a local minimum or maximum, as seen in Fig. 4.2, (C) and (D), and Fig. 4.4(B).

The discovery of a ZFP has led to an observation that a component in a multicomponent system can interdiffuse up its own thermodynamic activity gradient [29,31], which is not expected in a binary diffusion couple. In many ternary systems, relative extrema in concentration profiles are commonly observed. However, it is not necessary to develop a ZFP when a maximum or minimum is observed [29]. The formation of ZFPs depends on the strong diffusional interaction among different components. In the Cu–Ni–Zn ternary alloys, the compositions correspond to the intersections of its diffusion path and the isoactivity lines for the components drawn through the terminal alloy compositions on a ternary isotherm [14,29,31]. The diffusion path slope at the ZFP composition is uniquely defined and is correlated to the atomic mobility and thermodynamic data [33].

4.5 AVERAGE EFFECTIVE AND INTEGRATED DIFFUSION COEFFICIENTS IN MULTICOMPONENT SYSTEMS

Determination of the full matrix of interdiffusion coefficients is not trivial. To circumvent this problem, Dayananda and Behnke [34] introduced the idea of effective interdiffusion coefficient for each component for the analysis of multicomponent diffusion, and applied it to the determination of such coefficients for various components in a diffusion couple, IN 718 vs. Ta-modified IN-718. The effective interdiffusion coefficient provides a single diffusion coefficient for a compositional spectral even in multicomponent alloys. Such concepts were further explored by Dayananda and coworkers to other diffusion studies [35,36].

For a ternary alloy, Eq. (4.17) can be alternatively written as

$$\tilde{J}_i = -\tilde{D}_i^{eff} \frac{\partial C_i}{\partial x} \quad (i = 1, 2) \tag{4.34}$$

where $\tilde{D}_i^{eff}$ is the effective interdiffusion coefficients and can be expressed as

$$\tilde{D}_i^{eff} = \tilde{D}_{ii}^3 + \frac{\tilde{D}_{ij}^3 \partial C_j / \partial x}{\partial C_i / \partial x} \quad (j \neq i). \tag{4.35}$$

It is noted that $\tilde{D}_i^{eff}$ includes the diffusional interactions among the diffusing species since the cross ternary interdiffusion coefficients are included.

The idea of $\tilde{D}_i^{eff}$ is first applied to a single-phase diffusion couple, where an average value for effective interdiffusion coefficients is obtained on either side of the Matano plane, as expressed in the following equation:

$$\begin{aligned} &\int_{x-\infty}^{x_0} \tilde{J}_i dx = -\tilde{D}_{i,L}^{eff} \int_{C_i^-}^{C_i^0} dC_i = \tilde{D}_{i,L}^{eff}\left(C_i^- - C_i^0\right) \\ &\tilde{D}_{i,L}^{eff} = \frac{\int_{x-\infty}^{x_0} \tilde{J}_i dx}{C_i^- - C_i^0} \\ &\tilde{D}_{i,R}^{eff} = \frac{\int_{x_o}^{x+\infty} \tilde{J}_i dx}{C_i^0 - C_i^+} \end{aligned} \tag{4.36}$$

where C_i^0 is the composition at Matano plane. $\tilde{D}_{i,L}^{eff}$ and $\tilde{D}_{i,R}^{eff}$ are the average effective diffusion coefficients on the left-hand side and right-hand side of Matano plane, respectively. The concentration profiles can be generated using the error function solutions using $\tilde{D}_{i,L}^{eff}$ and $\tilde{D}_{i,R}^{eff}$ values and compared with the original concentration profile [36].

The procedure can be extended to the analysis of any component in a multicomponent system over a certain compositional range. The average effective interdiffusion coefficients of component i over the distance from x_1 to x_2 can be expressed as

$$\tilde{D}_i^{eff} = \frac{\int_{x_1}^{x_2} \tilde{J}_i dx}{C_i^{x1} - C_i^{x2}} \tag{4.37}$$

where C_i^{x1} and C_i^{x2} are the compositions of component i at location x_1 to x_2, respectively.

Therefore Eq. (4.37) can be employed to multiphase diffusion couples [35]. In this case, the location can be set up at the planar interphase boundaries of any phase formed during interdiffusion. Then, the average effective interdiffusion coefficients of that particular phase can be estimated and compared.

However, if the variation of the concentration over the phase layer is negligible (e.g., stoichiometric intermetallic compound), Eq. (4.37) becomes invalid because the denominator becomes zero. In this case, an integrated interdiffusion coefficient, $\tilde{D}_i^{\text{int}}$, over any compositional range between C_i^{x1} and C_i^{x2} can be used, which is defined as

[35,37]

$$\tilde{D}_i^{\text{int}} = \int_{x_1}^{x_2} \tilde{J}_i(x)dx. \tag{4.38}$$

The right-hand side of Eq. (4.38) involves integrating the interdiffusion flux over the distance from x_1 to x_2, which is schematically equivalent to the area S_2 shown in Fig. 4.1(B). This is different from the case of a binary alloy where only one integrated interdiffusion coefficient exists. Each component has its own integrated interdiffusion coefficient in a multicomponent alloy.

It then becomes clear that the average effective interdiffusion coefficient and integrated interdiffusion coefficient are related according to Eqs. (4.37) and (4.38) by [37]

$$\tilde{D}_i^{eff} = \frac{\tilde{D}_i^{\text{int}}}{C_i^{x1} - C_i^{x2}}. \tag{4.39}$$

4.6 AVERAGE TERNARY INTERDIFFUSION COEFFICIENTS

Another analysis of the ternary interdiffusion coefficients has been proposed by Dayananda and Sohn [25], where an average value of the ternary interdiffusion coefficients has been obtained, however, from only a single diffusion couple. In this analysis, mathematical treatments are applied to the interdiffusion flux in Eq. (4.18) so that multiple independent equations can be obtained to solve for the ternary interdiffusion coefficients.

Multiplying by $(x - x_0)^n$ on both sides of Eq. (4.18) and integrating over the range between x_1 and x_2 yields

$$\begin{aligned}\int_{x_1}^{x_2} \tilde{J}_i (x - x_0)^n dx &= -\int_{C_1^{x1}}^{C_1^{x2}} \tilde{D}_{i1}^3 (x - x_0)^n dC_1 - \int_{C_2^{x1}}^{C_2^{x2}} \tilde{D}_{i2}^3 (x - x_0)^n dC_2 \\ &= -\bar{\tilde{D}}_{i1}^3 \int_{C_1^{x1}}^{C_1^{x2}} (x - x_0)^n dC_1 - \bar{\tilde{D}}_{i2}^3 \int_{C_2^{x1}}^{C_2^{x2}} (x - x_0)^n dC_2\end{aligned} \tag{4.40}$$

where $\bar{\tilde{D}}_{i1}^3$ and $\bar{\tilde{D}}_{i2}^3$ correspond to the average main and cross ternary interdiffusion coefficients and are treated as constants over the compositional range. The average ternary interdiffusion coefficients can be defined as

$$\bar{\tilde{D}}_{ij}^3 = \frac{\int_{C_j^{x1}}^{C_j^{x2}} \tilde{D}_{ij}^3 dC_j}{\int_{C_j^{x1}}^{C_j^{x2}} dC_j} \quad (i = 1, 2). \tag{4.41}$$

If the exponent n in Eq. (4.40) is taken as 0, Eq. (4.40) is then simplified to

$$\begin{aligned}\int_{x_1}^{x_2} \tilde{J}_i dx &= -\bar{\tilde{D}}_{i1}^3 \int_{C_1^{x_1}}^{C_1^{x_2}} dC_1 - \bar{\tilde{D}}_{i2}^3 \int_{C_2^{x_1}}^{C_2^{x_2}} dC_2 \\ &= \bar{\tilde{D}}_{i1}^3 \left(C_1^{x_1} - C_1^{x_2}\right) + \bar{\tilde{D}}_{i2}^3 \left(C_2^{x_1} - C_2^{x_2}\right)\end{aligned} \tag{4.42}$$

If the exponent n is taken as 1, Eq. (4.40) becomes

$$\begin{aligned}\int_{x_1}^{x_2} \tilde{J}_i (x - x_0) dx &= -\bar{\tilde{D}}_{i1}^3 \int_{C_1^{x_1}}^{C_1^{x_2}} (x - x_0) dC_1 - \bar{\tilde{D}}_{i2}^3 \int_{C_2^{x_1}}^{C_2^{x_2}} (x - x_0) dC_2 \\ &= 2t\left\{\bar{\tilde{D}}_{i1}^3 \left[\tilde{J}_1(x_1) - \tilde{J}_1(x_2)\right] + \bar{\tilde{D}}_{i2}^3 \left[\tilde{J}_2(x_1) - \tilde{J}_2(x_2)\right]\right\}.\end{aligned} \tag{4.43}$$

The equations can be schematically understood from Fig. 4.1, where the profile of the interdiffusion flux $\tilde{J}_i$ is plotted in panel (B), and $\tilde{J}_i(x - x_0)$ vs. x plot is plotted in panel (C). The $\int_{x_1}^{x_2} \tilde{J}_i dx$ term in Eq. (4.42) is equivalent to the area S_2 shown in Fig. 4.1(B), while the $\int_{x_1}^{x_2} \tilde{J}_i (x - x_0) dx$ in Eq. (4.43) is equivalent to the area S_3 shown in Fig. 4.1(C). Although n can be set up to be higher than 1 in Eq. (4.40), the interpretation becomes much more difficult in terms of concentration and interdiffusion flux [25].

By applying Eqs. (4.42) and (4.43) to any two components in a ternary alloy, a total of four equations can be obtained. Therefore, the four average ternary interdiffusion coefficients, $\bar{\tilde{D}}_{11}^3$, $\bar{\tilde{D}}_{12}^3$, $\bar{\tilde{D}}_{21}^3$, and $\bar{\tilde{D}}_{22}^3$, can be determined.

For the Ni–Cr–Al alloys presented in Fig. 4.2, this approach was applied, and the segments on either side of Matano plane were analyzed. The calculated average ternary interdiffusion coefficients are listed in Table 4.2. Compared to the ternary interdiffusion coefficients listed in Table 4.1 that are calculated at the compositional intersection point, the signs are the same and the values are quite close. Furthermore, the concentration profile can be generated using the average interdiffusion coefficients and compared with the original concentration to validate the values [25].

To ensure that the four equations are independent, nonlinear segments of the profiles need to be selected [25]. If the concentration profile is close to linear, the calculated $\bar{\tilde{D}}_{ij}^3$ will be susceptible to large errors and not reliable. Furthermore, the moment $(x - x_0)^n$ also enlarges the uncertainty, especially towards the terminal end of the diffusion couples [38].

This approach provides an average value of ternary interdiffusion coefficients over certain compositional ranges. Cermak and Rothova [39] suggest, in an extension of the method, that the average values $\bar{\tilde{D}}_{ij}^3$ approach the true values of $\tilde{D}_{ij}^3$ if the interval between x_1 and x_2 is sufficiently small. The composition-dependent ternary interdiffusion coefficients have then been determined for NiAlX (X = Cr, Fe, Nb, Ti, and Mo) alloys [39,40]. However, it has been recently argued that the small interval between x_1

Table 4.2 Average ternary interdiffusion coefficients for diffusion couples of $Ni_{84.2}Cr_{8.0}Al_{7.8}$ vs. $Ni_{81.1}Cr_{11.6}Al_{7.3}$ and $Ni_{84.2}Cr_{9.6}Al_{6.2}$ vs. $Ni_{81.0}Cr_{9.5}Al_{9.5}$ on either side of Matano plane

Diffusion couple	Average ternary interdiffusion coefficients for composition range of left side of Matano plane (10^{-15} m^2/s)				Average ternary interdiffusion coefficients for composition range of right side of the Matano plane ($10^{-15} m^2/s$)			
	$\tilde{D}^{Ni}_{CrCr}$	$\tilde{D}^{Ni}_{CrAl}$	$\tilde{D}^{Ni}_{AlCr}$	$\tilde{D}^{Ni}_{AlAl}$	$\tilde{D}^{Ni}_{CrCr}$	$\tilde{D}^{Ni}_{CrAl}$	$\tilde{D}^{Ni}_{AlCr}$	$\tilde{D}^{Ni}_{AlAl}$
$Ni_{84.2}Cr_{8.0}Al_{7.8}$ vs. $Ni_{81.1}Cr_{11.6}Al_{7.3}$	15.4	9.3	9.8	27.7	14.1	8.3	9.4	26.3
$Ni_{84.2}Cr_{9.6}Al_{6.2}$ vs. $Ni_{81.0}Cr_{9.5}Al_{9.5}$	15.0	8.8	17.4	24.6	14.0	8.4	16.4	22.9

and x_2 is not applicable for obtaining the true composition-dependent interdiffusion coefficients.

There have been many efforts to develop either theoretical or experimental approaches to determine the diffusion coefficients in multicomponent alloys. For example, a square root diffusivity approach has been developed by Thompson and Morral, assuming constant diffusivity [41]. This method has been applied to obtain the square root diffusivity matrix $[r]$ and diffusivity matrix $[D]$ from diffusion profiles for ternary and quaternary alloys [42–44]. Bouchet and Mevrel proposed a numerical inverse method based on minimizing the difference between calculated and experimental concentrations. This method was created to extract composition-dependent interdiffusion coefficients in a single diffusion couple [38]. However, the process is time consuming and the quality of the procedure strongly depends on the data smoothing method [38]. Paul also developed a pseudo-binary method to analyze the interdiffusion coefficients as a function of composition in a multicomponent alloy [45,46], which requires experimentally that only two elements in the alloy can diffuse while other elements do not develop any significant interdiffusion fluxes. The method has been applied to Ni–Pt–Al alloys [47] and Cu–Sn–Si(Ga) alloys [48].

In summary, although many different approaches have been developed, each method has its own limitations and restrictions to be applied to understand the diffusional interactions in multicomponent alloys. Proper selection from these approaches is needed in order to obtain correct and meaningful diffusion coefficients in a multicomponent alloy.

4.7 A TRANSFER MATRIX ANALYSIS OF MULTICOMPONENT DIFFUSION

4.7.1 Derivation of Transfer Matrix Methodology

Diffusion problems in multicomponent systems with n components have been examined through a transfer matrix methodology (TMM) developed by Ram-Mohan and Dayananda [49,50]. The TMM methods enable the calculation of interdiffusion fluxes of all components at any section x in the diffusion zone from a given set of initial values of either the fluxes or concentration gradients at a given section x_s. Interdiffusion coefficients are determined as average values over selected composition ranges by the Method of Moments. This method is analogous to the approach proposed by Dayananda and Sohn [25], and is utilized in the calculations. General expressions for the concentrations are also obtained from initial conditions on the interdiffusion fluxes or the concentration gradients. Integration of the interdiffusion fluxes over time or distance can also be carried out with the transfer matrix. The TMM provides a generalization of the error function solutions available for ternary diffusion to n-component systems. Applications of TMM for the generation of concentration profiles are illustrated for two couples – a ternary and a quaternary diffusion couple.

The flux gradients for an isothermal solid–solid diffusion couple are expressed by [17,29]

$$\frac{\partial \tilde{J}_i(x,t)}{\partial x} = \frac{(x-x_0)}{2t}\frac{\partial C_i(x,t)}{\partial x}. \tag{4.44}$$

The concentration profiles $C_i(x, t)$ may be divided into small regions with a region s ranging over $x_s \leq x \leq x_{s+1}$. Onsager's formalism of Fick's law expressed in the matrix form is given by

$$\tilde{\boldsymbol{J}}^{(n)} = -\tilde{\boldsymbol{D}}^{(n)}\frac{\partial}{\partial x}\boldsymbol{C}^{(n)}(x) \tag{4.45}$$

where the vectors of concentration gradients and fluxes are of dimension $(n - 1)$, and the diffusion coefficient matrix $\tilde{\boldsymbol{D}}^{(n)}$ is assumed to be invertible. Hence, Eq. (4.44) can be written [50] in the matrix notation as

$$\frac{\partial}{\partial x}\tilde{\boldsymbol{J}}^{(n)}(x,t) = -\frac{(x-x_0)}{2t}\left(\tilde{\boldsymbol{D}}^{(n)}\right)^{-1}\cdot\tilde{\boldsymbol{J}}^{(n)}(x,t). \tag{4.46}$$

If the interdiffusion fluxes, $\tilde{J}_i(x_s)$, are given in section x_s, the $(n - 1)$ coupled linear first order differential equations in Eq. (4.46) can be solved for the interdiffusion fluxes at any other section x in the range $x_s \leq x \leq x_{s+1}$. For decoupling the equations, the matrix $\tilde{\boldsymbol{D}}^{(n)}$ is diagonalized by using a similarity transformation $\boldsymbol{P}$ such that

$$\boldsymbol{P}^{-1}\cdot\tilde{\boldsymbol{D}}^{(n)}\cdot\boldsymbol{P} = \Delta^{(n)}. \tag{4.47}$$

In Eq. (4.47), $\Delta^{(n)}$ is a diagonal matrix with elements $\Delta_{ij}^{(n)} = d_i\delta_{ij}$, where d_i represents the eigenvalues and δ_{ij} is the Kronecker tensor. Similarly, $\boldsymbol{P}$ also diagonalizes $(\tilde{\boldsymbol{D}}^{(n)})^{-1}$ so that $\boldsymbol{P}^{-1}\cdot(\tilde{\boldsymbol{D}}^{(n)})^{-1}\cdot\boldsymbol{P} = (\Delta^{(n)})^{-1}$. On the diagonal basis, the flux vector $\hat{\boldsymbol{J}}(x,t)$ denoted with a hat is defined to be

$$\boldsymbol{P}^{-1}\cdot\tilde{\boldsymbol{J}}^{(n)}(x,t)=\hat{\boldsymbol{J}}^{(n)}(x,t). \tag{4.48}$$

Multiplying Eq. (4.46) by $\boldsymbol{P}^{-1}$ from the left, one gets the $(n-1)$ decoupled equations, expressed in vector notation, by

$$\frac{\partial}{\partial x}\hat{\boldsymbol{J}}^{(n)}(x,t)=-\frac{(x-x_0)}{2t}\left(\Delta^{(n)}\right)^{-1}\cdot\hat{\boldsymbol{J}}^{(n)}(x,t). \tag{4.49}$$

Upon integrating Eq. (4.49) and converting the fluxes back into the original basis, one obtains

$$\tilde{\boldsymbol{J}}^{(n)}(x,t)=\boldsymbol{T}\left(x,x_s,t,\tilde{\boldsymbol{D}}^{(n)}\right)\cdot\tilde{\boldsymbol{J}}^{(n)}(x_s,t),\quad x_s\le x\le x_{s+1}, \tag{4.50}$$

where the matrix $\boldsymbol{T}(x, x_s, t, \tilde{\boldsymbol{D}}^{(n)})$ is called the *transfer matrix* [50] and is given by

$$\begin{aligned}\boldsymbol{T}\left(x,x_s,t,\tilde{\boldsymbol{D}}^{(n)}\right)&\equiv\mathbf{exp}\left(-\frac{(x-x_0)^2-(x_s-x_0)^2}{4t}\left(\tilde{\boldsymbol{D}}^{(n)}\right)^{-1}\right)\\&=\boldsymbol{P}\cdot\begin{bmatrix}\ddots & & 0\\ & \exp\left(-\frac{(x-x_0)^2-(x_s-x_0)^2}{4t}\left(d_i^{-1}\right)\right) & \\ 0 & & \ddots\end{bmatrix}\cdot\boldsymbol{P}^{-1}.\end{aligned} \tag{4.51}$$

The transfer matrix evolves the flux from its initial value $\tilde{\boldsymbol{J}}^{(n)}(x_s,t)$ at $x=x_s$ to its value at any x in the interval $x_s\le x\le x_{s+1}$.

Designating the diagonal form of the transfer matrix by $\boldsymbol{\tau}$, we have

$$\begin{aligned}\boldsymbol{\tau}\left(x,x_s,t,\tilde{\boldsymbol{D}}^{(n)}\right)&=\begin{bmatrix}\ddots & & 0\\ & \exp\left(-\frac{(x-x_0)^2-(x_s-x_0)^2}{4t}\left(d_i^{-1}\right)\right) & \\ 0 & & \ddots\end{bmatrix}.\\&=\frac{\mathbf{exp}\left(-\frac{(x-x_0)^2}{4t}\left(\Delta^{(n)}\right)^{-1}\right)}{\mathbf{exp}\left(-\frac{(x_s-x_0)^2}{4t}\left(\Delta^{(n)}\right)^{-1}\right)}\end{aligned} \tag{4.52}$$

In compact form Eq. (4.51) becomes $\boldsymbol{T}=\boldsymbol{P}\cdot\tau\cdot\boldsymbol{P}^{-1}$. Based on Eq. (4.45), Eq. (4.50) yields

$$\tilde{\boldsymbol{J}}^{(n)}(x,t)=-\boldsymbol{T}(x,x_s,t)\cdot\tilde{\boldsymbol{D}}^{(n)}\cdot\frac{\partial\boldsymbol{C}^{(n)}(x_s,t)}{\partial x}. \tag{4.53}$$

Thus, from the initial values of either the fluxes or the concentration gradients at x_s, the interdiffusion fluxes of all components can be determined at any x over each segment of the diffusion zone.

From the interdiffusion fluxes determined over x within the diffusion zone, one can determine the concentrations on the basis of Fick's law; thus,

$$\frac{\partial}{\partial x}\boldsymbol{C}^{(n)}(x,t) = -\left(\tilde{\boldsymbol{D}}^{(n)}\right)^{-1}\cdot\tilde{\boldsymbol{J}}^{(n)}(x,t). \tag{4.54}$$

Integrating both sides of Eq. (4.54) yields

$$\begin{aligned}\boldsymbol{C}^{(n)}(x,t) &= \boldsymbol{C}^{(n)}(x_s,t) - \left(\tilde{\boldsymbol{D}}^{(n)}\right)^{-1}\left[\int_{x_x}^{x} dx'\boldsymbol{T}\left(x',x_s,t,\tilde{\boldsymbol{D}}^{(n)}\right)\right]\tilde{\boldsymbol{J}}^{(n)}(x_s,t)\\ &= \boldsymbol{C}^{(n)}(x_s,t) - \left(\tilde{\boldsymbol{D}}^{(n)}\right)^{-1}\left[\int_{x_x}^{x} dx'\mathbf{exp}\left(-\frac{(x-x_0)^2-(x_s-x_0)^2}{4t}\left(\tilde{\boldsymbol{D}}^{(n)}\right)^{-1}\right)\right]\\ &\quad\times\tilde{\boldsymbol{J}}^{(n)}(x_s,t).\end{aligned} \tag{4.55}$$

Upon simplification [50], Eq. (4.55) becomes

$$\begin{aligned}\boldsymbol{C}^{(n)}(x,t) &= \boldsymbol{C}^{(n)}(x_s,t) + \boldsymbol{P}\cdot\begin{bmatrix}\ddots & & 0\\ & \phi_i & \\ 0 & & \ddots\end{bmatrix}\cdot\boldsymbol{P}^{-1}\cdot\left(\boldsymbol{C}^{(n)}(x_{s+1},t)-\boldsymbol{C}^{(n)}(x_s,t)\right)\\ &= \boldsymbol{C}^{(n)}(x_s,t) + \boldsymbol{P}\cdot\boldsymbol{\Phi}\left(x,x_s,x_{s+1},x_0,t,\Delta^{(n)}\right)\cdot\boldsymbol{P}^{-1}\cdot\left(\boldsymbol{C}^{(n)}(x_{s+1},t)-\boldsymbol{C}^{(n)}(x_s,t)\right),\end{aligned} \tag{4.56}$$

where ϕ_ι are given by

$$\phi_\iota\left(x,x_s,x_{s+1},x_0,t,d_i^{(n)}\right) = \frac{\left(\operatorname{erf}\left(\frac{x-x_0}{2\sqrt{td_i}}\right)-\operatorname{erf}\left(\frac{x_s-x_0}{2\sqrt{td_i}}\right)\right)}{\left(\operatorname{erf}\left(\frac{x_{s+1}-x_0}{2\sqrt{td_i}}\right)-\operatorname{erf}\left(\frac{x_s-x_0}{2\sqrt{td_i}}\right)\right)}. \tag{4.57}$$

Eqs. (4.56) and (4.57) can be considered as general expressions derived through TMM for the concentration profiles of diffusion couples with any number of components in terms of error functions; they yield expressions derived by Fujita and Gosting [51] for the concentration profiles of ternary diffusion couples.

Interdiffusion fluxes can also be integrated with respect to time. In terms of the Boltzmann parameter $\lambda = x/\sqrt{t}$, Eq. (4.20) for $\tilde{J}_i(x)$ can be written as [37]

$$\tilde{J}_i(x,t)2\sqrt{t} = \int_{C_i^{(+)}}^{C_i(\lambda)}(\lambda'-\lambda)dC_i(\lambda') \equiv K_i(\lambda), \quad (i=1,2,\ldots,n) \tag{4.58}$$

where $K_i(\lambda)$ is the specific value of the integral. The total interdiffusion flux $\boldsymbol{S}^{(n)}$ for a given value of λ is given by [37]

$$\boldsymbol{S}^{(n)} = \int_0^t \tilde{\boldsymbol{J}}^{(n)}(x,t)dt = 2t\tilde{\boldsymbol{J}}^{(n)}(x,t). \tag{4.59}$$

If the interdiffusion fluxes, $\tilde{\boldsymbol{J}}^{(n)}(x_0, t)$, at Matano plane are known, $\tilde{\boldsymbol{J}}^{(n)}(x, t)$ on the right-side of Eq. (4.59) can be expressed in terms of $\tilde{\boldsymbol{J}}^{(n)}(x_0, t)$ by utilizing Eq. (4.50). Hence, in terms of the transfer matrix $\boldsymbol{T}$, one gets

$$\boldsymbol{S}^{(n)} = 2t\boldsymbol{T}\left(x, x_0, t, \tilde{\boldsymbol{D}}^{(n)}\right) \cdot \tilde{\boldsymbol{J}}^{(n)}(x_0, t). \tag{4.60}$$

If the interdiffusion flux is integrated over a distance range from x_s to x, one obtains

$$\int_{x_s}^{x} dx' \tilde{\boldsymbol{J}}^{(n)}(x', t) = \int_{x_s}^{x} dx' \boldsymbol{T}\left(x, x_s, t, \tilde{\boldsymbol{D}}^{(n)}\right) \cdot \tilde{\boldsymbol{J}}^{(n)}(x_s, t). \tag{4.61}$$

If x_s is taken as x_o and $x \to \infty$, one can acquire from Eq. (4.61) the following expression:

$$\left(\boldsymbol{P}^{-1} \cdot \lim_{x\to\infty} \int_{x_0}^{\infty} dx' \tilde{\boldsymbol{J}}^{(n)}(x', t)\right) = \sqrt{\pi t}\left(\Delta^{(n)}\right)^{1/2} \cdot \left(\boldsymbol{P}^{-1} \cdot \tilde{\boldsymbol{J}}^{(n)}(x_0, t)\right). \tag{4.62}$$

4.7.2 Application to Ternary Diffusion

The transfer matrix analysis presented above is illustrated for the case of ternary diffusion characterized by the diffusion coefficient matrix, $\boldsymbol{D}^{(3)}$, expressed by

$$\boldsymbol{D}^{(3)} = \begin{pmatrix} \tilde{D}_{11}^{(3)} & \tilde{D}_{12}^{(3)} \\ \tilde{D}_{21}^{(3)} & \tilde{D}_{22}^{(3)} \end{pmatrix} \tag{4.63}$$

where component 3 is taken as the dependent component. Diagonalization of the diffusion matrix $\boldsymbol{D}^{(3)}$ has been addressed earlier by Toor [52,53], Gupta and Cooper [54], Taylor and Krishna [55], and Ram-Mohan and Dayananda [50]. The eigenvalue problem,

$$\tilde{D}_{ij}^{(3)} \Psi_j = d\mathbf{1}\Psi_i, \tag{4.64}$$

where the right side contains the unit matrix denoted by **1**, is initially solved and the eigenvalues d_i are given by the roots of the characteristic polynomial $\Xi(d)$ given by

$$\Xi(d) = \det \begin{vmatrix} D_{11}^{(3)} - d & D_{12}^{(3)} \\ D_{21}^{(3)} & D_{22}^{(3)} - d \end{vmatrix}. \tag{4.65}$$

The eigenvalues are

$$d_{1,2} = \frac{1}{2}\left[\left(\tilde{D}_{11}^{(3)} + \tilde{D}_{22}^{(3)}\right) \pm \sqrt{\Lambda}\right] \tag{4.66}$$

where the discriminant Λ is

$$\Lambda = \left(\tilde{D}_{11}^{(3)} - \tilde{D}_{22}^{(3)}\right)^2 + 4\tilde{D}_{12}^{(3)}\tilde{D}_{21}^{(3)}. \tag{4.67}$$

The corresponding eigenvectors, expressed by

$$\Psi^{(i)} = \begin{pmatrix} \alpha_i \\ \beta_i \end{pmatrix}, \quad (i = 1, 2) \tag{4.68}$$

are obtained by substituting d_1 into either of the two equations in Eq. (4.64) and solving for β_1 in terms of α_1, or by setting $\alpha_1 = 1$, assuming that the diagonal element in Eq. (4.64) is not zero. In a similar manner, we solve for α_2 in terms of $\beta_2 = 1$, invoking the traditional assumption of "diagonal dominance," without any loss of generality. Hence, one can show [50] that

$$\beta_1 = -\frac{(\tilde{D}_{11}^{(3)} - d_1)}{\tilde{D}_{12}^{(3)}} \tag{4.69}$$

and

$$\alpha_2 = -\frac{(\tilde{D}_{22}^{(3)} - d_2)}{\tilde{D}_{21}^{(3)}}. \tag{4.70}$$

Upon arranging the eigenvectors in columnar matrix form, a matrix $\boldsymbol{P}$ is defined by [50]

$$\boldsymbol{P} = \left(\begin{pmatrix} 1 \\ \beta_1 \end{pmatrix} \begin{pmatrix} \alpha_2 \\ 1 \end{pmatrix} \right) \tag{4.71}$$

(the parentheses around the columns are shown merely to describe the arrangement of the matrix elements in $\boldsymbol{P}$), together with its inverse

$$\boldsymbol{P}^{-1} = \frac{1}{(1 - \alpha_2\beta_1)} \begin{pmatrix} 1 & -\alpha_2 \\ -\beta_1 & 1 \end{pmatrix}. \tag{4.72}$$

Eq. (4.64) can now be written in a matrix form as

$$\tilde{\boldsymbol{D}}^{(3)} \cdot \boldsymbol{P} = \boldsymbol{P} \cdot \boldsymbol{\Delta}^{(3)} \tag{4.73}$$

where $\boldsymbol{\Delta}^{(3)}$ is the diagonal matrix

$$\boldsymbol{\Delta}^{(3)} = \begin{pmatrix} d_1 & 0 \\ 0 & d_2 \end{pmatrix}. \tag{4.74}$$

From Eq. (4.72) it follows that

$$\boldsymbol{P}^{-1} \cdot \tilde{\boldsymbol{D}}^{(3)} \cdot \boldsymbol{P} = \boldsymbol{\Delta}^{(3)}. \tag{4.75}$$

It can be seen from Eq. (4.75) that the matrix of eigenvectors, $\boldsymbol{P}$, is the similarity transformation matrix that diagonalizes matrix $\tilde{\boldsymbol{D}}^{(3)}$.

On the basis of Eq. (4.51), the transfer matrix $\boldsymbol{T}(x, x_s, t, \tilde{\boldsymbol{D}}^{(3)})$ over the interval $x_s \le x \le x_{s+1}$ is given by

$$\boldsymbol{T}\left(x, x_s, t, \tilde{\boldsymbol{D}}^{(3)}\right) = \boldsymbol{P} \cdot \begin{bmatrix} \exp\left[-\frac{\{(x-x_0)^2-(x_s-x_0)^2\}}{4td_1}\right] & 0 \\ 0 & \exp\left[-\frac{\{(x-x_0)^2-(x_s-x_0)^2\}}{4td_2}\right] \end{bmatrix} \cdot \boldsymbol{P}^{-1}. \tag{4.76}$$

On the basis of Eqs. (4.48) and (4.75), the equation $\boldsymbol{J}^{(3)} = -\boldsymbol{D}^{(3)} \cdot \frac{\partial \boldsymbol{C}^{(3)}}{\partial x}$ can be decoupled; the decoupled expressions [56,57] written in component form are as follows:

$$\begin{aligned} \tilde{J}_1 - \alpha_2 \tilde{J}_2 &= -d_1\left(\frac{\partial C_1}{\partial x} - \alpha_2 \frac{\partial C_2}{\partial x}\right) \\ -\beta_1 \tilde{J}_1 + \tilde{J}_2 &= -d_2\left(-\beta_1 \frac{\partial C_1}{\partial x} + \frac{\partial C_2}{\partial x}\right) \end{aligned} \tag{4.77}$$

where linear combinations of the interdiffusion fluxes of the independent components, 1 and 2, are directly proportional to a linear combination of their concentration gradients through eigenvalues.

An application of the TMM methodology is illustrated with an experimental Cu–Ni–Zn diffusion couple [17,29] assembled with Alloy I (30.1Cu–44.7Ni–25.2 Zn, at.%) and Alloy II (80.6Cu–19.4Ni, at.%) and annealed at 775°C for two days. From the experimental concentration profiles, two sets of ternary interdiffusion coefficients, $\tilde{D}_{ij}^3$, were determined, one for each composition range on either side of Matano plane, by employing the *Multi*DiFlux program [58] based on the methodology of Dayananda and Sohn [25] for the determination of interdiffusion coefficients. These coefficients are listed in Table 4.3 and considered as constants (average values) over the selected regions. From these sets of $\tilde{D}_{ij}^3$'s, Eq. (4.56) was utilized for the generation of concentration profiles on both sides of Matano plane. These generated profiles are presented in Fig. 4.5 along with those obtained on the basis of Fujita–Gosting solution [51] for ternary diffusion couples. The two sets of concentration profiles are found to be identical.

The TMM analysis of ternary diffusion has been recently extended by Ram-Mohan and Dayananda [57] to derive general expressions for the slope of the ternary diffusion path in terms of eigenvalue and eigenvector parameters at any section within the diffusion zone; additional constraints relating eigenvalues and concentration gradients of the individual components at selected sections have been derived.

Table 4.3 Average interdiffusion coefficients $\bar{\tilde{D}}_{ij}^3$ $(i, j = 1, 2)$ determined over two regions, one on each side of Matano plane, for the Cu–Ni–Zn ternary diffusion couple [17,29] with the aid of the MultiDiFlux software [58] (Zn = 1; Ni = 2; Cu = 3)

x coordinates of regions	$\bar{\tilde{D}}_{11}^3$ (m^2/s)	$\bar{\tilde{D}}_{12}^3$ (m^2/s)	$\bar{\tilde{D}}_{21}^3$ (m^2/s)	$\bar{\tilde{D}}_{22}^3$ (m^2/s)
0 to 198	$+1.36 \times 10^{-14}$	$+1.45 \times 10^{-15}$	-5.72×10^{-15}	$+1.60 \times 10^{-15}$
198 to 330	$+5.58 \times 10^{-15}$	$+2.40 \times 10^{-15}$	-3.22×10^{-18}	$+1.15 \times 10^{-15}$

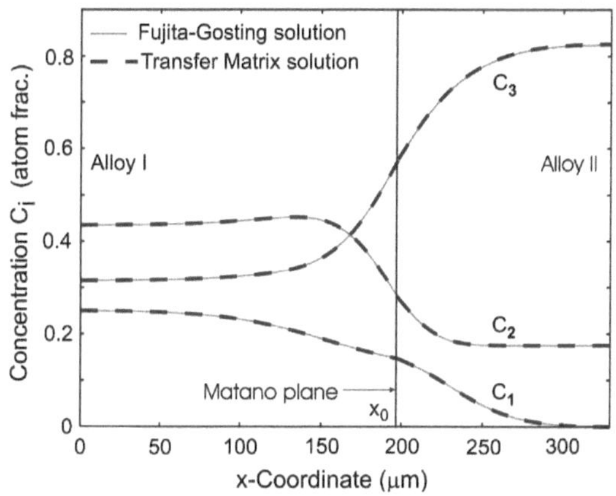

Figure 4.5 Concentration profiles generated by employing the Transfer Matrix Method (TMM) compared with those obtained from the Fujita–Gosting solution for the Cu–Ni–Zn diffusion couple characterized by $\bar{\tilde{D}}_{ij}^3$ reported in Table 4.3. (Zn = 1; Ni = 2; Cu = 3) [50].

4.7.3 Application to Quaternary Diffusion

For a diffusion couple involving four components, the matrix of diffusion coefficients $\tilde{\boldsymbol{D}}^{(4)}$ can be expressed by [59,60]

$$\tilde{\boldsymbol{D}}^{(4)} = \begin{bmatrix} \tilde{D}_{11}^4 & \tilde{D}_{12}^4 & \tilde{D}_{13}^4 \\ \tilde{D}_{21}^4 & \tilde{D}_{22}^4 & \tilde{D}_{23}^4 \\ \tilde{D}_{31}^4 & \tilde{D}_{32}^4 & \tilde{D}_{33}^4 \end{bmatrix} \tag{4.78}$$

The eigenvalues of the matrix $\tilde{\boldsymbol{D}}^{(4)}$ are d_1, d_2, and d_3, and the matrix $\boldsymbol{P}$ obtained from its eigenvectors is given by [60]

$$\boldsymbol{P} = [(\boldsymbol{p}_1), (\boldsymbol{p}_2), (\boldsymbol{p}_3)] = \left[\begin{pmatrix} 1 \\ \beta_1 \\ \gamma_1 \end{pmatrix} \begin{pmatrix} \alpha_2 \\ 1 \\ \gamma_2 \end{pmatrix} \begin{pmatrix} \alpha_3 \\ \beta_3 \\ 1 \end{pmatrix} \right]. \tag{4.79}$$

If $C_i^{(4)}(x_s)$ and $C_i^{(4)}(x_{s+1})$ represent the concentrations of components ($i = 1, 2, 3$) at locations x_s and x_{s+1}, respectively, the corresponding concentrations at any section x within x_s and x_{s+1} can be expressed in the matrix form by

$$\begin{bmatrix} C_1^{(4)}(x) \\ C_2^{(4)}(x) \\ C_3^{(4)}(x) \end{bmatrix} = \begin{bmatrix} C_1^{(4)}(x_s) \\ C_2^{(4)}(x_s) \\ C_3^{(4)}(x_s) \end{bmatrix} + \begin{bmatrix} K_{11} & K_{12} & K_{13} \\ K_{21} & K_{22} & K_{23} \\ K_{31} & K_{32} & K_{33} \end{bmatrix} \cdot \begin{bmatrix} \phi_1 \\ \phi_2 \\ \phi_3 \end{bmatrix} \tag{4.80}$$

where

$$\phi_i\left(x, x_s, x_{s+1}, x_0, t, d_i^{(4)}\right) = \frac{\operatorname{erf}\left(\frac{x-x_0}{2\sqrt{td_i}}\right) - \operatorname{erf}\left(\frac{x_s-x_0}{2\sqrt{td_i}}\right)}{\operatorname{erf}\left(\frac{x_{s+1}-x_0}{2\sqrt{td_i}}\right) - \operatorname{erf}\left(\frac{x_s-x_0}{2\sqrt{td_i}}\right)}. \tag{4.81}$$

The matrix elements K_{ij} in Eq. (4.80) are functions of eigenvalues, eigenvectors, and the concentrations, $C_i^{(4)}(x_{s+1}, t)$ and $C_i^{(4)}(x_s, t)$, as discussed in [60].

An application of Eq. (4.80) for the generation of concentration profiles is presented for a quaternary Cu–Ni–Zn–Mn diffusion couple investigated by Kansky and Dayananda [59] at 775°C for 48 hours. The couple identified by H/A was assembled with two terminal alloys, H (59Cu–22.3Ni–19Mn) and A (67.5Cu–23.3Ni–0.92 Zn, at%). From the experimental concentration profiles, a single set of average quaternary interdiffusion coefficients was calculated over the entire diffusion zone by employing the *Multi*DiFlux program [58], and the calculated $\tilde{\boldsymbol{D}}^{(4)}$ matrix is given by [61]

$$\tilde{\boldsymbol{D}}^{Cu} = \begin{pmatrix} 3.75 & 0.25 & -0.44 \\ 0.55 & 4.7 & 2.26 \\ -0.008 & -1.49 & 0.72 \end{pmatrix} \times 10^{-15}\ \text{m}^2\,\text{s}^{-1}. \tag{4.82}$$

In matrix $\tilde{\boldsymbol{D}}^{(4)}$ (see Eq. (4.78)), 1 = Zn, 2 = Mn, 3 = Ni, and 4 = Cu. The eigenvalues calculated for $\tilde{\boldsymbol{D}}^{(Cu)}$ are represented by:

$$\begin{bmatrix} d_1 \\ d_2 \\ d_3 \end{bmatrix} = \begin{bmatrix} 4.21 \\ 2.80 \\ 2.10 \end{bmatrix} \times 10^{-15}\ \text{m}^2\,\text{s}^{-1} \tag{4.83}$$

and the $\boldsymbol{P}$ matrix obtained from the eigenvectors is given by

$$\boldsymbol{P} = \left[\begin{pmatrix} 1 \\ 1.04 \\ -0.465 \end{pmatrix} \begin{pmatrix} -0.582 \\ 1 \\ -0.697 \end{pmatrix} \begin{pmatrix} 0.429 \\ -0.985 \\ 1 \end{pmatrix} \right]. \tag{4.84}$$

Concentration profiles regenerated from Eq. (4.80) for the quaternary couple are presented in Fig. 4.6 along with the experimental data points. It is apparent from Fig. 4.6

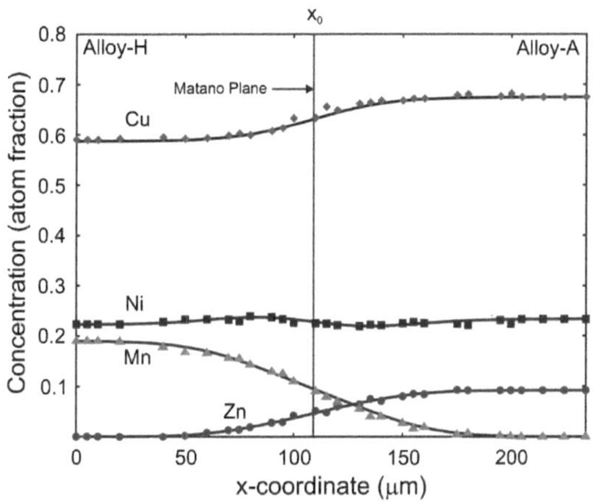

Figure 4.6 Concentration profiles generated for the quaternary diffusion couple H/A annealed at 775°C for 48 hours on the basis of evaluated quaternary interdiffusion coefficients. Experimental concentration data points are also shown [60].

that the TMM-profiles generated from the interdiffusion coefficients agree very well with the experimental concentration profiles.

4.8 ESTIMATION OF TRACER DIFFUSION COEFFICIENTS IN A TERNARY SYSTEM

A radioactive isotope of an element, considered identical to the stable isotope except for a small difference in mass, is generally used for the determination of impurity and tracer diffusion coefficients [62]. These coefficients are classified on the basis of the particular tracer–solvent system, composition, and temperature. Experimental determination of tracer diffusion coefficient D_i^* of component i at a given temperature generally involves solving Fick's law for diffusion of a thin film of tracer deposited on the solvent disk. For example, ^{59}Fe, ^{51}Cr and ^{57}Ni tracers were used in the determination of tracer diffusion coefficients in the ternary Fe–Cr–Ni alloys [63].

An alternative method of estimating tracer diffusion coefficients is to extrapolate the ternary interdiffusion coefficients. For a component i diffusing into a binary alloy with components j and k, the cross interdiffusion coefficient of component i goes to zero at infinite dilution in an alloy $j-k$. The tracer diffusion coefficient of component i, D_i^*, can then be estimated as

$$C_i \to 0 \qquad \tilde{J}_i = -\tilde{D}_{ii}^k \frac{\partial C_i}{\partial x} = -D_{i(jk)}^* \frac{\partial C_i}{\partial x} \tag{4.85}$$

where $\tilde{D}_{ii}^{k}$ is the main interdiffusion coefficient of component i in a binary alloy $j-k$ infinitely dilute in component i, and $D_{i(jk)}^{*}$ is the tracer diffusion coefficient of component i in a binary alloy $j-k$. This approach has been applied to the Fe–Ni–Al alloys, where D_{Ni}^{*} in Fe–Al is estimated from $\tilde{D}_{NiNi}^{Fe}$ and $\tilde{D}_{NiNi}^{Al}$ where Ni concentration approaches zero [3].

Another approach by Rönka et al. involves both thermodynamics and diffusion kinetics [64]. By combining the thermodynamics equation $\mu_i = \mu_i^0 + RT\ln a_i$ and Eqs. (4.10) and (4.29), the following equation is derived [65]:

$$\begin{aligned}\tilde{J}_1^M &= -\frac{C_1 D_1^*}{RTV_m}\frac{\partial \mu_1}{\partial x} + C_1\left(\frac{C_1 D_1^*}{RTV_m}\frac{\partial \mu_1}{\partial x} + \frac{C_2 D_2^*}{RTV_m}\frac{\partial \mu_2}{\partial x} + \frac{C_3 D_3^*}{RTV_m}\frac{\partial \mu_3}{\partial x}\right)\\ \tilde{J}_2^M &= -\frac{C_2 D_2^*}{RTV_m}\frac{\partial \mu_2}{\partial x} + C_2\left(\frac{C_1 D_1^*}{RTV_m}\frac{\partial \mu_1}{\partial x} + \frac{C_2 D_2^*}{RTV_m}\frac{\partial \mu_2}{\partial x} + \frac{C_3 D_3^*}{RTV_m}\frac{\partial \mu_3}{\partial x}\right)\end{aligned} \tag{4.86}$$

Eq. (4.86) relates the interdiffusion coefficients and tracer diffusion coefficients, and thus can be used to calculate the tracer diffusion coefficients at the intersection points of the two diffusion paths in the alloys. However, the thermodynamic parameters (chemical potential or activities) in the alloys must be known for this approach to work.

Traditional experimental measurements of tracer diffusion can be time consuming since only one value of the tracer diffusion coefficient for one component at one temperature at one given alloy composition can be obtained for multicomponent alloys. Recently, an efficient way of obtaining the composition-dependent tracer diffusion coefficients in multicomponent alloys has been developed by Belova et al. [66,67]. First, the authors provided a new isotopic interdiffusion formalism for obtaining tracer diffusion coefficients simultaneously with interdiffusion coefficients as a function of composition in a binary system [67]. Then, the isotopic interdiffusion formalism was expanded for the application in a multicomponent system, and then further refined to a case where the isotopes are not needed [66]. This method allows obtaining tracer diffusion coefficients in an interdiffusion condition. It is efficient in that only one diffusion couple/triple is needed to simultaneously obtain interdiffusion and tracer diffusion coefficients.

4.9 DETERMINATION OF EQUILIBRIUM PHASE DIAGRAM

The importance of phase diagrams to materials science and engineering needs no further attestation, particularly for multicomponent systems. Phase diagrams provide understanding of the properties of materials and help in selecting or designing materials. The traditional method of determining a phase diagram is based on equilibrium alloys and thermal analysis. This can be very time consuming due to the amount of alloys that need to be prepared and analyzed, particularly for multicomponent phase diagrams. However, with the help of diffusion couples or the diffusion multiples technique, the efficiency of determining a phase diagram can be greatly improved with more accuracy.

In the determination of phase diagrams through diffusion couple techniques, it is typically assumed that local equilibrium is established across the diffusion zone during the high temperature annealing. By quenching to room temperature, the high temperature phases are assumed to be retained. The compositions can be precisely measured through advanced characterization techniques (i.e., electron probe micro analyzer). The phase boundary composition and solubility range for different intermetallics and solid solutions can be evaluated. Therefore, the binary or isothermal ternary phase diagram can be constructed. For a binary phase diagram, two pure metals are needed for the terminal end of the diffusion couples. For a ternary phase diagram, multiple diffusion couples with different terminal ends and possibly with a few selected equilibrium alloys are necessary. Similarly, a diffusion triple involving ternary diffusion couple has been developed and applied to determine ternary phase diagram, where three pure metals are bonded to form a triple-junction [68]. Theoretically, only one diffusion triple is needed to determine one isothermal ternary phase diagram.

The diffusion couple technique has been further evolved to include the diffusion multiple technique, where an assembly of three or more metal blocks are held in contact [69]. These pure metal blocks are machined into proper shapes and fitted to a designed geometry. After cleaning the surfaces, the diffusion multiples are hot isostatic pressed to achieve an initial contact between different metal surfaces [70,71]. Then diffusion annealing can be carried out for the diffusion multiples. However, it needs to be pointed out that the analyses of diffusion coefficients using a diffusion multiple requires caution because certain boundary condition may not be satisfied. A detailed review of the diffusion multiple can be found in [69].

Obviously, the diffusion couples or multiples have proven to be an effective method to determine the phase diagram. However, for a complete phase diagram, equilibrium alloys may still be needed because not all phases will develop (i.e., nucleation and growth) in a diffusion couple. For example, in the binary Al–Zr diffusion couples, only two intermetallic phases were observed at 625°C [72], while the phase diagram showed a total of eight intermetallics at the same temperature [73]. This may be due to thermodynamics of some phases are not favored to nucleate and grow in the diffusion couple or in the presence of other phases. It is also possible that the diffusion kinetics of some phases is extremely slow and the phase layer is too thin to be properly analyzed. Furthermore, there may be phases that should not develop at high temperature but are present at room temperature in the diffusion couples. For example, precipitates may form during the cooling because of the saturation of the solutes [74] or martensitic phases may form because of martensitic transformation [75]. Therefore, when determining phase diagrams through the diffusion couples technique, it is important to exercise caution and careful observation.

4.10 EXAMPLES OF MULTICOMPONENT DIFFUSION

4.10.1 Diffusion in High Entropy Alloys

High entropy alloys (HEA) are defined as alloys containing at least five major elements, with concentration range between 5 and 35 at.% for each major element [76]. Since their discovery, HEAs have attracted much attention because of high hardness, outstanding high temperature strength, good oxidation and corrosion resistance. Interestingly, HEAs tend to form stable solid solutions with simple crystal structure rather than complicated intermetallics [77] because of their large mixing entropies. The diffusion kinetics in HEAs is important to understand their properties such as high temperature structural stability and the formation of nanostructure. Therefore, understanding the diffusion kinetics from diffusion experiments is important and has just been reported in recent years for HEAs.

Tsai et al. have designed diffusion couple experiments in Co–Cr–Fe–Mn–Ni HEA system and reported the diffusion coefficients [78]. The obtained diffusion coefficients were compared with diffusion in conventional systems with fewer components. It was concluded that the diffusion coefficients in the HEA Co–Cr–Fe–Mn–Ni alloy systems are the smallest compared to those in other systems. However, their method of calculating diffusion coefficients suffers from improper use of pseudo-binary approach, which is explained in Chapter 3 in the same volume.

Similar experiments have been carried out in another HEA system Al–Co–Cr–Fe–Ni recently [79]. The diffusion coefficients obtained from this study were again significantly lower than those in the case of conventional alloys. Therefore, both experiments have provided direct evidence for the sluggish diffusion effect in HEAs.

4.10.2 Diffusion in Amorphous Alloys

Amorphous alloys, also known as metallic glass, have practical applications, due to unique magnetic properties, good mechanical behavior, and high corrosion resistance. The production of amorphous alloys commonly requires a fast cooling rate, so that the atoms are frozen from liquid state. Therefore, there is no long-range order in amorphous alloys, which is different from typical metallic alloys.

The measurement of diffusivity in amorphous alloys has been limited because the amorphous alloys tend to crystallize above the glass transition temperature and the diffusion coefficients are small and usually less than 10^{-17} m^2/s below the crystallization temperature [80]. Due to the discovery of multicomponent bulk metallic glasses with higher resistance to crystallization, the diffusion in metallic glass in its supercooled state has been possible. In order to measure the diffusivity in amorphous alloys, tracer or self-diffusion coefficients have been extracted through the concentration depth profile obtained via ion-beam sputter-sectioning technique or ion mass spectrometry technique.

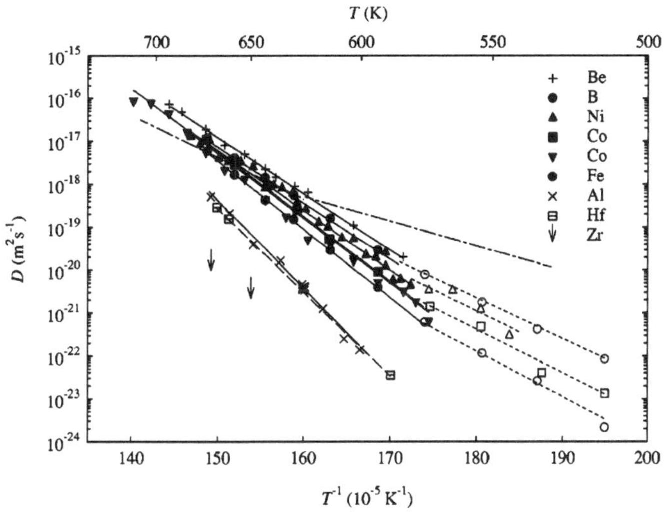

Figure 4.7 Arrhenius diagram for tracer or self-diffusion coefficients of Be, B, Ni, Co, Fe, Al, Hf, and Zr in $Zr_{46.75}Ti_{8.25}Cu_{7.5}Ni_{10}Be_{27.5}$ (Vitreloy 4) [81].

Fig. 4.7 presents the experimental tracer and self-diffusion coefficients of different elements in the amorphous $Zr_{46.75}Ti_{8.25}Cu_{7.5}Ni_{10}Be_{27.5}$ alloys (also known as Vitreloy 4) [81,82]. It is interesting to note that Arrhenius plot exhibits two different slopes at high and low temperature range. This indicates that the effective activation energy at high temperature is larger than that in the low temperature. By comparing the annealing parameters with the TTT diagram of the alloy, it was confirmed that the break point in Arrhenius plot corresponds to the transition from supercooled liquid to the amorphous state [82]. Therefore, it can be concluded that the diffusion mechanism is different between the supercooled and amorphous state [83]. A more detailed discussion on diffusion in amorphous alloys can be found in [84] and [81].

REFERENCES

[1] J.S. Kirkaldy, D.J. Young, Diffusion in the Condensed State, Taylor & Francis, 1987.

[2] T. Heumann, Zur Berechnung von Diffusions Koeffizienten bei Einund Mehrphasiger Diffusion in Festen Legierungen, Zeit Physik Chem 201 (1952) 168–189.

[3] Y.H. Sohn, M.A. Dayananda, Diffusion studies in the β(B2), β'(bcc), and γ(fcc) Fe–Ni–Al alloys at 1000°C, Metallurgical and Materials Transactions A 33 (2002) 3375–3392.

[4] J.R. Manning, L.J. Bruner, Diffusion kinetics for atoms in crystals, American Journal of Physics 36 (1968) 922–923.

[5] J.R. Manning, Cross terms in the thermodynamic diffusion equations for multicomponent alloys, Metallurgical and Materials Transactions B 1 (1970) 499–505.

[6] M.A. Dayananda, Atomic mobilities and vacancy wind effects in multicomponent alloys, Met Trans 2 (1971) 334–335.

[7] J.R. Manning, Diffusion and the Kirkendall shift in binary alloys, Acta Metallurgica 15 (1967) 817–826.
[8] P.T. Carlson, M.A. Dayananda, R.E. Grace, Atomic mobilities and vacancy wind effects for diffusion in ternary silver–zinc–cadmium solid solutions, Metallurgical Transactions A 6 (1975) 1245–1252.
[9] M.C. Petri, M.A. Dayananda, Vacancy wind contributions to intrinsic diffusion, Philosophical Magazine A 76 (1997) 1169–1185.
[10] L. Onsager, Reciprocal relations in irreversible processes. I, Physical Review 37 (1931) 405.
[11] L. Onsager, Reciprocal relations in irreversible processes. II, Physical Review 38 (1931) 2265.
[12] C. Matano, On the relation between the diffusion-coefficients and concentrations of solid metals (the nickel–copper system), Jpn J Phys 8 (1933) 109–113.
[13] L. Boltzmann, Zur integration der Diffusionsgleichung bei variabeln Diffusionscoefficienten, Annalen der Physik 289 (1894) 959–964.
[14] C.W. Kim, M.A. Dayananda, Zero-flux planes and flux reversals in the Cu–Ni–Zn system at 775°C, Metallurgical Transactions A 15 (1984) 649–659.
[15] J.R. Manning, Correlation factors for diffusion in nondilute alloys, Physical Review B 4 (1971) 1111.
[16] G.W. Roper, D.P. Whittle, Interdiffusion in ternary Co–Cr–AI alloys, Metal Science 14 (1980) 21–28.
[17] M.A. Dayananda, An analysis of concentration profiles for fluxes, diffusion depths, and zero-flux planes in multicomponent diffusion, Metallurgical Transactions A 14 (1983) 1851–1858.
[18] J. Philibert, A. Guy, Physique des metaux-diffusion dand les systems an constituants, Comptes Rendus Hebdomadaires Des Seances De L Academie Des Sciences 257 (1963) 2281.
[19] Y. Chen, B. Tang, G. Xu, C. Wang, H. Kou, J. Li, Y. Cui, Diffusion research in BCC Ti–Al–Mo ternary alloys, Metallurgical and Materials Transactions A 45 (2014) 1647–1652.
[20] S.K. Kailasam, J.C. Lacombe, M.E. Glicksman, Evaluation of the methods for calculating the concentration-dependent diffusivity in binary systems, Metallurgical and Materials Transactions A 30 (1999) 2605–2610.
[21] J.S. Kirkaldy, Diffusion in multicomponent metallic systems, Canadian Journal of Physics 35 (1957) 435–440.
[22] C. Kammerer, The Influence of Alloying Additions on Diffusion and Strengthening of Magnesium, University of Central, Florida Orlando, Florida, 2015.
[23] M.S. Thompson, J.E. Morral, A.D. Romig, Applications of the square root diffusivity to diffusion in Ni–Al–Cr alloys, Metallurgical Transactions A 21 (1990) 267–2685.
[24] G.R. Purdy, Fundamentals and applications of ternary diffusion, in: Proceedings of the International Symposium on Fundamentals and Applications of Ternary Diffusion, Hamilton, Ontario, Canada, August 27–28, 1990, Elsevier, 2013.
[25] M.A. Dayananda, Y.H. Sohn, A new analysis for the determination of ternary interdiffusion coefficients from a single diffusion couple, Metallurgical and Materials Transactions A 30 (1999) 535–543.
[26] T.O. Ziebold, Ternary Diffusion in Copper–Silver–Gold Alloys, Massachusetts Institute of Technology, 1965.
[27] V.D. Divya, U. Ramamurty, A. Paul, Interdiffusion and solid solution strengthening in Ni–Co–Pt and Ni–Co–Fe ternary systems, Philosophical Magazine 93 (2013) 2190–2206.
[28] M.A. Dayananda, Zero-flux planes and flux reversals and diffusion paths in ternary and quaternary systems, in diffusion in solids: recent developments, in: M.A. Dayananda, G.E. Murch (Eds.), TMS-AIME Conference Proceedings, 1984, pp. 195–230.
[29] M.A. Dayananda, C.W. Kim, Zero-flux planes and flux reversals in Cu–Ni–Zn diffusion couples, Metallurgical Transactions A 10 (1979) 1333–1339.
[30] Y.H. Sohn, M.A. Dayananda, A double-serpentine diffusion path for a ternary diffusion couple, Acta Materialia 48 (2000) 1427–1433.
[31] C.W. Kim, M.A. Dayananda, Identification of zero-flux planes and flux reversals in several studies of ternary diffusion, Metallurgical Transactions A 14 (1983) 857–864.

[32] J.G. Duh, M.A. Dayananda, Interdiffusion in Fe–Ni–Cr alloys at 1100°C, in: Defect Diffus Forum, vol. 39, Trans Tech Publ, 1985, p. 1.
[33] R. Venkatasubramanian, M.A. Dayananda, Estimation of diffusion path slopes at zero-flux plane compositions, Metallurgical and Materials Transactions A 17 (1986) 362–364.
[34] M.A. Dayananda, D.A. Behnke, Effective interdiffusion coefficients and penetration depths, Scripta Metall Mater 25 (1991) 2187–2191.
[35] M.A. Dayananda, Average effective interdiffusion coefficients in binary and multicomponent alloys, in: Defect Diffus Forum, vol. 95, Trans Tech Publ, 1993, pp. 521–536.
[36] M.A. Dayananda, Y.H. Sohn, Average effective interdiffusion coefficients and their applications for isothermal multicomponent diffusion couples, Scripta Materialia 35 (1996) 683–688.
[37] M.A. Dayananda, Average effective interdiffusion coefficients and the Matano plane composition, Metallurgical and Materials Transactions A 27 (1996) 2504–2509.
[38] R. Bouchet, R. Mevrel, A numerical inverse method for calculating the interdiffusion coefficients along a diffusion path in ternary systems, Acta Materialia 50 (2002) 4887–4900.
[39] J. Cermak, V. Rothova, Concentration dependence of ternary interdiffusion coefficients in Ni_3Al/Ni_3Al–X couples with X = Cr, Fe, Nb and Ti, Acta Materialia 51 (2003) 4411–4421.
[40] H. Wei, W. Gong, S. Huang, C. Zhou, Single-phase interdiffusion in the Ni_3Al–Mo ternary system, Scripta Materialia 64 (2011) 53–56.
[41] M.S. Thompson, J.E. Morral, The square root diffusivity, Acta Metallurgica 34 (1986) 2201–2203.
[42] M.K. Stalker, J.E. Morral, A.D. Romig, Application of the square root diffusivity to diffusion in Ni–Cr–Al–Mo alloys, Metallurgical Transactions A 23 (1992) 3245–3249.
[43] Y. Minamino, Y. Koizumi, N. Tsuji, T. Yamada, T. Takahashi, Interdiffusion in Co solid solutions of Co–Al–Cr–Ni system at 1423 K, Materials Transactions 44 (2003) 63–71.
[44] T. Takahashi, K. Hisayuki, T. Yamane, Y. Minamino, T. Hino, Interdiffusion in ALPHA. Solid solution of the Al–Cu–Mg–Ag system, Materials Transactions 44 (2003) 2252–2257.
[45] A. Paul, A pseudobinary approach to study interdiffusion and the Kirkendall effect in multicomponent systems, Philosophical Magazine 93 (2013) 2297–2315.
[46] A. Paul, T. Laurila, V. Vuorinen, S.V. Divinski, Thermodynamics, Diffusion and the Kirkendall Effect in Solids, Springer, 2014.
[47] P. Kiruthika, A. Paul, A pseudo-binary interdiffusion study in the β-Ni(Pt)Al phase, Phil Mag Lett 95 (2015) 138–144.
[48] S. Santra, A. Paul, Estimation of intrinsic diffusion coefficients in a pseudo-binary diffusion couple, Scripta Materialia 103 (2015) 18–21.
[49] L.R. Ram-Mohan, M.A. Dayananda, A transfer-matrix method for analysis of multicomponent diffusion with any number of components, J Phase Equilib Diff 27 (2006) 566–571.
[50] L.R. Ram-Mohan, M.A. Dayananda, A transfer matrix method for the calculation of concentrations and fluxes in multicomponent diffusion couples, Acta Materialia 54 (2006) 2325–2334.
[51] H. Fujita, L.J. Gosting, An exact solution of the equations for free diffusion in three-component systems with interacting flows, and its use in evaluation of the diffusion coefficients, Journal of the American Chemical Society 78 (1956) 1099–1106.
[52] H.L. Toor, Solution of the linearized equations of multicomponent mass transfer: I, AIChE Journal 10 (1964) 448–455.
[53] H.L. Toor, Solution of the linearized equations of multicomponent mass transfer: II. Matrix methods, AIChE Journal 10 (1964) 460–465.
[54] P.K. Gupta, A.R. Cooper, The [D] matrix for multicomponent diffusion, Physica 54 (1971) 39–59.
[55] R. Taylor, R. Krishna, Multicomponent Mass Transfer, John Wiley & Sons, 1993.
[56] M.A. Dayananda, An examination of a multicomponent diffusion couple, J Phase Equilib Diff 27 (2006) 572–581.

[57] L.R. Ram-Mohan, M.A. Dayananda, Ternary diffusion path in terms of eigenvalues and eigenvectors, Philosophical Magazine 96 (2016) 938–954.

[58] L.R. Ram-Mohan, M.A. Dayananda, "*Multi*DiFlux," software, https://engineering.purdue.edu/MSE/research/MultiDiFlux/index.html, Purdue University, 2005.

[59] K.E. Kansky, M.A. Dayananda, Quaternary diffusion in the Cu–Ni–Zn–Mn system at 775°C, Metallurgical Transactions A 16 (1985) 1123–1132.

[60] K.N. Kulkarni, A.M. Girgis, L.R. Ram-Mohan, M.A. Dayananda, A transfer matrix analysis of quaternary diffusion, Philosophical Magazine 87 (2007) 853–872.

[61] K. Kulkarni, M.A. Dayananda, A transfer matrix analysis of a quaternary Cu–Ni–Zn–Mn diffusion couple, in: Materials Science and Technology (MS&T) 2006: Fundamentals and Characterization, vol. 2, 2006, p. 155.

[62] P. Shewmon, Diffusion in Solids, second edition, The Minerals, Metals & Materials Society, 1989, p. 246 (Retroactive Coverage) (United States).

[63] S.J. Rothman, L.J. Nowicki, G.E. Murch, Self-diffusion in austenitic Fe–Cr–Ni alloys, Journal of Physics F: Metal Physics 10 (1980) 383.

[64] K.J. Rönkä, A.A. Kodentsov, P.J.J. Van Loon, J.K. Kivilahti, F.J.J. Van Loo, Thermodynamic and kinetic study of diffusion paths in the system Cu–Fe–Ni, Metallurgical and Materials Transactions A 27 (1996) 2229–2238.

[65] M.S. Thompson, J.E. Morral, The effect of composition on interdiffusion in ternary alloys, Acta Metallurgica 34 (1986) 339–346.

[66] I.V. Belova, Y.H. Sohn, G.E. Murch, Measurement of tracer diffusion coefficients in an interdiffusion context for multicomponent alloys, Phil Mag Lett 95 (2015) 416–424.

[67] I.V. Belova, N.S. Kulkarni, Y.H. Sohn, G.E. Murch, Simultaneous measurement of tracer and interdiffusion coefficients: an isotopic phenomenological diffusion formalism for the binary alloy, Philosophical Magazine 93 (2013) 3515–3526.

[68] J.C. Zhao, Z. Jin, Determination of phase equilibria in the Ti–Fe–Co system at 900°C, Z Metallkd 81 (1990) 247–250.

[69] J.C. Zhao, Combinatorial approaches as effective tools in the study of phase diagrams and composition–structure–property relationships, Prog Mater Sci 51 (2006) 557–631.

[70] J.C. Zhao, M.R. Jackson, L.A. Peluso, Mapping of the Nb–Ti–Si phase diagram using diffusion multiples, Materials Science and Engineering: A 372 (2004) 21–27.

[71] J.C. Zhao, Reliability of the diffusion-multiple approach for phase diagram mapping, J Mater Sci 39 (2004) 3913–3925.

[72] J. Dickson, L. Zhou, A.P. y Puente, M. Fu, D.D. Keiser, Y.H. Sohn, Interdiffusion and reaction between Zr and Al alloys from 425° to 625°C, Intermetallics 49 (2014) 154–162.

[73] J. Murray, A. Peruzzi, J.P. Abriata, The Al–Zr (aluminum–zirconium) system, Journal of Phase Equilibria 13 (1992) 277–291.

[74] Y. Park, K. Huang, A.P. y Puente, H.S. Lee, B.H. Sencer, J.R. Kennedy, Y.H. Sohn, Diffusional interaction between U-10 wt.% Zr and Fe at 903 K, 923 K, and 953 K (630°C, 650°C, and 680°C), Metallurgical and Materials Transactions A 46 (2015) 72–82.

[75] L. Zhou, C. Kammerer, A. Giri, K. Cho, Y.H. Sohn, Microstructural development and ternary interdiffusion in Ni–Mn–Ga alloys, Metallurgical and Materials Transactions A 46 (2015) 5572–5587.

[76] J.W. Yeh, S.K. Chen, S.J. Lin, J.Y. Gan, T.S. Chin, T.T. Shun, C.H. Tsau, S.Y. Chang, Nanostructured high-entropy alloys with multiple principal elements: novel alloy design concepts and outcomes, Advanced Engineering Materials 6 (2004) 299–303.

[77] J.W. Yeh, S.J. Lin, T.S. Chin, J.Y. Gan, S.K. Chen, T.T. Shun, C.H. Tsau, S.Y. Chou, Formation of simple crystal structures in Cu–Co–Ni–Cr–Al–Fe–Ti–V alloys with multiprincipal metallic elements, Metallurgical and Materials Transactions A 35 (2004) 2533–2536.

[78] K.Y. Tsai, M.H. Tsai, J.W. Yeh, Sluggish diffusion in Co–Cr–Fe–Mn–Ni high-entropy alloys, Acta Materialia 61 (2013) 4887–4897.

[79] J. Dąbrowa, W. Kucza, G. Cieślak, T. Kulik, M. Danielewski, J.W. Yeh, Interdiffusion in the FCC-structured Al–Co–Cr–Fe–Ni high entropy alloys: experimental studies and numerical simulations, Journal of Alloys and Compounds 674 (2016) 455–462.

[80] K. Hoshino, R.S. Averback, H.T. Hahn, S.J. Rothman, Tracer diffusion of ^{60}Co and ^{63}Ni in amorphous NiZr alloy, in: Defect Diffus Forum, vol. 59, Trans Tech Publ, 1988, pp. 225–232.

[81] F. Faupel, W. Frank, M.P. Macht, H. Mehrer, V. Naundorf, K. Rätzke, H.R. Schober, S.K. Sharma, H. Teichler, Diffusion in metallic glasses and supercooled melts, Reviews of Modern Physics 75 (2003) 237.

[82] K. Knorr, M.P. Macht, K. Freitag, H. Mehrer, Self-diffusion in the amorphous and supercooled liquid state of the bulk metallic glass $Zr_{46.75}Ti_{8.25}Cu_{7.5}Ni_{10}Be_{27.5}$, J Non-Cryst Solids 250 (1999) 669–673.

[83] P. Fielitz, M.P. Macht, V. Naundorf, G. Frohberg, Diffusion in ZrTiCuNiBe bulk glasses at temperatures around the glass transition, J Non-Cryst Solids 250 (1999) 674–678.

[84] H. Mehrer, Diffusion in Solids: Fundamentals, Methods, Materials, Diffusion-Controlled Processes, Springer Science & Business Media, 2007.

CHAPTER 5

Point Defects and Diffusion in Semiconductors

T.Y. Tan
Duke University, Durham, NC, USA

5.1 INTRODUCTION

Solid state electronic products have become an integral part of modern day life. Integrated circuit chips fabricated using Si as the substrate material are widely used in computers, communication devices such as smart phones, and entertaining machines, as well as sensors, as dedicated and general purpose measurements, and control instruments. Lasers and light emitting devices, which are optoelectronic devices fabricated using GaAs-based compound semiconductors as substrate materials, are used in communication devices, entertaining machines, instruments or just as lights. These solid-state devices are consisting of pn-junctions produced by diffusion of n- and p-type dopant atoms, and/or heterojunctions consisting of layers of different materials that are also doped. The n- and p-type dopants are specific kinds of substitutional impurity species producing the electric carriers electrons e and holes h, respectively, in a semiconductor. Diffusion processes are also involved in controlling detrimental metallic impurity contents, in silicide formation and in thermal SiO_2 growth in fabricating devices using Si. Diffusion in Ge will also be mentioned for the reasons that it is the simplest case among semiconductors on the one hand while on the other its importance as a material to modify Si for the future generations of devices is emerging.

Diffusion in semiconductors is considerably more complicated than in metals. In metals the dominant point defect species is vacancies, which lead to concentration profiles of self-diffusion and the diffusion of substitutional impurities of the erfc-function type with a diffusivity characterized by a single activation enthalpy over the whole temperature range of interests. In semiconductors diffusion profiles are more complex, they can have shapes substantially deviated away from that of the erfc-function family, and many sets of diffusivity values may exist. These features arise because of the following complicating factors. First, vacancies V and self-interstitials I both contribute to self-diffusion and to the diffusion of substitutional impurities. Second, the dominant point defect species are often consisting of the neutral and the electrically charged ones. The concentrations of charged point defect species are dependent upon the semiconductor Fermi level that is determined by doping conditions. This renders diffusivities of atom species utilizing these point defects also dependent upon the semiconductor Fermi level.

Handbook of Solid State Diffusion, Volume 1
DOI: 10.1016/B978-0-12-804287-8.00005-1

An interesting and complicated feature of a dopant is thus arrived at: its diffusivity can be dependent upon its own concentration. Third, a certain class of impurities is formed by interstitial–substitutional species in that the species predominantly occupies substitutional sites and with a small fraction on interstitial sites, which migrate rapidly and are hence responsible for the diffusion of the species. Fourth, compound semiconductors, e.g., GaAs, possess a range of thermodynamically allowed point defect concentrations, which coexists with an also thermodynamically allowed composition range. Hence, at a given temperature, any species utilizing point defects as diffusion vehicles cab exhibit a range of diffusivity values. Last but not least, semiconductor devices are consisting of pn-junctions and/or heterojunctions that are also doped. Segregation of dopant and other impurity species occurs concomitantly with diffusion at these junctions. Understanding, modeling, and quantitative treatment of the involved physical mechanisms of diffusion in Si, Ge, and GaAs, along with other associated features, are addressed in this chapter.

5.2 POINT DEFECT FUNDAMENTALS IN SEMICONDUCTORS

In semiconductors, hydrogen, noble gases, metal elements Li, Na, etc., and transition metal elements Fe, Ni, Cu, etc., dissolve on interstitial sites and migrate from one interstitial site to another not involving native point defects. On the other hand, self-diffusion and diffusion of substitutional species are governed by point defects. In crystals, a point defect is a thermal equilibrium species that lowers the crystal Gibbs free energy.

5.2.1 Point Defects in Si and Ge

Point defect species in the elemental semiconductors Si and Ge include both lattice vacancies V and self-interstitials I. In Si both I and V contribute to self-diffusion and the diffusion of substitutional impurity atoms, while in Ge only V contributes.

A vacancy has a formation Gibbs free energy, g_V^f, given by

$$g_V^f = h_V^f - Ts_V^f, \tag{5.1}$$

where h_V^f and s_V^f are respectively the enthalpy and entropy of formation of the vacancy. The values of h_V^f and s_V^f are over and above those of the host crystal atoms, i.e., positive. h_V^f is due to the creation of the atomic sized cavity, which possesses broken bonds, and s_V^f is due to lattice atom vibrational motion interruption by the vacancy with its value positive. The value of g_V^f is also positive because usually Ts_V^f is but a small fraction of h_V^f for all temperatures of interest, and hence $g_V^f \approx h_V^f$ holds for practical purposes. The thermal equilibrium concentration of V is

$$c_V^{eq} = \frac{C_V^{eq}}{C_o} = \exp\left(-\frac{g_V^f}{k_B T}\right), \tag{5.2}$$

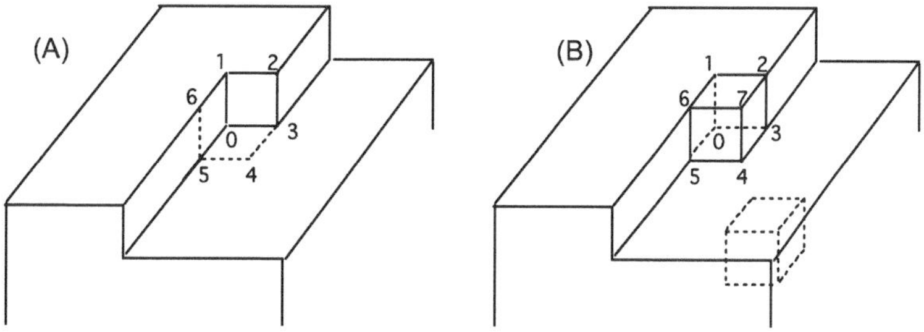

Figure 5.1 Schematic drawing showing, for an elemental crystal, (A) a surface kink site, and (B) the formation of a vacancy via the process of removing a crystal interior atom and placing it at the surface kink site.

where c_V^{eq} and C_V^{eq} are respectively the fractional and actual thermal equilibrium V concentrations, C_o the host crystal lattice atom density, k_B Boltzmann's constant, and T the absolute temperature.

The enthalpy of vacancy formation h_V^f is the crystal enthalpy increase occurred by removing an atom from a lattice site in the interior of the crystal to create the vacancy, which includes two parts. First, the enthalpy increase due to the removed atom, and second, that of the atomic sized cavity left behind, which is the vacancy. The removed atom may either be placed into vacuum an infinite distance away from the crystal, or placed onto the crystal surface. The latter case is the lower energy process, for in the first case the cohesive energy of one crystal atom must be included as part of h_V^f while in the latter case this energy is not present. On a crystal surface, the existence of atomistic steps having kinks in them is a thermal equilibrium phenomenon. Placing the atoms removed from the crystal interior at such kink sites is therefore the lowest energy process. Thus, V is created by removing an atom from a crystal interior lattice position and placing it at a crystal exterior surface kink site, see Fig. 5.1 for an illustration. The creation of I is the reverse. Placing the removed atom onto the crystal surface kink site results in no change in the number of the crystal bulk atoms, no change in the number of the crystal exterior surface atoms, and no change in the crystal exterior surface area, and hence also no energy change. These aspects are readily seen from Fig. 5.1. First, removing a crystal interior atom decreases the bulk atom number by one. Upon replacing the atom at the crystal surface kink site, Fig. 5.1(B), the original surface atom with its upward exposed surface marked 0345 now becomes a bulk atom. Therefore, there is no net bulk atom number change. The disappearance of the surface atom with its upward exposed surface marked 0345 is replaced by the new surface atom just placed at the kink site. Therefore, there is also no crystal exterior surface atom number change. As for the fact that there is no total surface area change, it is readily seen from the fact that the old unit surfaces marked 0123, 0345, and 1056 are now simply replaced by the new unit surfaces 4567, 1276, and 2347, respectively. For the creation of a large number of V, a crystal size change, and therefore also a surface area change, does occur. However, this change is extremely small. Consider a cube-shaped crystal containing 10^{21} atoms cm^{-3}. For 10^{14} cm^{-3} V created, the surface area change is $4 \times 10^7 a$, where a is the unit area depicted in Fig. 5.1 as 0123, etc. Therefore, the crystal surface area change associated with the formation of each vacancy is $4 \times 10^{-7} a$. Since the surface free energy associated with a shall in no case exceed g^f, it is seen that in determining the thermal equilibrium point defect concentrations the effect of this extremely small surface area change is negligible. This leaves only the energy associated with the existence of the cavity as the sole contributor to h_V^f. In a rough estimate, the upper bond of h_V^f shall just be the cohesive energy of one atom h_c, because nearest neighbor atoms surrounding the cavity now each possesses a severed bond with the total number of them just equal to that of the removed atom. Real values of h_V^f shall be smaller than h_c because bond reconstruction will occur to lower the energy of the severed bonds. For Si, the value of h_V^f deduced from experiments ranges from 3.5 to 4 eV [1,2], somewhat smaller than h_c value of ~4.63 eV, which is as expected. Quantum mechanically calculated h_V^f values for Si are in the same range [3–7]. For Ge, $h_c \sim 3.85$ eV, but there is no h_V^f value available. Much less is known for I in Si, but there are some indications that h_I^f is ~4 eV [1]. No corresponding work exists for Ge.

For V concentrations deviated away form C_V^{eq}, the crystal Gibbs free energy density expression is[1]

$$G_V = C_V k_B T \exp\left(\ln \frac{C_V}{C_V^{eq}} - 1\right), \tag{5.3}$$

where C_V is the actual V concentration, and the corresponding chemical potential is

$$\mu_V = k_B T \ln \frac{C_V}{C_V^{eq}}. \tag{5.4}$$

Eqs. (5.3) and (5.4) are plotted in Fig. 5.2 where it is seen that the crystal Gibbs free energy is the lowest for $C_V/C_V^{eq} = 1$, and $\mu_V \neq 0$ if $C_V \neq C_V^{eq}$, which tends to restore C_V value to C_V^{eq}.

A set of expressions analogous to Eqs. (5.2)–(5.4) holds for I.

5.2.2 Point Defects in GaAs

The compound semiconductor GaAs has a thermodynamically allowed equilibrium composition range around the $Ga_{0.5}As_{0.5}$ stoichiometric composition. In thermal equilibrium coexistence within a GaAs crystal, there are four vapor phase species: Ga_1, As_1, As_2, and As_4. In the crystal, there exist six single point defect species: vacancies of the

[1] For more than one species, the corresponding expression is $G = k_B T \sum_i C_i (\ln \frac{C}{C_i^{eq}} - 1)$.

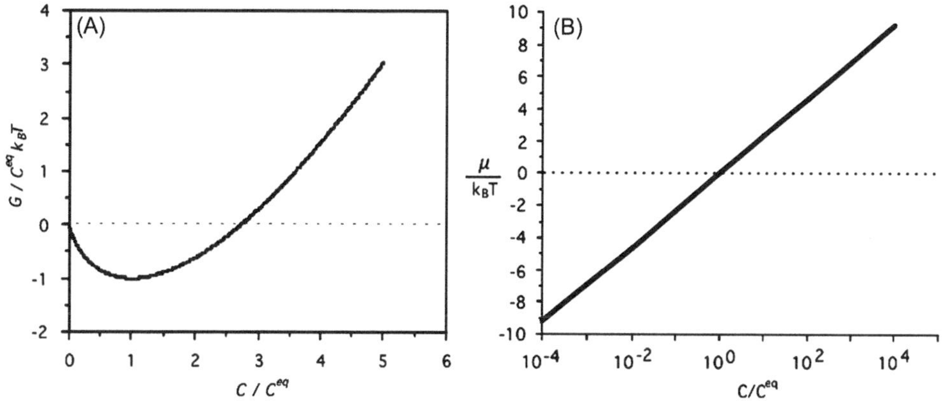

Figure 5.2 (A) Gibbs free energy of a crystal containing one point defect species. The curve is calculated according to Eq. (5.3) normalized by the factors $C^{eq}k_BT$ and C^{eq} for G and C, respectively. The subscript V is dropped because the same plot applies also to other point defect species. (B) The chemical potential of a point defect species as a function of its concentration according to Eq. (5.4).

Ga and As sublattices (V_{Ga} and V_{As}), self-interstitials of Ga and As (I_{Ga} and I_{As}), and antisite defects of a Ga atom on an As sublattice site (Ga_{As}) and of an As atom on a Ga sublattice site (As_{Ga}). Ignoring contributions from higher ordered point defect complexes (their concentrations shall be small), the *sum of the thermal equilibrium concentrations of the point defects constitutes the allowed GaAs crystal composition variation* within its thermodynamically allowed range. For instance, the excess As concentration (δC_{As}) is given by

$$\delta C_{As} = \left(C_{I_{As}}^{eq} + C_{As_{Ga}}^{eq} - C_{V_{As}}^{eq}\right) - \left(C_{I_{Ga}}^{eq} + C_{Ga_{As}}^{eq} - C_{V_{Ga}}^{eq}\right), \tag{5.5}$$

which is responsible for the compound crystal composition deviation from the $Ga_{0.5}As_{0.5}$ stoichiometry. Clearly, $\delta C_{Ga} = -\delta C_{As}$ is the excess Ga concentration. In Eq. (5.5), C_X^{eq} is the thermal equilibrium concentration of the point defect species X. Here the concentration of a point defect species includes those in all charge states.

Clearly, the following three entities form a mutual dependence, each of which may be regarded as the cause for the other two: (i) the vapor phase pressures; (ii) the GaAs crystal composition; and (iii) the point defect concentrations of the crystal.[2]

To analyze experiments, it is convenient to regard the vapor phase pressure as the cause and the other quantities as consequences. For III–V compounds, the group V element vapor phase pressures are large, e.g., P_{As_4} and/or P_{As_2}, and can be readily mea-

[2] It is noted that, for an elemental crystal, its own vapor pressure has no first-order effect on point defect concentrations. If the vapor pressure deviated away from its thermal equilibrium value, the crystal would grow or shrink accordingly.

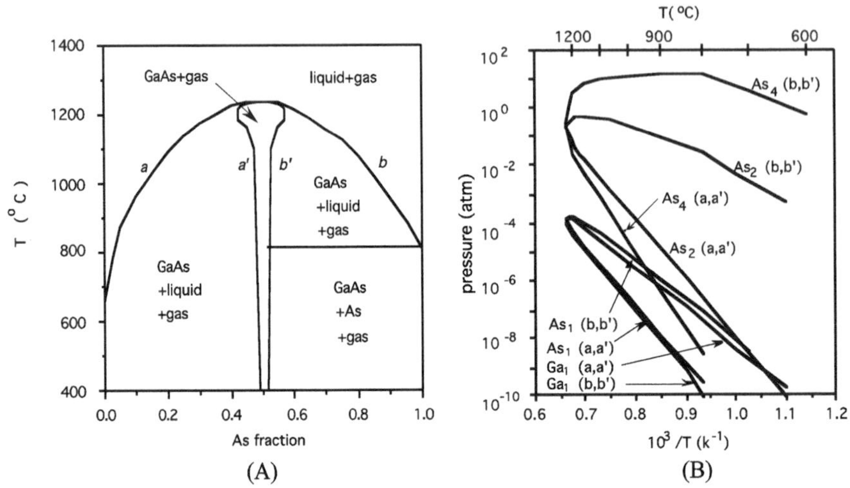

Figure 5.3 (A) Phase diagram of GaAs, with the thermodynamically allowed GaAs crystal composition range greatly exaggerated. (B) Partial pressures of the Ga and As vapor phases in equilibrium with the most gallium rich GaAs (a,a′) or the most arsenic rich GaAs (b,b′) [8].

sured. Fig. 5.3(A) shows the GaAs phase diagram, and Fig. 5.3(B) shows the vapor phase pressures [8].

Thus, for GaAs, according to the *pressure effect*, the expressions

$$C^{eq}_{I^o_{As}} \propto \frac{1}{C^{eq}_{V^o_{As}}} \propto \frac{1}{C^{eq}_{I^o_{Ga}}} \propto C^{eq}_{V^o_{Ga}} \propto P^{1/4}_{As4}, \tag{5.6a}$$

$$C^{eq}_{As^o_{Ga}} \propto \frac{1}{C^{eq}_{Ga^o_{As}}} \propto P^{1/2}_{As4} \tag{5.6b}$$

hold for the six point defect species in the neutral state. Explicit expressions for the thermal equilibrium concentrations of these neutral single point defects have been obtained [9], with that of the neutral As sublattice vacancy given by

$$c^{eq}_{V^o_{As}} = \frac{C^{eq}_{V^o_{As}}}{C_{As(GaAs)}} = \left(\frac{P_{As4}}{B_{As4}}\right)^{-1/4} \exp\left[-\frac{(g^f_{V^o_{As}})^{eff}}{k_B T}\right]. \tag{5.7}$$

In Eq. (5.7), B_{As_4} is the ideal gas pressure constant of As_4 given by

$$B_{As_4} = \left(\frac{2\pi m_{As4}}{h^2}\right)^{3/2} (k_B T)^{5/2}, \tag{5.8}$$

Table 5.1 Summary of the expressions of the thermal equilibrium concentrations of the six electrically neutral point defect species (X^o) in GaAs in accordance with the As_4 dominating assumption. These expressions are of the general form $c_{X^o}^{eq} = A_{X^o} \exp[-(g_{X^o}^f)^{eff}/k_BT]$.$A_{X^o}$ and $(g_{X^o}^f)^{eff}$ are listed for each case

X^o	**Kinetic reaction**	A_{X^o}	$(g_{X^o}^f)^{eff}$
V_{As}^o	$As_{As} \Leftrightarrow \frac{1}{4}As_4 + V_{As}^o$	$(P_{As_4}/B_{As_4})^{-1/4}$	$g_{V_{As}^o}^f - \delta g_{As}$
I_{As}^o	$\frac{1}{4}As_4 \Leftrightarrow I_{As}^o$	$(P_{As_4}/B_{As_4})^{1/4}$	$g_{I_{As}^o}^f + \delta g_{As}$
V_{Ga}^o	$\frac{1}{4}As_4 \Leftrightarrow GaAs + V_{Ga}^o$	$(P_{As_4}/B_{As_4})^{1/4}$	$g_{V_{Ga}^o}^f + \delta g_{As}$
I_{Ga}^o	$GaAs \Leftrightarrow \frac{1}{4}As_4 + I_{Ga}^o$	$(P_{As_4}/B_{As_4})^{-1/4}$	$g_{I_{Ga}^o}^f - \delta g_{As}$
Ga_{As}^o	$Ga_{Ga} + V_{As}^o \Leftrightarrow Ga_{As} + V_{Ga}^o$	$(P_{As_4}/B_{As_4})^{-1/2}$	$g_{Ga_{As}^o}^f - 2\delta g_{As}$
As_{Ga}^o	$As_{As} + V_{Ga}^o \Leftrightarrow As_{Ga} + V_{As}^o$	$(P_{As_4}/B_{As_4})^{1/2}$	$g_{Ga_{As}^o}^f + 2\delta g_{As}$

$\delta g_{As} = g_{As}(GaAs) - g_{As}(As_4)$, $B_{As_4} = (2\pi m_{As_4}/h)^{3/2}(k_BT)^{5/2}$.

where m_{As_4} is the mass of an As_4 molecule, h Plank's constant, and $(g_{V_{As}^o}^f)^{eff}$ is the effective Gibbs free energy of formation of V_{As}^o given by

$$(g_{V_{As}^o}^f)^{eff} = g_{V_{As}^o}^f - \delta g_{As}^f. \tag{5.9}$$

In Eq. (5.9), $g_{V_{As}^o}^f$ is the V_{As}^o Gibbs free energy of formation of the cavity, with its definition identical to that of the elemental crystal; δg_{As}^f is that due to the placement of the removed As atom, which turned out to be the difference of the As atom free energies of binding defined as

$$\delta g_{As}^f = g_{As(GaAs)}^f - g_{As(As_4)}^f, \tag{5.10}$$

where $g_{As(GaAs)}^f$ is the Gibbs free energy of formation of an As atom in the GaAs crystal and $g_{As(As_4)}^f$ is that of an As atom in the As_4 molecule. Expressions for all six neutral point defect species in GaAs, of the form $c_{X^o}^{eq} = A_{X^o} \exp[-(g_{X^o}^f)^{eff}/k_BT]$, are summarized in Table 5.1 by listing the preexponential factor A_{X^o} and the effective diffusivity $(g_{X^o}^f)^{eff}$. These expressions should be applicable to other III–V and to also II–VI compound semiconductors.

Although the thermal equilibrium point defect concentration expressions for elemental and compound semiconductors, e.g., Eqs. (5.2) and (5.7), respectively, are of the same form, the two cases differ considerably. That of the elemental semiconductors (also applicable to other elemental crystals), Eq. (5.2), is characterized by a preexponential factor of 1 with a single activation Gibbs free energy, $g_{X^o}^f$, which, for practical purposes, is always approximated by the formation enthalpy $h_{X^o}^f$. At a given temperature there is only one $c_{X^o}^{eq}$ value. Plotting $\log(c_{X^o}^{eq})$ as a function of $1/T$, one obtains the well-known Arrhenius plot, Fig. 5.4(A). On the other hand, that of GaAs (also applicable

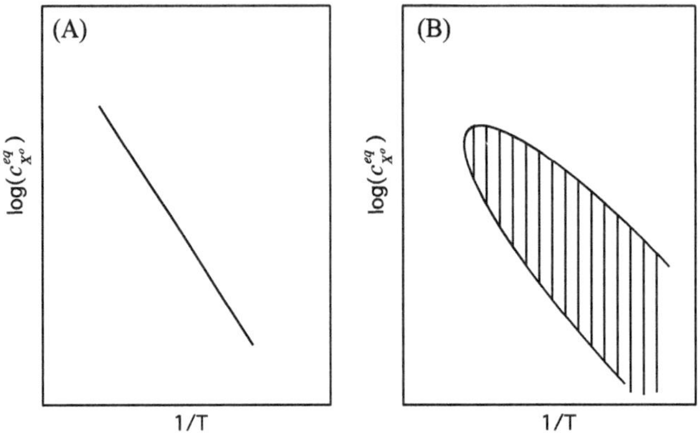

Figure 5.4 Schematic plots of point defect thermal equilibrium concentration $c^{eq}_{X^o}$ for an elemental crystal (A), and for a compound crystal (B). For a compound crystal, at a given temperature, $c^{eq}_{X^o}$ has a range of possible values, as indicated by the bounded vertical lines.

to other compound crystals), Eq. (5.7), is characterized by the fact that the preexponential factor is not 1 but instead a function of the pressure of a vapor phase coexisting in thermal equilibrium with the crystal. Corresponding to the allowed thermal equilibrium composition range and the allowed vapor phase pressure range, Fig. 5.3, the thermal equilibrium point defect concentration also has a range at a given temperature, Fig. 5.4(B). Moreover, the activation Gibbs free energy is more complex. In addition to that of the defect local formation energy, $g^f_{X^o}$, the term δg^f_{As}, which accounts for the energy difference of an As atom in GaAs and in As_4, is also present. This occurs because in forming V_{As}, unlike for the elemental crystal case, the removed As atom cannot be placed on a surface kink site without disturbing equilibrium among surface kinks. Refer to Fig. 5.5(A), wherein it is seen that there are two kinds of surface kink, the Ga kink that is surrounded by all Ga atoms and the As kink that is surrounded by all As atoms. Placing the As atom into a Ga kink will eliminate a Ga kink and create an As kink, Fig. 5.5(B), while placing an As atom into an As kink will create a surface As antisite defect, Fig. 5.5(C). In either case, thermal equilibrium on the surface is perturbed. To restore thermal equilibrium conditions of the surface, this atom must evaporate into the ambient to become part of a vapor species, e.g., As_4. The energetic situation associated with this process is shown in Fig. 5.6.

Using values of the known physical constants, the gas pressure constant of As_4 given by Eq. (5.8) reduces to $B_{As_4} = 135T^{5/2}$ atmosphere. Approximating the formation Gibbs free energy by the formation enthalpy and using that of GaAs and As_4, the difference of the As atom free energies of binding given by Eq. (5.10) reduces to $\delta g^f_{As} \approx h^f_{As(GaAs)} - h^f_{As(As_4)} = -0.69$ eV [9]. With these values, expressions of the six neutral point defect

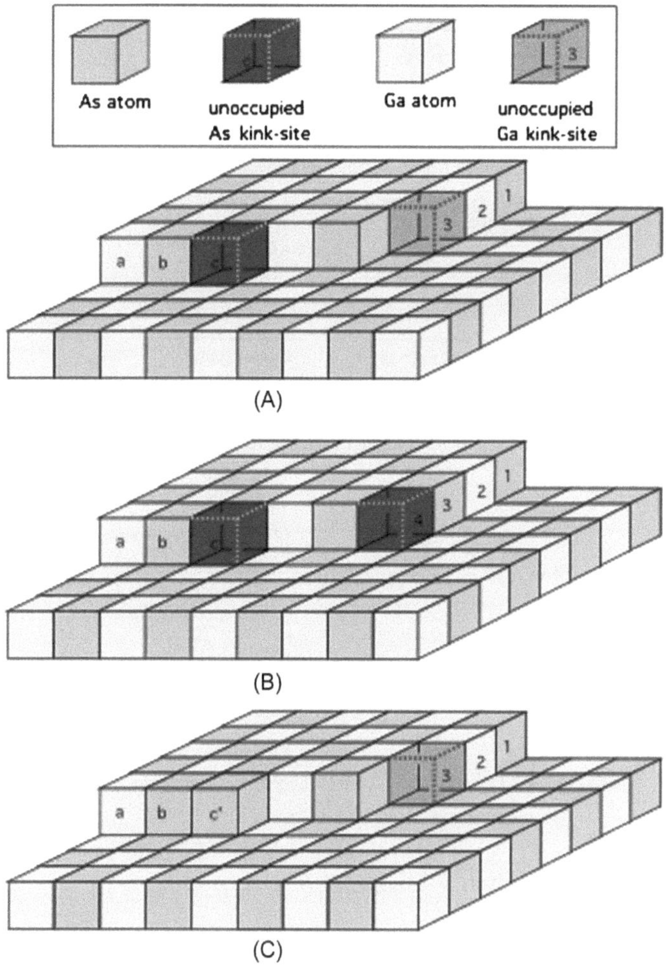

Figure 5.5 (A) Schematics of a Ga and an As surface step kink site, marked respectively by deep green and deep red colors and labeled as 3 and c. (B) Placing the As atom at the Ga kink site 3 results in the elimination of this Ga kink site and the creation of an additional As kink site colored deep red and labeled 4. (C) Placing the As atom at the As kink site c eliminates this kink site and creates the antisite defect labeled c′. (For interpretation of the references to color in this figure legend, the reader is referred to the web version of this chapter.)

species reduce to

$$c^{eq}_{V^o_{As}} \approx 3.4 T^{5/8} P_{As_4}^{-1/4} \exp\left(-\frac{g^f_{V^o_{As}} + 0.69eV}{k_B T}\right), \tag{5.11a}$$

$$c^{eq}_{I^o_{As}} \approx 0.293 T^{-5/8} P_{As_4}^{1/4} \exp\left(-\frac{g^f_{I^o_{As}} - 0.69eV}{k_B T}\right), \tag{5.11b}$$

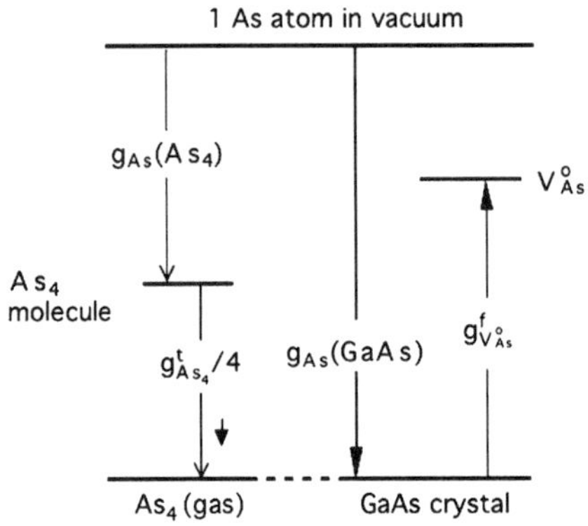

Figure 5.6 A schematic drawing showing the free energies of binding of an As atom in the various involved materials together with the free energy of formation of one V^o_{As}.

$$c^{eq}_{V^o_{Ga}} \approx 0.293T^{-5/8}P^{1/4}_{As4}\exp\left(-\frac{g^f_{V^o_{Ga}} - 0.69eV}{k_BT}\right), \tag{5.11c}$$

$$c^{eq}_{I^o_{Ga}} \approx 3.4T^{5/8}P^{-1/4}_{As4}\exp\left(-\frac{g^f_{I^o_{Ga}} + 0.69eV}{k_BT}\right), \tag{5.11d}$$

$$c^{eq}_{Ga^o_{As}} \approx 11.6T^{5/4}P^{-1/2}_{As4}\exp\left(-\frac{g^f_{Ga^o_{As}} + 1.38eV}{k_BT}\right), \tag{5.11e}$$

$$c^{eq}_{As^o_{Ga}} \approx 0.086T^{-5/4}P^{1/2}_{As4}\exp\left(-\frac{g^{eq}_{As^o_{Ga}} - 1.38eV}{k_BT}\right). \tag{5.11f}$$

For Si, although the formation enthalpies of both V and I are not precisely known, they are nonetheless both experimentally and theoretically determined to be within the relatively narrow range of ~3.5–4 eV. In contrast, the situation for GaAs is more perplexing. Experimental information are available from only two studies. Analyzing GaAs disordering results for superlattices made of Ga isotopes, Bracht et al. [10] obtained values 1.7–2.1 eV for the formation enthalpy of the neutral Ga vacancy. Using position annihilation, Gebauer et al. [11] obtained a formation enthalpy of (3.26 ± 0.5) eV and a formation entropy of $(9.661k_B)$ for the neutral Ga vacancy. These enthalpy values are in principle values of the effective enthalpy of the neutral Ga vacancy, $(h^f_{V^o_{Ga}})^{eff}$. Theoretically calculated formation enthalpy values of each of the six neutral point defect species, including those obtained from simpler considerations to those calculated using sophisticated quantum mechanical methods (ab initio or first principle calculations), are abundant, see Table 5.2 for some examples [12–18]. In Table 5.2 the values of van Vechten [12] and Wager [15] are $h^f_{X^o}$ while the rest are in principle $(h^f_{X^o})^{eff}$. It is seen

Table 5.2 Theoretical values of formation enthalpy of the six neutral point defect species in GaAs, $(h^f_{X^o})^{eff}$ (unmarked) or $h^f_{X^o}$ (denoted with *) (in eV)

	V^o_{Ga}	I^o_{Ga}	V^o_{As}	I^o_{As}	Ga_{As}	As_{Ga}
Van Vechten [12]*	2.59		2.59			
Baraff and Schlutter [13]	7.32			2.0	2.85	3.51
Janson and Shanky [14]	4.46		3.36		3.18	2.82
Wager [15]*	2.31		2.31		3.11	1.62
Cheong and Chang [16]			2.97	5.96	2.32	2.45
Schick [17]	2.85					1.8
El-Mellouhi and Mausseau [18]	2.95					

* $h^f_{X^o}$.

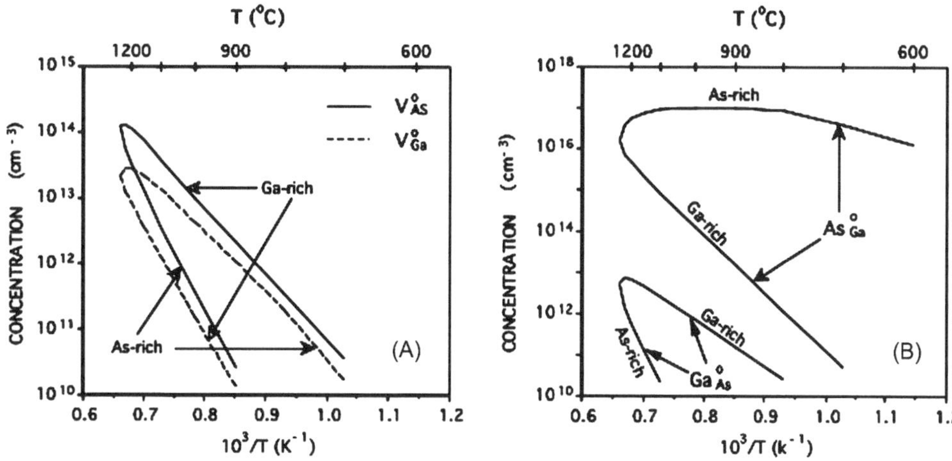

Figure 5.7 Calculated thermal equilibrium concentration values [9] of the two vacancies, V^o_{As} and V^o_{Ga}, (A), and of the two antisite defects, As^o_{Ga} and Ga^o_{As}, (B). It is interesting to note that the value of $C^{eq}_{As^o_{Ga}}$ is close to observed EL2 concentrations in As rich GaAs crystals. EL2 is an important deep (mid-gap) donor defect for producing semi-insulating GaAs crystals. It is believed that the dominant part of the defect EL2 is As^o_{Ga}.

from Table 5.2 that the reported values for each point defect species span such a wide range that it is not meaningful to further calculate the defect concentrations, if accuracy or reliability is sought after. However, in order to just demonstrate the physical features associated with Eq. (5.11), any $h^f_{X^o}$ value shown in Table 5.2 will suffice. For instance, Fig. 5.7 shows some numerical results obtained earlier [9] using the $h^f_{V^o_{Ga}}$ and $h^f_{V^o_{As}}$ values of van Vechten [12] and the $h^f_{Ga^o_{As}}$ and $h^f_{As^o_{Ga}}$ values of Wager [15]. It is seen from these results that the thermal equilibrium concentration of a point defect species span a range at a given temperature, just as that schematically shown in Fig. 5.4(B).

5.2.3 Charged Point Defects

Insofar as diffusion is concerned, in metals the question whether a point defect species is charged or not is purely academic. There is no convenient experiment that can be performed to either check its nature or put it to any practical use. This is because there is no practical way to alter the electron density of a metal crystal, which is from one to several times of the atom density. In contrast, charged point defects play a pivotal role in semiconductors. In semiconductors, both neutral and charged point defects serve as diffusion vehicles for self-diffusion and for the diffusion of substitutional impurities. Point defects also contribute and sometimes dominate electrical properties of the semiconductor: by changing charge states, they are recombination/generation centers of the electric carrier electrons and holes. Moreover, specific defects are deep donors, e.g., As_{Ga} (EL2). To build devices, a semiconductor can be doped to carrier concentrations many orders of magnitude of the intrinsic carrier concentration, which accordingly alter the charged point defect thermal equilibrium concentrations. Such altered charged point defect concentrations have huge effects on diffusion and on electrical properties of the semiconductor.

Charged point defect possesses energy transition (or ionization) levels in the semiconductor bandgap, see Fig. 5.8 for a schematic illustration wherein there are four energy levels, denoted respectively as $+/0$, $0/-$, $-/2-$, and $2-/3-$. The symbol to the left of the slash sign denotes the charge state of the defect before catching one more electron that is denoted by the symbol to the right of the slash sign. The packing order of the transition levels of this hypothetical defect is normal: it can become positively or negatively charged, and the more electrons the defect is charged with, the higher is its energy. This packing order resembles that of an atom in that the outer shell electrons are consecutively higher in energy. Thus, the charged point defect species is higher in energy then the neutral one. However, with a sufficiently high Fermi level, the defect will be negatively charged for the reason that the energy of the total system, consisting of the defect and electrons, will be lowered by the elimination of electrons that have the energy E_F, which is the energy of the Fermi level. On the other hand, if the Fermi level is lower than the donor level, the defect will lose electrons to become positively charged.

Irrespective of whether the semiconductor is elemental or compound, the charged point defect thermal equilibrium concentrations are given by the following expressions. For an acceptor type defect negatively charged by z electrons, the expressions

$$\frac{C_{X^{z-}}(n)}{C_{X^0}} = \exp\left(\frac{zE_F - \sum_{m=1}^{z} E_{am}}{k_B T}\right), \tag{5.12}$$

$$\frac{C_{X^{z-}}(n)}{C_{X^{z-}}(n_i)} = \left(\frac{n}{n_i}\right)^z, \tag{5.13}$$

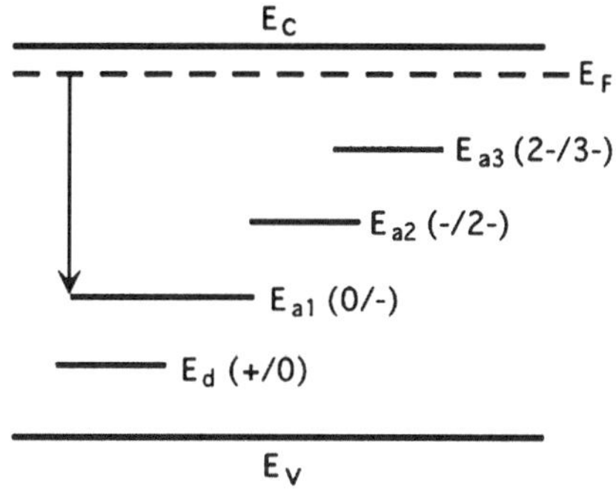

Figure 5.8 GaAs band diagram showing schematically the relation of the Fermi-level and the native point defect (*x*) deep charge states, in normal or positive-U order. The arrow indicates the total system (defect and electron systems) Gibbs free energy decrease of forming one x^- from x^o.

hold. Here n and n_i are respectively the actual and intrinsic electron concentrations, and E_{am} the energy of the mth acceptor level. For a donor type defect charged positively by having donated z electrons, the expressions

$$\frac{C_{x^{z+}}(p)}{C_{x^o}} = \exp\left(-\frac{zE_F - \sum_{m=1}^{z} E_{dm}}{k_B T}\right), \tag{5.14}$$

$$\frac{C_{x^{z+}}(p)}{C_{x^{z+}}(n_i)} = \left(\frac{p}{n_i}\right)^z, \tag{5.15}$$

hold. Here p is the actual hole concentration and E_{dm} the energy of the mth donor level.

Diffusion experiments in Si indicate the contributions of neutral as well as charged point defect species, but that of the neutral ones dominate. As reviewed by Watkins [19], for negatively charged Si vacancies, the $0/-$ and $-/2-$ transition levels are near midgap but in the upper part of the gap. On the other hand, ab initio calculations indicated that V_{Si}^- is much less stable than V_{Si}^{2-}, and the same holds also for the Si vacancy donor levels V_{Si}^+ and V_{Si}^{2+} [20]. A similar situation holds for the Si self-interstitials. Silicon has a relatively narrow bandgap (1.14 eV at room temperature) and consequently a fairly large intrinsic carrier concentration n_i throughout the temperature range of interests for diffusion experiments, resulting in that the Si crystal carrier concentration is usually close to n_i. Thus, whether the involved point defect species is charged or not is less important than in GaAs since in most experiments Si will be nearly intrinsic.

On the other hand, knowledge obtained from analyzing diffusion experiments showed that point defect species govern self-diffusion and diffusion of all substitutional impurities in GaAs are charged [21–23]. GaAs has a fairly large bandgap (1.45 eV at room temperature) and consequently a quite small intrinsic carrier concentration n_i throughout the temperature range of interests for diffusion processes, in that the GaAs

Table 5.3 Some available theoretically calculated Ga vacancy charge state energy levels together with the measured case of Brachet et al. [10] (in eV)

	$E_{V_{Ga}^{0/-}} - E_v$	$E_{V_{Ga}^{-/2-}} - E_v$	$E_{V_{Ga}^{2-/3-}} - E_v$
Baraff and Schlüter [13]	0.2	0.5	0.7
Jansen and Sankey [14]	0.1	0.35	0.5
Cheong and Chang [16]		0.49	0.69
Schick et al. [17]	0.09	0.13	0.2
El-Mellouhi and Mousseau [18]	0.05	0.4	0.55
Zhang and Northrup [24]	0.19	0.2	0.32
Pöykkö et al. [25]	0.11	0.22	0.33
Gorczyca et al. [26]	0.39	0.52	0.78
Janotti et al. [27]	0.13	0.15	0.18
Bracht et al. [10]*	0.42	0.6–0.72	1.33

* Experimentally extracted values.

crystal doping level can easily exceed n_i by orders of magnitude. Consequently, depending on the GaAs doping level, the diffusivity value of a diffusing species may span a range of several orders of magnitudes.

For GaAs, there is one experiment that allowed interpreting the nature of the charge states and their transition level positions for Ga vacancies [10]. On the other hand, there are many theoretical calculations predicting the nature of the charged states and their transition level positions for each of the six point defect species. Nonetheless, the situation is as perplexing as with the formation enthalpy values of the neutral point defects of GaAs discussed before, because the wide quantitative disagreement in both the charge state nature and their transition level positions. For example, Table 5.3 lists some of the available calculated results for V_{Ga} [13,14,16–18,24–27] together with the experimental results of Bracht et al. [10]. It is seen from Table 5.3 that each level position spans a wide energy range. A similar situation holds for all other defect species. Thus, it is not meaningful to further calculate the defect thermal concentrations if accuracy or reliability is sought after. However, in order to just demonstrate the physical features associated with Eqs. (5.12)–(5.15), any set of values in Table 5.3 may be chosen. For instance, Fig. 5.9 shows the results [28] obtained using the V_{Ga}^{o} concentration in Fig. 5.7 and the level positions of Baraff and Schlütter [13] listed in Table 5.3. For this case V_{Ga} is characterized by having 3 accepter states with their transition level positions all below the GaAs intrinsic level. Thus, in accordance with Eqs. (5.13) and (5.14), the concentrations of all 3 charged V_{Ga} species are higher than that of V_{Ga}^{o}, Fig. 5.9(A). Under n-doping, the concentration of each charged V_{Ga} species increases with an increase in dopant concentration, Fig. 5.9(B)–(D). Also seen from Fig. 5.9(B)–(D) is that the species' activation enthalpy decreases with the increase in dopant concentration. This effect is so pronounced that for V_{Ga}^{3-} its activation enthalpy becomes negative when very

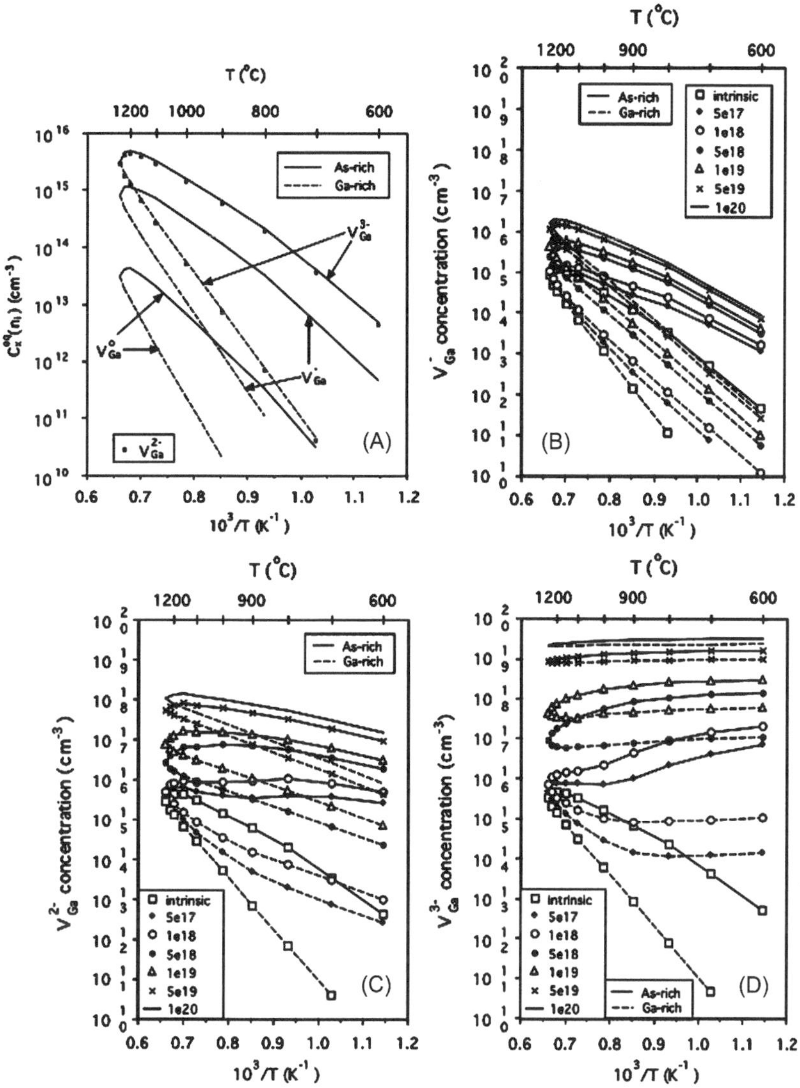

Figure 5.9 Calculated thermal equilibrium concentrations of charged V_{Ga} species in GaAs. (A) Thermal equilibrium concentration of all three charge states in intrinsic GaAs. (B)–(D) Thermal equilibrium concentrations of V_{Ga}^{-}, V_{Ga}^{2-}, and V_{Ga}^{3-}, respectively, in n-doped GaAs (from [28]).

heavily doped [28], Fig. 5.9(D). This means the concentration of V_{Ga}^{3-} will increase as the temperature if lowered, a feature that is quite contrary to intuitive expectations. Nonetheless, this feature has provided explanations to a large number of experimental phenomena occurring at low temperatures, including room temperature. The explained experimental results [28] include the formation of the so-called "dark-line defects" in

GaAs lasers, optical pumping damage in n-type GaAs, and enhanced Be diffusion in GaAs heterostructure bipolar transistor base regions under charge injection conditions. Indeed, in their positron annihilation experiment, Gebauer et al. [11] found that the concentration of V_{Ga}^{3-} increases as temperature is lowered.

5.3 DIFFUSION MECHANISM BASICS IN SEMICONDUCTORS

This section briefly introduces the mechanisms concerning diffusion in semiconductors. The details will be discussed where appropriate. The standard expressions for analyzing a diffusion profile is Fick's second and first law equations respectively given by

$$\frac{\partial C}{\partial t} = \frac{\partial}{\partial x}\left(D\frac{\partial C}{\partial x}\right), \tag{5.16a}$$

$$J = -D\frac{\partial C}{\partial x}. \tag{5.16b}$$

In Eq. (5.16), D is the diffusivity, C is the concentration of the species that generally both are functions of the spatial coordinate x, and J is the flux equation expressing Fick's first law.

Impurity atoms may dissolve in a semiconductor by occupying interstitial (*i*) or substitutional (*s*) sites, or both. Atoms of *i*-species migrate directly among the appropriate interstitial sites, which does not involve point defects. Self-diffusion, i.e., diffusion of the host crystal self-atoms, and diffusion of substitutional impurity species involve point defects in a variety of ways via a number of diffusion mechanisms. While the dominant native point defect species in metals is V, both V and I contribute in semiconductors. Self-diffusion proceeds by an atom directly *exchanging* its position with a neighboring I or V. An impurity *s* atom diffuses as a *pair* or *complex* with I or V. In each semiconductor, there exist a number of impurities that are interstitial–substitutional (*i–s*) species. An *i–s* species dissolve predominantly on substitutional sites, A_s, but diffuse via the rapid migration of those occupying interstitial sites, A_i, and the changeover between the A_i and A_s atoms mediated by I or V.

5.3.1 Interstitial Impurity Diffusion

Atoms of an impurity species dissolved in semiconductors or in metals by occupying interstitial sites, the *i*-species, migrate directly between the appropriate interstitial sites which does not involve point defects, Fig. 5.10(A). The diffusion process is described well by Fick's second law, Eq. (5.16). The diffusivity D of an *i*-species, a constant at a given temperature, is characterized by a Gibbs free energy of migration g_i^m, Fig. 5.10(B),

$$D = A\exp\left(-\frac{g_i^m}{k_B T}\right), \tag{5.17}$$

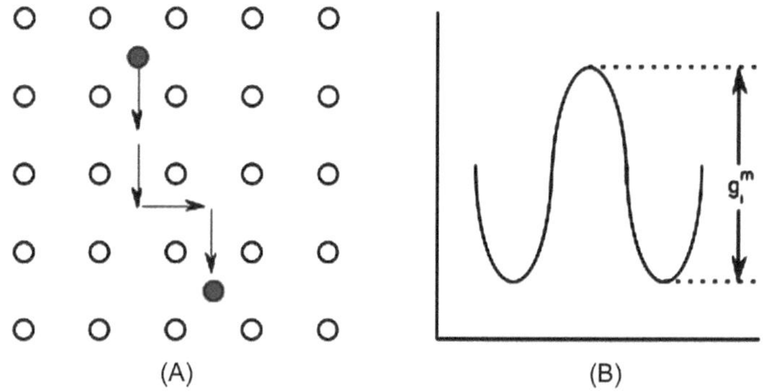

Figure 5.10 (A) Schematic diagram illustrating the diffusion of an interstitially dissolved impurity species in a crystal. (B) Schematic diagram illustrating the migration barrier to the diffusion of an atomic or point defect species in a crystal. The free energy difference between the unstable and stable positions of the atom constitutes g_i^m.

where the preexponential factor A is on the order of 10^{-3} to 10^{-2} $cm^2 s^{-1}$. The quantity g_i^m is usually approximated by an enthalpy term h_i^m. Strictly speaking, *i*-species are the only ones satisfying Eq. (5.16) precisely because no native point defects are involved.

5.3.2 The Exchange Mechanism

In this case the diffusing *s* atom simply exchanges its position with a neighboring point defect that will then migrate away. Schematically, Figs. 5.11(A)–(B) show what is appropriate for the elemental semiconductors Si and Ge (also applicable to metals), namely the atom exchanges its position with vacancy or self-interstitial located on one of its first nearest neighbor sites. Figs. 5.11(C)–(D) show that for the compound semiconductor GaAs (or other compounds) the atom can exchange its position with a self-interstitial located on a first nearest neighbor site. For the vacancy case, though, an atom can only exchange positions with a vacancy of the same sublattice by having the atom jump into that vacancy located only as close as on a second nearest neighbor position. For GaAs, it is highly improbably that the exchange process of vacancy and atom of the same sublattice can involve first nearest neighbor atom jumps. This is for the obvious reason that the first nearest neighbor atoms surrounding a vacancy of a given sublattice are all atoms of the other sublattice, and consequently a first nearest neighbor jump will lead immediately to the formation of an antisite defect plus a vacancy of the other sublattice. Thus, in order to finally result in an exchange of a vacancy and an atom of the given sublattice via a first nearest neighbor jump process, this will involve antisite defects of both sublattices and vacancy of the other sublattice as well as their complexes.

The exchange mechanism is responsible for self-diffusion. Occasionally diffusion of a substitutional impurity may resemble self-diffusion closely, e.g., Si in Ge. For this case Eq. (5.16) applies with a constant diffusivity D at a given temperature if the crystal

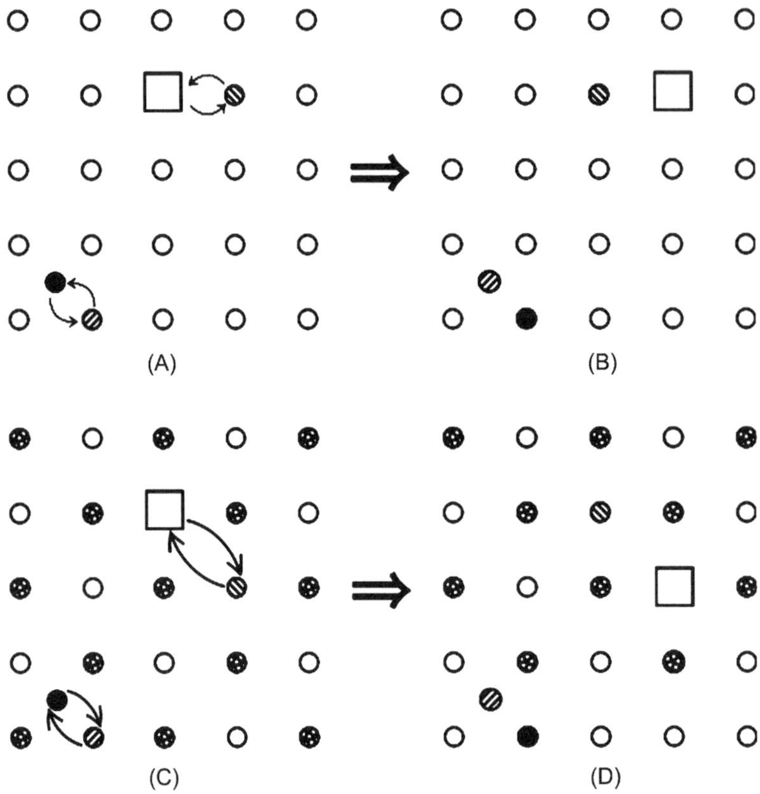

Figure 5.11 Schematic illustration of the direct exchange diffusion processes of a substitutional atom with a point defect. That of a vacancy (□) is shown in the upper center part and that of a self-interstitial (•) is shown in the lower left part of the diagram. For the elemental crystal case shown in (A) and (B), exchange takes place on the first nearest neighboring atom position bases for both I and V mediated diffusion. For the compound crystal case shown in (C) and (D), exchange takes place on the first nearest neighboring atom position basis for I mediated diffusion, while for the V mediated diffusion, exchange occurs on second nearest neighbor position basis.

surfaces are free and without a chemical reaction occurring, or internal point defect sources/sinks in the form of dislocations exist to a sufficient density. The s atom diffusive flux, e.g., that of the tracer atoms during a self-diffusion experiment, is balanced by a flux of the involved point defect species flowing to the opposite direction, which will in principle lead to an undersaturation of the point defect species in the diffused region. Since point defects diffuse much faster than s atoms, this undersaturation is efficiently alleviated by an inflow of point defects from the free surfaces or from dislocations inside the crystal. A point defect species is at its thermal equilibrium concentrations at these sources/sinks. The diffusion of a point defect species is described well by Eq. (5.16) with its diffusivity described also well by Eq. (5.17). Also described well by Eq. (5.16) is self-diffusion with the diffusivity D^{SD},

$$D^{SD} = D_V c_V^{eq} + D_I c_I^{eq}, \tag{5.18}$$

where D_V and c_V^{eq} are respectively the diffusivity and the fractional thermal equilibrium concentration of V, D_I and c_I^{eq} are respectively those of I. On account that both I and V contribute to diffusion in semiconductors, both are included in Eq. (5.18). In this equation, the V contribution is given by $D_V c_V^{eq}$ (and an analogous term for the I contribution) because, when exchange occurred, the s atom and V jumped at the same rate (but in opposite directions) and hence both are migrating with the diffusivity D_V. After this jump the V continuously jumps away, but for the s atom to make the next jump it must wait for the arrival of the another V to reach one of its nearest neighbor atom positions for which the probability is c_V^{eq}, resulting in $D_V c_V^{eq}$ being the V contribution to D^{SD}. The same holds for the I contribution.

Since both c_V^{eq} and c_I^{eq} are much smaller than 1, D^{SD} is much smaller than D_V and D_I.

5.3.3 The sV and sI Pairing Mechanisms and Point Defect Percolation Effect

With the exception of Si and B in Ge, all substitutional impurity species in all 3 considered semiconductors, Si, Ge, and GaAs, diffuse faster than self-diffusion and with a smaller activation enthalpy. The same phenomenon holds also for metals. This phenomenon prompted the notion of diffusion via the fast moving pairs sV and/or sI allowing the s species' diffusivity to be represented by

$$D_{sx} = D_x c_x^{eq} \exp\left(\frac{g_x^c}{k_B T}\right), \tag{5.19}$$

where x stands for V or I, and g^c is an effective coupling potential forming the pair. For g_x^c positive, which holds for almost all cases, the activation energy for diffusion of the s impurity is reduced by g_x^c and the diffusivity is increased. For Si and B in Ge, g^c is negative, which leads to diffusivity values lower than that of Ge self-diffusion together with a higher activation energy.

The coupling potential g^c is an effective one because it is not simply the binding energy of the pair, but a weighted (time averaged) value for V or I, occupying a few consecutively higher order nearest neighbor sites of the s atom. The point defect V or I of the pair actually wanders among these sites. If bonded tightly, the pair should move only slower or not move at all. Thus, for the pair to be mobile as one unit, it must be loosely bonded or coupled in such a way that *partial dissociation* and *reassociation* of the pair occurs continuously, i.e., V or I of the pair will move away from the s atom nearest neighbor position to the second, third, etc., neighbor positions and then wander back to an equivalent nearest position to allow the s atom to make one more jump. In this manner, it is said that the pair is migrating. When V or I is occupying second and still higher neighbor sites of the s atom, its energy becomes consecutively higher but still lower than that of the V or I in the lattice separated from the s atom by a long distance. Moreover, the activation barriers between the consecutive neighboring positions for the V or I to make the jump is not prohibitively high. In this manner, the probability for the diffusing pair to move along one direction becomes larger than that due the direct exchange mechanism because the probability of the same V or I, after wandering around, to reappear on the same side of the diffusing atom becomes larger than the probability of the arrival of another available I or V as in the exchange mechanism case. This idea is illustrated in Fig. 5.12 for a two-dimensional square lattice case, for which (A)–(D) schematically show the step by step sV partial-dissociation–association path for the V while (E) schematically shows the corresponding energetic situation, which is done by assuming that the s atom energy did not change in forming the pair and hence all the energy changes are reflected in that of V or I of the pair. While Fig. 5.12(E) is for cases for which s atom diffuses faster than host crystal self-atoms, Fig. 5.12(F) shall be that for B and to a much less extent Si in Ge, with two possibilities. In the first (α), it is depicted that the B atom is simply for some reason repelling V. In the second (β), since V is the dominant point defect species in Ge and B atom is under large tension because of its much smaller radius than that of an Si atom, it is reasonable to expect that the B atom would pair with V having a larger binding energy, as indicated by the deeper troughs of the few nearest neighbors in the curve β. Fig. 5.13 shows that of a real case, the E-center, which is a P^+V^- pair, identified by Watkins and Corbet more than 50 years ago [29]. For this case, Fig. 5.13(A) shows the P^+V^- partial-dissociation–association path while Fig. 5.13(B) is the corresponding energetic situation determined by Watkins and Corbet from EPR measurements [29].

The E-center diffuses very fast at high P concentrations, which is explained by the **V percolation effect** [30]. This effect is a special case of the sV pairing mechanism. When the s atom concentration is very high, V is always in the vicinity of one s atom, though not the same s atom at all times simply because there are so many s atoms around. Consider the hypothetical case that s concentration reached 10^{-3} of the host crystal atom density, then on average in any direction every tenth atom is an s atom. Then, instead of being dissociated away from a pair configuration, V is closely associated with one or another s atom for most of the time. In this way, the same V is used over and over for P migration to occur, leading to a higher P diffusivity than that of the low P concentration case. Using the sV exchange case as a basis for comparison, sV pairing results in an apparent increase in V concentration, while V percolation effect further enhances this apparent V concentration. Fig. 5.14 shows the energetics of a hypothetical case of E-center under V percolating conditions [30]. It is noted that the percolation effect obviously also applies to the sI case.

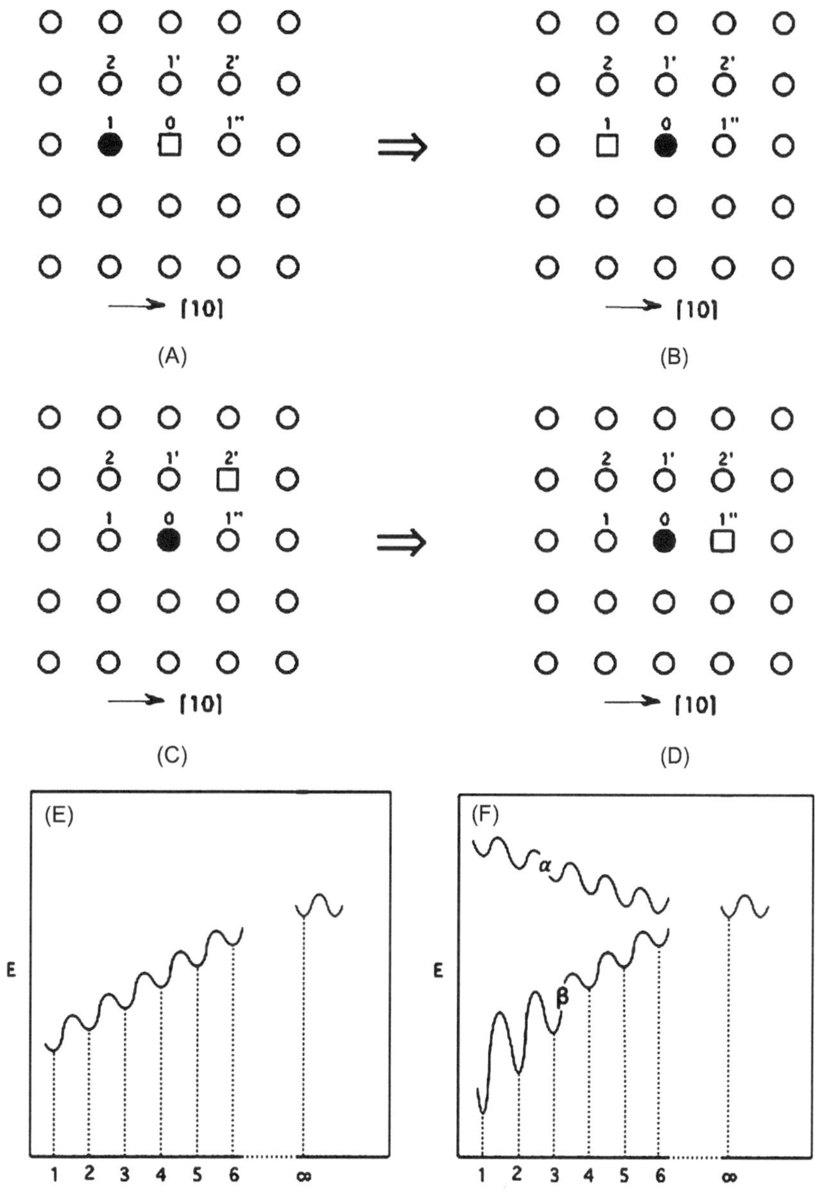

Figure 5.12 (A)–(D) Schematic drawing showing the migration of the substitutional impurity atom (•) and vacancy (□) pair in the [10] direction of a two-dimensional square lattice. In this process partial dissociation of the pair occurs, which allows the point defect to wondering around the impurity atom to lead to a directional flow of the pair. (E) Schematic energetic situation of the *sV* pair. Numerals denoting the first, second, etc., nearest neighbor positions *V* would occupy. Note that for *V* at ∞ the *sV* pair is effectively dissociated. (F) The energetic situation corresponds to that of B and to a much lesser extent Si in Ge, for which the atom and *V* repel each other.

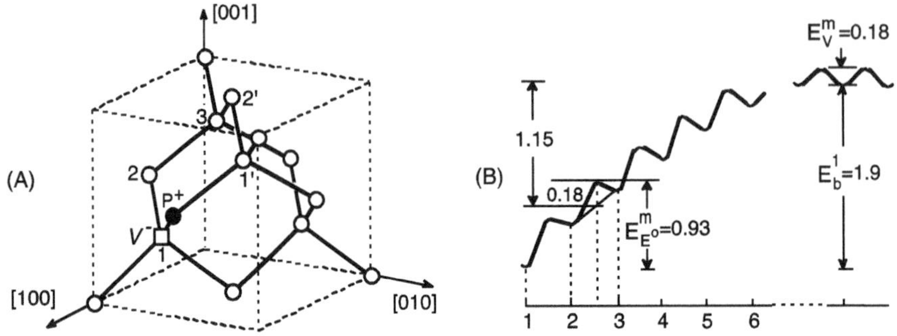

Figure 5.13 (A) The P^+ and V^- after a first change of their relative positions. The sequence of positions 1232′1′ denotes the path of nearest neighbors of the P^+ ion for V^- to migrate in facilitating the concurrent diffusion of the P^+V^- pair in the Si [001] direction. (B) Potential energy of V^- in the vicinity of P^+. The abscissa is the P^+V^- separation expressed by the consecutively numbered sites of V^-, corresponding to the first, second, third, etc., nearest neighboring positions of the P^+ ions.

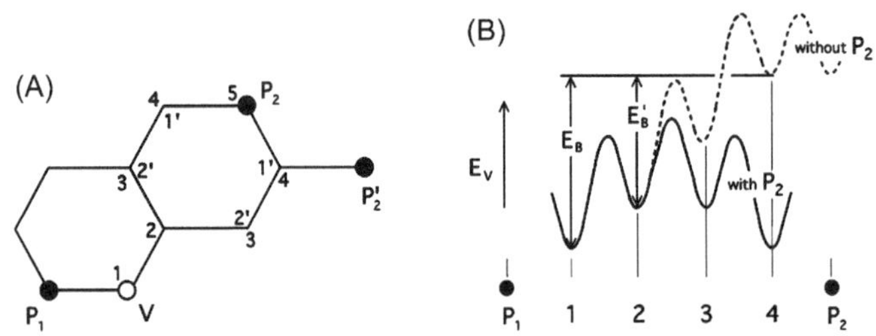

Figure 5.14 An E-center under V percolating conditions: (A) V percolation configuration of P atoms at high concentration; (B) corresponding energy of the vacancy. After Mathiot and Pfister [30].

The sV and sI pairing mechanisms are routinely called the vacancy and interstitialcy mechanisms, respectively, which, however, also include the exchange mechanism.

5.3.4 The Interstitial–Substitutional Species

A number of impurities are interstitial–substitutional (*i–s*) species that dissolve predominantly on substitutional sites, but their diffusion is actually due to the rapid migration of those occupying interstitial sites and the interchange between atoms occupying the two different types of sites. The *i–s* species diffuse in this manner because it is much faster than that of the A_s migration via either the vacancy or interstitialcy mechanism. The diffusivity in the Fick's second law description of the diffusion process of the *i–s* species is an *effective* diffusivity D^{eff} that can exhibit a strong dependence on its own concentration. The details of *i–s* diffusion are somewhat involved and will be deferred until discussing specific cases. Here it suffices to just briefly mention that there are two mechanisms for the diffusion of the *i–s* species, **Frank–Turnbull mechanism** [31] or

Longini mechanism [32], for which the A_i–A_s changeover is mediated by V, and the **kick-out mechanism,** for which the A_i–A_s changeover is facilitated by I [33,34]. These aspects will be discussed later on.

To explain why Cu diffuses in Ge extremely fast, Frank and Turnbull [31] suggested that Cu is an i–s species in Ge with the changeover of an atom between the i and s positions governed by V, which is called the **Frank and Turnbull mechanism** or **dissociative mechanism**, the details of which will be discussed in Section 5.4.2. According to this mechanism, the Cu profile should be one of the erfc type, which is correct, for V is the dominant point defect in Ge. To explain why Zn diffuses in GaAs extremely fast, Longini [32] similarly suggested that Zn is an i–s species in GaAs and V_{Ga} is the mediator for the exchange of Zn_i^- and Zn_s^+. His suggestion has since been referred to as the **Longini mechanism**. It turned out, however, for Zn diffusion in GaAs the Zn_i^-–Zn_s^+ changeover is actually facilitated by the Ga self-interstitial species I_{Ga}^{2+}.

In the **kick-out mechanism** [33,34] it is assumed that the point defect species governing the A_i–A_s changeover is the self-interstitial I, which is suitable for analyzing diffusion results of Au, Pt, and Zn in Si. Analogously, the kick-out mechanism holds for Zn and Be diffusion in GaAs.

5.3.5 Diffusion–Segregation

To fabricate solid-state electronic devices, dopant atoms are introduced into substrate materials both during crystal growth, and during device processing by diffusing dopant into the crystal using either an external source material or implanting the source material as ions into a shallow crystal surface layer. The device structures consist of electrical junctions and may also of heterojunctions, resulting in concurrent dopant diffusion and segregation when processing the device structures.

To treat the simultaneously occurring diffusion and segregation processes, a **diffusion–segregation equation (DSE)** and the corresponding flux equation have been derived [35]:

$$\frac{\partial C}{\partial t} = \frac{\partial}{\partial x}\left[D\left(\frac{\partial C}{\partial x} - \frac{C}{m}\frac{\partial m}{\partial x}\right)\right], \tag{5.20a}$$

$$J = -D\left(\frac{\partial C}{\partial x} - \frac{C}{m}\frac{\partial m}{\partial x}\right), \tag{5.20b}$$

where m is the segregation coefficient of the diffusing species defined with respect to a conveniently chosen reference host material located at x_o

$$m = \frac{C^{eq}(x)}{C^{eq}(x_o)}. \tag{5.21}$$

In Eq. (5.20) the quantities C, D, and m are all functions of the spatial coordinate x. Eq. (5.20) is as fundamental as Fick's second law expression, Eq. (5.16).

Before Eq. (5.20) became available, an empirical computational segregation flux criterion at an abrupt interface was being used in Si device processing simulation programs [36], which reads

$$F_s = h\left(C_1 - \frac{C_2}{m}\right), \tag{5.22}$$

where F_s is the *segregation-flux* across the abrupt interface, h is a phenomenological *mass transport coefficient*; C_1 is the diffusant concentration at the computational grid point 1, which is in one bulk region nearest to the interface, C_2 is the same in the other region. Eq. (5.22) is actually derivable with the mass transport coefficient identified to be $h = D^{eff}/\lambda$, where λ is the interatomic distance and D^{eff} denotes an effective diffusivity of the diffusant that may be conveniently chosen to be the average value of its diffusivities in the two different materials. Note that Eq. (5.20) is applicable to both abrupt and graded interface cases while Eq. (5.22) is basically restricted to abrupt interface cases. Applying Eq. (5.22) to cases with graded interfaces is possible but it will be extremely tedious.

5.3.6 Influence of Diffusion Source Conditions on Experiments

Interpretation of diffusion results relies upon solutions of the diffusion equation(s) subjected to the appropriate initial and boundary conditions. When solving the diffusion equation(s), it is implicitly assumed that the diffusant consists of dispersed atoms with a concentration at or below a well-defined thermal equilibrium concentration or solubility everywhere in the matrix crystal, so that clustering or precipitation process is not occurring.

The solubility of a diffusant is defined by the thermal equilibrium coexistence of the matrix crystal with a specific source material phase containing the diffusant. For either convenience or technology reasons, however, such a source material is rarely used. As an example, which is simpler, and sometimes much simpler, than real cases of diffusion in semiconductors, consider a binary eutectic system (of element A and B) with an intermediate phase that possesses the Gibbs free energy curves shown schematically in Fig. 5.16. This binary system involves 3 phases: two primary phases, α_A and α_B, and the intermediate phase β. To study B diffusion in A crystal, the appropriate source is the β phase material, but in real experiments almost always pure B material is used. Thus at the onset of the diffusion process the B atom concentration in A crystal surface layers first reaches the B atom solubility limit defined by the coexistence of the A crystal with the pure B crystal, as indicated by C_B^{eq} in Fig. 5.16. This B atom concentration decreases continuously until the value indicated by $C_{\alpha B}^{eq}$ is reached, when the surface region of the B crystal becomes the α_B material due to A atom counterdiffusion into the initially pure B crystal. For B atom concentration at either C_B^{eq} or $C_{\alpha B}^{eq}$ values, B atoms are supersaturated relative to the solubility defined by the coexistence of the α_A and β phase

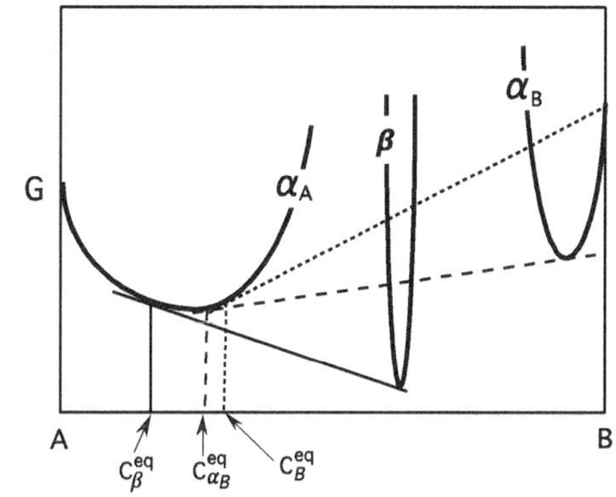

Figure 5.15 Schematic Gibbs free energy diagram illustrating the *B* source material strength evolution in diffusing *B* into *A* when using a pure *B* source material to start with. It is assumed that at the temperature of interest the stable source intermediate phase β will form. Initially *B* diffuses into *A* crystal to solubility C_B^{eq}, then to $C_{\alpha_B}^{eq}$, in accordance with the formation of the α_B phase due to counterdiffusion of A atoms into *B* crystal. The *B* atoms are supersaturated and hence forming the phase compound, which defines the stable *B* solubility in *A* crystal, C_β^{eq}.

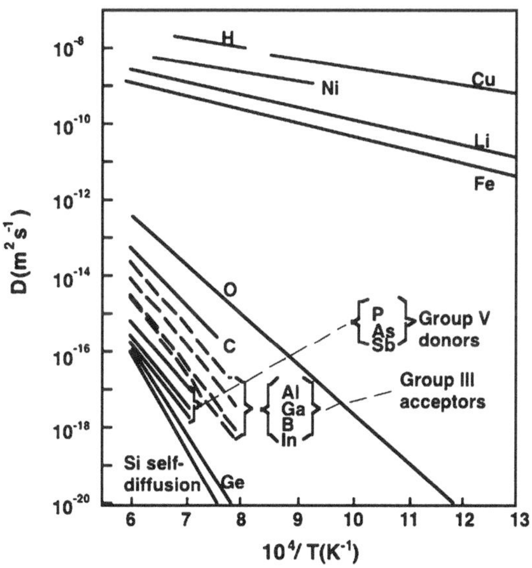

Figure 5.16 A survey of diffusivity values of the most important impurities in Si and of Si self-diffusion.

materials, i.e., C_β^{eq} in Fig. 5.15. Consequently, β phase precipitation can occur in the B diffused region of the A crystal. In the *B*-atom precipitated region, *B* atom diffusion will be apparently slower. This is the case of diffusion for a high concentration of P into Si using the vapor phase material PH_4 as a source.

In the prominent modern technology of manufacturing solid state electronic and optoelectronic devices, owing to technology needs, practically no diffusion process has been carried out using a thermal equilibrium source material in an inert ambient. Instead, either ion implanted sources or non-equilibrium source materials are used. The former involves non-equilibrium point defects due to damages incurred by the implantation process, and the latter obviously involves non-equilibrium phenomena due to the source material itself. In addition, the use of a reacting gas ambient adds another level of complexity to the results. Thus, it is seldom found that a simple description is satisfactory in interpreting diffusion results in Si, GaAs, and most other semiconductors.

5.4 DIFFUSION IN SILICON

Because of its use in fabricating integrated circuits, Si is the most important electronic material presently used and is likely to keep that position in the future. Diffusion processes are essential in the electronic device fabrication processes. Fig. 5.16 shows the diffusivities of the most important impurities and that of Si self-diffusion. Borrowing knowledge from metals, diffusion in Si had been first thought to be dominated by V-related mechanisms. Although in 1968 Seeger and Chik [37] already suggested that in Si both V and I contribute to self-diffusion and to dopant diffusion processes, the problem of the dominant point defect species in Si constituted a controversy for more than 20 years. Finally, during the 1980s [38–42], it was concluded and generally accepted that *both* I and V have to be taken into account, so as to consistently interpret the results of a large number of different experiments. The main indications for the contribution of I came from experiments performed under non-equilibrium native point defect conditions, such as the influence of surface oxidation or nitridation on dopant diffusion, and the diffusion of *i–s* species such as Au or Pt. What is still uncertain is the diffusivity and the thermal equilibrium concentration values of I and V, to be mentioned later.

With these developments, together with other available information, it appears that the framework of a global model of diffusion in Si is in place, which contains the following essential points: (i) both V and I participate in Si self-diffusion, the diffusion of dopants and diffusion of other substitutional impurities; (ii) neutral as well as charged V and I contribute to diffusion processes, known as the Fermi-level effect, which may or may not be prominent; (iii) there exist a special class of impurities, the *i–s* species, that will give rise to atypical diffusion profiles if dominated by I; (iv) chemical reactions at Si surfaces will alter concentrations of V and I. It is noted that this diffusion model is continuously improving quantitatively.

5.4.1 Silicon Self-Diffusion

Because I and V both contribute, the transport of Si self-atoms under thermal equilibrium conditions is governed by the uncorrelated self-diffusion coefficient

$$D^{SD} = D_I c_I^{eq} + D_V c_V^{eq}, \tag{5.23}$$

where D_I and D_V are respectively the diffusivities of I and V, and c_I^{eq} and c_V^{eq} are respectively the I and V thermal equilibrium concentrations normalized by the appropriate site densities into atomic fractions. As has been mentioned, native point defects may exist in several charge states. The observed doping dependence of group-III and V dopant diffusion (Section 5.4.3) indicates the contributions of neutral, positively charged, negatively and doubly negatively charged native point defects, but it is not known whether all these charge states occur for both I and V. Taking all observed charge states into account, we may write $D_I c_I^{eq}$ as

$$D_I c_I^{eq} = D_{I^o} c_{I^o}^{eq} + D_{I^-} c_{I^-}^{eq} + D_{I^{2-}} c_{I^{2-}}^{eq} + D_{I^+} c_{I^+}^{eq}, \tag{5.24}$$

and we have an analogous expression for V. The quantity $D_I c_I^{eq}$ comprises the sum of the concentrations of self-interstitials in the various charge states according to

$$c_I^{eq} = c_{I^o}^{eq} + c_{I^-}^{eq} + c_{I^{2-}}^{eq} + c_{I^+}^{eq}. \tag{5.25}$$

Therefore, the diffusivity D_I is actually an effective diffusion coefficient, that is, the weighted average of the diffusivities in the different charge states. The same holds analogously for c_V^{eq} and D_V.

Self-diffusion in Si is studied by measuring the diffusion of Si isotope tracer atoms D^T which differs slightly from D^{SD} of Eq. (5.23), since it contains the *correlation* factors f_I and f_V in the expression $D_T = f_I D_I c_I^{eq} + f_V D_V c_V^{eq}$. In the diamond cubic lattice, $f_V = 0.5$, and $f_I \leq 1$ (which depends on the unknown self-interstitial configuration). Measured results for D_T are usually fitted to the expression $D_T = D_o \exp(-Q/k_B T)$, where D_o is a preexponential factor, Q is the activation enthalpy. A summary of these results is shown in Fig. 5.17. Tracer measurements do not allow us to separate I and V contributions to self-diffusion. Such a separation became possible by investigating the diffusion of Au, Pt and Zn in Si, described in some detail in Section 5.4.2.

Earlier experiments, i.e., those listed in Fig. 5.18 but excluding that of Shimizu et al. (8 in the figure), allowed a fairly accurate determination of $D_I c_I^{eq}$ but only a crude estimate of $D_V c_V^{eq}$ derived from a combination of different types of experiments [40,42]. The results are:

$$D_I c_I^{eq} = 940 \exp(-4.84 eV/k_B T)\ \text{cm}^2\,\text{s}^{-1}, \tag{5.26}$$

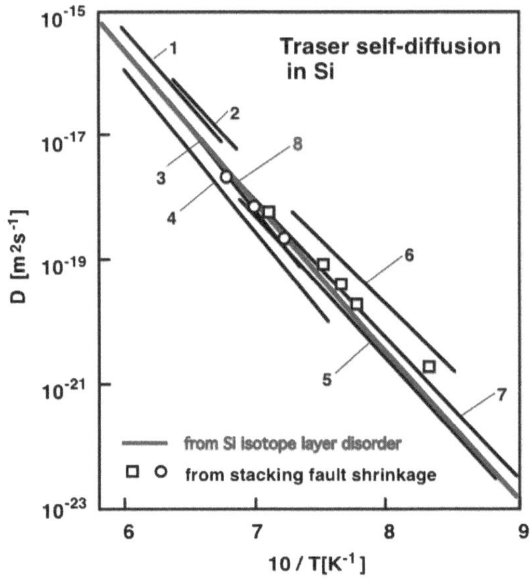

Figure 5.17 Tracer self-diffusion coefficients of silicon as a function of reciprocal absolute temperature. Indicated data are from: 1. R.F. Peart, *Phys. Stat. Sol. 15*, K119 (1966). 2. R.N. Ghostagore, *Phys. Rev. Lett. 16*, 890 (1966). 3. J.M. Fairfield and B.J. Masters, *J. Appl. Phys. 38*, 3148 (1967). 4. J.J. Mayer, H. Mehrer, and K. Maier, *Inst. Phys. Conf. Ser. 31*, 186K (1977). 5. L. Kalinowski and R. Seguin, *Appl. Phys. Lett. 35*, 171 (1980). 6. J. Hirvonen and A. Antilla, *Appl. Phys. Lett. 35*, 703 (1979). 7. F.J. Demond, S. Kalbitzer, H. Mannsperger, and H. Damjantschitsch, *Phys. Lett. 93A*, 503 (1983). 8. Y. Shimizu, M. Uematsu, and K.M. Ito, *Phys. Rev. Lett. 98*, 095901 (2007).

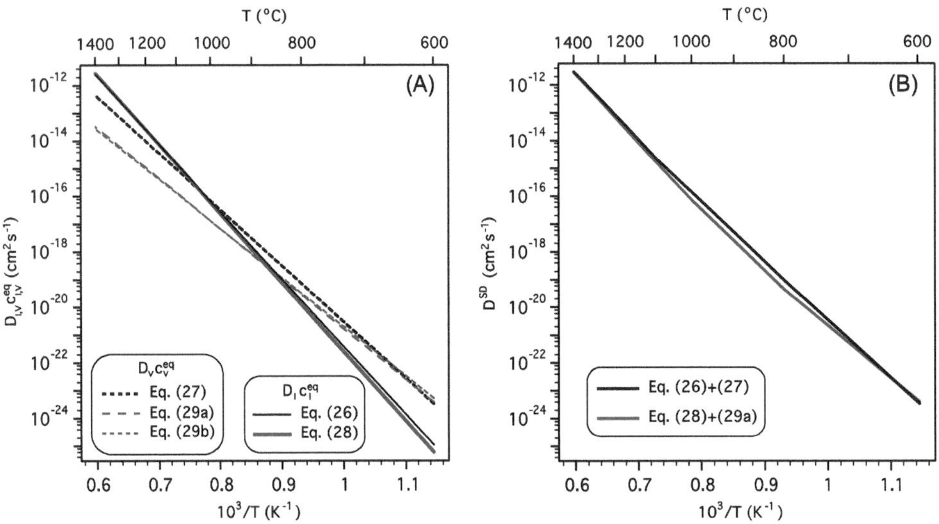

Figure 5.18 (A) Plots of the I and V contributions to Si self-diffusion, $D_I c_I^{eq}$ and $D_V c_V^{eq}$, respectively, and (B) the diffusivity of Si self-diffusion. The expressions in Eqs. (5.26) and (5.27) are from 30 years ago [40], while Eqs. (5.28) and (5.29) are more recent [47,48].

$$D_V c_V^{eq} = 0.6 \exp(-4.03eV/k_B T)\ \text{cm}^2\text{s}^{-1}. \tag{5.27}$$

The values in Eqs. (5.26) and (5.27) are within the experimental range of the measured Si tracer diffusion coefficient D^T. The doping dependence of Si self-diffusion [34] allows us to conclude that neutral, as well as positively and negatively charged, point defects are involved in self-diffusion, but the data are not accurate enough to determine the individual terms of Eq. (5.25) or the analogous expression for vacancies. Since D^T, as well as $D_I c_I^{eq}$ and $D_V c_V^{eq}$, each consist of various terms, their representation in terms of an expression as in Eqs. (5.26) and (5.27) can only be an approximation holding over a limited temperature range. In Section 5.4.5 we will discuss what is known about the individual factors D_I, c_I^{eq}, D_V, and c_V^{eq}.

Since it was determined that both I and V contribute in diffusion processes in Si self-diffusion and diffusion of dopants in the time frame of early to middle 1980s [38–42], including the assignment of the I and V contributions to self-diffusion given by Eqs. (5.26) and (5.27) and determination of the contributions to dopant diffusions by the interstitial component, Φ_I, to be discussed, diffusion studies in Si apparently slowed down considerably, until recently. The renewal of interest was apparently sparked by the need of more precise control in device fabrication and by the realization that stable isotopes ^{29}Si and ^{30}Si can be used in thin layered superlattice structures down to nanometer periods to study Si self-diffusion [43–48] and similarly for dopants. For self-diffusion, these studies provided further indications that both I and V contribute to Si self-diffusion in that the data covering a sufficiently large temperature range showed a slightly concave shape toward the upper right side on an Arrhenius plot, which indicates two different contributions to the total as described by Eq. (5.23) where I dominates at higher temperatures while V dominates at lower temperatures, just as Eqs. (5.26) and (5.27) predicted. Quantitatively, these works yielded some modifications to Eqs. (5.26) and (5.27). From the works of Shimizu et al. [47] and Kube et al. [48], contribution to self-diffusion by I became [47,48]

$$D_I c_I^{eq} = 2175.4 \exp\left(-\frac{4.95eV}{k_B T}\right)\ \text{cm}^2\,\text{s}^{-1}, \tag{5.28}$$

and that by V became [47]

$$D_V c_V^{eq} = 0.0023 \exp\left(-\frac{3.6eV}{k_B T}\right)\ \text{cm}^2\,\text{s}^{-1}, \tag{5.29a}$$

or [48]

$$D_V c_V^{eq} = 0.0011 \exp\left(-\frac{3.52eV}{k_B T}\right)\ \text{cm}^2\,\text{s}^{-1}. \tag{5.29b}$$

A fitting line in accordance with Eqs. (5.28) and (5.29a) has already been included in Fig. 5.18. As a direct comparison, computed values of the equation pairs (5.26), (5.27) and (5.28), (5.29) are shown in Fig. 5.18(A), while the corresponding D^{SD} values are shown in Fig. 5.18(B). In Fig. 5.18(A) it is seen that the I contribution to Si self-diffusion described by Eqs. (5.26) and (5.28) agrees well over the whole temperature range from 600 to 1400°C. For V contribution, the two newer descriptions, Eqs. (5.29a) and (5.29b), agree well with each other, but agree with that given in Eq. (5.27) fairly well only at low temperatures. At higher temperatures these values became smaller than that given by Eq. (5.27) by a fairly large factor, e.g., ~15 at 1400°C. On the other hand, it is seen from Fig. 5.18(B) that the two D^{SD} descriptions agree every well in both the low and high temperature range, but that given by Eqs. (5.26) and (5.27) is larger by a factor of 2 at ~850°C.

5.4.2 Interstitial–Substitutional Diffusion: Au, Pt, and Zn in Si

Both Au and Pt can reduce minority carrier lifetimes in Si because their energy levels are close to the middle of the bandgap. They are used in power devices to improve the device frequency behavior. In contrast, Au and, to a lesser extent, Pt are undesirable contaminants in integrated circuits and hence have to be avoided or removed. For these reasons, the behavior of Au and Pt has been investigated extensively. Zinc is not a technologically important impurity in Si, but scientifically it served as an example with diffusion behavior in between substitutional dopants and Au and Pt in Si.

These impurities are *i–s* species dissolved mainly on substitutional sites (A_s) but accomplish diffusion by switching over to an interstitial configuration (A_i) in which their diffusivity D_i is extremely high. The changeover from *i*-site to *s*-site and vice versa involves native point defects. For uncharged species, the two basic forms of the *i–s* changeover processes are the *kick-out* (KO) mechanism involving I [33],

$$A_i \Leftrightarrow A_s + I, \tag{5.30}$$

and the much earlier proposed *dissociative* mechanism or *Frank–Turnbull* (FT) mechanism involving V [31],

$$A_i + V \Leftrightarrow A_s. \tag{5.31}$$

Both mechanisms are schematically shown in Fig. 5.19. The KO mechanism is closely related to the interstitialcy mechanism. The main difference is that the foreign atom, once in an interstitial position, remains there for only one step in the interstitialcy mechanism and for many steps in the KO mechanism. In contrast, the FT mechanism and the vacancy exchange mechanism are qualitatively different. For the vacancy exchange mechanism an increase in V concentration enhances the diffusivity, whereas for the

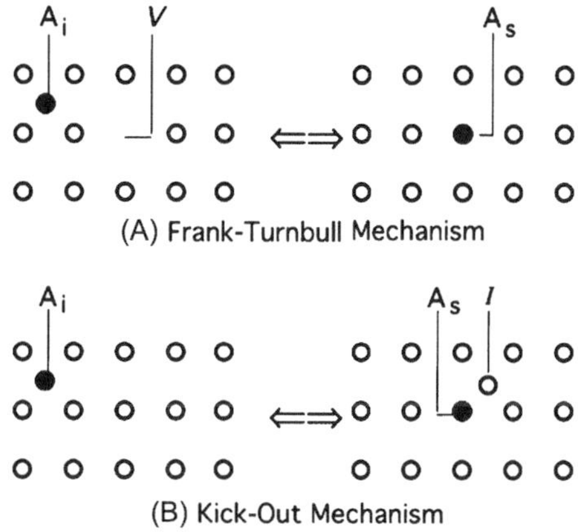

Figure 5.19 Schematic illustration of the interstitial–substitutional (*i–s*) impurity diffusion mechanisms. In both cases the *i–s* atom exchanges are involved. For the Frank–Turnbull mechanism shown in (A), a crystal vacancy is consumed when an interstitial impurity atom becomes substitutional, while for the kick-out mechanism shown in (B), a self-interstitial is generated when an interstitial impurity atom becomes substitutional.

FT mechanism an increase in V concentration rather decreases the diffusivity of the substitutional species.

For a detailed description of *i–s* atoms diffusing via the KO or FT mechanism, we need to solve a coupled system of three partial differential equations describing diffusion and reaction of A_i, A_s and the native point defect involved, V or I [33]. But for indiffusion from the surface, the A_s profile shapes are described well by an *effective* diffusivity D^{eff} obtained for use in the Fick's second law representation, Eq. (5.16). This requires assuming that the local equilibrium conditions in accordance with reactions (5.30) and (5.31),

$$C_s C_I / C_i = C_s^{eq} C_I^{eq} / C_i^{eq}, \tag{5.32}$$

$$C_i C_V / C_I = C_i^{eq} C_V^{eq} / C_I^{eq}, \tag{5.33}$$

hold for the KO and FT mechanisms, respectively. In Eqs. (5.32) and (5.33), C_s and C_i are respectively the concentrations of the s and i atoms of the *i–s* species. For the KO mechanism, if the indiffusion of A_i is slow enough to allow the Is generated to migrate out to the surface to keep their thermal equilibrium concentration, i.e., $D_i c_i^{eq} \ll D_I c_I^{eq}$ holds, then the effective diffusivity of A_s is given by

$$D_{(i)}^{eff} = D_i C_i / C_s^{eq}, \tag{5.34}$$

provided $C_s^{eq} \gg C_i^{eq}$ holds, which is generally the case. Here the subscript index in parentheses indicates the rate limiting species. If, on the other hand, the indiffusion of A_i is so fast that the generated I *cannot* escape quickly enough to the surface, i.e., $D_i c_i^{eq} \gg D_I c_I^{eq}$ holds, a *supersaturation* of I will develop and further incorporation of A_s will be limited by the outdiffusion of the generated I to the surface. This leads to an effective A_s diffusivity $D_{(I)}^{eff}$ given by

$$D_{(I)}^{eff} = \left(D_I C_I^{eq} / C_s^{eq}\right)\left(C_s^{eq} / C_s\right)^2. \tag{5.35}$$

For the FT mechanism, sufficiently slow indiffusion $D_i c_i^{eq} \ll D_V c_V^{eq}$ leads to the same $D_{(i)}^{eff}$ as given by Eq. (5.34). An effective A_s diffusivity $D_{(V)}^{eff}$ controlled by the indiffusion of V from the surface results if $D_i^{eff} c_i^{eq} \gg D_V c_V^{eq}$ holds:

$$D_{(V)}^{eff} = D_V C_V^{eq} / C_s^{eq}. \tag{5.36}$$

The strongly concentration-dependent effective diffusivity $D_{(I)}^{eff}$ of Eq. (5.35) leads to an A_s concentration profile so strongly deviated away from the erfc-type that it is actually concave upward in $\log C_s$ plotted as a function of 1/T. These profiles can easily be distinguished from the erfc-type profiles that are associated with $D_{(V)}^{eff}$. This macroscopic difference allows us not only to decide between different atomistic diffusion mechanisms of the specific foreign atom involved but also to obtain information on the *mechanism of self-diffusion*. The effective diffusivities given by Eqs. (5.35) and (5.36) have been derived under the assumption of dislocation-free crystals. The presence of a high density of dislocations in an elemental crystal maintains the equilibrium concentration of native point defects and thus an erfc-type profile characterized by the constant diffusivity $D_{(i)}^{eff}$ of Eq. (5.34) will result even if $D_i c_i^{eq} \gg D_I c_I^{eq}$ holds. For compound semiconductors, this statement does not hold in general, since the presence of dislocations does not necessarily guarantee native point defects to attain their thermal equilibrium concentrations. If I and V coexist, such as in the case of Si, the effective A_s diffusion coefficient in dislocation-free material for $D_i c_i^{eq} \gg (D_I c_I^{eq} + D_V c_V^{eq})$ is given by

$$D_{(I,V)}^{eff} = D_{(I)}^{eff} + D_{(V)}^{eff}. \tag{5.37}$$

The indiffusion profiles of both Au and Pt in dislocation-free Si show the concave profile shape typical for the KO mechanism [34,49,50]. Examples are shown in Figs. 5.20 and 5.21 respectively for Au and Pt. From profiles like these and from the measured solubility C_s^{eq} of Au_s and Pt_s in Si, the values of $D_I c_I^{eq}$ in Eq. (5.35) have been determined. Diffusion of Au into thin Si wafers leads to characteristic U-shaped profiles even if Au has been deposited on one side only. The increase of the Au concentration in the center of the wafer has also been used to determine $D_I c_I^{eq}$ [37].

In heavily dislocated Si the dislocations act as efficient sinks for Is to keep C_I close to C_I^{eq} so that the constant effective diffusivity $D_{(i)}^{eff}$ of Eq. (5.34) governs the A_s profile,

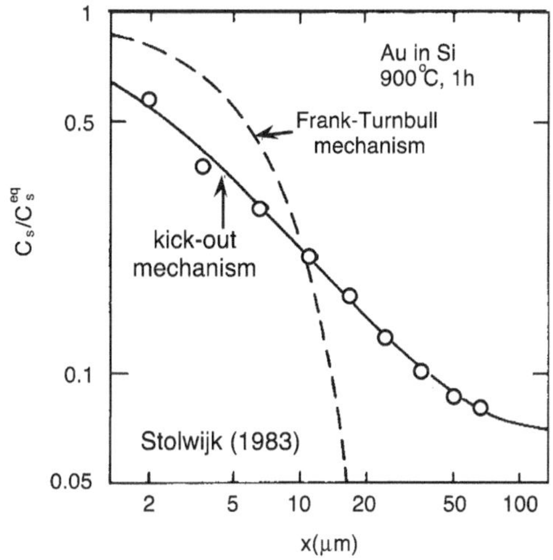

Figure 5.20 Experimental Au concentration profile in dislocation-free Si (circles) compared with predictions of the Frank–Turnbull and the kick-out mechanism. From Stolwijk et al. [49].

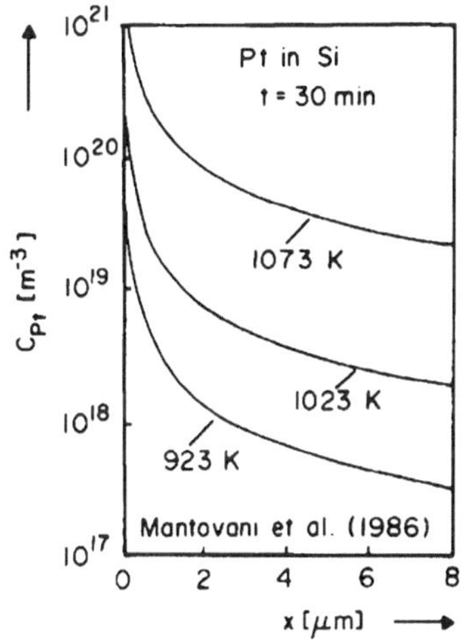

Figure 5.21 Platinum concentration profiles in dislocation-free Si. From Mantovani et al. [50].

which is erfc-shaped. This has been observed by Stolwijk et al. for Au [51]. Analysis of the resulting erfc-profiles allowed obtaining

$$D_i c_i^{eq} \approx 64 \exp(-3.93eV/k_B T) \text{ cm}^2\,\text{s}^{-1}. \tag{5.38}$$

This $D_i c_i^{eq}$ value turns out to be much larger than $D_I c_I^{eq}$ given by Eqs. (5.26) or (5.28), which is consistent with the observation that Au_s concentration profiles are governed by $D_{(I)}^{eff}$ in dislocation-free Si.

Zinc diffusion has also been investigated in Si [52]. In highly dislocated material, an erfc-profile develops as expected. In dislocation-free material only the profile part close to the surface shows the concave shape typical for the kick-out diffusion mechanism. For lower Zn concentrations, a constant diffusivity takes over. The reason for this changeover from one profile type to another is as follows. In contrast to the case of Au, the $D_i c_i^{eq}$ value determined for Zn is not much higher than $D_I c_I^{eq}$ so that even in dislocation-free Si only the profile close to the surface is governed by $D_{(I)}^{eff}$ of Eq. (5.35) which strongly increases with depth. For sufficiently large penetration depths, $D_{(I)}^{eff}$ finally exceeds $D_{(i)}^{eff}$ and a constant effective diffusivity begins to determine the concentration profile. A detailed analysis of this situation can be found elsewhere [42]. The changeover from a concave to an erfc-type profile has also been observed for the diffusion of Au either into very thick Si samples [53] or for short-time diffusions [54] into normal silicon wafers 300–800 μm in thickness.

5.4.3 Dopant Diffusion

5.4.3.1 Fermi Level Effect

Dopant diffusion has been studied extensively because of its importance in device fabrications. A detailed quantitative understanding of dopant diffusion is also a prerequisite for accurate and meaningful modeling in numerical process simulation programs. It is not the intention here to compile all available data on dopant diffusion in silicon, which may be found elsewhere (see [41,42] for references). Instead, let's concentrate on the diffusion mechanisms and native point defects involved in dopant diffusion, the effect of the Fermi level on dopant diffusion and on non-equilibrium point defect phenomena induced by high-concentration indiffusion of dopants.

The diffusivities D_s of all dopants in Si depend on the Fermi level. The experimentally observed doping dependencies may be described in terms of the expression

$$D_s(n) = D_s^o + D_s^+(n_i/n) + D_s^-(n/n_i) + D_s^{2-}(n/n_i)^2, \tag{5.39}$$

which reduces to

$$D_s(n_i) = D_s^o + D_s^+ + D_s^- + D_s^{2-} \tag{5.40}$$

for intrinsic conditions $n = n_i$. Depending on the specific dopant, some of the quantities in Eq. (5.40) may be negligibly small. $D_s(n_i)$ is an exponential function of inverse temperature as shown in Fig. 5.16. Values of these quantities in terms of preexponential factors and activation enthalpies are given in Table 5.4 [42,55,56]. It is seen from the

Table 5.4 Diffusion data of various dopants fitted to Eq. (5.40). Each term fitted to $D_o \exp(-Q/k_BT)$; D_o values in $10^{-4}\ m^2\ s^{-1}$ and Q values in eV [42,55,56]

Element	D_o^o	Q^o	D_o^+	Q^+	D_o^-	Q^-	D_o^{2-}	Q^{2-}
B	0.037	3.46	0.72	3.46	–	–	–	–
P	3.85	3.66	–	–	4.44	4.00	44.20	4.37
As	0.066	3.44	–	–	12.0	4.05	–	–
Sb	0.214	3.65	–	–	15.0	4.08	–	–

values listed in Table 5.4 that the charged point defect contributions amount to fractions of that of the neutral point defect species. More will be discussed in association with high concentration dopant diffusion studies.

The higher diffusivities of all dopants, as compared to self-diffusion, require faster moving complexes formed by the dopants and native point defects, i.e., the dopant–defect pairing mechanism, discussed in Section 5.3.3. The doping dependence of $D_s(n)$ is generally explained in terms of the various charge states of the native point defects carrying dopant diffusion. Since both I and V can be involved in dopant diffusion, each of the terms in Eq. (5.40) in general consists of an I and V related contribution, e.g.,

$$D_s^+ = D_s^{I^+} + D_s^{V^+}. \tag{5.41}$$

$D_s(n)$ may also be written in terms of I and V related contributions as

$$D_s(n) = D_s^I(n) + D_s^V(n) \tag{5.42}$$

with

$$D_s^I(n) = D_s^{I^o} + D_s^{I^+}(n_i/n) + D_s^{I^-}(n/n_i) + D_s^{I^{2-}}(n/n_i)^2 \tag{5.43}$$

and an analogous expression for $D_s^V(n)$.

The observed doping dependence expressed in Eq. (5.39) just shows that *charged point defects* are involved in the diffusion process, but nothing can be learned on the relative contributions of I and V in the various charge states. Strictly speaking, in contrast to the case of self-diffusion, the doping dependence of dopant diffusion does not necessarily prove the *presence of charged native point defects* but rather the *presence of charged point-defect/dopant complexes.* In Section 5.4.3.2 we will describe a way to determine the relative contribution of I and V to dopant diffusion by measuring the effect of non-equilibrium concentrations of native point defects on dopant diffusion.

In the interests of fabricating abrupt junctions, a number of high concentration dopant diffusion studies have been carried out, using ion implanted or layered grown-in source materials, under quasi-isoconcentration conditions using isotopes or dopants of the same type (n or p). The ion implanted source materials cases involve ion damage produced non-equilibrium point defects influenced by inadvertent factors. For the grown-in source materials cases it was found that in the regime Si is extrinsic at the diffusion temperature, the dopant diffusivities increase linearly with its concentration, irrespective of whether the dopant is n- or p-type [57–60], provided the dopant concentration is below a certain critical value believed to be the dopant solubility at the given temperature. This is clearly just the Fermi level effect, Eq. (5.43) or its analogous counter parts, to the first order charge-dependent term and in accordance with either I or V dominating. Above this critical concentration, the dopant diffusivity further increases but now with a much higher power dependence on its concentration, e.g., 3–5 [59–63]. This prompted a number of explanations, including the V percolation [30,64], discussed in Section 5.3.3, and the extended dopant–defect cluster model, e.g., Sb_2V [61].

5.4.3.2 High Concentration Dopant-Diffusion-Induced Non-equilibrium Effects

Non-equilibrium concentrations of native point defects may be induced by indiffusion of dopants starting from a high surface concentration. These non-equilibrium effects are most pronounced for high concentration P diffusion, but also present for other dopants such as B and to a lesser extend for Al and Ga. The source materials used in these experiments are non-equilibrium ones entering Si at concentrations exceeding the solubility of the species at the diffusion temperature, discussed in Section 5.3.6.

Phosphorus indiffusion profiles (Fig. 5.22) show a *tail* in which the P diffusivity is much higher (up to a factor of 100 at 900°C) than expected from isoconcentration studies. In n–p–n transistor structures in which high concentration P is used for the emitter diffusion, the diffusion of the base dopant B below the P diffused region is similarly enhanced, the so-called *emitter-push effect*. The diffusion of B, P, or Ga in buried layers many microns away from the P diffused region is also greatly enhanced. In contrast, the diffusion of Sb in buried layers is retarded under the same conditions. The enhanced and retarded diffusion phenomena are analogous to those occurring during surface oxidation, to be discussed in the next Section 5.4.3.3. As has also been confirmed by dislocation–climb experiments [65], all these phenomena are due to a *supersaturation* of I induced by high-concentration indiffusion of P, which is also associated with an *undersaturation* of V. The basic features of high concentration P diffusion are schematically shown in Fig. 5.23. Also shown in Fig. 5.23 is the presence of electrically neutral precipitates, which should be the compound SiP. This indicates P diffused into Si at concentrations exceeding the solubility limit defined by coexistence of Si and SiP at the diffusion temperature, which is a non-equilibrium source condition case discussed in Section 5.3.6. A much less pronounced supersaturation of I is generated by B starting from a high surface concentration as can be concluded from the B profiles and from the growth of interstitial-type stacking faults induced by B diffusion [57].

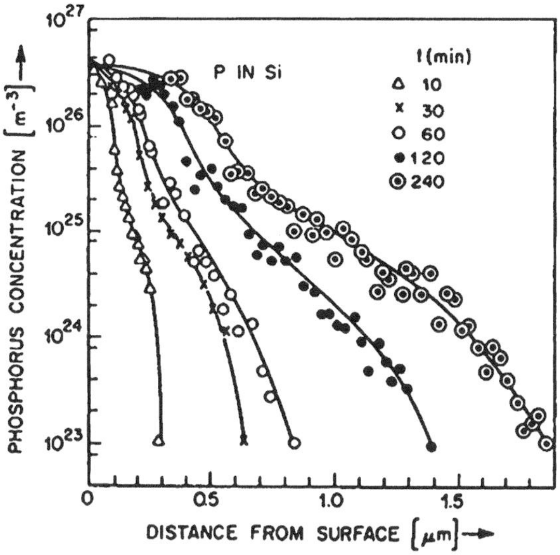

Figure 5.22 Concentration profiles of P diffused into Si at 900°C for the times t indicated. From Yoshida et al. [*J. Appl. Phys. 45*, 1498 (1974)].

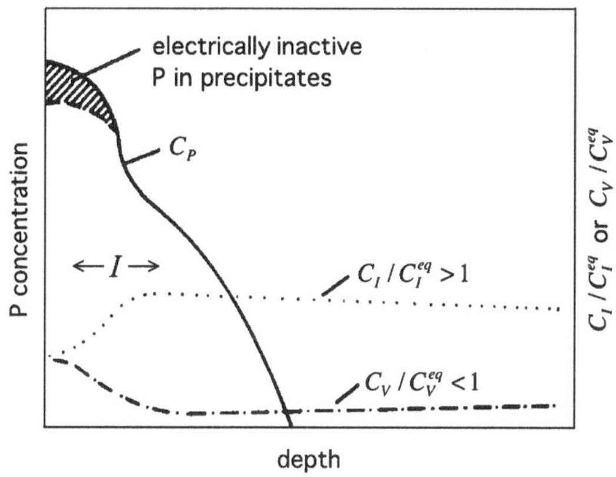

Figure 5.23 A schematic P concentration profile (C_P) and the normalized native point defect concentrations C_I/C_I^{eq} and C_V/C_V^{eq}. From Gösele [in: *Microelectronic Materials and Processes*: ed. R.A. Levy (Ed.). Dordrecht: Kluwer Academic, 1989, pp. 588–634].

Many models have been proposed to explain the phenomena associated with high concentration P diffusion. Earlier models are vacancy based and predict a P-induced V supersaturation, which contradicts the experimental results obtained in the meantime. In 1986, Morehead and Lever [58] presented a mathematical treatment of high-concentration dopant diffusion that is primarily based on the point defect species dominating the diffusion of the dopant, e.g., I for P and B and V for Sb. The concentration of the other native point-defect type is assumed to be determined by the dominating point defect via the local equilibrium conditions discussed in the next Section 5.4.3.3. The dopant-induced self-interstitial supersaturation s_I may be estimated by the influx of dopants which release part of the I involved in their diffusion process. These self-interstitials will diffuse to the surface where it is assumed that $C_I = C_I^{eq}$ holds, and also into the Si bulk. Finally, a quasi-steady-state supersaturation of self-interstitials will develop for which the dopant-induced flux of injected I just cancels the flux of Is to the surface. Fig. 5.23 shows schematically the situation described by Morehead and Lever [58].

5.4.3.3 Influence of Surface Reactions

Thermal oxidation is a standard process for forming field and gate oxides, or oxide masks for Si device fabrications. Oxidation of Si can lead to the formation of the so-called oxidation induced stacking faults, OSF [66,67], when nucleation sites of the OSF are present either due to mechanical damage or due to impurity contamination. The OSFs are extrinsic in nature and bounded by 1/3⟨111⟩ Frank partial dislocations [68]. In a growth kinetics study, Hu [69] found that the OSF exhibited a growth–retrogrowth phenomenon, which is dependent on the wafer orientation, temperature, and whether the ambient is dry or wet oxygen. On the other hand, oxidation also leads to the oxidation-enhanced-diffusion or oxidation-retarded-diffusion of dopants, OED or ORD, respectively. For oxidizing ⟨100⟩ Si wafers in dry oxygen below 1200°C for a few hours, OED were observed for B, P, As and Al [70–73], while ORD was observed for Sb [74]. The results varies for oxidizing Si in a wet oxygen ambient or for Si with ⟨111⟩ orientations. For oxidizing ⟨111⟩ in Si in dry oxygen, it was found that P exhibited OED and ORD, respectively, for shorter and longer oxidation times at a temperature of 1160°C [75], while under identical conditions just the reverse was found for Sb, i.e., ORD and OED, respectively, for short and long oxidations times [76].

The OSF growth/retrogrowth and dopant OED/ORD phenomena require the participation of both I and V in Si self-diffusion as well as in dopant diffusion. To consistently understand these phenomena, any attempt to ignore the role of either I or V is unequivocally ruled out. The OSF is of the extrinsic type, which consists of an extra {111} atom layer bonded by a 1/3⟨111⟩ Frank partial dislocation [68]. Thus, growth of OSF under oxidizing conditions clearly indicates that oxidation leads to the injection of I into Si. This injected I gives rise to an I supersaturation, which leads to growth of the OSF via the migration of these supersaturated I to reach the core of the Frank partial dislocation bonding the OSF. Migration of I is part of Si self-diffusion. The supersaturated I also gives rise to OED of dopants that diffuse primarily utilizing I as diffusion vehicles, e.g., B, P, As, Al. The supersaturated I will annihilate V, which leads to a V undersaturation that in turn leads to ORD of dopants primarily utilizing V as diffusion vehicles, e.g., Sb. V undersaturation is caused by the injected I via the

I–V recombination reaction $I + V \Leftrightarrow \phi$, where ϕ is a lattice atom, which leads to

$$C_I C_V = C_I^{eq} C_V^{eq}, \tag{5.44}$$

where C_I and C_V are respectively the actual I and V concentrations, and C_I^{eq} and C_V^{eq} are those under thermal equilibrium conditions.

Retrogrowth of OSF occurs with the diminishing of the oxide growth rate when the already grown oxide layer becomes so thick that transport of O_2 molecule to the SiO_2–Si interface through the oxide layer becomes ineffective. While this factor may appear to be sufficient to explain OSF retrogrowth with the driving force being the stacking fault energy, it is not sufficient to also explain OED/ORD reversal of P [75] and ORD/OED reversal of Sb [76]. Assuming that P diffuses mainly via I, P exhibits OED as the oxide is growing because of I injection. As soon as the oxide has grown so thick that oxidation is stopped, I injection also stops and P diffusion should return to its normal rate but not below it. Thus, just because of the diminishing oxidation rate at long times, OED should still be observed for much longer oxidation times, though to a lesser extent. Therefore, to explain the observed ORD of P, a mechanism that will lead to an I undersaturation concomitant with V supersaturation is needed. To this end, a V-injection mechanism has been suggested: the formation of SiO molecule at the SiO_2–Si interface at high temperatures [38]. The formation of the SiO molecule, which subsequently diffuses through the oxide to disappear into the ambient, consumes an Si atom and hence generates a V, or equivalently, consumes an I in Si at the interface, leading to V supersaturation and I undersaturation. This V-injection mechanism is in parallel with and independent of the I-injection mechanism due to oxidation. The effect of I-injection is larger for shorter oxidation times but diminishes afterwards, while the effect of V-injection is smaller but maintains a constant rate at all times, i.e., when I injection is stopped V injection is still operating. Assuming that Sb diffuses primarily via V, it exhibits ORD for shorter oxidation times but changes to OED for longer times [76], which is just the reverse situation compared to P [75].

The physical reason for I injection to occur during surface oxidation is as follows [77,78]. Oxidation occurs by the diffusion of oxygen through the oxide layer to react with the Si crystal atoms at the SiO_2–Si interface. The oxidation reaction is associated with a volume expansion of about a factor of 2.2, which is mostly accommodated by viscoelastic flow of the newly formed oxide but partly also by the injection of Si I into the Si crystal matrix, which leads to I supersaturation. Oxidation can also cause V injection, provided oxidation occurs at sufficiently high temperatures (typically 1150°C or higher) and the oxide is thick enough. Under these circumstances, Si, in the form of SiO [38], diffuses from the interface and reacts with oxygen in the oxide away from the interface. The resulting supersaturation of V associated with undersaturation of I gives rise to ORD of P diffusion [75] and OED of Sb [76]. Thermal nitridation of Si surfaces also causes supersaturation of V coupled with undersaturation of I, whereas oxynitridation (nitridation of oxides) behaves more like normal oxidation [41]. Silicidation reactions have also been found to inject native point defects and to cause enhanced dopant diffusion [79].

A simple quantitative formulation of oxidation- and nitridation-influenced diffusion is based on Eq. (5.42), which changes with perturbed native point-defect concentrations C_I and C_V approximately to

$$D_s^{per}(n) = D_s^I(n)\left[C_I/C_I^{eq}(n)\right] + D_s^V(n)\left[C_V/C_V^{eq}(n)\right]. \tag{5.45}$$

For long enough times and sufficiently high temperatures (e.g., one hour at 1100°C) local dynamical equilibrium between V and I according to Eq. (5.44) is established and Eq. (5.45) may be reformulated in terms of C_I/C_I^{eq}. Defining the normalized diffusivity enhancement as $\Delta_s^{per} = [D_s^{per}(n) - D_s(n)]/D_s(n)$, the fractional interstitialcy diffusion component as $\Phi_I(n) = D_s^I/D_s(n)$, and the I supersaturation ratio as $s_I(n) = [C_I - C_I^{eq}(n)]/C_I^{eq}(n)$, Eq. (5.45) may be rewritten as [38]

$$\Delta_s^{per}(n) = \left[2\Phi_I(n) + s_I\Phi_I(n) - 1\right]/(1 + s_I) \tag{5.46}$$

with Eq. (5.44) holding. The I supersaturation ratio s_I as derived from OSF growth data is

$$s_I = 6.6 \times 10^{-9}\exp(2.52eV/k_B T)t^{-1/4}, \tag{5.47}$$

where t is an instant of time into the oxidation. Normally Eq. (5.46) is for cases under intrinsic conditions and the dependence of Φ_I on n is not indicated. Eq. (5.46) is plotted in Fig. 5.24 for Φ_I values of 0.85, 0.5, and 0.2.

The left-hand side of Fig. 5.24, where $s_I < 0$ (associated with V supersaturation) has been realized by high-temperature oxidation and thermal nitridation of silicon surfaces, as mentioned above. Another possibility to generate vacancy supersaturation is the oxidation in an HCl containing atmosphere at sufficiently high temperatures and for sufficiently large HCl contents [40]. As expected, $s_I < 0$ results in enhanced Sb diffusion and retarded diffusion of P and B. Arsenic diffusion is enhanced as in the case of oxidation, which indicates that arsenic has appreciable components via both V and $I(\Phi_I \sim 0.5)$.

In Fig. 5.25 values of ϕ_I, which are the time averaged values of Φ_I over the oxidation period τ at 1100°C, are shown as a function of the atomic radius r_s of the various dopants for intrinsic doping conditions. For this purpose, Eqs. (5.46) and (5.47) respectively become [38]:

$$\bar{\Delta}_s^{per}(n) = \left[2\phi_I(n) + \bar{s}_I\phi_I(n) - 1\right]/(1 + \bar{s}_I), \tag{5.48}$$

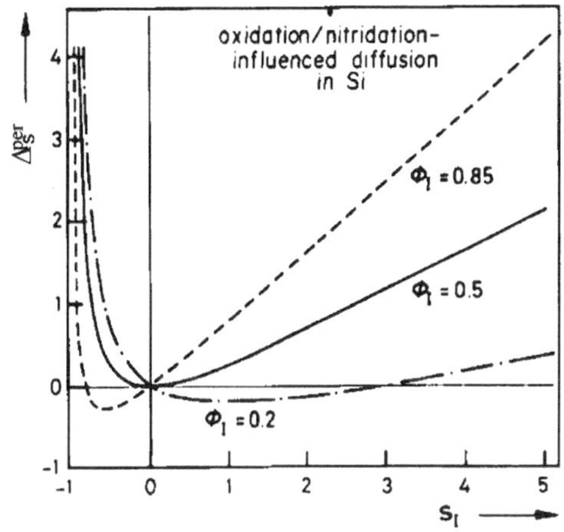

Figure 5.24 Normalized diffusion enhancement Δ_S^{per} versus self-interstitial supersaturation $s_I = (C_I - C_I^{wq})/C_I$ for different values of Φ_I (from Tan and Gösele [38]).

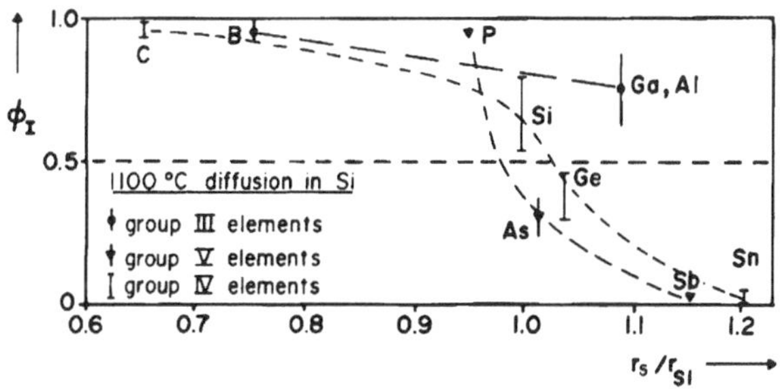

Figure 5.25 Interstitial-related fractional diffusion component Φ_I for group III, IV and V elements versus their atomic radius in units of the atomic radius r_{Si} of silicon. The values for carbon and tin are expected from theoretical considerations and limited experimental results. From Tan and Gösele [42].

$$\bar{s}_I = 8.8 \times 10^{-9} \exp(2.52eV/k_B T)\tau^{-1/4}. \tag{5.49}$$

Eqs. (5.47) and (5.49) are for oxidizing ⟨100⟩ Si wafers in dry oxygen. For ⟨111⟩ wafers, oxidation in wet ambient or ambient containing HCl, some modification to Eqs. (5.47) and (5.49) is needed [39,41].

Both the charge state (group III or V dopants) and the atomic size influence Φ_I. Φ_I has a tendency to increase with increasing temperature. Oxidation and nitridation experiments and extrinsic conditions indicate a decreasing value of Φ_I for P with increasing n-doping, but both P and B still remain dominated by $I(\Phi_I(n) > 0.5)$.

5.4.4 Diffusion of Carbon and Other Group IV Elements

The group IV elements C, Ge, and Sn dissolve in Si substitutionally, but knowledge on their diffusion mechanisms is incomplete. Ge and Sn diffusions are similarly slow as Si self-diffusion, whereas C diffusion is much faster (Fig. 5.16).

Germanium atoms are larger than Si atoms. Oxidation and nitridation experiments show a Φ_I value of Ge around 0.4 at 1100°C [80] which is slightly lower than that derived for Si self-diffusion. Diffusion of the much larger Sn atoms in Si is expected to be almost entirely due to the vacancy exchanged mechanism, similar as for the group V dopant Sb. Consistent with this expectation, a nitridation-induced supersaturation of V increases Sn diffusion [81], but no quantitative determination of Φ_I is available for Sn.

Indiffusion C profiles in Si are error function-shaped. Considering the atomic volume, it can be expected that the diffusion of C atoms, which are much smaller than Si, involves mainly Si Is. Based on EPR measurements, Watkins and Brower [82] proposed 50 years ago that C diffusion is accomplished by a highly mobile CI complex according to $C_s + I \Leftrightarrow CI$, where C_s denotes substitutional C. This is consistent with the experimental observation that I injected by oxidation or high-concentration P indiffusion enhance C diffusion [83]. Equivalently, we may regard C as an i–s impurity, just as Au. That is, we can regard the diffusion of C according to [84,85]

$$C_s + I \Leftrightarrow C_i, \tag{5.50}$$

where C_i denotes an interstitial carbon atom. Since whether C_s diffusion is actually carried by CI complexes or by C_i atoms has not yet been distinguished on a physical basis, and the mathematical descriptions for both cases are identical in form, we can regard C_s diffusion as being carried by C_i atoms in accordance with the KO mechanism of the i–s impurities. Under this assumption, when considering diffusion of C into Si for which the substitutional C concentration is at or below the solubility of the substitutional carbon atoms, C_s^{eq}, the substitutional C diffusivity D_s^{eff} is given by the effective diffusivity $D_i C_i^{eq}/C_s^{eq}$ where D_i is the diffusivity of the fast diffusing C_i atoms and C_i^{eq} is the solubility of the C_i atoms. Error function type C_s indiffusion profiles are obtained under indiffusion conditions because

$$D_s^{eff} C_s^{eq} = D_i C_i^{eq} < D_I C_I^{eq} \tag{5.51}$$

holds. Under this condition, C indiffusion induced Si Is migrated rapidly out to the Si surface and hence the C_I^{eq} condition was basically maintained, in agreement with experimental observations [86,87].

From the C indiffusion data, the solubility of C_s is given by [86,87]

$$C_s^{eq} = 4 \times 10^{24} \exp(-2.3eV/k_B T)\ \mathrm{cm}^{-3}, \tag{5.52}$$

and the diffusion coefficient of C_s is given by

$$D_s = 1.9\exp(-3.1eV/k_BT)\ \mathrm{cm}^2\,\mathrm{s}^{-1}. \qquad (5.53)$$

Interpreted in accordance with the *i–s* nature of *C*, we obtain

$$C_i^{eq} = 2 \times 10^{25}\exp(-4.52eV/k_BT)\ \mathrm{cm}^{-3}, \qquad (5.54)$$

$$D_i = 4.4 \times 10^8\exp(-0.88eV/k_BT)\ \mathrm{cm}^2\,\mathrm{s}^{-1}. \qquad (5.55)$$

For outdiffusion of C_s preintroduced to high concentrations, however, the situation is very different. For cases when the C_s concentration significantly exceeded its solubility, as pointed out by Scholz et al. [88],

$$D_i C_i^{eq} > D_I C_I^{eq} \qquad (5.56)$$

may be satisfied, leading to a severe undersaturation of I in the high C_s concentration region which significantly retards the outdiffusion of C_s atoms from the region. Indeed, such phenomena have been observed by Rücker et al. [89] and Werner et al. [90]. These experiments were performed using *molecular beam epitaxy* (MBE) grown Si layers containing regions with C_s concentrations in the 10^{25} to 10^{26} m^{-3} range, and hence tremendously exceeded the C_s solubility of the experimental temperature. A similar retardation of the diffusion of other impurity species diffusing via primarily I, e.g., B, in the same region is also expected. This is indeed the case of the experimental results of Rücker et al. [90], see Fig. 5.23. In order to highly satisfactorily fit the C_s profile and all the B spike-region profiles, Scholz et al. [88] found that additionally the contribution of Si V must also be included. Vacancy contributes a component to C_s diffusion via the dissociative or FT mechanism and a component to B diffusion via the vacancy-pairing mechanism. The V contribution to C_s diffusion is important in regions outside the initial C_s high concentration region and to B diffusion in all regions.

Using similarly grown samples containing C_s and B spikes, ion implantation induced Si I was found to be substantially attenuated in the C_s spike regions so that the diffusion of B buried beneath the C_s spikes was severely retarded when compared to cases of having no C_s spikes [91]. The phenomenon was interpreted by the authors as due to the reaction $C_s + I \Leftrightarrow CI$ but with the so-formed CI complexes assumed to be *immobile*, which is contrary to the suggestion of Watkins and Brower [82]. The assumption that immobile CI complexes are responsible for the retarded boron diffusion is not needed in the analysis of Scholz et al. [88]. It is expected that ion implantation or oxidation induced Si I supersaturation will enhance the diffusion of C and B with C in concentrations to a moderate level, e.g., in the range of 10^{17} cm^{-3}.

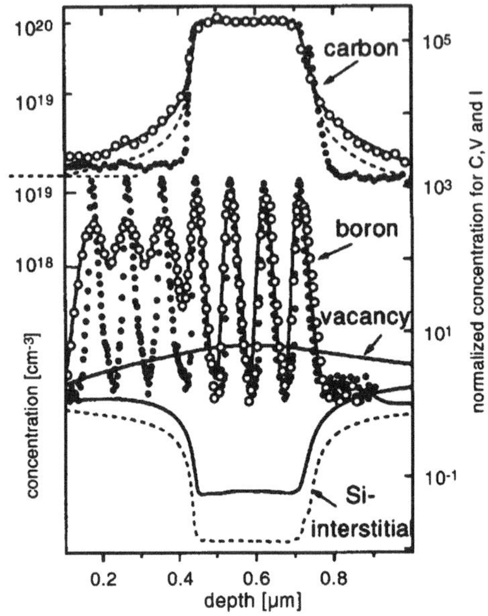

Figure 5.26 SIMS profiles of a 300 nm thick carbon layer with seven boron spikes [89]. Filled and open circles are respectively data of as-grown and annealed (900°C/45 min) cases. Dashed fitting lines are those with only the kick-out model, and solid lines are those with the dissociative mechanism also included. From Scholz et al. [88].

Another issue concerns the formation enthalpy of V in heavily C doped Si. A formation enthalpy of ~2.2 eV was extracted for V from the data of Scholz et al. [88] and other high concentration C diffusion data, in contrast to those from theoretical studies that are higher by 1 eV or more. While this discrepancy is attributed to influences such as pairing with O, etc., by Scholz et al., Pinacho et al. [92] suggested to consider C precipitation as an alternative responsible mechanism as merited by the C supersaturation and hence precipitation.

Regarding this issue, here we consider the contribution of an almost trivial and yet up to now ignored factor: the work term in the definition of enthalpy. Enthalpy is thermodynamically defined initially via studies of mechanical systems, e.g., an ideal gas system. The specific enthalpy expression reads $h = u + pv$, where u is the specific internal energy of the system constituent matter, p is pressure the system is subjected to, v is the specific volume of the system, and pv is the specific work associated with the flow of the constituent matter of the system. Applying the concept to point defects in a solid, e.g., enthalpy of formation of V, the pv term contribution is negligibly small. For instance, consider an otherwise perfect and pure Si crystal under the external pressure of 1 atmosphere and the point defect volume equals that of an atom, Ω, the pv work amounts to only 10^{-5} eV. With heavy doping by dopants or isoelectronic elements,

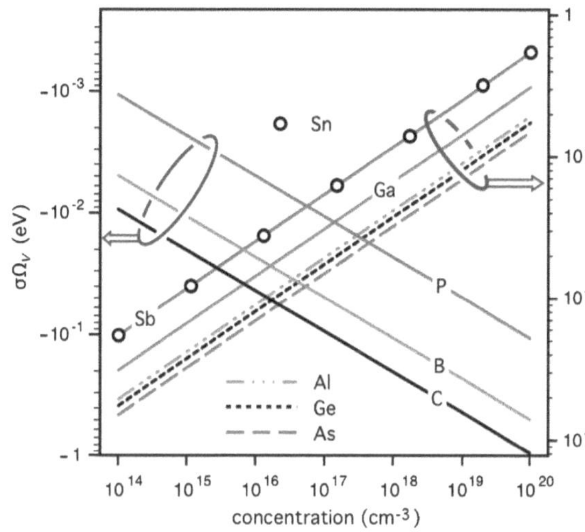

Figure 5.27 Calculated $\sigma\Omega_V$ work contribution to the enthalpy of formation of V for Si doped with p-type dopants B, Al, and Ga; p-type dopants P, As, and Sb; and isoelectronic elements C, Ge, and Sn. The positive and negative value cases correspond to compressive and tensile stress cases, respectively.

this term's contribution can no longer be ignored. Because the atom size difference between the impurity and Si, a substantial strain and a consequent internal stress σ that is also substantial renders the pv work term sizable. Atoms of the elements B and C are considerably smaller than that of the Si atom, which leads to a tensile stress, while Sb and Ga atoms are much larger than Si atom, which results in a compressive stress. For point defects, the term pv shall be written as $\sigma\Omega_x$, where σ is the stress, and Ω_x is the volume of the defect x, I or V. Keeping the thermodynamic definition of pv work, σ is regarded as positive for a compressive stress and negative for a tensile stress, with $\Omega_V > 0$ and $\Omega_I < 0$ holding. With the effect of $\sigma\Omega_x$ work, the point defect formation enthalpy becomes

$$h_x^f(C_y) = h_x^f(C_y = 0) + \sigma\Omega_x, \tag{5.57}$$

where $h_x^f(C_y)$ is the enthalpy of formation of the defect x doped by impurity species y to concentration C_y, h_x^f $(C_y = 0)$ is that without doping. In the simplest estimate, the term $\sigma\Omega_x$ is calculated using Hook's law expressed as

$$\sigma\Omega_x = Y\left(\frac{C_y}{C_o}\right)^{1/3}\frac{r_{Si} - r_y}{r_{Si}}\frac{1}{C_o}, \tag{5.58}$$

where Ω_x is taken to be $1/C_o$ with C_o being the density of Si, Y is Young's modulus of Si, $(C_x/C_o)^{1/3}$ is the linear density of the impurity, which, together with the atom radius difference ratio $(r_{Si} - r_y)/r_{Si}$, defines the linear strain in the Si crystal with the crystal assumed to be isotropic and homogeneous. Using Eq. (5.58), values of $\sigma\Omega_V$ for the p-type dopants B, Al, Ga, the n-type dopants P, As, Sb, and the isoelectronic elements C, Ge, and Sn are calculated as a function of the impurity concentration. Those obtained using Y = 190 GPa are shown in Fig. 5.27. The magnitude of the values will decrease by ~30% if Y = 130 GPa is used instead. It is seen from Fig. 5.27 for C doping to 10^{20} cm^{-3} $\sigma\Omega_V$ is about −1 eV, i.e., the V formation enthalpy in Si is reduced by $\sigma\Omega_V$ work by about 1 eV. While this result may appear to be coincidental, it nevertheless indicates that in heavily doped Si, for which the strain is large, $\sigma\Omega_x$ work contribution to the point defect formation enthalpy is no longer ignorable. The calculated values are also applicable to the I case but with the sign reversed, if it is assumed that the volumes of I and V are equal in magnitude. In general, impurities with covalent radii smaller than that of Si decrease and increase the formation enthalpy of V and I, respectively. The reverse holds for impurities with covalent radii larger than that of Si.

5.4.5 Diffusion of Si Self-Interstitials and Vacancies

For Si, although the product $D_I c_I^{eq}$ is known and estimates of $D_V c_V^{eq}$ are available, our knowledge of the individual factors, D_I, D_V, c_I^{eq} (or C_I^{eq}), and c_V^{eq} (or C_V^{eq}), is limited in spite of immense experimental efforts to determine these quantities.

The most direct way of measuring D_I is the injection of I (e.g., via surface oxidation) at one location of the Si crystal and the observation of its effect on dopant diffusion or on growth or shrinkage of stacking faults at another location as a function of time and of distance between the two locations. That is, the two locations may be the front- and the backside of an Si wafer. Extensive experiments on the spread of oxidation-induced I through wafers by Mizuo and Higuchi [93] have shown that supersaturation of I arrives at about the same time as corresponding undersaturation of V. Therefore, these kind of experiments at 1100°C just give information on an *effective* diffusivity of a perturbation in the I and V concentrations. This effective diffusivity may be expressed approximately by [40]

$$D_{(I,V)}^{eff} \approx \left(D_I c_I^{eq} + D_V c_V^{eq}\right)/\left(c_I^{eq} + c_V^{eq}\right) \tag{5.59}$$

and probably corresponds to the diffusivity values of about 3×10^{-13} m^2 s^{-1} in the experiments of Mizuo and Higuchi at 1100°C [93]. Many efforts had been expended on this approach but the results are inconsistent.

In most experiments aimed at determining D_I, it has not been taken into account that I may react with V according to the reaction $I + V \Leftrightarrow \phi$, which establishes local dynamical equilibrium condition given by Eq. (5.44). Based on experiments on oxidation-retarded diffusion of antimony [41,94], it has been estimated that an astonishingly long time, about one hour, is required to establish local dynamical equilibrium at

1100°C. This long recombination time indicates the presence of an energy or entropy barrier slowing down the recombination reaction. At lower temperatures much longer recombination times can be expected. These long recombination times hold for lightly doped material. There are indications that dopants or other foreign elements may act as *recombination centers* which can considerably speed up the recombination reaction, but no reliable data are available in this area. The D_I values so determined (and therefore indirectly also of c_I^{eq} via the known product $D_I c_I^{eq}$) were found to diverge over many orders of magnitude [42,95] and with I formation enthalpies from ~1 to 4 eV. This is clearly an unsatisfactory situation. The problem is further complicated by the observation that the measured effective diffusivity $D_{(I,V)}^{eff}$ depends on the type of Si material used. In the experiments of Fahey et al. [41] the transport of oxidation-induced Si I through epitaxially-grown Si layers was much faster than through equally thick layers of as grown float-zone (FZ) or Czochralski (CZ) Si. This difference has been attributed to the presence of vacancy-type agglomerates left from the crystal growth process, which might not be present in epitaxial Si layers. These vacancy agglomerates would have to be consumed by the injected I before further spread of interstitials can occur.

Nonetheless, considering development involving several different categories of studies, we can tentatively conclude that the migration enthalpies of vacancies and self-interstitials in silicon, h_V^m and h_I^m, respectively, are relatively small while their formation enthalpies, h_V^f and h_I^f, respectively, are large. This means that V and I are moving fairly fast while their thermal equilibrium concentrations are fairly small. The most probable value of h_V^m is ~0.5–1 eV while that of h_I^m is ~1 eV, and the corresponding most probable values of h_V^f is ~3–3.5 eV while that of h_I^f is ~4 eV. Sinno et al. [1] used values of 0.457 and 0.937 eV, respectively, for h_V^m and h_I^m to satisfactorily model the formation of swirl defects (interstitial-type dislocation loops and vacancy-type clusters) in FZ Si, including the defect location, density, size, and their dependence on the crystal growth rate and the thermal gradient. Plekhanov et al. [2] used an h_V^f value of ~3–3.4 eV to satisfactorily modeled the formation of voids in large diameter CZ Si. Moreover, in fitting the C and B diffusion results of Rücker et al. [89], as shown in Fig. 5.26, Scholz et al. [88] also needed to use h_I^m and h_V^m values smaller than 1 eV. This knowledge is consistent with quantum mechanical calculations which yielded fairly high h_I^f and h_V^f values and correspondingly low h_V^m and h_I^m values [3–7]. With the present estimates, it becomes also possible to connect in a reasonable and consistent way the fairly high diffusivities of native point defects found after low temperature electron irradiation [96] with the much lower apparent diffusivities which appear to be required to explain high temperature diffusion experiments.

5.4.6 Oxygen and Hydrogen Diffusion

Oxygen is the most important electrically inactive impurity element in Si. In CZ Si, O is incorporated from the quartz crucible and usually present in concentrations in the order

10^{18} cm^{-3}. An O atom in Si occupies the bond-centered interstitial position of two Si atoms and forms covalent bonds with the two Si atoms. Hence, its diffusion requires the breaking of bonds. The diffusivity of interstitial oxygen, O_i, has been measured between about 300°C and the melting point of Si and is fairly accurately described by

$$D_i = 0.13\exp(-2.53eV/k_BT)\ \text{cm}^2\,\text{s}^{-1}. \tag{5.60}$$

The solubility C_i^{eq} of interstitial O_i has been determined to be

$$C_i^{eq} = 9 \times 10^{22}\exp(-1.52eV/k_BT)\ \text{cm}^{-3}. \tag{5.61}$$

Since in most CZ Si crystals the grown-in O_i concentration exceeds C_i^{eq} at typical processing temperatures, O_i precipitation will occur in the interior but not the surface regions (because of O_i outdiffusion) of CZ Si. This leads to the important technological application *intrinsic gettering* [97] for improving the junction leakage and MOS capacitor charge holding time characteristics of integrated circuit devices fabricated using CZ Si, which is not available to FZ Si.

Around 450°C O_i forms electrically active agglomerates involving 4 O_i atoms, called *thermal donors* [98]. The formation kinetics of these agglomerates appears to require a fast diffusing species, for which both Si I [99] and molecular oxygen have been suggested [100]. Subsequent investigations favored an oxygen dimer coupled with I, or V, or an impurity atom [101–103].

Heavy doping in the range of 10^{18} to 10^{19} cm^{-3} influences O_i diffusion in Si. Heavy B and As doping reduces oxygen diffusivity at 800°C, while there is no influence at 1050°C [104]. For heavy Sb doping, no influence on O_i diffusion was found in the temperature range of 750–1150°C [104–106], but at still lower temperatures Sb (as well as B) strongly retards O_i diffusion [107]. Formation of immobile O complexes or clusters was suggested, which is possible when considering that 450°C is the thermal donor formation temperature and that above it until ~750°C is the SiO_2 precipitate nucleation temperature range. But a clear role played by doping is still not understood.

Hydrogen plays an important role in silicon device technology because of its capability to passivate electrically active defects. The passivation of dislocations and grain boundaries is especially important for inexpensive multicrystalline Si used for solar cells. Both acceptors and donors can be passivated by H, which is usually supplied to Si from a plasma. H in Si is assumed to diffuse as unbounded i atom in either a neutral or a positively charged form. The diffusivity of H in Si has been measured by Van Wieringen and Warmoltz [108] between 970 and 1200°C, see Fig. 5.16. Between room temperature and 600°C, H diffusivities are much lower than those extrapolated from the high temperature measured data. Pearton et al. [109] rationalized this observation by suggesting that atomic H may form interstitially dissolved, essentially immobile H_2 molecules. Apparently, these molecules can then form plate-like precipitates [110]. As in the case of oxygen, the existence of H molecules has not been proven experimentally.

5.5 DIFFUSION IN GERMANIUM

After its head start in 1947, Ge lost its leading role for solid state electronic device applications to Si in the early 1960s and became mainly used as a detector material. This is partly because of its lower bandgap and hence an inherently higher noise level in bipolar transistors, but more importantly because of its lack of a suitable oxide for MOSFET (metal–oxide–semiconductor field effect transistor) applications. The Ge oxide GeO_2 is a crystalline material prone to defects and hence to built-in charges in the GeO_2 bulk and at the GeO_2–Ge interface that is difficult to control. In contrast, thermally grown SiO_2 is an amorphous material that is highly stable and possesses few charges both in the bulk and at the SiO_2–Si interface after passivation by hydrogen. However, after nearly 50 years of continued miniaturization following Moore's scaling law, the Si MOSFET device has shrank to below 10 nm in its channel length and encountered a number of "*short channel*" problems. For example, carrier transport becomes limited by reduced mobility and source/drain parasitic resistance. Consequently, there is renewed interest in Ge because of its superior electron and hole low-field mobilities, its low-dopant activation temperatures and its small bandgap that could potentially reduce source/drain parasitic resistance. These factors, together with the development of the new high k (dielectric constant) gate insulator, would seem to favor an outright replacement of Si MOSFET by Ge FET with a high k gate insulator. Nonetheless, the current use of Ge is in Si/Ge superlattice or Si–Ge alloy layers or heterostructures, to modify the Si MOSFET channel region material by alloying with Ge or by the use of an alloy layer. Thus, from the early 1960s to 1990, publications of diffusion in Ge are sparse. That situation is improving.

Diffusion in Ge can be consistently explained in terms of V-related mechanisms and there is no need to invoke also I. Under intrinsic conditions at the diffusion temperature, profiles of all diffusants are erfc-shaped. Fig. 5.28 shows, in Ge, the diffusivities of group III and V dopants, some metals, and Ge as a function of inverse absolute temperature under intrinsic conditions. Under extrinsic conditions, doping dependence is observed [111–115]. Werner et al. [112] found that n-doping enhances self-diffusion while p-doping retards it; Südkamp et al. [113] found that p-doping by B retards Ge self-diffusion. Vainonen-Ahlgren et al. [114] and Bracht and Brotzmann [110] found that diffusivity of As in Ge depends on the electron concentration n quadratically. Naganawa et al. [115] found that Ge self-diffusion also depends on n quadratically. The observed doping dependence of dopant diffusion and self-diffusion can be explained by one kind of acceptor-type native point defect, doubly-negatively-charged, which turned out to be V^{2-}. The native point defects have been assumed to be V since the earliest studies of diffusion in Ge [37], but a convincing experimental proof was only given in 1985 by Stolwijk et al. based on the diffusion behavior of Cu in Ge [116].

Copper diffuses in Ge via an *i–s* mechanism [31]. In analogy to the case of Au and Pt in Si, its diffusion behavior may be used to check diffusion profiles for any indication of

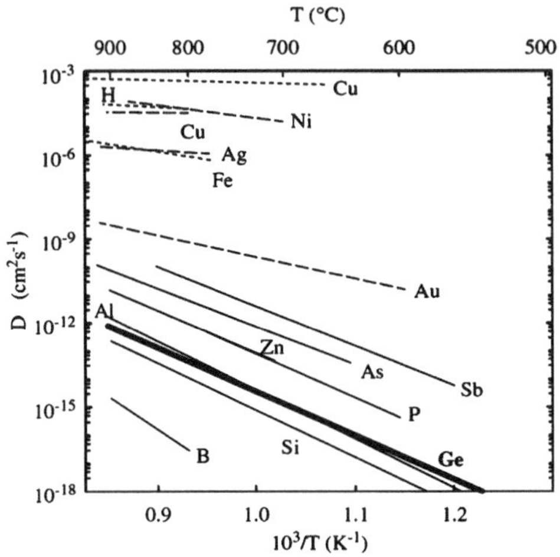

Figure 5.28 Diffusivities of various elements (including Ge) in Ge as a function of inverse absolute temperature (taken from [111]).

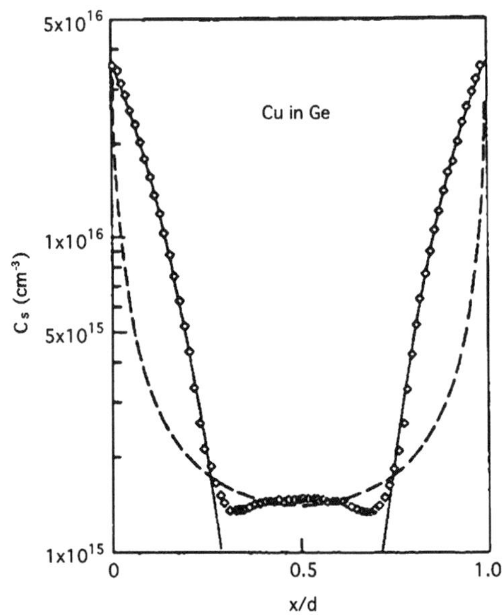

Figure 5.29 Concentration profiles of Cu into a dislocation-free Ge wafer after diffusion for 15 minutes at 878°C. The solid line is for the Frank–Turnbull and the dashed line for the kick-out mechanism. From Stolwijk et al. [116].

an *I* contribution via the KO mechanism. A concentration profile of Cu diffusion into a Ge wafer is shown in Fig. 5.29 [116]. The dashed U-shaped profile which is typical for the kick-out mechanism obviously does *not* fit the experimental data. In contrast, the experimental profiles may be well described by the constant diffusivity $D_{(I,V)}^{eff}$ given by Eq. (5.36). Values of the vacancy contribution to Ge self-diffusion,

$$D_V c_V^{eq} = 21.3 \exp(-3.11 eV/k_B T) \text{ cm}^2 \text{s}^{-1}, \tag{5.62}$$

as determined from Cu diffusion profiles, agree well with those measured from tracer self-diffusion in Ge [42,116] as well as the much later value of Hüger et al. [117] obtained using Ge isotope multilayer structures, $D_V c_V^{eq} = 25.4 \exp(-3.13 eV/k_B T)$ cm^2 s^{-1}. Excellent agreement shows that any Ge I contribution to the Ge self-diffusion process is negligible and hence Ge self-diffusion appears to be entirely governed by Vs. It is unclear why I play such an important role in diffusion processes in Si but induces no noticeable effect in Ge.

5.6 DIFFUSION IN GALLIUM ARSENIDE

Gallium arsenide is the most important base material used for optoelectronic applications with diffusion processes essential in fabricating the devices. Self-diffusion and diffusion of dopant and other important impurity species in GaAs are governed by native point defects, which is true also for other compound semiconductors. Compared to that in Si, diffusion in GaAs exhibits a much more prominent dependence on the Fermi-level effect, and it also shows a dependence on the pressure of an As vapor phase. Moreover, the number of participating point defect species is more than that in Si. Vacancy, interstitialcy, as well as *i–s* diffusion mechanisms are involved. See Section 5.2.2 and reviews given elsewhere [22,23,42,118].

Diffusions of many elements in GaAs have been investigated, with most of the studies focused on p-type dopants Zn and Be, on n-type dopants Si and Se, and on Cr, which is a deep acceptor used for producing semi-insulating GaAs. Since Zn, Be, Cr and a number of other elements diffuse via an *i–s* mechanism, this type of diffusion mechanism has historically received much more attention in GaAs than in Si and Ge. For each of these impurities, the diffusion results are laden with anomalies. That is, other than those obtained under isoconcentration diffusion conditions, the profiles strongly deviated away from that of a standard solution of the Fick's second law expression, Eq. (5.16), that are invariably of the error function family. Irrespective of this highly unusual circumstance, it had still been assumed for a long time that only vacancies need to be taken into account to understand diffusion processes in GaAs, e.g., see the review book of Tuck [119]. Considering that in Si the question whether V or I dominates diffusion processes was regarded as a controversy for more than 20 years, then one can conclude that for GaAs the situation can only be described as chaotic, which persisted until the late 1980s to early 1990s.

If we think that for Si the framework of a global model now exists for the diffusion of those atom species that utilize point defects, then it must be concluded that the same status is far from being reached for GaAs. It can at best be said that such a model is still in its prototyping stage. On the Ga sublattice side, though it is generally accepted that the roles of both I and V must be taken into account, and that the Fermi level effect dominates, the exact charge state of the responsible point defect is still debated. Even much less is known on the As sublattice side.

5.6.1 Fermi Level Effect and As_4 Pressure Effect

A compilation of earlier diffusion data in GaAs may be found elsewhere [119]. Only a few studies of self-diffusion in GaAs are available, but with advances in growing GaAs/AlAs-type superlattices using molecular beam epitaxy (MBE) or metalorganic chemical vapor deposition (MOCVD) methods, Al has served as an important foreign tracer element for elucidating Ga self-diffusion mechanisms. The observation that high concentration Zn diffusion into a $GaAs/Al_xGa_{1-x}As$ superlattice leads to a dramatic increase in the Al–Ga interdiffusion coefficient [120] opened up the possibility to fabricate laterally structured optoelectronic devices by locally disordering the superlattices. The dopant-enhanced superlattice disordering is a general phenomenon occurring for other p-type dopants, e.g., Mg, and for n-type dopants, e.g., Si, Se, and Te [121–127].

The dopant-enhanced superlattice disordering was the key to unravel the contribution of the Fermi level effect to self- and dopant diffusion processes in GaAs. In 1987 Mei et al. reported a set of quantitative data obtained using Si doped $GaAs/Al_xGa_{1-x}As$ superlattice structures [128]. Their data allowed a quantitative determination that the enhanced diffusivity of the Al–Ga interdiffusion is proportional to the cube of the electron concentration, n^3 [21]. This means that doping-enhanced diffusion is due to the Fermi level effect with the responsible point defect species triply-negatively charged. Subsequently it became clear that all substitutional species in GaAs, including self-atoms and dopants, diffuse via charged point defect species [22,23,42,118]. In GaAs, Fermi level effect plays a dominant role, which is different from the situation in Si for which neutral and charged point defects all play a role, but usually with the neutral ones dominating.

The dependence of diffusion processes on the As vapor pressure allowed us to distinguish the role of self-interstitials versus vacancies. The diffusivity of a substitutional species in GaAs generally shows a dependence on P_{As_4} because the concentration of the responsible point defect species is dependent on P_{As_4}, Eq. (5.11). Furthermore, non-equilibrium concentrations of native point defects may be induced by the indiffusion of dopants such as Zn starting from a high surface concentration.

Much less is known on the diffusion processes of atoms dissolved on the As sublattice, but some experiments indicate the dominance of As I on the diffusion of the isoelectronic group V element N [129], P and Sb [130–132], and the group VI n-type dopant S [132]. These results imply also the dominance of As Is on As self-diffusion, which is in contrast to the earlier interpretation of As self-diffusion results of Palfrey et al. [133] favoring the dominance of As Vs.

5.6.2 Gallium Self-Diffusion and Superlattice Disordering

5.6.2.1 Intrinsic and n-Type GaAs

The self-diffusion coefficient $D_{Ga}(n_i)$ of Ga in intrinsic GaAs has been measured by Goldstein [134] and Palfrey et al. [135] using radioactive Ga tracer atoms, and by Tan et al. [136] and Wang et al. [137] using ^{69}GaAs/^{71}GaAs superlattice structures. The radioactive tracer method allows measuring $D_{Ga}(n_i)$ down to about 10^{-15} cm^2/s. Measurements of mixing of the stable isotope atoms ^{69}Ga and ^{71}Ga extended the range to much lower values. The use of superlattices started in measurements of the interdiffusion of Ga and Al in $GaAs/Al_xGa_{1-x}As$ from superlattice structures [121,138–145]. It is noted that for all these results precautions were made so that they were obtained under nominally intrinsic and As-rich conditions. Lee et al. [139] noticed that in their work the results have a bimodal distribution, one characterized by a 6 eV activation enthalpy and the other by 4 eV. All 13 sets of data mentioned above [121,134–145] are plotted in Fig. 5.30 wherein it is clearly seen that the distribution is indeed bimodal. In Fig. 5.30 the fitting lines are

$$D^V_{Ga}(n_i, 1\text{-}atm, 6eV) \approx 2.9 \times 10^8 \exp(-6eV/k_BT)\ \text{cm}^2\,\text{s}^{-1} \tag{5.63}$$

for the 6 eV data, and

$$D^V_{Ga}(n_i, 1\text{-}atm, 4eV) \approx 7 \exp(-4eV/k_BT)\ \text{cm}^2\,\text{s}^{-1} \tag{5.64}$$

for the 4 eV data. Note that these expressions are also applicable to Al–Ga interdiffusions. A reasonable explanation of the data bimodal distribution is still lacking, let alone an acceptable one. The data bimodal distribution is apparently independent of the Fermi level effect because all results were obtained under nominally intrinsic conditions, and because the data cross-over each other at a temperature of ~1030°C instead of close to 1238°C (the GaAs melting point) and hence the argument that the 6 eV data may just happened to be due to the Ga-rich ones is not valid. It is apparently independent of the P_{As_4} effect, for in all cases some measure had been taken to ensure an As-rich ambient condition during annealing. It is apparently independent of whether it is Ga self-diffusion or Al–Ga interdiffusion, because both groups of data include those from Ga self-diffusion and Al–Ga interdiffusion. It is also apparently independent of the form of the starting materials, since bulk GaAs [134,135], superlattices of Ga isotopes [136,137] and superlattice structures of GaAs/AlAs are all involved. Could it be that the results are influenced by an inadvertent experimental condition, e.g., the introduction of hydrogen into the samples, during sample preparation and/or diffusion anneal? This question cannot be answered because of the lack of reported details. Corresponding to Eqs. (5.63) and (5.64), the D^V_{III} values for GaAs crystals at the Ga-rich boundary are respectively given by

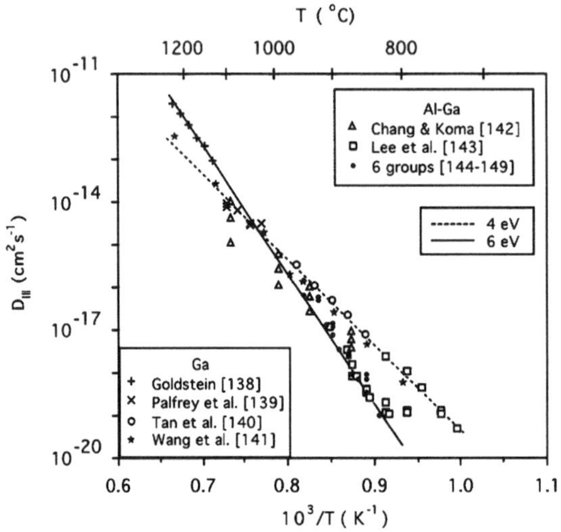

Figure 5.30 Plot of available data on Ga self-diffusion in GaAs and data on Ga/Al interdiffusion in GaAs/AlGaAs superlattices under intrinsic conditions. Fitting lines are given by Eqs. (5.65) and (5.66). Notice the clear data bimodal distribution.

$$D^V_{Ga}(n_i,\ Ga\text{-}rich,\ 6eV) \approx 3.93 \times 10^{12} \exp(-7.34eV/k_BT)\ \mathrm{cm^2\,s^{-1}}, \tag{5.65}$$

$$D^V_{Ga}(n_i,\ Ga\text{-}rich,\ 4eV) \approx 9.49 \times 10^{4} \exp(-5.34eV/k_BT)\ \mathrm{cm^2\,s^{-1}}. \tag{5.66}$$

The diffusivity D^V_{Ga} is due to the Ga sublattice V contribution, a conclusion arrived at because around 1 atm the disordering rate of the GaAs/$Al_xGa_{1-x}As$ superlattices increases as the ambient As_4 pressure is increased [146,147]. For Eqs. (5.63) and (5.64), it turned out that the responsible vacancy species is the triply-negatively-charged Ga vacancies, V^{3-}_{Ga}, discussed in the next paragraph.

In 1987 Mei et al. [128] published a set of Al–Ga interdiffusion data under Si doping conditions. Their data is associated with a 4 eV activation enthalpy, see Fig. 5.31. This set of data has been satisfactorily fitted by Tan and Gösele [21] to

$$\frac{D_{Al\text{-}Ga}(n)}{D^V_{Ga}(n_i,\ 1\text{-}atm,\ 6eV)} = \left(\frac{n}{n_i}\right)^3, \tag{5.67}$$

where the used $D^V_{Ga}(n_i,\ 1\text{-}atm,\ 6eV)$ value is that given by Eq. (5.63), i.e., the 6 eV value, see Fig. 5.32. Fig. 5.32(A) shows the Mei et al. data fitted according to Eq. (5.67), while Fig. 5.32(B) shows the extracted $D^V_{III}(n_i,\ 1\text{-}atm,\ 6eV)$ value data with Eq. (5.67) assumed to be valid. A power law fitting is not obtained using Eqs. (5.64)–(5.66), even accepting

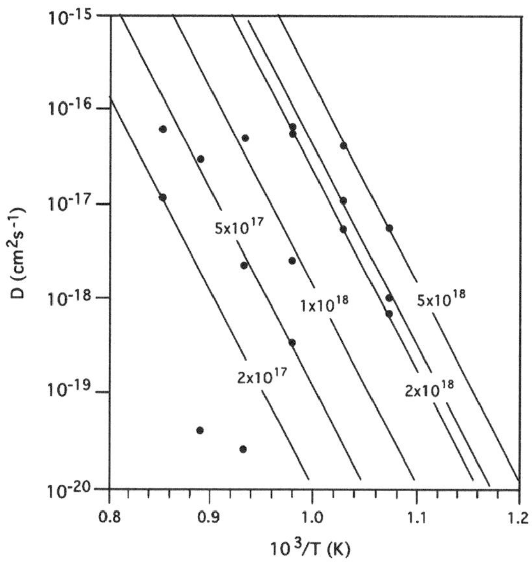

Figure 5.31 The Al–Ga interdiffusion data of Mei et al. [128] fitted to 4 eV activation enthalpy lines. Dopant (Si) concentrations are as indicated.

poor fitting. Eq. (5.67) shows that the responsible point defect species is the triply-negatively-charged Ga sublattice vacancy V_{Ga}^{3-}. This means that, under intrinsic and n-doping conditions, Ga self-diffusion and Al–Ga interdiffusion is governed by V_{Ga}^{3-}. This in turn means the dominance of the Fermi-level effect. It is seen from Fig. 5.32 that the diffusivity enhancement spans 7 orders of magnitude.

The fact that V_{Ga}^{3-} dominates Ga diffusion in GaAs under intrinsic and n-doping conditions is corroborated by almost all theoretically calculated results that showed the dominant presence of V_{Ga}^{3-} under intrinsic and n-doping conditions, see Section 5.2.3, in particular Table 5.3 and related references [13,14,16–18,24–27]. These results showed that the Ga vacancy transition level positions are in the normal packing order, i.e., the energies of the levels 0/−, −/2− and 2−/3− are successively higher but with the highest, that of 2−/3−, still below the Fermi level of intrinsic GaAs and hence the concentration of V_{Ga}^{3-} is the highest in intrinsic and in n-type GaAs. Nonetheless, by far the strongest support for the dominance of V_{Ga}^{3-} came from the positron measurement experiment of Gebauer et al. [11]. Using positron annihilation lifetime spectrometry, Gebauer et al. determined the charge state of Ga vacancies in n-type GaAs by directly probing into the bulk of GaAs to measure the vacancy concentration as a function of annealing temperature, arsenic vapor pressure, and doping concentration. They found that the detected vacancy concentration increases with arsenic vapor pressure to the 1/4 power and hence the detected vacancies must be V_{Ga} species. The concentration of this V_{Ga} species increases with an increase in dopant concentration, and, using equilibrium

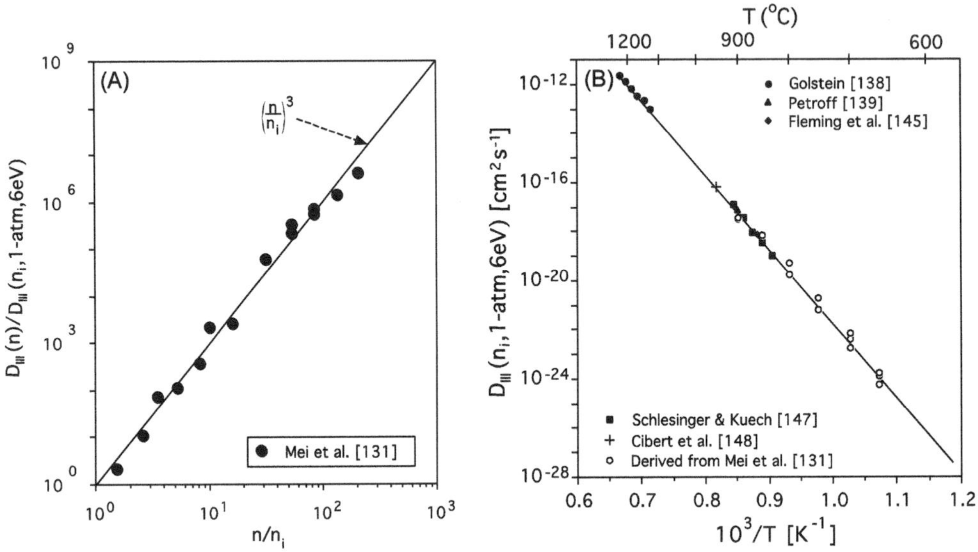

Figure 5.32 The data of Mei et al. [128] fitted to the power law $(n/n_i)^3$ according to Eq. (5.69): (A) the actual electron concentration n is computed from the dopant concentration subjected to the local charge neutrality condition and the intrinsic carrier concentration n_i is taken to be that of GaAs; (B) extracted $D^V_{Ga}(n_i, 1\text{-}atm, 6eV)$ data with the validity of Eq. (5.69) assumed.

thermodynamics, they obtained a 3e charge state of the Ga vacancy in n-doped GaAs, i.e., V_{Ga}^{3-}. Moreover, they also found that V_{Ga}^{3-} concentration decreases as the annealing temperature is increased, bearing out directly the prediction made in association with Fig. 5.9 for the concentration of V_{Ga}^{3-}, which is called the negative temperature dependence [28].

The data of Mei et al. [128] and the fitting of their data [21], Figs. 5.31, 5.32, together with the related expressions, Eqs. (5.63) and (5.69), formed the stepping stone for a systematic examination of diffusion in GaAs that has led to a consistent understanding of the seemingly chaotic situation on a qualitative to semi-quantitative basis. A fully satisfactory model is, however, still somewhat further away. As an example, just for materials discussed thus far, one must address the question of the bimodal data distribution in Fig. 5.30. As another example, one must ask the question why should Eq. (5.69) fit the Mei et al. data at all, let alone so nicely. In that fitting, the used n_i values are those of GaAs and thus ignore the influence of AlAs, which is a different material with a different bandgap and different n_i values. To make matters even more complicated, one notices that during disordering of the AlAs/GaAs superlattice structures Ga and Al concentrations are continuously changing with time.

The dominant role of V_{Ga}^{3-} has been challenged by Bracht et al. [10] who favor a V_{Ga}^{2-} dominance alternative. These authors performed a Ga self-diffusion experiment by disordering $^{71}GaAs/^{nat}GaAs$ isotope superlattice structures. Their result is described by

$$D_{Ga}(n_i, 1\text{-}atm) = 0.64 \exp(-3.71 eV/k_B T) \text{ cm}^2\,\text{s}^{-1}. \quad (5.68)$$

The values of Eq. (5.68) overlaps substantially with that given by Eq. (5.64), i.e., their results belong to the 4 eV group of results discussed in association with Eq. (5.64). They obtained better fits to their data with V_{Ga}^{2-} using Eq. (5.12). This is at odds with the theoretical predictions of the V_{Ga}^{3-} dominance [13,14,16–18,24–27], and, more importantly, with the measured V_{Ga}^{3-} dominance in n-type bulk GaAs by Gebauer et al. [11]. In an effort to resolve the dilemma between V_{Ga}^{3-} and V_{Ga}^{2-}, El-Mellouhi and Mousseau [148] developed a model predicting that V_{Ga}^{3-} dominates at lower temperatures while V_{Ga}^{2-} dominates at higher temperatures, with the two contributing equally at ~840°C. This crossover temperature appears to be too low as the positron measured dominance by V_{Ga}^{3-} included samples annealed up to 1100°C [11]. Nonetheless, since the experimental temperature of Bracht et al. is from 736°C upward, then, according to El-Mellouhi and Mousseau [148], some of their data would have contributions from both V_{Ga}^{3-} and V_{Ga}^{2-}. One indication of this possibility is the fit to the data of Mei et al. [128] by these authors: they found that $D_{Al-Ga}(n)/D_{Ga}(n_i) = 3.4(n/n_i)^{2.3}$. The n to the 2.3 power dependence can be interpreted as due to the combined contributions from V_{Ga}^{3-} and V_{Ga}^{2-}, and for some of the data points one can find that over 60% is due to V_{Ga}^{3-}. Notice that the work of Bracht et al. [10] is not a challenge to the validity of the Fermi level effect, but a search of refinement and accuracy within the framework of the Fermi level effect approach. A challenge to the validity of Fermi level effect per se would be on an entirely different level of significance.

5.6.2.2 Intrinsic and p-Type GaAs

No studies of Ga self-diffusion in p-doped bulk GaAs have been reported, but a wealth of data on Al–Ga interdiffusion in both n-type and p-type doped $GaAs/Al_xGa_{1-x}As$ superlattices is available. Moreover, diffusivity values can always be extracted from modeling dopant diffusion processes. Such extracted diffusivities are necessarily model-dependent and shall hence be cautiously regarded only as possible values.

The interdiffusion experiments were triggered by the observation of Zn indiffusion enhanced superlattice disordering due to Laidig et al. [120]. A number of disordering mechanisms were proposed [120,149–151] for a particular dopant at earlier times, but none is general enough to account for the occurrence of enhanced Al–Ga interdiffusion for also other dopants. The observed dopant enhanced interdiffusion appears to be due to two main effects [21,152]: (i) The thermal equilibrium concentration of appropriately charged point defects is enhanced by doping, i.e., the Fermi-level effect. In the case of the n-type dopant Si, mainly the *presence* of the dopant is of importance, but not its movement. Compensation doping, e.g., with Si and Be, should not lead to enhanced Al–Ga interdiffusion, in agreement with experimental results [126,153]; and (ii) For a dopant with high diffusivity and solubility, *non-equilibrium native point defects* are generated.

Depending on whether supersaturation or undersaturation of point defects develops, the enhanced disordering rate due to the Fermi level effect may be further increased or decreased. Irrespective of the starting material composition, such non-equilibrium native point defects drive the dopant diffused crystal region composition first toward an appropriate allowed GaAs crystal composition limits shown in Fig. 5.3(A). When the super- or undersaturation of point defects becomes so large that the crystal local region exceeds the allowed composition limit, extended defects form to bring the composition of the region back to that composition limit. Afterwards, this permits the diffusion processes to be described by an equilibrium point defect process appropriate for the crystal local region that is at an appropriate allowed composition limit. The crystal is in a non-equilibrium state because of the spatially changing composition. The diffusion of high concentration Zn, Be and Cr into GaAs [154–156] and their effects on $GaAs/Al_xGa_{1-x}As$ superlattices as analyzed by Tan et al. [157] appear to be such cases. Tan et al. [157] combined the Ga diffusivity used by Yu et al. to model indiffusion of Zn [154] and Cr [155] and the Al–Ga interdiffusion results of Hsieh et al. [145] to obtain

$$D^I_{Ga}(n_i,\ Ga\text{-}rich) \approx 4.46 \times 10^{-8} \exp\left(-\frac{3.37eV}{k_B T}\right)\ \mathrm{cm^2\,s^{-1}}, \tag{5.69}$$

see Fig. 5.33. The data of Hsieh et al. were obtained under intrinsic and Ga-rich ambient conditions while those of Zn and Cr were extracted values from the fittings [154, 155]. Both substitutional Zn and Cr atoms occupy Ga sites, and hence the diffused GaAs crystals should be the Ga-rich ones. The diffusivity given by Eq. (5.69) is that due to the contribution of Ga interstitial species under intrinsic and Ga-rich conditions, which turned out to be the doubly-positively-charged Ga self-interstitial, I^{2+}_{Ga}. With Eq. (5.69), the corresponding I^{2+}_{Ga} contribution to Ga self-diffusion under As-rich conditions is

$$D^I_{Ga}(n_i,\ 1\text{-}atm) \approx 6.05 \times 10^{-4} \exp\left(-\frac{4.71eV}{k_B T}\right)\ \mathrm{cm^2\,s^{-1}}. \tag{5.70}$$

It has also been noticed that the Al–Ga interdiffusion coefficient increases for *very low* arsenic vapor pressures [146,147], indicating that D_{Ga} is governed by Ga *Is* for sufficiently low As vapor pressures [22]. This Ga I is I^{2+}_{Ga} because it is the responsible point defect species used to fit the indiffusion profiles of Zn, Be, and Cr by Yu et al. [154,155], and because, as shown in Fig. 5.34, some of the available Al–Ga interdiffusion data in *p-type* GaAs based superlattices [126,159–161] were approximately fitted by Tan and Gösele [158] according to

$$D_{Al\text{-}Ga}(p,\ Ga\text{-}rich) = D^I_{Ga}(n_i,\ Ga\text{-}rich)(p/n_i)^2, \tag{5.71}$$

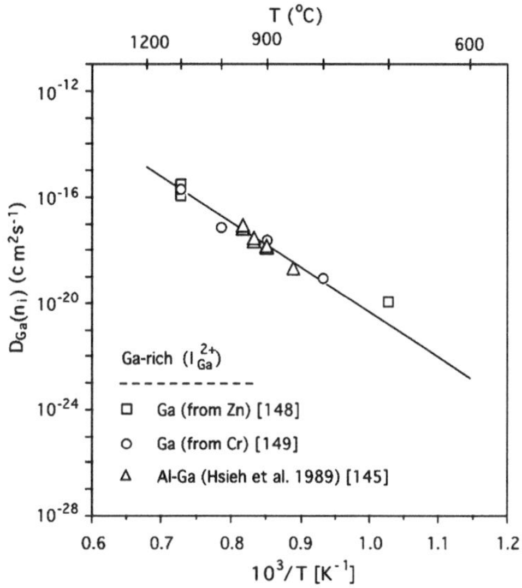

Figure 5.33 The I_{Ga}^{2+} contribution to Ga self-diffusion and to Al–Ga interdiffusion under intrinsic and Ga-rich conditions.

where $D_{Ga}^{I}(n_i, \textit{Ga-rich})$ is given by Eq. (5.69). In Eq. (5.71) the p^2 dependence of $D_{Al\text{-}Ga}(p, \textit{Ga-rich})$ shows that the I_{Ga} is doubly-positively-charged. The data used in Fig. 5.34 [126,159–161] are those under the dopant *indiffusion* conditions. Some data are obtained under the dopant *outdiffusion* conditions involving grown-in dopants without an outside dopant source. Under outdiffusion conditions, the dopant diffusivity values are too small to be reliably measured. The fit shown in Fig. 5.34 appears to be satisfactory but not as good as that for the Si-doping case (Fig. 5.32). Even *if* the fit were *perfect*, the essential native point defect equilibrium situation implied by Fig. 5.34 is only an *apparent* phenomenon, for it applies only to the p-dopant diffusion front region while the whole crystal is having a spatially varying composition. This point is most obvious in the data of Lee and Laidig [159] obtained in experiments using a high As_4 vapor pressure ambient.

The grossly different results for in- and outdiffusion conditions is due to non-equilibrium concentrations of native point defects induced by high-concentration diffusion of Zn or Be. Both Zn and Be diffuse via an *i–s* mechanism as will be discussed in more detail in the subsequent section. Historically, most authors [119] considered that diffusion of p-type dopants is governed by the Longini mechanism [32] involving Ga vacancies

$$A_i^+ + V_{Ga}^o \Leftrightarrow A_s^- + 2h, \tag{5.72}$$

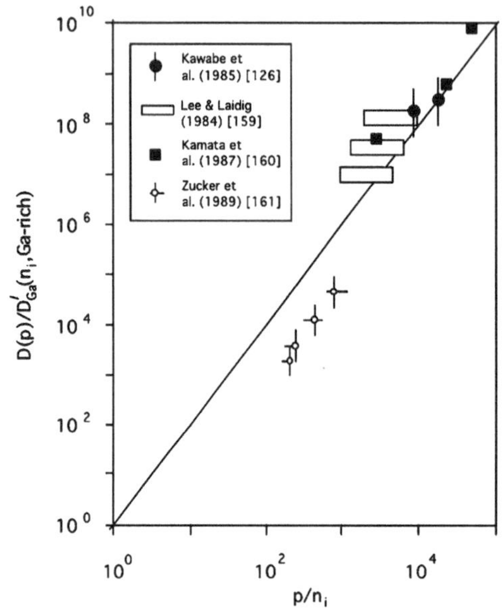

Figure 5.34 Fits of some available p-dopant enhanced Al–Ga interdiffusion data, with $D^I_{Ga}(\eta_i, Ga\text{-}rich)$ given by Eq. (5.71). The data exhibit an approximate quadratic dependence on p/n_i, indicating that the dominant native point defect is I^{2+}_{Ga}. From Tan and Gösele [158].

where h is a hole and the interstitial species of the dopant is assumed to be positively charged, A_i^+. Longini mechanism is the same as the FT mechanism, except that it deals with charged species. The superlattice disordering results indicate that for these dopants, the KO mechanism [33] involving Ga self-interstitials,

$$A_i^+ \Leftrightarrow A_s^- + I_{Ga}^{2+}, \tag{5.73}$$

is operating instead. This also indicates that Ga self-diffusion is governed by Ga *I*s under p-doping conditions. Within the framework of the kick-out mechanism, the dopant indiffusion generates a *supersaturation* of I_{Ga} with a corresponding *increase* of dopant diffusion and the Ga self-diffusion component involving Ga *I*s. Because of the I_{Ga} supersaturation, the dopant diffused region tends toward the Ga-rich composition. In the case of Zn indiffusion to very high concentrations, it will be discussed in detail that the I_{Ga} supersaturation is so large that in a small fraction of the diffusion time extended defect forms [156,162], resulting in that the Zn diffused region composition is at the thermodynamically allowed Ga-rich composition limit, and is associated with the appropriate thermal equilibrium point defect concentrations. This is the reason for the satisfactory fit shown in Fig. 5.34.

In the case of grown-in dopants without an outside source the KO mechanism involves the consumption of I_{Ga} which leads to an I_{Ga} *undersaturation* with a corresponding *decrease* in dopant diffusion [163–166] and the Ga self-diffusion component involving I_{Ga}. The results of the superlattice disordering experiments are consistent with the expectations based on the KO mechanism. In contrast, Longini mechanism predicts undersaturation of Vs for indiffusion conditions and supersaturation for outdiffusion conditions with a corresponding decrease and increase of a V dominated Ga self-diffusion component, respectively. Since the predictions based on Longini mechanism are just opposite to the observed superlattice disordering results, it can be concluded that: (i) Zn diffusion occurs via the KO mechanism, and (ii) Ga self-diffusion in p-type GaAs is governed by I_{Ga}.

In contrast to the group II acceptors Zn and Be, the group IV acceptor carbon (C) occupying the As sublattice sites diffuses slowly and should thus allow the native point defects to be maintained at their thermal equilibrium values. The effect of C on the disordering of $GaAs/Al_xGa_{1-x}As$ superlattices [167] is described well by

$$D^{eq}_{Al-Ga} = \alpha D^{I}_{Ga}(n_i)(p/n_i)^2, \tag{5.74}$$

where $D^{I}_{Ga}(n_i)$ is given by Eqs. (5.69) and (5.70), respectively, for data obtained under Ga-rich and As-rich ambient conditions, and α is a numerical factor of ~1 for the As-rich ambient results while it is ~0.01 for the Ga-rich ambient results. The presence of the factor α is attributed to a fairly thick intrinsic GaAs layer used to cap the C doped superlattice layers. This capping layer only allowed an inflow of I^{2+}_{Ga} from the wafer surface to reach the C doped region in accordance with $p = n_i$. For the As-rich ambient cases, the needed I^{2+}_{Ga} concentration in the C doped region is much smaller than that of the Ga-rich case.

The pressure dependence of disordering of p-doped superlattices confirms the predominance of Ga Is in Ga self-diffusion [118]. The magnitude of the enhancement effect, its restriction to the dopant-diffused region, and the implantation results of Zucker et al. [161] indicate that a Fermi level effect has to be considered *in addition* to non-equilibrium point defects.

Combining the results for the p-type and the n-type dopant induced disordering including an I supersaturation s_I defined as $s_I = C_I(n,p))/C^{eq}_I(n,p)$, and an analogous V supersaturation s_V defined as $s_V = C_V(n,p))/C^{eq}_V(n,p)$, where (n,p) indicates doping conditions, we may express the Ga self-diffusion coefficient approximately as

$$D_{Ga}\big((n,p), P_{As_4}\big) = D^{I}_{Ga}(n_i, 1\text{-}atm)(p/n_i)^2 P^{-1/4}_{As_4} s_I + D^{V}_{Ga}(n_i, 1\text{-}atm)(n/n_i)^3 P^{1/4}_{As_4} s_V. \tag{5.75}$$

In Eq. (5.75) the quantities $D^{V}_{Ga}(n_i, 1\text{-}atm)$ and $D^{I}_{Ga}(n_i, 1\text{-}atm)$ are given respectively by Eqs. (5.63) and (5.70). In writing down Eq. (5.75), the As-rich GaAs, designated by $P_{As_4} = 1$ atm, is chosen as the reference material state, and with GaAs crystals of all other compositions represented by an appropriate P_{As_4} value. Eq. (5.75) describes all presently known *essential effects* on GaAs/Al_xGa_{1-x}As superlattice disordering. In the case of non-equilibrium Ga Vs injected by an Si/As cap [168], $s_V > 0$ holds. In the case of ion-implantation, both $s_I > 0$ and $s_V > 0$ may hold and both quantities will be time dependent. In the case of diffusion-induced non-equilibrium point defects the presence of dislocations will allow local equilibrium between intrinsic point defects to establish in the two sublattices. In this way large supersaturation of I_{Ga} in the Ga sublattice may lead to undersaturation of I_{As} or supersaturation of V_{As} in the As sublattice.

5.6.3 Arsenic Self-Diffusion and Superlattice Disordering

Since there is only one stable As isotope, ^{75}As, As self-diffusion in GaAs cannot be studied using stable As isotopes. In intrinsic GaAs, however, three arsenic self-diffusion studies have been conducted using radioactive tracers [129,133,169]. In the experiment of Palfrey et al. [133], the As_4 pressure dependence of As self-diffusion indicated that V_{As} may be the responsible native point defect species. This is, however, in qualitative contradiction to the conclusion reached recently from a large number of studies involving As atoms and other group V and VI elements that the responsible native point defect species should be I_{As}. The latter studies include: (i) As–Sb and As–P interdiffusion in intrinsic GaAs/$GaSb_xAs_{1-x}$ and GaAs/GaP_xAs_{1-x} type superlattices for which x is small so as to avoid a large lattice mismatch [130,161,169]; (ii) P and Sb indiffusion into GaAs under appropriate P and As pressures so as to avoid extended defect formation which leads to complications [130,161,169]; (iii) an extensive analysis of the S indiffusion data in GaAs [132]; (iv) outdiffusion of N from GaAs [129]. A plot of the relevant data are shown in Fig. 5.35, from which the lower limit of the As self-diffusion coefficient, assigned to be due to the As self-interstitial contribution, is determined to be

$$D^I_{As}(n_i, 1\text{-}atm) \approx 6 \times 10^{-2} \exp(-4.8eV/k_BT)\ \mathrm{m^2\,s^{-1}}. \tag{5.76}$$

For P–As and Sb–As interdiffusion as well as P and Sb indiffusion cases [130,161, 169], the profiles are error function shaped. With P and Sb assumed to be *i–s* elements, such diffusion profiles are described by an effective diffusivity of the type

$$D_s^{eff} = D_i C_i^{eq} / C_s^{eq} \tag{5.77}$$

under native point defect equilibrium conditions, which is satisfied by either the KO mechanism involving I_{As} or Longini (or FT) mechanism involving V_{As}. The conclusion that I_{As} is the responsible species is reached for this group of experiments because the diffusion rate increases upon increasing the ambient As vapor pressure. I_{As} should be the responsible species in the N outdiffusion experiments [129] because the N profile is typical of that due to the KO reaction (5.73) but for the As sublattice under the

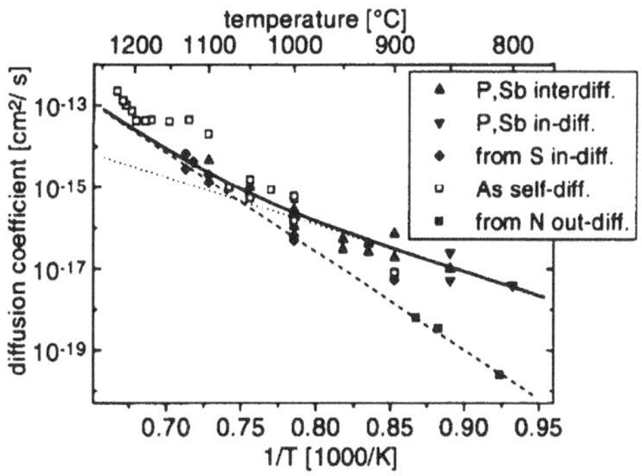

Figure 5.35 Data on As self-diffusion coefficient, obtained using radioactive As tracers (open squares), the group V elements N, P, and Sb and the group VI donor S (filled symbols). The dashed fitting line is given by Eq. (5.76), and the solid line is a better overall fitting. From Scholz et al. [130].

condition of I_{As} undersaturation, which is qualitatively different from those obtainable from the dissociative reaction (5.72). I_{As} should also be the responsible species in the S indiffusion experiments because the S profile [132] is typical of that due to the KO mechanism reaction $S_i^+ \leftrightarrow S_s^+ + I^o$ under the condition of I_{As} supersaturation, which is also qualitatively different from those obtained from any possible reactions of the FT or Longini mechanisms. It is seen from Fig. 5.35 that the available As self-diffusion data lie close to those deduced from the P, Sb, N, and S studies, and it may thus be inferred that As self-diffusion has a component contributed by I_{As}.

There is yet no doping dependence studies using the isoelectronic group V elements N, P, and Sb, and hence the charge nature of the involved I_{As} has not yet been so determined. However, S is a group VI donor occupying the As sublattice sites. In analyzing S indiffusion [132], it was necessary to assume that neutral I_{As} species were involved, which is therefore a most likely species responsible for As self-diffusion.

There is also a study on the disordering of GaAs/Al_xGa_{1-x}As superlattices by the group IV acceptor species C [167] which occupy the As sublattice sites. While there is no information obtained from this study on As self-diffusivity, satisfactory descriptions of the C diffusion profiles themselves were obtained also with the use of a kick-out reaction involving neutral As self-interstitials, discussed later. This lends further support to the interpretation that neutral As self-interstitials are responsible for As self-diffusion.

5.6.4 Impurity Diffusion in Gallium Arsenide

5.6.4.1 Silicon

For GaAs the main n-type dopant is Si. It is an amphoteric dopant mainly dissolved on the Ga sublattice but showing a high degree of self-compensation at high concentrations due to an increased solubility on the As sublattice.

The Si diffusivity shows a strong dependence on its own concentration, which had been modeled by several mechanisms [170–172]. Apparently, Si diffusion is dominated by negatively charged V_{Ga} and its apparent concentration dependence is actually a Fermi level effect. Results on Si diffusion into n-type (Sn-doped) GaAs confirm the Fermi level effect [172] and contradict the other models, e.g., the Si_{Ga}–Si_{As} pair-diffusion model of Greiner–Gibbons [171]. In the Fermi-level effect model, Yu et al. [172] used mainly V_{Ga}^{3-} to fit the Si indiffusion profiles, and Chen et al. [173] found that *only* V_{Ga}^{3-} is needed to fit the profiles obtained from heterostructures. This is consistent with the fact that V_{Ga}^{3-} dominates GaAs/AlAs superlattice disordering under n-doping conditions. In the work of Chen et al. [173] the Si source material and the GaAs crystal form a heterostructure so that electrical effects due to the heterojunction are also accounted for. In these analyses [172,173], the diffusivity of the Si donor species Si_{Ga}^{+} satisfies

$$D_{Si}(n) = D_{Si}(n_i)(n/n_i)^3 \tag{5.78}$$

which indicates that V_{Ga}^{3-} governs the diffusion of Si_{Ga}^{+}. Satisfactory fits of the experimental data of Greiner and Gibbons [171] and of Kavanagh et al. [168,170] were obtained using Eq. (5.78) with

$$D_{Si}(n_i) = 5.2\exp(-4.98eV/k_BT)\ \mathrm{m^2\,s^{-1}} \tag{5.79}$$

in the work of Yu et al. [172], while $D_{Si}(n_i)$ values 10 times larger than that given by Eq. (5.79) were needed in the analysis of Chen et al. [173].

In a set of Si outdiffusion experiments, You et al. [174] found that the Si profiles also satisfy Eq. (5.78) but the needed $D_{Si}(n_i)$ values are

$$D_{Si}(n_i, 1\text{-}atm) = 6.67 \times 10^4 \exp(-3.91eV/k_BT)\ \mathrm{cm^2\,s^{-1}}, \tag{5.80}$$

$$D_{Si}(n_i, Ga\text{-}rich) = 9.18 \times 10^8 \exp(-5.25eV/k_BT)\ \mathrm{cm^2\,s^{-1}}, \tag{5.81}$$

respectively, for experiments conducted under As-rich and Ga-rich ambient conditions.

The $D_{Si}(n_i)$ expressed by Eq. (5.80) and (5.81) are larger than those of Eq. (5.79) by several orders of magnitude at temperatures above ~800°C, which indicates the presence of undersaturation and supersaturation of V_{Ga}^{3-}, respectively, under the Si in- and outdiffusion conditions [174]. For the indiffusion case, the starting GaAs crystal contains V_{Ga}^{3-} and the neutral Ga vacancies V_{Ga}^{o} to the thermal equilibrium concentrations of those of the intrinsic material. Upon indiffusion of Si atoms, V_{Ga}^{3-} (and hence also V_{Ga}^{o}) becomes undersaturated relative to the thermal equilibrium V_{Ga}^{3-} concentration values appropriate for the n-doping conditions, which can only be alleviated via inflow of V_{Ga}^{3-} from the interface of the Si source material and the GaAs crystal. The reverse analogy holds for the Si out diffusion case. Since V_{Ga}^{3-} diffusion should be much faster than that of the Si_{Ga}^{+} atoms, in either case there shall be no substantial spatial variations in the distribution of the V_{Ga}^{o} species while the spatial distribution of V_{Ga}^{3-} follows the local n^3 value.

If Ga self-diffusion is regarded as dominated by V_{Ga}^{2-} instead of V_{Ga}^{3-}, as favored by Bracht et al. [10], Si diffusion will necessarily be also governed by V_{Ga}^{2-}. This will be in contrast to the successful fitting of the Si diffusion results in GaAs using V_{Ga}^{3-} [172–174].

5.6.4.2 Diffusion of Interstitial-Substitutional Species

5.6.4.2.1 Carbon

The group IV element carbon (C) occupies the As sublattice sites in GaAs and is a shallow acceptor, designated as C_s^- to emphasize that it is most likely an *i–s* species. By in situ doping during MBE crystal growth, C_s^- reaches high solubilities [175] and diffuses slowly [176], which are attractive features when compared to the main p-type dopants Zn and Be in GaAs. The measured C_s^- diffusivity values of a few groups obtained under As-rich annealing conditions [167,175–179] are fitted well by the expression

$$D_s(1\text{-}atm) = 4.79 \times 10^4 \exp(-3.13eV/k_BT)\ \text{cm}^2\,\text{s}^{-1}. \tag{5.82}$$

The corresponding D_s values under Ga-rich conditions should therefore be

$$D_s(Ga\text{-}rich) = 65 \times 10^4 \exp(-4.74eV/k_BT)\ \text{cm}^2\,\text{s}^{-1}. \tag{5.83}$$

These fits are shown in Fig. 5.36. In the work of You et al. [167], the C_s^- diffusivity data were obtained by the individual fittings of C_s^- profiles that are not quite error-function shaped. In order to fit these profiles well, together with a carbon precipitation process, it was also necessary to use the kick-out reaction

$$C_s^- + I_{As}^o \Leftrightarrow C_i^-, \tag{5.84}$$

where C_i^- is an interstitial C atom which is also assumed to be an acceptor, and I_{As}^o is a neutral As self-interstitial. Subsequently Moll et al. [180] identified the nature of the precipitation process as that of graphite formation. The As self-interstitials are maintained at their thermal equilibrium values during C diffusion because of the low diffusivity value of C.

5.6.4.2.2 Zinc and Beryllium

The main p-type dopants in GaAs based devices, Zn and Be, diffuse via an *i–s* mechanism in GaAs as well as in many other III–V compounds. In most works Zn and Be diffusion have been discussed in terms of the much earlier suggested FT or Longini mechanism [32], but only the KO mechanism involving I_{Ga} is quantitatively consistent with the superlattice disordering results as well as with Zn diffusion results [154,156].

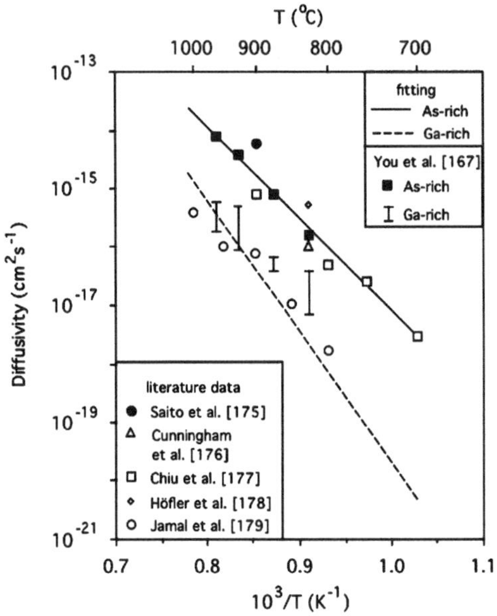

Figure 5.36 Available carbon diffusivity data and fittings in GaAs. From You et al. [167].

Isoconcentration diffusion of Zn isotopes in GaAs predoped by Zn showed error-function profiles [181,182] with the substitutional Zn diffusivity values of

$$D_s(p, 1\text{-}atm) = D_s(n_i, 1\text{-}atm)(C_s/n_i)^2 \tag{5.85}$$

for As-rich GaAs and an analogous expression for Ga-rich GaAs. At sufficiently high Zn concentrations, since the GaAs hole concentration p equals approximately to the Zn_s concentration ($p \sim C_s$), Eq. (5.85) shows that the responsible native point defect species can only be the doubly-positively-charged Ga self-interstitials or vacancies, I_{Ga}^{2+} or V_{Ga}^{2+}. Under high concentration Zn indiffusion conditions, the GaAs/$Al_xGa_{1-x}As$ superlattices disordering rates are tremendously high, indicating the presence of a high supersaturation of the responsible point defects. Thus, the native point defect species responsible for Zn diffusion, and also for Ga self-diffusion and Al–Ga interdiffusion under p-doping conditions, is I_{Ga}^{2+} according to reaction (5.73), and not V_{Ga}^{2+} according to reaction (5.72). In the latter case only undersaturation of V_{Ga}^{2+} can be incurred by Zn indiffusion which should then retard Al–Ga interdiffusion rates in superlattices, in contradiction with experimental results. In the Zn isoconcentration diffusion experiments, a non-equilibrium I_{Ga}^{2+} concentration is not involved. Similarly, for Zn diffusion to low concentrations below the n_i value, a non-equilibrium concentration of I_{Ga}^{2+} is also not present, and the Zn diffusivity values may be represented by that under the intrinsic conditions, $D_s(n_i)$. As analyzed by Yu et al. [154], Zn isoconcentration experiments and

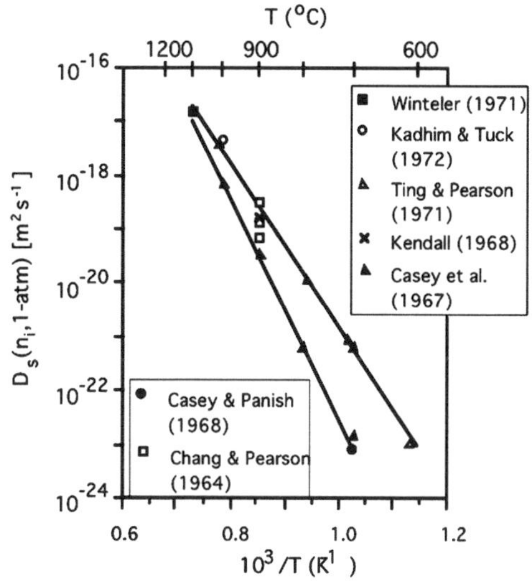

Figure 5.37 The substitutional Zn diffusivity values under intrinsic and 1-atm As_4 pressure conditions. From Yu et al. [154].

Zn indiffusion experiments at high concentrations yielded the value range of

$$D_s(n_i, 1\text{-}atm) = 1.6 \times 10^{-2} \exp(-2.98eV/k_B T)\ \mathrm{cm^2\,s^{-1}}, \tag{5.86a}$$

$$D_s(n_i, 1\text{-}atm) = 96.88 \exp(-4.07eV/k_B T)\ \mathrm{cm^2\,s^{-1}}. \tag{5.86b}$$

The two analogous expressions for Ga-rich materials are respectively

$$D_s(n_i, Ga\text{-}rich) = 1.18 \times 10^{-6} \exp(-1.64eV/k_B T)\ \mathrm{cm^2\,s^{-1}}, \tag{5.87a}$$

$$D_s(n_i, Ga\text{-}rich) = 7.14 \times 10^{-3} \exp(-2.73eV/k_B T)\ \mathrm{cm^2\,s^{-1}}. \tag{5.87b}$$

The values of Eqs. (5.86) and (5.87) and the associated data are plotted in Fig. 5.37.

The correspondingly deduced I_{Ga}^{2+} contribution to Ga self-diffusion has been included in Eqs. (5.69) and (5.70). Because of the lack of a proper Be source for indiffusion studies, and since in Be outdiffusion studies with Be incorporated using MBE or MOCVD methods Be diffusivity is too small, there are no reliable Be diffusivity data.

Outdiffusion of Zn or Be in GaAs doped to fairly high concentrations during crystal is associated with a high I_{Ga}^{2+} undersaturation, leading to Zn or Be outdiffusion rates orders of magnitude smaller than those under indiffusion conditions [164,165,177]. Indiffusion of high concentration Zn into GaAs induces an extremely large I_{Ga}^{2+} supersaturation because the condition

$$D_i C_i^{eq} \gg D_{Ga}^{SD}(p) \tag{5.88}$$

holds. As first noted by Winteler [162], this I_{Ga}^{2+} supersaturation leads to the formation of extended defects. In recent works three kinds of extended defects have been characterized and their formation process analyzed [22,156,183]: (i) interstitial-type dislocation loops, which degenerate into dislocation tangles in time; (ii) voids; and (iii) Ga precipitates coexisting with neighboring voids. For diffusing Zn into GaAs in a Ga-rich ambient, a Zn-diffused GaAs crystal region with compositions at the allowed Ga-rich boundary shown in Fig. 5.3(A) is obtained, irrespective of the GaAs starting composition. The fact that the Zn-diffused region is indeed rich in Ga is evidenced by the presence of Ga precipitates in the voids [156]. Formation of these defects ensures the Zn indiffusion profile to be governed by the thermal equilibrium concentrations of native point defects of the Ga-rich GaAs crystal, and the profile is box-shaped which reveals the p^2 (or C_s^2) dependence of the substitutional Zn diffusivity D_s. Such a profile is shown in Fig. 5.38 together with an illustration of the involved extended defects. It is, however, noted that the crystal is in a highly non-equilibrium state, for two reasons. First, extended defects are generated. Second, the starting material may not be rich in Ga and hence the crystal will now contain regions with different compositions which is, of course, a highly non-equilibrium crystal.

For diffusing Zn into GaAs in an As-rich ambient, the situation is more complicated. After a sufficient elapse of diffusion time, the crystal surface region becomes As-rich because of the presence of a high ambient As_4 pressure. But, since

$$D_i C_i^{eq} \gg D_{As}^{SD}(p) \tag{5.89}$$

holds in the Zn diffusion front region, it is Ga-rich. Thus, the high concentration Zn indiffusion profiles are of a kink-and-tail type resembling those of high concentration P indiffusion profiles in Si, see Fig. 5.39. The kink-and-tail profile develops because the Zn_s solubility value in the As-rich and Ga-rich GaAs materials are different [156]. In the high Zn concentration region the $D_s(n_i)$ values are those given by Eq. (5.86), while in the tail or Zn diffusion front region the $D_s(n_i)$ values are those given by Eq. (5.87). These profiles cannot be modeled with a high degree of self-consistency because the extended defect formation process cannot be modeled without the use of some phenomenological parameters [154]. The evolution of the extended defects, as suggested by Tan et al. [22] and Luysberg et al. [183], is as follows: (i) to reduce I_{Ga}^{2+} supersaturation, they form interstitial type dislocation loops containing extra GaAs molecules, with the needed As atoms taken from the surrounding As sites which generates a V_{As} supersaturation; (ii) the supersaturated V_{As} collapses to form voids, each of an initial volume about that of a neighboring Ga precipitate formed from Ga atoms lost the neighboring As atoms to the formation of the dislocation loops. The voids will be rapidly filled by subsequently generated Ga self-interstitials due to further Zn indiffusion. For cases of diffusing Zn into GaAs in a Ga-rich ambient, the voids contain Ga precipitates throughout the Zn indiffused region. For cases of diffusing Zn into GaAs in an As-rich ambient,

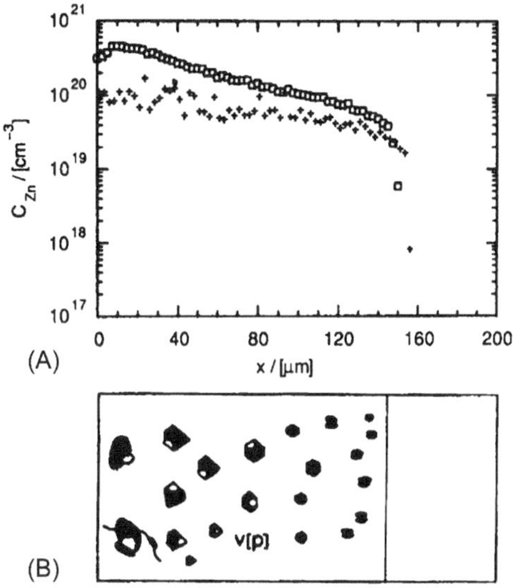

Figure 5.38 (A) Zn indiffusion profiles obtained at 900°C under Ga-rich ambient conditions. Squares are the total Zn concentration and crosses are the hole or Zn_S concentration. The higher total Zn concentration indicates formation of Zn containing precipitates caused by the use of a non-equilibrium Zn source material which diffused Zn into GaAs exceeding its solubility at 900°C. (B) A schematic diagram indicating the morphologies and distributions of voids (open) and Ga precipitates (filled), indicated by also v[p]. The presence of dislocations is not shown. From Jäger et al. [156].

the surface region voids are empty, which indicates that the surface region of the GaAs is As-rich but deeper inside the Zn-diffused region is Ga-rich. Thus, Zn diffusion into GaAs in an As-rich ambient produces a highly non-equilibrium crystal with its composition varying between the whole thermodynamically allowed limits.

The role of I_{Ga}^{2+} governing Zn diffusion and Ga self-diffusion under p-doping conditions have been questioned by some authors [184–186]. Bösker et al. [184] diffused Zn into GaAs to produce s-shaped Zn profiles that were fitted with the dominance of I_{Ga}^{2+} supplemented by also I_{Ga}^{3+}. It is possible that I_{Ga}^{3+} may have also contributed to Zn diffusion and to Ga self-diffusion under p-doping conditions to a minor extent, a point noted also by Wintler [162]. Bracht et al. [185] diffused Zn into ^{69}GaAs/^{71}GaAs superlattice layers and fitted their results with the contributions from V_{Ga}^{o}, V_{Ga}^{+}, and I_{Ga}^{o} for the diffusion of Zn and Ga self-diffusion. Bracht and Brotzmann [186] also produced s-shaped Zn indiffusion profiles and fitted them with the governing point defect species for Zn diffusion being I_{Ga}^{o} and I_{Ga}^{+}, which necessarily also then govern Ga self-diffusion. The findings of Bracht et al. [185] and Bracht and Brotzmann [186] are in contradiction with (i) each other; (ii) the isoconcentration Zn diffusion results as analyzed by Yu et al. [154], which clearly indicate a dominance by I_{Ga}^{2+} but not by other point defect species; and (iii) the p-doping enhanced GaAs/Al_xGa_{1-x}As superlattice disordering results of a

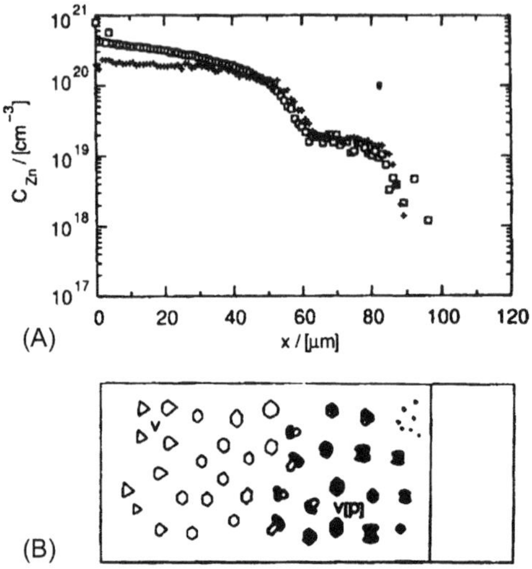

Figure 5.39 (A) Zn indiffusion profiles obtained at 900°C under As-rich ambient conditions. Squares are the total Zn concentration and crosses are the hole or Zn_S concentration. The higher total Zn concentration indicates formation of Zn containing precipitates caused by the use of a non-equilibrium Zn source material which diffused Zn into GaAs exceeding its solubility at 900°C. (B) A schematic diagram indicating the morphologies and distributions of voids (open) and Ga precipitates (filled). The presence of dislocations is not shown. From Jäger et al. [156].

few authors [126,159–161], for which the best fit is a p^2 dependence showing I_{Ga}^{2+} dominates the interdiffusion processes [158], see Fig. 5.34 and the associated discussions.

5.6.4.2.3 Chromium

Chromium is a deep acceptor occupying Ga sites and is used for fabricating semi-insulating GaAs. In undoped GaAs, diffusion of Cr involves no charge effects. Indiffusion profiles of Cr are characterized by a kick-out type profile from the crystal surface to a substantial depth and an erfc-type profile deeper in the material near the diffusion front [119,187]. Outdiffusion profiles are characterized by a constant diffusivity which is much lower than that for in-diffusion. The existence of the two types of profiles needs the description of the *i–s* diffusion mechanism in terms of the KO mechanism $Cr_i \Leftrightarrow Cr_s + I_{Ga}^{2+} + 2e$ and/or Longini mechanism $Cr_i + V_{Ga}^{3-} \Leftrightarrow Cr_s + 3e$, where e denotes an electron.

Tuck [119] and Deal and Stevenson [187] have discussed Cr diffusion in terms of Longini mechanism. Satisfactory treatment of the diffusion behavior of Cr in intrinsic GaAs [155], however, includes coexistence of V_{Ga} and I_{Ga}, dependence of C_s^{eq} and C_i^{eq} on the outside Cr vapor pressure, and dynamical equilibrium between the native point defects in the Ga and As sublattice at the crystal surface region. Indiffusion of Cr turned out to be governed by the effective diffusivity $D_{surf}^{eff} = (D_I C_I^{eq} / C_s^{eq})(C_s^{eq} / C_s)$ in the surface region, which is concentration dependent, and by the much faster constant diffusivity $D_{tail}^{eff} = D_i C_i^{eq} / C_s^{eq}$ in the tail region. The deduced $D_I C_I^{eq}$ values from Cr indiffusion profiles [155] were included in Eq. (5.69).

In the case of outdiffusion the Cr vapor pressure is so low that, similarly to the case of outdiffusion of Zn, a much lower diffusivity prevails. This slower outdiffusion turned out to be dominated either by the constant V-dominated diffusivity $D_s^{eff} = D_V C_V^{eq} / C_s^{eq}$ or the constant s_i dominated diffusivity $D_s^{eff} = D_i C_i^{eq} / C_s^{eq}$, whichever is larger for low outside chromium vapor pressure.

5.6.4.2.4 Sulfur

The group VI donor S occupies As sites. With lower surface concentrations, the S indiffusion profiles [132,188,189] resemble the erfc-function, but a concave shape develops in the surface region for higher concentration cases. The latter cases are indicative of the operation of the KO mechanism for an *i–s* impurity. The available S indiffusion profiles have been quantitatively explained [132] using the KO mechanism, assuming the involvement of the neutral As self-interstitials, I_{As}^{o}. The deduced $D_{As}^{I}(n_i, 1\text{-}atm)$ values were included in Eq. (5.76).

5.6.5 Diffusion in Other III–V Compounds

Gallium arsenide is certainly the one III–V compound in which self and impurity diffusion processes have been studied most extensively. The available results on self-diffusion in III–V compounds have been summarized by Willoughby [190]. The Group III and the Group V diffusivities appear to be so close in some compounds that a common defect mechanism involving multiple native point defects appears to be the case, although no definite conclusion has been reached. There are hardly any experimental results available which would allow us to draw conclusions on the type and charge states of the native point defects involved in self-diffusion processes. Zinc is an important p-type dopant also for other III–V compounds, and its diffusion behavior appears to be governed by an *i–s* mechanism as well. No information is available on whether the FT mechanism or the KO mechanism is operating. It can be expected that dopant diffusion induced superlattice disordering may rapidly advance the understanding of diffusion mechanisms in other III–V compounds similarly as has been accomplished in GaAs.

5.7 DIFFUSION–SEGREGATION: A SPECIAL SUBJECT

This section discusses a special subject usually not associated with a treatment on diffusion, the simultaneous diffusion and segregation process. This subject is especially relevant to semiconductor device fabrications and hence deserves special attention. To fabricate solid-state electronic devices, dopant atoms are introduced into substrate materials both during crystal growth, and during device processing by diffusing dopant into

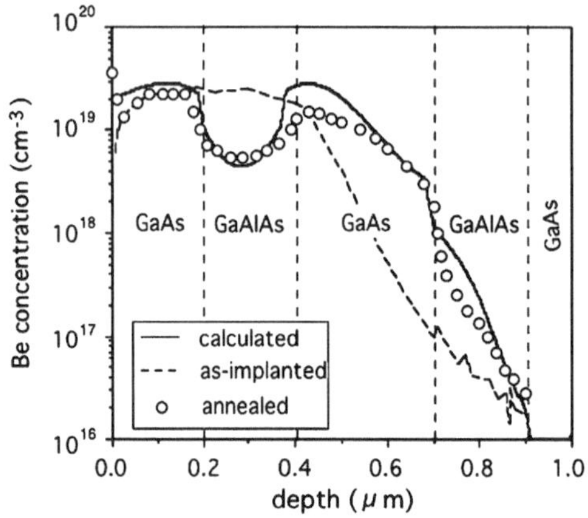

Figure 5.40 Be data of Humer-Hager et al. [192] in GaAs/AlGaAs SL obtained by ion implantation and annealing, together with the calculated fitting curve. The broken line is the as-implanted Be profile, the symbols are Be data after annealing, and the solid line is the fitting curve.

the crystal using either an external source material or implanting the source material as ions into a shallow crystal surface layer. The devices are compressing of electrical and/or heterojunctions, resulting in concurrent dopant diffusion and segregation.

In fabrications of traditional integrated circuit (IC) employing NMOS (n-channel metal–oxide–silicon) devices by diffusing the n-type dopant P or As into a p-type substrate typically doped with B to produce pn-junctions the segregation phenomenon is not prominent, since the as-grown p-type Si substrate is only lightly doped so that at the diffusion temperature the substrate is essentially intrinsic. The dopant diffused region becomes extrinsic wherein charged point defect concentrations, as determined by the dopant concentration, are not uniform and may lead to dopant profiles deviated away from that of the error function type. The dopant produces an advancing n–i (n-type to intrinsic) junction with a space charge region having an electric field which influences the carrier distribution that will be reflected upon the distribution of charged point defects and consequently also the dopant diffusion profile. For the more advanced CMOS (complementary MOS) circuit, each CMOS unit is a pair of side-by-side devices for which one is NMOS and the other PMOS (p-channel MOS). To fabricate the PMOS of the pair, n-type dopant is used to first produce an n-type pocket into the p-type substrate, and then a PMOS is fabricated by diffusion a p-type dopant, usually B, into the n-type pocket to produce the pn-junction. With the introduction of epitaxially grown GeSi strain layers into the more modern Si based IC devices, the dopant segregation phenomenon starts to play a more prominent role, because the GeSi layer, with either a graded or fixed Ge fraction, forms a heterojunction with Si.

It is obvious that the solubility of a dopant is different in different materials, which provides a chemical driving force for the dopant to segregate at an abrupt heterojunction

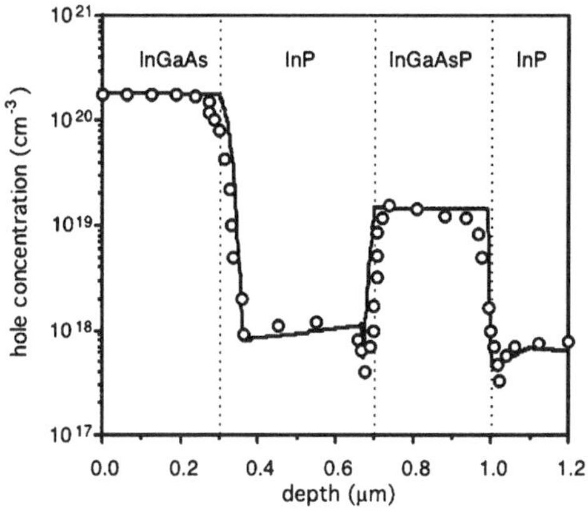

Figure 5.41 The hole data (open circles) in an InGaAs/InP/InGaAsP/InP SL obtained by Weber et al. [193] using Zn indiffusion, together with the calculated fitting curve (solid line).

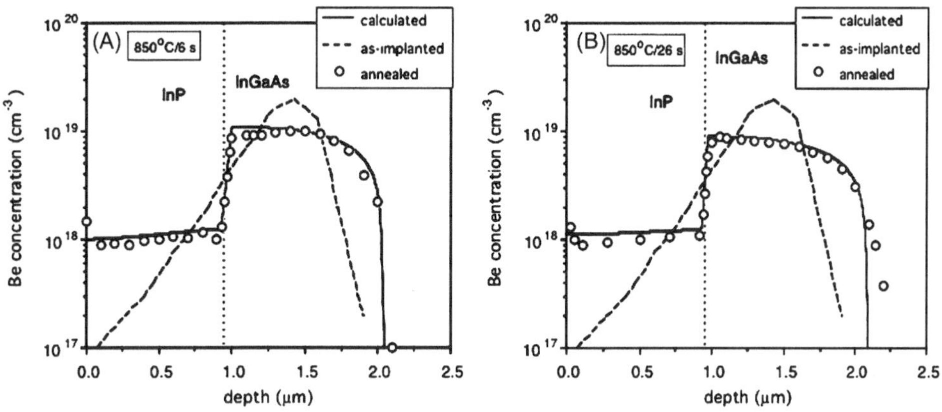

Figure 5.42 Be data of Häussler et al. [194] with Be implanted into the InGaAs layer and annealed at 850°C for (A) 6 s, and (B) 26 s. Dashed lines are the as-implanted data, open circles are those after annealing, and the solid lines are the calculated fitting curves.

or over a graded heterojunction region. Band offset and bending occur at a heterojunction with or without a diffusing dopant. In the presence of a diffusing dopant, a time changing band edge position shifting also occurs. These factors lead to a complicated time changing electrical junction that affects the electrical carrier concentration, which in turn affect the charged point defect concentrations, which in turn affects the dopant diffusion while the dopant's time changing concentration in turn also changes the carrier concentration. While for the GeSi/Si the effect is relative small but observable as the amount of Ge used is fairly small, for the compound semiconductor cases, however, the effect is quite large because the involved materials are fairly different. Optoelectronic devices are fabricated using GaAs based materials, to utilize the direct band gap of GaAs

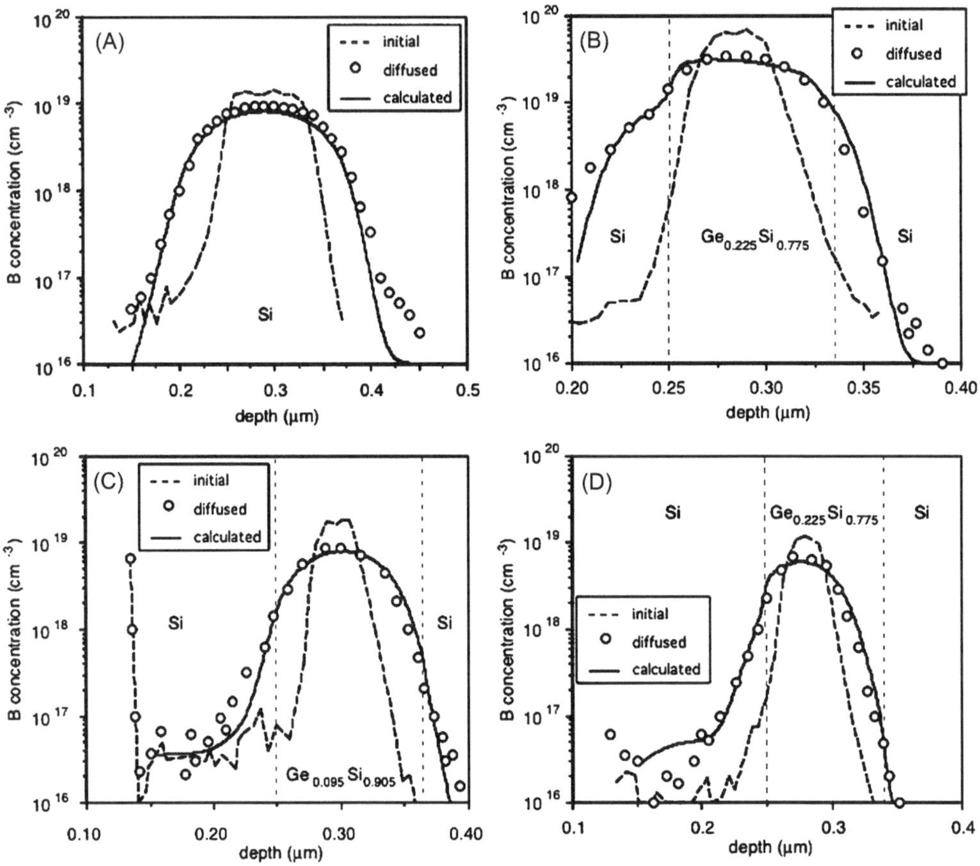

Figure 5.43 Experimental B_5^- distribution data of Fang et al. [196] at 850°C, fitted by the Fermi-level effect model with the effect of the junction carrier concentration also considered. The sample structural conditions and other experimental conditions are included in the drawing for each case.

for light emitting on the one hand and on the other the material need to be modified and confined for device reliability by adding other group III and/or group V elements to GaAs and by using other III–V compound material layers together with the GaAs layer.

To treat the concurrently occurring diffusion and segregation processes, the diffusion–segregation equation (DSE) of You et al. [35], Eq. (5.20), are simultaneously applied to three quantities: (i) the dopant, (ii) the point defect species governing the dopant diffusion, and (iii) the electrical carriers. In this way, Chen et al. [191] treated the ion implanted Be distribution results of Humer-Hager et al. [192] in a GaAs/AlGaAs superlattice structure, the hole data of Weber et al. [193] obtained by diffusing Zn into an InGaAs/InP/InGaAsP/InP superlattice structure, and the ion implanted Be

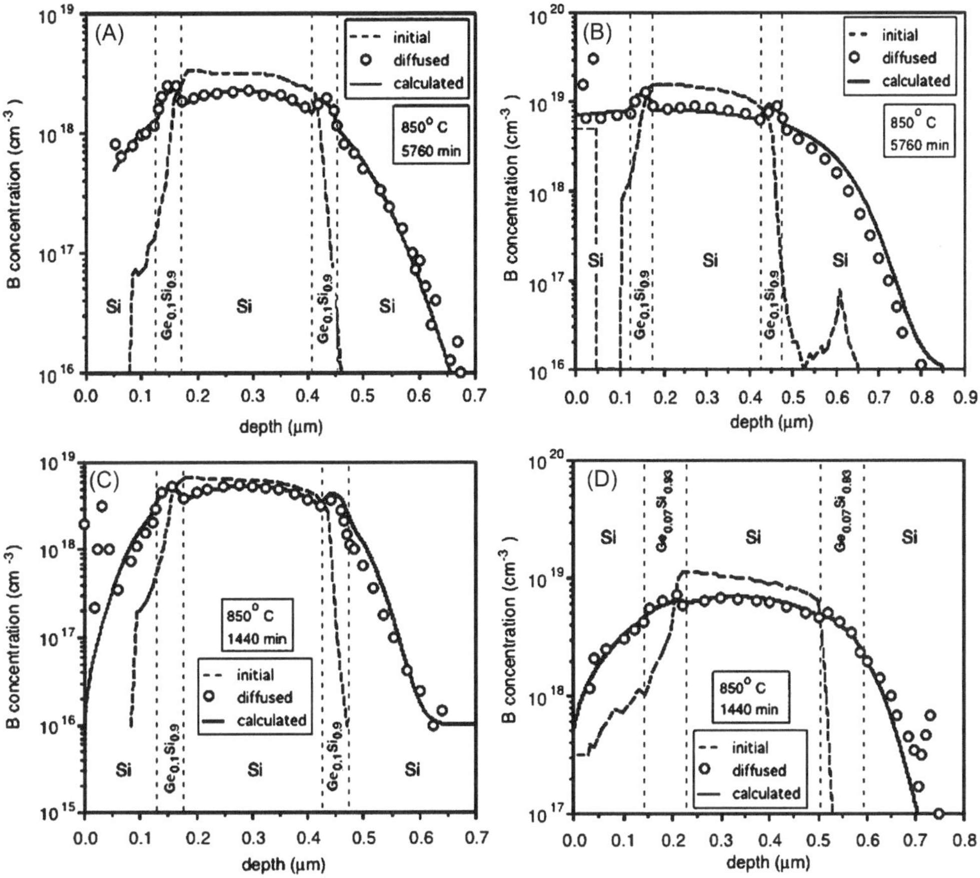

Figure 5.44 Experimental B_5^- distribution data of Lever et al. [197] at 850°C, fitted by the Fermi-level effect model with the effect of the junction carrier concentration also considered. The sample structural conditions and other experimental conditions are included in the drawing for each case.

distribution data of Häussler et al. [194] in a InP/InGaAs structure. These results are shown respectively in Figs. 5.40–5.42.

Moreover, Chen et al. [195] similarly treated two sets of implanted B redistribution data in GeSi/Si layer structures, that of Lever et al. [196] and that of Fang et al. [197], Figs. 5.43 and 5.44, respectively.

5.8 CONCLUDING REMARKS

In this chapter point defects, diffusion phenomena, and the responsible mechanisms in the semiconductors Si, Ge, and GaAs were discussed. Silicon and GaAs are the main materials used to fabricate electronic and optoelectronic devices, respectively. Diffusion in Ge is also mentioned for the reason that it is the simplest case among semiconductors and because of its emerging importance. Diffusion processes are used in doping a semiconductor to produce pn-junctions for device operations, and are involved also in controlling detrimental metallic impurity contents, in silicide formation and in thermal SiO_2 growth in fabricating devices using Si. Diffusion in semiconductors is considerably more complicated than in metals. This is because in semiconductors: (i) both vacancies and self-interstitials contribute to self-diffusion and to diffusion of substitutional impurities, (ii) the point defect species are consisting of neutral and charged ones, (iii) there exist a class of interstitial–substitutional impurities for which the atoms are predominantly dissolved on substitutional lattice sites but diffuses by the rapidly migrating interstitial species; (iv) compound semiconductors is characterized by having a range of thermodynamically allowed point defect concentrations and hence also a range of diffusivity values associated with self-diffusion and diffusion of substitutional impurity species; (v) segregation and diffusion proceed concomitantly at the device pn-junctions and heterojunctions.

REFERENCES

[1] T. Sinno, R.A. Brown, W. von Ammon, E. Dornberger, J. Electrochem. Soc. 145 (1998) 302.
[2] P.S. Plekhanov, U.M. Gösele, T.Y. Tan, J. Appl. Phys. 84 (1998) 718.
[3] L. Goodwin, A.J. Skinner, D.G. Pittfor, Europhys. Lett. 9 (1989) 701.
[4] C. Nichols, C.G. Van de Walle, S.T. Pantelides, Phys. Rev. Lett. 62 (1989) 1049.
[5] C.Z. Wang, C.T. Chan, K.M. Ho, Phys. Rev. Lett. 66 (1991) 189.
[6] J. Zhu, D. de la Rubia, L.H. Yang, C. Mailhiot, G.H. Gilmer, Phys. Rev. B 54 (1996) 4741.
[7] M. Tang, L. Colombo, J. Zhu, D. de la Rubia, Phys. Rev. B 55 (1997) 14279.
[8] J.R. Arthur, J. Phys. Chem. Solids 28 (1967) 2257.
[9] T.Y. Tan, Mat. Sci. Eng. B 10 (1991) 227.
[10] H. Bracht, M. Norseng, E.E. Haller, K. Eberl, M. Cardona, Solid State Commu. 112 (1999) 301.
[11] J. Gebauer, M. Lausmann, F. Redmann, R. Krause-Rehberg, H.S. Leipner, E.R. Weber, Ph. Ebert, Phys. Rev. B 67 (2003) 235207.
[12] J.A. van Vechten, J. Electrochem. Soc. 122 (1975) 419.
[13] G.A. Baraff, M. Schlutter, Phys. Rev. Lett. 12 (1985) 1327.
[14] R.W. Jansen, O.F. Sankey, Phys. Rev. B 39 (1989) 3192.
[15] J.F. Wager, J. Appl. Phys. 69 (1991) 3022.
[16] B.-H. Cheong, K.J. Chang, Phys. Rev. B 49 (1994) 17436.
[17] J.T. Schick, Phys. Rev. B 66 (2002) 195302.
[18] F. El-Mellouhi, N. Mausseau, Phys. Rev. B 71 (2005) 125207.
[19] G.D. Watkins, in: T. Diaz de la Rubia, S. Coffa, C.S. Rafferty, P.A. Stolk (Eds.), Defects and Diffusion in Silicon Processing, in: Mater. Res. Soc. Symp. Proc., vol. 469, Mater. Res. Soc., Pittsburg, PA, 1997, p. 139.
[20] R.M. Nieminen, M.J. Puska, 6.7 vacancy defects in c-Si: electronic and ionic structures, in: Robert Hull (Ed.), Properties of Crystalline Silicon, in: EMIS Data Review Series No. 20, INSPEC, 1999.
[21] T.Y. Tan, U. Gösele, Appl. Phys. Lett. 52 (1988) 1240.
[22] T.Y. Tan, U. Gösele, S. Yu, CRC Crit. Rev. Solid State and Mat. Sci. 17 (1991) 47.
[23] T.Y. Tan, U. Gösele, Diffusion in semiconductors, in: Paul Heitjans, Jörg Käger (Eds.), Diffusion in Condensed Matters: Methods, Materials, Models, Springer-Verlag, 2005, p. 165 (CH 4).

[24] S.B. Zhang, John E. Northrup, Phys. Rev. Lett. 2339 (1991).
[25] S. Pöykkö, M.J. Puska, R.M. Nieminen, Phys. Rev. B 54 (1996) 3813.
[26] I. Gorczyca, Phys. Rev. B 66 (2002) 075210.
[27] A. Janotti, Su-Huai Wei, S.B. Zhang, Sarah Kurtz, Phys. Rev. B 67 (2003) 161201.
[28] T.Y. Tan, H.-M. You, U. Gösele, Appl. Phys. A 56 (1993) 249.
[29] G.D. Watkins, J.W. Corbett, Phys. Rev. 134 (1964) A1359.
[30] D. Mathiot, J.C. Pfister, J. de Physique 43 (1982), L-453.
[31] F.C. Frank, D. Turnbull, Phys. Rev. 104 (1956) 617.
[32] R.L. Longini, Solid-State Electron. 5 (1962) 127.
[33] U. Gösele, W. Frank, A. Seeger, Appl. Phys. 23 (1980) 361.
[34] W. Frank, U. Gösele, H. Mehrer, A. Seeger, in: G.E. Murch, A. Nowick (Eds.), Diffusion in Crystalline Solids, Academic Press, New York, 1984, p. 31.
[35] H.-M. You, Ulrich M. Gösele, Teh Y. Tan, J. Appl. Phys. 74 (1993) 2461.
[36] D.A. Antoniadis, R. Dutton, IEEE Trans. Electron. Devices ED-26 (1979) 490.
[37] A. Seeger, C.P. Chik, Phys Stat. Sol. B 29 (1968) 455.
[38] T.Y. Tan, U. Gösele, Appl. Phys. Lett. 40 (1982) 616.
[39] T.Y. Tan, U. Gösele, F.F. Morehead, Appl. Phys. A 31 (1983) 97.
[40] T.Y. Tan, U. Gösele, Appl. Phys. A 37 (1985) 1.
[41] P. Fahey, P.B. Griffin, J.D. Plummer, Rev. Mod. Phys. 61 (1989) 289.
[42] T.Y. Tan, U. Gösele, Point defects, diffusion, and precipitation, in: K.A. Jackson, W. Schröter (Eds.), Handbook of Semiconductor Technology, vol. 1: Electronic Structure and Properties of Semiconductors, Wieley-VCH, New York, 2000, p. 231 (CH 5).
[43] S.R. Aid, T. Sakaguchi, K. Toyonaga, Y. Nakabayashi, S. Matumoto, M. Sakuraba, Y. Shimamune, Y. Hashiba, J. Murota, K. Wada, T. Abe, Mater. Sci. Eng. B 114–115 (2004) 330.
[44] H. Bracht, E.E. Haller, R. Clark-Phelps, Phys. Rev. Lett. 81 (1998) 393.
[45] A. Ural, P.B. Griffin, J.D. Plummer, Appl. Phys. Lett. 73 (1998) 1706.
[46] A. Ural, P.B. Griffin, J.D. Plummer, J. Appl. Phys. 85 (1999) 6440.
[47] Y. Shimizu, M. Uematsu, K.M. Ito, Phys. Rev. Lett. 98 (2007) 095901.
[48] R. Kube, H. Bracht, E. Hüger, H. Schmidt, J.L. Hansen, A.N. Larsen, J.W. Ager III, E.E. Haller, T. Geue, J. Stahn, Phys. Rev. B 88 (2013) 085206.
[49] N.A. Stolwijk, B. Schuster, J. Hölzl, H. Mehrer, W. Frank, Physica 115B (1983) 335.
[50] S. Mantovani, F. Nava, C. Nobili, G. Ottaviani, Phys. Rev. B 33 (1986) 5536.
[51] N.A. Stolwijk, M. Perret, H. Mehrer, Defect Diffus. Forum 59 (1988) 79.
[52] M. Perret, N.A. Stolwijk, L. Cohausz, J. Phys. Cond. Mater. 1 (1989) 6347.
[53] F.A. Huntley, A.F.W. Willoughby, Phil. Mag. 28 (1973) 1319.
[54] C. Boit, F. Lay, R. Sittig, Appl. Phys. A 50 (1990) 197.
[55] R.B. Fair, J. Electrochem. Soc. 128 (1981) 1360.
[56] C.P. Ho, J.D. Plummer, S.E. Hansen, R.W. Dutton, IEEE Trans. Electron. Devices ED-30 (1983) 1438.
[57] C.L. Claeys, G.J. DeClerck, P.J. van Overstraeten, Rév. Phys. Appliqueé 13 (1978) 797.
[58] F.F. Morehead, R.F. Lever, Appl. Phys. Lett. 48 (1986) 151.
[59] A. Nylandsted Larsen, K. Kyllesbech Larsen, P.E. Andersen, B.G. Svensson, J. Appl. Phys. 73 (1993) 691.
[60] A. Nylandsted Larsen, P. Kringho, J. Lundsgaard Hansen, S.Yu. Shiryaev, J. Appl. Phys. 81 (1997) 2173.
[61] S. Solmi, D. Nobili, J. Appl. Phys. 83 (1998) 2484.
[62] S. Mirabella, D. De Salvador, E. Napolitani, E. Bruno, F. Priolo, J. Appl. Phys. 113 (2013) 031101.
[63] D. Mathiot, J.C. Pfister, J. Appl. Phys. 66 (1989) 1970.
[64] S.T. Dunham, C.D. Wu, J. Appl. Phys. 78 (1995) 2362.

[65] H. Strunk, U. Gösele, B.O. Kolbesen, Appl. Phys. Lett. 34 (1979) 530.
[66] D.J.D. Thomas, Phys. Status Solidi 3 (1963) 2261.
[67] H.J. Queisser, P.G.G. Van Loon, J. Appl. Phys. 35 (1964) 3066.
[68] G.R. Booker, W.J. Tunstall, Phil. Mag. 13 (1966) 71.
[69] S.M. Hu, Appl. Phys. Lett. 27 (1975) 165.
[70] G.N. Wills, Solid-State Electron. 12 (1969) 133.
[71] G. Masetti, S. Solmi, G. Soncini, Solid-State Electron. 16 (1973) 1419.
[72] D.A. Antoniadis, A.M. Lin, R.W. Dutton, Appl. Phys. Lett. 33 (1979) 1030.
[73] S. Mizuo, H. Higuchi, Jpn. J. Appl. Phys. 21 (1982) 56.
[74] S. Mizuo, H. Higuchi, Jpn. J. Appl. Phys. 20 (1981) 739.
[75] R. Francis, P.S. Dobson, J. Appl. Phys. 50 (1979) 280.
[76] T.Y. Tan, B.J. Ginsberg, Appl. Phys. Lett. 42 (1983) 448.
[77] T.Y. Tan, U. Gösele, Appl. Phys. Lett. 39 (1981) 86.
[78] G.K. Celler, L.E. Trimble, Appl. Phys. Lett. 53 (1988) 2492.
[79] S.M. Hu, Appl. Phys. Lett. 51 (1987) 308.
[80] P. Fahey, S.S. Iyer, G.J. Scilla, Appl. Phys. Lett. 54 (1989) 843.
[81] B.P.R. Marioton, U. Gösele, Jpn. J. Appl. Phys. 28 (1989) 1274.
[82] G.D. Watkins, K.L. Brower, Phys. Rev. Lett. 36 (1976) 1329.
[83] L.A. Ladd, J.P. Kalejs, in: J.C. Mikkelsen Jr., S.J. Pearton, J.W. Corbett, S.J. Pennycook (Eds.), Oxygen, Carbon, Hydrogen and Nitrogen in Crystalline Silicon, in: Mater. Res. Soc. Proc., vol. 59, Mat. Res. Soc., Pittsburgh, 1986, p. 445.
[84] U. Gösele, A. Plössl, T.Y. Tan, in: Electrochem. Soc. Proc., vol. 96(4), Electcochem. Soc., Pennington, 1996, p. 309.
[85] R. Scholz, U. Gösele, J.-Y. Huh, T.Y. Tan, Appl. Phys. Lett. 72 (1998) 200.
[86] R.C. Newman, J. Wakefield, J. Phys. Chem. Solids 19 (1961) 230.
[87] F. Rollert, N.A. Stolwijk, H. Mehrer, Mater. Sci. Forum 38–41 (1989) 753.
[88] R.F. Scholz, P. Werner, U. Gösele, T.Y. Tan, Appl. Phys. Lett. 74 (1999) 392.
[89] H. Rücker, B. Heinemann, W. Röpke, D. Krüger, G. Lipert, H.J. Osten, Appl. Phys. Lett. 73 (1998) 2682.
[90] P. Werner, H.-J. Gossmann, D.C. Jacobson, U. Gösele, Appl. Phys. Lett. 73 (1998) 2465.
[91] P.A. Stolk, H.-J. Gossmann, D.J. Eaglesham, D.C. Jacobson, C.S. Raferty, G.H. Gilmer, M. Jaraiz, J.M. Poate, J. Appl. Phys. 81 (1997) 6031.
[92] R. Pinacho, P. Castrillo, M. Jaraiz, I. Martin-Bragado, J. Barbolla, J. Appl. Phys. 92 (2002) 1582.
[93] S. Mizuo, H. Higuchi, J. Electrochem. Soc. 130 (1983) 1942.
[94] D.A. Antoniadis, I. Moskowitz, J. Appl. Phys. 53 (1982) 6788.
[95] W. Taylor, B.P.R. Marioton, T.Y. Tan, U. Gösele, Rad. Eff. and Defects in Solids 111–112 (1989) 131.
[96] G.D. Watkins, in: I.A. Huntley (Ed.), Lattice Defects in Semiconductors 1974, in: Inst. Phys. Conf. Ser., vol. 23, 1975, p. 1, London.
[97] T.Y. Tan, E.E. Gardner, W.K. Tice, Appl. Phys. Lett. 30 (1977) 175.
[98] W. Kaiser, H.L. Frisch, H. Reiss, Phys. Rev. 112 (1958) 1546.
[99] R.C. Newman, J. Phys. C 18 (1985) L967.
[100] U. Gösele, T.Y. Tan, Appl. Phys. A 28 (1982) 79.
[101] S.A. McQuaid, M.J. Binns, C.A. Londos, J.H. Tucker, A.R. Brown, R.C. Newman, J. Appl. Phys. 77 (1995) 1427.
[102] S. Öberg, C.P. Ewels, R. Jones, T. Hallberg, J.L. Lindström, L.I. Murin, P.R. Briddon, Phys. Rev. Lett. 81 (1998) 2930.
[103] R.C. Newman, J. Phys.: Condensed Matter 12 (2000) R335.
[104] T. Ono, G.A. Rozgonyi, E. Asayama, H. Horie, H. Tsuya, K. Sueoka, Appl. Phys. Lett. 74 (1999) 3648.

[105] A.S. Oates, W. Lin, Appl. Phys. Lett. 53 (1988) 2659.
[106] M. Pagani, J. Appl. Phys. 68 (1990) 3726.
[107] H. Takeno, K. Sunakawa, M. Suezawa, Appl. Phys. Lett. 77 (2000) 376.
[108] A. Van Wieringen, N. Warmoltz, Physica 22 (1956) 849.
[109] S.J. Pearton, J.W. Corbett, T.S. Shi, Appl. Phys. A 43 (1987) 153.
[110] N.M. Johnson, F.A. Ponce, R.A. Street, R.J. Nemanich, Phys. Rev. B 35 (1987) 4166.
[111] H. Bracht, S. Brotzmann, Mater. Sci. in Semicon. Processing 9 (2006) 471.
[112] M. Werner, H. Mehrer, H.D. Hochheimer, Phys. Rev. B 32 (1985) 3930.
[113] T. Südkamp, H. Bracht, G. Impellizzeri, J. Lundsgaard Hansen, A. Nylandsted Larsen, E.E. Haller, Appl. Phys. Lett. 102 (2013) 242103.
[114] E. Vainonen-Ahlgren, T. Ahlgren, J. Likonen, S. Lehto, J. Keinonen, W. Li, J. Haapamaa, Appl. Phys. Lett. 77 (2000) 690.
[115] M. Naganawa, Y. Shimizu, M. Uematsu, K.M. Itoh, K. Sawano, Y. Shiraki, E.E. Haller, Appl. Phys. Lett. 93 (2008) 191905.
[116] N.A. Stolwijk, W. Frank, J. Hölzl, S.J. Pearton, E.E. Haller, J. Appl. Phys. 57 (1985) 5211.
[117] E. Hüger, U. Tietze, D. Lott, H. Bracht, D. Bougeard, E.E. Haller, H. Schmidt, Appl. Phys. Lett. 93 (2008) 162104.
[118] D.G. Deppe, N. Holonyak Jr., J. Appl. Phys. 64 (1988) R93.
[119] B. Tuck, Atomic Diffusion in III–V Semiconductors, Adam Hilger, Bristol, 1988.
[120] W.D. Laidig, N. Holonyak Jr., M.D. Camras, K. Hess, J.J. Coleman, P.D. Dapkus, J. Bardeen, Appl. Phys. Lett. 38 (1981) 776.
[121] M.D. Camras, N. Holonyak Jr., K. Hess, M.J. Ludowise, W.T. Dietze, C.R. Lews, Appl. Phys. Lett. 42 (1983) 185.
[122] J.J. Coleman, P.D. Dapkus, C.G. Kirkpatrick, M.D. Camras, N. Holonyak, Appl. Phys. Lett. 40 (1982) 904.
[123] M. Kawabe, N. Matsuura, N. Shimizu, F. Hasegawa, Y. Nannichi, Jpn. J. Appl. Phys. 23 (1984) L623.
[124] E.V.K. Rao, H. Thibierge, R. Brillouet, F. Alexandre, R. Azoulay, Appl. Phys. Lett. 46 (1985) 867.
[125] T. Nakamura, S. Komiya, T. Inata, S. Muto, S. Hiyamizu, I. Umebu, Proc. Mat. Res. Soc. 56 (1986) 339.
[126] M. Kawabe, N. Shimizu, F. Hasegawa, Y. Nannidi, Appl. Phys. Lett. 46 (1985) 849.
[127] D.R. Myers, R.M. Biefeld, I.J. Fritz, S.T. Picraux, T.E. Zipperian, Appl. Phys. Lett. 44 (1984) 1052.
[128] P. Mei, H.W. Yoon, T. Venkatesan, S.A. Schwarz, J.B. Harbison, Appl. Phys. Lett. 50 (1987) 1823.
[129] G. Bösker, N.A. Stolwijk, H. Mahrer, U. Södervall, J.V. Thordson, T.G. Anderson, A. Buchard, in: Y. Mishin, G. Vogl, N. Cowern, R. Catlow, D. Farkas (Eds.), Diffusion Mechanisms in Crystalline Materials, in: Mater. Res. Soc. Proc., vol. 527, Mat. Res. Soc., Pittsburgh, PA, 1998, p. 347.
[130] R. Scholz, U. Gösele, O. Breitenstein, U. Egger, T.Y. Tan, Diffus. Defect Data B 63–64 (1998) 183.
[131] M. Schultz, U. Egger, R. Scholz, O. Breitenstein, U. Gösele, T.Y. Tan, J. Appl. Phys. 83 (1998) 5295.
[132] M. Uematsu, P. Werner, M. Schultz, T.Y. Tan, U. Gösele, Appl. Phys. Lett. 67 (1995) 2863.
[133] H. Palfrey, M. Brown, A. Willoughby, J. Electr. Mater. 12 (1983) 863.
[134] B. Goldstein, Phys. Rev. 121 (1961) 1305.
[135] H.D. Palfrey, M. Brown, A.F.W. Willoughby, J. Electrochem. Soc. 128 (1981) 2224.
[136] T.Y. Tan, H.-M. You, S. Yu, U. Gösele, W. Jäger, D.W. Boeringer, F. Zypman, R. Tsu, S.-T. Lee, J. Appl. Phys. 72 (1992) 5206.
[137] L. Wang, L. Hsu, E.E. Haller, J.W. Erickson, A. Fischer, K. Eberl, M. Cardona, Phys. Rev. Lett. 76 (1996) 2342.
[138] L.L. Chang, A. Koma, Appl. Phys. Lett. 29 (1976) 138.
[139] J.-C. Lee, T.E. Schlesinger, T.F. Kuech, J. Vac. Sci. Technol. B 5 (1987) 1187.
[140] P.M. Petroff, J. Vac. Sci. Technol. 14 (1977) 973.
[141] R.M. Fleming, D.B. McWhan, A.C. Gossard, W. Weigmann, R.A. Logan, J. Appl. Phys. 51 (1980) 357.

[142] M.D. Camras, N. Holonyak Jr., R.D. Bumham, W. Streifer, D.R. Scifres, T.L. Paoli, C. Lindstrom, J. Appl. Phys. 54 (1983) 5637.
[143] T.E. Schlesinger, T. Kuech, Appl. Phys. Lett. 49 (1986) 519.
[144] J. Cibert, P.M. Petroff, D.J. Werder, S.J. Pearton, A.C. Gossard, J.H. English, Appl. Phys. Lett. 49 (1986) 223.
[145] K.C. Hsieh, Y.C. Lo, J.H. Lee, R.M. Kolbas, Inst. Phys. Conf. Ser. 96 (1988) 393.
[146] A. Furuya, O. Wada, A. Takamori, H. Hashimoto, Jpn. J. Appl. Phys. 26 (1987) L926.
[147] L.J. Guido, N. Holonyak Jr., K.C. Hsieh, R.W. Kalisiki, W.E. Plano, P.D. Burtham, R.L. Thornton, J.E. Epler, T.L. Paoli, J. Appl. Phys. 61 (1987) 1372.
[148] F. El-Mellouhi, N. Mousseau, J. Appl. Phys. 100 (2006) 083521.
[149] J.A. Van Vechten, J. Appl. Phys. 53 (1982) 7082.
[150] J.A. Van Vechten, J. Vac. Sci. Technol. B 2 (1984) 569.
[151] S.R. Tatti, S. Mitra, J.P. Stark, J. Appl. Phys. 65 (1989) 2547.
[152] T.Y. Tan, U. Gösele, Mat. Sci. Eng. B 1 (1988) 47.
[153] J. Kobayashi, M. Nakajima, T. Fukunagon, T. Takamori, K. Shida, H. Nakashima, K. Ishida, Jpn. J. Appl. Phys. 25 (1986), L 736.
[154] S. Yu, T.Y. Tan, U. Gösele, J. Appl. Phys. 69 (1991) 3547.
[155] S. Yu, T.Y. Tan, U. Gösele, J. Appl. Phys. 70 (1991) 4827.
[156] W. Jäger, A. Rucki, K. Urban, H.-G. Hettwer, N.A. Stolwijk, H. Mehrer, T.Y. Tan, J. Appl. Phys. 74 (1993) 4409.
[157] T.Y. Tan, S. Yu, U. Gösele, J. Appl. Phys. 70 (1991) 4823.
[158] T.Y. Tan, U.M. Gösele, Mater. Chem. Phys. 44 (1996) 45.
[159] J.W. Lee, W.D. Laidig, J. Electron. Mater. 13 (1984) 147.
[160] N. Kamata, K. Koboyashi, K. Endo, T. Sasudi, A. Misu, Jpn. J. Appl. Phys. 26 (1987) 1092.
[161] E.P. Zucker, A. Hasimoto, T. Fukunaga, N. Watanabe, Appl. Phys. Lett. 54 (1989) 564.
[162] H.R. Wintler, Helvetica Physica Acta 44 (1971) 451.
[163] K. Masu, M. Konagai, V. Takahoshi, Appl. Phys. Lett. 37 (1980) 182.
[164] B. Tuck, A.J.N. Houghton, J. Phys. D 14 (1981) 2147.
[165] P. Enquist, G.W. Wicks, L.F. Eastman, C. Hitzman, J. Appl. Phys. 58 (1985) 4130.
[166] P. Enquist, J.H. Hutchby, T.J. de Lyon, J. Appl. Phys. 63 (1988) 4485.
[167] H.M. You, T.Y. Tan, U.M. Gösele, S.-T. Lee, G.E. Höfler, K.C. Hsieh, N. Holonyak Jr., J. Appl. Phys. 74 (1993) 2450.
[168] K.L. Kavanagh, J.W. Mayer, C.W. Magee, J. Sheets, J. Tong, J.M. Woodall, Appl. Phys. Lett. 47 (1985) 1208.
[169] U. Egger, M. Schultz, P. Werner, O. Breitenstein, T.Y. Tan, U. Gösele, R. Franzheld, M. Uetmatsu, H. Ito, J. Appl. Phys. 81 (1997) 6056.
[170] K.L. Kavanagh, C.W. Magee, J. Sheets, J.W. Mayer, J. Appl. Phys. 64 (1988) 1845.
[171] M.E. Greiner, J.F. Gibbons, J. Appl. Phys. 57 (1985) 5181.
[172] S. Yu, U. Gösele, T.Y. Tan, J. Appl. Phys. 66 (1989) 2952.
[173] C.H. Chen, U.M. Gösele, T.Y. Tan, Appl. Phys. A 69 (1999) 313.
[174] H.-M. You, U.M. Gösele, T.Y. Tan, J. Appl. Phys. 73 (1993) 7207.
[175] K. Saito, E. Tokumitsu, T. Akatsuka, M. Miyauchi, T. Yamada, M. Konagai, K. Takahashi, J. Appl. Phys. 64 (1988) 3975.
[176] B.T. Cunningham, L.J. Guido, J.E. Baker, J.S. Major Jr., N. Holonyak Jr., G.E. Stillman, Appl. Phys. Lett. 55 (1989) 687.
[177] T.H. Chiu, J.E. Cunningham, J.A. Ditzenberger, W.Y. Jan, S.N.G. Chu, J. Crys. Growth 111 (1991) 274.
[178] G.E. Höfler, H.J. Höfler, N. Holonyak Jr., K.C. Hsieh, J. Appl. Phys. 72 (1992) 5318.

[179] Z. Jamal, P.J. Goodhew, in: C.R. Abernathy, C.W. Bates Jr., D.A. Bohling, W.S. Hobson (Eds.), Chemical Perspectives of Microelectronic Materials III, in: Mater. Res. Soc. Proc., vol. 282, Mat. Res. Soc., Pittsburgh, PA, 1993, p. 145.
[180] A.J. Moll, E.E. Haller, J.W. Ager III, W. Walukiewicz, Appl. Phys. Lett. 65 (1994) 1145.
[181] C.H. Ting, G.L. Pearson, J. Appl. Phys. 42 (1971) 2247.
[182] M.A.H. Kadhim, B. Tuck, J. Mater. Sci. 7 (1972) 68.
[183] M. Luysberg, W. Jäger, K. Urban, M. Schänzer, N. Stolwijk, H. Mehrer, Mat. Sci. Eng. B 13 (1992) 137.
[184] G. Bösker, N.A. Stolwijk, H.-G. Hettwer, A. Rucki, W. Jäger, U. Södervall, Phys. Rev. B 52 (1995) 11927.
[185] H. Brachta, M.S. Norseng, E.E. Haller, K. Eberl, Physica B 308 (2001) 831.
[186] H. Bracht, S. Brotzmann, Phys. Rev. B 71 (2005) 115216.
[187] M.D. Deal, D.A. Stevenson, J. Appl. Phys. 59 (1986) 2398.
[188] A.B.Y. Young, G.L. Pearson, J. Phys. Chem. Solids 31 (1970) 517.
[189] B. Tuck, R.G. Powell, J. Phys. D 14 (1981) 1317.
[190] A.F.W. Willoughby, in: S. Mahajan, J.W. Corbett (Eds.), Defects in Semiconductors II, North-Holland, New York, 1983, pp. 237–252.
[191] C.-H. Chen, U.M. Gösele, T.Y. Tan, Appl. Phys. A 68 (1999) 9–18.
[192] T. Humer-Hager, R. Treichler, P. Wurzinger, H. Tews, P. Zwicknagl, J. Appl. Phys. 66 (1989) 181.
[193] R. Weber, A. Paraskevopoulos, H. Schroeter-Janssen, H.G. Bach, J. Electrochem. Soc. 138 (1991) 2812.
[194] W. Häussler, J.W. Walter, J. Müller, in: N.W. Cheung, A.D. Marwick, J.B. Roberto (Eds.), Ion Beam Processing of Advanced Electronic Materials, in: Mat. Res. Soc. Symp. Proc., vol. 147, Mat. Res. Soc., Pittsburgh, PA, 1989, p. 333.
[195] C.-H. Chen, U.M. Gösele, T.Y. Tan, Appl. Phys. A 68 (1999) 19–24.
[196] W.T. Fang, P.B. Griffin, J.D. Plummer, in: J.C. Bean, K-Y.(N). Cheng, E.A. Fitzgerald, J. Hoyt (Eds.), Strained Layer Epitaxy–Materials, Processing, and Devise Applications, in: Mater. Res. Soc. Proc., vol. 379, Mater. Res. Soc., Pittsburg, PA, 1995, p. 379.
[197] R.F. Lever, J.M. Bonar, A.F.W. Willoughby, J. Appl. Phys. 83 (1998) 1988.

CHAPTER 6

CALPHAD-Type Modeling of Diffusion Kinetics in Multicomponent Alloys

Lijun Zhang*, Qing Chen†
*Central South University, Changsha, PR China
†Thermo-Calc Software AB, Stockholm, Sweden

Knowledge of both thermodynamic and diffusion characteristics in multicomponent alloys is of critical importance in various materials processes, such as solidification, heat treatment, recrystallization, protective coatings, and so on. In order to give an insight into those material processes, optimize the process parameters, and even accelerate the development of new alloys, an advanced computational technique, which uses fully integrating multicomponent thermodynamics and kinetics, should be an appropriate underlying tool. For the past decades, the CALPHAD (CALculation of PHAse Diagram) technique [1–3] and related software packages [4–6] have matured as powerful means to understand and predict phase relations and microstructures in multicomponent alloys of industrial relevance. Moreover, recent interest in Integrated Computational Materials Engineering (ICME) and Materials Genome Initiative (MGI) has put additional emphasis on the necessity and urgency of the CALPHAD-type computational tools as well as high-quality CALPHAD databases [7,8].

Owing to the pioneering work due to Andersson and Ågren [9], various diffusivities in multicomponent systems have also been modeled in the CALPHAD fashion, and diffusion-controlled phenomena in multicomponent alloys can be quantitatively simulated by computational tools, such as DIffusion-Controlled TRAnsformation (DICTRA) software package [4], which is based on the sharp interface and local equilibrium hypothesis. Based on high-quality thermodynamic and atomic mobility databases, DICTRA has been successfully utilized to simulate various diffusion-controlled phase transformation processes in different alloys [10,11].

This chapter starts from a brief introduction of the multicomponent diffusion theory, the concept of atomic mobility, as well as the relations between atomic mobility and various diffusion coefficients. After that, different formalisms for modeling atomic mobility in different phases are described, paying attention to the recent progress on model development. Subsequently, DICTRA software is introduced, followed by a demonstration of the general strategy to establish an atomic mobility database for multicomponent alloys. Then, examples of successful applications of DICTRA in different alloys during various materials processes are presented. Finally, the important roles of multicompo-

Handbook of Solid State Diffusion, Volume 1
DOI: 10.1016/B978-0-12-804287-8.00006-3

nent thermodynamic and atomic mobility databases in precipitation simulations using TC-PRISMA software and diffuse-interface phase-field modeling are pointed out.

6.1 MULTICOMPONENT DIFFUSION THEORY

In the presence of a concentration gradient, a net flux of the corresponding species can be established by using Fick's first law, which was initially proposed in the case of isothermal isobaric one-phase binary alloys. In real world, most alloys contain more than two species. For describing diffusion in a general isothermal isobaric one-phase multicomponent system, Onsager [12] postulated that the diffusion flux of component k in the lattice-fixed frame of reference was linearly related to the driving forces,

$$J_k = -\sum_{j=1}^{n} L_{kj} \nabla \mu_j, \tag{6.1}$$

where L_{kj}s are phenomenological parameters that relate the flux of k to all the driving forces. The μ_j terms are the chemical potentials for the various components, and may be assumed to be unique functions of the compositions, i.e., $\mu_j = f(c_1, c_2, c_3, \ldots, c_n)$. Transformation of diffusion flux to the volume-fixed frame of reference yields [9]

$$\tilde{J}_k = -\sum_{j=1}^{n} L'_{kj} \nabla \mu_j, \tag{6.2}$$

where

$$L'_{kj} = \sum_{i}^{n} \left[\delta_{ik} - x_k \frac{V_i}{V_m} \right] L_{ij}, \tag{6.3}$$

and δ_{ik} is the Kronecker delta, i.e., $\delta_{ik} = 1$ when $i = k$, and $\delta_{ik} = 0$ otherwise. x_k is the mole fraction of component k. V_m, the molar volume of the target phase, varies with the composition and can be expressed by

$$V_m = \sum_{k=1}^{n} x_k V_k, \tag{6.4}$$

where V_k is the partial molar volume defined by $(\partial V / \partial N_k)_{P,T,N_j}$ with the number of moles of component k, N_k. According to Andersson and Ågren's approximation [9], the components can be divided into two classes. The first class comprises those components that contribute to the volume of the phase, and those components are assumed to have the same partial molar volume, like substitutional solutes. While the components belonging to the other class do not contribute to the volume and their partial volume is thus zero, like interstitials. If the notation $k \in S$ is used to denote that a component belongs to the first class, and $k \notin S$ for the other class, one can have

$$\begin{matrix} V_k = V_S & \text{when } k \in S \\ V_k = 0 & \text{when } k \notin S \end{matrix}. \tag{6.5}$$

Then, the concentration c_k defined as moles of k per unit volume should be

$$c_k = \frac{x_k}{V_m} = \frac{x_k}{V_s \sum_{j \in s} x_j}. \tag{6.6}$$

In a volume-fixed frame of reference, where no net flow of volume appears, the flux obeys the following relation:

$$\sum_{k=1}^{n} \tilde{J}_k V_k = V_s \sum_{k \in s} \tilde{J}_k = 0. \tag{6.7}$$

In Eq. (6.2), the thermodynamic forces, $\nabla \mu_j$, are related by the Gibbs–Duhem equation, and thus not all of them are independent in an n-component system. Following Ågren's work [13], one can express the fluxes as linear functions of the independent set of driving forces, which reads

$$\tilde{J}_k = -\sum_{i=1}^{n} L''_{ki} \left[\nabla \mu_i - \frac{V_i}{V_n} \nabla \mu_n \right] \tag{6.8}$$

with n chosen to be the index of the dependent component. Here,

$$L''_{ki} = \sum_{j=1}^{n} \left[\delta_{ij} - x_i \frac{V_j}{V_m} \right] L'_{ki}. \tag{6.9}$$

Substituting Eq. (6.3) into Eq. (6.9), one can obtain

$$L''_{ki} = \sum_{j=1}^{n} \sum_{r=1}^{n} \left[\delta_{ij} - x_i \frac{V_r}{V_m} \right] \left[\delta_{jk} - x_k \frac{V_j}{V_m} \right] L_{jr}. \tag{6.10}$$

In practical calculations, it is usually more convenient to treat concentration gradients rather than potential gradients. In an n-component system, the driving force term, $\nabla \mu_i$, can be rewritten by applying the chain rule as

$$\nabla \mu_i = \frac{\partial \mu_i}{\partial c_1} \nabla c_1 + \frac{\partial \mu_i}{\partial c_2} \nabla c_2 + \cdots + \frac{\partial \mu_i}{\partial c_n} \nabla c_n = \sum_{j=1}^{n} \frac{\partial \mu_i}{\partial c_j} \nabla c_j. \tag{6.11}$$

Combining Eq. (6.2) and Eq. (6.11), we have

$$\tilde{J}_k = -\sum_{i=1}^{n} L'_{ki} \sum_{j=1}^{n} \frac{\partial \mu_i}{\partial c_j} \nabla c_j. \tag{6.12}$$

Similarly, combining Eq. (6.8) and Eq. (6.11), we get

$$\tilde{J}_k = -\sum_{i=1}^{n} L''_{ki} \left[\sum_{j=1}^{n} \left(\frac{\partial \mu_i}{\partial c_j} - \frac{V_i}{V_n} \frac{\partial \mu_n}{\partial c_j} \right) \nabla c_j \right]. \tag{6.13}$$

One expresses the diffusion flux as

$$\tilde{J}_k = -\sum_{j=1}^{n} D_{kj} \nabla c_j, \tag{6.14}$$

where the diffusivity D_{kj} can be given by [9]

$$D_{kj} = \sum_{i=1}^{n} L'_{ki} \frac{\partial \mu_i}{\partial c_j} = \sum_{i=1}^{n} L''_{ki} \left(\frac{\partial \mu_i}{\partial c_j} - \frac{V_i}{V_n} \frac{\partial \mu_n}{\partial c_j} \right). \tag{6.15}$$

In an n-component system, only $n-1$ concentrations are independent. Thus, one can write the diffusion flux shown in Eq. (6.14) as a function of the $n-1$ independent concentration gradients,

$$\tilde{J}_k = -\sum_{j=1}^{n-1} \tilde{D}^n_{kj} \nabla c_j, \tag{6.16}$$

with

$$\tilde{D}^n_{kj} = D_{kj} - D_{kn} \left(\frac{V_j}{V_n} \right). \tag{6.17}$$

Here, $\tilde{D}^n_{kj}$ is the so-called chemical diffusivity or interdiffusivity, which might be the most convenient quantity for practical use, and can be measured by means of the diffusion couple techniques, together with either the traditional Boltzmann/Kirkaldy–Matano method [14–17] or the more efficient numerical inverse method [18–20]. Moreover, in some actual numerical simulations, the choice of the dependent concentration variable may be varied frequently. By replacing the superscript n with l in Eq. (6.17), one can obtain

$$\tilde{D}^l_{kj} = \tilde{D}^n_{kj} - \tilde{D}^n_{kl} \left(\frac{V_j}{V_l} \right). \tag{6.18}$$

In fact, Fick's first law and its extensions are not very useful for diffusion kinetic simulation. However, when combined with the continuity equation,

$$\frac{\partial c_k}{\partial t} = -\nabla \tilde{J}_k, \tag{6.19}$$

it gives the fundamental differential equation of diffusion

$$\frac{\partial c_k}{\partial t} = \nabla \sum_{j=1}^{n-1} (\tilde{D}_{kj}^n \nabla c_j) \tag{6.20}$$

in a multicomponent system. Eq. (6.20) is usually referred as Fick's second law.

6.2 ATOMIC MOBILITY AND ITS RELATION WITH DIFFUSION COEFFICIENTS

For crystalline phases, the vacancy-exchange mechanism is predominant, i.e., diffusion occurs due to atoms jumping into neighboring vacant lattice sites. Based on a further assumption that there is random distribution of vacant sites at each instance, Andersson and Ågren [9] wrote the diffusional flux of a component in the lattice-fixed frame of reference as

$$J_k = -c_k y_{Va} \Omega_{kVa} (\nabla \mu_k - \nabla \mu_{Va}). \tag{6.21}$$

Here, y_{Va} is the fraction of vacant lattice sites on the sublattice where k is dissolved, and Ω_{kVa} is a kinetic parameter that gives the rate of exchange if there is a vacancy adjacent to a k atom. Assuming that the number of vacancies is everywhere adjusted according to equilibrium, one can have $\nabla \mu_{Va} = 0$, and thus Eq. (6.21) leads to

$$J_k = -c_k y_{Va} \Omega_{kVa} \nabla \mu_k. \tag{6.22}$$

Then, a concept of atomic mobility M_k is then introduced via

$$M_k = y_{Va} \Omega_{kVa}, \tag{6.23}$$

when k is substitutional, and

$$M_k = \Omega_{kVa}, \tag{6.24}$$

when k is interstitial.

The atomic mobility M_k in general depends on composition, temperature, and pressure. By comparing Eqs. (6.22)–(6.24) with Eq. (6.1), one has

$$L_{kk} = c_k M_k, \tag{6.25}$$

when k is substitutional, and

$$L_{kk} = c_k y_{Va} M_k, \tag{6.26}$$

when k is interstitial, and $L_{ki} = 0$ when $k \neq i$.

Transforming to the volume-fixed frame of reference, Andersson and Ågren [9] gave the following equation by combing Eqs. (6.3), (6.15), (6.25), and (6.26):

$$D_{kj} = \sum_{i \in S} (\delta_{ik} - u_k) u_i M_i \frac{\partial \mu_i}{\partial u_j} + \sum_{i \notin S} \delta_{ik} u_i y_{Va} M_i \frac{\partial \mu_i}{\partial u_j}. \tag{6.27}$$

Here, u_k, named as u-fraction of k, is a kind of concentration variable defined by

$$u_k = \frac{x_k}{\sum_{j \in S} x_j}. \tag{6.28}$$

In Eq. (6.27), the quantity $\partial \mu_i / \partial u_j$ is purely thermodynamic and corresponds to the well-known thermodynamic factor. This quantity can thus be evaluated from the thermodynamic description of the system. Moreover, based on Eq. (6.17), we have

$$\tilde{D}^n_{kj} = D_{kj} - D_{kn}, \tag{6.29}$$

when j is substitutional, and

$$\tilde{D}^n_{kj} = D_{kj}, \tag{6.30}$$

when j is interstitial.

In addition to interdiffusivity or chemical diffusivity, another kind of diffusivity is the tracer diffusivity, D^*_k, which can be experimentally measured by radioactive tracers. Assuming the mono-vacancy atomic exchange to be main diffusion mechanism, the tracer diffusivity D^*_k can be related to atomic mobility M_k via Einstein formula:

$$D^*_k = RTM_k. \tag{6.31}$$

In the theoretical treatment of diffusion kinetics, one usually works with diffusion coefficients, which can be evaluated from experimental measurements as mentioned in *Section 6.1*. However, in a realistic multicomponent system a large number of diffusion coefficients have to be evaluated. In a multicomponent system, the description of diffusion for n components requires an $(n - 1) \times (n - 1)$ interdiffusivity matrix, not to mention other diffusivities. Moreover, diffusivities are generally functions of alloy composition, temperature, and even pressure. A database of diffusivities would thus be very complex. A superior alternative is to store atomic mobilities in the database, rather than diffusivities. The number of parameters which need to be stored in a multicomponent system will then be substantially reduced, and the parameters are independent. For instance, only 10 mobilities need to be stored for a 10-component system, instead of a

9×9 interdiffusivity matrix. Various diffusivities, which may be used in the simulations, can be then obtained according to the above-mentioned relations (Eqs. (6.27)–(6.31)).

6.3 MODELS FOR ATOMIC MOBILITY IN DIFFERENT PHASES

6.3.1 Simple Phases

From absolute reaction rate theory arguments, the atomic mobility for an element B in a simple phase (referred to as disordered substitutional/substitutional–interstitial crystalline phase or disordered liquid), M_B, can be expressed as [9]

$$M_B = \frac{1}{RT}\exp\left(\frac{\Phi_B}{RT}\right), \tag{6.32}$$

where R is the gas constant, T the absolute temperature, and $\Phi_B = -Q_B + RT\ln M_B^0$ with Q_B as the activation enthalpy and M_B^0 as a frequency factor for simple phases. Φ_B is generally dependent on the composition, temperature, and pressure. In the spirit of the CALPHAD approach [21], the compositional dependency of Φ_B can be represented by a linear combination of the values at each end-point of the composition space in a Redlich–Kister expansion as

$$\begin{aligned}\Phi_B = &\sum_i x_i \Phi_B^i + \sum_i \sum_{j>i} x_i x_j \left[\sum_{r=0} {}^r\Phi_B^{i,j}(x_i - x_j)^r\right] \\ &+ \sum_i \sum_{j>i} \sum_{k>j} x_i x_j x_k \left[\sum_s v_{ijk}^s \, {}^s\Phi_B^{i,j,k}\right] + \cdots \quad (s = i, j, k)\end{aligned} \tag{6.33}$$

for disordered substitutional solution phase and liquid, while

$$\begin{aligned}\Phi_B = &\sum_i \sum_m y'_i y''_m \Phi_B^{i:m} + \sum_i \sum_{j>i} \sum_m y'_i y'_j y''_m \Phi_B^{i,j:m} \\ &+ \sum_j \sum_m \sum_{n>m} y'_j y''_m y''_n \Phi_B^{j:m,n} + \cdots\end{aligned} \tag{6.34}$$

holds for disordered solid solution phase with interstitial species. In Eq. (6.33), Φ_B^i is the value of Φ_B for pure element i, ${}^r\Phi_B^{i,j}$ is binary interaction parameter (if element i or j is component B) or ternary interaction value (if neither element i nor element j is component B), and ${}^s\Phi_B^{i,j,k}$ is ternary one. Parameter v_{ijk}^s is given by an equation of the form

$$v_{ijk}^s = x_s + (1 - x_i - x_j - x_k)/3. \tag{6.35}$$

While in Eq. (6.34), the subscripts $'$ and $''$ denote the first and second sublattices, respectively. y'_i is the site fraction of element i. $\Phi_B^{i:m}$ is the mobility parameter for endmember, while $\Phi_B^{i,j:m}$ and $\Phi_B^{j:m,n}$ are the interaction mobility parameters.

6.3.2 Phases With Ferromagnetic Ordering

According to Jönsson [22,23], the effect of ferromagnetic ordering on diffusion in fcc alloys can be negligible, while that in bcc alloys should be considered. Jönsson [22] suggested that the ferromagnetic effect on the atomic mobility of substitutional elements can be evaluated through

$$M_B = M_B^p \exp\left[a\xi\left(6 - \frac{Q_B^p}{RT}\right)\right], \tag{6.36}$$

where M_B is the atomic mobility for substitutional element B. M_B^p is the atomic mobility of element B under the paramagnetic condition, which can be evaluated by Eqs. (6.32) and (6.33). Q_B^p is activation energy of element B under the paramagnetic condition. α is a factor of proportionality which was determined by a fit to the experimental information [22], and equals 0.3 for substitutional elements in bcc alloys. ζ denotes the state of magnetic ordering ($0 < \zeta < 1$) defined as

$$\xi = \frac{^{mag}\Delta H(T)}{^{mag}\Delta H(0)}, \tag{6.37}$$

where $^{mag}\Delta H$ is the magnetic enthalpy.

While for the interstitial element (like carbon or nitrogen), a different expression for atomic mobility by taking the ferromagnetic effect into account follows [24]

$$M_I = M_I^p \exp\left[a_2 {}^{mag}\Delta H\left(\frac{6}{Q_I^p} - \frac{1}{RT}\right)\right], \tag{6.38}$$

where M_I^p denotes the atomic mobility of interstitial element under the paramagnetic condition, which can be evaluated on the basis of Eqs. (6.32) and (6.34). Q_I^p is the enthalpy part of M_I^p. Here, α_2 equals 1.8 for carbon while it equals 0.6 for nitrogen [24].

6.3.3 Phases With Chemical Ordering

Since the first experimental measurement of diffusivity in β brass by Kuper et al. [25], it has long been believed that a strong deviation from the usual linear behavior exists when plotting the temperature-dependent diffusivity in an Arrhenius diagram for alloys with a chemical order–disorder transition. Thus, for a chemically ordered phase, the composition dependence of the atomic mobility should also include the effect of chemical ordering. Based on Girifalco's conclusion [26] that the increase of the activation energy due to chemical ordering depends quadratically on the long-range order parameter, Helander and Ågren [27] first proposed a phenomenological model to describe diffusion in binary phases with the bcc-B2 ordering transformation modeled by a two-sublattice (2SL) model $(\mathbf{A}, \mathrm{B})^{\alpha}(\mathrm{A}, \mathbf{B})^{\beta}$ (here, the major constituent on each sublattice is highlighted in bold), where the activation energy can be expressed as

$$Q_{\mathrm{B}} = Q_{\mathrm{B}}^{\mathrm{dis}} + \Delta Q_{\mathrm{B}}^{\mathrm{ord}}, \tag{6.39}$$

where $Q_{\mathrm{B}}^{\mathrm{dis}}$ represents the contribution from the disordered state and can be expressed by Eqs. (6.32)–(6.34), while $\Delta Q_{\mathrm{B}}^{\mathrm{ord}}$ denotes the contribution from chemical ordering. This later quantity is given by an equation of the form:

$$\Delta Q_{\mathrm{B}}^{\mathrm{ord}} = \sum_{i} \sum_{i \neq j} \Delta Q_{\mathrm{B}ij}^{\mathrm{ord}} \left[y_i^{\alpha} y_j^{\beta} - x_i x_j \right], \tag{6.40}$$

where $\Delta Q_{\mathrm{B}ij}^{\mathrm{ord}}$ is a parameter describing the contribution of the component B as a result of the chemical ordering of the $i-j$ atoms on the two sublattices and y_i^{α} is the site fraction of component i on the α sublattice,

$$y_i^{\alpha} = \frac{N_i^{\alpha}}{N_{\mathrm{total}}^{\alpha}} = P_i^{\alpha}, \tag{6.41}$$

in which N_i^{α} is the number of sites on the α sublattice that are occupied by an i atom and $N_{\mathrm{total}}^{\alpha}$ is the total number of sites on the α sublattice. P_i^{α} is the probability of finding an i (i.e., A or B) atom on an α sublattice/site, which relates to the long-rang order parameter S via

$$S = P_A^{\alpha} - P_A^{\beta} = P_B^{\beta} - P_B^{\alpha}. \tag{6.42}$$

Based on Eqs. (6.41) and (6.42), together with the assumption that $\Delta Q_{\mathrm{B}ij}^{\mathrm{ord}} = \Delta Q_{\mathrm{B}ji}^{\mathrm{ord}}$ for the symmetric bcc-B2 model, Eq. (6.40) leads to

$$\Delta Q_{\mathrm{B}}^{\mathrm{ord}} = \frac{\Delta Q_{\mathrm{B}ij}^{\mathrm{ord}}}{2} S^2. \tag{6.43}$$

Eq. (6.43) indicates that the proposed phenomenological model by Helander and Ågren [27] is consistent with the conclusion derived by Girifalco [26]. Later, Helander and Ågren [28] generalized this phenomenological model to multicomponent systems. Moreover, Helander and Ågren [27,28] have also applied the phenomenological model to successfully evaluate the atomic mobility parameters in bcc-B2 alloys of the Fe–Al, Ni–Al, and Fe–Al–Ni systems, from which various diffusion properties were also simulated.

In 2008, Campbell [29] initiated the use of this phenomenological model to study also diffusion in the ordered fcc-$L1_2$ phase in Ni–Al and Ni–Al–Cr alloys, and the corresponding atomic mobility databases were successfully developed. Soon after that, Zhang et al. [30] tried the same phenomenological model to describe diffusion in the ordered fcc-$L1_2$ phase of the Ni–Al system by taking the homogeneity range and defect concentration into account, and as a result the authors were able to use fewer parameters. Though both sets of the resulted atomic mobility databases for the fcc-$L1_2$ phase in the Ni–Al system from Campbell [29] and Zhang et al. [30] can describe the experimental diffusion properties reasonably well, very different model parameters were employed. The seemingly arbitrary selection of model parameters triggered the very recent work by Zhang and Chen [31] on validating whether the phenomenological model developed for the ordered bcc-B2 phase by Helander and Ågren [27] is also applicable for other ordered phases, e.g., fcc-$L1_2$ and fcc-$L1_0$.

Based on a detailed literature survey, Zhang and Chen [31] found three publications by Girifalco and his colleague [32–34] which appeared after [26], from which the developed phenomenological model in 2SL formalism for describing the effect of chemical ordering on diffusion in bcc-B2 ordered alloys was carefully reanalyzed and validated to be applicable in ordered bcc-B2, fcc-$L1_0$, and fcc-$L1_2$ phases. Moreover, the phenomenological model in 2SL formalism was then extended into one in more physically based four-sublattice (4SL) formalism in a thermodynamically consistent way. In accordance with the 4SL thermodynamic model $(\mathrm{A,B})^{\alpha}_{0.25}(\mathrm{A,B})^{\beta}_{0.25}(\mathrm{A,B})^{\gamma}_{0.25}(\mathrm{A,B})^{\delta}_{0.25}$, the chemical ordering contribution to the activation energy of component B in ordered phases in 4SL formalism can be written as [31]

$$\Delta Q_{\mathrm{B}}^{\mathrm{ord\text{-}4SL}} = \sum_{i=A}^{B}\sum_{j=A}^{B}\sum_{k=A}^{B}\sum_{l=A}^{B}\left\{\Delta Q_{\mathrm{B}ijkl}^{\mathrm{ord\text{-}4SL}}\left[y_i^{\alpha} y_j^{\beta} y_k^{\delta} y_l^{\gamma} - x_i x_j x_k x_l\right]\right\}, \tag{6.44}$$

in which $\Delta Q_{\mathrm{B}ijkl}^{\mathrm{ord\text{-}4SL}}$ is the contribution to the activation energy for the diffusion of component B due to the chemical ordering of the *i-j-k-l* atoms. In principle, all the "endmembers" can be evaluated either based on the experimental information or directly from the first-principles calculations. When the first-principles calculations and/or sufficient experimental diffusivities are not available, a concept of "pair activation energy" similar to the thermodynamic treatment [35] was proposed by Zhang and Chen [31] to a first approximation of reasonable "endmembers" in 4SL model. The suggested phenomenological model in 4SL formalism (Eqs. (6.39) and (6.44)), together with the concept of "pair activation energy", was then applied to describe diffusion in the ordered fcc-$L1_2$ phase of the binary Ni–Al system by Zhang and Chen [31]. With the more physically based 4SL model, different ordered structures (e.g., fcc-$L1_0$ and fcc-$L1_2$) with respect to the same disordered phase (e.g., fcc-A1) can be simultaneously described. It was further found that an arbitrary selection of model parameters in the previous 2SL model can be nicely eliminated in the 4SL model.

6.3.4 Intermetallic Compounds

Most of the intermetallic compounds have their own specific crystal structures. In CALPHAD thermodynamic databases, the general sublattice model is commonly used to describe the Gibbs energy of the intermetallic compounds. For the compounds with noticeable homogeneity range, the general diffusion model, similar to that of interstitial solution phase shown in Eqs. (6.32) and (6.34), can be directly used, and thus not stated here.

There is a special class of compounds, for which the homogeneity range is negligible. They are usually named as either "stoichiometric compounds" or "line compounds". Because there is no concentration gradient in such stoichiometric compound and its thermodynamic factor is infinite (at least for the stoichiometric composition), Fick's law is not applicable. In order to circumvent this problem, Du and Schuster [36] introduced the concept of "average thermodynamic factor", and successfully described diffusion growth of various stoichiometric compounds in the Ni–Si system. Later on, Zhang et al. [37] extended the concept of average thermodynamic factor to the compounds with certain homogeneity range, and applied it to model the compounds in the Co–Si system.

For a binary stoichiometric compound A_aB_b, a very narrow homogeneity range can be first introduced by employing a sublattice model $(\mathbf{A}, B)_a(A, \mathbf{B})_b$ (again, the major constituent on each sublattice is highlighted in bold) to describe its Gibbs energy. One should be very careful when the thermodynamic parameters in the sublattice model are chosen. As suggested by Du and Schuster [36], the recalculated invariant reaction temperatures should be within 2 K of the temperatures computed earlier. For this new A_aB_b compound, assuming no volume change, its interdiffusion coefficient with narrow homogeneity range can be related to tracer self-diffusion coefficients via Darken equation

$$\tilde{D}_{A_aB_b} = \left(D^{A*}_{A_aB_b} \cdot x_A + D^{B*}_{A_aB_b} \cdot x_B\right) \cdot \phi, \tag{6.45}$$

where ϕ and x_i ($i = $ A or B) are the thermodynamic factor and mole fraction of component i, respectively. The term ϕ can be given by

$$\phi = 1 + \frac{\partial \ln \gamma_A}{\partial \ln x_A} = 1 + \frac{\partial \ln \gamma_B}{\partial \ln x_B}, \tag{6.46}$$

in which γ_i ($i = $ A or B) is the activity coefficient of component i in the A_aB_b compound. A typical plot for thermodynamic factor ϕ in the A_aB_b compound is presented in Fig. 6.1. As can be seen in the figure, ϕ changes significantly around its stoichiometric composition. In fact, for a strict stoichiometric compound, ϕ is infinite. In this case, the so-called average thermodynamic factor ϕ_{ave} can be proposed for the A_aB_b compound

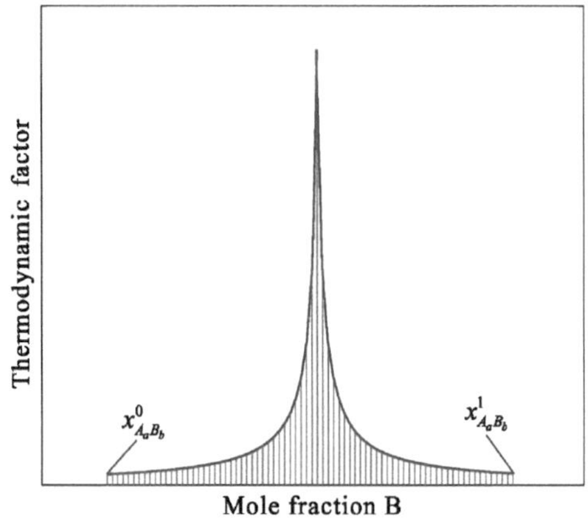

Figure 6.1 A typical plot for thermodynamic factor of A_aB_b compound at a certain temperature. $x^0_{A_aB_b}$ and $x^1_{A_aB_b}$ are the homogeneity limits of the A_aB_b compound due to thermodynamic calculations.

as

$$\phi_{ave} = \left(x^1_{A_aB_b} - x^0_{A_aB_b}\right)^{-1} \cdot \int_{x^0_{A_aB_b}}^{x^1_{A_aB_b}} \phi \cdot dx, \tag{6.47}$$

where $x^0_{A_aB_b}$ and $x^1_{A_aB_b}$ are the homogeneity limits of the A_aB_b compound and can be calculated directly from the thermodynamic parameters. The integral in Eq. (6.47) is dominated by the value of ϕ at the stoichiometric composition. Replacing ϕ in Eq. (6.45) by ϕ_{ave}, one gets the average interdiffusion coefficient of the A_aB_b compound:

$$\tilde{D}_{A_aB_b} = \left(D^{A*}_{A_aB_b} \cdot x_A + D^{B*}_{A_aB_b} \cdot x_B\right) \cdot \phi_{ave}. \tag{6.48}$$

With the known average interdiffusion coefficients, Fick's first and second laws can be solved numerically for the A_aB_b compound in DICTRA software [36,37].

6.3.5 Phases With Polycrystalline Structure

Most of real materials are not perfect single crystals, and contain various defects, like grain boundary, dislocation and surface, which may affect diffusion of species. For instance, the grain boundary usually provides a fast diffusion path. The grain-boundary and dislocation assisted diffusion model has also been implemented in DICTRA software based on an assumption that both mechanisms contribute to the diffusion by using the same frequency factor and a modified bulk activation energy. Grain boundaries and dislocations will accordingly contribute to the overall amount of diffusion according to their weighted factions. The effective atomic mobility including bulk, grain boundary, and dislocation diffusion is formulated as

$$M^{eff} = \frac{\delta}{d} M^{gd} + \rho b^2 M^{disl} + \left(1 - \frac{\delta}{d} - \rho b^2\right) M^{bulk}, \tag{6.49}$$

where δ is the grain boundary thickness (a typical value of 0.5×10^{-10} m is proposed by DICTRA), d the grain size as a function of time and temperature (a typical value of 25×10^{-6} m is proposed by DICTRA), and ρ the dislocation density as a function of time and temperature. b is Burgers vector (a typical value of 1×10^{-10} m is proposed by DICTRA). Here, δ/d can be regarded to be the fraction of grain boundaries in the bulk while ρb^2 the fraction of dislocations in the bulk. M^{bulk} is the atomic mobility in the bulk phase, which can be evaluated on the basis of equations displayed in *Sections 6.3.1–6.3.4*. M^{gd} is the atomic mobility for diffusion in grain boundary, and can be formulated as

$$M^{gd} = M_0^{bulk} \exp\left(\frac{F_{GB} Q^{bulk}}{RT}\right), \tag{6.50}$$

where M_0^{bulk} is the frequency factor in the bulk phase while Q^{bulk} activation energy in the bulk phase. F_{GB} is bulk diffusion activation energy multiplier or ratio between the activation energy for grain boundary and bulk diffusion. A typical value of 0.5 is proposed in DICTRA. M^{disl} is the atomic mobility for dislocation diffusion, and can be formulated as

$$M^{disl} = M_0^{bulk} \exp\left(\frac{F_{Disl} Q^{bulk}}{RT}\right), \tag{6.51}$$

where F_{Disl} is bulk diffusion activation energy multiplier or ratio between the activation energy for grain boundary and bulk diffusion. A typical value of 0.8 is proposed in DICTRA.

6.4 A SIMULATION TOOL FOR DIFFUSION-CONTROLLED TRANSFORMATION – DICTRA

DICTRA is a general software package released by Thermo-Calc Software AB (TC-SAB) for simulation of diffusion-controlled transformations in multicomponent systems. The DICTRA program is based on a numerical solution of the multicomponent diffusion equations in various regions of a material assuming that thermodynamic equilibrium holds locally at all phase interfaces. DICTRA simulations are both time- and space-dependent, but limited to be one-dimensional. Three simple geometries that can be reduced to one-dimensional space include planar, cylindrical, and spherical. Under these geometries, DICTRA can be used to solve certain well-defined problems very accurately. DICTRA is interfaced with Thermo-Calc which handles all the thermodynamic calculations needed by DICTRA. Provided that the necessary thermodynamic

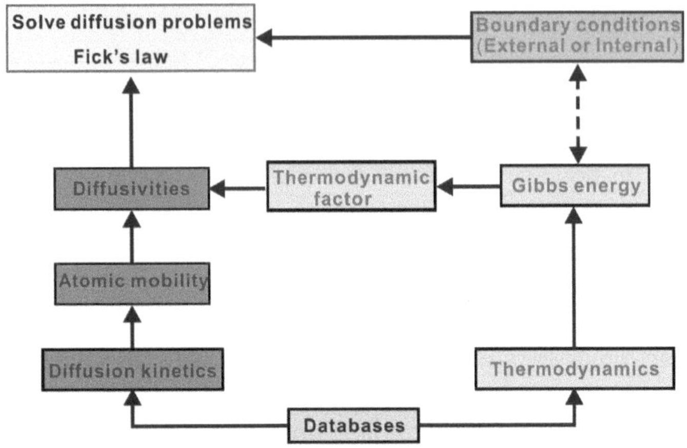

Figure 6.2 (Color online) Work flow for DICTRA simulation.

and atomic mobility databases are available, any number of components may be treated. DICTRA is particularly suitable for treating problems involving a moving boundary, e.g., Stefan problems. The boundary condition may be given in a rather flexible way which enables the user to treat problems of practical interest. The work flow for DICTRA simulation is schematically shown in Fig. 6.2.

The following lists some types of problem which have been nicely solved in DICTRA:

(i) *One-phase problems*, such as homogenization of alloys, carburizing and decarburizing of steel in austenitic state, and so on;

(ii) *Moving boundary problems*, for instance, growth or dissolution of precipitates, coarsening of precipitates, microsegregation during solidification, austenite-to-ferrite transformations, growth of intermediate phases in matrix, nitriding and nitrocarburization, etc.;

(iii) *Diffusion in dispersed systems*, like carburizing of superalloys, carburizing of high-speed steels;

(iv) *Long-range diffusion in multiphase systems*, for example, interdiffusion in compounds, e.g., coating systems;

(v) *Cooperative growth*, like growth of pearlite in alloyed steels;

(vi) *Deviation from local equilibrium*, e.g., calculations under paraequilibrium conditions.

DICTRA is essentially divided into 4 modules, namely, (i) solution of diffusion equations, (ii) calculation of thermodynamic equilibrium, (iii) solution of flux-balance equations, and (iv) displacement of phase interface positions and adjustment of grid points, etc. The detailed introduction of the modules and commands in DICTRA can be found in its manual from the website of TCSAB.

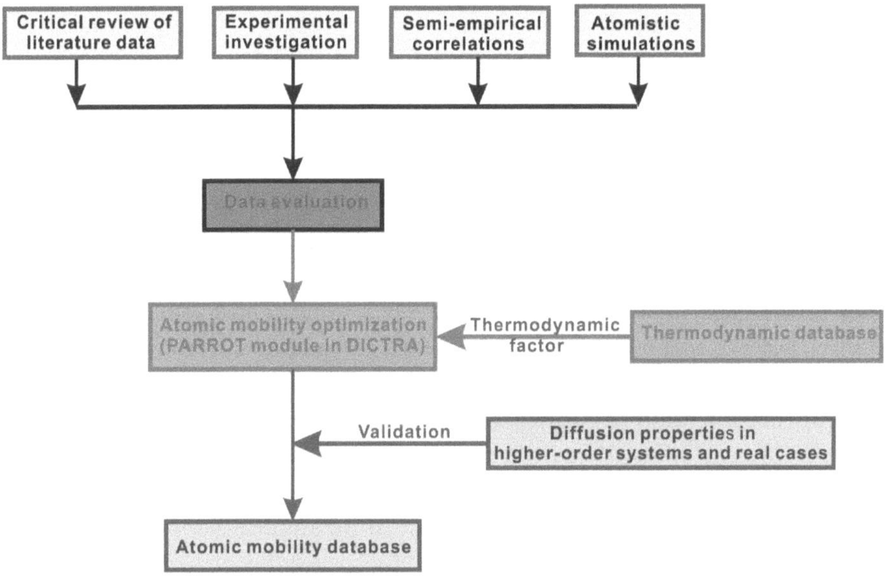

Figure 6.3 (Color online) General strategy for establishment of atomic mobility databases in multicomponent alloys.

In fact, the accuracy of DICTRA simulations is highly dependent on the accuracy of the thermodynamic and kinetic data used but also on different assumptions, such as choice of geometry. The best way to check the simulation accuracy is to compare the simulation results with the experimental data. The DICTRA software is commercial and also updated regularly with some new features. Throughout more than 20 year development, DICTRA now becomes an engineering tool used daily by engineers and scientists all over the world, allowing simulations to be performed with realistic conditions for alloys of practical importance [11]. It gives a possibility of bringing new insight into problems by linking fundamental models to critically assess thermodynamic and kinetic data.

6.5 GENERAL STRATEGY FOR ESTABLISHMENT OF ATOMIC MOBILITY DATABASE IN MULTICOMPONENT ALLOYS

As mentioned above, the accuracy of DICTRA simulations is highly dependent on the quality of the thermodynamic and atomic mobility databases. In order to establish high-quality atomic mobility databases in multicomponent alloys, one may refer to the following general strategies, as also schematically demonstrated in Fig. 6.3:

(i) *A high-quality thermodynamic database*: Before evaluation of any atomic mobility parameters, a high-quality thermodynamic database of the target system (or at least the reasonable thermodynamic descriptions for the target phases) should be available. This is because accurate thermodynamic factors need to be provided during the assessment of atomic mobilities.

(ii) *Critical review of the related literature data in all the subsystems*: A variety of diffusion properties like self-, impurity, tracer, intrinsic, and chemical diffusivities, as well as the composition–distance profiles, in all the subsystems of the target alloy should be collected and critically reviewed. At this step, suitable weighting should be assigned to the data from different sources. Thus, all the experimental factors which may affect the accuracy of the experimental data should be taken into account, such as the purity of initial materials, the accuracy of the experimental technique, single crystal or ploy crystal samples, and so on. For instance, the measured diffusivities in single crystal samples at low temperature should be more accurate than those in ploy crystal samples. Moreover, when different diffusivities from different sources are inconsistent, the following geometric relations [30] can be used as an important criterion: (I) the interdiffusion coefficient in a binary A–B system is equivalent to the impurity diffusion coefficient of A in B when $x_B \rightarrow 1$, and equivalent to the impurity diffusion coefficient of B in A when $x_B \rightarrow 0$; (II) the tracer diffusion coefficient of A is equivalent to the self-diffusion coefficient of A when $x_B \rightarrow 0$, and equivalent to the impurity diffusion coefficient of A in B when $x_B \rightarrow 1$; (III) the tracer diffusion coefficient of B is equivalent to the self-diffusion coefficient of B when $x_B \rightarrow 1$, and equivalent to the impurity diffusion coefficient of B in A when $x_B \rightarrow 0$; and (IV) the main interdiffusion coefficient in a ternary A–B–C system, like $\tilde{D}^C_{AA}$, is equivalent to the interdiffusion coefficient of the binary A–C system when $x_B \rightarrow 0$.

In addition, since some phases lack experimental diffusion properties, one needs to treat them carefully. If those phases are chemically stable, some experiments, e.g., diffusion couples, can be designed to obtain new experimental data. If those phases are metastable, atomistic simulations [38–40], like molecular dynamic (MD), Monte Carlo (MC), or first-principles, and some carefully-developed semiempirical relations [41,42] can be also utilized to provide some important diffusion properties.

(iii) *Optimization of atomic mobility parameters*: The optimization of mobility parameters can be then carefully carried out, e.g., within the PARROT module incorporated in the DICTRA software package by recurrent iteration based on minimization of the residuals between the calculated values and the critically-reviewed diffusion properties. Taking a quaternary system (e.g., A–B–C–D) for example, the evaluation of atomic mobilities can be conducted in general via the following steps:

For the first step, one needs to evaluate mobility parameters in all the 6 boundary binaries (i.e., A–B, A–C, A–D, B–C, B–D, and C–D). In this step, the mobility parameters for the end-members in binary systems are usually first assessed on the basis of the corresponding self- and impurity diffusivities. Then, the tracer, intrinsic and chemical diffusivities in binary alloys can be calculated by using the obtained atomic mobility parameters of the end-members, and directly compared with the literature data. Whether the binary interaction parameters need to be introduced is then determined by judging the degree of the fit between the calculations and the experimental data. After comprehensively comparing with the other diffusion properties, like concentration profiles, the mobility parameters in the binary systems can be temporarily fixed.

For the second step, the optimization of ternary interaction mobility parameters in all the 4 boundary ternaries (i.e., A–B–C, A–B–D, A–C–D, and B–C–D) may be evaluated. Before that, a direct extrapolation from the three binary systems into the ternary system is usually made first for each ternary system. Comparison between the calculated diffusion properties and the experimental data in the ternary system could suggest whether any ternary interaction parameters need to be utilized. Comprehensive comparison with the other diffusion properties, like concentration–distance profiles, should be also made. During this step, if large deviations between the calculated/simulated diffusion properties and the experimental data in the ternary systems are found, and such deviations cannot be made up by using the ternary interaction parameters, one may go back to adjust some mobility parameters in the related boundary binary systems. Sometimes one needs to balance the fit degrees to the experimental data in the binary and ternary systems by carefully adjusting mobility parameters. After that, the mobility parameters in the ternary systems can be temporarily fixed.

For the third step, a direct extrapolation from the four boundary ternaries should be made for the quaternary A–B–C–D system. Comprehensive comparison between the calculated diffusion properties and the experimental data in the quaternary system could suggest whether any quaternary interaction parameters need to be used. In most real cases, there is no need to introduce the quaternary interaction parameters. Again, if large deviations between the calculated/simulated diffusion properties and the experimental data in the quaternary system are observed, one needs to go back to slightly adjust some mobility parameters in the related boundary ternary systems or even binary systems. After balancing among the fit degrees to the experimental data in the all the binary, ternary, and quaternary systems, the mobility parameters in the quaternary system are then temporarily fixed.

(iv) *Validation of atomic mobility database in real cases and/or higher-order systems*: In order to establish a high-quality atomic mobility database in multicomponent alloys, further validation by model-predicting some diffusion properties in several possible real cases and/or in higher-order systems, together with the thermodynamic database, in a comparison with the experimental data is necessary. In some special cases, the mobility parameters in boundary binary or ternary systems might need to be modified to gain a better fit to the experimental data in real cases and/or in higher-order systems. After that, the atomic mobility parameters in the target multicomponent alloys can be fixed, and are ready for later diffusion-kinetic simulations.

6.6 APPLICATIONS OF DICTRA IN DIFFERENT MULTICOMPONENT ALLOYS

Through more than 30 years' effort within CALPHAD community, a variety of commercial and public/open-source thermodynamic databases for different multicomponent alloys have already been constructed, like TC-, TT- and Pan-series commercial databases from different companies [43–45], public unary database from Scientific Group Thermodata Europe (SGTE) [46], public solder database from National Institute of Standards and Technology (NIST), USA [47], etc. The related systems cover almost all the commonly-used alloys, such as Fe-, Al-, Mg-, Ni-, Ti-, TiAl-, Cu-, Si-, Zr-based alloys, solder, cemented carbides, and so on. As for atomic mobility databases, their development is slightly behind that of thermodynamic databases due to the historic reasons. So far, relatively complete atomic mobility databases for Fe-, Al-, Ni-based alloys [43] and cemented carbides [48] have been established, and frequently utilized together with the corresponding thermodynamic databases by engineers and scientists all over the world for computation simulations of material processes and even alloy design. In this section, some typical examples for application of DICTRA simulations in different multicomponent alloys will be presented with an aim to demonstrate how the CALPHAD-type thermodynamic and atomic mobility databases are applied in realistic material processes. A focus on Al alloys and cemented carbides is made, with an extension to a concise summary of other alloys, including Fe-, Ni-based alloys, solder, and oxides.

6.6.1 Al Alloys

The first completed atomic mobility database for Al alloys [49] was established by Prof. Yong Du's research group at Central South University (CSU), China, in cooperation with TCSAB, Sweden, and released by TCSAB as MOBAL-series databases [43] in 2012. Subsequent updates have been made regularly since then [50]. All the important additional elements in commercial aluminum alloys are incorporated in the MOBAL databases, in which the mobility parameters of liquid, fcc_A1, fcc_L1_2, bcc_A2, and bcc_B2 phases are critically assessed by using a combination of key experiments, first-principles calculations, semiempirical relations, and DICTRA modeling. The MOBAL databases should be used together with the TCAL-series thermodynamic databases in DICTRA and other kinetic simulations. It should be noted that during DICTRA simulations no diffusion is considered in the phases without mobility parameters while those phases are automatically entered in the simulations, provided a thermodynamic description for such phases has been retrieved prior to reading data from the mobility database.

The first example for demonstration here is the DICTRA simulation of the differential scanning calorimetry (DSC) curves of two commercial Al–Mg–Si alloys, i.e., A356.0

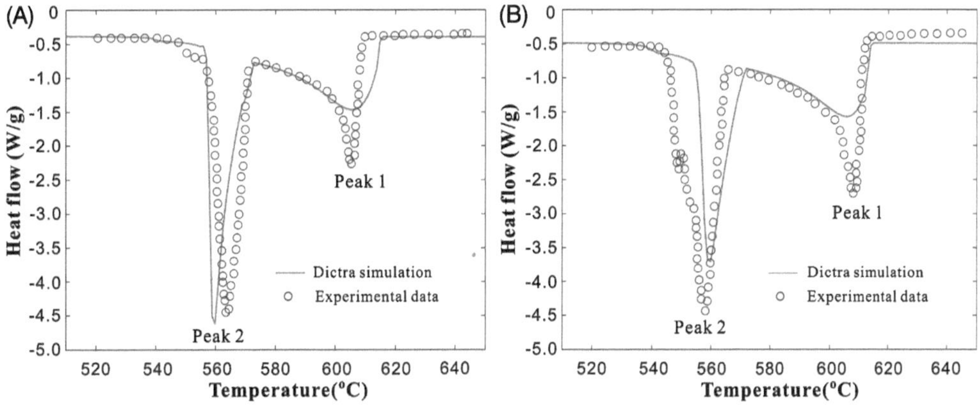

Figure 6.4 (Color online) DICTRA simulated DSC curves for (A) A356.0 (Al–0.3 wt.% Mg–7 wt.% Si), (B) A357.0 (Al–0.7 wt.% Mg–7 wt.% Si) at the cooling rate of 10 K/min, compared with the experimental data [51].

(Al–0.3 wt.% Mg–7 wt.% Si) and A357.0 (Al–0.7 wt.% Mg–7 wt.% Si), during solidification by Tang et al. [51] based on MOBAL mobility database together with TCAL thermodynamic database. According to the experimental information [52], starting solidification temperature and cooling rate for the two alloys are 650°C and 10°C/min, respectively. The Scheil simulation indicates that the major solidification sequence is: liquid → liquid + fcc-(Al) → liquid + fcc-(Al) + (Si) for both A356.0 and A357.0 alloys except that some minor phases (i.e., Al_5FeSi, $Al_8FeMg_3Si_6$ and Mg_2Si) also form at a later stage due to the relatively higher Mg content in A357.0 alloy. Only liquid, fcc-(Al) and (Si) phases were considered in the DICTRA simulation to simplify the simulation process. Moreover, no diffusion is considered in (Si) phase. Typical initial settings for eutectic solidification in DICTRA were established with the entire one-dimensional domain mostly occupied by the liquid phase region, and the other two regions, fcc-(Al) and (Si), with extremely small volume on the separated sides. In a heat flux DSC with pure Al as reference [52], the heat flow may be formulated as:

$$\text{heat flow(w/g)} = \frac{d(f_{\text{liq}} \cdot H_{\text{liq}} + f_{\text{fcc}} \cdot H_{\text{fcc}} + f_{(\text{Si})} \cdot H_{(\text{Si})} - H_{\text{pure-Al}})}{dT} \cdot \frac{\Delta T}{\Delta t}, \tag{6.52}$$

where f_{liq}, f_{fcc}, and $f_{(\text{Si})}$ are the volume fractions of liquid, fcc-(Al), and (Si) phases during the solidification, H_{liq}, H_{fcc}, $H_{(\text{Si})}$, and $H_{\text{pure-Al}}$ are the enthalpies of each phase in J/g, and $\Delta T/\Delta t$ is the cooling rate in K/s. The volume fraction and enthalpy for each phase may be obtained with the DICTRA simulation based on the thermodynamic database combined with the atomic database. The DICTRA simulated DSC curves of A356.0 and A357.0 alloys, along with the experimental data [52], are presented in Fig. 6.4. Primary fcc-(Al) phase solidifies and grows at the expense of the liquid phase as time increases (Fig. 6.4(A)) with the (Si) phase then forming at the eutectic reaction. Peak 1 corresponds to the formation of primary (Al) dendrites, while Peak 2 represents the eutectic reaction. Simulated temperatures for the two peaks are 605.5 and 560°C, very close to the experimental temperatures (605 and 563°C, respectively) [52]. Simulated starting temperature for primary fcc-(Al) phase formation is 616°C, which is 6°C higher than the experiment detected temperature (about 610°C) [52]. This could be explained as follows: (i) the starting temperature for primary fcc-(Al) phase formation is actually corresponding to the liquidus lines determined by the thermodynamic descriptions and nucleation was not considered in DICTRA simulation, and (ii) the driving force for nucleation in typical Al alloys is typically 6°C lower than the liquidus line. A similar situation is observed in A357.0 alloy (Fig. 6.4(B)) with a difference that some minor phases (i.e., Mg_2Si, Al_5FeSi) form just after Peak 2 in the experiment [52], leading to a wide peak temperature range due to peak overlaps, while in the simulation all minor phases were suspended.

The Al–Ag–Cu alloys, as an important type of eutectic alloys, provide an unusual opportunity for studying electronic, size effects, complex microstructure formation, and shape memory effect. The solidification behavior of such eutectic aluminum alloys can also be nicely predicted by using DICTRA simulations based on the thermodynamics and atomic mobility database. Here, two typical alloys are chosen: one is an univariant eutectic alloy with composition Al–3.93 at.% Ag–14.05 at.% Cu, and the other is an invariant eutectic alloy with composition Ag–69.1 at.% Ag–12.8 at.% Cu. For the univariant eutectic alloy, the thermodynamic calculation indicates that this alloy undergoes the univariant eutectic solidification, $L \Rightarrow \alpha\text{-(Al)} + \theta\text{-}Al_2Cu$. The choice of simulation domain and the phases presented in the DICTRA simulation at different stages are schematically shown in Fig. 6.5(A). The initial temperature in the simulation is set to be 1175 K, and the terminal temperature of the simulation is 772 K. A cooling rate of 0.038 K/s is chosen. Again, no diffusion was considered in the compound $\theta\text{-}Al_2Cu$ for simplification. The simulated solute concentration–distance profiles of Cu for different diffusion times at the cooling rate of 0.038 K/s for this Al–3.93 at.% Ag–14.05 at.% Cu univariant eutectic alloy are presented in Fig. 6.5(B), compared with the experimental element distribution along an energy-dispersive X-ray (EDX) line scan across the α-(Al) and $\theta\text{-}Al_2Cu$ lamellae just below the quenching front [53]. As can be seen in the plot, the simulated results agree quite well with the experimental data. As for the invariant eutectic alloy, the invariant reaction, $L \Rightarrow \alpha\text{-(Al)} + \theta\text{-}Al_2Cu + Ag_2Al$, occurs directly from liquid. For such an alloy, four phases are involved, resulting in the difficulty in the 1-D DICTRA simulation domain setting. In order to solve this problem, two liquid regions are set, and all the solid phases are kept in contact with liquid phase, as demonstrated in Fig. 6.6(A). Moreover,

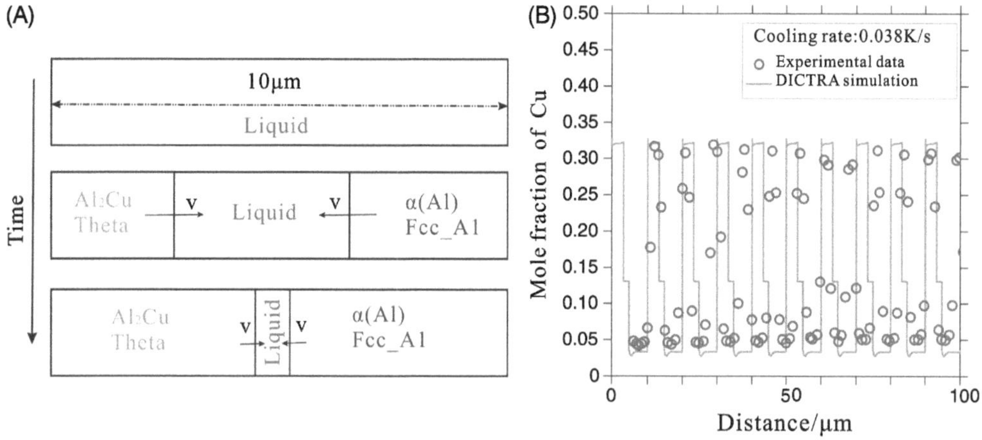

Figure 6.5 (Color online) (A) The choice of domain and the phases present in the simulation at different stages for the univariant eutectic solidification, (B) model-predicted solute concentration-distance profile of Cu at the cooling rate of 0.038 K/s in the Al–3.93 at.% Ag–14.05 at.% Cu univariant eutectic alloy, compared with the experimental element distribution along an EDX line scan across the α-(Al) and Al_2Cu lamellae just below the quenching front [53].

the initial concentrations for each phase are set to be the equilibrium ones according to thermodynamic calculation. The initial temperature in the simulation is set to be 772 K, which is lower by 2 K than the invariant eutectic temperature. The terminal temperature of the simulation is 767 K. The cooling rate for the simulation is also chosen to be 1 K/s. With the well-defined geometry and settings, a successful arrangement for four phases in this one-dimensional DICTRA simulation is then conducted for the first time. The simulated solute concentration profiles of Cu at 5 and 10 s are plotted in Fig. 6.6(B). As shown in the figure, the three solid phases grow at the expense of the liquid phase as time increases, indicating an invariant eutectic solidification.

One typical DICTRA simulation of solid-state transformation is displayed in Fig. 6.7, where a simulation of a diffusion couple for demonstrating diffusion of Si from an Al–0.88 wt.% Si into an Al–0.53 wt.% Fe–1.02 wt.% Mn was presented [54]. The simulation was performed by DICTRA on the basis of MOBAL mobility database together with TCAL thermodynamic database. As can be seen in the figure, the results show the phase fractions after heat treatment at 600°C for 4 days. The Al_6Mn phase has been transformed into alpha-phase ($Al_{15}Si_2(Mn,Fe)_4$) due to the diffusion of Si into the alloy. This is in very good agreement with the experiment by Alexander and Greer [55], who reported a micrograph of the interface where the alpha phase has been formed in the Al–0.53 wt.% Fe–1.02 wt.% Mn alloy due to Si diffusion. Moreover, the width of the alpha-phase zone is approximately 500 μm [55], which is also close to that predicted by the DICTRA simulation.

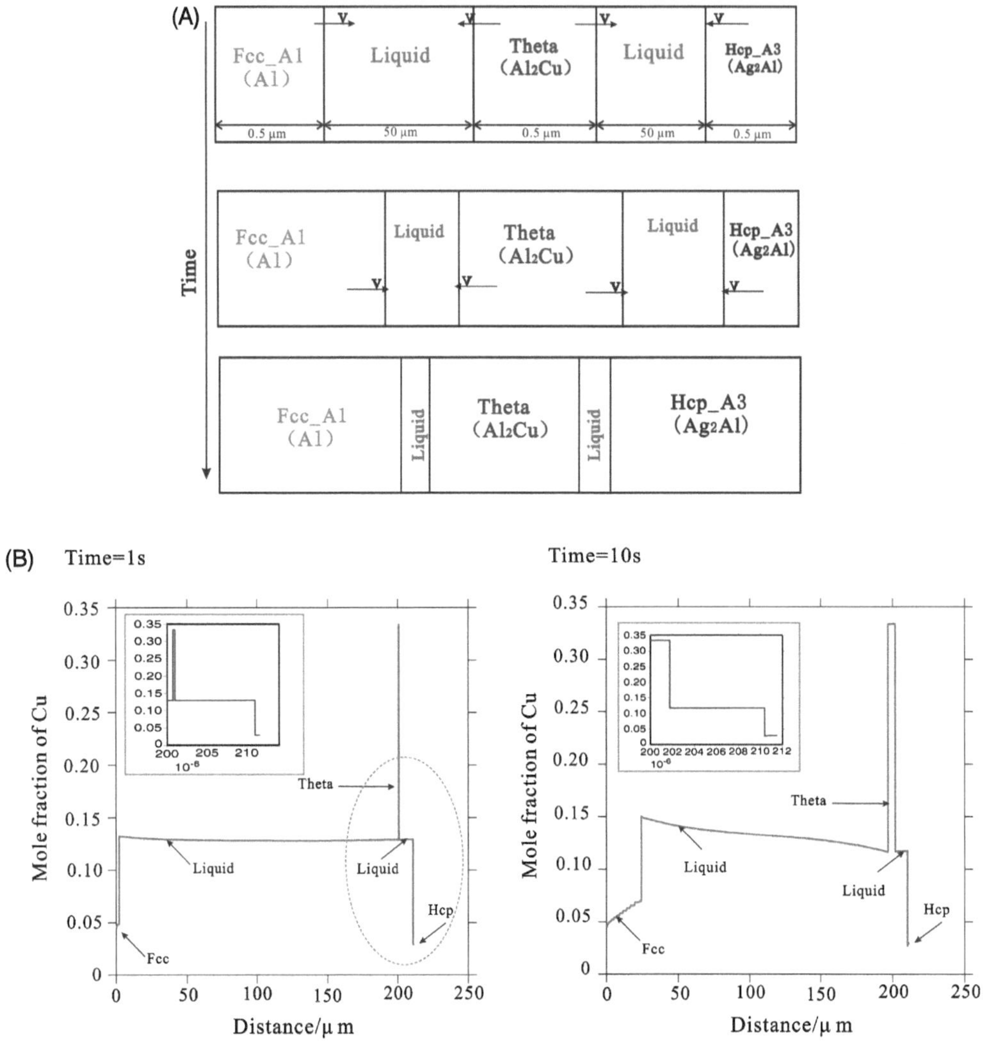

Figure 6.6 (Color online) (A) The choice of domain and the phases present in the simulation at different stages for the invariant eutectic solidification, (B) model-predicted solute concentration-distance profile of Cu for the Ag–69.1 at.% Ag–12.8 at.% Cu invariant eutectic alloy at the cooling rate of 1 K/s from DICTRA simulations with different simulation times (5 and 10 s).

6.6.2 Cemented Carbides

Cemented carbides are of great importance in the manufacturing industry, which have long been used in applications such as cutting, grinding, and drilling. Cemented carbide represents a class of composite materials, consisting of hard and wear resistant carbide and a tough and ductile metal binder, and thus owns a unique combination of hardness

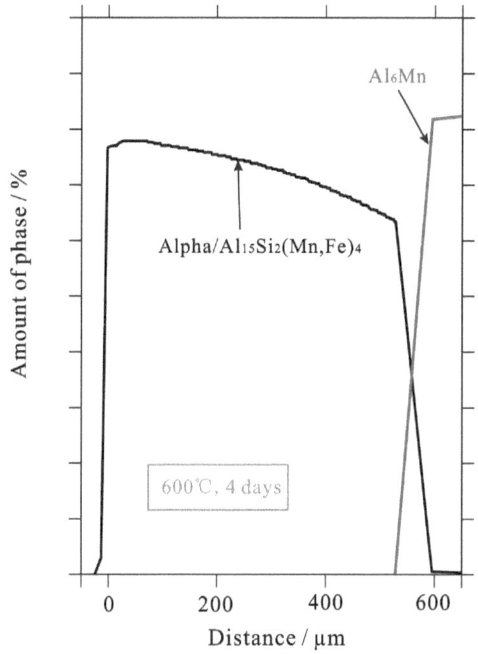

Figure 6.7 (Color online) DICTRA simulation of the phase fraction at different distances from the interface for a diffusion couple with Al–0.99 wt.% Si (below) and Al–0.53 wt.% Fe–1.02 wt.% Mn (above) after heat-treatment at 600°C for 4 days [54].

and toughness. In order to increase the performance and extend the service lifetime of the cemented carbides, graded cemented carbides have been developed. The formation of gradient zone during liquid phase sintering is a diffusion-controlled process, and computer kinetic simulations are indispensable to understand the formation mechanism, optimize technological parameters, and design a new type of graded cemented carbides. As a result of the cooperation between Prof. Yong Du's research group at CSU and Zhuzhou Cemented Carbide Cutting Tools Limited Company in China, a first version of atomic mobility database for cemented carbides, named as CSUDDCC1: Central South University Diffusion Database for Cemented Carbides – version-1, was established and released in 2014 [48]. In CSUDDCC1 mobility database, the core 8 elements in cemented carbides are included, i.e., C, Co, Cr, W, Ta, Ti, Nb, and N, and the mobility parameters of liquid and fcc_A1 are critically assessed by using a combination of key experiments, semi-empirical relations, modified Sutherland equation [56], and DICTRA modeling. The CSUDDCC1 mobility database in combination of CSUTDCC1 (Central South University Thermodynamic Database for Cemented Carbides – version-1) [57] thermodynamic database has been successfully applied to simulate various gradient sintering processes [48], e.g., carburization in the WC–Co cemented carbides, the effect of N partial pressure, Ti and N content on the WC–Ti(C, N)–Co ce-

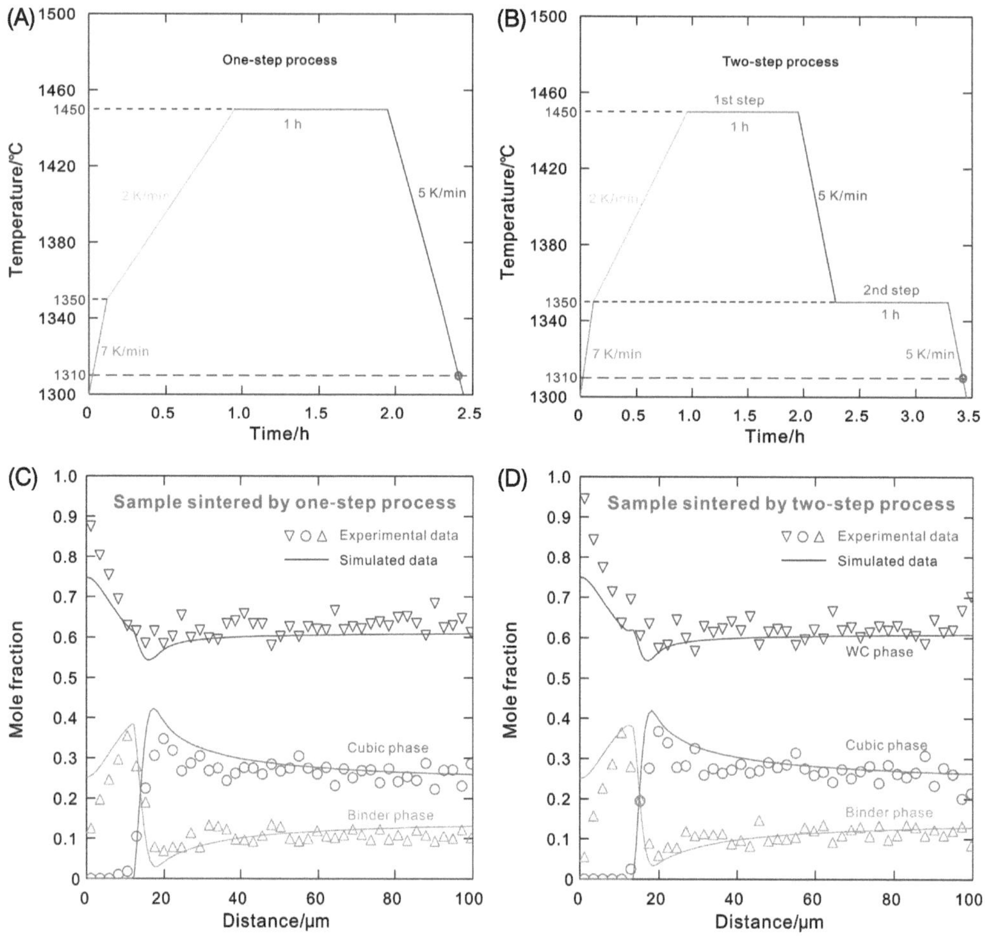

Figure 6.8 (Color online) Schematic diagram of the temperature profiles of the (A) one-step and (B) two-step processes. Model-predicted phase fractions in WC–Ti(C, N)–TaC–Co alloys after sintering by (C) one-step and (D) two-step processes, along with the experimental results [58].

mented carbides, sintering process on the WC–Ti(C, N)–TaC–Co cemented carbides, and Ta, Nb diffusion behavior in the WC–Ti(C, N)–(Ta, Nb)C–Co cemented carbides.

The formation of fcc-free surface layer on cemented carbide can minimize the propagation of thermal cracks formed between the coating and the substrate into the bulk, which is caused by the thermal expansion mismatch between the coating and the substrate. Here, one example on DICTRA simulation of fcc-free surface layer formation on WC–Ti(C, N)–TaC–Co cemented carbides during one-step and two-step sintering in vacuum performed by Chen et al. [58] is demonstrated. The temperature profiles for one-step and two-step sintering in vacuum are schematically shown in Figs. 6.8(A) and 6.8(B). The end-point temperature for the simulation was chosen to be 1310°C. The

CSUDDCC1 mobility database, together with CSUTDCC1 thermodynamic database, was used for the DICTRA simulations. Only diffusion in liquid phase (i.e., binder phase) was considered, while the other two phases (WC phase and cubic phase) acted as point sinks or sources of solute atoms during the simulation. Considering the presence of the dispersed hard phase in the cemented carbides, a model for long-range diffusion occurring in a continuous matrix with dispersed phases [59] was adopted to simulate the gradient layer formation. A labyrinth factor $\lambda(f)$ [60], which has been demonstrated by Mohammadpour et al. [61] to be linear in the volume fraction f of the binder phase, was employed. Thus, the effective interdiffusivity $\tilde{D}^{n}_{kjeff}$ in the binder phase can be given by

$$\tilde{D}^{n}_{kjeff} = \lambda(f)\tilde{D}^{n}_{kj} = f\tilde{D}^{n}_{kj}, \tag{6.53}$$

where $\tilde{D}^{n}_{kj}$ is the true interdiffusivity for binder phase. Thus, the effective interdiffusivity $\tilde{D}^{n}_{kjeff}$ instead of the true interdiffusivity $\tilde{D}^{n}_{kj}$ was then used in the DICTRA simulation. The model-predicted phase fractions in WC–Ti(C, N)–TaC–Co alloys after sintering by one-step and two-step processes along with the experimental results [58] are presented in Figs. 6.8(C) and 6.8(D). As can be seen in the figures, the model-simulated results are in good agreement with the corresponding experimental data. Actually, in addition to the volume fractions of different phases, the elemental concentration profiles in the WC–Ti(C, N)–TaC–Co cemented carbides can be also nicely predicted by means of DICTRA simulations, as displayed in the original publication by Chen et al. [58]. The results indicate that DICTRA simulation is an important tool for the prediction of the influence of compositions and process parameters (i.e., sintering temperature, sintering time, sintering atmosphere, heating rate, and cooling rate) in the industrial production of gradient surface layer on cemented carbides, from which the optimization of the compositions and process parameters and even design of the new cemented carbides become possible.

6.6.3 Miscellaneous

- Iron and steels

Historically, the MOBFE-series mobility databases for iron and steels, initially developed in Royal Institute of Technology (KTH) [43], are the first complete atomic mobility databases in the world. They have been successfully applied in a variety of diffusion-related cases in iron and steels together with TCFE-series thermodynamic databases [43], such as dissolution of cementite and other carbide precipitates in different kinds of steels [62,63], coarsening of carbides in austenite [64], and even when considering the effect of temperature gradient [65], and so on. Considering that most of those cases have been well summarized in the literature [11] for standard examples of DICTRA simulation, there is no need to repeat them in this section. Here, only one very recent work by Chen et al. [66] is given as an example to demonstrate the effect of grain boundary diffusion in the phase transformation from ferrite into austenite in the interface region between ferritic stainless steel interconnect (IC) plates and cell components in planar solid oxide fuel cell (SOEC) or electrolysis cell (SOEC) stacks.

Chromia-forming ferritic stainless steels are widely used as interconnecting materials in SOFC or SOEC stacks due to their high temperature stability and relatively low cost. In a planar stack design, a certain number of planar solid oxide cell (SOC) components are stacked together with shaped ICs in between. An intimate contact between the cell component and the IC plate is therefore essential to endure optimum cell and stack performance. During stack production and operation, interdiffusion across the cell–IC interface takes place, which may introduce adverse effects on the electrical, mechanical, and corrosion properties of the IC plates. For instance, nickel diffuses from the Ni/YSZ fuel electrode or from the Ni contact component into the IC plate, while iron and chromium from the steel diffuse in the reverse direction. Diffusion of Ni into the steel causes transformation of the ferritic bcc phase into the austenitic fcc phase in the interface region, accompanied with changes in volume and in mechanical and corrosion properties of the IC plates. In order to describe such problems, Chen et al. [66] designed a simple austenite (γ)/ferrite (α) diffusion couple, consisting of 13 μm Ni/150 μm $Fe_{0.77}Cr_{0.23}$, annealed at 800°C to mimic the realistic situation. Based on MOBFE-series mobility databases and TCFE-series thermodynamic databases, DICTRA simulations of the interdiffusion process between the two couples were performed. Two simulation cases were considered by Chen et al. [66]. In the first case, only bulk diffusion was considered in both γ and α phases, while both bulk and grain boundary diffusion were considered for the second case. In the second case, a grain size of 50 and 5 μm was adopted for the α and γ phase, respectively, according to Garcia-Fresnillo et al. [67]. The grain boundary thickness δ was set as 0.5×10^{-10} m, as recommended by DICTRA. F_{GB} in Eq. (6.50) was set to be 0.65 for both α and γ phases. According to Chen et al. [66], Ni diffuses into the steel while Fe and Cr diffuse in the reverse direction. During interdiffusion, the γ/α interface moves to α side, indicating the transformation of ferrite phase into the austenite phase. Fig. 6.9 presents the model-predicted composition profiles (in mass fraction) of the diffusion couple of $Ni/Fe_{0.77}Cr_{0.23}$ heat treated at 800°C for periods up to 2000 h, compared with the experimental data. As can be seen in the figure, only the simulation which takes the grain boundary diffusion into account can reproduce the experimental data.

- Ni-based superalloys

Accurate modeling of diffusion kinetics in Ni-based superalloys is important not only for predicting material properties for a given alloy after service or heat treatment, but also for determining coating lifetimes due to interdiffusion. In 2002, Campbell et al. from NIST, USA established a public atomic mobility database for 10-component Ni-

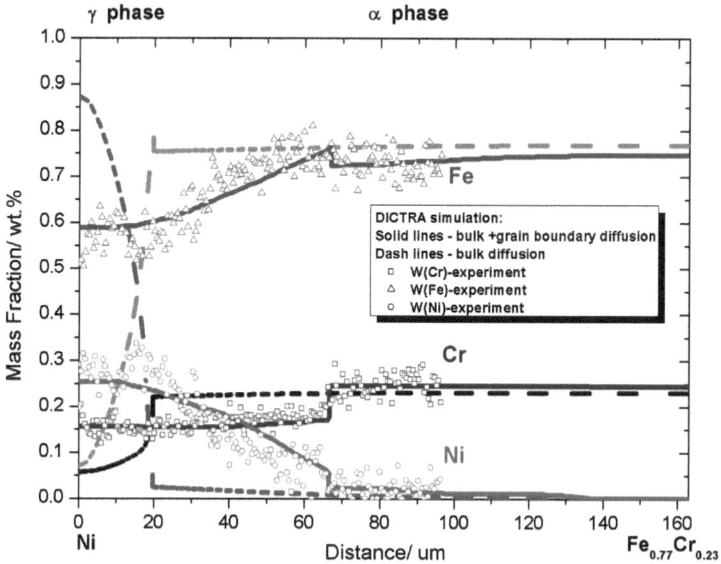

Figure 6.9 (Color online) DICTRA simulated composition profiles (in mass fraction) along the diffusion couple of Ni–$Fe_{0.77}Cr_{0.23}$ heat treated at 800°C for periods up to 2000 h [66]. The solid lines are due to the simulations considering both bulk and grain boundary diffusion, while the dash lines are due to the simulations considering only bulk diffusion.

based superalloys [68], which was later expanded to include Fe and C [69]. Its validation was made in a variety of single- and multiphase multicomponent Ni-based superalloy diffusion couples, like Ni/René-88, IN718/René-88 [69], René-95/René-88, René-95/IN718, IN100/IN718, and IN100/René-88 [70], etc. The concentration–distance profiles, interdiffusion coefficients, phase fraction profiles, as well as location of Kirkendall porosity, were model-predicted by using DICTRA simulations, and compared with the experimental data. However, only mobility parameters for fcc (γ) matrix phase was evaluated in Campbell's databases. This fact limits its wide application because diffusion in the strengthening γ' (ordered fcc-$L1_2$ structure) is also very important.

Recently, a relatively complete atomic mobility database, named MOBNI4, was released by TCSAB [43], in which as many as 25 elements were included, and mobility parameters for liquid, fcc-A1, fcc-$L1_2$, bcc-A2, bcc-B2 were evaluated. The MOBNI4 mobility database can be used together with TCNI8 thermodynamic database [43] for DICTRA and other kinetic simulations. However, it should be noted that diffusion in phases with chemical ordering (e.g., fcc-$L1_2$ and bcc-B2) in MOBNI4 was described by 2SL model, which may lead to arbitrary selection of model parameters as pointed out in *Section 6.3.3*. Therefore, in order to eliminate such an arbitrary selection of model parameters, it is highly recommended that the 4SL model as described in [31] should be used to establish the new-generation atomic mobility database for Ni-based superalloys.

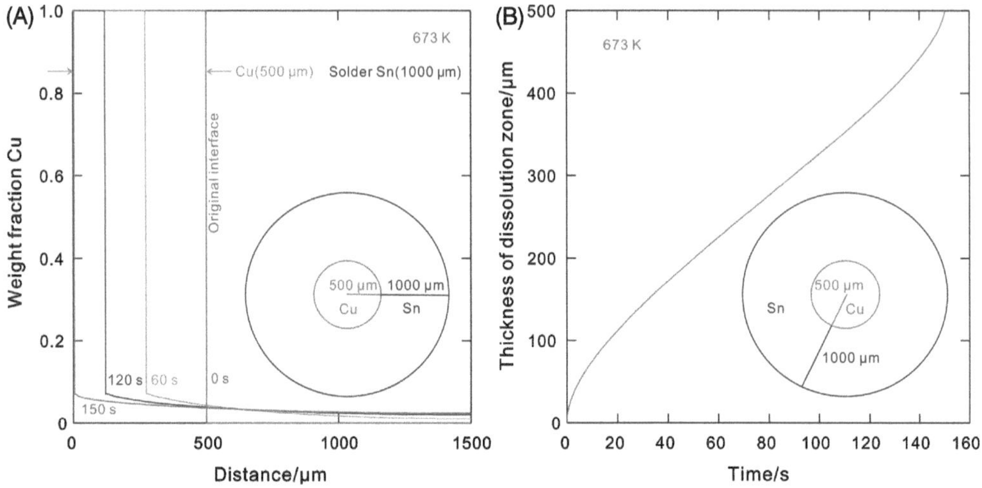

Figure 6.10 (A) Model-predicted composition profiles of Cu in a cylinder solid Cu/liquid solder Sn diffusion couple at 673 K, (B) the thickness of solid Cu dissolute into liquid Sn during reflow processing as according to DICTRA simulation.

- Solder

Very recently, Zhang and his colleagues from CSU, China, have established an atomic mobility database for solder alloys [20,71,72] in which 7 elements (Ag, Cu, Sn, Sb, Bi, Pb, and In) were included, and mobility parameters for liquid and fcc phases were critically reviewed. The developed atomic mobility database for solder can be used to model diffusion kinetics in the related solder alloys together with the pubic NIST solder thermodynamic database [47], such as the dissolution in liquid solder during the reflow process [71].

Fig. 6.10 shows the DICTRA simulation results on the dissolution of a cylinder solid Cu into liquid solder Sn at 673 K. It is assumed that there is no interfacial reaction in this simulation between the Cu-layer and the solder during heating in the solid state. The model-predicted composition profiles of Cu in a cylinder solid Cu/liquid solder Sn diffusion couple at 673 K are presented in Fig. 6.10(A), while the thickness of solid Cu dissolute into liquid Sn during reflow processing is shown in Fig. 6.10(B). As shown in Fig. 6.10(A), more solid Cu dissolute into liquid Sn at 673 K as the reflow time increases. While Fig. 6.10(B) displays that the thickness of dissolution zone is proportional to the square of the diffusion time, indicating that the dissolution behavior is diffusion-controlled.

- Oxides

Oxides usually widely act as functional materials in industry, e.g., in semiconductors, mixed electronic and ionic conductors, fuel cell components, substrates for thin films and thermal barrier coatings in Ni-based superalloys, and so on, and thus are of great

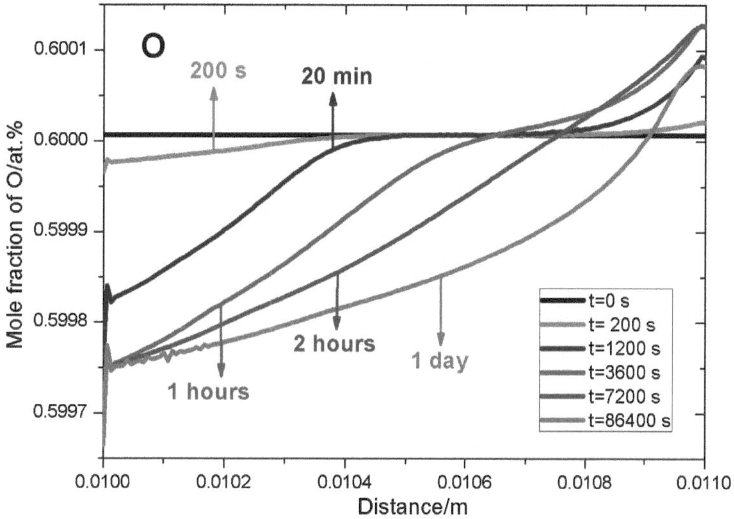

Figure 6.11 DICTRA simulated demixing profiles of O in $LaCoO_3$-δ oxygen membrane under 0.0001/0.21 bar oxygen partial pressure difference at 1073 K.

importance. Similar to metals, vacancy diffusion and interstitial diffusion are the primary diffusion mechanisms in oxides. For metallic systems, lattice defects, i.e., when atoms can jump to the adjacent vacant lattice sites, are important. While for oxides, different kinds of defects where the vacancies are of special importance exist, as postulated by Wagner [73]. Even for a stoichiometric oxide, there will be defects, e.g., Schottky and Frenkel defects.

The first report on DICTRA simulation on diffusion in oxides is from Hallström et al. [74], who successfully modeled diffusion in iron oxides (i.e., magnetite and hematite) by comparing with the experimental data. After that, Hallström et al. [75] studied the oxidation of pure metal Cr, and successfully modeled the Cr_2O_3 growth on pure Cr by using DICTRA. Very recently, Ta et al. due to the joint work between CSU and Technical University of Denmark (*unpublished work*) established the atomic mobility database in $LaCoO_3$-δ phase based on a critical review of various experimental diffusivities. Based on the established mobility database together with the thermodynamic database of the La–Co–O system, DICTRA simulations were performed to investigate the kinetic demixing and decomposition process in $LaCoO_3$-δ oxygen membrane, which was operated under a 0.0001/0.21 bar oxygen partial pressure difference at 1073 K. Fig. 6.11 presents the time evolution curves for demixing profiles of O from $t = 0$ h to $t = 24$ h according to the DICTRA simulation. As can be seen from the plot, the concentration gradient across the region gradually forms, indicating the kinetic demixing process at this stage. Though there is no complete atomic mobility database in multicomponent oxides yet, the above three examples demonstrate that DICTRA will play a more and more important role in diffusion kinetic simulation of oxides in the near future [76].

6.7 FURTHER EXTENSION TO COMPLEX PRECIPITATION AND MICROSTRUCTURE SIMULATION

Based on the reliable thermodynamic and atomic mobility databases, DICTRA has been successfully utilized to perform one-dimensional (1-D) simulation of various diffusion-controlled phase transformation processes in different alloys under the sharp interface and local equilibrium hypothesis, as already demonstrated in *Section 6.6*. All these successful applications make DICTRA as a necessary choice for the engineering-oriented MGI (Materials Genome Initiative) and ICME (Integrated Computational Materials Engineering) projects [7,8] nowadays. However, DICTRA software is still not a complete simulation tool because only some simple geometries, e.g., planar, cylindrical, and spherical ones, which may be transformed into one space variable, can be handled [11]. For the simulation of complex cases, like precipitation from a supersaturated solid matrix, or even two-/three-dimensional (2-/3-D) microstructure evolution during material processes, some other simulation tools/methods, for instance, TC-PRISMA based on sharp-interface assumption [77], and phase-field modeling based on the concept of diffuse interface [78–83], are needed.

6.7.1 Precipitation Simulation – TC-PRISMA

Precipitation hardening is probably the most important strengthening mechanism for various structural alloy materials. Subject to a suitable heat treatment schedule, these materials can be brought into a supersaturated state for small particles of second phases to nucleate from the supersaturated matrix phase and then grow and/or coarsen. The resulted dispersed second phase precipitates can impede the movement of dislocations and lead directly to the precipitation strengthening effect. At the same time, these particles can also pin grain boundaries and stop or slow down grain growth so that grain boundary strengthening effect can be enhanced. The precipitation microstructure is fully characterized by the particle size distribution (PSD), i.e., probability distribution of particle number densities over the particle size dimension. From a known PSD, the particle mean radius, number densities, volume fraction, and the inter-particle spacing can be calculated. As precipitation hardening provides such an effective means to improve material strength, it is highly desirable to have a computational tool to simulate and predict the temporal evolution of PSD in multicomponent industrial alloys.

The formation and evolution of the precipitate microstructure is a diffusion-controlled solid-state phase transformation process. Typically, it possesses three distinctive stages: nucleation, growth, and coarsening, but under certain conditions, these stages may overlap and happen simultaneously. As a widely used simulation tool for diffusion controlled phase transformation, DICTRA is capable of treating the growth or dissolution of one particle or at most several particles in several cells. Moreover, on the basis of the LSW theory, DICTRA can also predict the coarsening of precipitates at the stage when equilibrium volume fraction has almost been reached. While treating growth/dissolution or coarsening separately is of great value in many situations, DICTRA considers no nucleation, and as a result the system or cell size must be known in advance. Furthermore, the number of cells that can be dealt with is too small to represent a full PSD. Therefore, in order to be able to simulate the whole precipitation process involving concurrent nucleation, growth, and coarsening, a new computer program called TC-PRISMA has recently been developed by TCSAB [77].

The TC-PRISMA software is a general computational tool for simulating kinetics of diffusion controlled multiparticle precipitation process in multicomponent and multiphase alloy systems. It is based on the Langer–Schwartz theory [84] and adopts the Kampmann–Wagner numerical approach [85] to solve the continuity equation under the law of mass conservation for the evolution of particle size distribution function. An analytic growth rate model [86] for precipitate particles in multicomponent systems has been developed and implemented. In the aspect of numerical scheme, the continuous PSD is discretized into a sufficiently large number of size classes, and then integrated discretely in the time space. Following the method of characteristics, the number of particles in each size class is kept constant in TC-PRISMA unless particles of the same size have been nucleated.

With TC-PRISMA, the kinetics of concurrent nucleation, growth, and coarsening can be simulated by calculating the evolution of PSD. With the input of CALPHAD-based thermodynamic and kinetic data, interfacial energy, volume, elastic modulus, as well as initial conditions, variations with time of mean radius, number density, volume fraction, and size distribution of precipitate particles, can be simulated. Nucleation rate and precipitate composition can also be obtained during the simulation. The simulation results can be used to understand and guide how to obtain desirable precipitates with certain PSD or to avoid undesirable precipitations during heat treatments of alloys. If coupled with relevant microstructure-property models, the computer program could be even used to estimate the change of mechanical properties of alloys upon precipitation hardening treatment.

Up to now, TC-PRISMA has been successfully applied to simulate the kinetics of precipitation processes in steel, Ni-based superalloys, Al and Cu alloys. Some publications [54,77,87] can be referred to the interested readers. Here, two simple examples are presented below.

The first example is about the precipitation of Al_3Sc in the Al–0.18 at.% Sc alloy, as presented in Fig. 6.12. The evolution of average particle size at several temperatures has been well studied in experiments. For TC-PRISMA simulation, the thermodynamic database TCAL4 and the kinetic database MOBAL3 have been used. The molar

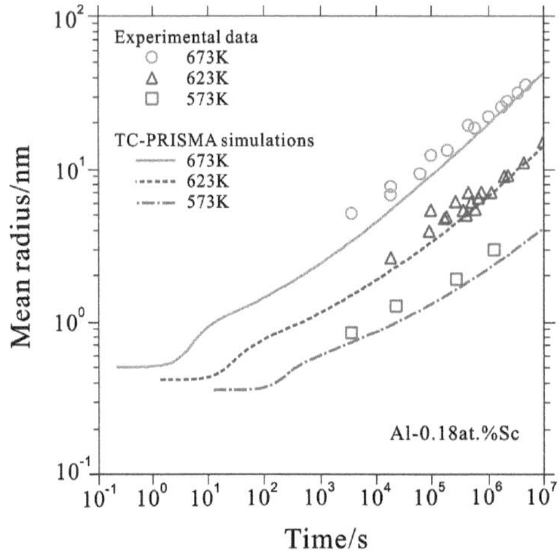

Figure 6.12 Comparison of calculated and experimental results for precipitation of Al_3Sc in Al–0.18 at.% Sc alloy at various temperatures.

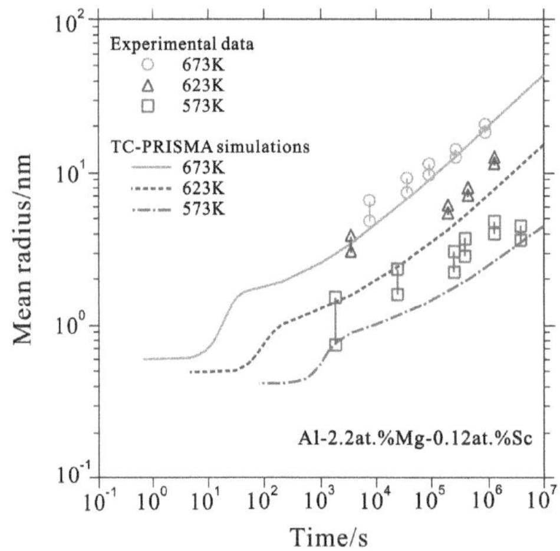

Figure 6.13 Prediction of Al_3Sc precipitation in the Al–2.2 at.% Mg–0.12 at.% Sc alloy [54].

volume data of both matrix and precipitate phases were also available in TCAL4. An experimental interfacial energy, 0.093 J/m^2, was directly adopted for all temperatures. The available nucleation sites were calculated by assuming homogeneous nucleation. As can be easily seen, the simulation results agree very well with experimental observations at all the temperatures.

The second example shown in Fig. 6.13 is the same as the first except that it is for a ternary Al-alloy. The same databases and interfacial energy value chosen in the first example were used in order to examine the prediction capability of the tool based on CALPHAD methodology. The precipitation of Al_3Sc in the Al–2.2 at.% Mg–0.12 at.% Sc alloy was simulated [54]. The agreement between modeling and measurement is excellent for the results at 673 K and acceptable at 623 and 573 K considering that the simulation is a direct extrapolation from the binary system and there were no fitting parameters involved.

6.7.2 Microstructure Simulation – Phase-Field Modeling

The phase-field method, being based on a diffuse representation of the phase boundaries, has emerged as a powerful tool to perform 2-D or 3-D simulation of the microstructural evolution in various materials processes during their lifetime [78–83]. Several widely recognized advantages of this approach include (i) the diffuse interface, with which the explicit tracking of the phase boundaries during complex phase transformations can be avoided, (ii) the description of non-equilibrium states in general, and (iii) the flexible coupling to different external fields, such as external stress, convective transport in liquid, temperature gradient, and so on. Considering the fact that a detail introduction of phase-field method will be made in *Chapter* 7 of this book, only a recent progress on the direct coupling between the phase-field model and the CALPHAD-type databases for a quantitative phase-field simulation of microstructure evolution during materials processes is highlighted here.

In general, in order for a quantitative phase-field simulation, three key points need to be fulfilled, i.e., (i) a quantitative phase-field model, (ii) input of the reliable thermophysical parameters, and (iii) key experimental validation [88]. Nowadays, a quantitative phase-field model is usually available in the literature, and a combination of classic (i.e., X-ray diffraction (XRD), metallography, Scanning Electron Microscope (SEM), Electron Probe Micro Analysis (EPMA)) and modern (i.e., Electron Backscattered Diffraction (EBSD), Transmission Electron Microscope (TEM), Three-dimensional Atom Probe (3DAP)) techniques also can meet the demands of validation of the simulated microstructure. While for input of the reliable thermophysical parameters, it is still a challenge for the phase-field community even today.

The thermophysical parameters needed in the phase-field simulations usually include bulk free energy and its derivatives (viz. diffusion potentials and thermodynamic factors), diffusion coefficients, interfacial energy, interfacial mobility, lattice parameters, elastic constants, viscosities, etc. As CALPHAD technique has been moving toward maturity, CALPHAD databases for thermodynamics, atomic mobility, lattice parameter, elastic constant, viscosity, thermal conductivity in different materials can be established nowadays, from which the respective thermophysical quantity at any composition, temperature and pressure can be retrieved. Thus, the coupling of phase-field model to the CALPHAD databases has become a good standard for dynamically providing the reasonable thermophysical information during the phase-field simulation [89,90].

It is well known that the parameters for each thermophysical property in CALPHAD-type databases are stored phase by phase in accordance with a certain rule. Therefore, the coupling of the phase-field model to CALPHAD databases can be in a straightforward way, for instance, that the complete description for each property in each phase is directly incorporated into the phase-field model. It is true for all the above-mentioned thermophysical properties like diffusivities except for the thermodynamic information. This is because the condition of a giving partitioning (i.e., local equilibrium hypothesis) [91,92] or the condition of equal diffusion potentials (i.e., quasi equilibrium hypothesis) [93,94] is needed in traditional phase-field models for retrieving the phase concentrations and potential over the diffuse interface, resulting in complex thermodynamic calculations. Hence, the traditional coupling approaches include (i) the link to a data file containing the necessary information on free energy, chemical potentials and phase equilibria calculated by an external CALPHAD software package [95,96], (ii) the approximation of Gibbs energy curve/surface due to the CALPHAD calculations via some simple functions [97,98], and (iii) the use of locally linearized phase diagrams by running a CALPHAD software package in parallel to the phase-field simulation via an interface, e.g., TQ interface for MICRESS (MICRostructure Evolution Simulation Software) [94,99]. The three traditional coupling approaches are categorized as the "indirect" method, and have some inherent drawbacks. For instance, the first and second coupling approaches are very difficult to be used in multicomponent alloys, while the dynamic interaction between phase-field code and CALPHAD software package in the third coupling approach is still a time-consuming process. One exception in the traditional phase-field models is the treatment of γ/γ' solid state transformation in, e.g., Ni-based superalloys, during phase-field simulation using the microscopic phase-field model [100]. For the γ/γ' order-disorder transition described by the 4SL model in CALPHAD formalism, the element site fraction in 4SL model can be converted into an order parameter (i.e., a kind of phase-field variable) in the microscopic phase-field model. Thus, the Gibbs energy description for γ/γ' phases can be directly incorporated into the phase-field model. This kind of coupling approach can not only be applicable in multicomponent alloys, but also avoid the complex equilibrium calculations and energy minimization during the simulation. Unfortunately, this coupling approach is only limited up to now to the γ/γ' order-disorder transition.

Very recently, a phase-field model with finite interface dissipation has been developed, and can provide description of various kinetic processes at the mesoscopic scale without restriction to the type of transformation ranging from the chemical equilibrium to strongly non-equilibrium ones [101–104]. The novel feature of this model is that each phase concentration is assigned by a kinetic equation to account for finite interface dissipation instead of applying an extra condition for solute partitioning between the phases as in traditional models. With such a novel feature, the external equilibrium calculations for the partitioning at the interface can be avoided in the phase-field simulations. For the substitutional and interstitial solution phases, their complete Gibbs energy descriptions can be directly incorporated into the phase-field model. While for the intermetallic compounds described by the general sublattice model, the internal minimization of sublattice occupancy and solute evolution during microstructure transformation can be simultaneously realized during the numerical simulation [104]. This treatment leads, in general, to a solution superior to the separate solution of the individual problems, and can be regarded to be a "direct" coupling method. Consequently, the phase-field model with finite interface dissipation opens a novel way to directly couple to the CALPHAD thermodynamic databases. The direct coupling to CALPHAD databases can provide precise thermophysical information for the phase-field simulation in an effective way. It is anticipated that the coupling between the phase-field model with finite interface dissipation and the CALPHAD databases may serve as a standard approach for quantitative phase-field simulation of microstructure evolution in materials processes [88].

Here, two representative examples are presented. One example is the phase-field simulation of microstructure evolution of primary (Al) in an industrial aluminum alloy, Al–1.12Si–0.43Mg–0.17Fe (in wt.%), during solidification. The cooling rate is set to be 0.6 K/s as done in experiment. The CALPHAD thermodynamic and atomic mobility databases, TCAL and MOBAL [43], are coupled with the phase-field modeling for retrieving reliable thermodynamic and diffusion information. The simulation domain is set to be 350 μm × 350 μm, which is consistent with the experimental scale. The finally simulated morphology of primary (Al) in the target alloy is presented in Fig. 6.14(B), which is in very good agreement with the experiment result (shown in Fig. 6.14(A)). The comparison between the phase-field simulated phase transition temperatures and those due to the Scheil simulation is made in Fig. 6.14(C). As can be seen in Fig. 6.14(C), the phase transition temperatures in the phase-field simulation are always lower than those in the Sheil simulation. This fact is reasonable because the nucleation and curvature undercooling are taken into account in the phase-field simulation, while not in the Scheil simulation. Moreover, the solute distributions are visualized in Fig. 6.14(D)–(F), from which the virtual EDX lines (shown in Fig. 6.14(G)–(I)) can be naturally generated. Based on Fig. 6.14(G)–(I), the microsegregation degree of different solutes in the primary (Al) dendrite can be nicely analyzed.

The other example is a 3-D phase-field simulation of precipitation process of metastable needle-like β''-Mg_5Si_6 phase from fcc-(Al) matrix in a Al–0.36% Mg–0.44% Si (in mole fraction) alloy during aging at 473 K by using the phase-field model with finite interface dissipation. In order to describe such solid-state phase transformation, the elastic contribution needs to be considered in the phase-field model with finite interface dissipation. Moreover, the thermodynamic and atomic mobility descriptions in the Al–Mg–Si system by Povoden-Karadeniz et al. [105] are directly coupled with the phase-field model. The simulation domain is set to be 250 × 250 × 250 grids with

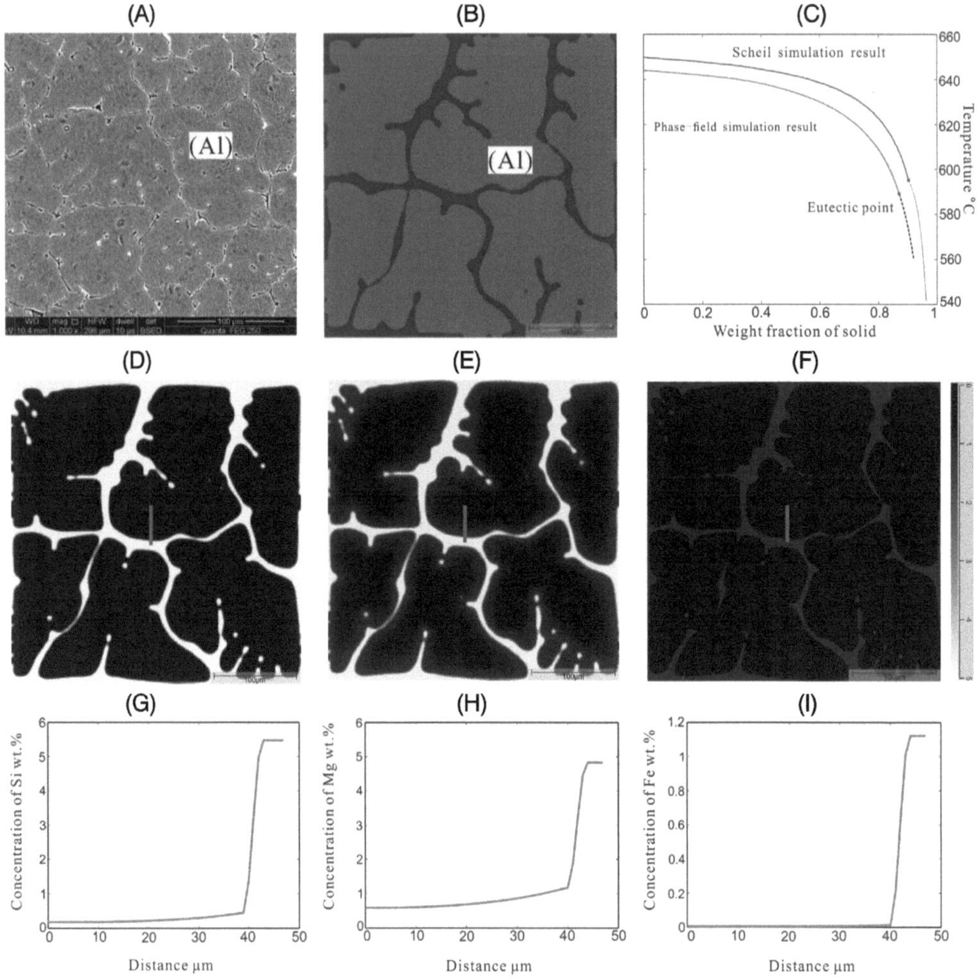

Figure 6.14 (Color online) Application of phase-field simulation in the solidification process of Al–1.12Si–0.43Mg–0.17Fe (in wt.%) alloy: (A) experimental result; (B) phase-field simulation result: red, primary (Al); purple, residual liquid; (C) phase-field simulated solidification process of primary (Al), compared with Scheil simulation result; the distributions of the solutes: (D) Si, (E) Mg, and (F) Fe; The composition–distance curves of (G) Si, (H) Mg, and (I) Fe, corresponding to the green lines in (D), (E), and (F).

interface width of 2.5 nm. The complete periodic boundary conditions were utilized. The 3-D display of the precipitation process of β''-Mg_5Si_6 phase in fcc-(Al) matrix due to the phase-field simulation can be seen in Fig. 6.15. There, 12 spherical particles of β'' are set in the domain with randomly radii and positions (see Fig. 6.15(A)). As time evolves, the β'' particles grow into a needle-like shape along the three $[100]_{Al}$ direc-

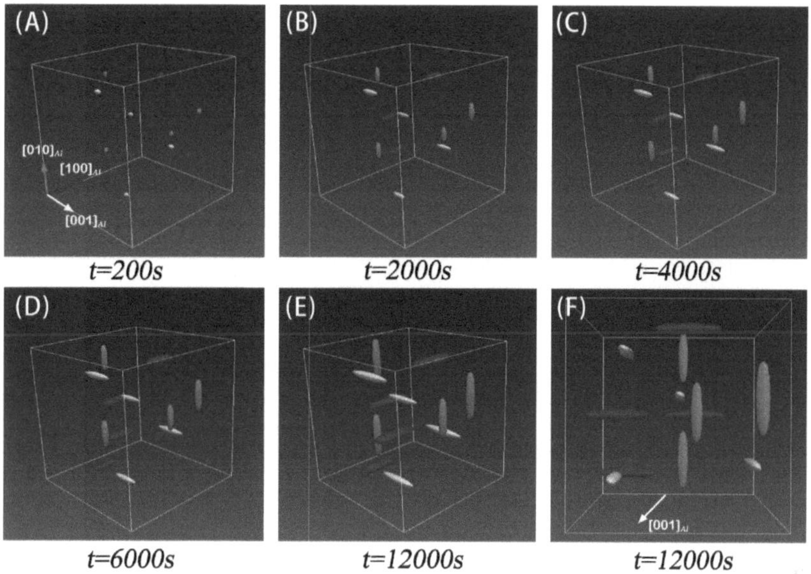

Figure 6.15 (Color online) (A)–(E) 3-D display of precipitation process of metastable β''-Mg_5Si_6 phase from fcc-(Al) matrix in a Al–0.36% Mg–0.44% Si (in mole fraction) alloy during aging at 473 K due to phase-field simulation; (F) $[001]_{Al}$ view of the simulated 3-D microstructure in (E).

tions (see Fig. 6.15(A)–(E)), and point-like in the $[001]_{Al}$ zone view (see Fig. 6.15(F)). The simulated results are in good agreement with the experimental observations in the literature [106].

The authors are grateful for the valuable help from Dr. Kaiming Cheng and doctoral students Mr. Weimin Chen, Ms. Na Ta, Ms. Rong Wang, Ms. Juan Chen, Ms. Huixia Xu, at State Key Laboratory of Powder Metallurgy, Central South University, China, as well as Dr. Ming Chen from Department of Energy Conversion and Storage, Technical University of Denmark, Denmark during preparation of this chapter.

REFERENCES

[1] L. Kaufman, H. Bernstein (Eds.), Computer Calculations of Phase Diagrams, Academic Press, New York, 1970.
[2] N. Saunders, A.P. Miodownik (Eds.), Calphad Calculations of Phase Diagrams: A Comprehensive Guide, Pergamon Materials Series, vol. 1, 1998.
[3] H.L. Lukas, S.G. Fries, B. Sundman (Eds.), Computational Thermodynamics: The Calphad Method, Cambridge University Press, 2007.
[4] Software @Thermo-Calc Software AB, http://www.thermocalc.com/products-services/software/, 2015 (accessed 15.12.15).
[5] Software @CompuTherm LLC, http://www.computherm.com/index.php?route=product/category&path=71, 2015 (accessed 15.12.15).

[6] Products @ FactSage, http://www.crct.polymtl.ca/factsage/fs_family.php, 2015 (accessed 15.12.15).

[7] L. Kaufman, J. Ågren, CALPHAD, first and second generation – birth of the materials genome, Scripta Mater. 70 (1) (2014) 3–6.

[8] G.B. Olson, C.J. Kuehmann, Materials genomics: from CALPHAD to flight, Scripta Mater. 70 (1) (2014) 25–30.

[9] J.O. Andersson, J. Agren, Models for numerical treatment of multicomponent diffusion in simple phases, J. App. Phys. 72 (1992) 1350–1355.

[10] J.O. Andersson, T. Helander, L. Höglund, P. Shi, B. Sundman, Thermo-Calc & DICTRA, computational tools for materials science, Calphad 26 (2) (2002) 273–312.

[11] Annika Borgenstam, A. Engstrom, Lars Hoglund, J. Ågren, DICTRA, a tool for simulation of diffusional transformations in alloys, J. Phase Equilib. Diff. 21 (2000) 269–280.

[12] L. Onsager, Reciprocal relations in irreversible processes. I, Phys. Rev. 37 (4) (1931) 405–426.

[13] J. Ågren, Numerical treatment of diffusional reactions in multicomponent alloys, J. Phys. Chem. Solids 43 (4) (1982) 385–391.

[14] L. Boltzmann, Zur integration der diffusionsgleichung bei variabeln diffusions coefficienten, Ann. Phys. 53 (1894) 959–964.

[15] C. Matano, Relation between diffusion coefficients and concentrations of solid metals, Jpn. J. Phys. 8 (1933) 109–113.

[16] J.S. Kirkaldy, Diffusion in multicomponent metallic systems, Can. J. Phys. 35 (1957) 435–440.

[17] J.S. Kirkaldy, J.E. Lane, G.R. Mason, Diffusion in multicomponent metallic systems, Can. J. Phys. 41 (1963) 2174–2186.

[18] K. Cheng, W. Chen, D. Liu, L. Zhang, Y. Du, Analysis of the Cermak–Rothova method for determining the concentration dependence of ternary interdiffusion coefficients with a single diffusion couple, Scripta Mater. 76 (2014) 5–8.

[19] W. Chen, L. Zhang, Y. Du, C. Tang, B. Huang, A pragmatic method to determine the composition-dependent interdiffusivities in ternary systems by using a single diffusion couple, Scripta Mater. 90–91 (2014) 53–56.

[20] H. Xu, W. Chen, L. Zhang, Y. Du, C. Tang, High-throughput determination of the composition-dependent interdiffusivities in Cu-rich fcc Cu–Ag–Sn alloys at 1073 K, J. Alloys Compd. 644 (2015) 687–693.

[21] J. Ågren, Calculation of phase diagrams: Calphad, Curr. Opin. Solid State Mater. Sci. 1 (3) (1996) 355–360.

[22] B. Jönsson, On ferromagnetic ordering and lattice diffusion—a simple model, Z. Metallkd. 83 (5) (1992) 349–355.

[23] B. Jönsson, Assessment of the mobility of carbon in fcc C–Cr–Fe–Ni alloys, Z. Metallkd. 85 (7) (1994) 502–509.

[24] B. Jönsson, Ferromagnetic ordering and diffusion of carbon and nitrogen in bcc Cr–Fe–Ni alloys, Z. Metallkd. 85 (7) (1994) 498–501.

[25] A.B. Kuper, D. Lazarus, J.R. Manning, C.T. Tomizuka, Diffusion in ordered and disordered copper–zinc, Phys. Rev. 104 (6) (1956) 1536–1541.

[26] L.A. Girifalco, Vacancy concentration and diffusion in order–disorder alloys, J. Phys. Chem. Solids 25 (3) (1964) 323–333.

[27] T. Helander, J. Ågren, A phenomenological treatment of diffusion in Al–Fe and Al–Ni alloys having B2-b.c.c. ordered structure, Acta Mater. 47 (4) (1999) 1141–1152.

[28] T. Helander, J. Ågren, Diffusion in the B2-b.c.c. phase of the Al–Fe–Ni system—application of a phenomenological model, Acta Mater. 47 (11) (1999) 3291–3300.

[29] C.E. Campbell, Assessment of the diffusion mobilites in the γ' and B2 phases in the Ni–Al–Cr system, Acta Mater. 56 (16) (2008) 4277–4290.

[30] L. Zhang, Y. Du, Q. Chen, I. Steinbach, B. Huang, Atomic mobilities and diffusivities in the fcc, L12 and B2 phases of the Ni–Al system, Int. J. Mater. Res. 101 (2010) 1461–1475.
[31] L. Zhang, Q. Chen, Revisiting the phenomenological model for effect of chemical ordering on diffusion, Intermetallics 70 (2016) 72–81.
[32] M. Schoijet, L.A. Girifalco, Theory of diffusion in ordered alloys of the β-brass type, J. Phys. Chem. Solids 29 (3) (1968) 481–495.
[33] M. Schoijet, L.A. Girifalco, Diffusion in the face centered cubic ordered alloys, J. Phys. Chem. Solids 29 (3) (1968) 497–508.
[34] M. Schoijet, L.A. Girifalco, Diffusion in order-disorder alloys. The face centered cubic AB3 alloy, J. Phys. Chem. Solids 29 (6) (1968) 911–922.
[35] N. Dupin, I. Ansara, B. Sundman, Thermodynamic re-assessment of the ternary system Al–Cr–Ni, Calphad 25 (2) (2001) 279–298.
[36] Y. Du, J.C. Schuster, An effective approach to describe growth of binary intermediate phases with narrow ranges of homogeneity, Metall. Mater. Trans. A 32A (2001) 2396–2400.
[37] L. Zhang, Y. Du, Y. Ouyang, H. Xu, X.-G. Lu, Y. Liu, Y. Kong, J. Wang, Atomic mobilities, diffusivities and simulation of diffusion growth in the Co–Si system, Acta Mater. 56 (2008) 3940–3950.
[38] M. Asta, D. Morgan, J.J. Hoyt, B. Sadigh, D. Althoff, D. de Fontaine, S.M. Foiles, Embedded-atom-method study of structural, thermodynamic, and atomic-transport properties of liquid Ni–Al alloys, Phys. Rev. B 59 (1999) 14271.
[39] M. Mantina, Y. Wang, R. Arroyave, L.-Q. Chen, Z.-K. Liu, First-principles calculation of self-diffusion coefficients, Phys. Rev. Lett. 100 (2008) 215901.
[40] D. Zhao, Y. Kong, A. Wang, L. Zhou, S. Cui, X. Yuan, L. Zhang, Y. Du, Self-diffusion coefficient of fcc Mg: first-principles calculations and semi-empirical predictions, J. Phase Equilib. Diff. 32 (2) (2011) 128–138.
[41] J. Askill (Ed.), Tracer Diffusion Data for Metals, Alloys and Simple Oxides, IFI/Plenum, New York, 1970.
[42] L. Zhang, W. Gong, J. Chen, Y. Du, Diffusivities and atomic mobilities in fcc Pt–Al alloys, Calphad 46 (2014) 118–123.
[43] Databases @ Thermo-Calc Software AB, http://www.thermocalc.com/products-services/databases/, 2015 (accessed 15.12.15).
[44] Databases @ CompuTherm LLC, http://www.computherm.com/index.php?route=product/category&path=59, 2015 (accessed 15.12.15).
[45] Databases @ FactSage, http://www.crct.polymtl.ca/fact/documentation/, 2015 (accessed 15.12.15).
[46] A.T. Dinsdale, SGTE data for pure elements, Calphad 15 (4) (1991) 317–425.
[47] Solder database (solder.tdb) available from NIST, USA, http://www.metallurgy.nist.gov/phase/solder/solder.html, 2015 (accessed 15.12.15).
[48] W. Zhang, Y. Du, W. Chen, Y. Peng, P. Zhou, S. Wang, G. Wen, W. Xie, CSUDDCC1 – a diffusion database for multicomponent cemented carbides, Int. J. Refract. Met. Hard Mater. 43 (2014) 164–180.
[49] Y. Du, L. Zhang, S. Cui, D. Zhao, D. Liu, W. Zhang, W. Sun, W. Jie, Atomic mobilities and diffusivities in Al alloys, Sci. China Tech. Sci. 55 (2012) 306–328.
[50] L. Zhang, D. Liu, W. Zhang, S. Wang, Y. Tang, N. Ta, M. Wei, Y. Du, A new diffusivity database for multi-component Al alloys: focusing on ternary systems and its effect on microstructure evolution during solidification, Mater. Sci. Forum 794–796 (2014) 611–616.
[51] Y. Tang, L. Zhang, Y. Du, Diffusivities in liquid and fcc Al–Mg–Si alloys and their application to the simulation of solidification and dissolution processes, Calphad 49 (2015) 58–66.
[52] Q.G. Wang, C.J. Davidson, Solidification and precipitation behavior of Al–Si–Mg cast alloys, J. Mater. Sci. 36 (2001) 739–750.

[53] A. Drevermann, U. Hecht, V. Witusiewicz, B. Böttger, S. Rex, Univariant eutetic growth in ternary Al–Cu–Ag-alloys, Microgravity Sci. Technol. 16 (1) (2005) 45–49.

[54] A. Markström, Y. Du, S.H. Liu, L.J. Zhang, P. Mason, L. Kjellqvist, J. Bratberg, A. Engström, Q. Chen, The development and validation of a new thermodynamic database for aluminum alloys, in: C.E. Suarez (Ed.), Light Metals 2012, TMS (The Minerals, Metals & Materials Society), 2012.

[55] D.T.L. Alexander, A.L. Greer, Nucleation of the Al6(Fe, Mn) to α-Al–(Fe, Mn)–Si transition in 3xxx aluminum alloys. I. Roll bonded diffusion couples, Philos. Mag. 84 (28) (2007) 3051–3070.

[56] W. Chen, L. Zhang, Y. Du, B. Huang, Viscosity and diffusivity in melts: from unary to multicomponent systems, Philos. Mag. 94 (14) (2014) 1552–1577.

[57] Y. Peng, Y. Du, P. Zhou, W. Zhang, W. Chen, L. Chen, Shequan Wang, G. Wen, W. Xie, CSUT-DCC1—a thermodynamic database for multicomponent cemented carbides, Int. J. Refract. Met. Hard Mater. 42 (2014) 57–70.

[58] W. Chen, W. Xie, L. Zhang, L. Chen, Y. Du, B. Huang, G. Wen, S. Wang, Diffusion-controlled growth of fcc-free surface layers on cemented carbides: experimental measurements coupled with computer simulation, Int. J. Refract. Met. Hard Mater. 41 (2013) 531–539.

[59] M. Ekroth, R. Frykholm, M. Lindholm, H.-O. Andrén, J. Ågren, Gradient zones in WC–Ti(C, N)–Co-based cemented carbides: experimental study and computer simulations, Acta Mater. 48 (9) (2000) 2177–2185.

[60] R. Fryholm, M. Ekroth, B. Jansson, A new labyrinth factor for modelling the effect of binder volume fraction on gradient sintering of cemented carbides, Acta Mater. 51 (2003) 1115–1121.

[61] M. Mohammadpour, P. Abachi, N. Parvin, K. Pourazarang, Study of cemented carbonitrides with nickel as binder: experimental investigations and computer calculations, Int. J. Refract. Met. Hard Mater. 31 (2012) 164–170.

[62] Z.-K. Liu, L. Höglund, B. Jönsson, J. Ågren, An experimental and theoretical study of cementite dissolution in an Fe–Cr–C alloy, Metall. Trans. A 22 (8) (1991) 1745–1752.

[63] J. Ågren, Computer simulations of diffusion-controlled precipitation and dissolution of carbide particles in steel, Mater. Sci. Eng. 55 (1) (1982) 135–141.

[64] J. Bratberg, J. Ågren, K. Frisk, Diffusion simulations of MC and M_7C_3 carbide coarsening in bcc and fcc matrix utilising new thermodynamic and kinetic description, Mater. Sci. Technol. 24 (6) (2008) 695–704.

[65] L. Höglund, J. Ågren, Simulation of carbon diffusion in steel driven by a temperature gradient, J. Phase Equilib. Diff. 31 (3) (2010) 212–215.

[66] M. Chen, S. Molin, L. Zhang, N. Ta, P.V. Hendriksen, W.R. Kiebach, Y. Du, Modeling of Ni diffusion induced austenite formation in ferritic stainless steel interconnects, ECS Trans. 68 (1) (2015) 1691–1700.

[67] L. Garcia-Fresnillo, V. Shemet, A. Chyrkin, L.G.J. de Haart, W.J. Quadakkers, Long-term behaviour of solid oxide fuel cell interconnect materials in contact with Ni-mesh during exposure in simulated anode gas at 700 and 800°C, J. Power Sources 271 (2014) 213–222.

[68] C.E. Campbell, W.J. Boettinger, U.R. Kattner, Development of a diffusion mobility database for Ni-base superalloys, Acta Mater. 50 (2002) 775–792.

[69] C.E. Campbell, J.C. Zhao, M.F. Henry, Comparison of experimental and simulated multicomponent Ni-base superalloy diffusion couples, J. Phase Equilib. Diff. 25 (1) (2004) 6–15.

[70] C.E. Campbell, J.-C. Zhao, M.F. Henry, Examination of Ni-base superalloy diffusion couples containing multiphase regions, Mater. Sci. Eng. A 407 (2005) 135–146.

[71] W. Chen, L. Zhang, Y. Du, B. Huang, Diffusivities and atomic mobilities of an Sn–Ag–Bi–Cu–Pb melt, Int. J. Mater. Res. 105 (9) (2014) 827–839.

[72] W. Chen, L. Zhang, Y. Du, B. Huang, Diffusivities and atomic mobilities of Sn–Ag and Sn–In melts, J. Electr. Mater. 43 (4) (2014) 1131–1143.

[73] C. Wagner, Equations for transport in solid oxides and sulfides of transition metals, Prog. Solid State Chem. 10 (1975) 3–16.
[74] S. Hallström, L. Höglund, J. Ågren, Modeling of iron diffusion in the iron oxides magnetite and hematite with variable stoichiometry, Acta Mater. 59 (1) (2011) 53–60.
[75] S. Hallström, M. Halvarsson, L. Höglund, T. Jonsson, J. Ågren, High temperature oxidation of chromium: kinetic modeling and microstructural investigation, Solid State Ionics 240 (2013) 41–50.
[76] S. Hallström, On High Temperature Oxidation Resistance: Towards the Materials Genome of High Temperature Alloys, PhD Thesis, 2014.
[77] Q. Chen, H.J. Jou, G. Sterner, TC-PRISMA Users Guide and Examples, Thermo-Calc Software AB, Stockholm, Sweden, 2011.
[78] W.J. Boettinger, J.A. Warren, C. Beckermann, A. Karma, Phase-field simulation of solidification, Annu. Rev. Mater. Res. 32 (1) (2002) 163–194.
[79] L.-Q. Chen, Phase-field models for microstructure evolution, Annu. Rev. Mater. Res. 32 (1) (2002) 113–140.
[80] N. Moelans, B. Blanpain, P. Wollants, An introduction to phase-field modeling of microstructure evolution, Calphad 32 (2) (2008) 268–294.
[81] I. Steinbach, Phase-field models in materials science, Modeling Simul. Mater. Sci. Eng. 17 (7) (2009) 073001.
[82] Y. Wang, J. Li, Phase field modeling of defects and deformation, Acta Mater. 58 (4) (2010) 1212–1235.
[83] I. Steinbach, Phase-field model for microstructure evolution at the mesoscopic scale, Annu. Rev. Mater. Res. 43 (1) (2013) 89–107.
[84] J.S. Langer, A.J. Schwartz, Kinetics of nucleation in near-critical fluids, Phys. Rev. A 21 (3) (1980) 948–958.
[85] R. Wagner, R. Kampmann, P.W. Voorhees (Eds.), Homogeneous Second-Phase Precipitation, Wiley-VCH Verlag GmbH & Co, KGaA, 2006.
[86] Q. Chen, J. Jeppsson, J. Ågren, Analytical treatment of diffusion during precipitate growth in multicomponent systems, Acta Mater. 56 (8) (2008) 1890–1896.
[87] L.J. Zhang, A. Markström, P. Mason, Y. Du, S. Liu, L. Kjellqvist, J. Bratberg, Q. Chen, A. Engström, TCAL1 and MOBAL2 – the development and validation of new thermodynamic and mobility databases for aluminium alloys, in: ICAA13: International Conference on Aluminum Alloys, 2012.
[88] L. Zhang, Y. Du, Phase-field model for finite interface dissipation: a novel way to directly couple with CALPHAD databases, J. Phase Equilib. Diff. 37 (3) (2016) 259–260.
[89] I. Steinbach, B. Böttger, J. Eiken, N. Warnken, S.G. Fries, CALPHAD and phase-field modeling: a successful liaison, J. Phase Equilib. Diff. 28 (1) (2007) 101–106.
[90] L. Zhang, I. Steinbach, Y. Du, Phase-field simulation of diffusion couples in the Ni–Al system, Int. J. Mater. Res. 102 (4) (2011) 371–380.
[91] A. Karma, Phase-field formulation for quantitative modeling of alloy solidification, Phys. Rev. Lett. 87 (11) (2001) 115701.
[92] J. Tiaden, B. Nestler, H.J. Diepers, I. Steinbach, The multiphase-field model with an integrated concept for modelling solute diffusion, Phys. D: Nonlinear Phenom. 115 (1–2) (1998) 73–86.
[93] S.G. Kim, W.T. Kim, T. Suzuki, Phase-field model for binary alloys, Phys. Rev. E 60 (6) (1999) 7186–7197.
[94] J. Eiken, B. Bottger, I. Steinbach, Multiphase-field approach for multicomponent alloys with extrapolation scheme for numerical application, Phys. Rev. E 73 (2006) 066122.
[95] M. Ode, T. Abe, H. Murakami, Y. Yamabe-Mitarai, H. Onodera, Prediction of microstructure evolution between Ir coating and Ni–Al alloy substrate using a phase-field model, J. Metals 59 (2) (2007) 209.
[96] T. Wang, G. Sheng, Z.K. Liu, L.Q. Chen, Coarsening kinetics of γ' precipitates in the Ni–Al–Mo system, Acta Mater. 56 (19) (2008) 5544–5551.

[97] K. Wu, Y.A. Chang, Y. Wang, Simulating interdiffusion microstructures in Ni–Al–Cr diffusion couples: a phase field approach coupled with CALPHAD database, Scripta Mater. 50 (8) (2004) 1145–1150.
[98] K. Wu, N. Zhou, X. Pan, J.E. Morral, Y. Wang, Multiphase Ni–Cr–Al diffusion couples: a comparison of phase field simulations with experimental data, Acta Mater. 56 (15) (2008) 3854–3861.
[99] U. Grafe, B. Böttger, J. Tiaden, S.G. Fries, Coupling of multicomponent thermodynamic databases to a phase field model: application to solidification and solid state transformations of superalloys, Scripta Mater. 42 (12) (2000) 1179–1186.
[100] J.Z. Zhu, Z.K. Liu, V. Vaithyanathan, L.Q. Chen, Linking phase-field model to CALPHAD: application to precipitate shape evolution in Ni-base alloys, Scripta Mater. 46 (5) (2002) 401–406.
[101] I. Steinbach, L. Zhang, M. Plapp, Phase-field model with finite interface dissipation, Acta Mater. 60 (6–7) (2012) 2689–2701.
[102] L. Zhang, I. Steinbach, Phase-field model with finite interface dissipation: extension to multicomponent multi-phase alloys, Acta Mater. 60 (6–7) (2012) 2702–2710.
[103] L. Zhang, E.V. Danilova, I. Steinbach, D. Medvedev, P.K. Galenko, Diffuse-interface modeling of solute trapping in rapid solidification: predictions of the hyperbolic phase-field model and parabolic model with finite interface dissipation, Acta Mater. 61 (11) (2013) 4155–4168.
[104] L. Zhang, M. Stratmann, Y. Du, B. Sundman, I. Steinbach, Incorporating the CALPHAD sublattice approach of ordering into the phase-field model with finite interface dissipation, Acta Mater. 88 (2015) 156–169.
[105] E. Povoden-Karadeniz, P. Lang, P. Warczok, A. Falahati, J. Wu, E. Kozeschnik, CALPHAD modeling of metastable phases in the Al–Mg–Si system, Calphad 43 (12) (2013) 94–104.
[106] S.J. Andersen, H.W. Zandbergen, J. Jansen, C. Treholt, U. Tundal, O. Reiso, The crystal structure of the β'' phase in Al–Mg–Si alloys, Acta Mater. 46 (9) (1998) 3283–3298.

CHAPTER 7

Phase-Field Modeling as a Method Relevant for Modeling Phase Transformation During Interdiffusion

Abhik Choudhury
Indian Institute of Science, Bangalore, India

7.1 INTRODUCTION

Phase-field models have become an efficient solution method for problems on phase boundary evolution as a result of phase transformations. Classically developed for modeling phase transformations, the models have now evolved for treating problems on electrochemical reactions, mechanical deformation and fracture processes, and also domain evolution in ferro-electric and magnetic materials. The elegance of the phase-field formulations for these moving boundary problems is due to the lack of explicit treatment of the boundary conditions at the moving interface, thereby complicated morphological evolution is easily generalizable in all dimensions.

In this chapter, I will briefly outline the history of the phase-field model development, followed by a description of a generic phase-field model which can be utilized for the modeling of diffusion coupled phase transformations. Subsequently, I will present some exemplary work related to the modeling of basic diffusional phenomena and growth of phases in diffusion couples. Finally, I will present the topic of coupling of the phase-field techniques with CALPHAD databases, which then places phase-field modeling as one of the important methods for simulating multicomponent diffusion driven growth in real alloys.

7.2 PHASE-FIELD MODELS

Phase-field modeling as a continuum method for simulating phase transformations can be traced back to the works of Cahn et al. [1] where the problem of modeling spinodal decomposition in a binary alloy is discussed and [2] where an electromagnetic problem related to the transition to a superconducting phase is described. In both of these formulations, an *order parameter* linked to a material property that shows variance between the different phases is used to create what are called *free energy functionals*, usually consisting of not only the spatially varying order-parameters, but also their gradients and higher order variations as well. Thereafter, equations describing the spatio-temporal

Handbook of Solid State Diffusion, Volume 1
DOI: 10.1016/B978-0-12-804287-8.00007-5

evolution of the order parameters are derived, ensuring monotonic minimization of the functional. The evolution equations have, as their equilibrium solutions, constant values in the bulk phases and a smooth variation of the order parameters over a finite width, which we will refer to as the *interface width*, between the two constant values in the bulk. For conditions out of equilibrium, the evolution equations will describe the motion of these *diffuse interfaces*, indicating phase transformation.

There are important differences here. In [1], the evolution of the order parameter, which is the local composition field, naturally follows from the mass-conservation equation, i.e., the local update of the composition field follows from the divergence of the flux, which is in turn determined by the gradients of the generalized *chemical potential.* In contrast, in [2] the order parameter evolves according to a non-conservation equation, which simply follows from the requirement that the rate of change of the order parameter is directly proportional to the variational derivative of the free-energy functional w.r.t. this parameter. The non-conserved parameter evolution equation is also used for order–disorder phase transition models in solids later in [3,4], where the extent of ordering is used as an order parameter. A description of the different applications of the phase-field model can be found in different reviews [5–7]. In the following, I will present two separate evolution chains of phase-field models for *diffusion-coupled phase transformations*, one pertaining to the solidification community and the other more for simulating solid-state phenomena.

7.2.1 Mesoscopic and Microscopic Formulations

In the context of diffusion-coupled transformations related to growth of phases in solidification, the most commonly used models are related to extensions of the model C of Halerpin and Hohenberg for the solidification of a binary alloy, as mentioned in the review [8], where they couple a non-conserved variable for the transition of the phases with the evolution of another variable that is conserved, such as the concentration or the internal energy. Extensions of this model can be found in Langer [9], Levine [10], and the first numerical simulations by Kobayashi [11]. The order parameter used in these models may be linked to the density field which is inherently varying across the solid–liquid interface, however, this is not essential; and the order parameters can in general be viewed just as distinguishing mathematical functions, which allows for the determination of the spatial distribution of the different phase entities.

Contemporary works in the solid-state community discuss models for precipitate growth [12–19], where the additional factor of elastic coherency stresses at interfaces is incorporated along with mass-diffusion. These models are inherently *microscopic* in their formulation, in that the phenomena that are investigated occur on length and time scales that are comparable to the dimensions of the physical interface. The goal of these models is to effectively capture the physics of the interaction of the different fields in the interface and its influence on growth of the phases. Therefore, the order parameters that are usually chosen in these models are related to the material properties that vary across the interface. The symmetry elements of these properties are directly utilized to formulate the free-energy functional and thereafter the evolution equations are described in a standard manner, following minimization of the functional, using either conserved or non-conserved formulations.

Till now, we have been able to follow the evolution of the phase-field models formulated in general for the modeling of phase evolution under varied conditions of diffusion-coupled phase transformations. It is important, however, to note a classification among the different model formulations. In one of them, which I will refer to as the *microscopic formulation*, the physical phenomena occurring at the scale of the real physical interface cannot be decoupled into the eventual evolution of the phases, thereby these models are generally formulated using a bottom-up approach, where essential details of atomic structure, distribution, and crystal symmetry are utilized in the construction of the free-energy functional. This approach is more prevalent in the solid-state community. The other class of models, which I will refer to as *mesoscopic formulations*, are more prevalent in the solidification community. In these models, formulated for the phase evolution at mesoscopic length scales, the focus is more on retrieving the sharp-interface free-boundary problem governing the interface motion rather than understanding the structure of the physical interface. Thereby, the relevant length scales captured in these formulations are often determined by the morphological length scale that is simulated rather than that of the physical interface. Thus, an additional difference between the two broad classes of formulations is that, while the interface width in the mesoscopic formulations is a degree of freedom which is determined by the morphological length scale to be simulated, in the microscopic formulation it is fixed by the variation of fundamental material properties chosen to differentiate the relevant phases. A commonality of both formulations (microscopic and mesoscopic), however, is that the properties of the interface, such as the interfacial energy and its effect on the interface evolution through the Gibbs–Thomson effect, are effectively captured. Additionally, mass/momentum/energy balances associated with the phase boundary evolution, which are generally referred to as Stefan conditions, are also implicitly satisfied at the interface position.

7.2.2 Mesoscopic Formulations and Quantitative Simulations

Given that the mesoscopic type of formulations are more amenable for simulations of realistic microstructures at the scale of micrometers, the discussion henceforth will be more in line with the structure of mesoscopic phase-field models. Of course, the inherent assumption is that there is complete scale separation between the effects at the scale of the physical interface and those at the scale of the morphological evolution, and thereby, those degrees of freedom are effectively integrated out and captured as quantities such as the interfacial energy and its mobility. In the event that this assumption does not hold, the models will need to be effectively modified. In practice, for performing reasonable simulations at the mesoscopic scale, it is important therefore to upscale the interface width. During the past two decades major developments in the phase-field models have progressed in the direction of making simulations *quantitative* i.e., retrieving the characteristics of the sharp-interface free-boundary problem while using interface widths that are much larger than the capillary length. An immediate implication of this is that the quantities derived from the phase-field simulations have to be independent of the interface width, since it does not figure as a parameter in the sharp-interface free-boundary problem. The challenges in the choice of larger interface widths for performing quantitative simulations arise as modifications to Gibbs–Thomson and Stefan conditions which are proportional to the interface width. These can be further separated into *equilibrium defects* and *kinetic defects*.

7.2.2.1 Kinetic Defects and Thin-Interface Asymptotics

The aggregate of the kinetic defects can be found in the literature related to pure material solidification [20,21] as well as alloys [22–24]. Solutions to these problems have also been proposed in [22–24]. Essentially, the answer to these problems relies on conducting a "thin-interface asymptotics" of the coupled problem of phase evolution and mass/heat diffusion, and determining the conditions such that the leading order error in the interface conditions is proportional to the square of the interface width (ϵ). Using these, simulations can be performed in a range of interface widths such that the results of the simulation are independent of the choice of ϵ (interface width), which is finite but much smaller than the macroscopic length scales, namely the diffusion and thermal lengths. An additional outcome of these solutions is that the relevant asymptotic limits such as the kinetic coefficients and their dependence on interface width can be derived, which allows for determining the conditions such that diffusion controlled phase transformations can be performed [22]. This is quite important for comparison not only with experiments but also with analytical formulations, which are in most cases written in the diffusion controlled regime.

7.2.2.2 Equilibrium Defects

The second problem of *equilibrium defects* has more to do with the appropriate formulation of the free-energy functionals with respect to the choice of the correct state variables. A review of this problem which is prevalent for alloys is present in [25–28, 24,29,30], which essentially states that, if the bulk free-energy density is written as an interpolation of the individual free-energies at a given composition, which is equivalent to choosing the composition as a state variable, there are two resultant problems. First, the equilibrium interface has a profile determined by the properties of the chemical free-energy density of the phases and, as a result, the interfacial energy and the interface widths, which were meant to be degrees of freedom for the *mesoscopic model formulations*, cease to be such. Second, the interfacial energies become functions of the interface width, and thereby quantitative simulations cannot be performed. An associated problem, which is also the most restrictive, is that the resultant interface width, which is now also a function of the chemical system, becomes far too small for certain alloys making the method computationally inefficient. The solution to this problem relates to the correct choice of the relevant state variable for alloy phase transformation problems, as described in [26,31,28]. Here, the resultant free-energy density is written as an interpolation of the individual free-energy densities of the relevant phases at their respective phase compositions. These phase compositions are related by the condition that they satisfy the local mass-balance condition and, secondly, are such that the diffusion potentials of the phases are equal at those compositions. This formulation reintroduces the interface energy and interface width as two independent parameters and degrees of freedom which can be chosen as per the simulation conditions, and thus retains the spirit of the mesoscopic phase-field formulation.

7.2.2.3 Grand-Potential Formalisms

A thermodynamically equivalent, more elegant and computationally useful formulation, which is presented in [29,30] as a grand-canonical formulation, is actually choosing the *diffusion potential* that relates the phase-composition fields in the previous formulation [26,31,28], as the state variable, and recasting the equation for the evolution of the concentration field into that of the temporal evolution of the diffusion potential. This reduces computational memory requirements as well as draws parallels with the solidification models for pure materials, where a similar evolution equation for the temperature field is formulated from the conservation equation for the internal energy. The analogy is not surprising, if one relates to the condition for equilibrium, i.e., equal free-energies that occur at the same temperature for both phases in pure materials, while the corresponding condition for the alloys is that of equal grand canonical potentials which appear at equal diffusion potentials in the phases. Further developments using the diffusion potentials as state variables for simulating local non-equilibrium conditions arising from finite energy dissipations at the interface have also been proposed in [32].

In the subsequent sections, therefore, I will base my discussions on the usage of phase-field models for problems in diffusion and in particular diffusion-coupled phase transformations with respect to the particular formulation presented in [30] and its variations. This formulation has all the features of *mesoscopic formulations* discussed above.

7.3 PHASE-FIELD MODEL: RELEVANT TO MODELING PHASE TRANSFORMATIONS IN DIFFUSION COUPLES

In this section we describe a multiphase, multicomponent model based on a grand-potential formalism for modeling generic diffusion-coupled phase transformations. We utilize a vector phase-field formulation, where the variable $\boldsymbol{\phi} = \{\phi_1 \dots \phi_N\}$ is the phase-field vector containing the volume fraction of the N phases in the system. The vector $\tilde{\boldsymbol{\mu}}^m$ is the molar diffusion potential vector consisting of the K independent diffusion potentials. These vectors are utilized for formulating the grand-potential functional Ψ as a function of temperature T as

$$\Psi\left(\tilde{\boldsymbol{\mu}}^m, T, \boldsymbol{\phi}\right) = \sum_{\alpha=1}^{N} \Psi^{\alpha}\left(\tilde{\boldsymbol{\mu}}^m, T\right) h_{\alpha}(\boldsymbol{\phi}), \tag{7.1}$$

where $h_{\alpha}(\boldsymbol{\phi})$ is the interpolation polynomial given as $h_{\alpha}(\boldsymbol{\phi}) = \phi_{\alpha}^2 (3 - 2\phi_{\alpha}) + 2\phi_{\alpha} \times \sum_{\beta \neq \alpha, \gamma \neq \alpha, \beta < \gamma}^{N,N} \phi_{\beta}\phi_{\gamma}$, which ensures that $\sum_{\nu=1}^{N} h_{\nu}(\boldsymbol{\phi}) = 1$. The phase concentration expressions can be derived in terms of the diffusion potential using the relation $c_i^{\alpha}\left(\tilde{\boldsymbol{\mu}}^m, T\right) = -\dfrac{\partial \Psi^{\alpha}\left(\tilde{\boldsymbol{\mu}}^m, T\right)}{\partial \tilde{\mu}_i^m}$, where V_m is molar volume (which is assumed equal for all the components). This completes the description of the bulk thermodynamics of the chemical systems under consideration.

The grand-potential functional Ω can then be written following [30] as

$$\Omega = \int a(\boldsymbol{\phi}, \nabla\boldsymbol{\phi}) + w(\boldsymbol{\phi}) + \Psi\left(\tilde{\boldsymbol{\mu}}^m, T, \boldsymbol{\phi}\right) dV, \tag{7.2}$$

where $a(\boldsymbol{\phi}, \nabla\boldsymbol{\phi})$ represents the gradient energy density and w is the surface potential density as a double obstacle function. Their respective forms are described in [33,34].

The evolution equations for the N phase-field variables $\phi_{\alpha}, \alpha = \{1, \dots, N\}$ can then be derived using the non-conserved dynamics [3] as

$$\begin{aligned} \tau\epsilon\frac{\partial \phi_{\alpha}}{\partial t} = \epsilon\left(\nabla \cdot \frac{\partial a(\boldsymbol{\phi}, \nabla\boldsymbol{\phi})}{\partial \nabla\phi_{\alpha}} - \frac{\partial a(\boldsymbol{\phi}, \nabla\boldsymbol{\phi})}{\partial \phi_{\alpha}}\right) \\ - \frac{1}{\epsilon}\frac{\partial w(\boldsymbol{\phi})}{\partial \phi_{\alpha}} - \frac{\partial \Psi\left(\tilde{\boldsymbol{\mu}}^m, T, \boldsymbol{\phi}\right)}{\partial \phi_{\alpha}} - \Lambda, \end{aligned} \tag{7.3}$$

where Λ is the Lagrange parameter to maintain the constraint $\sum_{\alpha=1}^{N} \phi_{\alpha} = 1$ (owing to the order parameter being viewed as a volume fraction) and the parameter τ is written as $\dfrac{\sum_{\alpha<\beta}^{N,N} \tau_{\alpha\beta}\phi_{\alpha}\phi_{\beta}}{\sum_{\alpha<\beta}^{N,N} \phi_{\alpha}\phi_{\beta}}$, where $\tau_{\alpha\beta}$ is the relaxation constant of the $\alpha\beta$ interface, ϵ is a parameter, which is a part of the functions a and w and has the units of length.

For a two-phase interface in one dimension, there are certain properties that can be defined. The bracketed and the following term on the right-hand side of Eq. (7.3) are the variational derivatives of the part of the functional which determines the properties of the equilibrium interface. At equilibrium the last term on the right-hand side of Eq. (7.3) is zero, and the equilibrated profile of the phase-fields of the phases in one dimension can be derived by solving the differential equation formed by equating the bracketed and the following term on the right-hand side of Eq. (7.3). The resultant equilibrated profile when inserted into the functional in Eq. (7.2) yields as a result the interfacial energy for a single interface between two phases. For the particular choice of the obstacle potential that is used in this formulation [33,34], the equilibrated profile is a truncated *sinus* profile, where the range is limited between 0 and 1. Further, the interface width corresponding to this variation can be derived as $\pi^2\epsilon/4$.

For a general multiphase, multicomponent system, the evolution equations for the components of the diffusion potential $\tilde{\boldsymbol{\mu}}^m$ can be expressed in vector–matrix form by Eq. (7.4), where {} represents a vector of dimension K while [·] denotes a matrix of dimension $K \times K$. Here, $M_{ij}(\boldsymbol{\phi})$ is the atomic diffusion mobility, where the individual phase mobilities are interpolated as

$$M_{ij}(\boldsymbol{\phi}) = \sum_{\alpha=1}^{N} M_{ij}^{\alpha} g_{\alpha}(\boldsymbol{\phi}).$$

Each of the M_{ij}^{α} is defined using the expression

$$M_{ij}^{\alpha} = \sum_{k=1}^{K} D_{ik}^{\alpha} \frac{\partial c_k^{\alpha}(\tilde{\boldsymbol{\mu}}^m, T)}{\partial \tilde{\mu}_j^m}.$$

The function $g_{\alpha}(\boldsymbol{\phi}) = \phi_{\alpha}$ interpolates the mobilities. D_{ij}^{α} are the interdiffusivities in each phase α, and $\dfrac{\partial c_k^{\alpha}(\tilde{\boldsymbol{\mu}}^m, T)}{\partial \tilde{\mu}_j^m}$ is the susceptibility. The evolution equations for the K diffusion potential vectors can thus be written as

$$\left\{\frac{\partial \tilde{\mu}_i^m}{\partial t}\right\} = \left[\sum_{\alpha=1}^{N} h_{\alpha}(\boldsymbol{\phi}) \frac{\partial c_i^{\alpha}(\tilde{\boldsymbol{\mu}}^m, T)}{\partial \tilde{\mu}_j^m}\right]_{ij}^{-1} \left\{\nabla \cdot \sum_{j=1}^{K} M_{ij}(\boldsymbol{\phi}) \nabla \tilde{\mu}_j^m - \right.$$
$$\sum_{\alpha}^{N} c_i^{\alpha}(\tilde{\boldsymbol{\mu}}^m, T) \frac{\partial h_{\alpha}(\boldsymbol{\phi})}{\partial t} -$$
$$\left. \frac{\partial T}{\partial t} \sum_{\alpha}^{N} \left(\frac{\partial c_i^{\alpha}(\tilde{\boldsymbol{\mu}}^m, T)}{\partial T}\right)_{\tilde{\boldsymbol{\mu}}^m} h_{\alpha}(\boldsymbol{\phi}) \right\}. \tag{7.4}$$

Some interesting points should be mentioned about the latter equation. In the braces {}, the first term determines the diffusional flux depending upon the diffusion potential gradients, the second term is the solute redistribution term due to interface migration, while the last term accounts for non-isothermal processes. For the discussion in this particular chapter, we are going to treat isothermal situations when this last term in

{} drops out. Following the same idea, we will also drop the temperature argument T from the functionals. At steady state in one dimension, during diffusion-coupled phase transformation, the asymptotic value of the diffusion potential at the interface is at equilibrium. At this state the integral of the diffusional flux across the interface balances the solute redistribution, which is the Stefan boundary condition in the sharp interface free-boundary problem.

Simultaneous solution of Eqs. (7.4) and (7.3) will describe the particular diffusion-coupled phase transformations. Appropriate conditions for the choice of the relaxation parameters such that diffusion-controlled phase transformations can be performed are derived using the results of the thin-interface asymptotics which has been performed for the model in [30].

In the subsequent sections, I will show applications of the phase model for modeling diffusional phenomena in diffusional couples, thus highlighting the usefulness of the method for such problems.

7.4 MODELING KIRKENDALL EFFECT IN A BINARY ALLOY

In the following I will present a brief description of the Kirkendall effect and thereafter depict how the phase-field model described above can be used to model phenomena in diffusion couples. I will consider modeling the following three different diffusion cases: (a) of a couple formed between the same phases, (b) of a couple formed between two different phases, and (c) of a multicomponent diffusion couple formed between two phases.

7.4.1 Kirkendall Effect in a Binary Alloy

Let us begin with a brief description of the phenomena of Kirkendall marker motion [35,36] in a binary alloy with A and B as the constituent elements, with D_A and D_B as the intrinsic diffusivities. In a diffusion couple created between an A-rich region and a B-rich region, diffusion of A and B occurs between the regions following the composition gradients of the alloy as:

$$J_A = -D_A \frac{dc_A}{dx}, \tag{7.5}$$

$$J_B = -D_B \frac{dc_B}{dx}, \tag{7.6}$$

where c_A and c_B represent the concentrations of A and B in (mol/m^3), with D_A and D_B as the intrinsic diffusivities, and J_A and J_B as the respective intrinsic fluxes, respectively. In the laboratory frame of reference, an unequal intrinsic flux of the components will bring about a displacement of the material (with no vacancies) with a velocity being a function of space and time. Therefore, in the laboratory frame, the fluxes can be expressed as:

$$\tilde{J}_A = J_A + v c_A, \tag{7.7}$$
$$\tilde{J}_B = J_B + v c_B, \tag{7.8}$$

where v is the rate of displacement of material. The rate of change of concentrations of A and B can then be expressed using mass-conservation equations as:

$$\frac{dc_A}{dt} = -\frac{d}{dx}\left(\tilde{J}_A\right), \tag{7.9}$$
$$\frac{dc_B}{dt} = -\frac{d}{dx}\left(\tilde{J}_B\right). \tag{7.10}$$

Given that the partial molar volume of A is V_A and that of B is V_B, together with the relations $X_A = c_A V_A$, $X_B = c_B V_B$, and $X_A + X_B = 1$, we can deduce from the previous relation that

$$\left(V_A \tilde{J}_A + V_B \tilde{J}_B\right) = 0, \tag{7.11}$$

which is derived after integrating $\frac{d}{dx}\left(V_A \tilde{J}_A + V_B \tilde{J}_B\right) = 0$, with the boundary conditions that the far-field fluxes go to zero.

Utilizing the relations in Eq. (7.8), we can thus derive the velocity of the hypothetical marker positions as

$$v = -\left(J_A V_A + J_B V_B\right), \tag{7.12}$$

which after further simplification gives

$$v = \left(D_A - D_B\right) \frac{dX_A}{dx}. \tag{7.13}$$

This is the fundamental result of Darken's analysis [36], meaning that a difference in the intrinsic fluxes of the two components gives rise to a velocity field of the markers. Since a relation exists between the extrinsic fluxes in Eq. (7.11), there is only one independent extrinsic flux. So, for example, the extrinsic flux for the component A reads

$$\tilde{J}_A = J_A - \left(J_A V_A + J_B V_B\right) c_A, \tag{7.14}$$

which can be simplified as

$$\tilde{J}_A = J_A c_B V_B - J_B c_A V_B \tag{7.15}$$

$$= -V_B\left(D_A c_B + D_B c_A\right)\frac{dc_A}{dx}, \tag{7.16}$$

where we have used the fact that the total number of moles at any position is constant $c = c_A + c_B$, and thereby the relation $\frac{dc_A}{dx} = -\frac{dc_B}{dx}$ holds. For a system with equal partial molar volumes of the elements, an interdiffusivity $\tilde{D}$ can be derived as

$$\tilde{D} = \left(D_A X_B + D_B X_A\right). \tag{7.17}$$

In the following, we will model this phenomena of Kirkendall motion in binary alloys using the construct of the phase-field model. Similar phase-field work can be found in [37] and [38].

For this, we will assume that the partial molar volumes of both the elements A and B are equal to V_m, which reduces the relation for the velocity of the markers as $v = -\left(J_A + J_B\right)V_m$ and the extrinsic fluxes follow as $\tilde{J}_A = -\tilde{J}_B$. In order to model the motion of Kirkendall markers in a binary alloy, we construct a three-component formulation, wherein A and B represent the major components and C represents the hypothetical marker/vacancy concentrations.

Unequal diffusion of the major components leads to motion of the species C (marker/vacancies) in a direction opposite to the resultant sum of the fluxes of the other two components, thereby leading to the relation $J_C = -\left(J_A + J_B\right)$. Note the similarity between the actual velocity obtained from the previous analysis which reads as $(v/V_m) = -\left(J_A + J_B\right)$. However, in this formulation, there is no deformation of the material, and hence the extrinsic and intrinsic fluxes are equal. The result of this is that the relation $\tilde{J}_A = -\tilde{J}_B$ is no longer true. The difference of the intrinsic fluxes is, however, successfully captured in the behavior of the flux of the marker composition, which is given by J_C. Thus the sum of the fluxes of the three species is $J_A + J_B + J_C = 0$ at each point in space. Also, as a result of this formulation, one can derive a variation of the equivalent Kirkendall marker velocity in the entire domain, which can be computed as $V_m J_C$. The formulation can be seen as that of a lattice with vacancies, where their inherent creation or annihilation is determined by the fluxes of the majority components. While the presence of markers is certainly not essential for the treatment of Kirkendall effect, their presence leads to calculation of the marker positions and velocity that allows for a clear visualization of the consequences of Kirkendall motion.

7.4.2 Same Phase Couples

Among the first cases are diffusion couples formed between the same two phases which have different compositions. The model ingredients required for this construction are described as follows. The compositions of the two ends of the diffusion couple are $\alpha \equiv (c_A, c_B) : (0.78, 0.21)$ on the left end and $\beta \equiv (c_A, c_B) : (0.23, 0.76)$ on the right. To define their intrinsic diffusivities, we set different values for the diagonal entries, while the cross-diffusivities are set to zero. Since, in the phase-field model, we derive the driving force for diffusion on the basis of the gradients of the diffusion potentials, we give a brief description of the thermodynamics of the system. We create a simple system which is meant for illustrating the phenomena, while more complicated descriptions in real alloys could also be accounted for. In this, only the diffusion potential as a function of the composition is essential, which reads

$$\tilde{\mu}_A^m = 2A_{AA}c_A + A_{AB}c_B, \tag{7.18}$$

$$\tilde{\mu}_B^m = 2A_{BB}c_B + A_{AB}c_A, \tag{7.19}$$

where A_{ij}'s are thermodynamic constants taken as

$$A_{ij} = \begin{bmatrix} \frac{1}{2}\left(\frac{1}{0.77} + \frac{1}{0.01}\right) & \frac{1}{0.01} \\ \frac{1}{0.01} & \frac{1}{2}\left(\frac{1}{0.22} + \frac{1}{0.01}\right) \end{bmatrix}. \tag{7.20}$$

Once the chemical potentials are described, the diffusion of the two species occurs according to the equations described in the phase-field model section, Eq. (7.4), with the mobility matrix

$$M_{ij} = \begin{bmatrix} D_{AA} & 0.0 \\ 0.0 & D_{BB} \end{bmatrix} \begin{bmatrix} 2A_{AA} & A_{AB} \\ A_{AB} & 2A_{BB} \end{bmatrix}^{-1}. \tag{7.21}$$

The phase-field vector is set to $\boldsymbol{\phi} = \phi_1$, constituting of a single phase, which makes the evolution rate of the order parameter zero. Thus, there is only evolution of the diffusion potentials.

In the following, we present three different situations, corresponding to the choice of the intrinsic diffusivities. Firstly, we have the case of equal diffusivities of the components A and B, the results of which are depicted in Fig. 7.1, where $D_{AA} = 1.0$ and $D_{BB} = 1.0$. The marker concentrations are derived out of the condition that $c_A + c_B + c_C = const.$, therefore, knowing the updated c_A and c_B, the corresponding values of c_C can be derived.

In Fig. 7.1, we see that the diffusion profiles of the components A and B homogenize, while the concentration of the markers stays at the equilibrium concentration throughout. This would in turn imply that the velocity of Kirkendall markers in Darken's description would be zero.

Consequently, the results of two choices of diffusivities, $D_{AA} = 1.0$ and $D_{BB} = 0.1$ as well as $D_{AA} = 0.1$ and $D_{BB} = 1.0$, are depicted in Fig. 7.2(A) and Fig. 7.2(B), respectively. In Fig. 7.2(A) we see that the vacancy concentration to the left of the diffusion

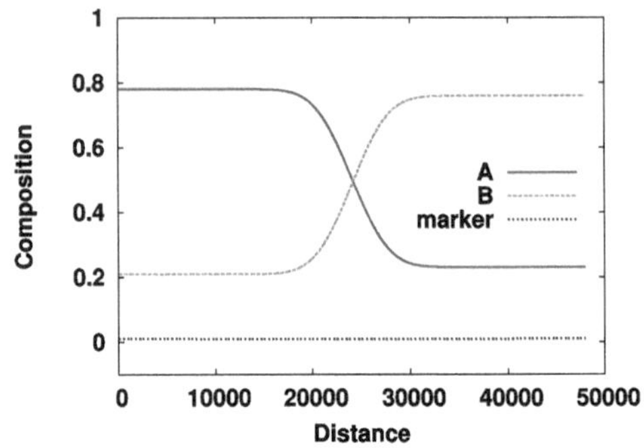

Figure 7.1 Interdiffusion in a diffusion couple formed between the same phases. Given equal diffusivity of the components, there is no vacancy flux and the composition stays at the equilibrium value.

couple boundary increases while it decreases to the right, which indicates a net local motion of the material to the right. The converse is true for the opposing case, Fig. 7.2(B), of higher diffusivity for the component B (D_{BB}) where the net local motion of the material occurs to the left. In both Figs. 7.2(A) and 7.2(B), a negative composition for vacancies implies a concentration of vacancies lesser than equilibrium, while a positive one implies something larger than the equilibrium composition.

7.4.3 Different Phase Couples

In the previous studies diffusion couples were created with the two ends of a couple being of the same phase. In many situations, a couple consists of two phases, where the end-compositions are away from the value at which thermodynamic equilibrium exists. The phase compositions at the interface assume equilibrium values, giving rise to a concentration gradient between the bulk and the interfacial compositions, on either side of the interface. The eventual motion of the interface is then determined by Stefan condition, relating the diffusion fluxes on either side of the interface, which for a binary alloy with two independent intrinsic fluxes reads as

$$D^{\alpha}_{AA}\frac{\partial c^{\alpha}_{A}}{\partial x} - D^{\beta}_{AA}\frac{\partial c^{\beta}_{A}}{\partial x} = \left(c^{\beta}_{A} - c^{\alpha}_{A}\right) v, \tag{7.22}$$

$$D^{\alpha}_{BB}\frac{\partial c^{\alpha}_{B}}{\partial x} - D^{\beta}_{BB}\frac{\partial c^{\beta}_{B}}{\partial x} = \left(c^{\beta}_{B} - c^{\alpha}_{B}\right) v. \tag{7.23}$$

In this situation, there are several interesting cases related to the combination of the different diffusivities in the two phases, which can result in vacancy fluxes in different directions in the two phases.

We simulate this situation using the phase-field model described before with the following thermodynamics of the phases α and β. The free-energies of the phases are

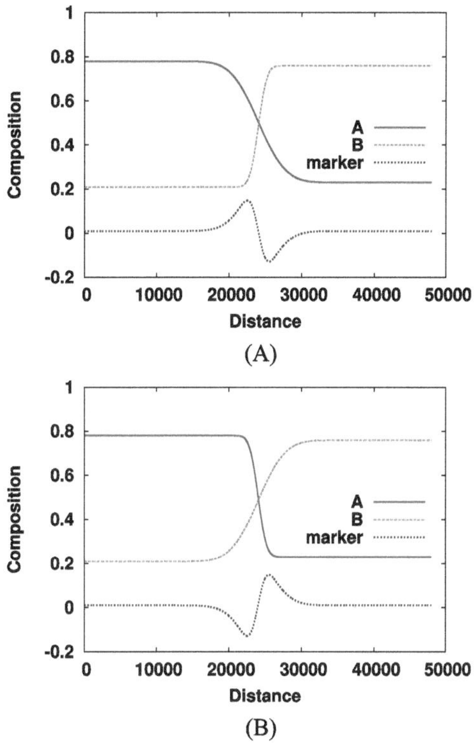

Figure 7.2 Diffusion couple for two situations: (A) $D_{AA}=1.0$ and $D_{BB}=0.1$, (B) $D_{BB}=1.0$ and $D_{AA}=0.1$. The blue lines depict the variation of the vacancy concentration due to motion of the major components A and B. (For interpretation of the references to color in this figure legend, the reader is referred to the web version of this chapter.)

described by the following relations:

$$f^{\alpha} = A^{\alpha}_{AA}c_A^2 + A^{\alpha}_{BB}c_B^2 + A^{\alpha}_{AB}c_Ac_B, \tag{7.24}$$
$$f^{\beta} = A^{\beta}_{AA}c_A^2 + A^{\beta}_{BB}c_B^2 + A^{\beta}_{AB}c_Ac_B. \tag{7.25}$$

The matrix A^{ν}_{ij}, $\nu=\alpha, \beta$, is

$$A^{\alpha}_{ij} = \begin{bmatrix} \frac{1}{2}\left(\frac{1}{0.77}+\frac{1}{0.01}\right) & \frac{1}{0.01} \\ \frac{1}{0.01} & \frac{1}{2}\left(\frac{1}{0.22}+\frac{1}{0.01}\right) \end{bmatrix}, \tag{7.26}$$

$$A_{ij}^{\beta} = \begin{bmatrix} \frac{1}{2}\left(\frac{1}{0.22}+\frac{1}{0.01}\right) & \frac{1}{0.01} \\ \frac{1}{0.01} & \frac{1}{2}\left(\frac{1}{0.77}+\frac{1}{0.01}\right) \end{bmatrix}. \tag{7.27}$$

Using these matrices, the diffusion potential of each of the phases can be computed as

$$\tilde{\mu}_i^{m,\nu} = 2A_{ii}^{\nu}c_i + A_{ij}^{\nu}c_j \quad \forall i,j \in \{A,B\} \quad \& \quad \nu \in \{\alpha,\beta\}, \tag{7.28}$$

and thereby the matrix $\left[\dfrac{dc_i^{\nu}}{d\tilde{\mu}_j^{m}}\right]$ can be derived as the inverse of $\left[\dfrac{d\tilde{\mu}_i^{m,\nu}}{dc_j}\right]$ which reads

$$\left[\frac{d\tilde{\mu}_i^{m,\nu}}{dc_j}\right] = \begin{bmatrix} 2A_{AA}^{\nu} & A_{AB}^{\nu} \\ A_{AB}^{\nu} & 2A_{BB}^{\nu} \end{bmatrix}. \tag{7.29}$$

The phase compositions $c_i^{\nu}(\tilde{\boldsymbol{\mu}}^{m})$ can thereafter be derived and used for determining the Legendre transform of the free-energy densities $\Psi^{\alpha} = f^{\alpha} - \sum_i^{K-1} \tilde{\mu}_i{}^{m} c_i^{\alpha}$, which will be utilized for determining the thermodynamic driving forces for α–β transitions in the phase-field model. A detailed description of this can be found in [30] and related papers [39,40].

For the diffusivity matrix, we consider two cases with the following choices: the first for the phases α and β reads as

$$[D]^{\alpha} = \begin{bmatrix} 1.0 & 0.0 \\ 0.0 & 0.1 \end{bmatrix}, \tag{7.30}$$

$$[D]^{\beta} = \begin{bmatrix} 1.0 & 0.0 \\ 0.0 & 0.1 \end{bmatrix}. \tag{7.31}$$

The equilibrium compositions of the phase α are $c_{A,eq}^{\alpha} = 0.77$ and $c_{B,eq}^{\alpha} = 0.22$, while for the β phase they are $c_{A,eq}^{\beta} = 0.22$ and $c_{B,eq}^{\beta} = 0.77$. The end-compositions of the phases are initialized as $\alpha \equiv (c_A, c_B) : (0.78, 0.21)$ and $\beta \equiv (c_A, c_B) : (0.23, 0.76)$. Under such conditions, the flux of B atoms occurs away from the α–β interface in the β phase while A atoms arrive at the α–β interface. On the α side, A atoms arrive at the interface while B atoms leave the interface. Since the A atoms diffuse faster on either side of the interface, this implies that there should be annihilation of the vacancy concentration at the interface, due to excess A atoms accumulating at the interface. This is depicted in the composition profiles plotted in Figs. 7.3(B) and 7.3(C). Along with the annihilation of the vacancies at the interface, further away from the interface on either side, the reverse must hold true, i.e., there must be vacancy generation due to fewer A atoms

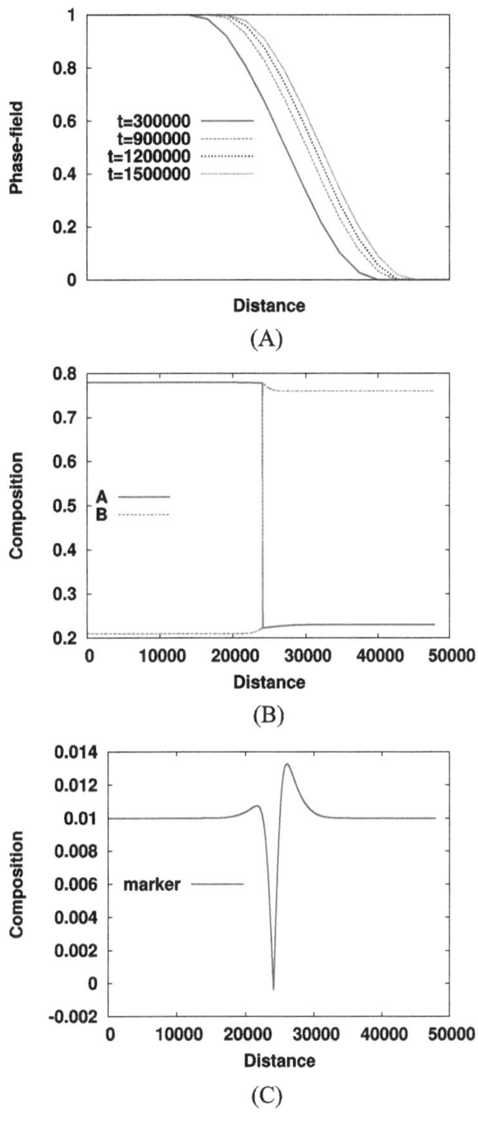

Figure 7.3 (A) Diffuse interface profile positions of the order parameter ϕ_α at various times, indicating α–β transition, as the diffusion of *A* and *B* components progresses in both phases. For this particular situation, β phase transforms to α phase, which is indicated by the rightward movement of the phase interface ϕ_α with time. (B) The composition profiles of *A* and *B* with time, along with the vacancy compositions in (C).

than *B* atoms to take the missing atomic positions. This is reflected by the accumulating vacancy concentration next to the interface on either side, which is unequal, on account of the unequal fluxes. Along with this, there is also migration of the phase interface, as

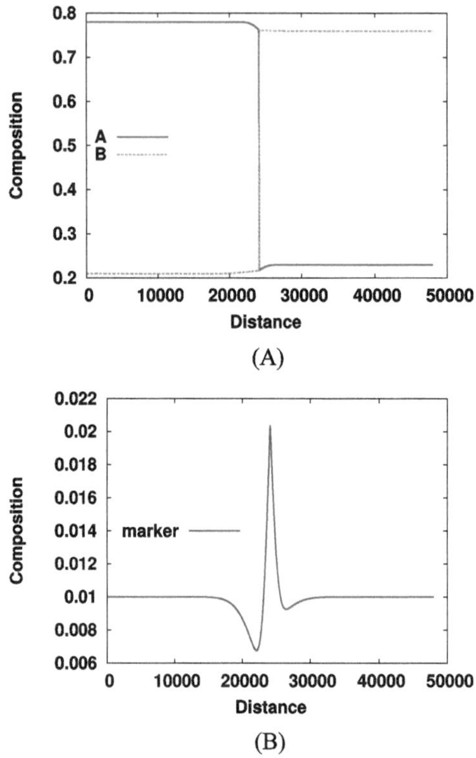

Figure 7.4 Composition profiles of the major components *A* and *B* in (A), along with the marker concentrations in (B), which indicate a vacancy flux towards the α–β interface on its either side.

shown in Fig. 7.3(a), which shows β–α phase transition. The converse case, with the diffusivities reversed in both phases, can be seen in Fig. 7.4. Now more B atoms leave the interface on both sides of the it than equivalent A atoms can take their place. This implies that there is a reversal of the vacancy flux direction at the interface, which is now towards the α–β interface in both phases, leading to accumulation of the vacancies at the interface. In the Darken picture, this implies an expansion of the material.

The other two possibilities are with the inverse diffusivities for the components on either side of the interface. For the first of these cases, we have:

$$[D]^{\alpha} = \begin{bmatrix} 1.0 & 0.0 \\ 0.0 & 0.1 \end{bmatrix}, \tag{7.32}$$

$$[D]^{\beta} = \begin{bmatrix} 0.1 & 0.0 \\ 0.0 & 1.0 \end{bmatrix}. \tag{7.33}$$

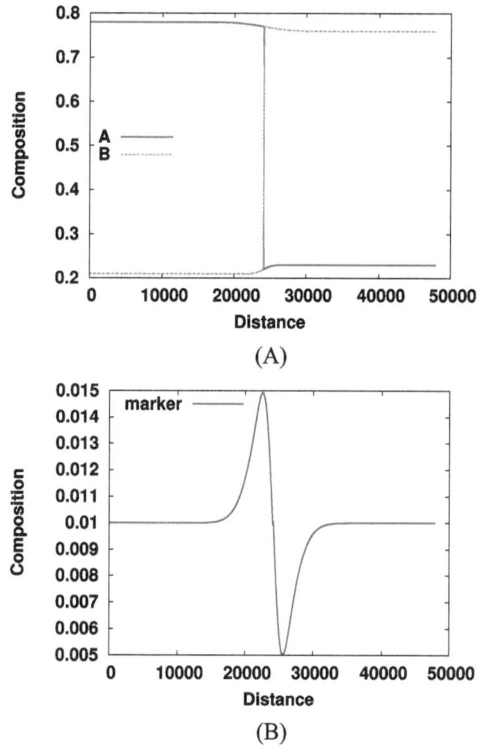

Figure 7.5 Composition profiles of the major components *A* and *B* in (A), along with the marker concentrations (B), which indicates vacancy accumulation on the α side of the α–β interface, while vacancy depletion occurs on the β side of the α–β interface, leaving the interface vacancy composition in equilibrium.

Now, while there is annihilation of vacancies on the α side of the interface, the reverse occurs on the β side, with vacancies accumulating at the interface. Both of these influences cancel out at the interface, implying that the vacancy composition at the interface remains in equilibrium, which is depicted in the composition profiles presented in Fig. 7.5. Next to the interface on both sides, however, we have accumulation of vacancies on the α side, while on the β side there is depletion. In the absence of any vacancies, the same would occur with a net motion of the material to the right.

Finally, for the other possibility of reversing the possible diffusivity matrices, i.e.,

$$[D]^{\alpha} = \begin{bmatrix} 0.1 & 0.0 \\ 0.0 & 1.0 \end{bmatrix}, \tag{7.34}$$

$$[D]^{\beta} = \begin{bmatrix} 1.0 & 0.0 \\ 0.0 & 0.1 \end{bmatrix}, \tag{7.35}$$

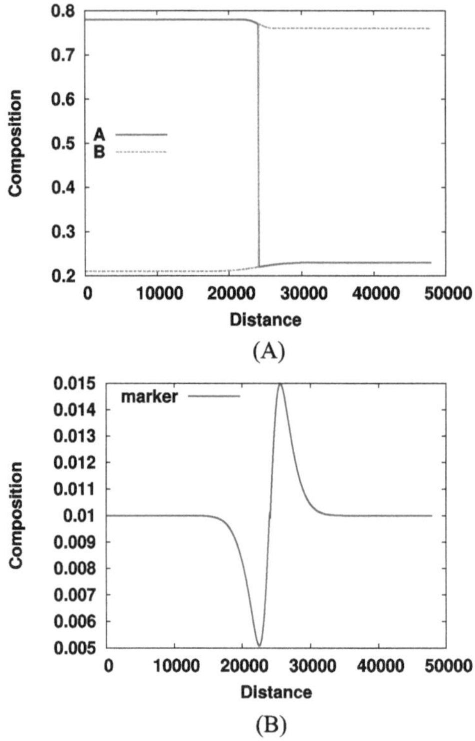

Figure 7.6 Composition profiles of the major components *A* and *B* in (A), along with the marker concentrations in (B), which indicates vacancy annihilation on the α side of the α–β interface, while vacancy generation occurs on the β side of the α–β interface, leaving the interface vacancy composition in equilibrium.

the composition profiles are depicted in Fig. 7.6. Here again, due to annihilation of vacancies on the β side of the interface and accumulation on the α side, the vacancy concentration at the interface remains unaltered. However, next to the interface now there is depletion on the α side and concomitant accumulation of vacancies next to the interface on the β side.

7.5 MULTICOMPONENT COUPLES (NO VACANCIES)

In the following section a diffusion couple in a ternary alloy is treated, where there is a two-phase couple formed with arbitrary compositions. In this situation, the composition at the interface is dictated by the choice of the inter-diffusivity of the elements in the two phases and thereby also the composition pathway between the chosen matrix compositions. This is of interest in the materials community since the composition pathway may help determining the nucleation of particular phases as well as the growth dynamics of the phase boundary between the two phases forming the diffusion couple.

In this context, there are already analytical studies of Maugis et al. [41,42] for diffusion couples and previous examples for precipitate growth [43–45], which essentially can be seen as generalizations of the Zener and Frank theories for binary alloys [46,47]. The problem lies in the derivation of the composition profiles during phase transformation in a diffusion couple, assuming that the composition profiles on either side of the interface decay exponentially to their values given by the boundary condition. The problem is well-posed and the governing equation for the composition fields in either of the phases can be written for the set of K independent elements as

$$\left\{\frac{\partial c_i^{\nu}}{\partial t}\right\} = \left[D_{ii}^{\nu}\right]\left\{\frac{\partial^2 c_i^{\nu}}{\partial x^2}\right\} \quad \forall \nu \in \{\alpha, \beta\}, \tag{7.36}$$

where the equation is written in vector–matrix notation, with $\{\cdot\}$ denoting a vector and $[\cdot]$ denoting a matrix. For this particular study, we will assume a diagonal interdiffusivity matrix, which is also indicated by the repeated index D_{ii}. For the case of the off-diagonal matrices, one can use the same formulation as described below, after performing diagonalization of the diffusivity matrices and repeating the analysis in the transformed coordinate system. The governing equations have Stefan conditions applicable at the interface boundary as their boundary conditions. These can again be written in generic fashion for all the K components as

$$v\left\{c_i^{\alpha} - c_i^{\beta}\right\} = -\left[D_{ii}^{\alpha}\right]\left\{\frac{\partial c_i^{\alpha}}{\partial x}\right\} + \left[D_{ii}^{\beta}\right]\left\{\frac{\partial c_i^{\beta}}{\partial x}\right\}. \tag{7.37}$$

As is usual in such diffusion controlled growth problems, there exists a scaling relationship between the growth rate of the interface and the time, which can be encapsulated in the growth coefficient $\eta^* = x_i/\sqrt{t}$, where x_i is the position of the interface and t is the time. The procedure henceforth involves a transformation of the variables in terms of the variable $\eta = x/\sqrt{t}$, resulting in the governing equations written as

$$-\frac{\eta}{2}\left\{\frac{\partial c_i^{\nu}}{\partial \eta}\right\} = \left[D_{ii}^{\nu}\right]\left\{\frac{\partial^2 c_i^{\nu}}{\partial \eta^2}\right\} \quad \forall \nu \in \{\alpha, \beta\}, \tag{7.38}$$

and the Stefan conditions being transformed to

$$-\frac{\eta^*}{2}\left\{c_i^{\alpha} - c_i^{\beta}\right\} = -\left[D_{ii}^{\alpha}\right]\left\{\frac{\partial c_i^{\alpha}}{\partial \eta}\right\}|_i + \left[D_{ii}^{\beta}\right]\left\{\frac{\partial c_i^{\beta}}{\partial \eta}\right\}|_i, \tag{7.39}$$

where η^* is the value of η at the interface, and the derivatives w.r.t. to η are evaluated at the interface denoted by the subscript i. Integrating the transformed governing equations in Eq. (7.38) once, we derive

$$\left\{\frac{\partial c_i^\nu}{\partial \eta}\right\} = \left\{\lambda_i^\nu \exp\frac{-\eta^2}{4D_{ii}^\nu}\right\}, \tag{7.40}$$

where λ_i^ν are constants of integration independent of η.

Substituting the derivatives of the preceding relation computed at the interface $\eta = \eta^*$ into Eq. (7.39), we obtain the relations

$$-\frac{\eta^*}{2}\left\{c_i^\alpha - c_i^\beta\right\} = -\left[D_{ii}^\alpha\right]\left\{\lambda_i^\alpha \exp\frac{-\eta^{*2}}{4D_{ii}^\alpha}\right\} + \left[D_{ii}^\beta\right]\left\{\lambda_i^\beta \exp\frac{-\eta^{*2}}{4D_{ii}^\beta}\right\}. \tag{7.41}$$

Integrating Eq. (7.40) and applying the far-field boundary conditions for the compositions of the components, we arrive at the equations for the constants λ_i^ν for each component in the respective phase, namely

$$\left\{\lambda_i^\nu\right\} = \left\{\frac{c_i^{\nu,\infty} - c_i^\nu}{\int_{\eta^*}^{\infty} \exp\left(\frac{-\eta^2}{4D_{ii}^\nu}\right) d\eta}\right\} \quad \forall \nu \in \{\alpha, \beta\}, \tag{7.42}$$

where we have introduced $c_i^{\nu,\infty}$ as the far-field boundary condition for the concentration of each component in phase ν. Thus, we will use the preceding relation as a function of the growth coefficient, $\lambda_i^\nu(\eta^*)$. Also note that since λ_i is an integration constant, it cannot vary with space or time, thereby indicating that η^* itself must be a constant.

Substituting these relations into Eq. (7.41), we derive

$$-\frac{\eta^*}{2}\left\{c_i^\alpha - c_i^\beta\right\} = -\left[D_{ii}^\alpha\right]\left\{\lambda_i^\alpha(\eta^*) \exp\frac{-\eta^{*2}}{4D_{ii}^\alpha}\right\} + \left[D_{ii}^\beta\right]\left\{\lambda_i^\beta(\eta^*) \exp\frac{-\eta^{*2}}{4D_{ii}^\beta}\right\}, \tag{7.43}$$

which are a set of nonlinear equations in terms of the growth coefficient η^*. These equations need to be simultaneously solved for both the growth coefficient and equilibrium compositions c_i^ν. Altogether, there are $2K$ unknown compositions for two-phase equilibrium and the growth exponent η^*. Invoking the local thermodynamic equilibrium condition, which is equating the K diffusion potentials among the phases, along with the condition of the same grand-potential, there are K independent equations required, which is exactly the number of available Stefan conditions. Therefore, both the equilibrium compositions and growth coefficient are fixed by the thermodynamics and the diffusivities of the elements in the respective phases, which is different from the case of binary alloys where it is only possible to fix the growth coefficient from the given Stefan condition.

Finding the growth coefficient completes the solution procedure for the composition profiles in both phases which reads

$$\left\{c_i^{\nu}(\eta)\right\} = \left\{ c_i^{\nu,\infty} - \left(c_i^{\nu,\infty} - c_i^{\alpha}\right) \frac{\int_{\eta}^{\infty} \exp\left(\frac{-\eta^2}{4D_{ii}^{\nu}}\right)}{\int_{\eta^*}^{\infty} \exp\left(\frac{-\eta^2}{4D_{ii}^{\nu}}\right)} \right\}. \tag{7.44}$$

7.5.1 Representative Phase-Field Simulation

We utilize the phase-field model and the thermodynamics as described in the previous sections for simulating a multicomponent diffusion couple. A similar study can be found in the literature [38]. Here we will choose a ternary alloy with components A and B as the independent elements. The chosen tie-line equilibrium compositions of the phase α are $c_{A,eq}^{\alpha} = 0.4$ and $c_{B,eq}^{\alpha} = 0.4$, while the β phase has the compositions $c_{A,eq}^{\alpha} = 0.1$ and $c_{B,eq}^{\alpha} = 0.1$. The end compositions are chosen as $\alpha \equiv (c_A, c_B) = (0.48, 0.28)$ and $\beta \equiv (c_A, c_B) = (0.08, 0.1)$. The free-energy densities of the phases are described as

$$f^{\nu} = A_{AA}^{\nu} c_A^2 + A_{BB}^{\nu} c_A c_B + B_A^{\nu} c_A + B_B^{\nu} c_B + C^{\nu}. \tag{7.45}$$

The matrix A_{ij} is assumed to be such that the curvature of the free-energy densities at the equilibrium compositions would be the same as that from an ideal solution model, thereby the matrix A_{ij}^{ν}, $\nu = \alpha, \beta$, reads

$$A_{ij}^{\nu} = \begin{bmatrix} \frac{1}{2}\left(\frac{1}{c_{A,eq}^{\nu}} + \frac{1}{\left(1 - c_{A,eq}^{\nu} - c_{B,eq}^{\nu}\right)}\right) & \frac{1}{\left(1 - c_{A,eq}^{\nu} - c_{B,eq}^{\nu}\right)} \\ \frac{1}{\left(1 - c_{A,eq}^{\nu} - c_{B,eq}^{\nu}\right)} & \frac{1}{2}\left(\frac{1}{c_{B,eq}^{\nu}} + \frac{1}{\left(1 - c_{A,eq}^{\nu} - c_{B,eq}^{\nu}\right)}\right) \end{bmatrix}. \tag{7.46}$$

The constants B_A^{β} and B_B^{β} are assumed to be zero, while B_A^{α} and B_B^{α} are determined using the chosen values of the equilibrium compositions $\mathbf{c}_{i,eq}^{\nu}$, $\nu = \alpha, \beta$, as

$$B_i^{\alpha} = 2A_{ii} c_{i,eq}^{\beta} + A_{ij} c_{j,eq}^{\beta} - \left(2A_{ii} c_{i,eq}^{\alpha} + A_{ij} c_{j,eq}^{\alpha}\right). \tag{7.47}$$

Correspondingly, the constant C^{β} is assumed to be zero, while C^{α} is determined according to

$$C^{\alpha} = -A_{ii}\left(c_{i,eq}^{\beta}\right)^2 - A_{ij} c_{i,eq}^{\beta} c_{j,eq}^{\beta} + \left(A_{ii}\left(c_{i,eq}^{\alpha}\right)^2 + A_{ij} c_{i,eq}^{\alpha} c_{j,eq}^{\alpha}\right). \tag{7.48}$$

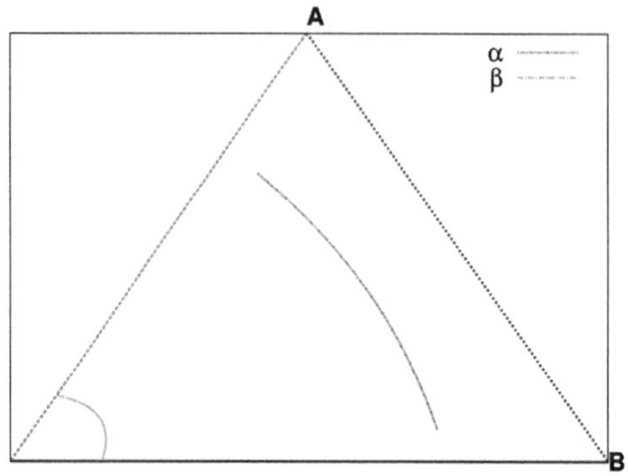

Figure 7.7 Coexistence lines showing equilibrium compositions of the two phases.

The constructed free energy has one of the equilibrium compositions given by the chosen tie-line values $c^{\nu}_{i,eq}$, $\nu = \alpha, \beta$. The complete phase-diagram spanning the entire coexistence lines of the two phases are plotted in Fig. 7.7.

Thus all the required diffusion potentials and the compositions as functions of the diffusion potentials, along with the driving force for transformation, that are required for the evolution equations of the diffusion potential and the phase-field variables can be determined [30,39].

In the following we perform phase-field simulations for the chosen diffusion couple with different diffusivity matrices, which are defined as

$$D^{\alpha}_{PathI} = \begin{bmatrix} 1.0 & 0.0 \\ 0.0 & 0.1 \end{bmatrix}, \tag{7.49}$$

$$D^{\beta}_{PathI} = \begin{bmatrix} 1.0 & 0.0 \\ 0.0 & 0.1 \end{bmatrix}, \tag{7.50}$$

$$D^{\alpha}_{PathII} = \begin{bmatrix} 0.1 & 0.0 \\ 0.0 & 1.0 \end{bmatrix}, \tag{7.51}$$

$$D^{\beta}_{PathII} = \begin{bmatrix} 0.1 & 0.0 \\ 0.0 & 1.0 \end{bmatrix}, \tag{7.52}$$

and

$$D^{\alpha}_{PathIII} = \begin{bmatrix} 1.0 & 0.0 \\ 0.0 & 1.0 \end{bmatrix}, \tag{7.53}$$

$$D^{\beta}_{PathIII} = \begin{bmatrix} 1.0 & 0.0 \\ 0.0 & 1.0 \end{bmatrix}. \tag{7.54}$$

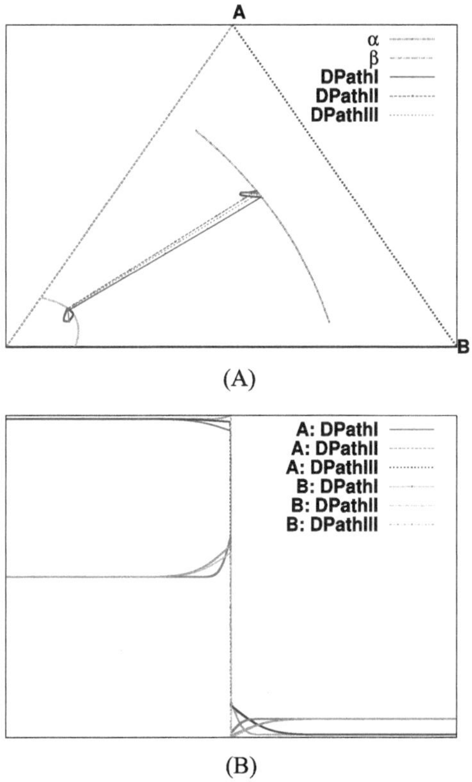

Figure 7.8 Diffusion paths represented on the simplex (A) and the corresponding composition profiles in the diffusion couple (B). Both figures show the three different composition pathways that are possible with different combinations of the diffusivity matrix.

The composition profiles from the phase-field simulations are plotted on the simplex as highlighted in Fig. 7.8. We see that there are three possible solutions corresponding to the different choices of the diffusion matrices. While in this chapter we do not show a one-to-one comparison with theory, it can be found in another work [48]. Thus phase-field models can be shown to be accurate and elegant methods for describing the interfacial compositions as well as the growth rates of the phases occurring during interdiffusion in diffusion couples. Their generality with respect to incorporation of the thermodynamic and kinetic coupling and elegance of treating multidimensional moving boundary problems makes them ideal choices for modeling phase transformations, particularly in multicomponent alloys, where the corresponding sharp interface methods become increasingly complex numerically.

7.6 INCORPORATING DATABASES

Given the discussion in the previous sections, it is clear that phase-field models are quite efficient schemes for treating problems in diffusion-coupled phase transformations. Therefore, they also become ideal choices for modeling phenomena occurring in real materials. To do so, it is important that the phase-field models can be coupled to thermodynamics as well as mobility databases.

The CALPHAD (CALculation of PHAse Diagrams) method [49,50] was developed to compute phase diagrams in multicomponent alloys based on the Gibbs energies of the bulk phases in a specific alloy. The functional expressions of the Gibbs energies in a multicomponent system are constructed on the basis of thermodynamical solution models, such as substitutional as well as sublattice models. This approach uses an extrapolation from binary (lower-order) subsystems including suitable higher-order interaction terms. The required thermodynamical data rely on experimental measurements of the material properties, phase relationships, and phase equilibria. By extrapolation, experimentally unaccessible concentration regions can also be addressed. Therefore, the CALPHAD method provides Gibbs energy functions of all bulk phases depending on the temperature and composition. By introducing a standardized data format (.tdb / .tdc format), massive thermodynamic databases have been established for a broad range of industrial alloy systems. Software packages such as ThermoCalc [51] and Pandat [52] build upon these thermodynamic databases and contain algorithms for computing the phase diagrams and the optimization of parameters in the Gibbs energies.

This progress also enabled extensive developments in phase-field modeling of multicomponent systems over the last 15 years by coupling with thermodynamical databases. In addition, the DICTRA (DIffusion Controlled TRAnsformations) [53] software calculates diffusion controlled phase transformations in multicomponent alloys based on a sharp interface model. The package provides temperature- and concentration-dependent expressions for the atomistic mobilities in order to calculate diffusion matrices in multicomponent systems. Due to the lack of experimental measurements of diffusion coefficients in alloys, the mobility database DICTRA possesses fewer data sets than the thermodynamical CALPHAD database.

In 2000, Grafe et al. [54] kicked off the progress of integrating thermodynamics databases and applied them to phase transformations in ternary Ni–Al–Cr superalloys. The software ThermoCalc [51] was employed as an interface (TQ-interface) for the CALPHAD method to calculate the driving forces for the phase transformations from the molar Gibbs energies and from the chemical potentials. The diffusion coefficients have been determined from a subroutine of the software DICTRA containing standardized kinetic databases. Several other applications have followed thereafter [55,56]. Phase-field models have also progressed with the coupling between thermodynamics databases also being enabled, with models incorporating finite interface dissipation [32] which can be utilized for studying phase transformations in steels. Other possibilities using approximations for the driving forces in the region of interest have also been utilized [39], which also share the same phase-field formulation that has been used in this chapter. Therefore, it is now possible to formulate phase-field models which can incorporate thermodynamics and mobility databases, and thereby modeling of real alloys is enabled.

7.7 CONCLUSIONS

In conclusion, we have elaborated on a brief history of the phase-field formulations and their different origins and applications. Thereafter, we have seen the applications of the phase-field model for simulating different phenomena in diffusion couples, namely diffusion in couples formed out of the same phase as well as of different phases. For the case of the diffusion couple formed out of different phases, we have considered the cases of the material expanding and contracting on the basis of accumulation or depletion of vacancies at the phase boundary. Finally, we have also highlighted the complexity in a diffusion couple formed in generic multicomponent alloys, where the growth coefficient as well as the compositions at the interface depend not only on the thermodynamics but also on the diffusivity matrices in both phases, using both analytical calculations as well as phase-field simulations. These example investigations clarify that the phase-field model can be an ideal choice for simulating such phenomena involving moving boundaries coupled with mass-diffusion of multiple components.

REFERENCES

[1] J.W. Cahn, J.E. Hilliard, J. Chem. Phys. 28 (1958) 258.
[2] V.L. Ginzburg, L.D. Landau, Sov. Phys. JETP 20 (1950) 1064.
[3] J.W.C. Samuel, M. Allen, A microscopic theory for antiphase boundary motion and its application to antiphase domain coarsening, Acta Metallurgica 27 (1979) 1085–1095.
[4] A.J. Bray, Theory of phase-ordering kinetics, Advances in Physics 43 (1994) 357–459.
[5] L. Chen, Annual Review in Materials Research 32 (2002) 113–140.
[6] I. Steinbach, Modeling and Simulation in Mat. Sci. and Engg. 17 (2009) 073001, 31 p.
[7] B. Nestler, A. Choudhury, Current Opinion in Solid State and Materials Science 15 (2011) 93.
[8] P.C. Hohenberg, B.I. Halerpin, Theory of dynamic critical phenomena, Rev. Mod. Phys. 49 (1977) 435–479.
[9] J.S. Langer, Directions in Condensed Matter, World Scientific, Singapore, 1986, p. 165.
[10] J.B. Collins, H. Levine, Diffuse interface model of diffusion-limited crystal growth, Phys. Rev. B 31 (1985) 6119–6122.
[11] R. Kobayashi, Physica D 63 (1993) 410.
[12] Y. Wang, L.Q. Chen, A.G. Khachaturiyan, Acta Metallurgica et Materialia 41 (1993) 279.
[13] V. Vaithyanathan, L.Q. Chen, Acta Materialia 50 (2002) 4061.
[14] Y. Wang, D. Banerjee, C.C. Su, A.G. Khachaturiyan, Acta Materialia 46 (1998) 2983.
[15] Y. Wang, J. Li, Acta Materialia 58 (2010) 1212.
[16] J.Z. Zhu, Z.K. Liu, V. Vaithyanathan, L.Q. Chen, Scripta Materialia 46 (2002) 401.

[17] M.P. Gururajan, T.A. Abinandanan, Phase-field study of precipitate rafting under a uniaxial stress, Acta Materialia 55 (2007) 5015–5026.
[18] R. Mukherjee, T.A. Abinandanan, M.P. Gururajan, Phase-field study of precipitate growth: effect of misfit strain and interface curvature, Acta Materialia 57 (2009) 3947–3954.
[19] S. Bhattacharya, T.A. Abinandanan, Evolution of multivariant microstructures with anisotropic misfit: a phase-field study, Acta Materialia 57 (2009) 646–656.
[20] R.F. Almgren, SIAM J. Appl. Math 59 (1999) 2086.
[21] G.B. McFadden, A.A. Wheeler, D.M. Anderson, Physica D 144 (2000) 154.
[22] A. Karma, W.J. Rappel, Phys. Rev. E 53 (4) (1996) R3017.
[23] A. Karma, Phys. Rev. Lett. 87 (2001) 115701.
[24] B. Eschebaria, R. Folch, A. Karma, M. Plapp, Phys. Rev. E 70 (2004) 061604.
[25] A.A. Wheeler, W.J. Boettinger, G.B. McFadden, Phys. Rev. E 45 (1992) 7424.
[26] J. Tiaden, B. Nestler, H.J. Diepers, I. Steinbach, Physica D 115 (1998) 73.
[27] S.G. Kim, W.T. Kim, T. Suzuki, Phys. Rev. E 58 (1998) 3316.
[28] S.G. Kim, W.T. Kim, T. Suzuki, Phys. Rev. E 60 (1999) 7186.
[29] M. Plapp, Phys. Rev. E 84 (2011) 031601.
[30] A. Choudhury, Quantitative Phase-Field Model for Phase Transformations in Multi-component Alloys, vol. 21, KIT Scientific Publishing, 2012.
[31] I. Steinbach, B. Boettger, J. Eiken, N. Warnken, S.G. Fries, ASM International 28 (2007) 101–106.
[32] I. Steinbach, L. Zhang, M. Plapp, Phase-field model with finite interface dissipation, Acta Materialia 60 (2012) 2689–2701.
[33] H. Garcke, B. Nestler, B. Stinner, SIAM J. Appl. Math. 64 (2004) 775.
[34] B. Nestler, H. Garcke, B. Stinner, Phys. Rev. E 71 (2005) 041609.
[35] A.D. Smigelskas, O.E. Kirkendall, Zinc diffusion in alpha brass, Trans. AIME 171 (1947) 130–142.
[36] L.S. Darken, Diffusion, mobility and their interrelation through free energy in binary metallic systems, Trans. AIME 175 (1948) 184–201.
[37] W.J. Boettinger, J.E. Guyer, C.E. Campbell, G.B. McFadden, Computation of the Kirkendall velocity and displacement fields in a one-dimensional binary diffusion couple with a moving interface, The Royal Society 463 (2007) 3347–3373.
[38] M. O, G. Vakanas, N. Moelans, M. Kajihara, W. Zhang, Formation of compounds and Kirkendall vacancy in the Cu–Sn system, Microelectronic Engineering 120 (2014) 133–137.
[39] A. Choudhury, M. Kellner, B. Nestler, A method for coupling the phase-field model based on a grand potential formalism to databases, Current Opinion in Solid State and Materials Science 19 (2015) 287–300.
[40] A. Choudhury, Pattern formation during self-organization in three-phase eutectic solidification, Transactions of the Indian Institute of Metals 68 (2015) 1137–1143.
[41] P. Maugis, W.D. Hopfe, J.E. Morral, J.S. Kirkaldy, Degeneracy of diffusion paths in ternary, two-phase diffusion couples, Journal of Applied physics 79 (1996) 7592.
[42] P. Maugis, W.D. Hopfe, J.E. Morral, J.S. Kirkaldy, Multiple interface velocity solutions for ternary bi-phase infinite diffusion couples, Acta Materialia 45 (1997) 1941–1954.
[43] D.E. Coates, Diffusion-controlled precipitate growth in ternary systems I, Metallurgical Transactions 3 (5) (1972) 1203–1212.
[44] D.E. Coates, Diffusion controlled precipitate growth in ternary systems: II, Metallurgical Transactions 4 (4) (1973) 1077–1086.
[45] J.P. Bourne, C. Atkinson, R.C. Reed, Diffusion-controlled growth in ternary systems, Metallurgical and Materials Transactions A 25 (12) (1994) 2683–2694.
[46] C. Zener, Theory of growth of spherical precipitates from solid solution, Journal of Applied Physics 20 (10) (1949).

[47] F.C. Frank, Radially symmetric phase growth controlled by diffusion, Proceedings of the Royal Society of London A: Mathematical, Physical and Engineering Sciences 201 (1067) (1950) 586–599.
[48] A. Lahiri, T.A. Abinandanan, A. Choudhury, M.S. Bhaskar, Theoretical and numerical determination of phase concentrations during growth in ternary alloys, arXiv:1602.0534.
[49] N. Saunders, A. Miodownik, Elsevier Science, Amsterdam, 1998.
[50] J.O. Saunders, J. Agren, J. Appl. Phys. 72 (1992) 1350.
[51] B. Sundman, B. Jansoon, J.O. Anderson, CALPHAD 9 (2) (1985) 153.
[52] S.L. Chen, S. Daniel, F. Zhang, Y.A. Chang, X.Y. Yan, R.S. Fetzer, W.A. Oates, CALPHAD 26 (2002) 175.
[53] A. Engstroem, L. Hoeglund, J. Agren, Metall. Trans. 25A (1994) 1127.
[54] U. Grafe, B. Böttger, J. Tiaden, S.G. Fries, Scripta Mat. 42 (2000) 1179.
[55] S.G. Fries, B. Boettger, J. Eiken, I. Steinbach, Int. Journal of Materials Research 100 (2009).
[56] B. Boettger, M. Apel, J. Eiken, P. Schaggnit, I. Steinbach, Steel Research Isnt. 79 (2008) 608.

CHAPTER 8

Thermodynamic Treatment of Diffusive Phase Transformation (Reactive Diffusion)

Franz Dieter Fischer*, Jiří Svoboda†
*Montanuniversität Leoben, Leoben, Austria
†Academy of Sciences of the Czech Republic, Brno, Czech Republic

8.1 INTRODUCTION

The microstructural development in material systems can be in most cases considered as a thermodynamically non-equilibrium process approaching to equilibrium. The evolution of the material system is described by kinetic equations for internal state variables. These relations are usually established as so-called constitutive equations taken from experiments. However, in some material systems such constitutive equations are not available and must be "invented." Diffusion of an interstitial element like carbon in iron can be considered as an example. The phenomenological equation relating carbon flux to the gradient of its concentration is given by well-known Fick's first law. Via the mass conservation equation, the flux of the component can be related to its concentration rate and inserted into Fick's first law. This provides the constitutive equation, the well-known Fick's second law, relating the rate of the concentration to the second space derivatives of concentration. Fick's second law represents a partial differential equation in time and space for the concentration field.

The situation is much more complicated in the case of diffusion in multicomponent multiphase systems, where diffusive transformation occurs. Moreover, one must distinguish between interstitial and substitutional components, as well as account for existence and role of vacancies. For finding kinetic (evolution) equations of such systems it is reasonable to look for physically rational concepts. Such a handy tool was invented several decades ago and is represented by Thermodynamic Extremal Principle (TEP), allowing an effective treatment of non-equilibrium material systems. Here some remarks concerning the TEP may be useful. This principle was first formulated by Onsager [1] in 1931 for the heat conduction problem. The second pioneering step was done by Ziegler 30 years later (see [2,3]), who extended the results established by Onsager to a general variational principle, applicable to non-equilibrium thermodynamic system; for an overview see [4]. It may be of interest that several thermodynamic variational principles were formulated in the meantime, see the discussion by Hackl and

Handbook of Solid State Diffusion, Volume 1
DOI: 10.1016/B978-0-12-804287-8.00008-7

Fischer [5]. We concentrate here on the TEP, which has the potential to provide, e.g., the diffusion equations for multicomponent systems, the evolution equations for grains in polycrystals or for precipitates embedded in the matrix; diffusive phase transformations in multicomponent multiphase system can also be treated by the TEP.

Although different formulations of TEP are available for many decades now, they were scarcely used in materials science after their original formulation. One may think about Onsager (for details see later sections) as the pioneer in the field of TEPs, who used it only once for diffusion [6], and this paper from 1945 has been nearly forgotten. But, as a matter of fact, different versions of the TEP have become an efficient tool in materials science since more than the last two decades, see Cocks et al. [7], Svoboda et al. [8,9], Petryk [10], Hackl et al. [11,12], Hillert and Ågren, [13,14], and recent contributions from the Chinese community, e.g., Yang et al. [15] and Wang et al. [16,17] as examples; there is no claim for any completeness of this list. However, one may still have the impression that much more literature was published on the principles themselves, see papers by Martyushev et al. [18–20], Attard [21,22], Lucia [23], Rajagopal et al. [24,25] or the recent book by Haslach [26].

If the characterization of a system is performed with discrete (thermodynamic) parameters, the TEP formulated directly in discrete parameters (CPs) delivers the evolution equations for the rates of these CPs. Examples for discrete CPs are the radii of grains or precipitates in a grain growth and coarsening process (see, e.g., [27]), the shape parameters of precipitates of given fixed chemical composition (see, e.g., [28]), or the hole radius in a spherical nanoparticle (see [29]). If a material system must be described by fields, e.g., by concentration profiles, then it is possible to replace the fields with a sum of shape functions multiplied by discrete "intensity factors", a practice used in the finite element method. Then the "intensity factors" overtake the role of characteristic parameters (CPs), and their evolution equations can be derived by the TEP. A typical example is the evolution of a segregation parameter during transient solute drag; for details see [30]. Thinking about multiparticle, multicomponent and multiphase systems, such as modern steel, the TEP has proven to be an extremely efficient tool, providing thousands of parallel evolution equations solvable by an extremely quick time integration procedure (see, e.g., [27] or [28] and their implementation in MATCALC (http://matcalc.tuwien.ac.at)). One can hardly imagine that the evolution of such realistic material systems can be simulated with other concepts in a comparable time with a comparable accuracy and comparable statistics.

In the first part of this chapter we introduce and describe the TEP formulated in discrete CPs. Further parts of the chapter are devoted to the treatment of diffusive phase transformations in multiphase systems. In the second part the concept of reactive diffusion allowing an effective treatment of diffusional transformation in binary multiphase systems is introduced. Application of this concept to simulation of a two-dimensional problem is demonstrated in the third part. The fourth part presents the application of TEP formulated in CPs to diffusive transformation (formation of single crystalline reactive zones) in a binary multiphase system. This treatment is then generalized to formation of polycrystalline reactive zones in the fifth part.

8.2 FORMULATION OF TEP IN DISCRETE CHARACTERISTIC PARAMETERS

The classical way of determining the evolution of a system is represented by the solution of phenomenological equations supplemented by conservation laws and proper boundary and contact, as well as initial, conditions. By setting some assumptions on the system geometry, a certain degree of idealization of the system can be achieved, so that the phenomenological equations can be solved and the time evolution of the CPs can be extracted from the solution. However, in many practical cases the level of necessary idealization may completely miss the physical substance of the problem. Obviously, the solution of standard phenomenological equations need not be the most efficient way. This problem can be overcome by the application of the TEP formulated directly in CPs, which provides a systematic way of derivation of evolution equations for the CPs. It is necessary to point out that the approach by means of TEP, expressed in discrete CPs, provides the rates of the CPs without the need to solve partial differential equations. Therefore, applying the TEP often drastically simplifies the problem solution.

The systems treated in materials science are usually rather complex. In most cases, however, only a limited number of CPs need to be used for the description of the state and evolution of the system, and only a limited number of CPs are of interest, implying a certain degree of idealization of the system. The number of CPs depends on the required accuracy of the description and on the complexity of the system. Moreover, constraints of local or global character amongst the CPs must often be taken into account.

The TEP formulated in CPs with constraints provides a direct way for determination of the evolution equations for the CPs.

Let us assume that Gibbs energy G of the system can be expressed as a function $G = G(q_1, q_2, \ldots, q_N)$ of the set of the CPs denoted by q_i $(i = 1, \ldots, N)$ at constant absolute temperature T and pressure p. The time evolution of the system is caused by the diffusive fluxes $\mathbf{j}_k$ $(k = 1, \ldots, n)$, n is the number of diffusing components in the system, and by migration of the interfaces moving with the normal velocity v. The time evolution results in the change of CPs described by the rates of the CPs denoted by $\dot{q}_i$ $(i = 1, \ldots, N)$, which were designated as "kinetic variables".

We expect that, using the mass conservation law, proper simplifications and assumptions on the geometry of the system, linear relations[1]

$$\mathbf{j}_k = \sum_{i=1}^{N} \tilde{\mathbf{j}}_{ik}(q_1, q_2, \ldots, q_N)\dot{q}_i, \quad (k = 1, \ldots, n), \tag{8.1}$$

exist for diffusive fluxes $\mathbf{j}_k$. For a distinct phase interface, the normal interface velocity v can be formulated as

$$v = \sum_{i=1}^{N} \tilde{v}_i(q_1, q_2, \ldots, q_N)\dot{q}_i. \tag{8.2}$$

Within the frame of linear non-equilibrium thermodynamics, the dissipation function Q for fixed T and p in an isotropic system is a quadratic function in the fluxes and interface velocities, given as

$$Q = \int_V \sum_{k=1}^{n} \frac{RT\mathbf{j}_k^2}{c_k D_k} \mathrm{d}V + \int_{\text{Interfaces}} \frac{v^2}{M} \mathrm{d}A, \tag{8.3}$$

where R is the gas constant, c_k the concentration of component k, D_k the tracer diffusion coefficient of component k, and M is the interface mobility.

After inserting Eqs. (8.1) and (8.2) into Eq. (8.3) and computing the integrals, one obtains the dissipation function Q in the form

$$Q = \sum_{i=1}^{N} \sum_{j=1}^{N} U_{ij}(q_1, q_2, \ldots, q_N)\dot{q}_i\dot{q}_j, \quad U_{ij} = \frac{1}{2}\frac{\partial^2 Q}{\partial \dot{q}_i \partial \dot{q}_j}. \tag{8.4}$$

The matrix U_{ij} is positive definite and takes into account the material properties and geometry of the system. The rate $\dot{G}$ of the Gibbs energy G can be calculated as

$$\dot{G} = \sum_{i=1}^{N} \frac{\partial G}{\partial q_i}\dot{q}_i. \tag{8.5}$$

The dissipation D follows (considering all q_i as internal variables) as

$$D = -\dot{G}. \tag{8.6}$$

Assume m constraints amongst the CPs of the form

$$F_l(q_1, q_2, \ldots, q_N) \equiv 0, \quad (l = 1, \ldots, m). \tag{8.7}$$

[1] Remark: Material time derivatives (rates) are marked by a dot.

Differentiating Eq. (8.7) with respect to time, one obtains the constraints in terms of the rates $\dot{q}_i$ as

$$\sum_{j=1}^{N} a_{jl}(q_1, q_2, \ldots, q_N)\dot{q}_j = 0, \quad (l = 1, \ldots, m), \; a_{jl} = \frac{\partial F_l}{\partial q_j}. \tag{8.8}$$

The TEP asserts that *"A closed system under constant temperature and pressure evolves in a way corresponding to the maximum dissipation D constrained by the balance equation Q = D and by further constraints Eq. (8.8)."*

The necessary condition for the maximum of dissipation D constrained by $Q = D$ and by Eq. (8.8) is given as

$$\frac{\partial}{\partial \dot{q}_i}\left[D - \lambda(Q - D) + 2\sum_{l=1}^{m} \beta_l \sum_{j=1}^{N} a_{jl}\dot{q}_j\right] = 0, \quad (i = 1, \ldots, N), \tag{8.9}$$

with λ and β_l $(l = 1, \ldots, m)$ being Lagrange multipliers. After inserting D and Q from Eqs. (8.6) and (8.3), respectively, and evaluating the derivatives, one obtains

$$-\frac{\partial G}{\partial q_i} - \lambda\left(2\sum_{j=1}^{N} U_{ij}\dot{q}_j + \frac{\partial G}{\partial q_i}\right) + 2\sum_{l=1}^{m} \beta_l a_{il} = 0, \quad (i = 1, \ldots, N). \tag{8.10}$$

If Eq. (8.10) is multiplied by $\dot{q}_i$, summed over i $(i = 1, \ldots, N)$, and Eqs. (8.4)–(8.6) are used, one obtains

$$D - \lambda(2Q - D) = 0, \tag{8.11}$$

which, together with the constraint $Q = D$, allows the evaluation of λ as $\lambda = 1$. Then Eq. (8.10) can be rewritten in a final form as

$$\sum_{j=1}^{N} U_{ij}\dot{q}_j + \sum_{l=1}^{m} a_{il}\beta_l = -\frac{\partial G}{\partial q_i} = -\frac{\partial \dot{G}}{\partial \dot{q}_i}, \quad (i = 1, \ldots, N). \tag{8.12}$$

It is advantageous to note that matrix U_{ij} can be calculated from the dissipation function Q as

$$U_{ij} = \frac{1}{2}\frac{\partial^2 Q}{\partial \dot{q}_i \partial \dot{q}_j}, \quad (i, j = 1, \ldots, N). \tag{8.13}$$

Eqs. (8.12), together with the constraints Eq. (8.8), can always be solved with respect to $\dot{q}_i$ $(i = 1, \ldots, N)$ and integrated with respect to time; then one obtains $q_i = q_i(t)$ $(i = 1, \ldots, N)$.

8.3 TREATMENT OF REACTIVE DIFFUSION IN BINARY SYSTEMS WITH MULTIPLE STOICHIOMETRIC PHASES BY ADAPTED DIFFUSION EQUATIONS

One-Dimensional Simulation

Reactive diffusion is understood in the literature, see, e.g., theoretical treatments in [31–33] or experimental reports in [34,35], as the formation of one or more new (product) phases between two old (parent) phases in contact under the assistance of diffusion. Let us assume as the simplest example a binary diffusion couple with one phase of pure A-component on the left side and another phase of pure B-component on the right side of an interface as a one-dimensional starting configuration (coordinate x in the actual configuration). After a certain nucleation process along the interface, see, e.g., the studies by Gusak and co-workers [36,37], one or more stable layers of new phases can develop between the parent phases. Moreover, also a certain molar fraction of B-component can be admitted in the left phase and a certain molar fraction of A-component can be admitted in the right phase, depending on the temperature, see the numerous phase diagrams for A–B-systems for solubilities.

There are several approaches to the treatment of reactive diffusion. Dybkov [38] uses a standard chemical approach in his book and introduces for each system a certain number of "reaction constants" describing the reactions at the interfaces and diffusion of components across the layers. In the case of interface reaction control, linear growth occurs; for diffusion control, the growth is parabolic. The links amongst the reaction constants, interface mobilities, activity of sources and sinks for vacancies at the interface, and bulk diffusion coefficients are not directly pursued.

A further step in understanding has been done by van Loo and coworkers [39–41] or later by Paul and coworkers, see, e.g., [42], who presented the physicochemical approach. They introduced an average interdiffusion coefficient for each phase, and the ratios of the intrinsic diffusion coefficients of the species in each phase were considered as fitting parameters. This approach can treat not only the parabolic growth of the layers controlled by diffusion but also the splitting and motion of Kirkendall planes, which can serve for the determination of the ratios of the intrinsic diffusion coefficients.

A crucial relation to be met in a solution concept is the mass balance at an interface. If a mathematical (sharp) interface is considered, the transport theorem of continuum mechanics teaches that the velocities of the interface relative to the lattice are generally different on both sides of the interface and that they must be introduced into the mass balance for the actual configuration; for derivation see, e.g., [43] and its transfer to the reference configuration in [44, Sect. 4]. The difference in the interface velocity with respect to adjacent lattices is due to generation/annihilation of vacancies at the interface and/or due to different partial molar volumes of different components in different phases. Svoboda et al. published several papers on reactive diffusion, addressing

Kirkendall effect by engaging the TEP, meeting the above mentioned mass balance as a necessary constraint for the individual fluxes, see [45,46], and [47]. These papers deal with stoichiometric intermetallic phases, taking into account the activity of sources and sinks for vacancies either at the interfaces only or in the bulk and at the interfaces. A recent contribution [48] deals also with the interaction of bulk diffusion and grain boundary/interface diffusion.

Most recently Erdelyi and Schmitz presented a solution concept in [31], where they avoided working with the interface and replaced the coupled problem of diffusion and interface migration by just the solution of a single modified diffusion equation expressed in gradients of the site fractions and using thermodynamic factors. This concept was previously applied only for intermixing of binary one-phase systems (and not reactive diffusion with moving interfaces); see recent works by Gusak et al. [49], and Beke, Erdelyi et al. [50]. If this idea were to work also in multiphase systems, this would be a significant simplification of the treatment in the field of reactive diffusion. This concept has been successfully modified by the authors and works directly with the gradients of the chemical potentials and tracer diffusion coefficients, see [51].

The Reactive Diffusion Process in a One-Dimensional Setting

Choose a simple binary system of substitutional components A and B, as used in [31], where y describes the site fraction of component A. We work with the coordinate x in the actual configuration. One phase is supposed to be a solid solution (superscript SS), forming an ideal solution with the molar Gibbs energy given by

$$g^{SS} = RT\left[y \ln y + (1-y)\ln(1-y)\right]; \tag{8.14}$$

the second phase is supposed to be an intermetallic phase (superscript IMP) with the molar Gibbs energy approximated by

$$g^{IMP} = g_0 + W(y - y_m)^2 \tag{8.15}$$

with g_0 being the minimum value of the molar Gibbs energy of the IMP for y_m, and W characterizing the width of solubility in the IMP. Then the chemical potentials of individual components of the individual phases are given by

$$\begin{aligned} &\mu_A^{SS} = RT\ln y, \qquad \mu_B^{SS} = RT\ln(1-y), \\ &\mu_A^{IMP} = g_0 + W\left(2y - y^2 + y_m^2 - 2y_m\right), \qquad \mu_B^{IMP} = g_0 + W\left(-y^2 + y_m^2\right). \end{aligned} \tag{8.16}$$

The solubility limits y_1^{eq} and y_4^{eq} in the SS and y_2^{eq} and y_3^{eq} in the IMP are given by the local equilibrium condition, solving the equations below for y_1^{eq}, y_2^{eq} and y_3^{eq}, y_4^{eq},

$$\mu_A^{SS}\left(y_1^{eq}\right) = \mu_A^{IMP}\left(y_2^{eq}\right), \qquad \mu_B^{SS}\left(y_1^{eq}\right) = \mu_B^{IMP}\left(y_2^{eq}\right),$$

$$\mu_A^{SS}\left(y_4^{eq}\right) = \mu_A^{IMP}\left(y_3^{eq}\right), \qquad \mu_B^{SS}\left(y_4^{eq}\right) = \mu_B^{IMP}\left(y_3^{eq}\right), \tag{8.17}$$

which can be determined numerically, see Fig. 8.1 for the data used in simulations.

We suppose that the molar volumes of both components and of vacancies in both phases are the same and denoted by Ω. The bulk and the interfaces act as ideal sources and sinks for vacancies, which is in accordance with the so-called Darken concept. Furthermore, we assume that the equilibrium vacancy site fraction is negligible with respect to one, the influence of the internal stress state plays a negligible role, the mobility of the interfaces is infinite (which means that there is no dissipation due to the interface motion), and the tracer diffusion coefficients D_A and D_B are independent of y.

We formulate the fluxes j_A and j_B of both substitutional components A and B, respectively, according to [52, Sect. 4], for the geometrical correlation factor $f = 1$ and for ideal sources and sinks for vacancies (i.e., the chemical potential of vacancies is assumed as 0) as

$$\Omega j_A = -\frac{y D_A}{RT}\frac{\partial \mu_A}{\partial x}, \qquad \Omega j_B = -\frac{(1-y) D_B}{RT}\frac{\partial \mu_B}{\partial x}. \tag{8.18}$$

The vacancy flux j_V is given by

$$j_V = -j_A - j_B, \tag{8.19}$$

and the vacancy generation rate (for details see [52, Sect. 2, Eq. (8.15)]) as

$$\alpha = \Omega \frac{\partial j_V}{\partial x}. \tag{8.20}$$

The local swelling due to generation of vacancies (supposed to be only in the x-direction) reads according to [52, Sect. 2] as

$$\dot{\varepsilon} = \alpha. \tag{8.21}$$

The Mass Balance at the Interface and in the Bulk

Assume a migrating sharp interface, acting as an ideal source and sink for vacancies. Both chemical potentials are continuous across the interface, and the site fractions y^L and y^R to the left and right from the interface, respectively, hence the superscripts L, R, are given by the values of y_1^{eq} and y_2^{eq} or y_3^{eq} and y_4^{eq}. The solution of the diffusion equation provides the values of fluxes j_A^L and j_B^L just left from the interface and j_A^R and j_B^R just right from the interface. If we denote the velocity of the interface relative to the left lattice as U and the velocity of the motion of the interface relative to the right lattice as V, then the mass conservation of both components, written with respect to the actual configuration, reads as

$$\Omega\left(j_A^L - j_A^R\right) = U y^L - V y^R,$$

$$\Omega\left(j_B^L - j_B^R\right) = U\left(1 - y^L\right) - V\left(1 - y^R\right); \tag{8.22}$$

for derivation see [43]. Then the values of U and V can be solved from Eqs. (8.21)–(8.22) and expressed by the fluxes and site fractions as

$$U = \frac{\Omega(j_B^L - j_B^R)y^R - \Omega(j_A^L - j_A^R)(1 - y^R)}{y^R(1 - y^L) - y^L(1 - y^R)}, \tag{8.23.1}$$

$$V = \frac{\Omega(j_B^L - j_B^R)y^L - \Omega(j_A^L - j_A^R)(1 - y^L)}{y^R(1 - y^L) - y^L(1 - y^R)}. \tag{8.23.2}$$

The reader may be interested in the fact that the mass conservation of both components, written with respect to the reference configuration, contains only one velocity, i.e., the interface velocity, with respect to the reference configuration, see [44].

The mass balance in the bulk can be taken from [52, Sect. 2, Eq. (8.14)], and reads with Eqs. (8.19)–(8.21) as

$$\dot{y} = -\Omega\frac{\partial j_A}{\partial x} - \dot{\varepsilon}y. \tag{8.24}$$

Numerical Treatment of the One-Dimensional Problem

The present one-dimensional problem described by the concept of reactive diffusion can be treated numerically by means of finite differences on a non-equidistant grid in the actual configuration. We suppose that the system is closed and divided into N regions (intervals) Δ_i, $i = 1, \ldots, N$, of initially the same length, which, however, may change their length during the system evolution as outlined below. The fluxes $j_{A,i}$ and $j_{B,i}$, $i = 0, \ldots, N$, are defined at the nodal points, and $j_{A,0} = j_{B,0} = j_{A,N} = j_{B,N} = 0$ holds at both ends of the system for a closed system.

Bulk Equations

The diffusive fluxes between regions i and $i+1$ are then calculated by application of the TEP, as discussed in the previous section, according to [53, Eq. (8.6.1,2)]:

$$\Omega j_{A,i} = -\frac{2}{RT}\left(\frac{\Delta_i}{y_i D_A} + \frac{\Delta_{i+1}}{y_{i+1} D_A}\right)^{-1}(\mu_{A,i+1} - \mu_{A,i}), \tag{8.25.1}$$

$$\Omega j_{B,i} = -\frac{2}{RT}\left(\frac{\Delta_i}{(1 - y_i)D_B} + \frac{\Delta_{i+1}}{(1 - y_{i+1})D_B}\right)^{-1}(\mu_{B,i+1} - \mu_{B,i}). \tag{8.25.2}$$

Recall that the quantities y_i are the site fractions of A-component in the region i.

The rate of y_i follows from the mass conservation Eq. (8.24) as

$$\dot{y}_i = \frac{\Omega(j_{A,i-1} - j_{A,i})}{\Delta_i} - \dot{\varepsilon}_i y_i. \tag{8.26}$$

The quantity $\dot{\varepsilon}_i$ represents now the strain rate $\dot{\varepsilon}$ in the region i given by Eqs. (8.19)–(8.21) as

$$\dot{\varepsilon}_i = -\Omega \frac{j_{A,i} + j_{B,i} - j_{A,i-1} - j_{B,i-1}}{\Delta_i}. \tag{8.27}$$

The regions change their lengths according to

$$\dot{\Delta}_i = \dot{\varepsilon}_i \Delta_i. \tag{8.28}$$

Interface Equations

Special attention must be paid to the regions being in contact with the interface. The values of y_i in these regions are fixed quantities corresponding to y_1^{eq}, y_2^{eq}, y_3^{eq} or y_4^{eq}, and they coincide with the values of y^L and y^R, which appear in Eqs. (8.22)–(8.23). Thus, no evolution equations are necessary for y in these regions. Assume that the interface is situated between the regions l and $l+1$. The fluxes j_A^L, j_B^L (j_A^R, j_B^R) can be obtained by linear extrapolation from the two nearest nodal points on the left (right) side of the interface.

The region left from the interface changes its length, accounting also for the relative velocity U of the interface, Eq. (8.23.1), as

$$\dot{\Delta}_l = \dot{\varepsilon}_l \Delta_l + U \tag{8.29}$$

with $\dot{\varepsilon}_l$ given by Eqs. (8.19)–(8.21) as

$$\dot{\varepsilon}_l = -\Omega \frac{j_A^L + j_B^L - j_{A,l-1} - j_{B,l-1}}{\Delta_l}. \tag{8.30}$$

Analogously, the region right from the interface changes its length, accounting also for the relative velocity V of the interface, Eq. (8.23.2), as

$$\dot{\Delta}_{l+1} = \dot{\varepsilon}_{l+1} \Delta_{l+1} - V, \tag{8.31}$$

with $\dot{\varepsilon}_{l+1}$ given by Eqs. (8.19)–(8.21) as

$$\dot{\varepsilon}_{l+1} = -\Omega \frac{j_{A,l+1} + j_{B,l+1} - j_A^R - j_B^R}{\Delta_{l+1}}. \tag{8.32}$$

Since the values of Δ_i evolve in time during simulations, it is necessary to keep reasonable values of all Δ_i to ensure sufficient accuracy and stability of computation. Thus, if Δ_i obtains less than 60% of its initial length, the smaller neighbor region and the actual region are unified to one new region with a weighted value of the according data there. For the case that Δ_i obtains more than 150% of its initial length, the actual region is divided in two new regions of half the length of the actual region.

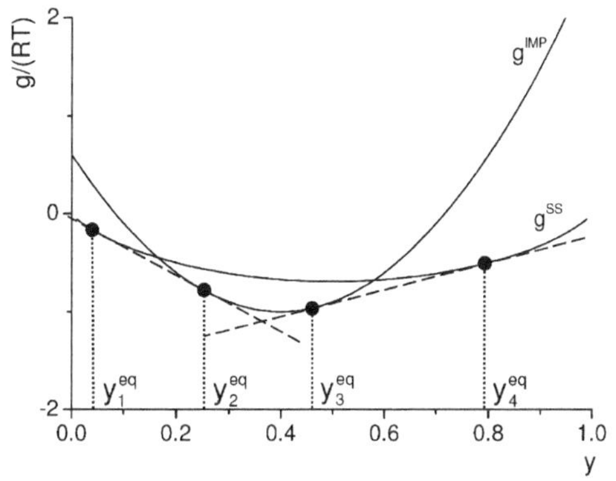

Figure 8.1 Molar Gibbs energy diagram for SS and IMP phases plotted for the used values of thermodynamic quantities at the interfaces. The solubility limits y_1^{eq} to y_4^{eq} are given by the double-tangent construction. Reprinted from Computational Material Science, 78, Svoboda J. & Fischer F.D., A new computational treatment of reactive diffusion in binary systems, 39–46, 2013 with permission from Elsevier.

Eqs. (8.25)–(8.32) represent a complete set of evolution equations for the coupled problem of diffusion and interface migration in the sense of the rigorous treatment of reactive diffusion in the system with ideal sources and sinks for vacancies and infinite interface mobility. This treatment accounts also for proper site fraction jumps at the interfaces.

Simulations

We use a system with the total initial length of $l_0 = 10^{-6}$ m, and different initial discretizations, represented by N_{start}, are selected.

The following values of thermodynamic parameters are used: $RT = 10^4\ \mathrm{J\,mol^{-1}}$, $g_0 = -RT$, $W = 10RT$, $y_m = 0.4$, $\Omega = 10^{-5}\ \mathrm{m^3\,mol^{-1}}$, $D_A = 2 \cdot 10^{-14}\ \mathrm{m^2\,s^{-1}}$ and $D_B = 5 \cdot 10^{-14}\ \mathrm{m^2\,s^{-1}}$. The selected thermodynamic data lead with Eqs. (8.14)–(8.17) to the solubility limits of $y_1^{eq} = 0.05297$, $y_2^{eq} = 0.25581$, $y_3^{eq} = 0.46827$, and $y_4^{eq} = 0.79664$. A graphical representation of the Gibbs energies can be taken from Fig. 8.1. If the total Gibbs energy is represented by the convex hull (envelope), then double-tangents replace g^{SS} and g^{IMP} in the intervals $y_1^{eq} \leq y \leq y_2^{eq}$ and $y_3^{eq} \leq y \leq y_2^{eq}$. In those intervals, the chemical potentials are constant. Under this assumption, the site fraction y may range also in one of the intervals (y_1^{eq}, y_2^{eq}) and (y_3^{eq}, y_4^{eq}). Then only Eqs. (8.25)–(8.28), formulated for the bulk, provide the kinetics of the whole system, if the chemical potentials are calculated according to Eq. (8.16).

The results of simulations are presented in Fig. 8.2. The jumps in site fractions y are represented by very steep site fraction gradients, and always only one nodal point

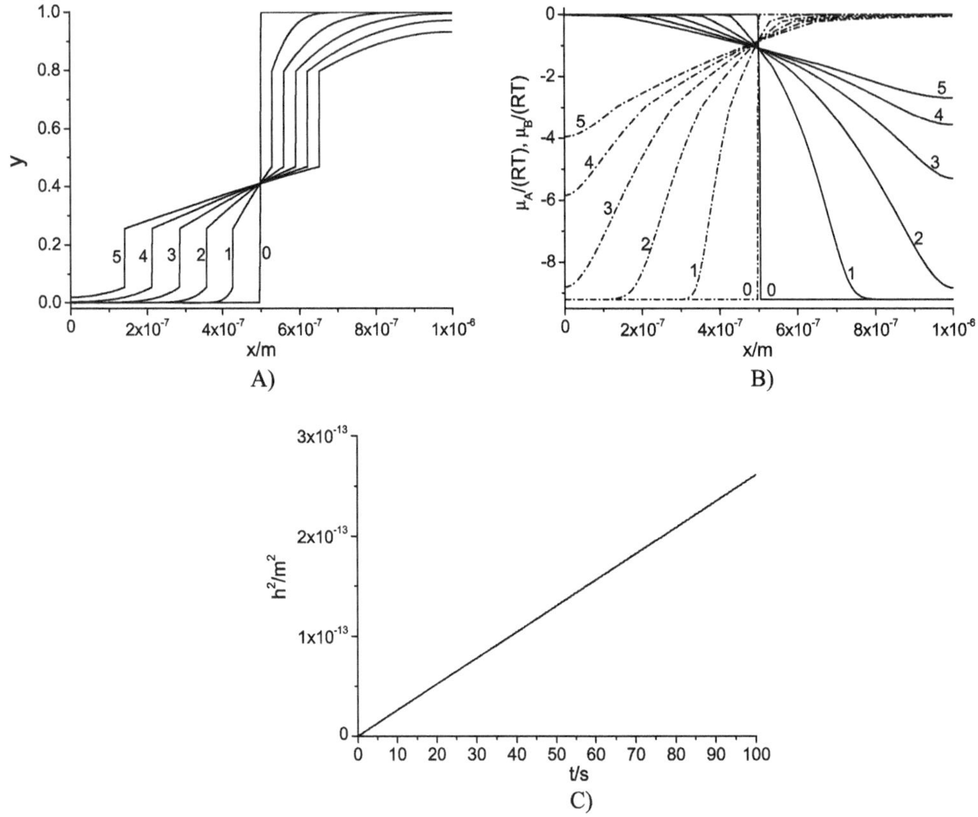

Figure 8.2 Evolution of (A) site fraction profiles y, (B) chemical potentials μ_A (solid) and μ_B (dashed), and (C) of IMP thickness h. The individual curves in (a) and (b) correspond to times $0-t=0$ s, $1-t=4$ s, $2-t=16$ s, $3-t=36$ s, $4-t=64$ s, $5-t=100$ s, $N_{start}=1000$. Reprinted from Computational Material Science, 78, Svoboda J. & Fischer F.D., A new computational treatment of reactive diffusion in binary systems, 39–46, 2013 with permission from Elsevier.

corresponds to a value of y from the interval $y_1^{eq} < y < y_2^{eq}$ or $y_3^{eq} < y < y_4^{eq}$. The method automatically localizes the interface precisely by an extremely steep site fraction gradient, which is limited only by the discretization used.

Discussion

One must highly appreciate the simplicity and accuracy of the present reactive diffusion concept. It has a high potential to get a widely used method for simulations of reactive diffusion in binary systems and to replace the much more complicated phase field method in two- and three-dimensional simulations.

The Reactive Diffusion Process in a Two-Dimensional Setting

Let us repeat that the reactive diffusion concept, based on a modified form of a proposal by Erdelyi and Schmitz [31], has the following advantages:

(i) it works only with thermodynamic quantities (e.g., chemical potentials, diffusion coefficients, diffusive fluxes, etc.) without the necessity to handle explicitly the individual phases;

(ii) it is very simple since this concept does not need to engage a jump condition at interfaces between phases; the according algorithm automatically finds the positions of the phase interfaces as steep concentration gradients.

A research task is now to simulate reactive diffusion processes also in multidimensional configurations. Details can be found in [54].

We consider again a binary system of substitutional components A and B, which may form several intermetallic phases A_yB_{1-y}. Mainly to avoid any deformation state as a consequence of the diffusion process, we assume that no sources and sinks for vacancies exist in the system and the role of vacancies is restricted only to mediation of diffusion, and that the molar volume Ω is independent of chemical composition and phase. The site fractions of A and B are denoted as y and $1-y$, respectively. Chemical potentials μ_A and μ_B, respectively, are assigned to both components. Furthermore, the components diffuse with the fluxes $\mathbf{j}_A$ and $\mathbf{j}_B$ through the lattices of individual phases and interfaces. Two conservation equations exist, namely for

(i) the mass as $$\dot{y} = -\Omega\,\mathrm{div}(\mathbf{j}_A), \tag{8.33}$$

(ii) the lattice positions as $$\mathbf{j}_A + \mathbf{j}_B = \mathbf{0}. \tag{8.34}$$

By grad(·) and div(·) we denote the gradient and divergence operators, respectively, applied to a quantity (·) with respect to the actual configuration. The constitutive equations for the fluxes can be formulated in a common way, relating the fluxes to the gradients of the chemical potentials as

$$\mathbf{j}_A = -\frac{y(1-y)}{RT\Omega}\frac{D_AD_B}{(yD_A + (1-y)D_B)}\,\mathrm{grad}(\mu_A - \mu_B). \tag{8.35}$$

The second term on the right-hand side of Eq. (8.35) represents a so-called "interdiffusion" coefficient D^{NP}, see the Appendix in [54] with respect to its derivation and the superscript "NP" (related to Nernst and Planck), yielding the diffusion equation in the form

$$RT\Omega\mathbf{j}_A = -y(1-y)D^{NP}\,\mathrm{grad}(\mu_A - \mu_B). \tag{8.36}$$

For ideal solutions of components A and B, one finds the classical Fick's first law as

$$\Omega\mathbf{j}_A = -D^{NP}\,\mathrm{grad}(y). \tag{8.37}$$

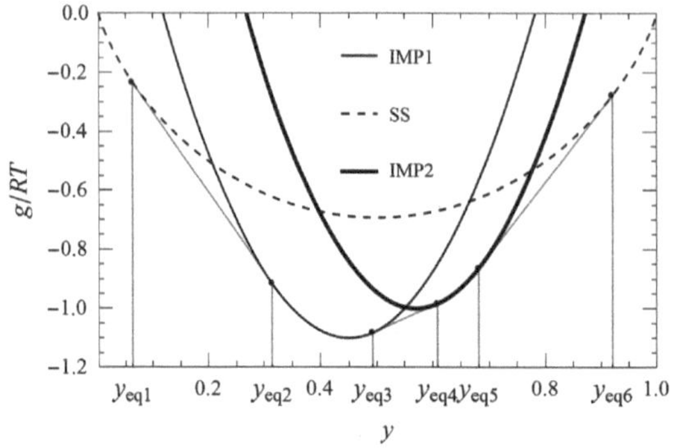

Figure 8.3 Molar Gibbs energy diagram for SS, IMP1, and IMP phases plotted for the used values of thermodynamic quantities. The solubility limits y_{eq1} to y_{eq6} are given by the double-tangent construction. Reprinted from Computational Material Science, 95, Svoboda J., Stopka J. & Fischer F.D., Two-dimensional simulation of reactive diffusion in binary systems, 309–315, 2014, with permission from Elsevier.

Combination of Eq. (8.33) and Eq. (8.37) allows formulating the diffusion equation as

$$\dot{y} = \mathrm{div}\left(D^{NP}\,\mathrm{grad}(y)\right). \tag{8.38}$$

The "interdiffusion" coefficient D^{NP}, however, depends on y, see Eq. (8.35), and, therefore, a nonlinear partial differential equation must be dealt with. Only for (nearly) identical tracer diffusion coefficients the problem formulation becomes linear in y.

We consider a quadratic region with a periodic boundary condition and a given distribution of the components A (site fraction y) and B (site fraction $(1-y)$) in this region as initial condition. The real demanding task, however, is to meet correctly the separation of the total region into subregions with different chemical composition and phases. As contact condition at the interfaces, having obtained their shapes as rather complicated curves, the equality of chemical potentials must be taken into account, see Eq. (8.41) below.

In the following context we assume as initial phase a solid solution of component A in component B, which allows formulating the chemical potential in the frame of ideal solution as Phase SS:

$$\mu_A^{SS} = RT\ln(y), \qquad \mu_B^{SS} = RT\ln(1-y). \tag{8.39}$$

Moreover, we consider two intermetallic phases A_yB_{1-y} denoted as IMP1 and IMP2. The molar Gibbs energy diagram for all phases SS, IMP1, and IMP2 is depicted schematically in Fig. 8.3 by using parameters explained later. The chemical potentials are defined according to the concept presented in [54] for both intermetallic phases

IMP1 and IMP2 as

$$\begin{aligned}
&\text{(Phase IMP1)} \quad \mu_A^{IMP1} = g_{01} + W_1(2y - y^2 + y_{m1}^2 - 2y_{m1}),\\
&\qquad\qquad\qquad \mu_B^{IMP1} = g_{01} + W_1(-y^2 + y_{m1}^2); \qquad (8.40.1)\\
&\text{(Phase IMP2)} \quad \mu_A^{IMP2} = g_{02} + W_2(2y - y^2 + y_{m2}^2 - 2y_{m2}),\\
&\qquad\qquad\qquad \mu_B^{IMP2} = g_{02} + W_2(-y^2 + y_{m2}^2). \qquad (8.40.2)
\end{aligned}$$

The quantities g_{01} and g_{02} represent the minima of the molar Gibbs energies of IMP1 and IMP2 for $y = y_{m1}$ and $y = y_{m2}$, respectively; see Fig. 8.3. The width of the solubility of each intermetallic phase is characterized by W_1 and W_2. The solubility limits y_{eq1}, y_{eq6} in the SS, y_{eq2}, y_{eq3} in the IMP1, and y_{eq4}, y_{eq5} in the IMP2, see Fig. 8.3, are given by the local equilibrium conditions as

$$\begin{aligned}
&\mu_A^{SS}(y_{eq1}) = \mu_A^{IMP1}(y_{eq2}), \quad && \mu_B^{SS}(y_{eq1}) = \mu_B^{IMP1}(y_{eq2}),\\
&\mu_A^{IMP1}(y_{eq3}) = \mu_A^{IMP2}(y_{eq4}), && \mu_B^{IMP1}(y_{eq3}) = \mu_B^{IMP2}(y_{eq4}), \qquad (8.41)\\
&\mu_A^{IMP2}(y_{eq5}) = \mu_A^{SS}(y_{eq6}), && \mu_B^{IMP2}(y_{eq5}) = \mu_B^{SS}(y_{eq6}),
\end{aligned}$$

and they can be determined numerically.

Numerical Treatment of the Two-Dimensional Problem

We assume as the initial two-dimensional configuration only the SS phase. Two kinds of pattern with periodic boundary conditions are selected by means of a random generator. The first case starts with impinging circles of phase SS with 36% of its area having the chemical composition $y = 0.99$ (marked black in Fig. 8.4 for $t = 0$) and with 64% of the remaining area having the chemical composition $y = 0.01$ (marked white in Fig. 8.4 for $t = 0$). The second case starts with impinging squares of phase SS with 41% of its area having the chemical composition $y = 0.99$ (marked black in Fig. 8.5 for $t = 0$) and with 59% of the remaining area having the chemical composition $y = 0.01$ (marked white in Fig. 8.5 for $t = 0$).

Bulk Equations

The problem is treated on a grid with 280×280 cells (pixels) with a uniform site fraction $y(x_1, x_2, t)$ inside; x_1 and x_2 are the coordinates of the pixel, $x_1, x_2 \in \{1, 2, \ldots, 280\}$. The flux $\mathbf{j}_A$ has component $j_1(x_1, x_2, t)$, representing the flux from the pixel with the coordinates $[x_1, x_2]$ to the pixel with the coordinates $[x_1 + 1, x_2]$, and component $j_2(x_1, x_2, t)$, representing the flux from the pixel with the coordinates $[x_1, x_2]$ to the pixel with the coordinates $[x_1, x_2 + 1]$. The periodic boundary conditions are set as

$$j_1(0, x_2, t) = j_1(280, x_2, t), \qquad j_2(x_1, 0, t) = j_2(x_1, 280, t). \qquad (8.42)$$

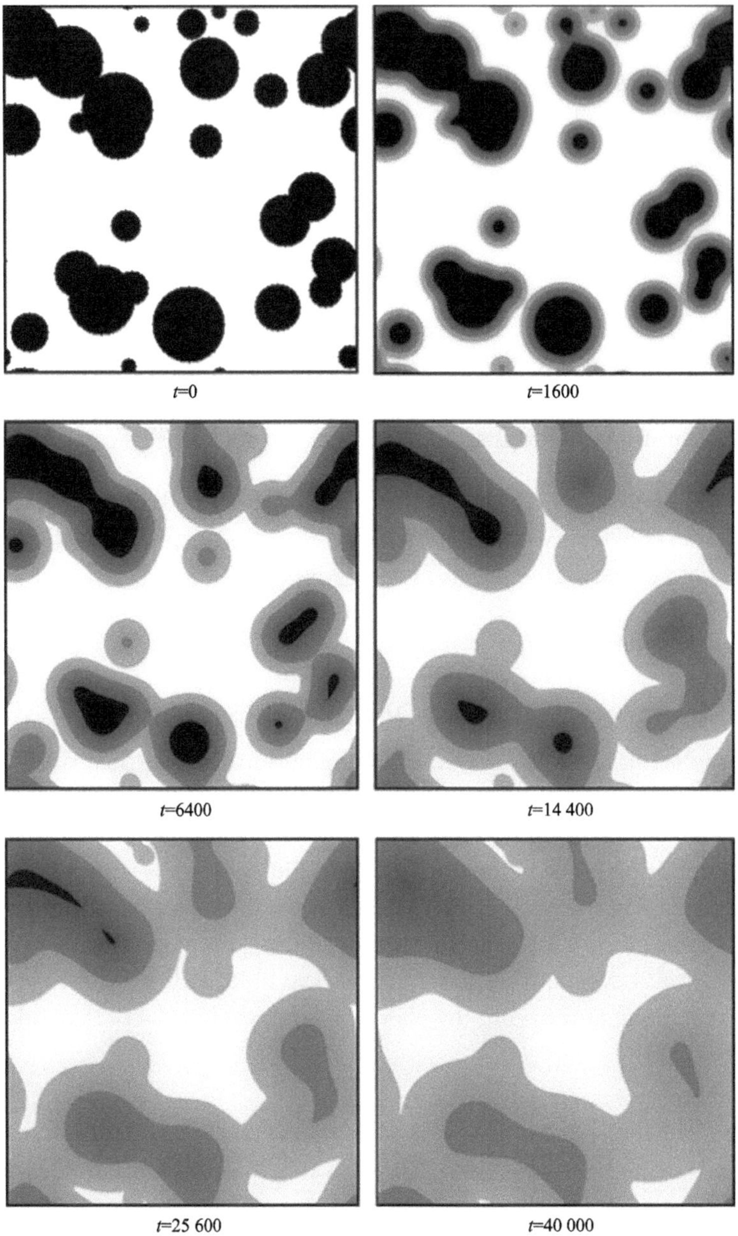

Figure 8.4 Two-dimensional simulation of the evolution of patterns of impinging circles with periodic boundary conditions for different times. At time $t = 0$ the circles of SS phase cover 33% of the total area with the chemical composition $y = 0.99$ (marked black) and are embedded in SS phase of chemical composition $y = 0.01$ (marked white). Reprinted from Computational Material Science, 95, Svoboda J., Stopka J. & Fischer F.D., Two-dimensional simulation of reactive diffusion in binary systems, 309–315, 2014, with permission from Elsevier.

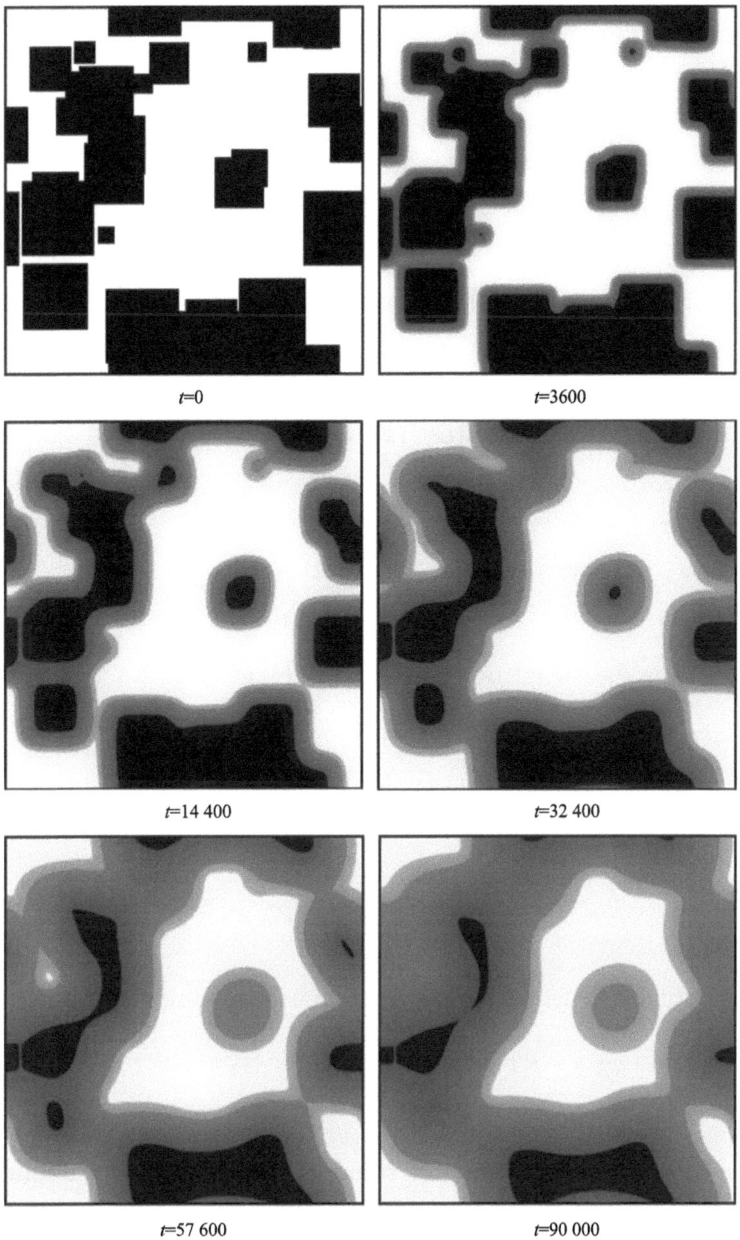

Figure 8.5 Two-dimensional simulation of the evolution of patterns of impinging squares with periodic boundary conditions. At time $t = 0$ the squares of SS phase cover 48% of the total area with the chemical composition $y = 0.99$ (marked black) and are embedded in SS phase of chemical composition $y = 0.01$ (marked white). Reprinted from Computational Material Science, 95, Svoboda J., Stopka J. & Fischer F.D., Two-dimensional simulation of reactive diffusion in binary systems, 309–315, 2014, with permission from Elsevier.

The fluxes interacting with a pixel at coordinates $[x_1, x_2]$ are calculated according to Eq. (8.36) as

$$\begin{aligned} &RT\Omega j_1(x_1-1,x_2,t) \\ &\quad = -y(x_1,x_2,t)\left[1-y(x_1,x_2,t)\right]D^{NP}(x_1,x_2,t) \\ &\qquad \times \left[\mu_A(x_1,x_2,t)-\mu_B(x_1,x_2,t)-\mu_A(x_1-1,x_2,t)+\mu_B(x_1-1,x_2,t)\right], \end{aligned} \tag{8.43.1}$$

$$\begin{aligned} &RT\Omega j_1(x_1,x_2,t) \\ &\quad = -y(x_1,x_2,t)\left[1-y(x_1,x_2,t)\right]D^{NP}(x_1,x_2,t) \\ &\qquad \times \left[\mu_A(x_1+1,x_2,t)-\mu_B(x_1+1,x_2,t)-\mu_A(x_1,x_2,t)+\mu_B(x_1,x_2,t)\right], \end{aligned} \tag{8.43.2}$$

$$\begin{aligned} &RT\Omega j_2(x_1,x_2-1,t) \\ &\quad = -y(x_1,x_2,t)\left[1-y(x_1,x_2,t)\right]D^{NP}(x_1,x_2,t) \\ &\qquad \times \left[\mu_A(x_1,x_2,t)-\mu_B(x_1,x_2,t)-\mu_A(x_1,x_2-1,t)+\mu_B(x_1,x_2-1,t)\right], \end{aligned} \tag{8.43.3}$$

$$\begin{aligned} &RT\Omega j_2(x_1,x_2,t) \\ &\quad = -y(x_1,x_2,t)\left[1-y(x_1,x_2,t)\right]D^{NP}(x_1,x_2,t) \\ &\qquad \times \left[\mu_A(x_1,x_2+1,t)-\mu_B(x_1,x_2+1,t)-\mu_A(x_1,x_2,t)+\mu_B(x_1,x_2,t)\right]. \end{aligned} \tag{8.43.4}$$

Then the change of the site fraction y in each pixel is given by Eq. (8.33) as

$$\begin{aligned} &y(x_1,x_2,t+\Delta t) \\ &\quad = y(x_1,x_2,t)-\Omega\left[j_1(x_1,x_2,t)-j_1(x_1-1,x_2,t)+j_2(x_1,x_2,t)-j_2(x_1,x_2-1,t)\right]\Delta t \end{aligned} \tag{8.44}$$

with Δt being the time step.

Simulations

The following values of thermodynamic quantities are used similarly as in [54]:

$$\begin{aligned} &g_{01}=-RT, \quad g_{02}=-1.1RT, \quad W_1=10RT, \\ &y_{m1}=0.45, \quad W_2=11RT, \quad y_{m2}=0.57, \end{aligned}$$

which leads, according to the numerical solution of Eq. (8.41), to the solubility limits $y_{eq1}=0.0624588$, $y_{eq2}=0.314562$, $y_{eq3}=0.492346$, $y_{eq4}=0.608496$, $y_{eq5}=0.681392$, $y_{eq6}=0.920607$. In the first case the values of diffusion coefficients $D_A=0.02$, $D_B=0.05$ are selected as independent of the phase and chemical composition. The values of the diffusion coefficients are adapted with respect to their dimensions, expressed by the pixel size = 1, and the time step is selected as $\Delta t=1$. In the second case the values of diffusion coefficients are assumed to be dependent on the phase and are selected as $D_A^{SS}=D_A$, $D_B^{SS}=D_B$, $D_A^{IMP1}=D_A/3$, $D_B^{IMP1}=D_B/4$, $D_A^{IMP2}=D_A/10$, $D_B^{IMP2}=\mathrm{D}_B/10$.

The values of RT and Ω are irrelevant, since they are eliminated in the final evolution equations.

The evolution of the systems is depicted for the first case in Fig. 8.4 and for the second case in Fig. 8.5. One can observe a spontaneous separation of the intermetallic phases by rather sharp interfaces after several time steps. The pixels with site fractions between the solubility limits of adjacent phases form very thin regions never exceeding the thickness of one pixel. An example demonstrating only the interface regions is presented in Fig. 8.6, which clearly shows that there is at most one and sometimes even no pixel at the interface. The simulation allows also calculating the evolution of areal fractions f of individual phases as depicted in Fig. 8.7. The scattering of the diagrams results from a rather coarse mesh compared to the thicknesses of the product phases. Note that the current concept needs no impinging correction factor as in [55], since it handles impinging automatically.

Discussion

The authors have already shown that the Reactive Diffusion Concept (RDC), working with chemical potentials and tracer diffusion coefficients (instead with chemical diffusion coefficients only), perfectly simulates the phase transformation in binary one-dimensional systems. However, the shape of an interface and the interface energy after nucleation of a phase do not play any role for a one-dimensional configuration. Therefore, one may be inclined to expect that two- and three-dimensional configurations behave in a similar way, if the change of chemical energy dominates the change of the interface energy during the transformation process in the system. Or in other words, the RDC can be used successfully, if the volume fractions of the individual phases change significantly. Of course, one cannot expect a successful application of the RDC to coarsening, driven only by the change of the interface energy. This fact motivated us to apply the RDC to the treatment of chemically driven growth of new phases for a two-dimensional configuration. As the results reported above show the RDC works also successfully in a two-dimensional configuration, providing the microstructure evolution as well as the kinetics of the fractions of the individual phases.

It may be of interest to compare the RDC with the Cahn–Hilliard Phase-Field Method applied to a binary system; we refer here to original works [56,57]. The interface is treated in the Cahn–Hilliard concept by addition of a weighted squared gradient term of the site fraction to the total Gibbs energy function. This term determines the properties of the interface as its specific interface energy and its thickness. Comparing both concepts shows clearly that the RDC is a special case of the Cahn–Hilliard concept. Reduction of the role of the gradient term to a zero value (switching from the Cahn–Hilliard concept to the RDC) has the consequence of zero interface energy and zero interface thickness (i.e., sharp interfaces) and infinite interface mobility.

Figure 8.6 Chart of the interface pixels only, corresponding to Fig. 8.4 and $t = 14400$. Reprinted from Computational Material Science, 95, Svoboda J., Stopka J. & Fischer F.D., Two-dimensional simulation of reactive diffusion in binary systems, 309–315, 2014, with permission from Elsevier.

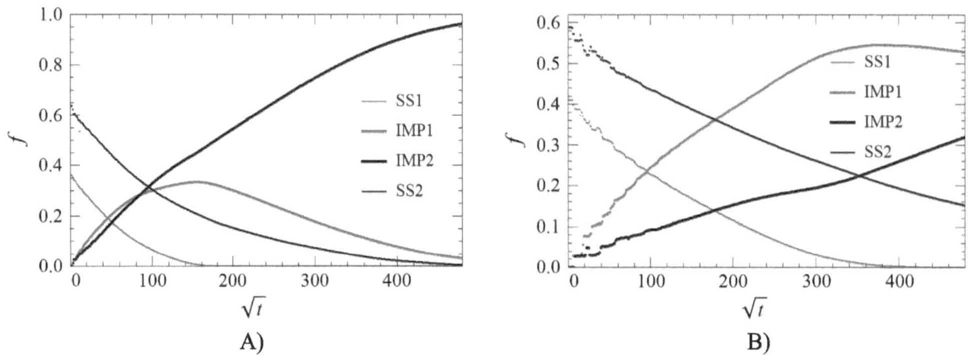

Figure 8.7 Time evolution of the areal fractions f of individual phases: SS phase with $y \geq y_{eq6}$ (SS1), IMP1, IMP2, and SS2 phase with $y \leq y_{eq1}$. (A) Starting configuration are impinging circles, corresponding to Fig. 8.4; (B) Starting configuration are impinging squares, corresponding to Fig. 8.5. Reprinted from Computational Material Science, 95, Svoboda J., Stopka J. & Fischer F.D., Two-dimensional simulation of reactive diffusion in binary systems, 309–315, 2014, with permission from Elsevier.

8.4 TREATMENT OF REACTIVE DIFFUSION IN BINARY SYSTEMS WITH MULTIPLE STOICHIOMETRIC PHASES BY APPLICATION OF TEP

Understanding the mechanism of diffusion has been the topic of research since more than 50 years. Kirkendall experiments provide a basis for that research; see the report on Kirkendall's seminal work by Nakajima [58]. Later the experiments have been concentrated on systems of nearly stoichiometric phases, providing the surprising result that the originally single Kirkendall plane furcates and is localized in various new phases.

Kirkendall plane is understood as a plane with markers deposited at the contact of the original diffusion couple. Several authors, e.g., Paul et al. [59], used a physicochemical approach for the problem description. This approach utilizes the theory of reactive diffusion [60–65] for several newly nucleated phases. Kinetic parameters such as the integrated interdiffusion coefficient in the newly nucleated phases and the ratio of intrinsic fluxes are used in the treatment.

The theory of reactive diffusion tacitly assumes ideal sources and sinks for vacancies at the interfaces and also infinite interface mobility, which is a realistic assumption for incoherent interfaces. The sources and sinks for vacancies in the bulk can be considered as ideal or inactive, which is the assumption of the analysis [59] for the bulk of the newly nucleated phases with stoichiometric composition.

The models in [45] and later in [46] represent an alternative to the model in [59] of the diffusive phase transformation for binary systems treating multiple stoichiometric phases. These new models are purely thermodynamic. They are more general, allow for arbitrary many newly nucleated phases, and utilize standard thermodynamic data as molar Gibbs energies, molar volumes in the stoichiometric phases, and the tracer diffusion coefficients. In [46] it is shown that such a treatment enables also solution of the inverse problem, i.e., determination of tracer diffusion coefficients of components in the newly nucleated phases by evaluation of experiments.

The present model is demonstrated on an example of a binary system forming four stoichiometric phases. The motion of Kirkendall planes depends on the values of the Gibbs energies of all phases and on the tracer diffusion coefficients in the newly nucleated phases.

The Model

Suppose that a binary system consists of components A and B, forming n stable stoichiometric phases $A_{r_i}B_{1-r_i}$, where r_i is the mole fraction of component A in phase i, $0 \leq r_i \leq 1$, $i = 1, \ldots, n$, and r_i increases with i. The partial molar volumes Ω_{Ai} and Ω_{Bi} correspond to components A and B, in individual phases i; they are assumed known, and the molar volume Ω_i of phase i is given by

$$\Omega_i = r_i\Omega_{Ai} + (1 - r_i)\Omega_{Bi}. \tag{8.45}$$

Moreover, g_i, $i = 1, \ldots, n$ denote the molar Gibbs energies of the phases.

As starting configuration we assume a diffusion couple in a one-dimensional setting with unit cross-section which consists of two phases: $A_{r_1}B_{1-r_1}$ at the left side and $A_{r_n}B_{1-r_n}$ at the right side of the contact with inert markers on it. We assume that very small nuclei of all new phases are formed at the original interface within a negligible time interval. Then the nuclei grow into a sequence of phases as schematically shown in Fig. 8.8. Some markers are present in all newly nucleated phases. All phases are considered as defect-free

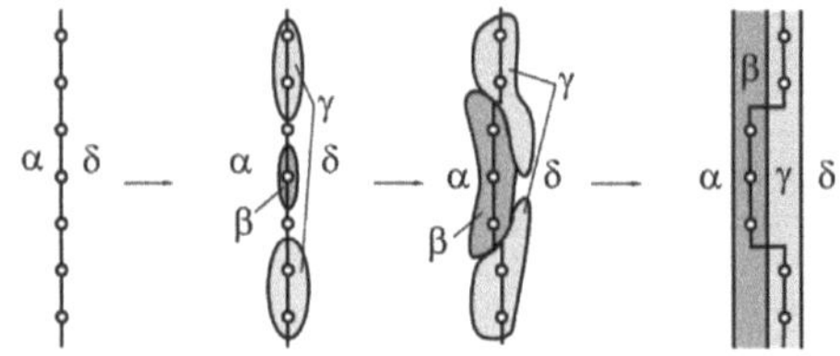

Figure 8.8 Nucleation stages at the interface leading to bi-furcation of the plane with deposited markers. With kind permission from Springer Science + Business Media: Journal of Phase Equilibria and Diffusion, Modeling of kinetics of diffusive phase transformation in binary system with multiple phases, 27, 2006, 622–628, Svoboda J., Gamsjäger E., Fischer F.D. & Kozeschnik E.

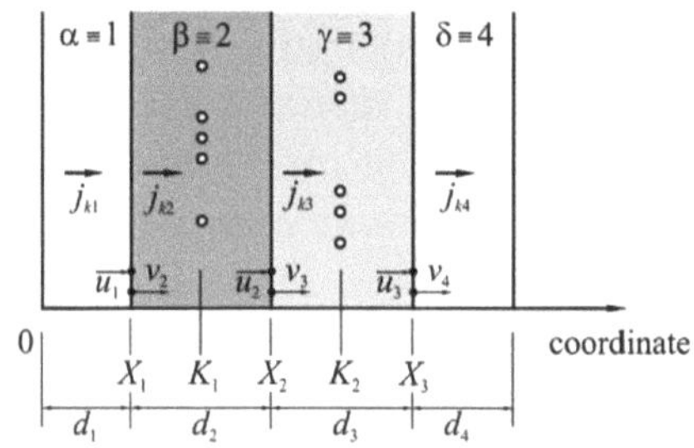

Figure 8.9 Schematic plot of the system geometry together with the definition of the kinetic variables. With kind permission from Springer Science + Business Media: Journal of Phase Equilibria and Diffusion, Modeling of kinetics of diffusive phase transformation in binary system with multiple phases, 27, 2006, 622–628, Svoboda J., Gamsjäger E., Fischer F.D. & Kozeschnik E.

single crystals, and thus no sources and sinks for vacancies are active in the bulk. The stoichiometricity of phases results in a constant molar volume and a constant chemical composition in each phase. No sources and sinks for vacancies assumed in the bulk of each phase imply that both diffusive fluxes j_{Ai} and j_{Bi}, as well as the material velocity, are constant in the whole region occupied by each individual phase. The existence of fluxes is due to a fact that a very small (negligible) alteration of the mole fraction in each phase causes a large change in the chemical potentials of both components. The model assumes that interfaces between the phases are incoherent, freely mobile, and act as ideal sources and sinks for vacancies.

At an interface between the phases i and $i+1$, the diffusive fluxes j_{Ai}, j_{Bi} and j_{Ai+1}, j_{Bi+1} occur, see Fig. 8.9 for an advanced stage of the system. The fluxes at each interface are balanced against interface migration and deposition or removal of layers of new atoms at the interface, which cause relative motion of the lattices of the individual phases.

Denoting by u_i the velocity of the interface between the phases i and $i+1$, relative to the lattice of the phase i, and by v_{i+1} the velocity of the interface relative to the lattice of the phase $i+1$, the mass conservation for A and B components is expressed as (see

[43, Sect. 3])

$$\frac{u_i r_i}{\Omega_i} - \frac{v_{i+1} r_{i+1}}{\Omega_{i+1}} = j_{Ai} - j_{Ai+1}, \quad i = 1, \ldots, n-1, \tag{8.46}$$

$$\frac{u_i(1-r_i)}{\Omega_i} - \frac{v_{i+1}(1-r_{i+1})}{\Omega_{i+1}} = j_{Bi} - j_{Bi+1}, \quad i = 1, \ldots, n-1. \tag{8.47}$$

Solving Eqs. (8.46) and (8.47) with respect to the velocities u_i and v_i provides

$$u_i = \frac{\Omega_i}{r_i - r_{i+1}}\left[(j_{Ai} - j_{Ai+1})(1 - r_{i+1}) - (j_{Bi} - j_{Bi+1})r_{i+1}\right], \quad i = 1, \ldots, n-1, \tag{8.48}$$

$$v_{i+1} = \frac{\Omega_{i+1}}{r_i - r_{i+1}}\left[(j_{Ai} - j_{Ai+1})(1 - r_i) - (j_{Bi} - j_{Bi+1})r_i\right], \quad i = 1, \ldots, n-1. \tag{8.49}$$

Considering a closed system with no deposition of matter at the surface leads to boundary conditions

$$j_{A1} = 0, \quad j_{B1} = 0, \quad j_{An} = 0, \quad j_{Bn} = 0, \quad v_1 = 0 \quad \text{and} \quad u_n = 0. \tag{8.50}$$

Then the rate $\dot{d}_i$ of the width d_i of the region of phase i (see Fig. 8.9) is given by

$$\dot{d}_i = u_i - v_i, \quad i = 1, \ldots, n. \tag{8.51}$$

The rate $\dot{d}_i$ can be expressed by means of the independent fluxes j_{Ai} and j_{Bi}, $i = 2, \ldots, n-1$ by using Eqs. (8.48)–(8.50).

For the left end of the specimen fixed to the origin of a coordinate system, the velocity w_i of the lattice of the phase i in this coordinate system can be calculated as

$$w_{i+1} = w_i + u_i - v_{i+1}, \quad i = 1, \ldots, n-1, \tag{8.52}$$

and with $w_1 = 0$ as

$$w_{i+1} = \sum_{k=1}^{i}(u_k - v_{k+1}), \quad i = 1, \ldots, n-1. \tag{8.53}$$

Using Eqs. (8.45), (8.48)–(8.49) in Eq. (8.53), one obtains after some algebra

$$w_i = -\Omega_{Ai} j_{Ai} - \Omega_{Bi} j_{Bi}, \quad i = 2, \ldots, n-1, \tag{8.54}$$

which is a commonly used equation. Velocity w_i can be determined by using Eqs. (8.48) and (8.49) in Eq. (8.53) for the case when the partial molar volumes of components A and B in the individual phases i are not explicitly known. This may be the case for stoichiometric phases, where only the molar volumes Ω_i are known. Then velocity w_i is given by a more complicated expression.

One can also calculate velocity V_i of interface i between the phases i and $i+1$ as

$$V_i = \sum_{k=1}^{i} \dot{d}_k = \sum_{k=1}^{i} (u_k - v_k), \quad i = 1, \ldots, n-1, \tag{8.55}$$

and then position X_i of interface i at time t is obtained by integration of Eq. (8.55) in time as

$$X_i = X_{0i} + \int_0^t V_i dt, \quad i = 1, \ldots, n-1, \tag{8.56}$$

with X_{0i} being the position of the interface at time $t = 0$.

Several experimental studies of Kirkendall effect in multiphase systems, as, e.g., [66, 67], show an evident polyfurcation of Kirkendall plane. The inert markers, positioned originally at the contact plane of the diffusion couple, are found at different positions in the new phases nucleated at the contact plane and grown during a certain time of annealing. Thus, it is reasonable to assume that after nucleation at the contact region, the lattice of the new phase carries some of the markers with velocity w_i. Velocity w_i is given by Eq. (8.53) or (8.54). However, a specific interface may pass Kirkendall plane by migration, and then Kirkendall plane with markers originally positioned in a newly nucleated phase i becomes fixed at the migrating interface or remains embedded in another phase. This slightly complicates the treatment of calculation of the positions of several Kirkendall planes. In the model any drag of the markers by moving or any pinning of the interface by markers is not considered.

Let the position of Kirkendall plane k be x_{0k} at time $t = 0$ and let its actual correspondence to a phase l be known. Then w_l is the actual velocity of Kirkendall plane k, and the position of Kirkendall plane k at time t can be calculated as

$$x_k = x_{0k} + \int_0^t w_l dt. \tag{8.57}$$

At any time, the corresponding phase index l must be calculated from the actual position of Kirkendall plane x_k and the actual positions of interfaces X_i. In multiphase Kirkendall experiments starting with a diffusion couple, as shown, e.g., in [59], all the positions x_{0k} and X_{0i} coincide, and they resemble the original position of the contact plane occupied by markers. After contacting both parts of the diffusion couple and heating, the new phases nucleate, and each individual marker becomes situated in the bulk of one of the new phases. This scenario is schematically depicted in Fig. 8.9, and the final stage represents the starting configuration in our model with already furcated Kirkendall plane. Then also no singularity occurs due to zero widths of new phases.

The independent diffusive fluxes j_{Ai} and j_{Bi}, $i = 2, \ldots, n-1$, are free kinetic parameters of the system, unambiguously determining the kinetics of the system. The values of the fluxes j_{Ai} and j_{Bi}, $i = 2, \ldots, n-1$, can be calculated by using TEP.

The Total Gibbs Energy and Its Rate

The total Gibbs energy G of the system can be calculated as

$$G = \sum_{i=1}^{n} \frac{g_i}{\Omega_i} d_i, \tag{8.58}$$

with g_i being the molar Gibbs energies. These quantities are supposed to be known and fixed at a given fixed temperature. Then one can calculate the rate $\dot{G}$ of the total Gibbs energy by using Eqs. (8.48)–(8.51) as

$$\begin{aligned}\dot{G} &= \sum_{i=1}^{n} \frac{g_i}{\Omega_i} \dot{d}_i \\ &= \sum_{i=1}^{n} g_i \left\{ (j_{Ai} - j_{Ai+1}) \frac{1 - r_{i+1}}{r_i - r_{i+1}} - (j_{Bi} - j_{Bi+1}) \frac{r_{i+1}}{r_i - r_{i+1}} \right. \\ &\quad \left. - (j_{Ai-1} - j_{Ai}) \frac{1 - r_{i-1}}{r_{i-1} - r_i} + (j_{Bi-1} - j_{Bi}) \frac{r_{i-1}}{r_{i-1} - r_i} \right\}.\end{aligned} \tag{8.59}$$

The partial derivatives of the rate of the total Gibbs energy with respect to diffusive fluxes are calculated as

$$\frac{\partial \dot{G}}{\partial j_{Ai}} = g_i \left(\frac{1 - r_{i+1}}{r_i - r_{i+1}} + \frac{1 - r_{i-1}}{r_{i-1} - r_i} \right) - g_{i-1} \frac{\Omega_{i-1}}{\Omega_i} \frac{1 - r_i}{r_{i-1} - r_i} - g_{i+1} \frac{\Omega_{i+1}}{\Omega_i} \frac{1 - r_i}{r_i - r_{i+1}} = -F_{Ai}, \quad i = 2, \ldots, n-1, \tag{8.60}$$

$$\frac{\partial \dot{G}}{\partial j_{Bi}} = -g_i \left(\frac{r_{i+1}}{r_i - r_{i+1}} + \frac{r_{i-1}}{r_{i-1} - r_i} \right) + g_{i-1} \frac{\Omega_{i-1}}{\Omega_i} \frac{r_i}{r_{i-1} - r_i} + g_{i+1} \frac{\Omega_{i+1}}{\Omega_i} \frac{r_i}{r_i - r_{i+1}} = -F_{Bi}, \quad i = 2, \ldots, n-1. \tag{8.61}$$

The quantities F_{Ai} and F_{Bi} are the driving forces conjugated to the fluxes j_{Ai} and j_{Bi}.

The Dissipation Function

We assume that the Gibbs energy dissipates by the bulk diffusion only, and no dissipation is associated to the deposition or removal of new atomic layers at the interface and the migration of interfaces. Then the total dissipation function, see [9], can be calculated as

$$Q = \sum_{i=2}^{n-1} RT\Omega_i d_i \left(\frac{j_{Ai}^2}{r_i D_{Ai}^*} + \frac{j_{Bi}^2}{(1 - r_i) D_{Bi}^*} \right) \tag{8.62}$$

with D_{Ai}^* and D_{Bi}^* being the tracer diffusion coefficients of components A and B, respectively, in phase i.

The partial derivatives of Q with respect to the independent fluxes are calculated as

$$\frac{\partial Q}{\partial j_{Ai}} = \frac{2RT\Omega_i d_i j_{Ai}}{r_i D^*_{Ai}}, \quad i=2,\ldots,n-1, \tag{8.63}$$

$$\frac{\partial Q}{\partial j_{Bi}} = \frac{2RT\Omega_i d_i j_{Bi}}{(1-r_i)D^*_{Bi}}, \quad i=2,\ldots,n-1. \tag{8.64}$$

Equations for the System Evolution

Application of the TEP [9] allows determination of the independent diffusive fluxes j_{Ai} and j_{Bi}, $i=2,\ldots,n-1$ given by a set of linear algebraic equations for the fluxes as

$$-\frac{\partial \dot{G}}{\partial j_{Ai}} = \frac{1}{2}\frac{\partial Q}{\partial j_{Ai}} \quad \text{and} \quad -\frac{\partial \dot{G}}{\partial j_{Bi}} = \frac{1}{2}\frac{\partial Q}{\partial j_{Bi}}, \quad i=2,\ldots,n-1. \tag{8.65}$$

The set of linear equations has a diagonal matrix, which can easily be solved as

$$j_{Ai} = F_{Ai}\frac{r_i D^*_{Ai}}{RT\Omega_i d_i} \quad \text{and} \quad j_{Bi} = F_{Bi}\frac{(1-r_i)D^*_{Bi}}{RT\Omega_i d_i}, \quad i=2,\ldots,n-1. \tag{8.66}$$

If the values of d_i, $i=2,\ldots,n-1$, are nearly zero at time $t=0$, then integration of differential equations Eq. (8.51) provides, after insertion of Eqs. (8.48), (8.49), and (8.66), a solution of the parabolic type, namely

$$d_i = \tilde{d}_i\sqrt{t} \quad \text{and} \quad \dot{d}_i = \tilde{d}_i/(2\sqrt{t}) \quad \Rightarrow \quad \dot{d}_i = d_i/(2t), \quad i=2,\ldots,n-1 \tag{8.67}$$

with $\tilde{d}_i$ being time-independent parameters. After eliminating time from Eqs. (8.48), (8.49), (8.51), (8.63), and (8.66), the problem is described by a set of nonlinear equations for $\tilde{d}_i$, $i=2,\ldots,n-1$. One can show that also all other kinetic parameters j_{Ai}, j_{Bi}, u_i, v_i, w_i, and V_i scale with $1/\sqrt{t}$, and thus the changes of the position of each interface $X_i - X_{0i}$, $i=1,\ldots,n-1$, (see Eq. (8.56)) scale with $\sqrt{t}$. This holds also for the changes of the positions of Kirkendall planes $x_k - x_{0k}$ for the case that Kirkendall plane k always remains in the same phase. Under this assumption, the velocities V_i are given by

$$V_i = (X_i - X_{0i})/(2t), \quad i=1,\ldots,n-1. \tag{8.68}$$

Moreover, for Kirkendall plane k remaining always in the same phase i, velocity w_i is given by

$$w_i = (x_k - x_{0k})/(2t). \tag{8.69}$$

Results of Simulations

The model is demonstrated for $n=4$, and integration of Eqs. (8.56) and (8.57) in time is performed numerically. The molar volumes of all phases are chosen as $\Omega_i =$

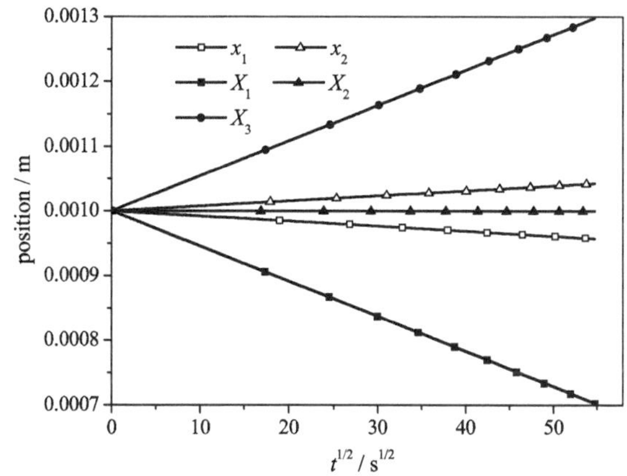

Figure 8.10 Time dependence of Kirkendall planes and of the positions of the interfaces for the sets $\boldsymbol{D}_A^0 = \{10^{-5}, 10^{-5}, 2\cdot10^{-5}, 10^{-5}\}\ \mathrm{m^2/s}$ and $\boldsymbol{D}_B^0 = \{10^{-5}, 2\cdot10^{-5}, 10^{-5}, 10^{-5}\}\ \mathrm{m^2/s}$. With kind permission from Springer Science + Business Media: Journal of Phase Equilibria and Diffusion, Modeling of kinetics of diffusive phase transformation in binary system with multiple phases, 27, 2006, 622–628, Svoboda J., Gamsjäger E., Fischer F.D. & Kozeschnik E.

$7\cdot10^{-6}\ \mathrm{m^3\,mol^{-1}}$ in all phases. Activation energies of all tracer diffusion coefficients are chosen as $E_{Ai} = E_{Bi} = 1.5\cdot10^5\ \mathrm{J\,mol^{-1}}$ and used in $D_{ki}^* = D_{ki}^0 \exp[-E_{ki}/(R_g T)]$, $k = A, B$. The temperature $T = 1000$ K is chosen. The values of g_i are taken from the set $\boldsymbol{g} = \{10^5, 0, 0, 10^5\}\ \mathrm{J\,mol^{-1}}$, the values of r_i from the set $\boldsymbol{r} = \{0.2, 0.4, 0.6, 0.8\}$, and the initial values of d_i are given by $\boldsymbol{d}_0 = \{10^{-3}, 10^{-8}, 10^{-8}, 10^{-3}\}$ m. The initial Kirkendall planes are positioned in the center of the phases 2 and 3. The sets $\boldsymbol{D}_A^0$ and $\boldsymbol{D}_B^0$ of the pre-exponential factors D_{Ai}^0 and D_{Bi}^0 of the tracer diffusion coefficients are taken as the only free parameters.

The first simulation is calculated for a "symmetric" case given by $\boldsymbol{D}_A^0 = \{10^{-5}, 10^{-5}, 2\cdot10^{-5}, 10^{-5}\}\ \mathrm{m^2/s}$ and $\boldsymbol{D}_B^0 = \{10^{-5}, 2\cdot10^{-5}, 10^{-5}, 10^{-5}\}\ \mathrm{m^2/s}$, which is depicted in Fig. 8.10. From this figure it is evident (as is from the following figures) that the parabolic time law holds. The selected symmetric diffusion coefficients ensure the deposition of atoms at the central interface, leading to symmetric bifurcation of Kirkendall plane. When destroying symmetry of the problem by choosing $\boldsymbol{D}_A^0 = \{10^{-5}, 10^{-5}, 3\cdot10^{-5}, 10^{-5}\}\ \mathrm{m^2/s}$ and $\boldsymbol{D}_B^0 = \{10^{-5}, 2\cdot10^{-5}, 10^{-5}, 10^{-5}\}\ \mathrm{m^2/s}$ (i.e., increasing the value of D_{A3}^0), the results do not change qualitatively; compare Fig. 8.10 and Fig. 8.11. Further increasing of D_{A3}^0 ($\boldsymbol{D}_A^0 = \{10^{-5}, 10^{-5}, 5\cdot10^{-5}, 10^{-5}\}\ \mathrm{m^2/s}$ and $\boldsymbol{D}_B^0 = \{10^{-5}, 2\cdot10^{-5}, 10^{-5}, 10^{-5}\}\ \mathrm{m^2/s}$) leads, however, to the effect that Kirkendall plane, originally positioned in phase 3, very quickly becomes positioned in phase 2. Then both practically unseparated Kirkendall planes remain in the bulk of phase 2 as shown in Fig. 8.12 and, thus, no bifurcation of Kirkendall plane is observed. The sets $\boldsymbol{D}_A^0 = \{10^{-5}, 2\cdot10^{-5}, 10^{-5}, 10^{-5}\}\ \mathrm{m^2/s}$ and $\boldsymbol{D}_B^0 = \{10^{-5}, 10^{-5}, 2\cdot10^{-5}, 10^{-5}\}\ \mathrm{m^2/s}$ pro-

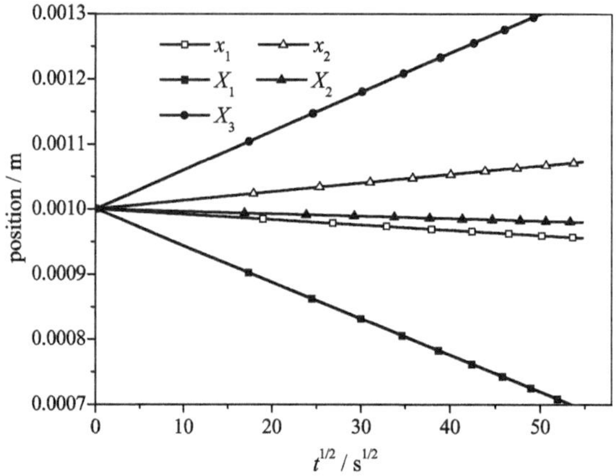

Figure 8.11 Time dependence of Kirkendall planes and of the positions of the interfaces for the sets $D_A^0 = \{10^{-5}, 10^{-5}, 3 \cdot 10^{-5}, 10^{-5}\}$ m^2/s and $D_B^0 = \{10^{-5}, 2 \cdot 10^{-5}, 10^{-5}, 10^{-5}\}$ m^2/s. With kind permission from Springer Science + Business Media: Journal of Phase Equilibria and Diffusion, Modeling of kinetics of diffusive phase transformation in binary system with multiple phases, 27, 2006, 622–628, Svoboda J., Gamsjäger E., Fischer F.D. & Kozeschnik E.

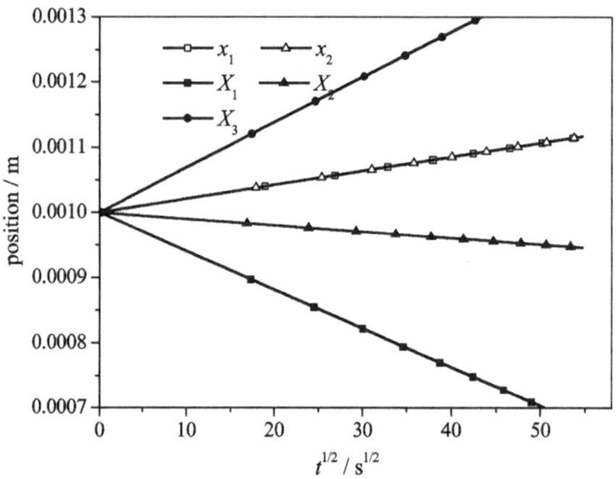

Figure 8.12 Time dependence of the Kirkendall planes and of the positions of the interfaces for the sets $D_A^0 = \{10^{-5}, 10^{-5}, 5 \cdot 10^{-5}, 10^{-5}\}$ m^2/s and $D_B^0 = \{10^{-5}, 2 \cdot 10^{-5}, 10^{-5}, 10^{-5}\}$ m^2/s. With kind permission from Springer Science + Business Media: Journal of Phase Equilibria and Diffusion, Modeling of kinetics of diffusive phase transformation in binary system with multiple phases, 27, 2006, 622–628, Svoboda J., Gamsjäger E., Fischer F.D. & Kozeschnik E.

voke collection of atoms at the central interface. Then both originally slightly separated Kirkendall planes move together, and they become fixed at the central interface, as it is shown in Fig. 8.13. This configuration remains qualitatively unchanged for increasing D_{A2}^0, as is shown in Fig. 8.14.

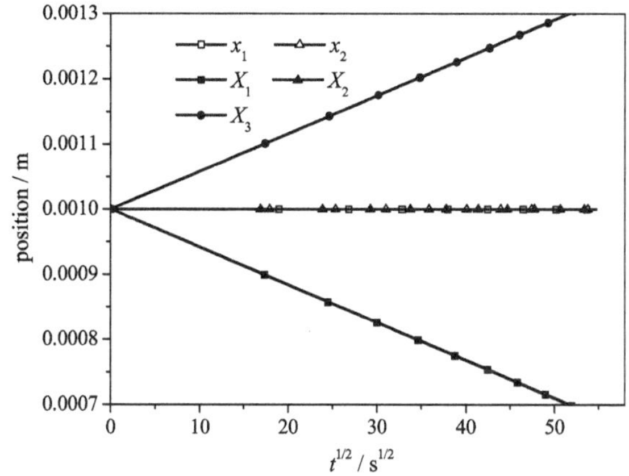

Figure 8.13 Time dependence of the Kirkendall planes and of the positions of the interfaces for the sets $D_A^0 = \{10^{-5}, 2 \cdot 10^{-5}, 10^{-5}, 10^{-5}\}$ m^2/s and $D_B^0 = \{10^{-5}, 10^{-5}, 2 \cdot 10^{-5}, 10^{-5}\}$ m^2/s. With kind permission from Springer Science + Business Media: Journal of Phase Equilibria and Diffusion, Modeling of kinetics of diffusive phase transformation in binary system with multiple phases, 27, 2006, 622–628, Svoboda J., Gamsjäger E., Fischer F.D. & Kozeschnik E.

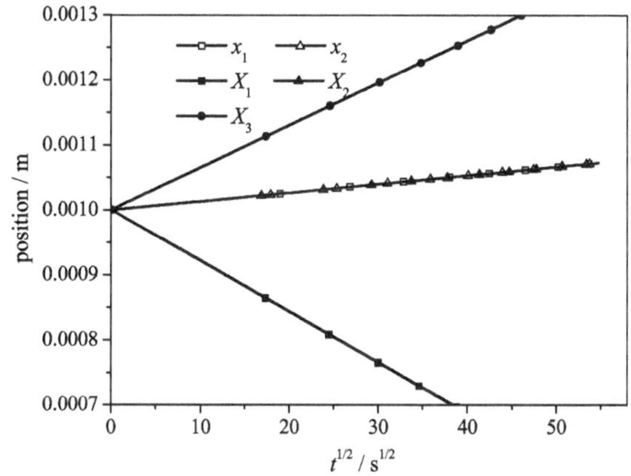

Figure 8.14 Time dependence of Kirkendall planes and of the positions of the interfaces for the sets $D_A^0 = \{10^{-5}, 5 \cdot 10^{-5}, 10^{-5}, 10^{-5}\}$ m^2/s and $D_B^0 = \{10^{-5}, 10^{-5}, 2 \cdot 10^{-5}, 10^{-5}\}$ m^2/s. With kind permission from Springer Science + Business Media: Journal of Phase Equilibria and Diffusion, Modeling of kinetics of diffusive phase transformation in binary system with multiple phases, 27, 2006, 622–628, Svoboda J., Gamsjäger E., Fischer F.D. & Kozeschnik E.

Discussion of Results of Simulations

The simulations clearly demonstrate that the number of furcation of Kirkendall planes cannot increase during the "computer experiment." This indicates that, if no markers are already present in a newly nucleated phase at the beginning of the simulation, then the phase remains free of markers, unless an interface does not pass a Kirkendall plane. This can be understood so that a Kirkendall plane enters the phase from another phase. Such a new phase without markers can be created by a late nucleation at a newly created marker-free interface. Thus, one can deduce that furcating of Kirkendall plane cannot occur, unless the nucleation of the new phases ensures positioning of the markers into at least two new phases as, e.g., according to the scenario shown in Fig. 8.8. Such a nucleation process is most probably tacitly assumed also by Paul et al. [59], meaning that the nucleation automatically redistributes the markers into the bulks of all newly nucleated phases. Then the phases with no markers are those which loose the markers by interface migration. From this point of view, the model in [59] and the current model are in accordance.

8.5 FORMATION OF MULTIPLE STOICHIOMETRIC PHASES IN BINARY SYSTEMS BY COMBINED BULK AND GRAIN BOUNDARY DIFFUSION – EXPERIMENTS AND MODELING BY THE TEP

Nearly in all modeling studies of reactive diffusion only bulk diffusion is accounted for. A typical example is a solid Cu–Sn couple, as it was studied in detail by Paul in his Thesis [68], by Paul et al. in [69,70], and most recently by Kumar et al. [71]. However, special soldering systems exist, where the nucleation and growth of the intermetallic phases is extremely fast. The phases have rather the microstructure of fine polycrystals than that of a single crystal, and grain boundary diffusion interacts strongly with bulk diffusion. The effect of interacting bulk and grain boundary diffusion occurs predominantly in the η-phase Cu_6Sn_5. The η-phase shows a "scalloped" morphology, see, e.g., [68,72,73] and later [74,75], which can only be explained by intensive nucleation of the η-phase and the cooperative diffusion through the bulk and along the grain boundaries.

A model is presented below, where the macroscopic diffusive fluxes of the components A, B in some of the newly formed phases between the parent phases consist of bulk and grain boundary contributions. The TEP, as used in [45,46], is engaged to analyze the individual contributions to the fluxes as well as the kinetics of the system.

The Model

We shortly outline the problem setting as it was already presented in [45,46]. Therefore, we avoid repeating corresponding figures and refer the reader to [45,46]. Details with respect to the applied equations (kinematics, mass balance) can also be taken from [45, 46].

Let us assume a binary system of components A and B, forming n stable stoichiometric phases $A_{r_i}B_{1-r_i}$ with r_i being the mole fraction of the component A in the phase i, $0 \le r_i \le 1$, $i = 1, \ldots, n$, and r_i increasing with i. The system is described in detail in the text following Eq. (8.45). However, we assume that the macroscopic diffusive fluxes

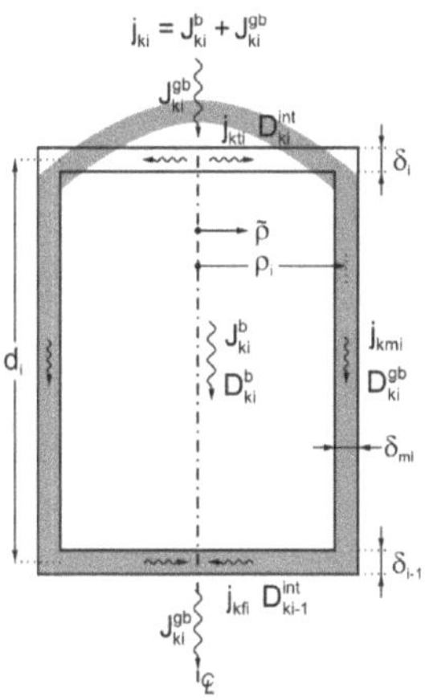

Figure 8.15 Sketch of a scalloped grain with diffusive fluxes ($k = A, B$) in the interior (bulk) of the grain (phase i) and along the grain surface. The layers with thickness $\delta_i, \delta_{i-1}, \delta_{mi}/2$ are assigned to one grain "unit cell." The apparent layer thickness of phase i is d_i, for details see [48]. Reprinted from Acta Materialia, 61, Svoboda, J., Fischer F.D. & Schillinger W., Formation of multiple stoichiometric phases in binary systems by combined bulk and grain boundary diffusion – experiments and model, 32–39, 2013 with permission from Elsevier.

consist of two contributions, namely the flux through the bulk of grains and the flux along a combined path consisting of grain boundaries and interfaces.

Let us analyze the behavior of an interface between the phases i and $i+1$ where the macroscopic diffusive fluxes j_{Ai}, j_{Bi} and j_{Ai+1}, j_{Bi+1} occur. The velocity of the interface between phases i and $i+1$ relative to the lattice of phase i is denoted by u_i. The velocity of the interface between phases i and $i+1$ relative to the lattice of phase $i+1$ is denoted by v_{i+1}. Respective treatment accounting for the mass conservation for both the A and B components is given by Eqs. (8.46)–(8.57). Diffusive fluxes are assumed to furcate into a bulk contribution and a grain boundary/interface contribution. The balance between the fluxes at each interface is coupled with both the interface migration and the deposition or removal of new atomic layers at the interface, causing a motion of the lattices of the individual phases relative to each other.

Furcating of Diffusive Fluxes

The relevance of grain boundary diffusion in the context of the Kirkendall effect was already shown by Thornton and coworkers in [76] and [77] applying the phase field method.

We introduce now the furcating of the macroscopic diffusive fluxes j_{Ai}, j_{Bi} into bulk and grain boundary/interface contributions, specially adapted for an intermetallic layer (with label "i") consisting of a columnar grain as sketched in Fig. 8.15. The scallops are approximated by cylinders of a height being identical to the layer thickness d_i and a circular top with radius ρ_i representing a unit cell. Insofar the model is similar to that used by researchers in surface science; see, e.g., [78]. However, in our case we consider different flow paths and furcate the macroscopic total fluxes j_{Ai}, j_{Bi} into macroscopic fluxes J^b_{Ai}, J^b_{Bi} through the bulk (superscript b) of the grain in axial direction and in macroscopic diffusive fluxes J^{gb}_{Ai} and J^{gb}_{Bi} (superscript gb) along the grain boundary layer of the unit cell represented by layers of finite thicknesses δ_i on the top, $\delta_{mi}/2$ on the mantle, and δ_{i-1} on the foot of the cylinder (δ_i is the thickness of interface i), see Fig. 8.15. Consequently, the macroscopic diffusive fluxes j_{ki} are given by the sum

$$j_{ki} = J^b_{ki} + J^{gb}_{ki}, \quad k = A, B. \tag{8.70}$$

The fluxes along the boundary layer are correspondingly denoted as j_{kti} at the top, j_{kmi} at the mantle, and j_{kfi} at the foot for $k = A, B$. Mass conservation yields for j_{kti}, j_{kmi}, j_{kfi}:

$$j_{kti}(\tilde{\rho}) = J^{gb}_{ki}\tilde{\rho}/(2\delta_i) \quad \text{for } 0 \le \tilde{\rho} \le \rho_i, \tag{8.71}$$

$$j_{kmi} = J^{gb}_{ki}\tilde{\rho}/(\delta_{mi}) \quad \text{at } \tilde{\rho} = \rho_i, \tag{8.72}$$

$$j_{kfi}(\tilde{\rho}) = J^{gb}_{ki}\tilde{\rho}/(2\delta_{i-1}) \quad \text{for } 0 \le \tilde{\rho} \le \rho_i. \tag{8.73}$$

Different concepts exist in the literature to describe the interaction of bulk diffusion, grain boundary diffusion, and finally surface segregation, e.g., a modified Darken approach, the so-called "Darken–du Plessis" concept, developed by du Plessis et al. (see, e.g., [79,80]), and a recent application [79] with the concentration rate directly related to the gradient of the chemical potential. In contrast, we prefer to avoid some rather artificial assumptions and will engage the TEP to find the evolution equations for combined bulk and grain boundary/interface diffusion. A first step towards this goal was already done by Kozeschnik et al. [81] by providing the evolution of precipitates at grain boundaries. The task is now to describe the kinetics and Kirkendall effect in a diffusion couple, where the individual stoichiometric phases may grow by combined bulk and grain boundary/interface diffusion. The kinetics of the system is unambiguously determined by the diffusive fluxes J^b_{Ai}, J^{gb}_{Ai} and J^b_{Bi}, J^{gb}_{Bi}, $i = 2, \ldots, n-1$, which represent independent kinetic variables. The actual values of these fluxes are determined by application of the TEP in a similar way to [45,46].

Total Gibbs Energy and Its Rate

The total Gibbs energy per unit cross-section is given as

$$G = \sum_{i=1}^{n}\left(\frac{g_i}{\Omega_i} + \frac{\gamma_{mi}}{\rho_i}\right)d_i + \sum_{i=1}^{n-1}\gamma_{tfi} \tag{8.74}$$

with the molar Gibbs energies g_i and the molar volumes Ω_i supposed as fixed given quantities at a given temperature. The specific grain boundary energy γ_{mi} and the interface energy γ_{tfi} represent fixed quantities, too; from now on, we introduce $\tilde{g}_i = g_i/\Omega_i + \gamma_{mi}/\rho_i$ with the dimension J/m^2. Then the rate of the total Gibbs energy can be calculated by using Eqs. (8.48)–(8.51) and (8.70) as

$$\dot{G} = \sum_{i=1}^{n} \tilde{g}_i \left\{ \begin{array}{l} (J_{Ai}^b + J_{Ai}^{gb} - J_{Ai+1}^b - J_{Ai+1}^{gb}) \dfrac{1 - r_{i+1}}{r_i - r_{i+1}} - (J_{Bi}^b + J_{Bi}^{gb} - J_{Bi+1}^b - J_{Bi+1}^{gb}) \dfrac{r_{i+1}}{r_i - r_{i+1}} - \\ (J_{Ai-1}^b + J_{Ai-1}^{gb} - J_{Ai}^b - J_{Ai}^{gb}) \dfrac{1 - r_{i-1}}{r_{i-1} - r_i} + (J_{Bi-1}^b + J_{Bi-1}^{gb} - J_{Bi}^b - J_{Bi}^{gb}) \dfrac{r_{i-1}}{r_{i-1} - r_i} \end{array} \right\}. \tag{8.75}$$

The partial derivatives with respect to diffusive fluxes follow as

$$\frac{\partial \dot{G}}{\partial J_{Ai}^b} = \frac{\partial \dot{G}}{\partial J_{Ai}^{gb}} = \tilde{g}_i \left(\frac{1 - r_{i+1}}{r_i - r_{i+1}} + \frac{1 - r_{i-1}}{r_{i-1} - r_i} \right) - \tilde{g}_{i-1} \frac{\Omega_{i-1}}{\Omega_i} \frac{1 - r_i}{r_{i-1} - r_i} - \tilde{g}_{i+1} \frac{\Omega_{i+1}}{\Omega_i} \frac{1 - r_i}{r_i - r_{i+1}} = -F_{Ai},$$
$$i = 2, \ldots, n-1, \tag{8.76}$$

$$\frac{\partial \dot{G}}{\partial J_{Bi}^b} = \frac{\partial \dot{G}}{\partial J_{Bi}^{gb}} = -\tilde{g}_i \left(\frac{r_{i+1}}{r_i - r_{i+1}} + \frac{r_{i-1}}{r_{i-1} - r_i} \right) + \tilde{g}_{i-1} \frac{\Omega_{i-1}}{\Omega_i} \frac{r_i}{r_{i-1} - r_i} + \tilde{g}_{i+1} \frac{\Omega_{i+1}}{\Omega_i} \frac{r_i}{r_i - r_{i+1}} = -F_{Bi},$$
$$i = 2, \ldots, n-1. \tag{8.77}$$

The quantities F_{Ai}, F_{Bi} can be understood as the driving forces conjugated to the fluxes J_{Ai}^b and J_{Ai}^{gb}, J_{Bi}^b and J_{Bi}^{gb}.

Dissipation Function

In this section we study the dissipation due to diffusion in the bulk and along the grain boundaries/interfaces and calculate all quantities per unit area.

In phase i the dissipation function Q_{it}^{gb} due to interface diffusion along the top follows as

$$Q_{it}^{gb} = \frac{2\pi \delta_i \Omega_i RT}{\pi \rho_i^2} \int_0^{\rho_i} \left(\frac{(j_{Ati})^2}{r_i D_{Ai}^{int}} + \frac{(j_{Bti})^2}{(1 - r_i) D_{Bi}^{int}} \right) \tilde{\rho} \, d\tilde{\rho}, \tag{8.78}$$

with D_{ki}^{int} being the tracer diffusion coefficient of component k in interface i. Inserting Eq. (8.71) into Eq. (8.78) yields the dissipation function Q_{it}^{gb} as

$$Q_{it}^{gb} = RT\Omega_i \frac{\rho_i^2}{8\delta_i} \left(\frac{(J_{Ai}^{gb})^2}{r_i D_{Ai}^{int}} + \frac{(J_{Bi}^{gb})^2}{(1 - r_i) D_{Bi}^{int}} \right). \tag{8.79}$$

In an analogous way one calculates the dissipation function Q_{if}^{gb} due to interface diffusion along the foot layer as

$$Q_{if}^{gb} = RT\Omega_i \frac{\rho_i^2}{8\delta_{i-1}} \left(\frac{(J_{Ai}^{gb})^2}{r_i D_{Ai-1}^{int}} + \frac{(J_{Bi}^{gb})^2}{(1-r_i) D_{Bi-1}^{int}} \right). \tag{8.80}$$

For the dissipation function Q_{im}^{gb} due to interface diffusion along the mantle one finds

$$Q_{im}^{gb} = RT\Omega_i \frac{\rho_i d_i}{\delta_{mi}} \left(\frac{(J_{Ai}^{gb})^2}{r_i D_{Ai}^{gb}} + \frac{(J_{Bi}^{gb})^2}{(1-r_i) D_{Bi}^{gb}} \right). \tag{8.81}$$

Note that D_{ki}^{gb} is the tracer diffusion coefficient of component k in the mantle in phase i.

The dissipation function Q_i^b due to bulk diffusion in the phase i follows with the tracer diffusion coefficient D_{ki}^b in the bulk for component k as

$$Q_i^b = RT\Omega_i d_i \left(\frac{(J_{Ai}^b)^2}{r_i D_{Ai}^b} + \frac{(J_{Bi}^b)^2}{(1-r_i) D_{Bi}^b} \right). \tag{8.82}$$

Then the total dissipation function Q in the system is given by

$$Q = \sum_{i=2}^{n-1} (Q_{it}^{gb} + Q_{im}^{gb} + Q_{if}^{gb} + Q_i^b). \tag{8.83}$$

The partial derivatives of the total dissipation function Q with respect to the independent kinetic variables read for $2 \le i \le n-1$ as

$$\frac{1}{2}\frac{\partial Q}{\partial J_{Ai}^b} = \frac{RT\Omega_i d_i}{r_i D_{Ai}^b} J_{Ai}^b, \qquad \frac{1}{2}\frac{\partial Q}{\partial J_{Bi}^b} = \frac{R_g T\Omega_i d_i}{(1-r_i) D_{Bi}^b} J_{Bi}^b, \tag{8.84}$$

$$\frac{1}{2}\frac{\partial Q}{\partial J_{Ai}^{gb}} = \frac{RT\Omega_i \rho_i}{r_i} \left(\frac{\rho_i}{8\delta_i D_{Ai}^{int}} + \frac{d_i}{\delta_{mi} D_{Ai}^{gb}} + \frac{\rho_i}{8\delta_{i-1} D_{Ai-1}^{int}} \right) J_{Ai}^{gb}, \tag{8.85}$$

$$\frac{1}{2}\frac{\partial Q}{\partial J_{Bi}^{gb}} = \frac{RT\Omega_i \rho_i}{1-r_i} \left(\frac{\rho_i}{8\delta_i D_{Bi}^{int}} + \frac{d_i}{\delta_{mi} D_{Bi}^{gb}} + \frac{\rho_i}{8\delta_{i-1} D_{Bi-1}^{int}} \right) J_{Bi}^{gb}. \tag{8.86}$$

Kinetic Equations of the System

The evolution of the system is given according to the TEP (see, e.g., [45,46]) as

$$\frac{1}{2}\frac{\partial Q}{\partial J_{ki}^b} = -\frac{\partial \dot{G}}{\partial J_{ki}^b} = F_{ki}, \qquad \frac{1}{2}\frac{\partial Q}{\partial J_{ki}^{gb}} = -\frac{\partial \dot{G}}{\partial J_{ki}^{gb}} = F_{ki}, \qquad i = 2, \ldots, n-1, \; k = A, B. \tag{8.87}$$

Solving the equations with respect to the independent kinetic variables J_{Ai}^{b}, J_{Ai}^{gb} and J_{Bi}^{b}, J_{Bi}^{gb}, $i = 2, \ldots, n-1$, provides

$$J_{Ai}^{b} = \frac{r_i D_{Ai}^{b} F_{Ai}}{RT\Omega_i d_i}, \qquad J_{Bi}^{b} = \frac{(1-r_i) D_{Bi}^{b} F_{Bi}}{RT\Omega_i d_i}, \tag{8.88}$$

$$J_{Ai}^{gb} = \frac{r_i F_{Ai}}{RT\Omega_i \rho_i}\left(\frac{\rho_i}{8\delta_i D_{Ai}^{int}} + \frac{d_i}{\delta_{mi} D_{Ai}^{gb}} + \frac{\rho_i}{8\delta_{i-1i} D_{Ai-1}^{int}}\right)^{-1}, \tag{8.89}$$

$$J_{Bi}^{gb} = \frac{(1-r_i) F_{Bi}}{RT\Omega_i \rho_i}\left(\frac{\rho_i}{8\delta_i D_{Bi}^{int}} + \frac{d_i}{\delta_{mi} D_{Bi}^{gb}} + \frac{\rho_i}{8\delta_{i-1i} D_{Bi-1}^{int}}\right)^{-1}. \tag{8.90}$$

Then the macroscopic diffusive fluxes j_{Ai} and j_{Bi} are given by

$$j_{Ai} = J_{Ai}^{b} + J_{Ai}^{gb} = \frac{r_i F_{Ai}}{RT\Omega_i d_i}\left[D_{Ai}^{b} + \frac{d_i}{\rho_i}\left(\frac{\rho_i}{8\delta_i D_{Ai}^{int}} + \frac{d_i}{\delta_{mi} D_{Ai}^{gb}} + \frac{\rho_i}{8\delta_{i-1} D_{Ai-1}^{int}}\right)^{-1}\right], \tag{8.91}$$

$$j_{Bi} = J_{Bi}^{b} + J_{Bi}^{gb} = \frac{(1-r_i) F_{Bi}}{RT\Omega_i d_i}\left[D_{Bi}^{b} + \frac{d_i}{\rho_i}\left(\frac{\rho_i}{8\delta_i D_{Bi}^{int}} + \frac{d_i}{\delta_{mi} D_{Bi}^{gb}} + \frac{\rho_i}{8\delta_{i-1} D_{Bi-1}^{int}}\right)^{-1}\right]. \tag{8.92}$$

The expressions in square brackets of Eqs. (8.91) and (8.92) can be considered as effective diffusion coefficients. As the values of d_i evolve with time, the effective diffusion coefficients must be applied in an updated form. Moreover, the values of ρ_i may develop according to any established grain growth law; see, e.g., [27]. The evolution of the system is then given by time integration of equations (8.48)–(8.57) as presented in [45,46].

Here a short discussion on effective diffusion coefficients seems to be useful. We can base this discussion on an elementary study by Belova and Murch [82]. They study different arrangements of grains and grain boundaries in relation to the diffusion direction (parallel, series, grains in a square pattern). In our case we have columnar grains with both types of interfaces, see Fig. 8.15, namely a parallel arrangement of the mantle grain boundary and a series arrangement of the top and foot interfaces. In addition the fluxes in these interfaces are coupled by the mass conservation, Eqs. (8.71)–(8.73). Insofar the configuration dealt with in this paper cannot directly be related to one of the three arrangements discussed above. However, if one assumes that the diffusion coefficients of the top and foot interfaces tend to infinity, $D_{ki}^{\text{int}}, D_{ki-1}^{\text{int}} \to \infty$, $k = A, B$, then the bulk and the mantle grain boundary form a parallel arrangement, which Hart [83] investigated already in 1957, concluding that the effective diffusion coefficient D_{eff} follows as

$$D_{eff} = fD^{gb} + (1-f)D^{b}, \tag{8.93}$$

where f is the volume fraction of the grain boundary. In the actual case f can be calculated as $f = \delta_{mi}/\rho_i \ll 1$. With the above assumptions we find for D_{eff} from Eqs. (8.91) and (8.92)

$$D_{eff,ki} = \frac{\delta_{mi}}{\rho_i} D_{ki}^{gb} + D_{ki}^{b}, \tag{8.94}$$

representing a Hart-type relation.

Examples

Let us consider the Cu–Sn couple with the phase sequence Cu–Cu_3Sn (ε-phase)–Cu_6Sn_5 (η-phase)–Sn as it was studied by Paul [68] and by Paul et al. in [69,70] and most recently by Kumar et al. [71]. Although material data can be found in those publications, we will collect them below.

The molar volumes of solid Cu and solid Sn are calculated as the ratio of standard atomic weight and the density (http://en.wikipedia.org/Copper and Tin) as $\Omega_{Cu} = 7.11 \cdot 10^{-6}$ m^3/mol, $\Omega_{Sn} = 16.31 \cdot 10^{-6}$ m^3/mol and for liquid Sn as $\Omega_{Sn}^{L} = 16.98 \cdot 10^{-6}$ m^3/mol. The molar volumes of the ε- and η-phase are calculated from atomic weights of Cu and Sn and the densities of the ε- and η-phase. These quantities are taken from a Research Publication of the Materials Performance Group as 8.90 g/cm^3 and 8.28 g/cm^3, respectively. Then the molar volumes follow as $\Omega_\varepsilon = 8.69 \cdot 10^{-6}$ m^3/mol, $\Omega_\eta = 10.70 \cdot 10^{-6}$ m^3/mol.

The Gibbs energies g_i for the pure phases, as well as the ε- and η-phase, can be taken from Shim et al. [84] and can easily be programmed. Here we would like to mention that a contribution due the hydrostatic stress state σ_H, i.e., $-\Omega_i\sigma_H$, is not added to the chemical potentials. As Klinger and Rabkin have shown in several papers (see, e.g., [85, 86]), diffusion of components from a liquid layer into the grain boundaries of a solid phase beneath the liquid layer may cause a wedge-like formed second phase, whose shape is controlled by the local stress state. However, as one can see from Fig. 8.17, such wedges do not appear in the grain boundaries of the solid phase beneath the liquid (tin) layer. Therefore, it is justified to omit the hydrostatic stress term in the Gibbs energy for the system under investigation.

The following relations for the tracer diffusion coefficients in the bulk are used:

$$\begin{aligned}
(\varepsilon\text{-phase})\quad & D^b_{Cu\varepsilon} = 1.8 \cdot 10^{-8} \exp\left(-78.8\ \text{kJ mol}^{-1}/RT\right)\ \text{m}^2/\text{s},\\
& D^b_{Sn\varepsilon} = 7.9 \cdot 10^{-10} \exp\left(-79.7\ \text{kJ mol}^{-1}/RT\right)\ \text{m}^2/\text{s};\\
(\eta\text{-phase})\quad & D^b_{Cu\eta} = 6.2 \cdot 10^{-8} \exp\left(-80.5\ \text{kJ mol}^{-1}/RT\right)\ \text{m}^2/\text{s},\\
& D^b_{Sn\eta} = 5.9 \cdot 10^{-7} \exp\left(-85.2\ \text{kJ mol}^{-1}/RT\right)\ \text{m}^2/\text{s}.
\end{aligned}$$

The data above are taken from Paul et al. [70, Fig. 6]. Here we would like to refer also to the possibility to calculate the a.m. diffusion coefficients in the intermetallics by density functional theory, see Ulrich et al. [87].

The tracer (i.e., self-) diffusion coefficients of Sn and Cu in liquid Sn are taken from Smithells et al. [88] for Sn and from [89] for Cu as

$$D^L_{\mathrm{Cu}} = 3.66 \cdot 10^{-8} \exp\left(-11.6\ \mathrm{kJ\,mol^{-1}}/RT\right)\ \mathrm{m^2/s},$$
$$D^L_{\mathrm{Sn}} = 3.18 \cdot 10^{-8} \exp\left(-11.6\ \mathrm{kJ\,mol^{-1}}/RT\right)\ \mathrm{m^2/s}.$$

Finally, let us report the grain boundary tracer diffusion coefficient for Cu from [90] as

$$D^{gb}_{\mathrm{Cu}}\delta = 8.8 \cdot 10^{-14} \exp\left(-120\ \mathrm{kJ\,mol^{-1}}/RT\right)\ \mathrm{m^3/s}$$

with δ being the grain boundary thickness. Unfortunately, grain boundary and interface diffusion coefficients for Cu and Sn in the ε- and η-phase are not at our disposal.

First let us compare two very recent investigations [70,71] on Cu–Sn solid diffusion couples performed at 200 °C over approximately 10 days. Paul and co-workers [70] report a thickness of (9 ± 3.6) μm for the ε-phase and (16.4 ± 2.1) μm for the η-phase and Kumar et al. [71], (4.1 ± 1.8) μm for the ε-phase and (7.9 ± 3.4) μm for the η-phase. Although the results of both groups differ by a factor of approximately 5 in their kinetics (if a parabolic growth law is assumed), both groups confirm their experimental results by own simulations.

Irritated, but also motivated by this discrepancy, we used our model with the tracer diffusion coefficients and assumed no grain boundary/interface diffusion ($\rho_i \to \infty$ in the effective diffusion coefficients in Eqs. (8.90)–(8.91)). We found nearly perfect agreement for the thickness of the ε-phase and η-phase and the positions of the furcated Kirkendall planes relative to the position of the interfaces with the results of Paul and coworkers, see Fig. 10 in [70], and our Fig. 8.16, where the positions of the interfaces as well of Kirkendall planes are marked by vertical lines along the length coordinate axis ($0 \le x \le 45$ μm) in the reference configuration.

Any information about the absolute position of the interfaces, as well the shrinkage of the specimen, cannot be deduced from Fig. 10 in [70]. Our results are referred to the spatially fixed left end of the specimen and show shrinkage in the transformed region of nearly 8%. This effect is discussed by Kumar et al. [71], whose results we can reproduce only if the tracer diffusion coefficients in our simulations are reduced by a factor of 5. Moreover, Kumar et al. [71] do not report any furcating of Kirkendall plane, which is obviously situated in their results in the η-phase.

Finally, let us report a further experiment performed with a strip of Cu–4 wt.% Sn alloy of thickness 0.8 mm in contact at both surfaces with the liquid Sn–1 wt.% Cu at 250 °C for 5 s. The liquid is saturated with respect to the neighboring η-phase. Thus, no flux of Cu into the liquid exists, which means that the η-phase does not dissolve in the liquid phase. In this case the liquid behaves as a stoichiometric phase with a fixed chemical composition, which is in agreement with the preposition of the model. This situation differs to the experimental settings in several works as Lord et al. [91],

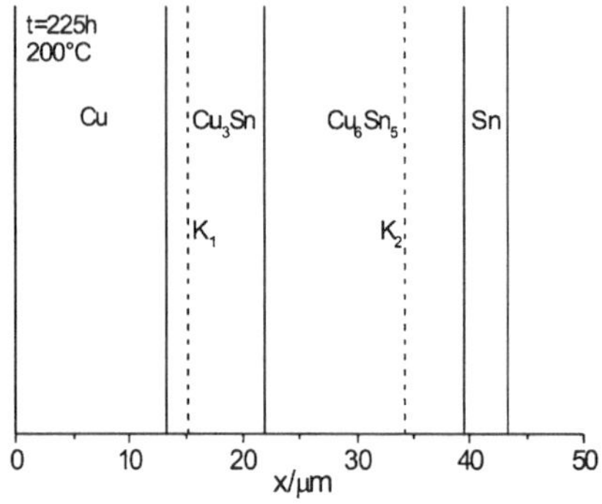

Figure 8.16 Sketch of the phase arrangement in a Cu–Sn diffusion couple as investigated by Paul et al. [70] and Kumar et al. [71]. Calculated positions of the interfaces and Kirkendall planes after 225 h, marked by vertical lines along the length coordinate axis ($0 \leq x \leq 45$ μm) in the reference configuration. The position of the interface in the starting configuration was $x = 25$ μm. Reprinted from Acta Materialia, 61, Svoboda, J., Fischer F.D. & Schillinger W., Formation of multiple stoichiometric phases in binary systems by combined bulk and grain boundary diffusion – experiments and model, 32–39, 2013 with permission from Elsevier.

Gong et al. [92], and Hsiao et al. [93], where a Cu-free solder was used, giving rise to an interaction between the η-phase and the solder. The temperature in the strip is equilibrated during a fraction of a second (a study not mentioned here estimates this time as $4 \cdot 10^{-3}$ s). Thus, the experiment can be treated as isothermal. During the time of 5 s an 80 nm thick ε-phase layer and a 1000 nm thick η-phase layer grow at the Cu–4 wt.% Sn strip (see Fig. 8.17). The ε-phase exhibits an extremely fine grain structure characterized by a grain size of $2\rho_\varepsilon = 50$ nm, and the η-phase exhibits a scallop morphology characterized by the grain size of $2\rho_\eta = 600$ nm. The scallops are formed according to the dihedral angle given by the local equilibrium of specific energies of the interface between the η-phase and the liquid phase Sn–1 wt. %Cu and of the η-phase grain boundary.

If the growth of the layers is simulated by assuming no grain boundary/interface diffusion, i.e., by using the bulk diffusion ($\rho_i \to \infty$ in the effective diffusion coefficients in Eqs. (8.90)–(8.91)), the calculated kinetics of the system is nearly two orders of magnitude slower than the observed one (after 5 s the ε-phase is 114 nm thick and the η-phase is 55 nm thick). This fact indicates that it is necessary to engage the grain boundary/interface diffusion to meet the experimental observation. Thus, a radius ρ_ε of 25 nm is used for the ε-phase, and a radius ρ_η of 300 nm is used for the η-phase. The grain boundary/interface thicknesses δ_ε, δ_η are selected as 0.3 nm. In the first attempt to meet the experimental observation, all grain boundary/interface diffusion coefficients are approximated by the diffusion coefficients D^L_{Cu}, D^L_{Sn} for the liquid phase except the diffusion coefficients of Cu and Sn at the interface between the η-phase and the liquid Sn–1 wt.% Cu layer, which are taken as infinite. The kinetics of the system in this case

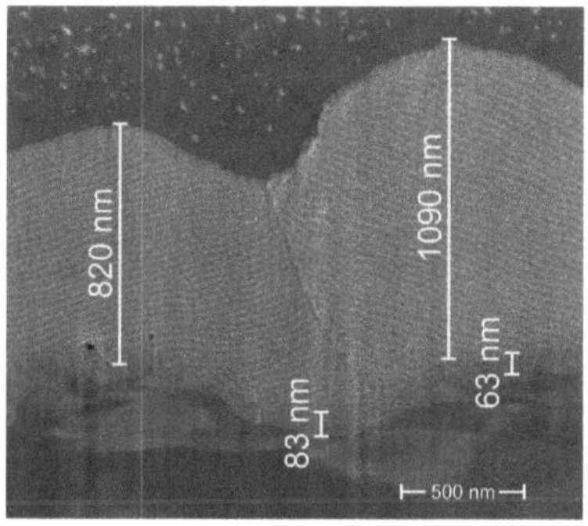

Figure 8.17 Cross-section through a strip of Cu–4 wt.% Sn in contact with liquid Sn–1 wt.% Cu at 250 °C for 5 s. The ε-layer is marked by its thickness of 63 to 83 nm, and the η-layer is marked by its thickness of 820 to 1090 nm. Below the ε-layer the Cu–4 wt.% Sn solid layer is situated. Above the η-layer an Sn–1 wt.% Cu originally liquid layer (but now a solid layer with tiny η-precipitates (gray)) is situated, for details see [48]. Focused Ion Beam image Courtesy of Chair of Functional Materials, Department Materials Science and Engineering, Saarland University, D-66123 Saarbrücken, Germany. Reprinted from Acta Materialia, 61, Svoboda, J., Fischer F.D. & Schillinger W., Formation of multiple stoichiometric phases in binary systems by combined bulk and grain boundary diffusion – experiments and model, 32–39, 2013 with permission from Elsevier.

is much faster than the observed one (after 5 s the ε-phase is 623 nm thick and the η-phase is 36550 nm thick). To meet the experimental observation, it is obviously necessary to reduce the diffusions coefficients. Compared to the former case, the diffusion coefficients (for Sn and Cu, abbreviated as X) are set at the Cu/ε-phase interface as $D_X^L/1.2 \cdot 10^6$ at the mantle as $D_X^L/15$ and as infinite at the η-phase/liquid Sn–1 wt.% Cu interface. The mentioned diffusion coefficient $D_{\mathrm{Cu}}^{int} = 2 \cdot 17 \cdot 10^{-15}$ m^2 s^{-1} ranges between $D_{\mathrm{Cu}}^{gb} = 3.04 \cdot 10^{-16}$ m^2 s^{-1} and $D_{\mathrm{Cu}}^{L} = 2.54 \cdot 10^{-9}$ m^2 s^{-1} for $T = 250\,°\mathrm{C}$. So the value of D_{Cu}^{int} can be considered as an acceptable assumption. Then the simulation provides results (depicted in Fig. 8.18), being in agreement with the experiment and predicting that after 5 s the ε-phase is 81 nm thick and η-phase is 965 nm thick compared to the experimental observations of 80 and 1000 nm, respectively.

The kinetics in Fig. 8.18 shows for the interfaces a slight deviation from the $\sqrt{t}$-behavior. This is due to the fact that the effective diffusion coefficients given by Eqs. (8.90)–(8.91) depend on the phase thickness d_i, i.e., the height of the grains, for fixed values of grain radii ρ_i. In Fig. 8.18 also the positions of Kirkendall planes K_1 and K_2, calculated according to Eq. (8.57), are depicted.

Finally, let us shortly discuss the motion and location of Kirkendall planes. If Kirkendall plane is located in the bulk, it moves with the lattice at a velocity given by Eq. (8.53). In the case that interface i between phases i and $i+1$ crosses Kirkendall plane, the following situations can occur after crossing. If $u_i \leq 0$ and $v_{i+1} \geq 0$, Kirkendall plane remains located at the interface. If $u_i \geq 0$, Kirkendall plane enters either phase i

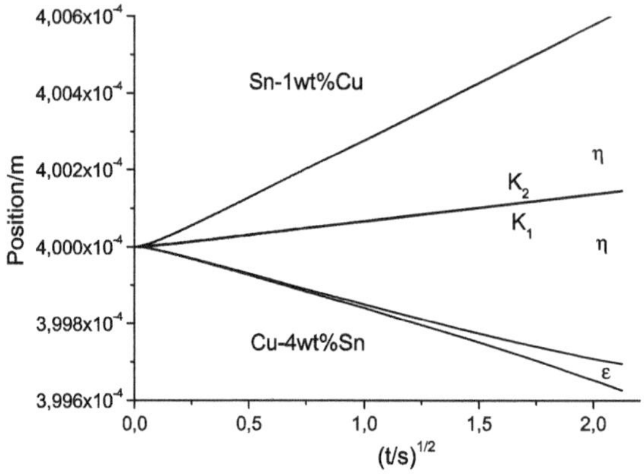

Figure 8.18 Simulated positions of the interfaces measured as the distances from the midplane of the strip with initial thickness of 8×10^{-4} m. Only the interface between Cu–4 wt.% Sn phase and the Sn–1 wt.% Cu phase exists at $t = 0$ coinciding with Kirkendall plane $K_1 \equiv K_2$. Kirkendall planes K_1 and K_2 coincide during the entire system evolution. Reprinted from Acta Materialia, 61, Svoboda, J., Fischer F.D. & Schillinger W., Formation of multiple stoichiometric phases in binary systems by combined bulk and grain boundary diffusion – experiments and model, 32–39, 2013 with permission from Elsevier.

for $u_i > -v_{i+1}$ or phase $i+1$ for $u_i < -v_{i+1}$, furcates and enters both i and $i+1$ phases if $u_i = -v_{i+1}$ (rather rare case). Since we work with effective bulk diffusion coefficients, we do not distinguish between the bulk and the grain boundary in the phase. Consequently, we assume that the marker defining Kirkendall plane located in the grain boundary moves with the same velocity as the marker located in the bulk. This is also in accordance with the assumption that the adjacent grains do not move relatively and that the atoms in the grain boundaries have no reason to move relative to the grains.

ACKNOWLEDGEMENTS

Financial support by the Austrian Federal Government (in particular from the Bundesministerium für Verkehr, Innovation und Technologie and the Bundesministerium für Wirtschaft und Arbeit) and the Styrian Provincial Government, represented by Österreichische Forschungsförderungsgesellschaft mbH and by Steirische Wirtschaftsförderungsgesellschaft mbH, within the research activities of the K2 Competence Centre on "Integrated Research in Materials, Processing and Product Engineering", operated by the Materials Center Leoben Forschung GmbH in the framework of the Austrian COMET Competence Centre Programme, Project A1.17 is gratefully acknowledged.

REFERENCES

[1] L. Onsager, Reciprocal relations in irreversible processes I, Phys. Rev. 37 (1931) 405–426.

[2] H. Ziegler, Some extremum principles in irreversible thermodynamics with applications to continuum mechanics, in: I.N. Sneddon, R. Hill (Eds.), Progress in Solid Mechanics, North-Holland, Amsterdam et al., 1963 (chapter 15).

[3] H. Ziegler, An Introduction to Thermomechanics, North-Holland, Amsterdam et al., 1977 (chapter 15).
[4] F.D. Fischer, J. Svoboda, H. Petryk, Thermodynamic extremal principles for irreversible processes in materials science, Acta Mater. 67 (2014) 1–20.
[5] K. Hackl, F.D. Fischer, On the relation between the principle of maximum dissipation and inelastic evolution given by dissipation potentials, Proc. R. Soc. A 464 (2008) 117–132.
[6] L. Onsager, Theories and problems of liquid diffusion, Ann. New York Acad. Sci. 46 (1945) 241–265.
[7] A.C.F. Cocks, S.P.A. Gill, J.-Z. Pan, et al., Modelling microstructure evolution in engineering materials, in: E. van der Giessen, T.Y. Wu (Eds.), Advances in Applied Mechanics, vol. 36, Academic Press, San Diego, 1998, pp. 81–162.
[8] J. Svoboda, I. Turek, V. Sklenička, Unified thermodynamic treatment of cavity nucleation and growth in high temperature creep, Acta Metall. Mater. 38 (1990) 573–580.
[9] J. Svoboda, I. Turek, F.D. Fischer, Application of the thermodynamic extremal principle to modeling of thermodynamic processes in material sciences, Philos. Mag. 85 (2005) 3699–3707.
[10] H. Petryk, Incremental energy minimization in dissipative solids, CR Mecanique 331 (2003) 469–474.
[11] K. Hackl, R.A. Heinen, A micromechanical model for pretextured polycrystalline shape-memory alloys including elastic anisotropy, Cont. Mech. Thermodyn. 19 (2008) 499–510.
[12] K. Hackl, F.D. Fischer, J. Svoboda, A study on the principle of maximum dissipation for coupled and non-coupled non-isothermal processes in materials, Proc. Royal Soc. A 467 (2011) 1186–1196. Addendum, Proc. Royal Soc. A 467 (2011) 2422–2426.
[13] M. Hillert, J. Ågren, Extremum principles for irreversible processes, Acta Mater. 54 (2006) 2063–2066.
[14] M. Hillert, Phase Equilibria, Phase Diagrams and Phase Transformations, Cambridge University Press, Cambridge, 2008 (chapter 17).
[15] Q. Yang, J.Q. Bao, Y.R. Liu, Asymptotic stability in constrained configuration space for solids, J. Non-Equilib. Thermodyn. 34 (2009) 155–170.
[16] H. Wang, F. Liu, H. Zhai, K. Wang, Application of the maximal entropy production principle to rapid solidification: a sharp interface model, Acta Mater. 60 (2012) 1444–1454.
[17] H. Wang, F. Liu, G.J. Ehlen, D.M. Herlach, Application of the maximal entropy production principle to rapid solidification: a multi-phase-field model, Acta Mater. 61 (2013) 2617–2627.
[18] L.M. Martyushev, V.D. Seleznev, Maximum entropy production principle in physics, chemistry and biology, Phys. Rep. 426 (2006) 1–45.
[19] L.M. Martyushev, A.S. Nazarova, V.D. Seleznev, On the problem of the minimum entropy production in the nonequilibrium stationary state, J. Phys. A 40 (2007) 371–380.
[20] L.M. Martyushev, The maximum entropy production principle: two basic questions, Phil. Trans. Royal Soc. B 365 (2010) 1333–1334.
[21] P. Attard, Theory for non-equilibrium statistical mechanics, Phys. Chem. Chem. Phys. 8 (2006) 3585–3611.
[22] P. Attard, Statistical mechanical theory for steady state systems. VI. Variational principles, J. Chem. Phys. 125 (2006) 214502.
[23] U. Lucia, Probability, ergodicity, irreversibility and dynamical systems, Proc. Royal Soc. A 464 (2008) 1089–1104.
[24] K.R. Rajagopal, A.R. Srinivasa, On thermomechanical restrictions of continua, Proc. Royal Soc. Lond. A 460 (2004) 631–651.
[25] K.R. Rajagopal, A.R. Srinivasa, On the thermomechanics of materials that have multiple natural configurations Part I: viscoelasticity and classical plasticity, Z. Angew. Math. Phys. 55 (2004) 861–893.
[26] H.W. Haslach Jr., Maximum Dissipation Non-equilibrium Thermodynamics and Its Geometric Structure, Springer, New York et al., 2011.
[27] F.D. Fischer, J. Svoboda, P. Fratzl, A thermodynamical approach to grain growth and coarsening, Philos. Mag. 83 (2003) 1075–1093.

[28] J. Svoboda, F.D. Fischer, P. Fratzl, E. Kozeschnik, Modelling of kinetics in multi-component multi-phase systems with spherical precipitates I. – Theory, Mat. Sci. Engng. A 385 (2004) 166–174.
[29] J. Svoboda, F.D. Fischer, D. Vollath, Modeling of formation of binary-phase hollow nanospheres from metallic solid nanospheres, Acta Mater. 57 (2009) 1912–1919.
[30] J. Svoboda, F.D. Fischer, M. Leindl, Transient solute drag in migrating grain boundaries, Acta Mater. 59 (2011) 6556–6562.
[31] Z. Erdélyi, G. Schmitz, Reactive diffusion and stresses in spherical geometry, Acta Mater. 60 (2012) 1807–1817.
[32] M. Kajihara, Analysis of kinetics of reactive diffusion in a hypothetical binary system, Acta Mater. 52 (2004) 1193–1200.
[33] V. Buscaglia, U. Anselmi-Tamburini, On the diffusional growth of compounds with narrow homogeneity range in multiphase binary systems, Acta Mater. 50 (2002) 525–535.
[34] H.L. Wang, Z.B. Wang, K. Lu, Enhanced reactive diffusion on Zn in a nanostructured Fe produced by means of surface mechanical attrition treatment, Acta Mater. 60 (2012) 1762–1770.
[35] C. Milanese, V. Buscaglia, F. Maglia, U. Anselmi-Tamburini, Reactive diffusion in the system vanadium–silicon, Acta Mater. 50 (2002) 1393–1403.
[36] A.M. Gusak, F. Hodaj, G. Schmitz, Flux-driven nucleation at interfaces during reactive diffusion, Philos. Mag. Letters 91 (2011) 610–620.
[37] F. Hodaj, A.M. Gusak, Suppression of intermediate phase nucleation in binary couples with metastable solubility, Acta Mater. 52 (2004) 4305–4315.
[38] V.I. Dybkov, Reaction Diffusion and Solid State Chemical Kinetics, 2nd ed., Trans Tech Publications, Durnten-Zurich, 2010.
[39] F.J.J. van Loo, Multiphase diffusion in binary and ternary solid-state systems, Prog. Solid State Chem. 20 (1990) 47–99.
[40] F.J.J. van Loo, M.C.L.P. Pleumeerkers, A.A. Kondentsov, M.J.H. van Dal, Diffusion studies and re-examination of the Kirkendall effect in the Au–Ni system, J. Alloy Compd. 309 (2000) 132–140.
[41] M.J.H. van Dal, M.C.L.P. Pleumeerkers, A.A. Kondentsov, F.J.J. van Loo, Intrinsic diffusion and Kirkendall effect in Ni–Pd and Fe–Pd solid solutions, Acta Mater. 48 (2000) 385–396.
[42] S. Santra, A. Mondal, P. Aloke, Interdiffusion in the Fe–Pd system, Metall. Mater. Trans. A 43 (2012) 791–795.
[43] F.D. Fischer, N.K. Simha, Influence of material flux on the jump relations at a singular interface in a multicomponent solid, Acta Mech. 171 (2004) 213–223.
[44] F.D. Fischer, T. Waitz, D. Vollath, N.K. Simha, On the role of surface energy and surface stress in phase-transforming nanoparticles, Prog. Mater. Sci. 53 (2008) 481–527.
[45] J. Svoboda, E. Gamsjäger, F.D. Fischer, E. Kozeschnik, Modeling of kinetics of diffusive phase transformation in binary systems with multiple stoichiometric phases, J. Phase Equilibria Diff. 27 (2006) 622–628.
[46] J. Svoboda, F.D. Fischer, R. Abart, Modelling of diffusional phase transformation in multi-component systems with stoichiometric phases, Acta Mater. 58 (2010) 2905–2911.
[47] J. Svoboda, F.D. Fischer, R. Abart, Modeling the role of sources and sinks for vacancies on the kinetics of diffusive phase transformation in binary systems with several stoichiometric phases, Philos. Mag. Letters 92 (2012) 67–76.
[48] J. Svoboda, F.D. Fischer, W. Schillinger, Formation of multiple stoichiometric phases in binary systems by combined bulk and grain boundary diffusion – experiments and model, Acta Mater. 61 (2013) 32–39.
[49] A.M. Gusak, N.V. Tyutyunnyk, K.-N. Tu, Models of interdiffusion in a poly-crystalline alloy: Kirkendall effect versus non-equilibrium vacancies and backstress, Defect and Diffusion Forum 309–310 (2011) 135–142.

[50] D.L. Beke, Z. Erdélyi, B. Parditka, Effect of diffusion induced driving forces on interdiffusion-stress development, relaxation and kinetics of diffusion processes, Defect and Diffusion Forum 309–310 (2011) 113–120.
[51] J. Svoboda, F.D. Fischer, A new computational treatment of reactive diffusion in binary systems, Comput. Mater. Sci. 78 (2013) 39–46.
[52] J. Svoboda, F.D. Fischer, P. Fratzl, Diffusion and creep in multi-component alloys with non-ideal sources and sinks for vacancies, Acta Mater. 54 (2006) 3043–3053.
[53] J. Svoboda, E. Gamsjäger, F.D. Fischer, P. Fratzl, Application of the thermodynamic extremal principle to the diffusional phase transformations, Acta Mater. 52 (2004) 959–967.
[54] J. Svoboda, J. Stopka, F.D. Fischer, Two-dimensional simulation of reactive diffusion in binary systems, Comput. Mater. Sci. 95 (2014) 309–315.
[55] B. Rheingans, E.J. Mittemeijer, Phase transformation kinetics: advanced modelling strategies, JoM 65 (2013) 1145–1154.
[56] J.W. Cahn, J.E. Hilliard, Free energy of a nonuniform system: I. Interfacial free energy, J. Chem. Phys. 28 (1958) 258–267.
[57] J.W. Cahn, Free energy of a nonuniform system: II. Thermodynamic basis, J. Chem. Phys. 30 (1959) 1121–1124.
[58] H. Nakajima, The discovery and acceptance of the Kirkendall effect: the result of a short research career, JoM 6 (1997) 15–19.
[59] A. Paul, M.J.H. van Dal, A.A. Kodentsov, F.J.J. van Loo, The Kirkendall effect in multiphase diffusion, Acta Mater. 52 (2004) 623–630.
[60] C. Wagner, The evaluation of data obtained with diffusion couples of binary single-phase and multiphase systems, Acta Metall. 17 (1969) 99–107.
[61] S.R. Shatynski, J.P. Hirth, R.A. Rapp, A theory of multiphase binary diffusion, Acta Metall. 24 (1976) 1071–1078.
[62] D.S. Williams, R.A. Rapp, J.P. Hirth, Multilayer diffusional growth in silver–zinc alloys, Metall. Mater. Trans. A 12 (1981) 639–652.
[63] G.X. Li, G.W. Powell, Theory of reaction diffusion in binary systems, Acta Metall. 33 (1985) 23–31.
[64] E. Metin, O.T. Inal, A.D. Romig Jr., Solutions to multiphase diffusion in binary metal interstitial systems, Metall. Mater. Trans. A 36 (2005) 1407–1415.
[65] M. Kajihara, Relationship between temperature dependence of interdiffusion and kinetics of reactive diffusion in a hypothetical binary system, Mat. Sci. Engng. A 403 (2005) 234–240.
[66] M.J.H. van Dal, D.G.G.M. Huibers, A.A. Kodentsov, F.J.J. van Loo, Formation of Co–Si intermetallics in bulk diffusion couples. Part I. Growth kinetics and mobilities of species in the silicide phases, Intermetallics 9 (2001) 409–421.
[67] M.J.H. van Dal, A.A. Kodentsov, F.J.J. van Loo, Formation of Co–Si intermetallics in bulk diffusion couples. Part II. Manifestations of the Kirkendall effect accompanying reactive diffusion, Intermetallics 9 (2001) 451–456.
[68] A. Paul, The Kirkendall Effect in Solid State Diffusion, PhD Thesis, Techn. Univ. Eindhoven, The Netherlands, ISBN 90-386-2646-0, 2004.
[69] A. Paul, A.A. Kodentsov, F.J.J. van Loo, Intermetallic growth and Kirkendall effect manifestations in Cu/Sn and Au/Sn diffusion couples, Z. Metallkd. 95 (2004) 913–920.
[70] A. Paul, C. Ghosh, W.J. Boettinger, Diffusion parameters and growth mechanism of phases in the Cu–Sn system, Metall. Mater. Trans. A 42 (2011) 952–963.
[71] S. Kumar, C.A. Handwerker, M.A. Dayananda, Intrinsic and interdiffusion in Cu–Sn system, J. Phase Equilibria Diff. 32 (2011) 309–319.
[72] R.A. Gagliano, M.E. Fine, Thickening kinetics of interfacial Cu_6Sn_5 and Cu_3Sn layers during reaction of liquid tin with solid copper, J. Electron. Mater. 32 (2003) 1441–1447.

[73] N. Dariavach, P. Callahan, J. Liang, R. Fournelle, Intermetallic growth kinetics for Sn–Ag, Sn–Ag, Sn–Cu, and Sn–Ag–Cu lead-free solders on Cu, Ni, and Fe–42Ni substrates, J. Electron. Mater. 35 (2006) 1581–1592.
[74] M.S. Park, R. Arróyave, Early stages of intermetallic compound formation and growth during lead-free soldering, Acta Mater. 58 (2010) 4900–4910.
[75] M.S. Park, R. Arróyave, Concurrent nucleation, formation and growth of two intermetallic compounds (Cu_6Sn_5 and Cu_3Sn) during the early stages of lead-free soldering, Acta Mater. 60 (2012) 923–934.
[76] H.C. Yu, A. Van der Ven, K. Thornton, Theory of grain boundary diffusion induced by the Kirkendall effect, Appl. Phys. Letters 93 (2008) 091908.
[77] K.R. Elder, K. Thornton, J.J. Hoyt, The Kirkendall effect in the phase field crystal model, Philos. Mag. 91 (2011) 151–164.
[78] F. Christien, R. Le Gall, Ultra-fast grain boundary diffusion and its contribution to surface segregation on a martensitic steel. Experiments and modeling, Surface Science 605 (2011) 1711–1718.
[79] J. du Plessis, G.N. van Wyk, A model for surface segregation in multicomponent alloys – part III: the kinetics of surface segregation in a binary alloy, J. Phys. Chem. Solids 50 (1989) 237–245.
[80] J.Y. Wang, J. du Plessis, J.J. Terblans, G.N. van Wyk, Kinetics near the discontinuous surface transition in the Cu(Ag)(111) binary segregating system, Surface Science 423 (1999) 12–18.
[81] E. Kozeschnik, J. Svoboda, R. Radis, F.D. Fischer, Mean-field model for the growth and coarsening of stoichiometric precipitates at grain boundaries, Model. Simul. Mater. Sci. Eng. 18 (2011) 015011.
[82] I.V. Belova, G.E. Murch, Analysis of the effective diffusivity in nanocrystalline materials, J. Metastable Nanocryst. Mater. 19 (2004) 25–34.
[83] E.W. Hart, On the role of dislocations in bulk diffusion, Acta Metall. 5 (1957) 597.
[84] J.-H. Shim, C.S. Oh, B.J. Lee, D.N. Lee, Thermodynamic assessment of the Cu–Sn system, Z. Metallkd. 87 (1996) 205–212.
[85] L. Klinger, E. Rabkin, Grain boundary interdiffusion and stresses in thin polycrystalline films, J. Mater. Sci. 46 (2011) 4343–4348.
[86] L. Klinger, E. Rabkin, Theory of the Kirkendall effect during grain boundary interdiffusion, Acta Mater. 59 (2011) 1389–1399.
[87] C.M. Ulrich, A. Hashibon, J. Svoboda, C. Elsässer, D. Helm, H. Riedel, Diffusion kinetics in aluminium–gold bond contacts from first-principles density functional calculations, Acta Mater. 59 (2011) 7634–7644.
[88] C.J. Smithells, E.A. Brandes (Eds.), Metals Reference Book, 5th ed., Butterworths, London et al., 1976, p. 939.
[89] C.H. Ma, R.A. Swalin, A study of solute diffusion in liquid tin, Acta Metall. 8 (1960) 388–395.
[90] H. Mehrer (Ed.), Diffusion in Solid Metals and Alloys, Landolt-Börnstein, New Series, vol. 26, Springer-Verlag, Berlin et al., 1990, p. 641 (Sect. 12.2.1).
[91] R.A. Lord, A. Umantsev, Early stages of soldering reactions, J. Appl. Phys. 98 (2005) 063525.
[92] J. Gong, C. Liu, P.P. Conway, V.V. Silberschmidt, Evolution of CuSn intermetallics between molten SnAgCu solder and Cu substrate, Acta Mater. 56 (2008) 4291–4297.
[93] H.-Y. Hsiao, C.-C. Hu, M.-Y. Gua, C. Chen, K.N. Tu, Inhibiting the consumption of Cu during multiple reflows of Pb-free solder on Cu, Scripta Mater. 65 (2011) 907–910.

CHAPTER 9

Monte Carlo Methods in Solid State Diffusion

Graeme E. Murch, Irina V. Belova
The University of Newcastle, Callaghan, NSW, Australia

9.1 INTRODUCTION

The Monte Carlo method is a unique numerical method that makes use of computer-generated random numbers to solve scientific or engineering problems when a regular mathematical procedure or numerical analysis is difficult or essentially impossible to use. The name "Monte Carlo" was used as a code name in the Manhattan Project in World War II for an investigation to model neutron trajectories. Since that time, the Monte Carlo method has broadened enormously and undergone countless developments and refinements. It has been applied in numerous areas right across science, engineering and finance.

Simulation of diffusion by the Monte Carlo method was first suggested in 1951 by King [1] who used it for simulating random walks of particles. Because of the relative slowness of the computers at the time, little progress was made until 1960. In that year, Flinn and McManus [2] reported on more substantial Monte Carlo simulations of solid state diffusion. Over the next decade, that paper was followed by quite a few others reporting on Monte Carlo simulations of various diffusion-related phenomena such as the kinetics of ordering in binary alloys and radiation damage annealing [3–8]. Perhaps surprisingly, even by 1970, no Monte Carlo calculations had been made of any fundamental quantities in solid state diffusion, by which we mean the tracer diffusion coefficient, the ionic conductivity, the chemical diffusion coefficient and tracer correlation factors, etc. In 1971, essentially as an offshoot to their main investigation of what they called persistency effects in diffusion in hard-sphere gases, Bennett and Alder [9] reported on Monte Carlo calculations of geometric tracer correlation factors for the single vacancy and divacancy mechanisms in the f.c.c. lattice. The values obtained were in good agreement with analytically determined results for those mechanisms. That paper probably marks the start of a new era in which fundamental solid state diffusion properties are routinely calculated by the Monte Carlo method.

At this point, we need to differentiate between several types of Monte Carlo method as applied to solid state diffusion. First, there are Monte Carlo calculations of defect formation energies and investigation of the jump dynamics in continuous space systems. These calculations are now quite rare. Next, there are Monte Carlo calculations

Handbook of Solid State Diffusion, Volume 1
DOI: 10.1016/B978-0-12-804287-8.00009-9

by way of discrete hopping models. This is by far the largest category of the Monte Carlo method in solid state diffusion and there are a great many variations within it. These calculations are often now known under the name Kinetic Monte Carlo (KMC). However, this name is of comparatively recent origin, roughly since about 1990. Furthermore KMC is considerably broader in concept and application than simply solid state diffusion. It should be pointed out too that a good few calculations since that time have not explicitly used the term KMC. Finally, there are Monte Carlo calculations that use random walks of particles to solve the Diffusion Equation (Fick's Second Law). When these calculations use a hypothetical lattice overlay (usually a simple cubic lattice) and discrete particles make jumps on it, then they are usually referred to as Lattice Monte Carlo (LMC) calculations, to draw a distinction in process from not only the quite different KMC calculations but also the continuous random walk methods that have long been used in phenomenological heat conduction problems [10].

In the present contribution, we will restrict ourselves to solid state KMC and LMC calculations: specifically how they work and how the non-specialist reader might undertake a very basic calculation of each type. At this juncture, we need to point out the quite sizable number of reviews that have been published in the last 30 years on KMC and LMC. The following is by no means an exhaustive list. Much of the earlier work by way of KMC calculations of basic diffusion quantities was reviewed in 1984 by one of the present authors [11]. A shorter review coupled with a tutorial-like description of a solid state diffusion KMC calculation along with a description of a basic calculation of the geometric tracer correlation factor (complete with a listing of what is now obsolete FORTRAN 77 code) was also published, again by one of the authors [12]. Several shorter pedagogical papers on KMC and solid state diffusion have also appeared [13,14]. An exhaustive review of KMC calculations in lattice gases has also been published [15]. A more general introduction to KMC with an emphasis on addressing radiation damage has also appeared [16]. The rather more recent LMC method was reviewed almost ten years ago by the present authors [17]. There has been a very useful recent compilation of the abstracts of papers published on Monte Carlo calculations in solid state diffusion [18]. In its 516 entries it captures much of the Monte Carlo activity in the area of solid state diffusion since the 1960s. This should be consulted by the beginner wishing to do Monte Carlo calculations.

9.2 SOLID STATE DIFFUSION AND KINETIC MONTE CARLO

9.2.1 Tracer Diffusion

The most basic measurable quantity in solid state diffusion is almost certainly the tracer diffusion coefficient or diffusivity, D^*. This is sometimes referred to loosely as the self-diffusion coefficient. But often the self-diffusion coefficient is defined as the tracer diffusion coefficient divided by the tracer correlation factor (see further below) or the collective diffusion coefficient. The tracer diffusion coefficient can be measured using radioisotopes or stable isotopes and SIMS [19]. The tracer diffusion coefficient can be defined as a proportionality factor between the tracer flux J^* and the tracer concentration gradient $\partial C^*/\partial x$ as expressed in Fick's First law,

$$J^* = -D^* \frac{\partial C^*}{\partial x}. \tag{9.1}$$

Equivalently, D^* can also be defined in the Einstein–Smoluchowski Equation, given here for three dimensions:

$$D^* = \frac{< r^2 >}{6t} \tag{9.2}$$

where r is the displacement of a given tracer or marked atom in time t. In a situation where discrete hopping of atoms to a nearest neighboring vacant site is assumed, the tracer diffusion coefficient can be expanded for isotropic lattices as:

$$D^* = \frac{na^2}{6t} f = \Gamma \frac{a^2}{6} f \tag{9.3}$$

where n is the (average) number of atomic jumps of the tracer atom of length a in time t, (n/t is the atomic jump frequency Γ). The factor f is the tracer, sometimes called the single particle, correlation factor. This factor captures correlations in the directions of successive jumps of a given tracer or marked atom. When there are no correlations, then $f = 1$ and the walk of the atom (or even a defect if it were to be followed) would be a completely random one. This would be the case for an isolated interstitial atom such as a C atom in a metal such as Fe. When all atomic jumps are immediately reversed, then $f = 0$. In general, correlations can come about because of the proximity of the defect (generally this is a vacancy) to a tracer or marked atom. But they can also arise, for example, as a result of vacancy–vacancy interactions, competition between several atomic species for a vacancy, and jump reversals for intersublattice jumps in ordered structures. In some of these cases, the tracer correlation factor is quite temperature dependent and can carry perhaps as much as 20% of the observed total activation energy for the entire tracer diffusion process [11,19].

There are quite a few different ways of using Monte Carlo to calculate the tracer correlation factor. These are listed in [11]. But the method based on the combination of Eqs. (9.2) and (9.3), i.e.,

$$f = \lim_{n \to \infty} \frac{< r^2 >}{na^2}, \tag{9.4}$$

is certainly the widest known, the easiest to use and probably the most adaptable to new situations. Furthermore, it is very easily extended to collective correlations, see Section 9.2.2.

Let us consider tracer diffusion in a binary random alloy AB and how a simple Monte Carlo calculation might be programmed. We will deal with this in some detail and make a few comments about more elaborate calculations as required. In a typical Monte Carlo program to calculate f, an integer array of reasonably large size, e.g., $30 \times 30 \times 30$, is first of all defined. If the array is too small, historical effects about the atomic configuration may be perpetuated. For example, a vacancy may continue to approach a given atom from an unphysical direction for a bulk system. This initially simple cubic array can be adapted to another lattice such as f.c.c. by simply reordering the array elements and tagging them appropriately. In the case of f.c.c., this would leave half of the array elements unused for the calculation. A vacant site can be introduced into the array by tagging one array element differently. At this stage, the array now contains three types of integer: one for a lattice site, one for an unused site (if it is a non-simple cubic lattice being simulated), and finally, one for the vacancy. In order to calculate the displacement r of each atom, we will also need to keep account of the coordinates of these numbered atoms as if each atom started at $(0, 0, 0)$. We also need jump counters. We can now introduce a second type of atomic species B. Since we are dealing with the random alloy model, these B atoms are introduced at randomly chosen locations. Those lattice sites left can be assigned as the A atoms. In models with atomic interactions, we will need to bring the system to thermodynamic equilibrium first. There are many ways to achieve this, including by way of diffusion which is obviously convenient, but it may well be the slowest. We refer to the reviews [11,15]. Now we can number the lattice sites representing the species of atoms of type A starting from unity and up. Each of these numbered atoms will represent a (different) tracer atom of A. The B atoms will be numbered as well and other information initialized as was done for A. Perhaps negative integers can be used for the numbering purpose to distinguish these atoms from A, or the first species A might use integers from 1 up to 100,000 and the second species B from 100,000 and above and so on.

We are now in a position to start the simulation itself. Although it would be more physically correct to attempt to move a randomly chosen atom in a random direction, this would be, of course, very inefficient since there will rarely be a vacancy as a nearest neighbor to a chosen atom. Instead, we attempt to move the vacancy itself in a random direction. If this is an f.c.c. lattice, there will be 12 directions. To make this choice, a random number is generated uniformly on the interval $[0, 1]$ and is converted to a random integer from 1 to 12. A random choice of direction is now made. There are two atomic species present. We have two fixed atom–vacancy exchange frequencies, one for A, w_A, and one for B, w_B. These frequencies are rescaled so that the larger one is put equal to unity. They will be treated as probabilities in the calculation. Once the type of atom (A or B) is identified in the randomly chosen direction, we now generate another random number to determine the outcome of the attempt. If the identified atom is an A atom and the A species is the faster moving one (its rescaled frequency is unity), then the vacancy and the *A* atom are interchanged forthwith and the coordinates of that particular A atom are updated. If the identified atom was a *B* atom, then a further random number is generated. If this number is less than the exchange frequency of *B* then the vacancy jump attempt is accepted and the vacancy and the *B* atom are interchanged. The jump counters and the stored atom coordinates are updated. If not, the vacancy jump attempt is abandoned entirely and a random direction is chosen once again. This process is repeated until the average number of jumps for an atom of the slower species is about 20. Then we can calculate the correlation factors of both species according to Eq. (9.4).

Typically in most such calculations, the correlation factors converge fairly quickly if the exchange frequencies differ by not more than about a factor of ten. This is not the case, however, when the frequencies differ by large amounts, say 100 to 1000. Then the average number of jumps per (slower) atom must be greatly increased, perhaps to several thousand [20]. It is important to ensure that enough jumps have occurred so that the correlation factors have converged. It should be noted that the lack of a sufficient number of jumps seriously marred the results of quite a number of early KMC calculations of the tracer correlation factor, as was pointed out in [20]. Periodic boundaries are generally used in these simulations in order to avoid surface effects. Thus the system appears to be infinite in the sense that it is surrounded by periodic images of itself. However, although the system has no surfaces, it should be noted that the system is still a relatively small system from a thermodynamic perspective. In the simulation, a vacancy or atom that leaves the first or last plane of the array is simply inserted on the other side of the array. However, the running coordinates of the tracer atoms are always updated after a jump as if no periodic boundary exists. There have been a large number of calculations of tracer correlation factors by KMC. The reader is directed to the compilation [18].

In some calculations, it is a (reduced) effective diffusion coefficient that is really required rather than a tracer correlation factor. Then the Einstein–Smoluchowski Equation (Eq. (9.2)) is used directly, with real time being represented by the vacancy jump attempt counter. This is, of course, the diffusion coefficient per unit vacancy concentration. The calculation of an effective diffusion coefficient is often performed explicitly with LMC and we will delay discussion of that until Section 9.3. When one of the jump frequencies is very small compared with the other(s), there can be a large computational inefficiency with the basic algorithm described above. This can be avoided, or at least minimized, by making use of straightforward vacancy residence time algorithms; see, for example, [21–24]. However, there is inevitably some computational overhead that often significantly reduces the gains made.

9.2.2 Collective Diffusion

Collective diffusion refers to such diffusion processes as chemical or interdiffusion and ionic conductivity. The Onsager formalism of irreversible thermodynamics provides a convenient means of expressing the various types of collective diffusion, and the Onsager phenomenological coefficients provide the foundation for expressing the collective correlation factors, which are the correlated parts of those coefficients in a very similar way to the way that the tracer correlation factor is the correlated part of the tracer diffusion coefficient in Eq. (9.3). The interdiffusion coefficient $\tilde{D}$ in a binary alloy AB can be expressed in terms of the phenomenological coefficients as [19]

$$\tilde{D} = \left(c_B \frac{L_{AA}}{c_A} + c_A \frac{L_{BB}}{c_B} - 2L_{AB} \right) \frac{\phi kTV}{N} \tag{9.5}$$

where the L_{ij} are the phenomenological (Onsager) coefficients, c_i is the mole fraction of species i, N is the total number of lattice sites, V is the volume, ϕ is the thermodynamic factor, k is the Boltzmann constant, and T is the absolute temperature. As an aside, it should be mentioned that a direct KMC calculation of the interdiffusion coefficient in a binary alloy couple, whilst seemingly straightforward, is in fact quite difficult because of the formal need to keep the vacancies at equilibrium during the simulation [25]. In real materials, this is handled by dislocation climb to create and annihilate vacancies, i.e., processes that are on a different spatial scale altogether from a simulation in a periodic box of typical dimensions 30 × 30 × 30.

The dc ionic conductivity σ_A of an ionic conductor (with one mobile sublattice and one type of atom A) can be expressed in terms of the phenomenological coefficients as

$$\sigma_A = q_A^2 L_{AA} \tag{9.6}$$

where q_A is the charge on species A. If there are two contributions to the ionic conductivity from two types of atom (A and B) on the same sublattice, then the individual contributions are written as

$$\sigma_A = q_A (q_A L_{AA} + q_B L_{AB}), \tag{9.7}$$

and similarly for the B component.

The Onsager phenomenological coefficients L_{ij} can be expressed in terms of collective displacements using the well-known Green–Kubo formulae as

$$L_{ij} = \frac{< \Delta R_i \cdot \Delta R_j >}{6tkTV} \tag{9.8}$$

which can be decomposed into [26]

$$L_{ii} = \Gamma_i \frac{a^2 N c_i}{6kTV} f_{ii},$$

$$L_{ij} = \Gamma_i \frac{a^2 N c_i}{6kTV} f_{ij}^{(i)} = \Gamma_j \frac{a^2 N c_j}{6kTV} f_{ij}^{(j)} \tag{9.9}$$

where the f_{ii} and $f_{ij}^{(i)}$ are the collective correlation factors [26,27]. These factors can be expressed in Einsteinian forms quite analogous to the tracer correlation factor of Eq. (9.4):

$$f_{ii} = \lim_{N_i \to \infty} \frac{< \Delta R_i^2 >}{N_i a^2}, \tag{9.10}$$

$$f_{ij}^{(i)} = \lim_{N_i \to \infty} \frac{< \Delta R_i \cdot \Delta R_j >}{N_i a^2}, \tag{9.11}$$

$$f_{ij}^{(j)} = \lim_{N_j \to \infty} \frac{< \Delta R_i \cdot \Delta R_j >}{N_j a^2}, \tag{9.12}$$

and the collective displacement ΔR_i is the displacement of the center of mass of all atoms of type i. KMC provides a very useful way to gain access to the collective correlation factors. Their calculation differs slightly in detail from the calculation above in Section 9.2.1 of the tracer correlation factor. Each run provides only a single contribution to the collective displacements of the A and B species. Thus the run needs to be repeated many times to build up a suitable average, 10,000 times is a typical number. However, the number of jumps per atom does not need to be as large as in the case for the tracer correlation factor, see above in Section 9.2.1. Experience has shown that several jumps per atom in a given run are generally sufficient. But in any new simulation one should always check that convergence has been achieved. To gather the various contributions for the average, the runs should have the same number of attempts to jump by the vacancy, in other words, the time should be constant.

9.2.3 The Jump Frequency

The atomic jump frequency Γ is generally decomposed into the expression

$$\Gamma = Z p_{iV} w_i \tag{9.13}$$

where Z is the coordination number, p_{iV} is the vacancy availability factor and is a generalization of the usual vacancy fraction c_V, and w_i is the atom i-vacancy exchange frequency. The vacancy availability factor captures the fact that because of the interactions between different atomic species, each atomic species may not "see" the vacancies that are present in an equal way. Sometimes the quantity $\overline{p}_{iV}$ is introduced as a prefactor to c_V, i.e., $p_{iV} = \overline{p}_{iV} c_V$. There will be a small temperature dependence of $\overline{p}_{iV}$. In the case of the random alloy model that was mentioned above, where the atom–vacancy exchange frequencies were fixed everywhere, p_{iV} is simply equal to c_V. In many Monte Carlo calculations, there are imposed interactions between the different atomic species, and w_i will, in effect, vary from site to site. In a KMC calculation the highest atom–vacancy exchange frequency encountered would still be scaled to unity since the actual exchange frequency would generally be much too small to do KMC calculations. An overall (scaled frequency) Γ' can easily be determined using a vacancy attempt to jump as the basic "clock" counter as the denominator and the average number of actual jumps for each species n_i as the numerator. In order to obtain the components w_i and p_{iV} of the frequency, one must also calculate the inverse quantity p_{Vi} the atom i availability to the vacancy. It is simply related to p_{iV} by

$$p_{iV} = p_{Vi} c_V c_i. \tag{9.14}$$

In the early years, the well-known Ising magnet model, in its rigid lattice binary alloy guise, was frequently used to write the site-dependent exchange frequency w_i in KMC calculations; for a review, see [11]. There are, in principle, an infinite number of possibilities for writing such an exchange frequency. The only formal restriction for equilibrium calculations is that the exchange frequencies satisfy the requirement for detailed balance [11]. A choice that has a reasonable semblance to reality is the so-called bond-breaking one first introduced by Sato and Kikuchi [28] in their Path Probability analytical calculations and exemplified here for an A atom:

$$w_A = \nu \exp\left(\frac{-E_A + Z_{AA}E_{AA} + Z_{AB}E_{AB}}{kT}\right) \tag{9.15}$$

where E_{AA}, E_{BB} and E_{AB} are the nearest neighbor interaction energies between AA, BB and AB nearest neighbors, Z_{AA} is the number of like bonds at the initial site of the jumping A atom, Z_{AB} is the number of unlike bonds at the site of the jumping A atom, ν is the (constant) prefactor, and E_A is the migration or saddle point energy of a hypothetical A atom with no interactions. A refinement to this Ising alloy model introduces modification to the saddle point itself. With extension to next nearest neighbors and input to the E_{ij} from embedded atom method potentials [29] such rigid lattice calculations have continued up to the present day.

A more realistic approach avoids the rigid lattice approximation altogether and attempts to calculate the jump frequency on-the-fly [30]. In this, the exchange frequency of the atom chosen to jump is calculated as required in a separate calculation. However, unless all the frequencies are calculated beforehand and stored together with the local environments of the atoms (which then effectively reduces it to the next approach), this kind of calculation has proved to be much too slow to use routinely, especially if tracer or collective correlation factors are to be determined.

The next approach, chronologically appearing a little earlier than the on-the fly method, attempts to calculate beforehand the site-dependent frequencies by calculating the possible migration energies and therefore exchange frequencies for all possible environments in close proximity to the atom that has been selected to jump. Like the rigid lattice Ising-type calculations the attempt frequency ν was first effectively made constant. The first such attempt at such a calculation was performed in two parts [31]. In the first part, lattice relaxation (sometimes called lattice statics or molecular statics) was used to calculate the migration energy by moving the atom towards the vacancy in a series of small steps and relaxing the system at each step. This was performed for a large number of different local environments of the atom neighboring to a vacancy using a classical potential. The second part, which was the KMC calculation itself, was largely concerned with "pattern recognition" of the local environment of the atom neighboring to the vacancy and, once identified, then looking up the relevant exchange frequency for that atom. This approach has now been greatly extended and refined over the last 30 years; see, for example, the reviews [34,35].

In more recent calculations of this type, the exchange frequency is written according to transition state theory as [32,33]:

$$w = \frac{kT}{h}\exp(-\Delta G_m/kT) \tag{9.16}$$

where ΔG_m is the difference between the Gibbs free energy of the transition state after neglecting the unstable vibration mode and the Gibbs free energy of the initial equilibrium state which has contributions from all vibrational degrees of freedom, h is Planck's constant. ΔG_m is the free energy barrier for the jumping atom and is associated with the saddle point in the potential energy surface. The identification of the saddle point is now routinely obtained in a first principles calculation using software such as VASP with the nudged elastic band method in which a potential pathway of migration is changed so that its barrier height in minimized with a spring force acting on the jumping atom [34]. Eq. (9.16) can be rewritten in the harmonic approximation into the well-known form:

$$w = \nu^* \exp(-\Delta H_m/kT) \tag{9.17}$$

where ΔH_m is the enthalpy of migration. The prefactor ν^* is given by:

$$\nu^* = \frac{\prod_{i=1}^{3N-3}\nu_i}{\prod_{j=1}^{3N-4}\nu_j'} \tag{9.18}$$

where ν_i and ν_j' are the normal vibrational frequencies at the initial and transition states respectively for N atoms and a vacancy. Accumulated errors when using Eq. (9.18) mean that the direct Eq. (9.16) is now often preferred. For further information on the subject of calculations of jump frequencies in KMC calculations in ionic compounds, metallic alloys and grain boundaries we refer to the recent reviews [35–37].

It can also be mentioned that KMC can be combined with the simulation method of molecular dynamics. In this, short-time molecular dynamics is used to determine the frequencies or probabilities for various atomic jumps and these are fed into a separate KMC calculation for the required long-time calculation of the tracer correlation factor or effective diffusion coefficient. This works well for cases where the jump frequency depends on position such as in a grain boundary [38].

9.3 SOLID STATE DIFFUSION AND LATTICE MONTE CARLO

9.3.1 The Effective Diffusion Coefficient

Many phenomenological problems in solid state diffusion, and indeed diffusion problems generally, require the calculation of an effective diffusion coefficient of diffusants in the system. The effective diffusion coefficient is the (tracer) diffusion coefficient at the long time limit in the material. Experimentally, the effective diffusion coefficient might be the diffusivity measured in a tracer diffusion experiment made on a polycrystalline material in the so-called Harrison Type-A kinetics regime. In this regime, the effective diffusion coefficient of the diffusant is some (usually unknown) weighted average of the bulk (i.e., lattice) and grain boundary diffusion coefficients. It will depend on the geometry of the grain arrangements. Other examples include the effective ionic conductivity of a composite solid electrolyte or the effective diffusion coefficient of a diffusant in a two-phase material. In these examples, the value of the diffusion coefficient of the diffusant depends on the types of "phase" and their spatial arrangement. At very long diffusion times, the material overall has an average or effective diffusion coefficient for that diffusant.

The Lattice Monte Carlo (LMC) method was developed to address these problems by overlaying the system with a hypothetical very fine-grained lattice upon which particles can diffuse on random walks, see the review [17]. It should be noted that these particles are not atoms: they should be considered as mass packets that represent equal amounts of the diffusant species. As an example, let the model system consist of a two-phase system of spheres of equal size embedded in a matrix where the diffusant has a different diffusion coefficient in the spheres compared with the matrix. We wish to know the effective diffusion coefficient of the diffusant in this material as a whole. A very fine-grained simple cubic lattice, say $100 \times 100 \times 100$, is overlaid on the system. A plane of this structure is shown schematically in Fig. 9.1. The requirement here is that the lattice is sufficiently fine grained so that it can capture reasonably well the shape of the spherical inclusions in this case. The lattice has periodic boundaries, similar to what was described in Section 9.2.1. In this sense, the lattice acts like a unit cell of the complete system. Particles are released, one by one, from randomly chosen lattice sites and each particle is permitted to explore the entire lattice. The jump frequency Γ that is chosen for a particle diffusing on the lattice in each of the two different phases is linked directly to their corresponding diffusion coefficients for that diffusant. (Strictly speaking, we deal with rescaled or reduced diffusion coefficients.)

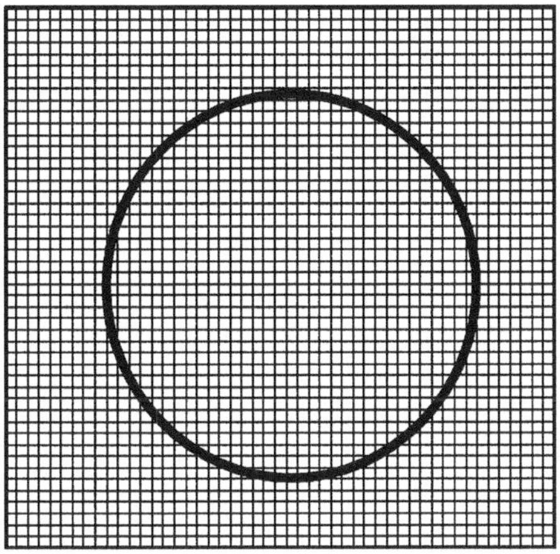

Figure 9.1 Schematic of a plane of a model composite with regularly spaced spherical inclusions in a simple cubic arrangement overlaid by a simple cubic lattice. The lattice spacing is shown larger than would be used in an LMC calculation.

Thus

$$\Gamma = 6Da^2, \tag{9.19}$$

see Eq. (9.3).

There is no correlation factor in Eq. (9.19) (f is trivially equal to unity) because each particle explores the empty lattice with no interactions whatsoever with other particles, not even site blocking, see further below. The jump distance a is the lattice spacing and, accordingly, this can be very conveniently scaled to the actual spatial dimensions of the problem. The effective diffusion coefficient in the long time limit is given by the Einstein–Smoluchowski Equation, Eq. (9.2). The time t is scaled to a counter that counts attempts to jump by a particle. Typically, it might be 10^5 attempts. It is important that the particles explore the structure adequately. As a reasonable estimate for what would be a sufficiently long time, it is best to estimate the diffusion length $(6D_{eff}t)^{1/2}$ which should be at least half the length of the lattice. The process is repeated with a large cohort of particles, say 10^6, all diffusing for the same length of time. This would give an error for the effective diffusion coefficient of about 0.1%.

In principle, all particles of the cohort can be released at the same time rather than one by one. If this is done, multiple occupancy by particles of a given lattice site is specifically permitted. This still means that the correlation factor is trivially equal to unity. A particle chosen to jump must be chosen randomly from the cohort before a jump attempt is made.

In many materials, the diffusants will exhibit segregation among the various phases that are present. For two phases, this is conveniently expressed through a segregation factor s which is defined by way of Henry's Law as

$$s = \frac{c_2}{c_1}\left(= \exp\left(-\frac{E_s}{kT}\right)\right) \tag{9.20}$$

where for convenience $c_1 < c_2$ where c_i is the equilibrium concentration of the diffusant in phase i. In Eq. (9.20) E_s is the segregation energy of the diffusant between the two phases. In the implementation of segregation in the LMC method, the jump frequency Γ_{12} from phase 1 to phase 2 and the jump probability Γ_{21} for the reverse jump are simply related by

$$\Gamma_{12}/\Gamma_{12} = s. \tag{9.21}$$

When segregation is present in the model, the method for the initial release of the particles must also be adjusted to reflect this. This is done by weighting the phases according to Eq. (9.20) and their relative fractions [17].

9.3.2 Concentration Profiles

On some occasions, it is useful to determine the time-dependent concentration profiles of a diffusant that has been released from one region (the source) for diffusion into the rest of the material. Obtaining the concentration profiles using the LMC method is equivalent to solving the Diffusion Equation (Fick's Second law) for the problem once initial and boundary conditions have been supplied. The most common example in solid state diffusion is tracer diffusion in a material containing grain boundaries and this has attracted the most attention of the LMC method; see, for example, [39].

The basic LMC procedure to produce concentration profiles is quite similar to that already described in Section 9.3.1. In the region of the lattice where the particles are released, i.e., the source, the particles are released one by one from sites chosen randomly in that region or they can be released all at once from randomly chosen sites in that region. Either method corresponds to a finite supply of particles of diffusant. If we assume for convenience the simultaneous release of particles, the particles are then permitted to diffuse independently of one another for some time t, at which time the final position of each particle is recorded. The final positions of all of the particles are then simply assembled from these individual final positions to simply form the concentration profile.

In some diffusion problems, the diffusant has an infinite supply at the source, for example, it could be the gas phase, in which case the surface concentration of the diffusant is effectively kept constant. The LMC method simply implements this by keeping the total number of particles at the surface constant. A large number of particles, e.g., 10^5, are allowed to be released simultaneously. As each particle makes a jump away from the source plane it is immediately replaced by a new one, which is again generated at a random site at the surface. On the other hand, if a particle returns to the source plane, and therefore the (constant) number of particles at the source is exceeded, then that particle is permanently removed from the system. Over a period of time, the number of particles within the system naturally increases as more particles diffuse away from the source region than return to it. Accordingly, the diffusion time t must now be rescaled since the diffusion time is required to be proportional to the number of jump attempts per particle.

ACKNOWLEDGEMENTS

This work was supported by the Australian Research Council under its Discovery Grants Scheme.

REFERENCES

[1] G.W. King, Ind. Eng. Chem. 43 (1951) 2475.
[2] P.A. Flinn, G.M. McManus, Phys. Rev. 124 (1961) 54.
[3] J.R. Beeler, Phys. Rev. A 134 (1964) 1396.
[4] J.R. Beeler, J.A. Delaney, Phys. Rev. 130 (1963) 962.
[5] A. Scholz, Phys. Stat. Solidi 14 (1969) 169.
[6] W.M. Young, E.W. Elcock, Proc. Phys. Soc. 89 (1966) 735.
[7] H. Mehrer, Z. Naturforsch 24a (1969) 358.
[8] H. Mehrer, Z. Naturforsch 241 (1969) 367.
[9] C.H. Bennett, B.J. Alder, J. Phys. Chem. Solids 32 (1971) 2111.
[10] A. Haji-Sheikh, E.M. Sparrow, J. Heat Transfer (Trans. ASME) 89 (1967) 121.
[11] G.E. Murch, in: G.E. Murch, A.S. Nowick (Eds.), Diffusion in Crystalline Solids, Academic Press, 1984, Chapter 7.
[12] G.E. Murch, S.J. Rothman, in: G.E. Murch, H.K. Birnbaum, J.R. Cost (Eds.), Nontraditional Methods in Diffusion, The Metallurgical Society of AIME, 1984, pp. 281–305.
[13] G.E. Murch, Am. J. Phys. 47 (1979) 78.
[14] G.E. Murch, Am. J. Phys. 47 (1979) 958.
[15] K.W. Kehr, K. Binder, in: K. Binder (Ed.), Applications of the Monte Carlo Method in Statistical Physics, Springer, 1984, pp. 181–221.
[16] A.F. Voter, in: K.E. Sickafus, et al. (Eds.), Radiation Effects in Solids, Springer, 2007, pp. 1–23.
[17] I.V. Belova, G.E. Murch, T. Fiedler, A. Öchsner, Diffusion Fundamentals 4 (2007) 15.1–15.23.
[18] D.J. Fisher (Ed.), Monte Carlo Diffusion Studies, Defect and Diffusion Forum, vol. 358, 2015.
[19] H. Mehrer, Diffusion in Solids; Fundamentals, Methods, Materials, Diffusion Controlled Processes, Springer, 2007.
[20] I.V. Belova, G.E. Murch, Phil. Mag. A 80 (2000) 1469–1479.
[21] J.P. Lavine, D.L. Losee, J. Appl. Phys. 59 (1984) 924.
[22] J.P. Lavine, in: K. Board, D.R.J. Owen (Eds.), Simulation of Semiconductor Devices and Processes, Pineridge Press, 1984.
[23] P. Metsch, F.H.M. Spit, H. Bakker, Phys. Stat. Solidi, (a) 93 (1986) 543.
[24] Y. Limoge, J.L. Bocquet, Acta Met. 36 (1988) 1717–1722.
[25] J. Zenisek, J. Svoboda, F.D. Fischer, Phil. Mag. 85 (2005) 2363–2389.
[26] A.R. Allnatt, J. Phys. C 15 (1982) 5603.
[27] I.V. Belova, G.E. Murch, Phil. Mag. A 80 (2000) 599–607.

[28] R. Kikuchi, H. Sato, J. Chem. Phys. 51 (1969) 161.
[29] P. Sowa, R. Kozubski, A. Biborski, E.V. Levchenko, A.V. Evteev, I.V. Belova, G.E. Murch, V. Pierson-Bohnes, Phil. Mag. 93 (2013) 1987–1998.
[30] J.L. Bocquet, C. Schmidt, Defect Diffus. Forum 194–199 (2001) 145–150.
[31] A.D. Murray, G.E. Murch, C.R.A. Catlow, Solid State Ionics 18–19 (1986) 196–202.
[32] H. Eyring, J. Chem. Phys. 3 (1935) 107–115.
[33] G.H. Vineyard, J. Phys. Chem. Solids 3 (1957) 121–127.
[34] A. Van der Ven, J.C. Thomas, Q. Xu, Phys. Rev. B 78 (2008) 104306.
[35] B.C. Zhou, Z.K. Liu, Diffusion Foundations 8 (2016) 1–30.
[36] A. Van der Ven, H.C. Yu, G. Ceder, K. Thornton, Prog. Materials Science 55 (2010) 61–105.
[37] Y. Mishin, M. Asta, Ju Li, Acta Materialia 58 (2010) 117.
[38] M. Guillpe, G. Ciccotti, V. Pontikis, Surface Science 144 (1984) 67.
[39] I.V. Belova, N.J. Kulkarni, G.E. Murch, Phil. Mag. 92 (2012) 1742–1763.

CHAPTER 10

Defects and Diffusion in Ordered Compounds

Sergiy Divinski
University of Münster, Münster, Germany

10.1 INTRODUCTION

The physical processes which determine the long-time stability of materials are known to be related to the rate of atomic transport. The processes of random migration of atoms of main components, solutes, and of impurities are termed as diffusion. Depending on the specific features of the atomic jumps and on the space limitations (confinement), different phenomena like bulk diffusion (i.e., atomic motion in bulk of individual crystals), diffusion on grain (interface) or phase (interphase) boundaries, dislocation diffusion, surface diffusion, etc., have to be considered. Generally, macroscopic diffusion is a response on non-zero gradient of a chemical potential and it results in flattening of such gradient with the course of the time. The diffusion rate is described with a help of diffusion coefficients and the following diffusion coefficients are important in the present content:

- D_X^* or D_X in this chapter, the tracer diffusion coefficient of the species X in the given compound. This is one of the fundamental parameters which represents directly the migration rate of atoms in the given compound or pure element. The value of D^* is determined in direct experiments with isotopes (e.g., using radioisotopes) and it corresponds to the transport rate of marked atoms in dilute regime (small concentrations). This value for various elements is in focus of the present analysis.
- $\widetilde{D}$, the interdiffusion or chemical diffusion coefficient (see corresponding chapter in this book for further reading). It determines the rate of diffusion transport between two different phases or components. In metals, the interdiffusion coefficients and the tracer coefficients (for the ideal vacancy sinks and sources, i.e., when the vacancy concentration corresponds to equilibrium values) are related via so-called Darken–Manning relation, Eq. (10.18), see below. The investigation of interdiffusion does not only provide additional and valuable information on the long-time stability of multicomponent and multiphase materials. By combing the data on $\widetilde{D}$ and the tracer diffusion coefficient of one of the components in a binary alloy, the other diffusion coefficient can be derived, if the necessary thermodynamic data are available (see Eq. (10.20) below).

Handbook of Solid State Diffusion, Volume 1
DOI: 10.1016/B978-0-12-804287-8.00010-5

The present chapter will focus on tracer diffusion in intermetallics, where the corresponding tracer diffusion coefficients, D_X^*, are of main importance. Both self- (i.e., transport of one of the main components of the alloy) and solute tracer diffusion will be considered. The interdiffusion coefficients, $\widetilde{D}$, will be analyzed only as a way to access the tracer diffusion coefficient of Al in aluminides, see below.

The intermetallics are compounds of two and more metals with structures which differ from those of the constituent metals [1]. The intermetallic compounds have different lattice structures: from the simple cubic, e.g., *B*2 (NiAl, FeAl, NiTi), and slightly tetragonally distorted $L1_0$ (TiAl, CuAu) to very sophisticated configurations like quasicrystalline *i*-AlCuFe alloys. The short- and long-range order are important phenomena of the intermetallic compounds which significantly affect their mechanical and physical properties [2]. Moreover, *sublattices* and *preferential site occupation* are important terms which describe microscopic features of the atomic distribution in these compounds.

The technological application of intermetallic compounds requires the knowledge of the relevant bulk- and grain boundary (GB) self- and solute diffusion data. However, diffusion phenomena in intermetallics attract considerable attention not only from the point of view of accumulating of experimental data, but also with the purpose of gaining a deeper insight into the underlying microscopic diffusion mechanisms, especially in highly ordered alloys.

The history of development and practical use of the intermetallics is, e.g., described in the collective monograph published under edition of Westbrook [2]. The intermetallic are used from the very beginning of metallurgy. Their application was determined by enhanced hardness and friction resistance of typical intermetallics in combination with metallic properties. The intermetallic compounds drifted to the focus of scientific studies with the development of physical metallurgy. Detailed studies of structure and properties of intermetallics were initiated by Tammann in Germany [4] and by Kurnakov in Russia [5]. Early studies on intermetallics included examination of phase stability, phase composition, and phase reaction in order to establish the corresponding phase diagrams and to investigate the physical properties [6]. Simultaneously, it was established that the high hardness and strength of intermetallics are typically combined with their increased brittleness.

From the very beginning, the intermetallics were used as functional materials. For example, such a material as SENDAST based on $Fe_3(Si,Al)$ reveals high magnetic properties and enhanced resistance to frictional load. As a result, this material was widely used in recording magnetic heads. Other examples for initial application of intermetallics represent semiconductor materials and shape-memory alloys, just to name a few.

Initially, the intermetallic compounds were not used as structure materials due to increased brittleness, although they were widely used as second-phase inclusions to base materials. Titanium aluminides Ti_3Al and TiAl represent the first intermetallics which were commercially used [7]. Subsequent research was devoted to nickel aluminide

Ni_3Al, which was used as secondary phase inclusion in nemonics and superalloys [8]. The commercialization of intermetallics was intensified after the so-called boron effect was discovered, when small amounts of added boron resulted in significant improvement of intermetallic ductility [9]. This discovery contributed notably to the general interest to intermetallics [2,3]. Nowadays intermetallic compounds are used for structural, heat-resistant and corrosion-resistant applications [10], and especially for thermal or wear coatings [11]. As an emerging class of intermetallic alloys one has to mention **complex metal alloys** which comprise a huge group of (largely unknown) alloys and compounds with the crystal structures based on very large (100 and even more atoms) unit cells including atom clusters [12]. Another current example represent the so-called **high-entropy alloys** which are multicomponent alloys in equiatomic or nearly equiatomic composition [13].

In the present analysis, we will focus on aluminum-based ordered intermetallic compounds, namely nickel, titanium, and iron aluminides which are used in high-temperature structural applications [14]. These compounds were the subject of intensive research due to their high mechanical properties and low specific weight and they represent interesting subjects from a didactic point of view, too, especially in view of an intensive research and deeply understood diffusion properties. Diffusion in these intermetallic compounds was analyzed in several reviews which could be proposed for further reading; see, e.g., [15–18]. Specific systems were also comprehensively analyzed in dedicated research papers, e.g., in the $L1_2$-structures in the Ni–Al, Ni–Ge, and Ni–Ga systems [19], in the Ti–Al system [20,21], in the Fe–Al system [22], just to mention some of them.

Diffusion in pure materials is known to be mostly mediated by a random motion of vacancies. Due to an ordered structure of intermetallic compounds and different probabilities to find a vacancy on a particular sublattice (see below) it was recognized that diffusion in intermetallics occurs through more sophisticated jump sequences involving several atoms. We have to highlight that even in a completely ordered intermetallic compound at a stoichiometric composition a given amount of thermal defects (vacancies and antistructure atoms) exists at the given temperature $T \neq 0$ K under equilibrium conditions. This fact implies that the relevant order parameter is not unity. This feature has an important consequence. While the given state of order has to be maintained on average, local deviations are possible. Thus, the relevance of a *specific diffusion mechanism* in an intermetallic compound as a well-defined sequence of atom jumps becomes somewhat vague, especially for compounds and compositions with a large deviation from the perfectly ordered state.

However, there were a lot of efforts in classification of the vacancy motion in terms of a particular diffusion mechanism which maintains the local order. Such diffusion mechanisms were suggested for the ordered structures, as the well-known 6-jump cycle [23] (almost *the* diffusion mechanism for ordered structures), the triple defect mechanism [24], the antistructure bridge (ASB) mechanism [25], or 4-jump cycle [26]. The concept of a particular diffusion mechanism is attractive not only because it provides a simplified physical pattern of a complex diffusion behavior, but also since it allows calculating the effective activation energies and entropies for these processes, including the correlation effects; see, e.g., the analysis for the 6-jump cycle diffusion mechanism in NiAl [27]. Using elaborate approaches (*ab initio* calculations, molecular static and dynamic calculations) one can compare the theoretical predictions with experimental data and select the most plausible diffusion mechanism. These and some other specific diffusion mechanisms, which are relevant for aluminides of transition metals, will be analyzed in detail in this chapter. Such analysis covers almost all known mechanisms discussed for ordered intermetallic compounds.

The ordered binary aluminides of Ni, Ti, and Fe form different structures, such as $B2$ (NiAl, FeAl), $L1_2$ (Ni_3Al), $D0_{19}$ (α_2-Ti_3Al), $L1_0$ (γ-TiAl), and $D0_3$ (Fe_3Al). Thus, our analysis will provide some fundamental insight into the interdependence of diffusion behavior and diffusion mechanisms on structure and ordering.

In ordered aluminides with XAl and X_3Al compositions (here X = Ni, Ti, or Fe), the ordered phases exist in differently wide compositional intervals of the corresponding phase diagrams, Figs. 10.1, 10.2, and 10.3. Whereas the TiAl phase field extends predominantly on the Al-rich side of the stoichiometric composition and FeAl on the corresponding Fe-rich side, NiAl can accommodate a remarkable excess of both Al and Ni atoms (Fig. 10.1). The Ni_3Al phase field also exists on both sides of the stoichiometric composition, but in a narrower compositional interval. In contrast, both Ti_3Al and Fe_3Al exist as Al-rich ordered phases in limited temperature intervals.

Note that the composition–temperature interval for the $B2$ phase FeAl is divided into three areas where different types of dominant point defects were found [32] (for a detailed analysis see the corresponding section below).

10.2 POINT DEFECTS IN INTERMETALLIC COMPOUNDS

The ideally ordered crystalline structures of Ni-, Ti-, and Fe-aluminides are shown in the corresponding phase diagrams, Figs. 10.1, 10.2 and 10.3, respectively. These atomic arrangements correspond to the structures at zero temperature and at a perfect stoichiometric compositions. As the temperature increases and/or the composition deviates from stoichiometry, point defects are inevitably generated. Four types of point defect can be introduced in two-atomic intermetallics AB, namely the vacancies on both sublattices V_A and V_B and the atoms on unlike sublattices A_B and B_A (the antistructure atoms), see Fig. 10.4 where these defects are illustrated for the B2 structure of NiAl.

In addition to these elementary defects, different defect agglomeration or clusters could be found in compounds. Most important are di-vacancies (typically two vacancies on the same sublattice in neighboring positions, see Fig. 10.5(A)) or even triple defects,

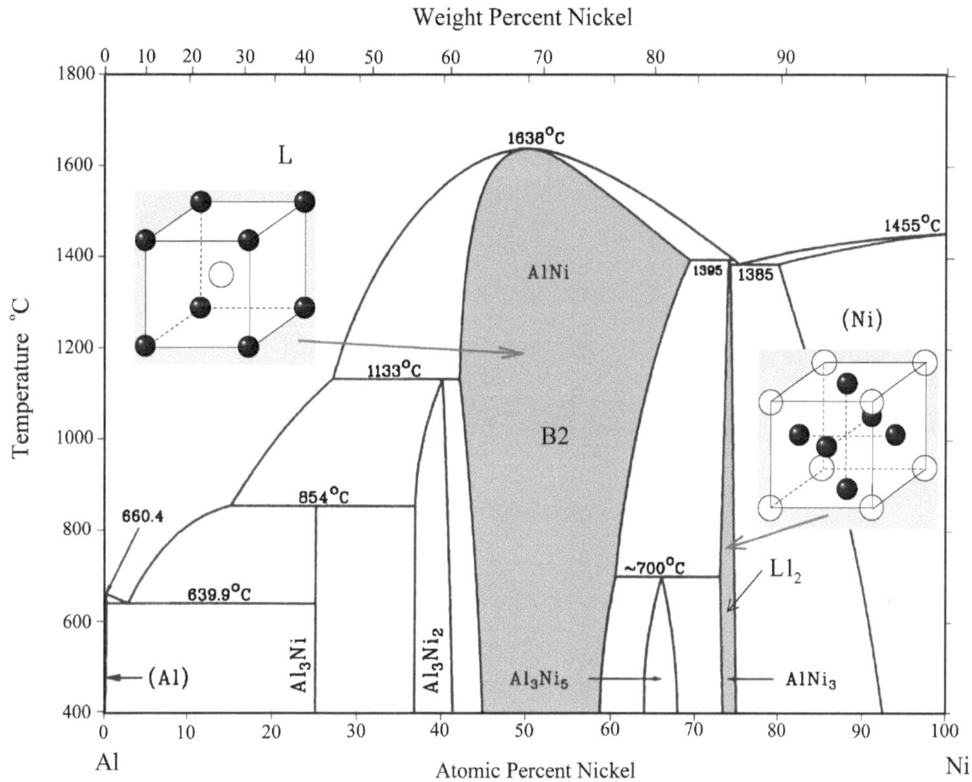

Figure 10.1 Phase diagram of Ni–Al system [28]. The structures of ordered *B*2 NiAl and *L*1$_2$ Ni$_3$Al phases are schematically shown.

Fig. 10.5(B). These triple defects will be shown are important for understanding of diffusion in the B2 structure of NiAl.

The existence of both structural (constitutional) and thermal point defects is recognized for an off-stoichiometric intermetallic compound. After Korzhavy et al. [33], the structural defects are those defects which remain in thermal equilibrium in the intermetallic compound even at $T = 0$ in its maximally ordered state to accommodate the deviation from a stoichiometric composition. Accordingly, the difference between a real concentration of defects at $T \neq 0$ and the concentration of so-defined structural defects represents the concentration of thermal defects. In such a definition the *concentration of thermal defects can even be negative*, as it was proposed, e.g., for the Al-rich phase NiAl [33]. A further difference between the structural and thermal defects is related to the fact that while one type of structural point defect is generally sufficient to accommodate a deviation from the stoichiometric composition, at least two types of thermal point defect have to be simultaneously created to satisfy the mass–balance conditions and to

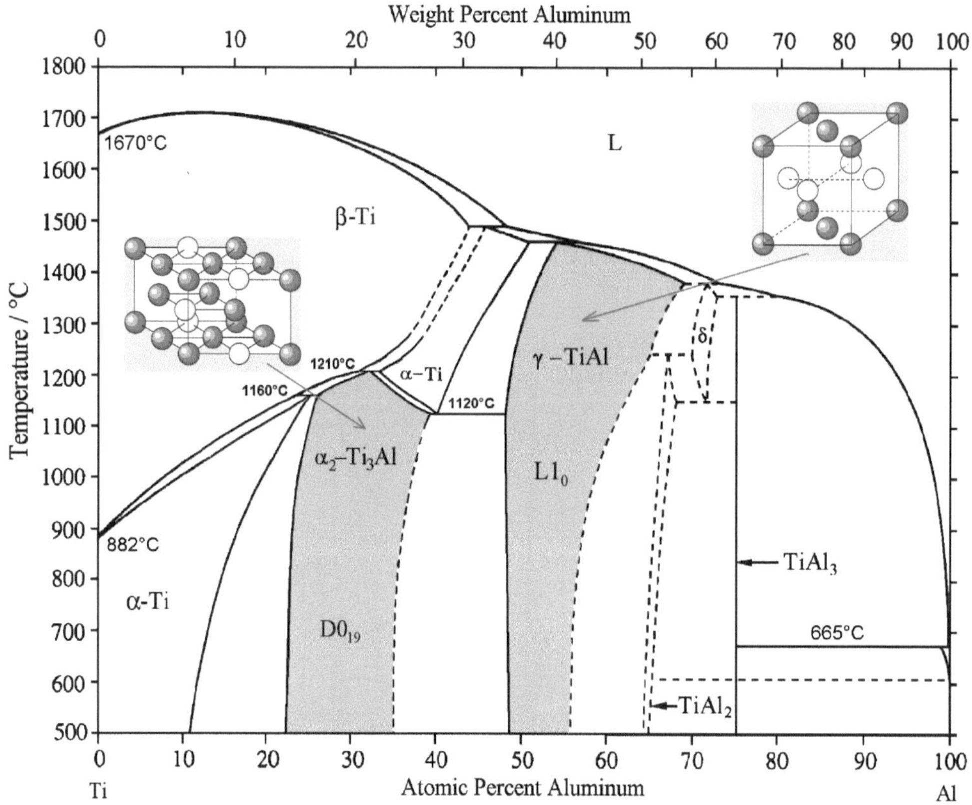

Figure 10.2 Phase diagram of Ti–Al system [29,30]. The structures of ordered $L1_0$ TiAl and $L1_2$ Ti_3Al phases are schematically shown.

preserve the given composition, i.e., the ratio between the constitutional elements in the compound, see the next section.

Moreover, the point defects are typically not uniformly distributed over different sublattices in an ordered intermetallic compound, see the next section.

10.2.1 Concentration of Point Defects

An approach described in Chapter 2 can be applied to determine the given concentrations of point defects in ordered intermetallics. To be specific, let us demonstrate the calculations for the NiAl compound with the B2 structure.

Eqs. (1.21)–(1.23) from Chapter 1 will be read as

$$\varepsilon_{td} + kT \ln X_{V_{Ni}}^2 X_{Ni_{Al}} = 0, \tag{10.1}$$

$$\varepsilon_{Ni_{Al}} + \varepsilon_{Al_{Ni}} + kT \ln X_{Ni_{Al}} X_{Al_{Ni}} = 0, \tag{10.2}$$

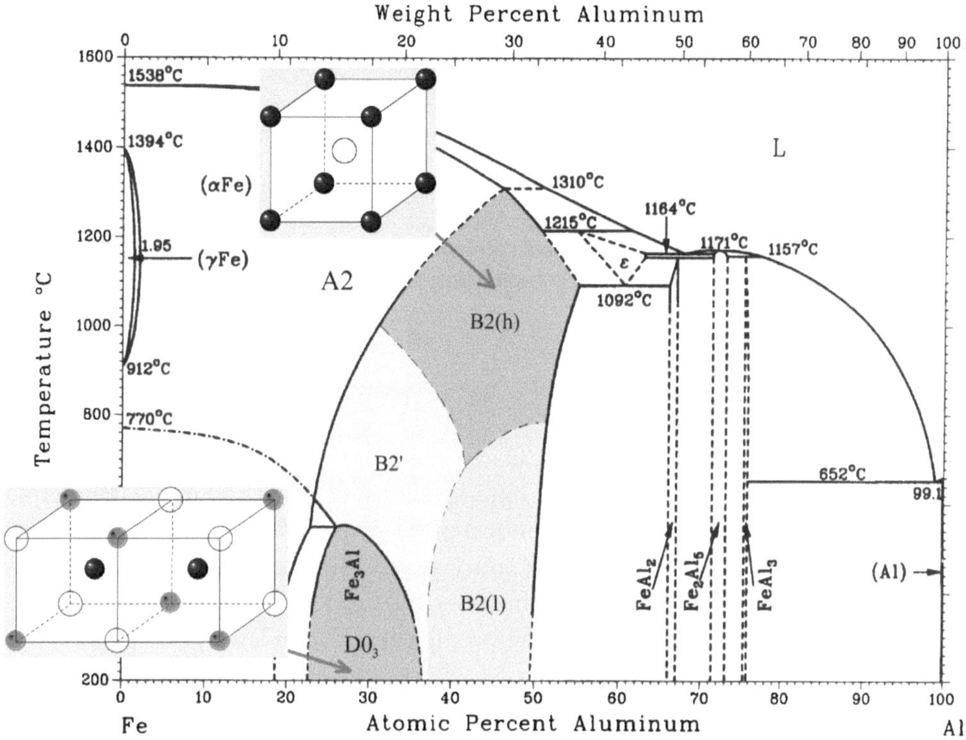

Figure 10.3 Phase diagram of Fe–Al system [31]. The structures of ordered *B*2 FeAl and *D*0$_3$ Fe_3Al phases are schematically shown. In the *D*0$_3$ structure of Fe_3Al (e), the two sublattices for Fe atoms are distinguished (black and gray spheres). The *B*2 area of the phase diagram is divided into three domains after [32].

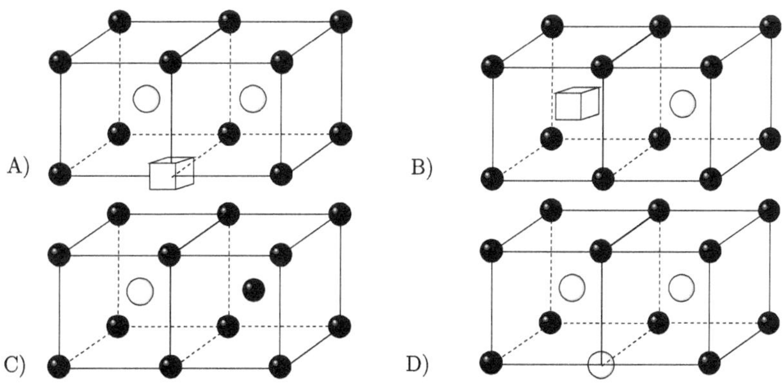

Figure 10.4 Point defects in ordered compounds: vacancy on A sublattice (A), vacancy on B sublattice (B), atom A on B sublattice, the A_B anti-cite atom (C), and atom B on A sublattice, the B_A anti-cite atom (D). As an example the ordered B2 lattice is shown. The A and B atoms are shown as black and white spheres, and cubes represent the unoccupied positions (vacancies).

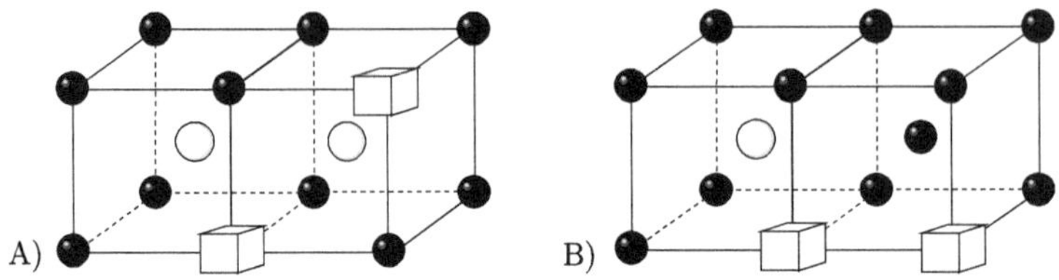

Figure 10.5 Defect clusters in ordered compounds: divacancy, i.e., two vacancies on A sublattice (A) and a triple defect – two A vacancies and one A antisite (B). The A and B atoms are shown as black and white spheres, and cubes represent the unoccupied positions (vacancies).

$$2\varepsilon_{V_{\mathrm{Al}}} - \varepsilon_{\mathrm{Ni_{Al}}} + 2\varepsilon_0 + kT\ln\left(\frac{X^2_{V_{\mathrm{Al}}}}{X_{\mathrm{Ni_{Al}}}}\right) = 0. \tag{10.3}$$

Here $\varepsilon_{\mathrm{td}} = 2\varepsilon_0 + 2\varepsilon_{V_{\mathrm{Ni}}} + \varepsilon_{\mathrm{Ni_{Al}}}$ is the so-called triple-defect energy, i.e., the energy of the defect shown in Fig. 10.5(B); $2\varepsilon_0$ is the binding energy of a unit cell in the compound; and X_i are the concentrations (mole fractions) of the given defects. These equations can be solved numerically for the four defect concentrations – vacancies on both sublattices, V_{Ni} and V_{Al}, and antisites on the two sublattices, $\mathrm{Ni_{Al}}$ and $\mathrm{Al_{Ni}}$ – in dependence on the composition of the alloy applying in addition the conservation law,

$$\xi\left(1 - 2X_{V_{\mathrm{Al}}} - 2X_{V_{\mathrm{Ni}}}\right) = \frac{1}{4}\left(X_{V_{\mathrm{Al}}} - X_{V_{\mathrm{Ni}}}\right) + \frac{1}{2}\left(X_{\mathrm{Ni_{Al}}} - X_{\mathrm{Al_{Ni}}}\right) \tag{10.4}$$

where

$$\xi = c_{\mathrm{Ni}} - \frac{1}{2} \tag{10.5}$$

is the deviation of the alloy composition (mole fraction of Ni, c_{Ni}) from the ideal stoichiometry ($c_{\mathrm{Ni}} = 0.5$). Neglecting the terms like ξ^2 in Eq. (10.4) for small deviations from the stoichiometric composition and small concentrations of the point defects, the above equation can be presented as

$$\xi = \frac{1}{4}\left(X_{V_{\mathrm{Al}}} - X_{V_{\mathrm{Ni}}}\right) + \frac{1}{2}\left(X_{\mathrm{Ni_{Al}}} - X_{\mathrm{Al_{Ni}}}\right). \tag{10.6}$$

Now, we have four equations with four unknown concentrations which are to be solved for. The effective formation energies of the given type of defects can be estimated by linear fitting in the logarithm of concentration *vs.* inverse temperature in a given temperature interval. The analysis shows that the temperature dependence of the point defect concentrations can be adequately treated with a single formation energy.

In Fig. 10.6, the concentrations of different point defects in the given intermetallic compounds are compared at $T = 0.8T_m$. This is typical temperature for bulk tracer diffusion measurements, and it corresponds to $T = 1335$ K for $\mathrm{Ni_3Al}$, 1530 K for NiAl,

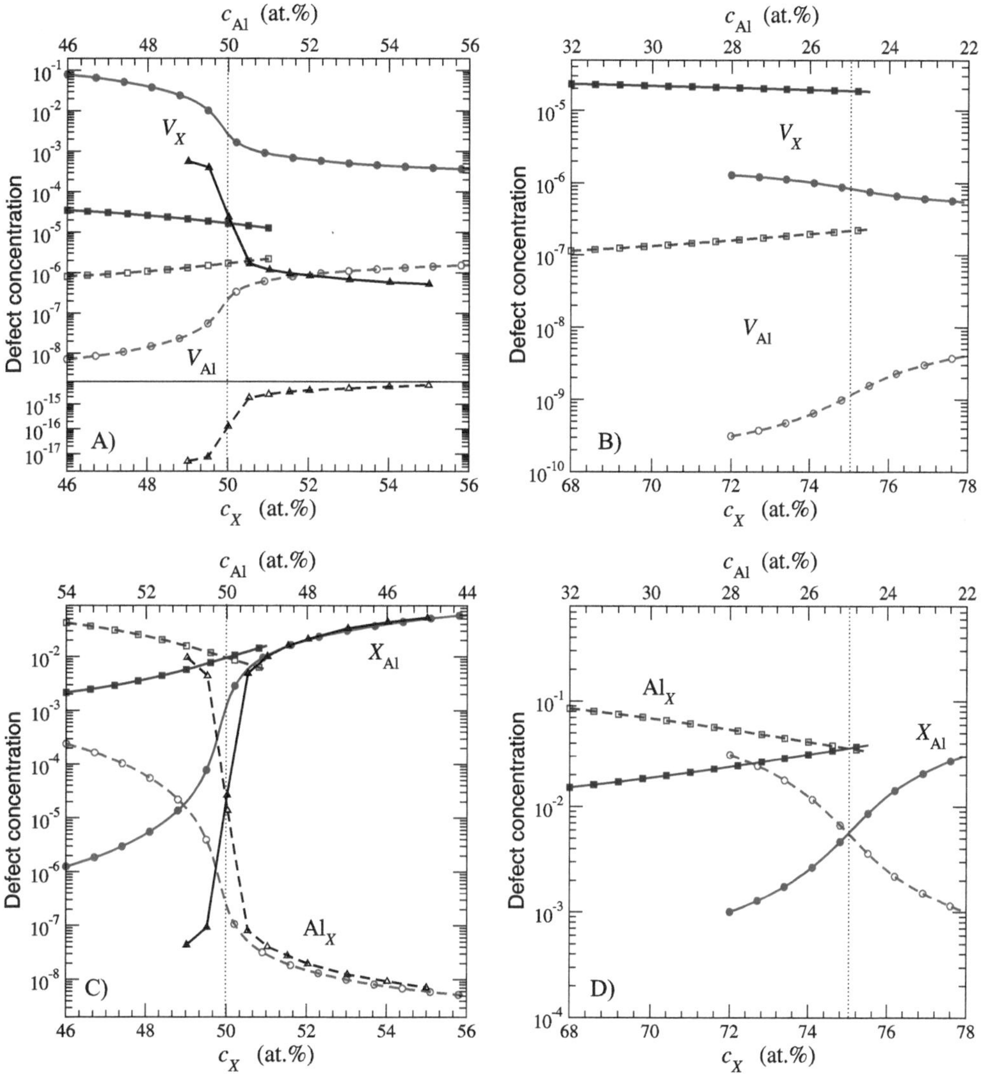

Figure 10.6 Vacancy (A)–(B) and antistructure atom (C)–(D) concentrations on transition metal (solid lines) and Al (dashed lines) sublattices in *X*Al (A) and (C), as well as in X_3Al (B) and (D) (X = Ni, Ti, and Fe), as functions of the composition at $T = 0.8T_m$. The defect concentrations are shown by red circles, blue squares, and black triangles for Ni, Ti, and Fe aluminides, respectively. Filled and open symbols represent the defect concentrations on transition metals and Al sublattices, respectively. (For interpretation of the references to color in this figure legend, the reader is referred to the web version of this chapter.)

1554 K for Ti_3Al, 1380 K for TiAl, and 1275 K for FeAl. T_m is the melting temperature of the stoichiometric composition of the given compound. The defect concentrations were calculated according to the above-outlined chemical-reaction approach [21]. Since in an intermetallic compound point defects are created in a correlated manner in order to preserve the given composition, the concentration of point defects depends on the formation energies of all four types of defects. The defect formation energies for Ti_3Al and TiAl calculated in [20,21] were directly used. The EAM potentials developed by Voter and Chen [34] were used in the calculations for Ni_3Al, and those of Mishin et al. [35] were applied for NiAl. The results of *ab initio* calculations of defect formation energies of Meyer and Fähnle [36,37] were used in the case of *B*2-FeAl. Formation entropy effects were neglected (the calculations of the formation entropies are outlined in the next section).

Fig. 10.6 demonstrates a few important features of the defect behavior in these intermetallics:

- TiAl, Ti_3Al, and Ni_3Al belong to the *antistructure-defect type* of intermetallic compounds, since the antistructure atoms are predominantly generated to accommodate a deviation from the stoichiometric composition (compare the concentrations of vacancies and antisites in these compounds; in X_3Al compounds, e.g., the antisite concentrations exceed 10^{-3} and those of vacancies are smaller than 3×10^{-5} at $T = 0.8T_m$ in the given concentration intervals, Figs. 10.6(B), (D); the same holds for TiAl, too, Fig. 10.6(A)).
- NiAl reveals a *triple-defect-type* of point defect disorder and constitutional Ni vacancies exist in NiAl on the Al-rich side and $C_{V_{Ni}} > 10^{-2}$, Fig. 10.6(A), and the concentration of Ni vacancies does practically not depend on temperature in Al-rich compositions. In Ni-rich compositions, however, Ni antistructure atoms dominate the disorder, cf. Figs. 10.6(A), (C).
- In the ordered aluminides the vacancies are mainly concentrated on the transition-metal sublattice and their concentration amounts to about 10^{-6} to 10^{-5} at $T = 0.8T_m$ (excluding NiAl alloys, in which Ni vacancy concentration is significantly larger); see Figs. 10.6(A)–(B). These are also the typical vacancy concentrations in close-packed pure metals at the same reduced temperature. The vacancy concentrations on the Al sublattice are remarkably smaller, especially in FeAl, Fig. 10.6(A).
- Ni vacancy concentration in NiAl is very large also on the Ni-rich side, e.g., $C_{V_{Ni}} \sim 10^{-4}$ at $T = 0.8T_m$.
- The antisite type of disorder is demonstrated by the fact that the concentrations of antistructure atoms of the two sublattices are almost symmetrical with respect to the stoichiometric composition $c_{Al} = 50$ at.%; see, e.g., Fig. 10.6(D).

Alternatively to Ti- and Ni-aluminides, the defects in FeAl are more complex. According to [36,37], *B*2-FeAl is neither a compound with a pure antisite disorder nor a compound with pure triple-defect disorder. FeAl demonstrates a hybrid behavior in which the relation between the Fe vacancy concentration and that of the antistructure atoms depends crucially on temperature.

The concentration of Ti antistructure atoms in the Ti aluminides (blue squares) is generally larger than that of Ni antistructure atoms in the Ni aluminides (red circles) of the same composition, Fig. 10.6(C)–(D). This fact corresponds to a higher degree of thermal disorder inherent in Ti aluminides at similar reduced temperatures and it plays a decisive role in the analysis of the respective self-diffusion behavior.

10.2.1.1 Formation Entropy of Point Defect

The effective formation entropy of point defects can be calculated within the canonical ensemble formalism, too. Again, the procedure is here outlined for the B2 ordered NiAl compound, with Ni atoms being at origin of the coordinate axes [27]. First, the "raw" formation entropies of point defects are computed (the vibrational contribution to the formation entropy), e.g., in a quasiharmonic approximation [38]. The Ni atoms with the coordinates $R_i < R_N$ and Al atoms with the coordinates $\|R_i - r_0\| < R_N$ are allowed to vibrate. Here $r_0 = (\frac{a}{2}, \frac{a}{2}, \frac{a}{2})$ and a is the lattice parameter. Such a choice of the vibration block guarantees its perfect stoichiometric composition in the absence of defects.

The eigenfrequencies ν_i of normal vibrations of atoms can be calculated within both the local harmonic approximation (LHA) [39] and the quasiharmonic approximation (QHA) [40]. The vibrational free energy F_v is calculated as [38]

$$F_v = kT \sum_{i=1}^{3N} \ln 2 \left\{ \sinh \frac{h\nu_i}{2kT} \right\}. \tag{10.7}$$

The entropies can be calculated according to the scheme proposed in [40]. At first, the equilibrium atomic positions at 0 K are to be determined by energy minimization. Then, this configuration is uniformly expanded and the thermal expansion factor α of the perfect block is established by minimizing the corresponding ansatz for Gibbs energy G with respect to volume V. This gives us the entropy $S_0^{(N)}$ of the perfect block calculated within the quasiharmonic approximation as [38]

$$S_0^{(N)} = -\left(\frac{\partial F_v}{\partial T}\right)_V = k \sum_i \left\{ \frac{h\nu_i}{2kT} \coth\left(\frac{h\nu_i}{2kT}\right) - \ln\left(2\sinh\frac{h\nu_i}{2kT}\right) \right\}. \tag{10.8}$$

Here superscript (N) indicates that S_0 was evaluated for the cluster of N vibrating atoms.

The entropy of the block with a given defect is determined similarly. At first, the defect is introduced and the structure is relaxed at 0 K by energy minimization. Then, using α calculated for the perfect block, the relaxed structure is expanded and the entropy of an N-atom cluster around the defect is determined by Eq. (10.8).

Table 10.1 Calculated "raw" formation entropies s_d of point defects in NiAl as a function of temperature and the effective formation entropy $S_{V_{Ni}}$ of Ni vacancies for selected stoichiometric and Ni-rich compositions. $2s_0$ is the entropy per unit cell in the perfect lattice. The entropy values are given in the units of Boltzmann constant k

Approximation	T **[K]**	$s_{V_{Ni}}$	$s_{V_{Al}}$	$s_{Ni_{Al}}$	$s_{Al_{Ni}}$	$2s_0$	$S_{V_{Ni}}$	
							$c_{Ni}=0.5$	$c_{Ni}=0.52$
LHA	1000	−4.69	−2.19	0.76	9.83	12.78	0.92	3.00
	1200	−5.40	−3.06	0.67	9.74	14.07	0.85	2.89
	1500	−6.16	−3.59	0.70	10.69	15.80	0.93	3.01
	1700	−6.63	−3.81	0.71	11.60	16.91	0.99	3.10
QHA	1200	−6.02	−5.29	1.02	10.5	14.91	0.83	2.86

Note that in the case of vacancies (Ni or Al) the N block contains $N-1$ atoms and the number of vibrational modes is $3(N-1)$. The "raw" defect entropies, s_d, are calculated as $s_d = S_d^{(N)} - S_0^{(N)}$ at different N and the values are extrapolated to $N \to \infty$ on the plots of s_d *vs.* $1/N$ [41]. In this scheme we neglect the inhomogeneity of thermal expansion around the defect. Note that this effect was established to be small for pure metals [40]. The results of our calculations at $T = 1200$ K for NiAl are presented in Fig. 10.7 and the resulting values of the "raw" formation entropies are summarized in Table 10.1. Fig. 10.7 suggests the importance of collective vibrational modes. The "raw" entropy values computed by QHA differ on average by 1 k from the corresponding values calculated by LHA, see Table 10.1.

Note that in the case of intermetallics it is not possible to use $\delta S = S_d^{(N-1)} - \frac{N-1}{N} S_0^{(N)}$ as the vacancy formation entropy (exactly as it is the case for formation energies). Instead, the effective formation entropies have to be calculated. Similar to the effective formation energies, the effective formation entropies can be considered as an Arrhenius approximation of "true" formation entropies which are temperature dependent. Following the general approach of Chapter 1 and defining the defect concentrations C_d as $C_d = \exp\left\{\frac{S^f}{k}\right\} \exp\left\{-\frac{E^f}{kT}\right\}$, the effective formation entropy $S^f_{V_{Ni}}$ of the Ni vacancies in dependence on composition is:

$$S^f_{V_{Ni}} = k \ln(2\xi), \qquad \xi = 0.5 - c_{Ni} > 0,$$

$$S^f_{V_{Ni}} = \frac{1}{3}\left(2s_0 + 2s_{V_{Ni}} + s_{Ni_{Al}}\right) - \frac{2}{3} k \ln 2, \quad c_{Ni} = 0.5, \tag{10.9}$$

$$S^f_{V_{Ni}} = \frac{1}{2}\left(2s_0 + 2s_{V_{Ni}} + s_{Ni_{Al}}\right) - \frac{1}{2} k \ln(8\xi), \ \xi = c_{Ni} - 0.5 > 0.$$

Here c_{Ni} is the atomic concentration of Ni atoms in the NiAl alloy. Similarly, the formation entropies of all constitutional defects can be determined.

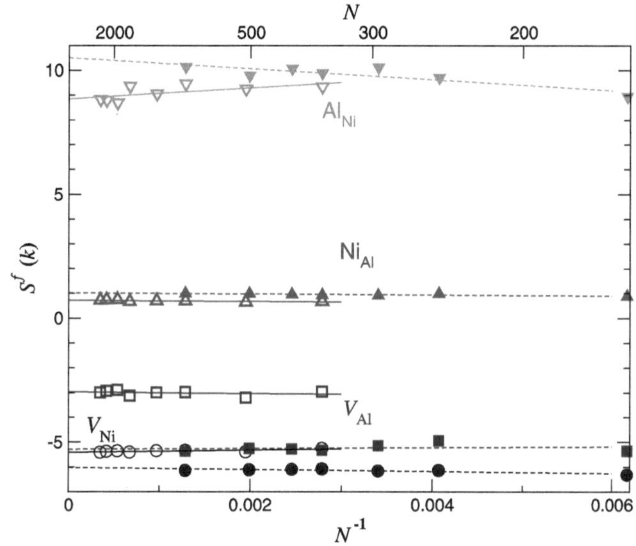

Figure 10.7 Calculated "raw" formation entropies S^f (in units of the Boltzmann constant k) of different point defects – Ni vacancies (circles), Al vacancies (squares), Ni antistructure atoms (triangles up), and Al antistructure atoms (triangles down) – in NiAl as functions of the number of atoms N in the vibration block. Both LHA (open symbols and solid lines) and QHA (full symbols and dotted lines) results are presented. $T = 1200$ K.

Table 10.1 indicates that the formation entropies could be non-constant for a given defect and depend on temperature. Nevertheless, the effect here is not appreciably large. Recently, the temperature dependence of the Gibbs energy of vacancy formation in Al and Cu from $T = 0$ K up to the melting temperature was determined from ab initio thermodynamics, fully taking into account anharmonic contributions [42]. Their results convincingly demonstrated that the formation entropy of vacancies is not constant but increases almost linearly with temperature [42]. As a result, deviations from perfect linear Arrhenius-type temperature dependence of the diffusion coefficients can also be expected.

10.3 DIFFUSION MECHANISMS IN ORDERED INTERMETALLICS

Now we are confronted with an issue of how does a particular crystal structure of the given intermetallic compound affect the self-diffusion properties. Self-diffusion in close-packed structures is generally accepted to occur via nearest-neighbor jumps of vacancies. Since random vacancy jumps between different sublattices would inevitably introduce a disorder and since there is a strong tendency to accomplish a reverse, ordering jump after a given disordering jump, the correlated jumps of vacancies play a decisive role in the long-range diffusion process. These jump sequences of vacancies where a given level of disorder is completely restored after their completion is termed here as a diffusion mechanism.

When comparing the atomic structures sketched in Figs. 10.1, 10.2, and 10.3, an important feature can be realized which, to a large extent, determines the self-diffusion properties of these compounds. It is obvious that in Ni_3Al (Fig. 10.1), Ti_3Al and TiAl (Fig. 10.2), and Fe_3Al (Fig. 10.3) the transition metal sublattice represents an interconnected network for nearest-neighbor jumps of vacancies, whereas the nearest neighbor jumps in the B2 structures of NiAl and FeAl (Figs. 10.1 and 10.3, respectively) are exclusively the jumps between the different sublattices. Since vacancies are concentrated on the transition-metal sublattices, see Fig. 10.6(A)–(B), the *sublattice* diffusion mechanism may be proposed for Ni_3Al, Ti_3Al, TiAl, and Fe_3Al from this general consideration.

One may expect that structural point defects significantly affect self-diffusion (temperature-driven jumps of atoms to nearby vacancies) in intermetallic compounds and produce a compositional dependence of the diffusivity. Thus, do structural vacancies represent the defects which most remarkably enhance self-diffusion? To what extent do antistructural atoms affect the diffusivity? How are these effects related to the given crystalline structure? Does the absence of the connected network for nearest-neighbor jumps of Ni vacancies slow down self-diffusion in NiAl in comparison with the presence of such a network in the more close structure of Ni_3Al? These and other issues will be analyzed below, and an additional insight into the interdependence of the diffusion behavior and the diffusion mechanisms on the structure and the type of disorder in the Ni-, Ti-, and Fe-aluminides will be provided. However, before to proceed we present a short overview of main diffusion mechanisms in ordered intermetallics.

The ordered structure of intermetallic compounds imposes certain limitations on geometrically possible vacancy-mediated diffusion mechanisms. Below, the most important diffusion mechanisms which are relevant to the diffusion behavior of the aluminides under consideration are discussed.

The following probable diffusion mechanisms will be considered in the present analysis:

1. Next nearest neighbor (nnn) jumps
2. 6-jump cycle mechanism
3. 4-jump cycle mechanism (antisite-assisted 6-jump cycle)
4. Antistructure bridge mechanism (ASB)
5. Divacancy mechanism
6. Triple defect mechanism

10.3.1 Next Nearest Neighbor Jumps

The next nearest neighbor (nnn) jumps are intra-sublattice jumps, i.e., jumps within the same sublattice. As an example, in Fig. 10.8 this mechanism is illustrated for a B2 structure (NiAl). For any Ni vacancy, there are 6 directions of such jumps (along

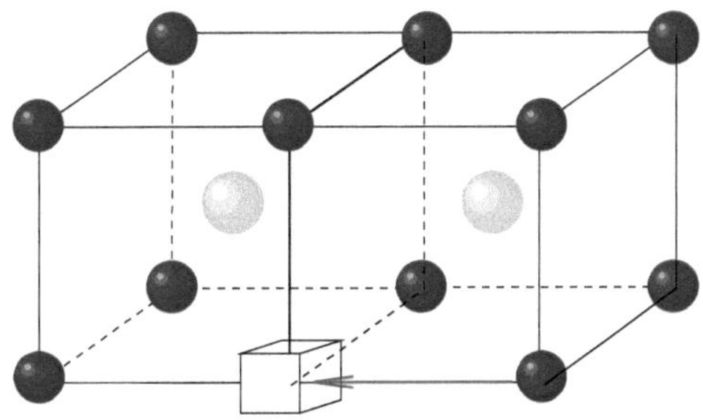

Figure 10.8 A scheme of the next nearest neighbor jump mechanism in a *B*2 compound *AB*.

edges of the corresponding cubic cells). It was predicted that this might be a dominant mechanism in NiAl, at least on the Al-rich side of composition [43].

These nnn jumps are jumps within a given sublattice, thus, they preserve the given ordered state of the lattice. Such a mechanism is probably not possible in a dense lattice (like FCC or HCP) due to energy penalty which is imposed at a saddle-point configuration. However, in a more open structure (BCC, e.g., B2) the activation barriers for these jumps may be reasonably small.

Note that the tracer diffusion coefficients of the components will be unbounded if the nnn jumps dominate diffusion of one of the components.

10.3.2 Six-Jump Cycle Mechanism

Originally, this mechanism was proposed for the *B*2 compounds [23], but later it was elaborated for other ordered structures, like $L1_2$ [44] or even icosahedral quasicrystals [45,46], too.

For a B2 structure, this mechanism is sketched in Fig. 10.9. As it is directly seen, having accomplished six successive jumps, a vacancy moved to a neighboring position (second or third neighbor shell). If the initial state represents a ground state of the alloy (with the given concentration of equilibrium vacancies), the local order increases during the first three jumps, reaching its maximum with two B antisites, one A antisite, and a B vacancy, Fig. 10.9. Then, the successive vacancy jumps sequentially restore the initial order. By inspection of the lattice structure, one can recognize that three different cycles are possible in a B2 structure, Fig. 10.10, left panel, namely one flat and two bend cycles with [011] and [010] net displacements of the vacancy.

The correlated jumps of atoms during these cycles impose certain limitations on the diffusion coefficient of two species, which can be measured in a diffusion experiment. Firstly, in a highly ordered state the ratio of tracer diffusivities of both components,

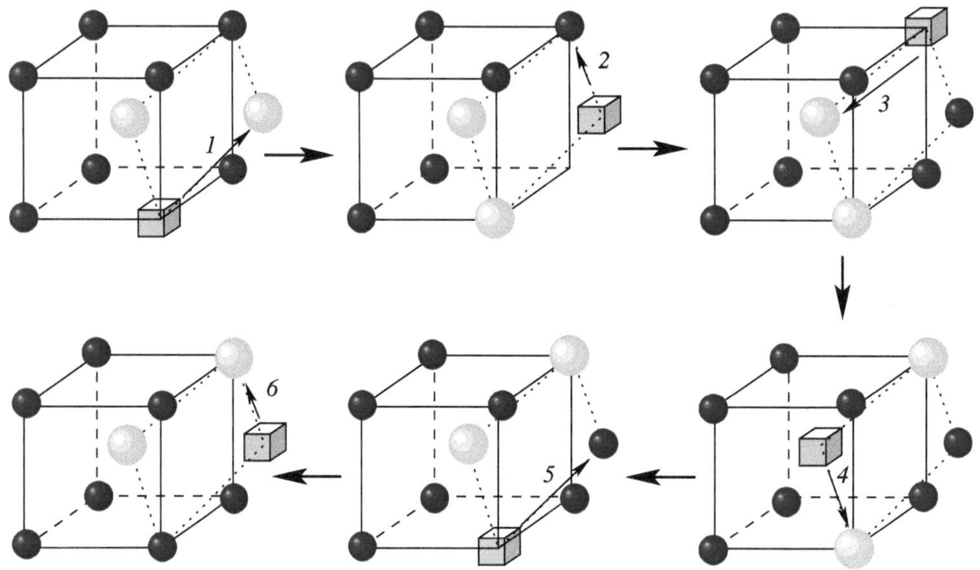

Figure 10.9 A scheme of the six-jump cycle mechanism in a *B*2 compound *AB*. The A and B atoms are shown by blue and yellow spheres, respectively. (For interpretation of the references to color in this figure legend, the reader is referred to the web version of this chapter.)

D_A^*/D_B^*, can adopt only the values within the limits

$$\frac{1}{q} \le \frac{D_A^*}{D_B^*} \le q, \tag{10.10}$$

where q was first determined to be 2 [47] and later corrected to $q \cong 2.034$ by including the correlation effects [48]. The ratio of diffusivities of Ag and Mg in β-AgMg [47], Zn and Au in β'-AuZn [49], and Cd and Au in β-AuCd [50] fall into these limits, and this was considered as a strong support of the six-jump cycle diffusion mechanism in these compounds.

As it was shown in previous sections, with a deviation of the composition from a stoichiometric one, many constitutional antistructure defects typically appear. Including these interactions remarkably changes the limits in Eq. (10.10), and q could even attain a value of 10 [51]. Thus, in a less ordered state experimental values of D_A/D_B for a B2 compound being larger than 2 can no longer be considered as an indication that the six-jump cycle mechanism does not operate.

Since the ratio of the tracer diffusion coefficients of the two components is inevitably coupled by Eq. (10.10) and the same vacancy facilitates motion of both components, one may argue that the ratio of the corresponding correlation factors is also highly coupled. Since atom diffusion through the six-jump cycles represents in its turn a highly

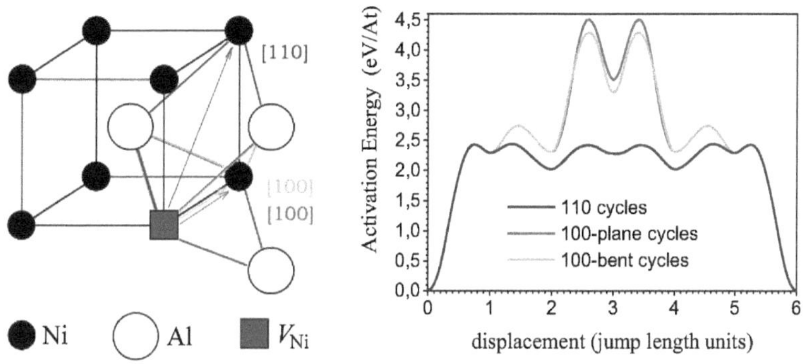

Figure 10.10 System energy (right) of a function of vacancy position in the course of jumps according to the 6-jump-cycle mechanism (left) for NiAl.

correlated process, the corresponding correlation factor is supposed to be rather small (calculation of the correlation factors is outlined below in a separate section). However, one should generally distinguish two types of correlations which characterize the six-jump cycles. One can consider the individual 6-jump cycles as one effective vacancy jump which occurs with the given frequency. Then, analyzing long-range diffusion via the 6-jump cycles, the resulting correlation factors $\widehat{f_A}$ and $\widehat{f_B}$ can be computed [48]. For example, for *B*2 NiAl, the Monte Carlo (MC) calculations according to Eq. (10.14) resulted in $\widehat{f_{Ni}} \cong 0.782$ and $\widehat{f_{Al}} \cong 0.860$ [27]. In contrast, the tracer correlation factors (including all individual jumps of atoms) for Ni and Al atoms in that case were determined to be $f_{Ni} \cong 0.445$ and $f_{Al} \cong 0.022$, respectively. We highlight that f_{Ni} is not so small as it was usually anticipated for a highly-correlated cycle mechanism.

An experimental approach to determine the correlation factor is the usage of the isotope effect when diffusion of two isotopes of the same chemical species with different masses is simultaneously measured [52]. The ratio of the determined diffusion coefficients determines the isotope effect, *E* [52]. In [53], the isotope effect was measured for both Au and Zn in the *B*2-ordered β'-AuZn alloys. It was shown that E_{Au} (and correspondingly f_{Au}) is considerably larger than E_{Zn} (f_{Zn}) in Zn-rich alloys ($E_{Au} = 0.35$ and $E_{Zn} = 0.05$ in the Au–51.85at.%Zn alloy). This fact resembles the above relation between f_{Ni} and f_{Al} in *B*2-NiAl for the six-jump cycle mechanism [27] and can be explained by the predominant vacancy concentration in the Au-sublattice, and an increased probability of a reverse jump of a Zn atom which has initiated a six-jump cycle. We conclude that the relationship $f_A \gg f_B$ in an *AB* compound cannot be considered as an argument against the six-jump cycle mechanism.

As an example, Fig. 10.10 represents a calculation of the migration energy for a 6-jump-cycle mechanism in B2 NiAl. It is seen that the 110 cycles dominate in NiAl. Here the cycles starting from Ni vacancy are considered, as the calculations of the defect concentrations in NiAl suggest. We conclude that although several distinct types

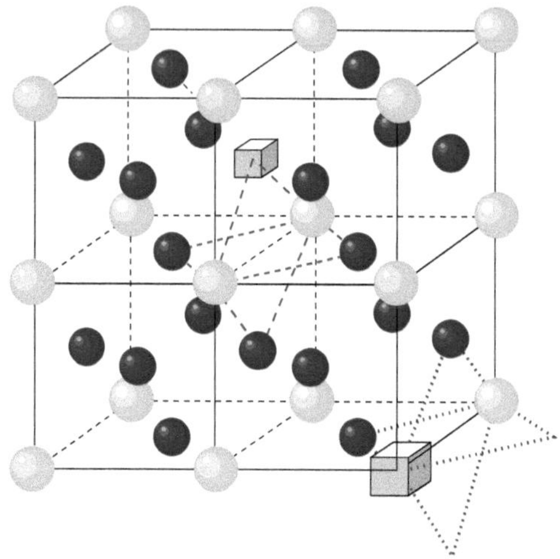

Figure 10.11 A scheme of the 6-jump cycle mechanism for the $L1_2$ structure. Two subsets of the mechanism are shown: starting with A (red dashed lines) and B (blue dotted lines) atoms. The A and B atoms are shown by blue and yellow spheres, respectively. (For interpretation of the references to color in this figure legend, the reader is referred to the web version of this chapter.)

of 6-jump cycles are generally possible, in some specific situations a given type of cycles may dominate.

As it was mentioned, the 6-jump cycle mechanism was extended for other ordered intermetallics. In Figs. 10.11 and 10.12, it is illustrated for the $L1_2$ and icosahedral quasicrystalline structures, respectively.

Fig. 10.11 reveals a high similarity of 6-jump cycles in the $L1_2$ structure to those in the $B2$ one. Again, flat and bent cycles are possible and atomistic calculations are required to establish the dominant process, if this mechanism does contribute to long-range diffusion. One recognizes that the geometry of the cycles is different for the cases with starting A or B vacancy, red and blue lines, respectively. Indeed, due to a different number of nearest neighbors of the opposite type for a vacancy on a given sublattice (12 for B and 4 for A), the ratio of the diffusion coefficients will be different from Eq. (10.10).

In Fig. 10.12 the sequence of atom jumps according to the 10-cycle mechanism [46] for the icosahedral *i*-AlMnSi phase is shown. Hereto, an outward part of Mackay icosahedron is considered and its plane projection is presented in Fig. 10.12(A). This structure corresponds to a spherical triangle formed by three Mn atoms (in Fig. 10.12 one Mn atom is replaced by a vacancy) and with three Al/Si atoms on the edges between them (Al and Si atoms share this sublattice and Al atoms dominate). Diffusion of the Mn atoms can occur by a correlated sequence of ten jumps without a loss of the order after completing the whole circle. This sequence is shown in Fig. 10.12(B)–(G) a by a set of

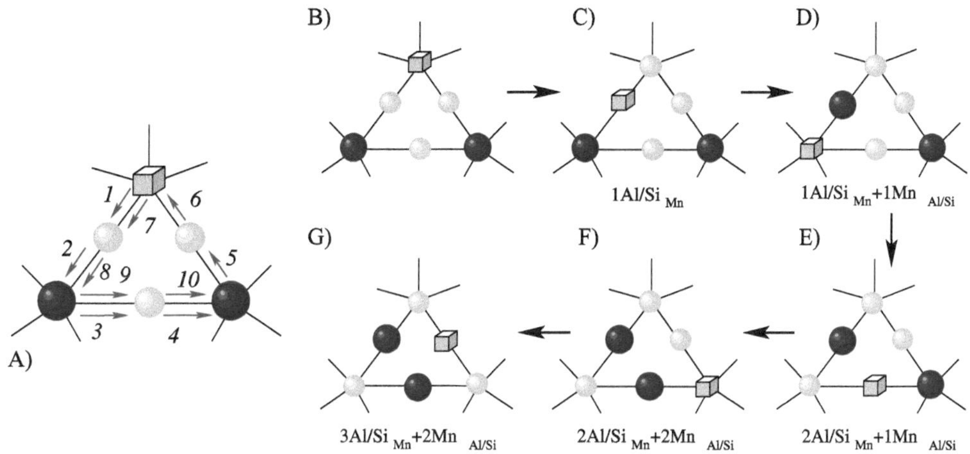

Figure 10.12 Local atomic configurations and the ten-jump-cycle mechanism for the *i*-AlMnSi structure (A), and appearance of antistructural defects after the first five jumps (B)–(G). Mn and Al(Si) atoms are sketched by blue and yellow spheres, and the vacancy is shown by a box. (For interpretation of the references to color in this figure legend, the reader is referred to the web version of this chapter.)

snap-shots. As a result, the vacancy on the Mn sublattice is shifted by one position in its own sublattice, and two Mn atoms and three Al atoms jump into the nearest neighboring sites on their particular sublattices. To demonstrate the loss of order during this cycle jumps, the first five vacancy jumps are shown by the sequence of Fig. 10.12(B)–(G). The atomic order is seen to be locally destroyed by the appearance of antistructural atoms. In Fig. 10.12 the antistructural atoms are designated by the $Mn_{Al/Si}$ or Al/Si_{Mn} symbols. One has to take into account that the diffusion mechanism of Al atoms is mixed, it is mediated both through the ten-jump-cycles and by directed jumps within the Al sublattice [46].

Taking into account all possibilities to perform the ten-jump-cycles as well the direct jumps of Al atoms within their own Al/Si sublattice, one can determine the ratio of diffusion coefficients of Al, D_{Al} and Mn, D_{Mn} atoms. The calculation have shown [46] that this ratio is (by neglecting the correlation effects),

$$\frac{D_{Al}}{D_{Mn}} = \frac{n_1\omega_1(4h^2 + l^2) + 3n_2\omega_2 h^2 + \gamma n_2\omega_2^* a_R}{n_1\omega_1(2h^2 + l^2) + n_2\omega_2(5h^2 + l^2/2)}. \tag{10.11}$$

Here $h = \tau a_R/2$ is the distance between the icosahedron center and the center of the outer edges; $l = a_R\sqrt{3-\tau}$ is the length of these outer edges ($\tau = (1+\sqrt{5})/2$ is the golden ratio); a_R is the distance between the Mackay icosahedron center and the Mn atoms. Here n_1 and n_2 are the vacancy concentrations on the Mn and Al/Si sublattices, and ω_1 and ω_2 are the frequencies of the ten-jump cycles with a starting vacancy on Mn and Al sites, respectively. The value of the geometric factor γ is about 1/4 and

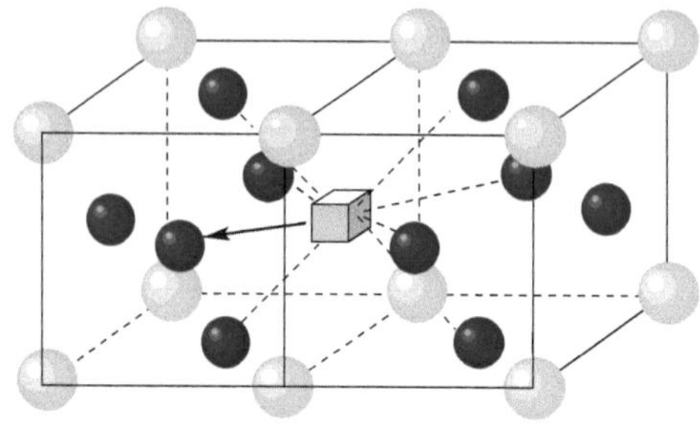

Figure 10.13 A scheme of the sublattice diffusion mechanism in the $L1_2$ structure of an A_3B compound.

takes into account the decrease in the number of Al nearest neighbors of an Al atom in the intermetallic quasicrystal in comparison with pure Al. Finally, ω_2^* is the frequency of the nearest neighbor jumps of Al atoms within their sublattice. One sees that this ratio approaches 2 if $n_1 \gg n_2$ and it is $(3\tau^2 + 4\gamma\omega_2^*/\omega_2)/(3\tau^2 + 8)$ in the reverse case. If we assume that ω_2^* is larger than ω_2 by at least an order of magnitude, the ratio $D_{Al}/D_{Mn} > 0.5$ if $n_1 \ll n_2$.

The tracer diffusion of Mn and Zn (as Al substitute) was measured in quasicrystalline Al–Pd–Mn alloy [54] and it was found that the Zn diffusion rate exceeds that of Mn atoms by more than four orders of magnitude and there is a large difference of the corresponding activation enthalpies. This fact suggests that in icosahedral Al–Pd–Mn alloy direct jumps of Al-substitutes over Al-sublattice most probably dominate Zn (as well as Ga or Ge diffusion [54]) and diffusion rates of Mn atoms are more correlated.

10.3.3 Sublattice Diffusion Mechanism

As it is seen from Figs. 10.1–10.3, there is a number of cases in which one of the components forms a lattice structure which enables nearest-neighbor jumps through the respective sublattice, e.g., the $L1_2$ structure of Ni_3Al. In this case, random jumps of a vacancy on this sublattice do not affect the order in the compound. This mechanism is exemplary sketched in Fig. 10.13 for the $L1_2$ structure. One has to recognize that this mechanism can dominate diffusion not only of the majority component (A atoms in Fig. 10.13), but also of the minority component (B atoms in Fig. 10.13). In the latter case, a minority atom jumps into the "wrong" sublattice and migrates further through this sublattice as an antistructure atom. The sublattice diffusion mechanism was extensively analyzed, e.g., in [55,56].

An analysis reveals immediately that the diffusion coefficients of both components are not coupled by a relation similar to Eqs. (10.10) or (10.11), if the sublattice diffusion mechanism operates.

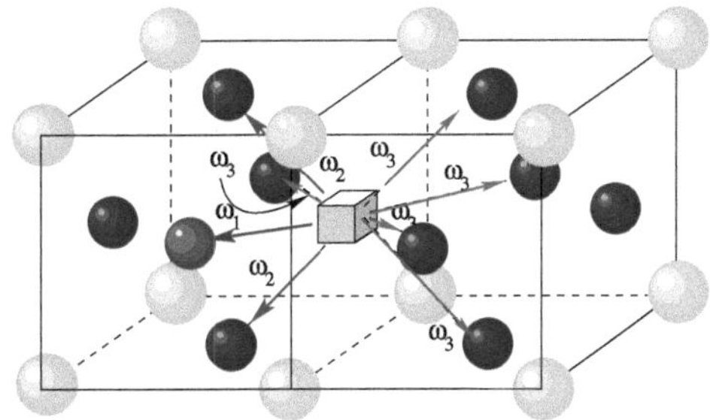

Figure 10.14 The jump frequencies for an A vacancy (box) neighboring a B (or solute *X*) atom B_A (red sphere) which occupies a position on the A sublattice. The ω_4 frequency corresponds to a jump which is just the reverse of a ω_3 jump. ω_0 is the frequency of all other jumps of the B vacancy in the B sublattice. ω_2 are rotational jumps (the vacancy remains in a nearest-neighbor position of the red atom). (For interpretation of the references to color in this figure legend, the reader is referred to the web version of this chapter.)

For the $L1_2$ structure of Ni_3Al, the correlation factors for the sublattice diffusion mechanism in were calculated in [56], and it was found that $f_{Ni} = 0.689$ (is a given constant) and f_{Al} can be expressed in a usual way via the vacancy-Al atom exchange frequency ω_2 and the vacancy escape frequency H [56]:

$$f_{Al} = \frac{H}{\omega_2 + H}. \tag{10.12}$$

Modifying the standard five-frequency model [56], the escape frequency H becomes

$$H = \frac{10\omega_3 F^z(2\omega_1 + 5\omega_3 F^x)}{2\omega_1 + 5\omega_3(F^x + F^z)}. \tag{10.13}$$

The modified five-frequency model for the sublattice mechanism is schematically shown in Fig. 10.14. The situation corresponds to the case when a B atom jumped to the A sublattice (it is an antisite atom, red sphere) and it becomes a neighbor of an A vacancy (box). For any A vacancy there are 8 nearest-neighbor jumps within the A sublattice: one ω_1 jump (exchange with the antisite), two ω_2 jumps (rotational jumps when a vacancy rotates around the antisite remaining at a nearest-neighbor position), and 5 ω_3 jumps (the A vacancy occupies a position from second, third or fourth coordination shell of the antisite). The frequency ω_0 corresponds to "normal" jumps of the A vacancy over the A sublattice (exchanges with A atoms far away from the antisite) and ω_4 is the reverse of ω_3 jump. In Eq. (10.13), F^x and F^z are the functions of ω_4/ω_0 and they are changed over the range from 0.2 to 1, if ω_4/ω_0 increases from 0 to 1 [56].

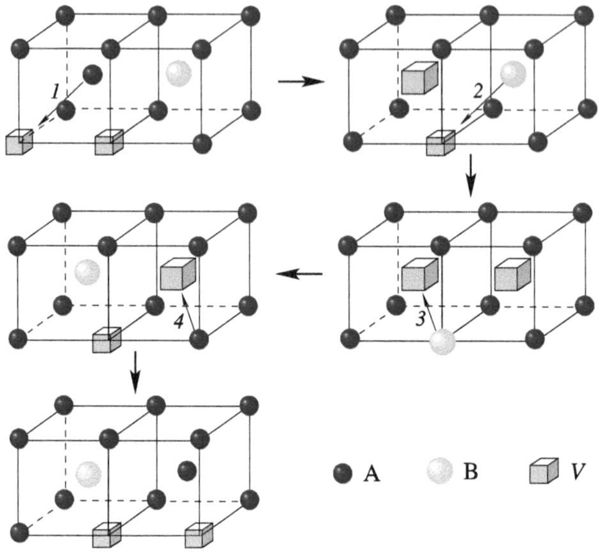

Figure 10.15 A scheme of the triple defect diffusion mechanism in a *B*2 structure *AB*.

10.3.4 Triple Defect Diffusion Mechanism

Originally, this mechanism was invented by Bakker and Stolwijk for the *B*2 compound CoGa [24]. It describes migration of a triple defect (Fig. 10.5(B)), i.e., a bounded entity composed of two transition metal (A) vacancies and one transition metal atom in an antistructure position (A_B) as nearest-neighbor defects. The triple defect mechanism in CoGa was proposed to correspond to two nearest-neighbor jumps of a Co atom and one next-nearest-neighbor jump of a Ga atom [24]. Generally, a B atom may perform two nearest-neighbor jumps instead of one next-nearest-neighbor jump as it was proposed to be the case of NiAl [57]. Fig. 10.15 shows a general triple defect mechanism, where an inverse triple defect ($2V_B + B_A$) exists as an intermediate stage. As a result of the indicated sequence of four jumps of atoms, the triple defect moves leaving the order in the compound unchanged.

Due to a correlation between the successive atomic jumps, the diffusion coefficients of both components in a perfectly ordered state are coupled by equation like Eq. (10.10) with $q = 13.3$ [24].

As an example, the results of energy barrier calculations for the triple defect mechanism in NiAl are presented in Fig. 10.16 for the sequence of jumps shown in Fig. 10.15. This sequence turned out to reveal a lowest energy barriers for the steps *3* and *4*.

The correlation factors for the triple defect diffusion mechanism were calculated for NiAl and $f_{Ni} \cong 0.05$ at $T = 1300$ K [57]. Remarkably, f_{Ni} was found to depend significantly on temperature, and the contribution of this temperature dependence to

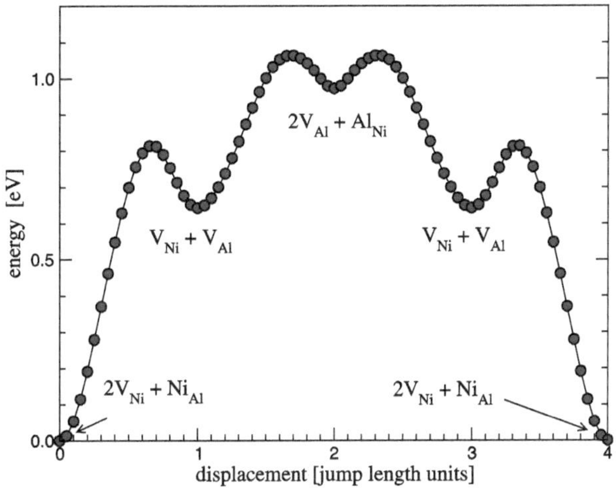

Figure 10.16 Energy vs. displacement curves for the triple defect diffusion mechanism in NiAl presented in Fig. 10.15.

the overall effective activation enthalpy of Ni diffusion by the triple defect mechanism amounts to 17 kJ/mol [57]. To the best of author's knowledge, this is one of the largest contributions to the effective activation energy of diffusion stemming from correlation effects in an ordered intermetallic compound.

It will be shown below that the triple defect mechanism is closely related to a divacancy diffusion mechanism, Fig. 10.23, which in its sequence corresponds to the configurations appearing after the jumps *1* or *3* in Fig. 10.15.

10.3.5 Antistructure Bridge Mechanism

The antistructure bridge (ASB) mechanism was originally proposed by Kao and Chang for the *B*2 structure [25] and was later applied to other structures like the $L1_2$ structure [58]. This mechanism for the *B*2 structure is sketched in Fig. 10.17(A). The vacancy and the antistructure atom effectively exchange their positions as a result of the two indicated jumps. Since the vacancy can in fact reach the fourth or fifth coordination shell from its initial position (depending on the lattice structure [58]), the resulting large geometrical factor of the ASB mechanism increases its contribution to the diffusivity, in parallel to typically low value of the effective activation energy.

A Ni antistructure atom and an Al vacancy participate, e.g., in the ASB mechanism in NiAl which results in an exchange of the positions between the initial vacancy and the anti-structure atom, Fig. 10.17(A). The migration energy was computed for NiAl and it was found that it is very low, about 0.19 eV, see Fig. 10.18. This small value of the migration energy suggests that this mechanism can be very plausible.

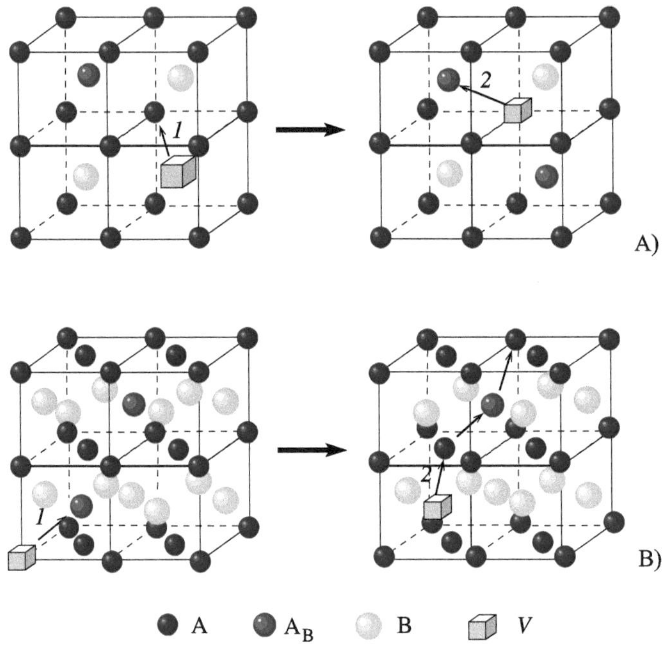

Figure 10.17 Schemes of the anti-structure bridge mechanism in a $B2$ structure (A) and $L1_0$ (B) structures. Typically, a vacancy V_A moves along the "bridges" formed by the anti-structure atoms A_B.

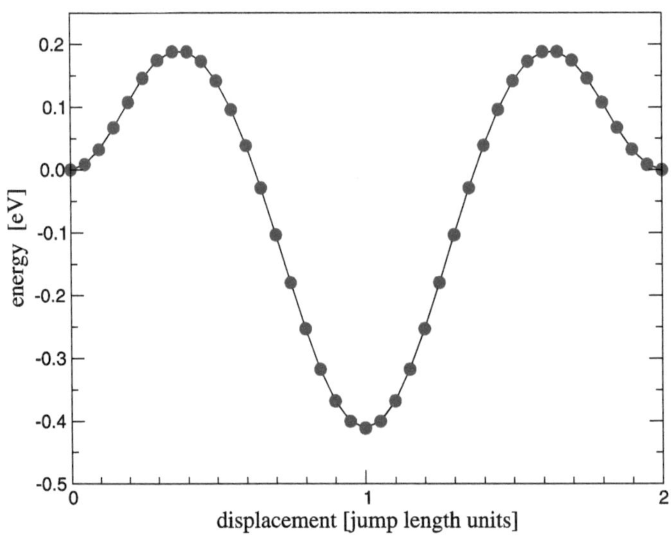

Figure 10.18 Energy as a function of displacement for the antistructure bridge (ASB) mechanism in NiAl.

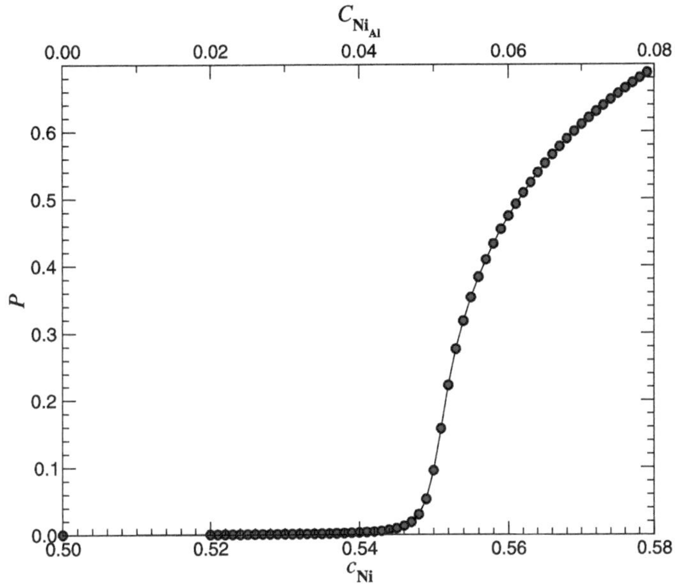

Figure 10.19 Probability P that an Al vacancy belongs to a percolative cluster of Ni antistructure atoms as a function of their concentration.

However, it is important to take into account that the contribution of the ASB mechanism has a percolation nature in the sense that long-range diffusion solely via this mechanism occurs only if the concentration of the anti-structure atoms is larger than a percolation threshold. The Monte Carlo simulation of vacancy jumps according to the ASB mechanism established that the percolation threshold is about $\sim 4.9(2)\%$ in the $B2$ structure [58], see Fig. 10.19. Later on this value was refined to 4.88225(5)% [59,60] that represents probably a more academic interest for description of experimental findings. Such an antistructure atom concentration can indeed exist in intermetallics, and the ASB mechanism becomes important for explaining the observed diffusion behavior in the NiAl and Ni_3Al phases [57,61], see also below.

Long-range diffusion of Al vacancies through a network of "bridges" from the Ni anti-structure atoms will give no contribution to the Al self-diffusion, but an appreciable contribution to the Ni self-diffusion will be obtained. It should be noted that within the framework of the ASB mechanism it is not important which type of defects (Al vacancies or Ni anti-structure atoms) prevails. The defect which dominates has to form a connected network of bridges for the long-range movement of the other type of defect. Referring to the defect structure of the NiAl compound on the Ni-rich side of composition, Fig. 10.6(A) and (C), the concentration of Ni antistructure atoms will determine the percolation threshold for the diffusion by the ASB mechanism. The value of the activation energy suggests that a remarkable contribution of this mechanism can be achieved after passing the percolation threshold.

Let P be the probability that an Al vacancy belongs to a percolative cluster of the Ni anti-structure atoms. Such a vacancy can perform a long-range diffusion by the ASB mechanism and appreciably contribute to the diffusivity of Ni atoms. Fig. 10.19 presents the dependence of P on the concentration of Ni antistructure atoms. The bottom X-axis in Fig. 10.19 presents the composition of NiAl which corresponds to the given concentration of Ni antistructure atoms (upper X-axis) at $T = 1300$ K. It is seen that the percolation threshold does occur at about $c_{Ni} \approx 0.549(2)$. Therefore, the experimentally observed notable increase of the diffusion coefficient of Ni in the $Ni_{56.8}Al_{43.2}$ alloy along with a pronounced decrease of the activation energy is very likely to result from the direct contribution of the ASB mechanism, see Section 10.4.2.2.

Below the percolation threshold, Ni atom diffusion by the ASB mechanism is confined to finite clusters isolated one from another. Nevertheless, since the diffusion in such a cluster may proceed very fast with a low activation energy, its appearance will result in some enhancement of the overall diffusion rate. The effective activation energy of the resulting diffusion kinetics will be determined by the slowest diffusion mechanism, but the pre-exponential factor can profit from the large geometric factor corresponding to the ASB mechanism, see [58,62]. With approaching the percolation threshold, the contribution of the ASB mechanism increases, see Fig. 10.19, and the diffusion will be notably enhanced.

The ASB mechanism is also of importance in other ordered structures, e.g., in the $L1_0$ structure of the phase TiAl [20,21]. One of ASB variants for the $L1_0$ structure is presented in Fig. 10.17(B). Having performed the indicated two jumps (*1+2*), the A vacancy moves perpendicular to the A atom layers using an antistructure A atom as a "bridge." If a further antistructure atom is available in a suitable nearest-neighbor position, the next ASB sequence may start, see Fig. 10.17(B). The Monte Carlo calculations of the percolation threshold for the long-range diffusion by this variant of the ASB mechanism yield the critical concentration of the antistructure atoms at about 11%.

It is important to recall that during the ASB sequence of jumps only one type of atoms moves, see Fig. 10.17. Therefore, the diffusivities of the two components are not coupled when this mechanism operates.

In a strict sense, the genuine ASB mechanism operates only after the percolation threshold is reached. However, as it was indicated above, in combination with another mechanism (usually the sublattice diffusion mechanism) the ASB mechanism (as $1 \rightarrow 2$ in Fig. 10.17(B)) can substantially contribute already to long-range diffusion without any percolation threshold [20,21,63]. This fact allows to use the term "ASB mechanism" also in such cases, referring to the specific sequence presented in Figs. 10.17(A)–(B).

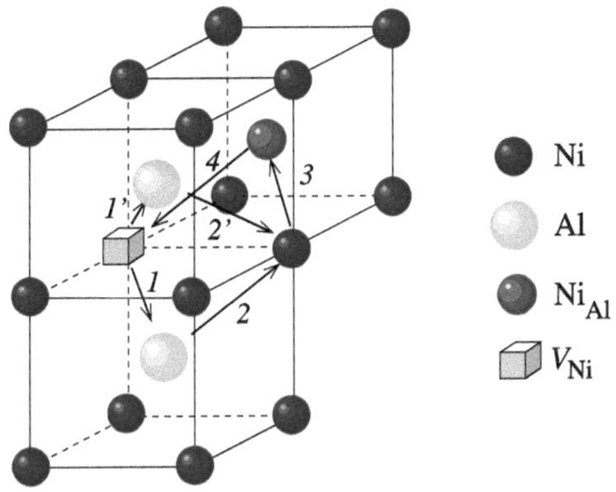

Figure 10.20 A scheme of 4-jump cycle. The flat (*1–2–3–4*) and bent (*1′–2′–3–4*) cycles are shown.

10.3.6 4-Jump Cycle Mechanism (6-Jump Cycles Assisted by an Antistructure Atom)

In the presence of antisites the 6-jump cycle mechanism can be modified and it will be reduced to 4-jump cycles. These feature will be demonstrated for B2 NiAl below.

The nearest neighboring pair of a Ni vacancy and a Ni antistructure atom is to be considered as a starting configuration, see Fig. 10.20. Then, the Ni vacancy may jump into the adjacent Al site, producing an Al antistructure atom and an Al vacancy, jump *1* in Fig. 10.20. After accomplishing the jumps *1–4*, the vacancy returns to its initial location and the Ni antistructure atom and Al atoms exchange effectively their positions. The term "effectively" should be used, since, in fact, the initial and final Ni antistructure atoms are indeed different atoms, as a close inspection of Fig. 10.20 reveals. The given degree of the order is not changed during the cycle. As a result, there is no movement of the starting Ni vacancy, and two Ni atoms participate in this correlated sequence, producing an effective [100] or [110] "jump" of the initial antistructure Ni atom within the Al sublattice. In fact, this modified cycle corresponds to steps *2*, *3*, *4*, and *5* in the common 6-jump cycle when the vacancy is initially located on the Al sublattice. Indeed, after the exchange of the Al vacancy with a neighboring Ni atom, a Ni antistructure atom and a Ni vacancy will be produced; this is just the starting configuration for the 4-jump cycle mechanism, see Fig. 10.20.

Since after the cycle the vacancy occupies the same initial position, this mechanism cannot provide a long-range diffusion of the Ni antistructural atoms, unless the concentration of the Ni vacancies is unreasonably high, allowing for the Ni antistructure atoms to meet another, "fresh" Ni vacancy at its nearest-neighbor position. However, Fig. 10.21(A) suggests that two successive nearest-neighbor jumps of the Ni vacancy (into the initial position of the Ni antistructure atom and then into either next nearest-neighbor Ni site) will result into a total jump of this Ni vacancy towards its second,

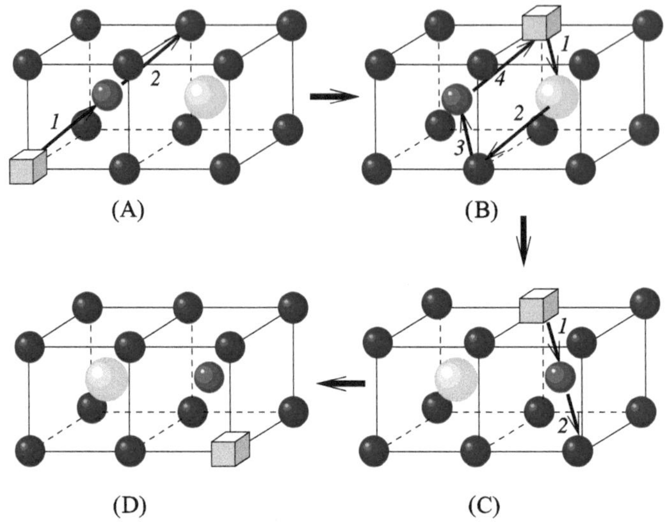

Figure 10.21 A scheme of jumps of a pair of Ni vacancy and Ni anti-structure atom: 2-jump sequences (*1–2*) resulting in the movement of the Ni vacancy (A), (C) and 4-jump cycles (*1–2–3–4*) giving rise to the displacement of the Ni anti-structure atom (B). Stage (B) reveals highest energy barriers and presents a critic point of the mechanism.

third, or fifth coordination shell, jumps *1* and *2* in Fig. 10.21(A) and (C). The final position of the Ni antistructure atom will be the same as initial, but two nearest-neighbor jumps of Ni atoms will be performed, giving a contribution to the Ni self-diffusion. This consideration predicts that a nearest-neighbor pair of a Ni vacancy and a Ni antistructure atom can diffuse over large distances in B2 lattice by two different sets of nearest-neighbor jumps, see Fig. 10.21.

For the estimation of the effective activation energy for diffusion by this mechanism in NiAl, the jumps can be artificially divided into two steps: one resulting in a displacement of the Ni vacancy (two-jump sequences in Fig. 10.21(A) and (C)), and the other giving rise to the migration of the Ni antistructure atom (4-jump sequence in Fig. 10.21(B)). The dependence of the energy on the displacement of the Ni vacancy from its initial to final position is presented in Fig. 10.22 for both sets of jumps. It is seen that the 4-jump cycle involves higher energy barrier and presents a critical point of the mechanism under consideration.

The analysis shows that the 4-jump multi-barrier process corresponding to the net movement of the Ni anti-structure atom (Fig. 10.22(B)) can be adequately thought as a one-barrier process which is characterized by a single effective migration energy.

A close inspection indicates that three different types of the 4-jump cycles are to be distinguished, the plane (*1–2–3–4*) and bent (*1'–2'–3–4*) [100]-cycles, see Fig. 10.20, and [110]-ones. The energy vs. displacement curve for the [100] cycles is presented

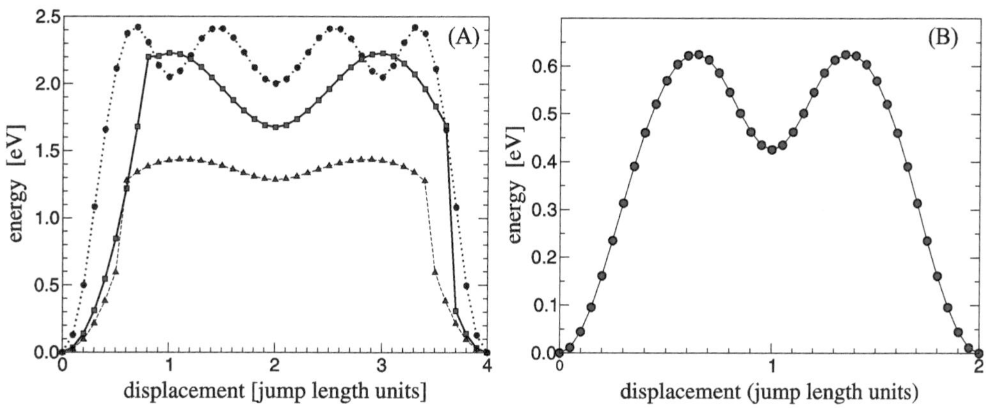

Figure 10.22 Energy vs. displacement curve for the plane (black circles) and bent [100] (red squares) and [110] (blue triangles) 4-jump cycles (A) and the 2-jump sequences from Fig. 10.22(A). (For interpretation of the references to color in this figure legend, the reader is referred to the web version of this chapter.)

in Fig. 10.22(A). The situation with the bent [100] and [110] cycles is more delicate. Simulation shows that the configuration corresponding to the jump *1'* in Fig. 10.20 is not stable and the Al atom returns to its initial position on the Al sublattice. The same holds for the [110] cycles. However, if we suppose that the jumps *1'* and *2'* have been accomplished, the vacancy will perform the jumps *3* and *4* in one step: the Al atom will jump to its sublattice immediately when the vacancy will appear at its nearest neighbor position without any additional energy barrier. Thus, the 4-jump sequence *1'–2'–3–4* (and similarly for the [110] cycles) is indeed a sequence of two composed jumps, *1'+2'* and *3+4*. To perform the *1'+2'* step, Al and Ni atoms have to move in a correlated manner. Corresponding energy curves are presented in Fig. 10.22(A), where only two steps can be distinguished.

In order to establish the operating type of the 4-jump cycles, the correlation factors are to be analyzed. A Monte Carlo analysis of the diffusivity by the 4-jump cycle mechanism has been performed, and the temperature dependence of the correlation factor f_{4j} is presented in Fig. 10.25. Only [110] 4-jump cycles were found to operate in B2 NiAl. Note that f_{4j} is found to depend remarkably on the temperature, and gives $\delta E_{corr} \approx 0.77$ eV contribution to the effective diffusion activation energy.

10.3.7 Divacancy Diffusion Mechanism

Again, the analysis will be done for the *B2* ordered structure of NiAl. Consider a Ni divacancy, namely [100]-, [110]-, or [111]-pair of Ni vacancies in terms of the Ni sublattice. These positions correspond to second, third, and fifth coordination shells in terms of the *B2* structure. The sequence of four jumps *1–4* indicated in Fig. 10.23 will

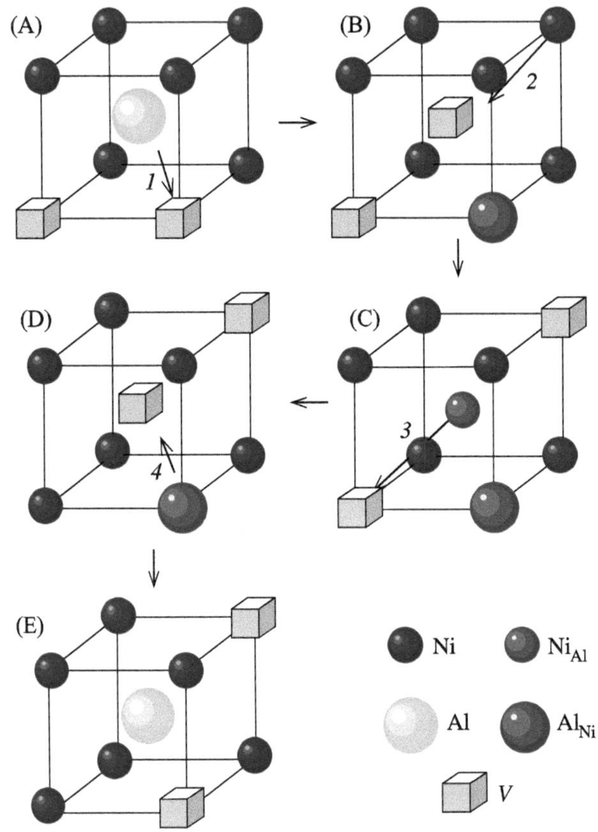

Figure 10.23 A scheme of atomic jumps corresponding to [100] → [011] jump of a Ni divacancy.

provide a long-range diffusion of the divacancy, without introducing additional disorder in the lattice of the off-stoichiometric NiAl compound. As a result, a Ni atom performs a jump to the second, third, or fifth coordination shell, and there is no long-range diffusion for the Al atoms.

In B2 NiAl, EAM calculations [26] show that there is a small binding energy of the divacancy in the [100] position, about −0.1 eV. On the other hand, the [110]- and [111]-divacancies show a slight repulsive interaction, excess energies are about +0.08 and +0.03 eV, respectively. The energy of the system versus displacement is presented in Fig. 10.24 for the [110]→[101] reorientation of the initial divacancy.

The analysis shows that the sophisticated 4-jump sequence of the Ni divacancy motion can be thought as an effective one-barrier jump with the migration energy being in the range of 2.6÷3.1 eV in NiAl, depending on the particular configuration of the divacancy. Note that the migration of the most stable configuration, [100], requires the

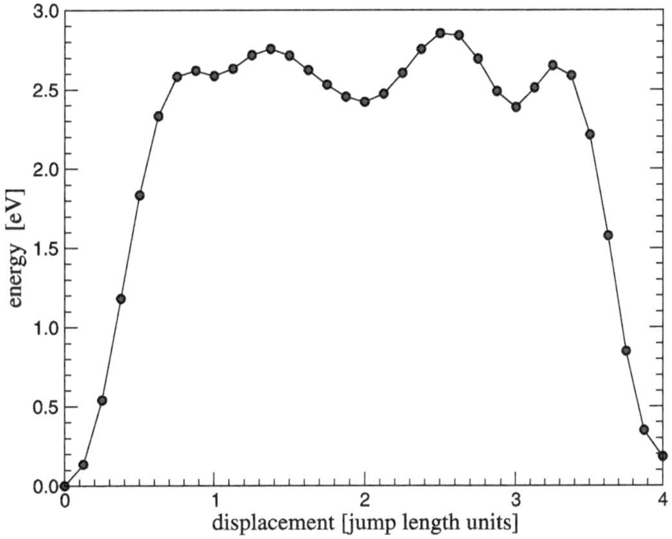

Figure 10.24 Energy vs. displacement curves for the [100]→[101] jumps of a Ni divacancy.

highest barrier, 3.1 eV. Alternatively, the configuration [110], which was found to reveal a positive binding energy, has the lowest migration energy of about 2.6 eV.

The correlation factor according to the divacancy diffusion mechanism is presented in Fig. 10.25 as a function of the inverse temperature. It is seen that it practically does not depend on the temperature and has a value of about 0.15.

Note that this mechanism has common features with the triple defect mechanism of diffusion [24], however, the participation of the Ni antistructure atoms is not necessary. Moreover, the triple defect configuration refers to a $V_{Ni}-V_{Al}$ divacancy, while the $V_{Ni}-V_{Ni}$ divacancies are considered here. The configurations corresponding to the Ni divacancies were really observed in experiments on the Al-rich side of the composition of NiAl alloys by perturbed angular correlation of gamma rays [64].

10.3.8 Calculation of Correlation Factors

Having established the diffusion mechanism and calculated the transition probabilities for all possible vacancy jumps in the given structure, the correlation factor (Chapter 2) can be determined, e.g., by a numerical Monte Carlo approach.

Let us illustrate this approach for the case of B2 NiAl. First, a simulation block is created. A Ni vacancy is centered in the block and the vacancy is allowed to perform nearest-neighbor jumps. The probabilities p_i of these jumps are calculated according to the transition frequencies ν_i estimated by accounting for the migration energies and

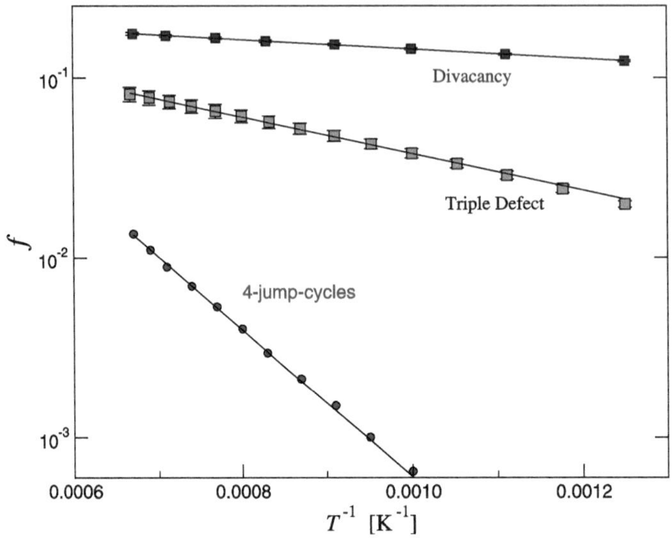

Figure 10.25 Temperature dependence of correlation factors for the divacancy f_{2d} and triple defect f_{td} mechanisms. The contributions of the correlation factors to the effective activation energy of diffusion by the given mechanisms are also given.

entropies, and so $p_i = \nu_i / \sum_{j=1}^{z} \nu_j$, where $z = 8$ is the coordination number of the $B2$ lattice. The time-residence algorithm is applied. Each jump of the vacancy is successful and the time variable is increased by $\tau_i = 1/\sum_{j=1}^{z} \nu_j$.

The atomic correlation factors for species k (= Ni or Al) are calculated from the equation

$$f_k = \frac{4\sum_{i=1}^{n_k} R_i^2}{3m_k a^2} \tag{10.14}$$

where R_i is the net displacement of the ith atom of type k, n_k the total number of atoms of type k, m_k the total number of vacancy exchanges with atoms of type k, $\frac{3}{4}a^2$ the elementary jump length, and a the lattice parameter.

During one MC run, the average number of vacancy jumps has to be larger than at least 1 to 10 per atom, and the calculated correlation factors are averaged usually over 10^4 Monte Carlo runs.

As an example, the calculated temperature dependencies of the tracer correlation factor for divacancy (Fig. 10.23), 4-jump cycles (Fig. 10.21), and triple defect diffusion mechanisms (Fig. 10.15) are presented in Fig. 10.25.

10.3.9 Effective Activation Energy of a Complex Vacancy-Mediated Mechanism

In terms of the transition-state theory [65,66], the tracer diffusion coefficient D^* for a vacancy-mediated mechanism, which includes a single jump of the tracer atom, is simply given by

$$D^* = \frac{1}{2} f \nu_0 \exp\left\{-\frac{S^f + S^m}{k}\right\} \exp\left\{-\frac{E^f_{V_i}}{kT}\right\} \exp\left\{-\frac{E^m}{kT}\right\} \sum_j x_j^2 \quad (10.15)$$

where f is the tracer correlation factor (a constant for the given lattice), ν_0 the attempt frequency, x_j the jump length projection on the X axis for the jth jump, and the summation proceeds over all possible jumps. $E^f_{V_i}$ (S^f) and E^m (S^m) are the energies (entropies) of the V_i vacancy formation and migration, respectively.

This expression holds for pure metals where a single value of E^m is present. In the case of intermetallic compounds, the situation is more sophisticated. As we have shown above, there are specific sequences of nearest-neighbor jumps of a vacancy, resulting in full restoration of the given state of order. Thus, we have a multibarrier process and, generally, Eq. (10.15) will be violated. However, as it was shown above, in many cases the energy barrier for the first jump exceeds substantially the energy barriers of the subsequent jumps of the vacancy along the given path. Since the energy barrier enters exponentially into the expression for the diffusion coefficient, this results in a difference by many orders of magnitude between the frequencies of corresponding jumps. The analytical treatment and Monte Carlo runs suggest that in such cases the multibarrier process may be presented as an *effective* one-barrier process with some effective value of $E^m_{(eff)}$ which is of the order of magnitude of the difference between the energy of the initial configuration and *the highest energy* of the saddle-point configurations occurring in the given sequence of diffusion jumps.

Within this approximation and assuming that a second defect γ ($\gamma = V_{Ni}$, V_{Al}, Ni_{Al}, or Al_{Ni}) also participates in the multibarrier process (in addition to the vacancy V_i) Eq. (10.15) will be modified as

$$D^* = \frac{1}{2} f(T) \nu_0 \exp\left\{-\frac{E^f_{V_i} + E^f_{\gamma} + E^b}{kT}\right\} \exp\left\{-\frac{E^m_{(eff)}}{kT}\right\} \sum_j x_j^2. \quad (10.16)$$

Here E^f_γ is the formation enthalpy of the defect γ, and E^b is the binding energy of the composed defect $V_i - \gamma$, and the correlation factor $f(T)$ is temperature dependent now. Eq. (10.16) suggests that the effective activation energy Q^{eff} associated with the given mechanism can be presented as

$$Q^{eff} = E^f_{V_i} + E^f_\gamma + E^b + E^m + \delta E_{corr} \quad (10.17)$$

where δE_{corr} presents the contribution to the activation energy resulting from the temperature dependence of the correlation factor $f(T)$. Usually, this value is quite small, but in some cases, see Fig. 10.25, it can substantially affect the Q^{eff} value.

10.4 MEASUREMENTS OF AL DIFFUSION IN ALUMINIDES

As a matter of fact, the direct tracer measurements of Al self-diffusion in aluminides are practically absent. The problem is due to very high price and low specific activity of the only available ^{26}Al radioisotope, on the one hand, and due to the difficulty of avoiding oxidation of this isotope during a diffusion study, on the other. The only available tracer Al-diffusion data in NiAl and TiAl were measured by Larikov and coworkers [15].

Two approaches were typically used to bypass this difficulty:

- a combination of interdiffusion and tracer diffusion measurements;
- application of Al-substituting solutes which mimic Al tracer diffusion, e.g., Ga, In, or Ge.

According to the Darken–Manning formalism [67,68], the interdiffusion coefficient, $\widetilde{D}$, in AB compound (A = transition metal, B = Al) can be written as (neglecting by the volume effects):

$$\widetilde{D} = (N_B D_A^* + N_A D_B^*) \cdot \Phi \cdot S. \tag{10.18}$$

Here N_A and N_B are the mole fractions of elements, and D_A^* and D_B^* are the corresponding diffusion coefficients, Φ is the thermodynamical factor, and S is the vacancy wind factor. Φ can be determined by independent activity measurements or estimated from thermodynamic calculations, and S is usually assumed to be about unity (only several investigations were published where S in intermetallic compounds was estimated; see, e.g., [69]). For the random alloy model, it was shown [68] that

$$1 \leq S \leq \frac{1}{f_0}, \tag{10.19}$$

where f_0 is the geometrical correlation factor for the given lattice.

Since Eq. (10.18) was experimentally verified for several compounds [70,71], it can be applied to interdiffusion in aluminides, too. Then, the (mostly unknown) tracer diffusion coefficient of Al is estimated as

$$D_B^* = \frac{\frac{\widetilde{D}}{\Phi \cdot S} - N_B D_A^*}{N_A}. \tag{10.20}$$

Eq. (10.20) suggests that if D_B^* is remarkably smaller than D_A^*, the reliability of such determination of D_B^* becomes questionable due to unavoidable experimental uncertainties related in the determination of $\widetilde{D}$ and Φ, since the terms $\frac{\widetilde{D}}{\Phi \cdot S}$ and $N_B D_A^*$ are almost identical within the experimental uncertainties.

A comment is due here. Murch and Belova in a series of careful studies have shown that the Darken–Manning approach is applicable to intermetallic compounds with the $B2$, $L1_2$, and $D0_3$ lattices at lower levels of order [72,73]. They found that the correlated diffusion mechanisms in ordered intermetallic compounds result in a deviation of the S value from the limits predicted by Eq. (10.19). In the $B2$ compounds, e.g., S can be 0.42 for the six-jump cycle mechanism [74]. Likewise the upper value of S is not bound above by f_0^{-1} in the $L1_2$ structures [75].

10.5 DIFFUSION IN ORDERED BINARY ALUMINIDES

In this section the experimental data on bulk diffusion in binary Ti, Ni, and Fe ordered binary aluminides are reviewed. Since self- and solute diffusion in binary aluminides was analyzed in detail by Christian Herzig and the present author in a book chapter long time ago [18], these data will be shown only shortly here, and the main attention is paid to findings concerning the alloying effects and tracer diffusion in ternary aluminides. Note that the available grain boundary diffusion data in binary aluminides (not discussed here) were also analyzed in [18] and the reader is referred to that work.

10.5.1 Diffusion in Ti Aluminides

10.5.1.1 Diffusion in Ti_3Al

Self-diffusion. For a most comprehensive review of self- and solute tracer diffusion in Ti_3Al, the reader is referred to [21]. Ti self-diffusion in the phase α_2-Ti_3Al was in detail investigated in [76]. Within the experimental errors, Ti tracer diffusion was almost the same in compositions with 75, 72, 68, and 65 at.% Ti, although a marginal tendency towards an increase of D^*_{Ti} with increasing Al content at the lowest temperature of the investigation may be indicated [76]. As a result, one "master" Arrhenius dependence for Ti self-diffusion in α_2-Ti_3Al was suggested (Fig. 10.26) with the Arrhenius parameters given in Table 10.2.

Tracer diffusion of Ti in Ti_3Al occurs most probably via the Ti-sublattice diffusion mechanism [76]. The experimental data suggest that the compositional Ti or Al antistructure atoms (for the concentrations see Fig. 10.6(D)) only slightly change the energy barriers for vacancy jumps on the Ti sublattice. This fact correlates with almost constant Ti vacancy formation enthalpies reported for the stoichiometric- and Al-rich alloys [21,77].

Tracer diffusion of Al was not measured in Ti_3Al. However, interdiffusion in the α_2-Ti_3Al-phase field was studied [76,78]. Applying Eq. (10.20), assuming $S = 1$, and using the relevant thermodynamic data [79], the Al tracer diffusion coefficient D^*_{Al} can be determined as a function of temperature. The results are presented in Fig. 10.26. An analysis of the experimental data suggests that the large value of Q_{Al}, about 400 kJ/mol

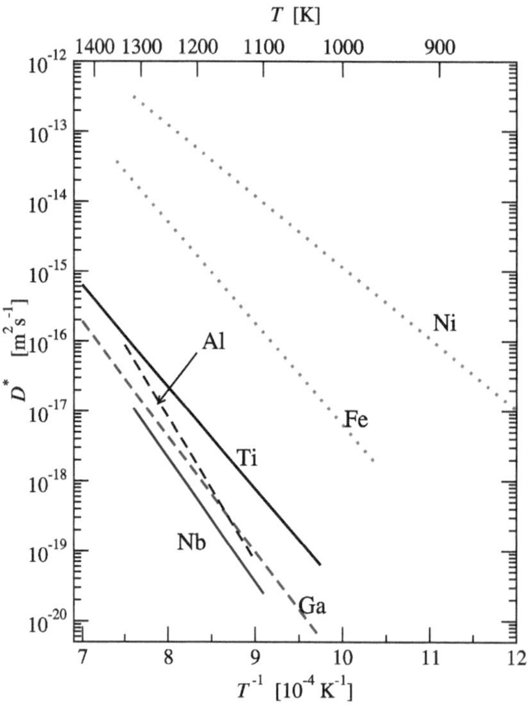

Figure 10.26 Temperature dependence of self- (black lines) and solute (color lines) diffusion in Ti_3Al. The diffusion data for Ti [76] (solid black line), Al [78] (dashed black line), Ga [80] (dashed red line), and Nb (solid blue line), and Fe and Ni (dotted magenta lines) [82] diffusion are compared. (For interpretation of the references to color in this figure legend, the reader is referred to the web version of this chapter.)

(Table 10.2), is probably related to uncertainties in the D^*_{Al} determination via the Darken relation Eq. (10.18), especially at low temperatures [80].

Solute Diffusion. In Fig. 10.26 the results on solute diffusion in Ti_3Al are differentiated depending on the relevant type of site occupation, with red color used for Al substitutes and blue one applied to solutes which dissolve on the Ti sublattice. Fe and Ni solute represents a special case, see below.

Ga is commonly accepted to occupy the Al sublattice in Ti_3Al [81], and the measured Ga diffusion coefficients are presented in Fig. 10.26 as a function of temperature. Although the absolute values of D^*_{Ga} are similar to those of D^*_{Al}, the activation enthalpy of Al diffusion is distinctly larger, see Table 10.2. Ga (as well as Al) diffuses in Ti_3Al according to the Ti-sublattice diffusion mechanism of the minority component [80]. In this case, the excess of Q_{Ga} with respect to Q_{Ti} is related to the formation enthalpy of Ga atoms in antistructure positions on the Ti-sublattice [80]. This fact is fully supported

Table 10.2 Arrhenius parameters of self- and solute diffusion in Ti aluminides

Tracer	D_0 (m^2s^{-1})	Q (kJ/mol)	Reference
	$D0_{19}$ phase Ti_3Al		
Ti	2.24×10^{-5}	288	[76]
Al*	2.3×10^{-1}	394	[76]
Ga	6.3×10^{-5}	315	[80]
Fe	2.0×10^{-3}	277	[82]
Ni	1.8×10^{-5}	195	[82]
Nb	3.2×10^{-4}	339	[82]
	$L1_0$ phase TiAl		
Ti	3.2×10^{-6}	261	[20]
Al*	2.1×10^{-2}	358	[20]
Ga	4.4×10^{-5}	293	[80]
Co	1.1×10^{-3}	318	[87]
Cr	4.4×10^{-3}	350	[88]
Fe**	2.7×10^{-6}	252	[63]
Ni**	1.8×10^{-2}	340	[63]
Nb	1.5×10^{-4}	324	[63]
Zr	6.4×10^{-3}	348	[63]
Cr	5.2×10^{-3}	332	[63]

* Determined applying the Darken–Manning equation.
** Activation enthalpies of Ti and Fe diffusion in TiAl were calculated by fitting experimental points below $T = 1470$ K.

by the calculation of the concentrations of point defects in the compound, Fig. 10.6(A) and (C).

Unfortunately, there is only quite limited information on diffusion of Ti-substitutes in α_2-Ti_3Al. In Fig. 10.26 diffusion of Nb in Ti_3Al [82] is presented (blue line); it was found that the diffusion rate does not practically depend on the alloy composition within the composition range from 27 to 33 at.% Al. Since Nb atoms are considered to occupy Ti sites in Ti_3Al [81], their diffusion is dominated by exchanges with vacancies on the Ti sublattice (Ti sublattice diffusion mechanism).

As it was mentioned, diffusion of Fe and Ni in α_2-Ti_3Al presents a special case. These transition metals are ultra-fast diffusers in α-Ti where their diffusion coefficients are higher than those of Ti tracer by 7 to 8 orders of magnitude [83–85]. Fig. 10.26 suggests that Fe and Ni are fast diffusers in Ti_3Al, too, though diffusion enhancement is within two to three orders of magnitude with respect to the rate of Ti self-diffusion. The diffusion mechanism is briefly discussed in the next section together with the data on TiAl.

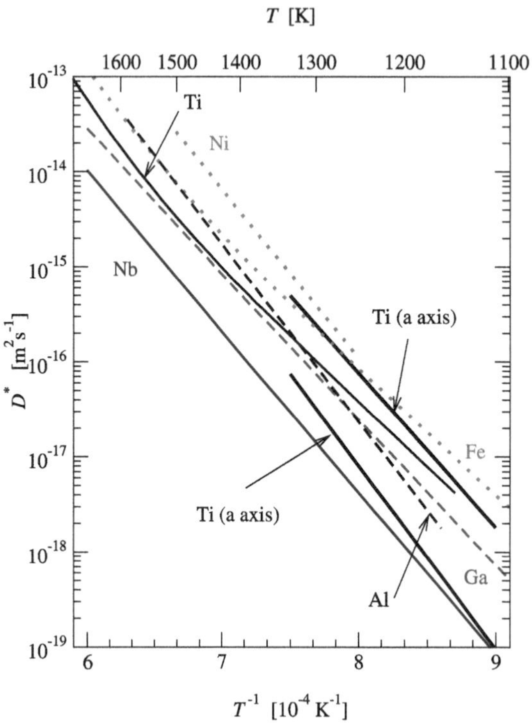

Figure 10.27 Self- and solute tracer diffusion in TiAl. Ti self-diffusion was measured in polycrystalline TiAl alloys [20] (solid black line) and compared with Ti diffusion measured along the c and a axes in the $L1_0$ structure of TiAl [86]. The estimation of Al diffusion coefficient from interdiffusion data [20] is shown (dashed black line). Solute diffusion is shown according to the type of preferential solubility on the Ti (Nb [63], blue line) or Al (Ga [80], dashed red line) solubility. The diffusion data for Fe and Ni [63] are plotted by dotted lines. (For interpretation of the references to color in this figure legend, the reader is referred to the web version of this chapter.)

10.5.1.2 Diffusion in TiAl

Self-diffusion. Ti self-diffusion in γ-TiAl revealed clearly a non-Arrhenius character [20], see Fig. 10.27, it follows an Arrhenius temperature dependence up to about $T = 1470$ K with a subsequent upward deviation to larger D^*_{Ti} values at higher temperatures. This effect is almost independent on the composition in TiAl [20]. Such a behavior of the temperature dependence of tracer diffusion in an ordered compound is atypical, for the majority of intermetallic phases linear Arrhenius-type temperature dependencies were usually reported. However, having recognized the existence of a variety of point defects and a number of possible diffusion mechanisms in ordered intermetallics, see preceding sections, linear Arrhenius behavior should probably be unexpected and nonlinear dependencies have to be typically observed. However, experimental data suggest that in many intermetallic phases there is a given diffusion mechanism which dominates self-diffusion over a wide temperature interval, and a single effective activation enthalpy can be determined.

The pronounced curvature of Ti self-diffusion is explained by a simultaneous operation of two diffusion mechanisms: the Ti-sublattice mechanism and the ASB mechanism [20]. Fig. 10.6(C) suggests that the concentration of Ti antistructure atoms is large even in Al-rich compositions that facilitates the antistructure bridge mechanism (Fig. 10.17(B)). The layered $L1_0$ structure of TiAl, see Fig. 10.2, results in a slight tetragonal distortion of the cubic lattice, and an anisotropy of self-diffusion may be expected. Indeed, the anisotropy of Ti self-diffusion in single crystalline TiAl alloys was observed [86], see Fig. 10.27. Ti diffusion along Ti layers in the $L1_0$ structure (along the a axis) is faster than that perpendicular to this direction (i.e., along the c axis).

By a simple inspection of the geometry of the $L1_0$ structure, Fig. 10.2, one can conclude that the Ti-sublattice diffusion mechanism should result in a strong anisotropic contribution with a dominant contribution to diffusion along the *a* axis (along the Ti layers), whereas the ASB mechanism, which occurs by intersublattice jumps of the Ti atoms, corresponds to an almost isotropic mass-transport. As the temperature increases, the contribution of the ASB mechanism increases. This increase in the contribution of the ASB mechanism manifests itself as a nonlinear Arrhenius dependence of Ti self-diffusion [20]. Again, we conclude that the basically different rates and effective activation energies for diffusion parallel to the *a* and *c* axes are caused by different diffusion mechanisms.

Briefly, the current view on Ti diffusion in γ-TiAl can be summarized as follows. According to the atomistic calculations, the vacancy concentration on the Ti sublattice exceeds that on the Al sublattice by orders of magnitude [20,89]. Moreover, a relatively high concentration of Ti anti-structure atoms, i.e., Ti atoms on the Al sublattice in antistructure positions was predicted, especially at higher temperatures [20]. The lowest activation energy was calculated for the jumps of Ti atoms into a nearest-neighbor vacancy within the same Ti layer. The diffusion jumps between the layers (i.e., parallel to the *c* axis) involve higher activation energies. Therefore, the Ti sublattice diffusion mechanism dominates Ti diffusion at lower temperatures. At higher temperatures, the antistructure bridge mechanism benefits from the large concentration of Ti antistructure atoms and enhances significantly the Ti diffusivity [20]. These features result in a nonlinear temperature dependence of Ti diffusion as experimentally observed.

The diffusivity of the polycrystalline material is seen to lie just in-between the diffusivities of the single crystalline alloys for the two extreme orientations, Fig. 10.27. At lower temperatures the Ti diffusivity of the polycrystalline material approaches the diffusion rate of Ti along the Ti layers in the $L1_0$ structure, whereas it is the increased diffusivity perpendicular to these layers which additionally enhances Ti diffusion at higher temperatures and results in a curvature of the temperature dependence in the polycrystalline binary γ-phase. The activation enthalpy of Ti diffusion along Ti layers is smaller than that of Ti diffusion perpendicular to them [86].

As it was mentioned, Al tracer diffusion was not directly measured in TiAl. Interdiffusion in γ-TiAl was measured in [90] using single-phase diffusion couples, and these data were applied in [20] to deduce the Al tracer diffusion coefficients D^*_{Al}. The corresponding Arrhenius dependence is presented in Fig. 10.27 (black dashed line). The resulting activation enthalpy of Al diffusion is fairly large, $Q_{Al} = 358$ kJ/mol, Table 10.2.

Solute Diffusion. We will analyze the dependence of the solute diffusion parameters in TiAl on the character of preferential solubility in the compound. Diffusion of Ga as an Al-substitute was measured [80]. The corresponding diffusion data are presented in Fig. 10.27 (red dashed line) and it is seen that $D^*_{Ga} < D^*_{Ti}$ in the whole temperature interval of the investigations, whereas D^*_{Al} turns out to be smaller than D^*_{Ti} only at lower temperatures. However, with respect to the absolute values, D^*_{Al} and D^*_{Ga} are very similar. It was established that Ga occupies exclusively the Al sublattice in TiAl [81], but Ti vacancies dominate, see Fig. 10.6(A). Thus, we conclude that Ga jumps over Ti sublattice dominate Ga diffusion and the difference between Q_{Ga} and Q_{Ti} in the low-temperature interval is mainly attributed to the formation enthalpy of Ga atoms as antistructure atoms on the Ti sublattice.

On the other hand, Nb is a typical Ti substitute, it occupies preferentially Ti sites and is a slow diffuser in TiAl, Fig. 10.27. The slow diffusivity of Nb atoms most likely is related to a certain repulsive vacancy-Nb atom interaction. Fig. 10.27 suggests that diffusion of the studied solutes, at maximum, deviates by one order of magnitude from Ti self-diffusion. This behavior corresponds to the substitutional character of these solutes and to a vacancy mechanism of diffusion.

10.5.1.3 Fe and Ni Diffusion in Ti Aluminides

Fe and Ni diffusion in Ti aluminides is of particular interest, since these transition metals are ultra-fast diffusers in α-Ti and it was already shown that they remain to be fast diffusers in α_2-Ti_3Al (though, not "ultra-fast"). These two lattices are hexagonal and question arises whether Fe and Ni are fast diffusers in cubic γ-TiAl, too.

It turned out that Fe diffusion in TiAl is very interesting indeed, in view of its strong nonlinear Arrhenius dependence, see Fig. 10.27 (dotted magenta line). Most importantly, the magnitude of the Fe diffusion coefficients is similar to that of Ti self-diffusion and the curvatures of Ti and Fe Arrhenius plots are also quite similar. This specific behavior was explained [63] by accounting for an equal preference of Fe atoms to occupy both Ti and Al sites in TiAl [81]: Fe atoms located on the Ti sublattice mainly dominate Fe diffusion at lower temperatures and the ASB mechanism becomes progressively important at higher temperatures. As a result, the Arrhenius dependence attains an upward curvature [63].

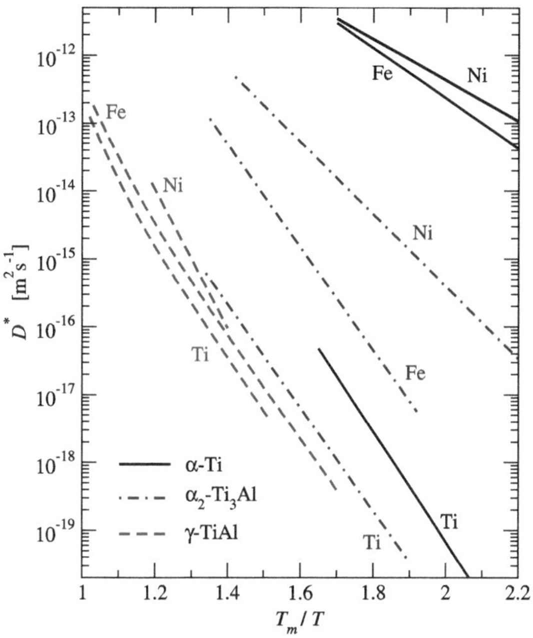

Figure 10.28 Fe and Ni diffusion in α-Ti [83–85], α_2-Ti_3Al [82,76], and γ-TiAl [63,20] when plotted in reduced coordinates T_m/T in comparison to Ti self-diffusion. T_m is the melting temperature of corresponding material.

Ni diffusion in TiAl shows almost linear temperature dependence and its magnitude is about that of Ti self-diffusivity. The findings for γ-TiAl are in contrast to Fe and Ni diffusion behavior in Ti_3Al. We explain this fact by a change of the diffusion mechanism from interstitial to dissociative, and then finally to substitutional diffusion when changing from α-Ti to α_2-Ti_3Al and to γ-TiAl. Simultaneously, there is a remarkable decrease in the tracer diffusion rates: while Fe and Ni mobilities are by seven to eight orders of magnitude greater than Ti in α-Ti [83,84], they exceed Ti self-diffusion in α_2-Ti_3Al only by two to four orders of magnitude, Fig. 10.26(B), and all of them diffuse with similar magnitudes in γ-TiAl, Fig. 10.28.

This behavior strongly correlates with the different type of atomic arrangements which form the interstitial sites in these materials [63], Fig. 10.29. Naturally, only Ti atoms form interstitial sites in α-Ti, Fig. 10.29(A). In the $D0_{19}$ structure of Ti_3Al, there are two different interstitial sites, Fig. 10.29(B): one of them is built exclusively by Ti atoms, whereas the second type is formed by two Al and four Ti atoms. In the $L1_0$ structure of TiAl, all interstitial sites have both Al and Ti atoms as environment, Fig. 10.29(C). Thus, the availability of Al atoms as a nearest neighbor strongly decreases the interstitial solubility of Fe and Ni atoms at such positions [63].

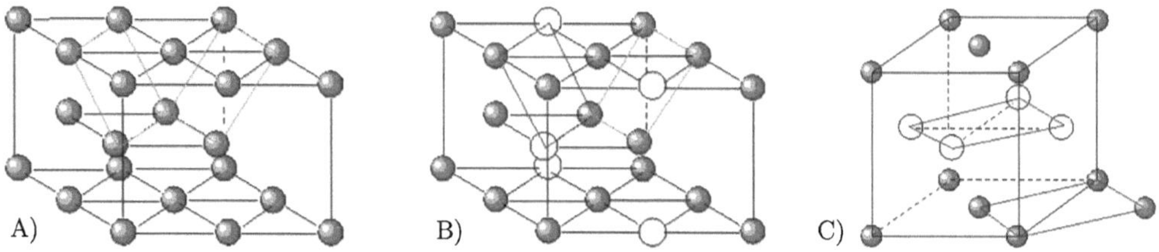

Figure 10.29 Atomic environments of interstitial site in α-Ti (A), α_2-Ti_3Al (B), and γ-TiAl (C). The environments composed only from Ti atoms are sketched in green, and those containing Al atoms are shown in red. (For interpretation of the references to color in this figure legend, the reader is referred to the web version of this chapter.)

If solutes occupy both interstitial and substitutional sites, which corresponds to the dissociative diffusion mechanism, the effective diffusion coefficient D is given by

$$D = \frac{c_i}{c_i + c_s} D_i + \frac{c_s}{c_i + c_s} D_s \cong \frac{c_i}{c_s} D_i. \tag{10.21}$$

Here D_i (D_s) is the interstitial (substitutional) diffusion coefficient and c_i (c_s) is the interstitial (substitutional) solubility of the solute atoms. Eq. (10.21) is simplified under the plausible assumptions that $c_i \ll c_s$ and $c_i D_i \gg c_s D_s$ which typically hold for metals. Since c_i and c_s are temperature dependent, the activation enthalpy of the dissociative diffusion can be remarkably larger than that of pure interstitial diffusion. This conclusion agrees with the observation that the activation enthalpies of Fe and Ni diffusion in α_2-Ti_3Al (277 and 195 kJ/mol, respectively) in comparison with the Ti self-diffusion activation enthalpy (288 kJ/mol) are not as small as expected for the pure interstitial diffusion mechanism. For comparison, the activation energy of Fe and Ni diffusion in α-Ti is about 110 kJ/mol [83,84].

10.5.2 Diffusion in Ni Aluminides

10.5.2.1 Diffusion in Ni_3Al

Ni Self-diffusion. Ni self-diffusion in Ni_3Al was intensively investigated in several experimental studies, e.g., [91–94]. Most reliable data for the single crystalline stoichiometric Ni_3Al [94] are shown in Fig. 10.30.

The Ni data indicate that only one diffusion mechanism operates in the whole temperature interval, and this mechanism is identified with the Ni sublattice diffusion mechanism, e.g., [93] and [94]. Correspondingly, a Ni vacancy performs random jumps in the Ni sublattice with the coordination number 8, see Fig. 10.13. Monte Carlo simulations of the correlation effects demonstrated that the correlation factor for the Ni sublattice diffusion mechanism is marginally changed across the stoichiometry and equals about 0.689.

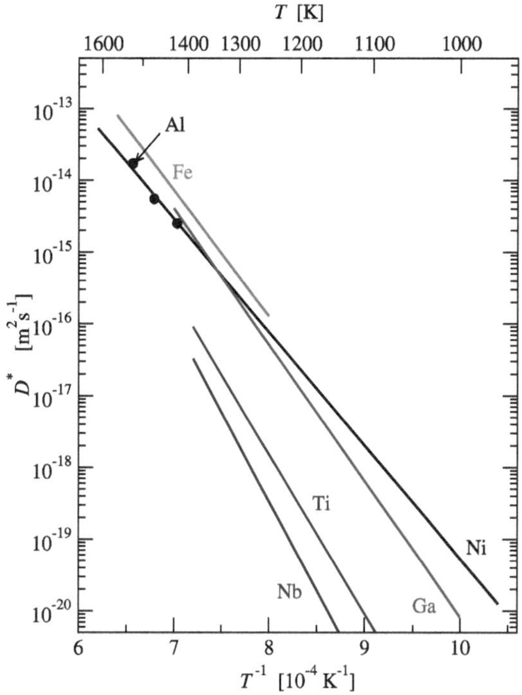

Figure 10.30 Self- (black line and symbols) and solute (colored lines) diffusion in stoichiometric Ni_3Al. The calculated D^*_{Al} (full circles [97]) are shown in comparison to Ni diffusion (black line [94]). Solute diffusion is differentiated with respect to Al substitutes (Ga, red line [61]), Ni substitutes (Nb and Ti, blue lines [101]), and a mixed solubility (Fe, magenta line [102]). (For interpretation of the references to color in this figure legend, the reader is referred to the web version of this chapter.)

Al Self-diffusion. ^{26}Al radiotracer diffusion in nearly stoichiometric Ni_3Al (together with the ^{63}Ni tracer) was measured in a single study [15] at temperatures between 1273 and 1473 K. The data are shown in Fig. 10.31. It is seen that Al and Ni diffuse with similar rates in Ni_3Al in this temperature interval.

Having performed interdiffusion experiments [95,96] and analyzed the relative positions of Kirkendall and Matano planes, a conclusion was drawn that the ratio of Al and Ni tracer diffusivities is very close to unity in the investigated temperature interval of 1450 to 1550 K [97]. Thus the direct experimental findings of Larikov and coworkers [15] are fully supported. The estimated Al diffusion coefficients are compared with the Ni diffusion coefficients in Fig. 10.30.

Solute Diffusion. Since Ga occupies exclusively the Al sublattice in Ni_3Al [99], it is used to mimic Al diffusion [61,100]. The diffusion data are presented in Fig. 10.30 in comparison to Ni and Al diffusion coefficients. One can see that Ga diffusion follows a straight Arrhenius dependence with an activation enthalpy of about 60 kJ/mol larger

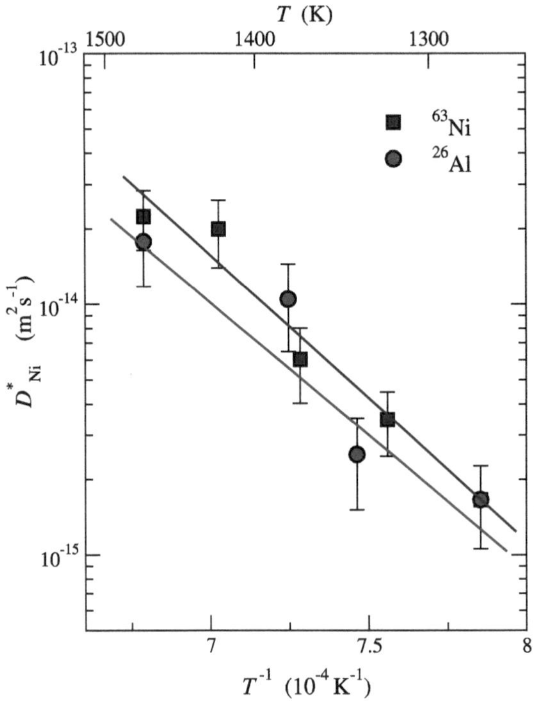

Figure 10.31 Tracer diffusion of Ni and Al measured in Ni_3Al [15].

than that of Ni self-diffusion, see also Table 10.3. The Ga diffusion data in Ni_3Al are very close to the Al diffusivity measured directly [15] or estimated from the interdiffusion data [97], Fig. 10.30.

Ti and Nb are known to occupy preferentially Ni sites [99] and these species reveal a distinctly smaller diffusivity with a larger activation enthalpy, when compared to that of Ni diffusion in Ni_3Al, Fig. 10.30. The difference in the activation enthalpies of Ni and solute diffusion is related to the formation energy of the solute atom at an anti-structural position in the Ni sublattice [94].

Since Fe shows a mixed solubility in Ni_3Al, its diffusion behavior, Fig. 10.30, corresponds to an interplay of the intra-sublattice (the Ni sublattice mechanism) and inter-sublattice (the ASB mechanism) jumps of Fe atoms [100]. The Arrhenius parameters of self- and solute diffusion in Ni_3Al are summarized in Table 10.3.

10.5.2.2 Diffusion in NiAl

Ni Self-diffusion. Alternatively to considered Ni- and Ti-aluminides, NiAl (i) exhibits a triple defect type of disorder and (ii) all nearest-neighbor jumps of a vacancy are the jumps between different sublattices in this structure, see Fig. 10.1. A triple de-

Table 10.3 Arrhenius parameters of self- and solute diffusion in Ni aluminides

Tracer	Composition (at.%Ni)	D_0 (m^2s^{-1})	Q (kJ/mol)	Reference
		$L1_2$ Ni_3Al		
Ni	75.0	1.5×10^{-2}	347	[92]
Ni	74.7	1.0×10^{-4}	303	[98]
Ni	75.9	3.1×10^{-4}	302	[94]
Al*	75.0	8.0×10^{-3}	340	
Ga	75.9	7.8×10^{-2}	363	[101]
Ge	75.9	1.0×10^{-1}	369	[101]
Ge	75.0	3.6×10^{-1}	379	[100]
Nb	75.9	2.6×10^{1}	476	[101]
Ti	75.9	8.6×10^{-1}	425	[101]
Mo	75.1	2.3×10^{2}	493	[100]
Fe	75.1	1.3×10^{-2}	335	[102]
		$B2-NiAl$		
Ni	46.8	2.3×10^{-5}	287	[57]
	50.0	3.5×10^{-5}	290	
	51.8	4.8×10^{-5}	288	
	54.6	4.4×10^{-5}	278	
	56.6	1.0×10^{-6}	231	
In	47.7	1.6×10^{-4}	271	[108]
	49.5	4.6×10^{-4}	311	
	50.6	2.0×10^{-3}	336	
	52.0	9.8×10^{-4}	322	
	55.3	1.1×10^{-3}	311	

* Estimation from the Al intrinsic diffusion data of [97].

fect (two Ni vacancies and one Ni antistructure atom) has the lowest formation energy between the point defects, the thermal formation of which does not change the composition. Moreover, there exist so-called structural Ni vacancies, which are available on the Al-rich side, Fig. 10.6(A). One may expect that the structural vacancies (which concentration approaches several percent!) should significantly enhance Ni self-diffusion in this compound, as it was expected in [98,103].

The temperature dependence of Ni diffusion in single crystalline NiAl alloys of different compositions [57] is shown in Fig. 10.32.

In Fig. 10.32(C) the results of diffusion measurements for different nearly stoichiometric NiAl alloys of different origin are presented, and good consistency of the data is pronounced. We note that at $T > 1500$ K an upward deviation from the otherwise straight Arrhenius dependencies was observed in all compositions in [57] and a strong dependence of Ni diffusion coefficient on the composition is seen, Fig. 10.32.

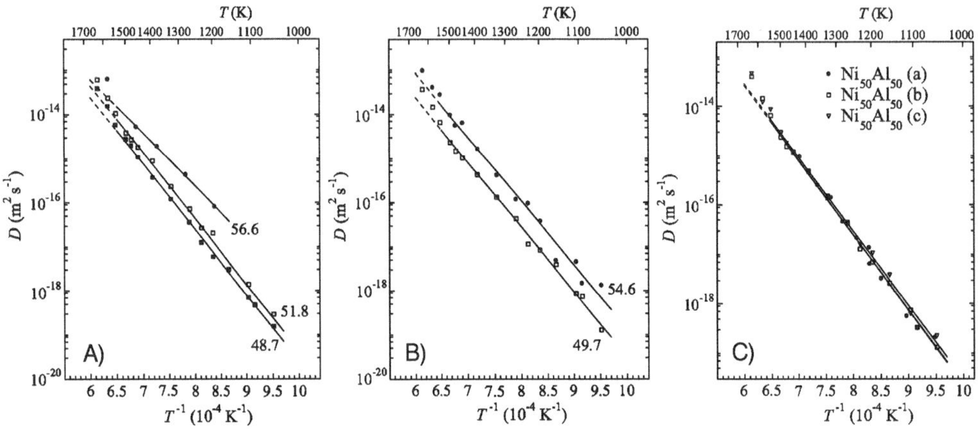

Figure 10.32 Temperature dependencies of Ni diffusion in single crystalline Ni-rich and Ni-poor (A)–(B) and stoichiometric (C) NiAl alloys [57].

The activation enthalpy Q of Ni self-diffusion in NiAl calculated by the Arrhenius fit in the interval 1050 K $< T <$ 1500 K [57] and it is given in Fig. 10.33(B) as a function of composition. In Fig. 10.33(A) the composition dependence of Ni tracer diffusion coefficients, D^*_{Ni}, measured in different alloys [57] is plotted for two selected temperatures. As a matter of fact, both D^*_{Ni} and Q values are almost constant for compositions $c_{Ni} < 53$ at.%, and significantly increase and decrease, respectively, in alloys with larger Ni concentrations. The Arrhenius parameters of Ni self-diffusion in representative Al-rich, stoichiometric, and Ni-rich NiAl alloys are summarized in Table 10.3.

Diffusion mechanisms in NiAl with respect to composition were analyzed in [26, 57] and it was proposed that it is the triple defect mechanism, which is characterized by a constant activation enthalpy of about 300 kJ/mol independent on the composition, Fig. 10.15, which most probably dominates the diffusion process in NiAl [57], in perfect agreement with the experiments, see Fig. 10.33(B). The observed increase of D^*_{Ni}, Fig. 10.33(A), and the decrease of Q_{Ni}, Fig. 10.33(B), at $c_{Ni} > 54$ at.% is explained by the activation of the ASB mechanism [57].

It is important that the upward deviation of D^*_{Ni} at $T >$ 1500 K from the otherwise linear Arrhenius dependence, Fig. 10.32, correlates with the recent results of differential dilatometry measurements in NiAl alloys [32], which reveal a nonlinearity in the Arrhenius plot of the vacancy concentration in this high temperature range.

Diffusion of Al in NiAl. So far Al tracer diffusion was not directly measured in NiAl. Interdiffusion in NiAl was investigated in several studies, and a notable compositional dependence of the interdiffusion coefficient $\widetilde{D}$ was established [104,105]. Again, using the thermodynamic data of [106] and the Darken–Manning equation, the contribution of Al tracer diffusion to the interdiffusion coefficient can be estimated by applying the

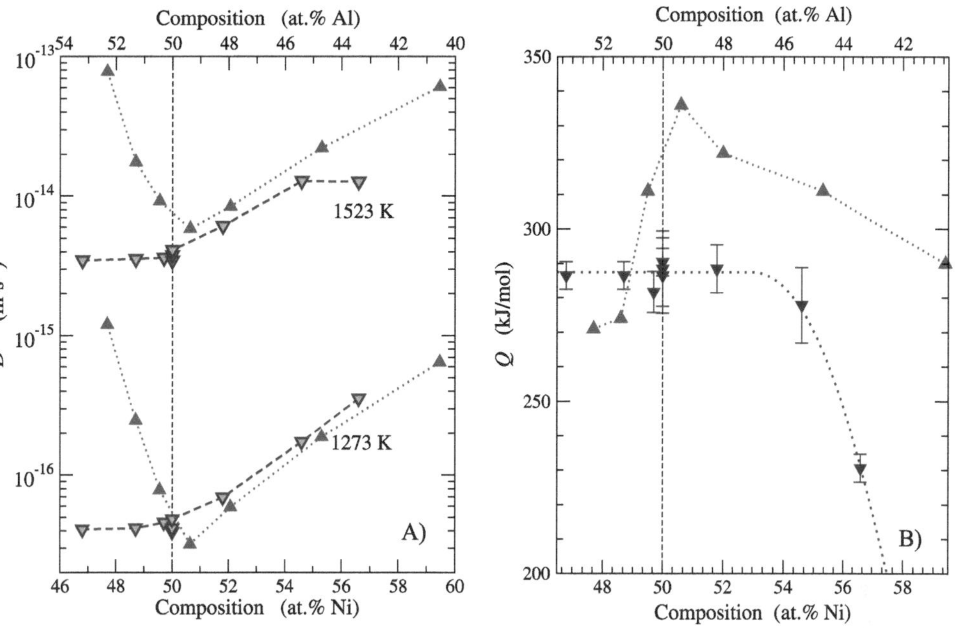

Figure 10.33 Composition dependence of the diffusion coefficients (A) and of the effective activation enthalpies (B) of Ni (triangles down [57]) and In (triangles up [109]) diffusion in NiAl. The diffusion coefficients are shown for two selected temperatures.

usual assumption $S = 1$. Still, the data can be unreliable in view of pronounced deviation of the vacancy wind S from unity for the triple defect mechanism [107]. We conclude that the usage of Al-substituting solute to mimic Al diffusion is preferred.

Solute Diffusion in NiAl. Indium is considered as Al-substituting solute and its diffusion in NiAl was measured in [108,109]. In Fig. 10.33(A) and (B), the compositional dependence of In diffusivity D^*_{In} and the determined activation enthalpy [109] are respectively compared with the Ni tracer diffusion data. We see directly that In is faster than Ni at higher temperatures in Ni-rich compositions and this relation is reversed at lower temperatures. However, the absolute values of D^*_{In} and D^*_{Ni} are almost similar.

On the other hand, In diffuses remarkably faster than Ni in Al-rich compositions. Such behavior of In diffusion can be explained by the triple defect diffusion mechanism and an additional activation of the ASB mechanism (which involves Ni vacancies and Al antistructure atoms) in Al-rich compositions. The appearance of Al atoms in antistructure positions in Al-rich compositions of NiAl is seen in Fig. 10.6(C).

We conclude that the diffusion behavior in NiAl is somehow paradoxical, a large amount of structural Ni vacancies does not enhance Ni self-diffusion, whereas they do facilitate diffusion of the Al component. The reason is the triple defect character of thermal disorder.

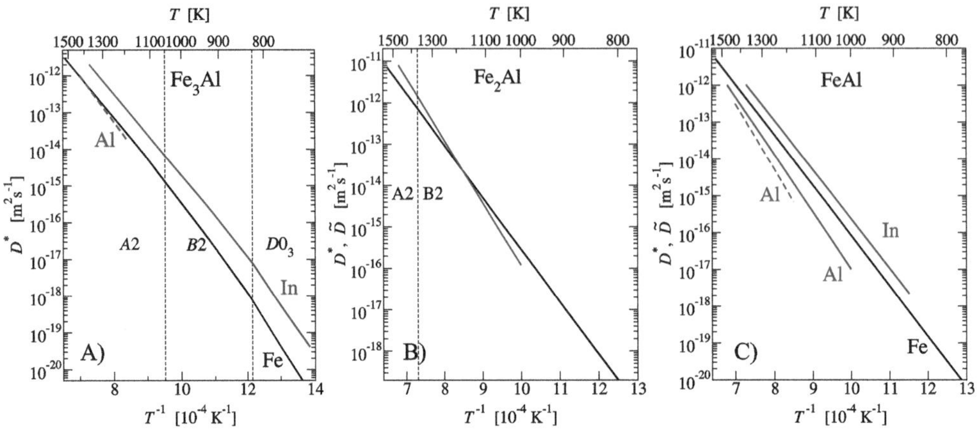

Figure 10.34 Fe (solid black lines) [110] and Al (red lines) [69] diffusion in Fe_3Al (A), Fe_2Al (B), and FeAl (C). Al tracer diffusion data measured directly [15] are plotted, too (red dashed lines). In solute diffusion (red lines) [110] are shown as well. The vertical dashed lines differentiate existing phases. (For interpretation of the references to color in this figure legend, the reader is referred to the web version of this chapter.)

10.5.3 Fe–Al System

10.5.3.1 Self-diffusion of Fe

Most extensive tracer measurements of self-diffusion in the Fe–Al system were performed by Mehrer with coworkers [110], and the results for nominal compositions of Fe_3Al, Fe_2Al, and FeAl are shown in Figs. 10.34(A)–(C), respectively.

The phase Fe_3Al reveals sequentially $A2$, $B2$, and $D0_3$ ordered structures with decreasing temperature, Fig. 10.3. In [110] Fe self-diffusion was measured in a very extended temperature interval which comprises all three possible structures, Fig. 10.34(A), and it was convincingly demonstrated that the increase of the order is accompanied by an increase in the activation enthalpy, i.e., the following relationships are valid, see Table 10.4,

$$Q_{\rm Fe}^{(A2)} < Q_{\rm Fe}^{(B2)} < Q_{\rm Fe}^{(D0_3)}. \tag{10.22}$$

According to Girifalco's model [111], increasing order slows down diffusion, and the following formula was proposed:

$$D_v = D_0 \times \exp\left\{-\frac{Q(1+\gamma S^2)}{RT}\right\}. \tag{10.23}$$

Here, D_v is the bulk tracer diffusion coefficient, D_0 and Q are Arrhenius parameters of diffusion in the disordered state, S is the long-range order parameter, and γ is a

Table 10.4 Arrhenius parameters of self- and solute diffusion in FeAl

Approximate composition	Tracer	Structure	D_0 (m^2s^{-1})	Q (kJ/mol)	Reference
Fe_3Al	Fe	$A2$	8.1×10^{-5}	217	[110]
	Fe	$A2$	3.2×10^{-5}	204	[113]
	Fe	$B2$	3.8×10^{-4}	232	[110]
	Fe	$D0_3$	3.3×10^{-1}	278	[110]
	In	$A2/B2$	1.9×10^{-4}	214	[110]
	Zn	$A2/B2$	1.8×10^{-4}	221	[110]
	Ni	$A2$	4.7×10^{-4}	240	[113]
	Ni	$A2/B2$	2.9×10^{-1}	310	[118]
	Co	$A2/B2$	5.4×10^{-2}	290	[118]
	Cr	$A2/B2$	8.6×10^{-4}	233	[118]
	Mn	$A2/B2$	1.1×10^{-3}	240	[118]
Fe_2Al	Fe	$A2/B2$	1.1×10^{-3}	241	[110]
	In	$A2/B2$	2.1×10^{-3}	239	[110]
	Zn	$A2/B2$	4.3×10^{-3}	240	[110]
FeAl	Fe	$B2$	4.1×10^{-3}	262	[110]
	In	$B2$	5.4×10^{-3}	257	[110]
	Zn	$B2$	1.2×10^{-2}	251	[110]

constant. Note that, however, in the Girifalco approach the change in the correlation factor below the transition temperature is neglected [112].

Fe self-diffusion in Fe_2Al and FeAl is investigated mainly in the $B2$ phase region and perfectly linear Arrhenius dependencies are established, Figs. 10.34(B)–(C) [110]. However, generally, Eq. (10.22) holds for Fe_2Al, too.

At a given absolute temperature the Fe tracer diffusion coefficient obeys the following relations:

$$D_{Fe}^{(FeAl)} < D_{Fe}^{(Fe_2Al)} < D_{Fe}^{(Fe_3Al)}. \qquad (10.24)$$

Simultaneously, the activation enthalpies increase consistently with a decreased content of Fe (increased amount of Al):

$$Q_{Fe}^{(FeAl)} > Q_{Fe}^{(Fe_2Al)} > Q_{Fe}^{(Fe_3Al)}. \qquad (10.25)$$

The dominating diffusion mechanisms in FeAl are not completely understood so far [110]. A large activation volume of Fe self-diffusion has been reported [114], which indicates diffusion contribution of composed defects. Results of Mössbauer spectroscopy and quasielastic neutron scattering measurements suggest nearest-neighbor jumps of Fe atoms [115,116]. Recent atomic-scale X-ray Photon Correlation Spectroscopy study [117] directly reveals dominance of the effective <111> jumps of atoms, too.

10.5.3.2 Al Diffusion

^{26}Al diffusion was measured in several FeAl alloys [15] and these data are plotted in Figs. 10.34(A) and (C), too. The main result is that the Al and Fe diffusivities have the same order of magnitude.

Interdiffusion in the Fe–Al system was intensively investigated [119,69], and the deduced tracer diffusion coefficients of Al are shown in Figs. 10.34(B) and (C), too. By measuring Kirkendall shift, it was found that the ratio of intrinsic diffusivities, D^*_{Al}/D^*_{Fe}, is always larger than unity and is about 1.2 to 1.5 in *B*2 FeAl and about 1.8 in the *A*2 phase region of the Fe_3Al alloy [120].

10.5.3.3 Solute Diffusion

Solute diffusion was most extensively measured in Fe_3Al, and a significant effect of the *A*2–*B*2 transition on Ni diffusion in Fe_3Al was determined [121]. The In solute was used as Al-substitute [110], and it was found that In diffuses faster than Fe (Fig. 10.34(A)). Whereas the *A*2–*B*2 transition does not influence diffusion of Al substitutes; a small effect from the *B*2–*D*0$_3$ transition is indicated in the case of In diffusion, Fig. 10.34(A).

10.5.4 General Trends of Diffusion in Binary Ni, Ti, and Fe Aluminides

An analysis of tracer diffusion behavior of Ni-, Ti-, and Fe-aluminides indicates a number of similarities and fundamental differences in the properties which can be tracked back to pertinent crystalline structures and the type of atomic disorder.

- Typically, linear Arrhenius-type dependencies are observed for tracer diffusion of both components in binary alloys. At first glance, this fact may be considered as surprising in view of diversity of atomistic mechanisms of diffusion and different types of point defects existing in these alloys. However, usually one diffusion mechanism dominates in the given temperature interval[1] and, in view of inevitable experimental uncertainties, a single effective activation enthalpy is determined.
- However, the aluminides of transition metals provide two characteristic examples of strongly nonlinear Arrhenius dependencies. These features are clearly seen for TiAl (Fig. 10.27) and NiAl (Fig. 10.32). The deviations from otherwise linear (low-temperature) Arrhenius behavior starts at high temperatures (above 1470 K in TiAl [20] and above 1500 K in NiAl [57]).
- The nonlinear Arrhenius plots for Ti and Ni diffusion in γ-TiAl and B2 NiAl, respectively, seem to have different origins. One of them is explained by a contribution of a specific ASB mechanism, Fig. 10.17(B), which provides an additional contribution to the Ti sublattice diffusion mechanism at higher temperatures. The

[1] Note that standard methods limit the accessible tracer diffusion coefficients to the range of 10^{-21} to 10^{-12} m^2/s, which correspondingly shrinks the temperature range of these measurements.

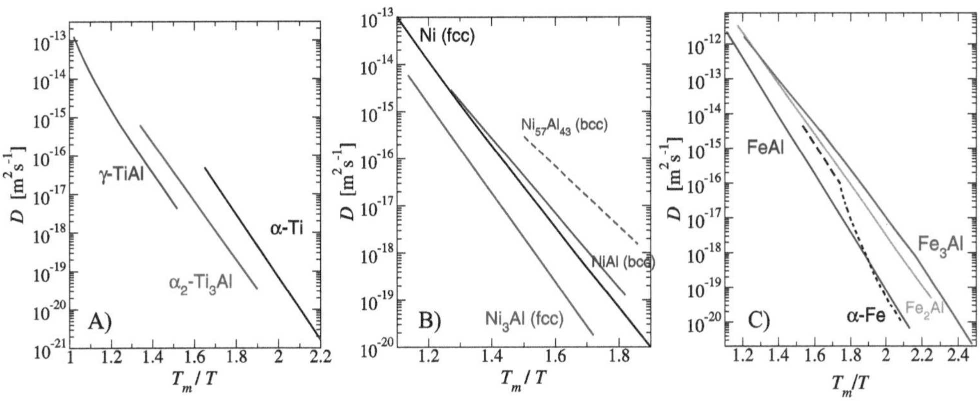

Figure 10.35 Ti (A), Ni (B), and Fe (C) tracer self-diffusion in the corresponding pure metal and its aluminum alloys as a function of the inverse homologous temperature T_m/T. Here T_m is the melting point of the corresponding material. Ni diffusion in both stoichiometric as well as Ni-rich B2 NiAl phase is shown in (B).

ASB mechanism is related to the large concentration of thermal Ti antistructure atoms at high temperatures even in the Al-rich TiAl alloys [21,20].

- Presently, there is no adequate explanation of the observed upward deviation of D^*_{Ni} from the otherwise straight Arrhenius line in NiAl at $T > 1500$ K – a feature which is almost independent on the composition of the B2 NiAl phase [57]. We highlight, however, that there is a direct correlation between the nonlinear tracer diffusivity and the measured vacancy concentration in NiAl alloys [32].
- It is remarkable that the crystalline phases with an antistructure type of disorder on both sides from the stoichiometric composition (i.e., Ni_3Al, Ti_3Al, and TiAl), reveal a marginal compositional dependence of the tracer self-diffusion coefficient of the transition metal. This behavior is supposed to be induced by the following reasons: (i) the dominant sublattice diffusion mechanism for the transition metal component and (ii) the dominant concentration of vacancies on the transition-metal sublattice.
- Both tracer diffusion rates and the effective activation enthalpies in the B2 phase NiAl reveal a pronounced compositional dependencies. A striking feature is that the Ni antistructure atoms remarkably enhance Ni tracer diffusion on the Ni-rich side, whereas Ni atoms do not profit from abandoned structure vacancies on the Ni-poor side. These features are explained by a dominant contribution of the triple defect diffusion mechanism with an additional contribution of the ASB mechanism at larger Ni content.

Fig. 10.35 represents the tracer diffusion coefficients of Ti, Ni, and Fe in corresponding pure metals in comparison to the transition metal diffusion rates in aluminides when plotted using the homologous temperature, T/T_m. Here T_m is the melting point of the corresponding stoichiometric compound. Whereas systematic changes are observed in the diffusion behavior of Ti-aluminides – Ti diffusion decreases gradually from pure α-Ti to Ti_3Al, and finally to TiAl – the diffusion behavior in Ni and Fe aluminides is more sophisticated.

The diffusion systematics for Ti-aluminides is most probably caused by the following reasons:

- vacancy-mediated diffusion mechanisms in all materials;
- densely-packed crystalline structures (HCP and slightly tetragonally distorted FCC lattices) in these materials;
- the sublattice diffusion mechanism in pure α-Ti (for a trivial reason) and Ti aluminides;
- prevailing vacancy concentrations on the transition metal sublattices;
- a systematic decrease of the coordination number of the transition metal sublattice from pure α-Ti to TiAl.

The latter feature is directly supported by Fig. 10.35(A) which reveals the following relationship for the tracer diffusion coefficients:

$$D^*_{\mathrm{Ti}} > D^*_{\mathrm{Ti_3Al}} > D^*_{\mathrm{TiAl}}, \tag{10.26}$$

which is in a line with the coordination number of the transition metal sublattices in the corresponding structures:

$$z^{\mathrm{Ti}}_{\mathrm{Ti}} = 12 > z^{\mathrm{Ti}}_{\mathrm{Ti_3Al}} = 8 > z^{\mathrm{Ti}}_{\mathrm{TiAl}} = 4. \tag{10.27}$$

On the other hand, the tracer diffusion coefficients of Ni in pure Ni and Ni aluminides at a given homologous temperature, Fig. 10.35(B), follow the relationship

$$D^*_{\mathrm{NiAl}} \geq D^*_{\mathrm{Ni}} > D^*_{\mathrm{Ni_3Al}}. \tag{10.28}$$

The relatively high diffusion rates in B2 NiAl can be explained by its relatively more open structure (BCC) comparing to those of pure Ni or Ni_3Al (fcc).

Fe tracer diffusion in iron aluminides, Fig. 10.35(C), nicely follows the same trend established for Ti aluminides, namely

$$D^*_{\mathrm{Fe_3Al}} > D^*_{\mathrm{Fe_2Al}} > D^*_{\mathrm{FeAl}}, \tag{10.29}$$

which have almost similar crystalline structures (bcc). The specific features of Fe tracer diffusion in α-Fe are probably explained by ferromagnetic properties of the metal, especially its pronounced curvature in the Arrhenius coordinates below the Curie temperature [122,123].

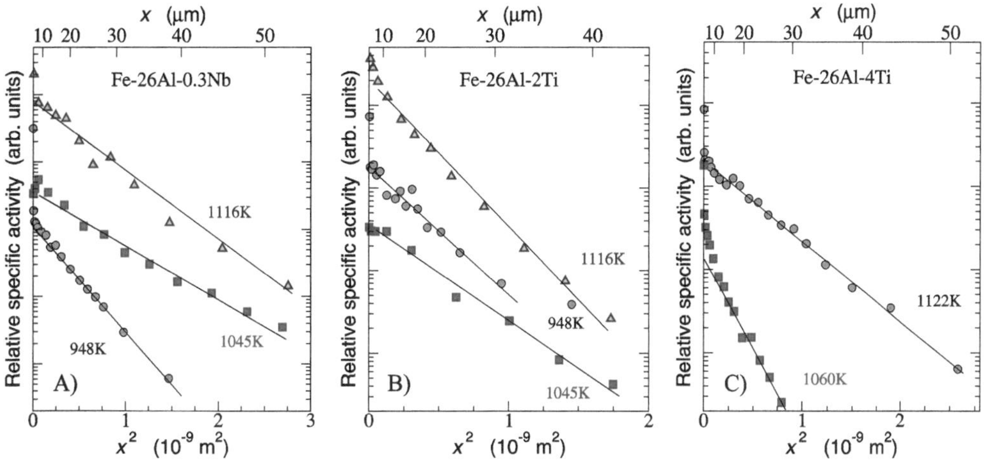

Figure 10.36 Penetration profiles of Fe diffusion in Fe–26Al–0.3Nb (A), Fe–26Al–2Ti (B), and Fe–26Al–4Ti (C) alloys.

10.5.5 Effect of Alloying on Diffusion in Ordered Intermetallics

Alloying of binary aluminides is commonly used to improve their mechanical and functional properties. In this section, we will analyze the effect of alloying by a third component on diffusion and diffusion mechanisms of main components and alloying elements for the FeAl, TiAl, and NiAl alloys.

10.5.5.1 Effect of Alloying by Ti and Nb on Fe Diffusion in Fe_3Al

The effect of alloying on diffusion was investigated for the Fe_3Al phase, and diffusion of Fe atoms was followed by the radiotracer method [124]. Diffusion was measured in Fe–26Al–0.3Nb, Fe–26Al–2Ti, and Fe–26Al–4Ti alloys, and the precision mechanical grinding was utilized. The penetration profiles are presented in Fig. 10.36. Although Gaussian solution is seen to be generally fulfilled, in some cases the error-function solution has to be used.

The resulting temperature dependence of Fe diffusion in Fe_3Al-based alloys is summarized in Fig. 10.37 and it is compared to the data on Fe diffusion in pure Fe_3Al [110]. Since the composition of Fe_3Al investigated in [110] slightly deviates from 26% of Al, Fe diffusion was remeasured in Fe–26Al. The result, Fig. 10.37, does not practically deviate from the literature data, confirming a minor effect of Al content on Fe diffusion in the Fe_3Al phase.

The main result is that the alloying by Nb and Ti does not practically change Fe diffusion in ternary alloys [124]. Therefore, a strong change of mechanical damping in these alloys [125] is not related to the Fe diffusion rate. It is known that both Nb and Ti are strong carbide-forming elements and the amount of C is strongly affected by the

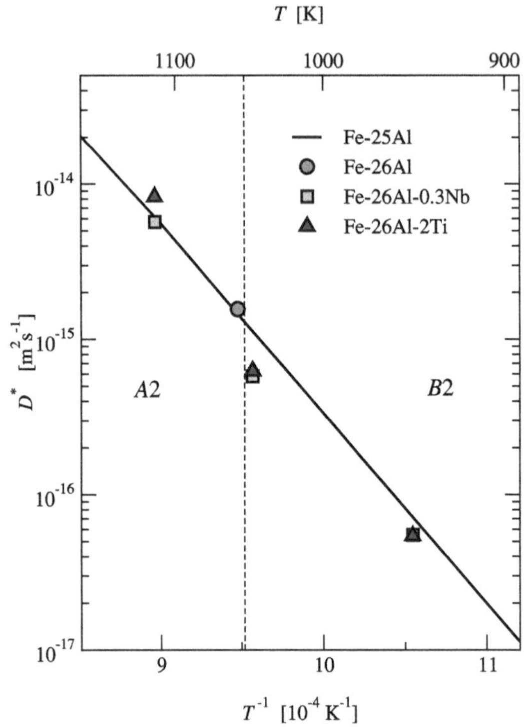

Figure 10.37 The effect of alloying on Fe diffusion in Fe_3Al-based alloys. The limits of the existence of A2 and B2 ordered phases in Fe_3Al are indicated.

alloying of Fe_3Al, and this feature was concluded to affect the damping properties of the FeAl alloys.

10.5.5.2 Effect of Alloying by Nb on Diffusion in γ-TiAl

Nb Diffusion. Nb diffusion in the binary Ti–56Al and ternary Ti–54Al–10Nb alloys was investigated in Refs. [126,127], and the determined tracer diffusion coefficients are plotted in Figs. 10.38(A)–(B), respectively, as functions of the inverse temperature (circles). Both mechanical sectioning (full symbols) and ion-beam sputtering (open symbols) techniques were utilized to record the penetration profiles, and good agreement of the data is seen. In both cases Nb diffuses slower than Ti, though Nb alloying level ups the two diffusivities, Fig. 10.38.

Ti Diffusion. Typical penetration profiles measured by mechanical sectioning technique are presented in Fig. 10.39(A). Gaussian-like concentration profiles are usually observed with the penetration depths ranging from five to several tens of micrometers. In some cases the Ti penetration profiles, corresponding to the constant-source solution of the diffusion problem, are recorded, see Fig. 10.39(B).

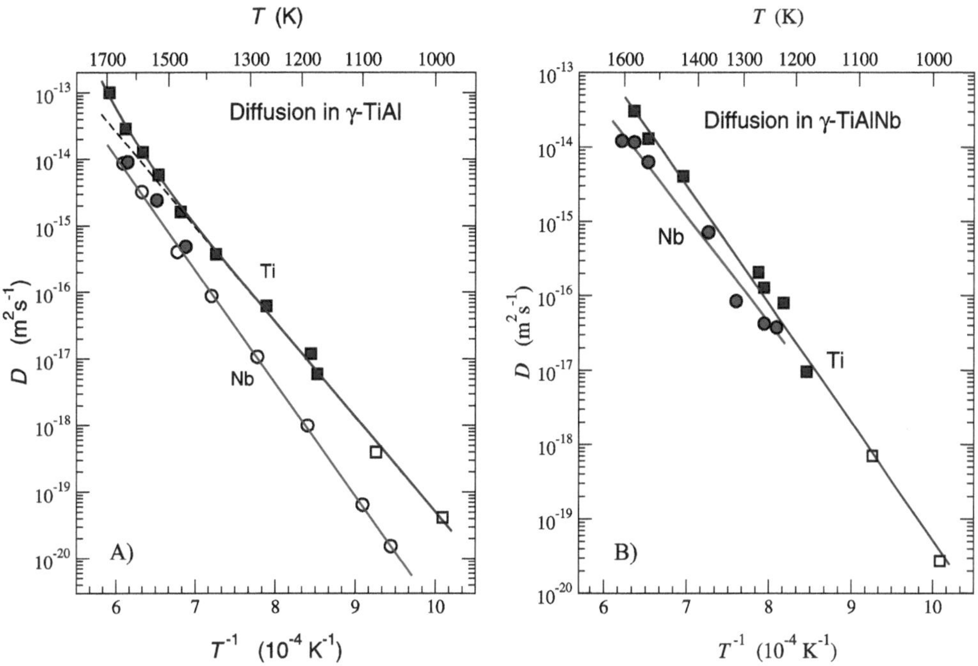

Figure 10.38 Arrhenius plots of Ti (squares) and Nb (circles) tracer diffusion in Ti–56Al (A) and Ti–54Al–10Nb (B) alloys. The results of the mechanical sectioning (full symbols) and ion-beam sputtering (open symbols) are shown.

At lower temperatures, the penetration profiles were measured by ion-beam sputtering technique which allows reliably determining the small diffusivities, see Fig. 10.39(C). The penetration depths which are even smaller than one micrometer are accessible and can be reliably measured in this case.

The ion-beam sputtering inevitably comprises several physical processes which affect the resulting penetration profile shape. Atomic mixing, atom redeposition, and other effects limit the resolution of this sectioning method. As a result, the measured penetration profile would be not only broader than the real one, but an artificial deep-penetration tail may additionally appear. The measured profile is then represented by a convolution of a real one and the device (sputter) functions.

We have evaluated the sputter function by measuring a penetration profile which corresponds to the nominal diffusion length of about 30 nm. The tracer material was evaporated onto the specimen surface and short annealing was performed at $T = 1080$ K for 340 s. The resulted sputtering profile is shown in Fig. 10.39(C) (triangles). The profile consists of a sharp Gaussian-like near-surface part (less than 100 nm) and a deep low-activity tail. Analysis shows that the distortion of the measured penetration profiles

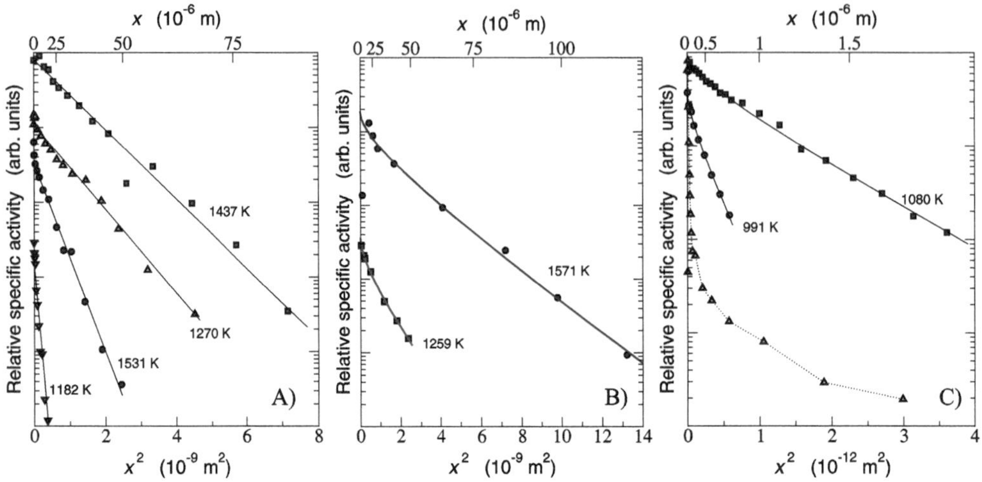

Figure 10.39 Penetration profiles of ^{44}Ti diffusion measured by mechanical sectioning (A)–(B) and ion-beam sputtering (C) techniques in the Ti–54Al–10Nb alloy. x is the penetration depth. In (B) the results of codiffusion experiments with Nb are presented, see text. In (C) the resolution function of the sputter device, which was measured after a very short diffusion anneal at $T = 1080$ K for 340 s, is also shown (triangles).

(squares and circles in Fig. 10.39(C)) induced by such a sputter function is very small and can be totally neglected in all cases under consideration.

The temperature dependence of the Ti tracer diffusion in the Ti–54Al–10Nb alloy is presented in Fig. 10.38(B) (squares) and compared to that in Ti–56Al alloy, Fig. 10.38(A). The results of mechanical sectioning and ion-beam sputtering are highly consistent with each other, confirming the reliability of the diffusion measurements. Both Ti and Nb isotopes are seen to follow almost linear temperature dependencies in the Arrhenius coordinates in the ternary Ti–54Al–10Nb alloy. Nb diffuses slower than Ti by a factor of 2 to 3. The activation enthalpies of Ti and Nb diffusion in the Ti–54Al–10Nb alloy are similar within the experimental uncertainties.

The ternary high niobium-containing γ-Ti–54Al–10Nb alloy has a layered $L1_0$ structure, which is schematically presented in Fig. 10.2 for the binary γ-TiAl alloy. Nb was shown to occupy the Ti sublattice in γ-TiAl [128–130]. Heavy Nb alloying increases both the lattice parameters a and c as well as their ratio c/a [129,131].

As it was discussed, Ti self-diffusion was extensively studied in binary γ-TiAl [20] and α_2-Ti_3Al [76] alloys. In Fig. 10.40 the Ti and Nb tracer diffusion coefficients are compared in the binary and ternary titanium aluminides. One of the most striking features is a strongly curved temperature dependence of Ti diffusion in γ-TiAl, if it is measured in an extended temperature interval [20].

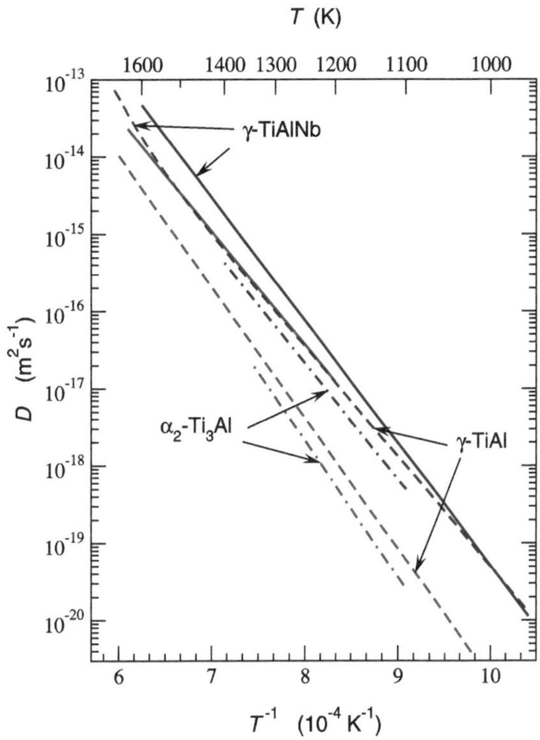

Figure 10.40 Ti (blue lines) and Nb (red lines) diffusion in the γ-Ti–54Al–10Nb ternary alloy (solid lines) in comparison to their diffusivities in the binary γ-Ti–56Al [20,133] (dashed lines) and α_2-Ti–33Al [76, 132] (dashed-dotted lines) alloys. (For interpretation of the references to color in this figure legend, the reader is referred to the web version of this chapter.)

Different to the γ-TiAl phase, a linear temperature dependence of Ti diffusion was observed in α_2-Ti_3Al when plotted in the Arrhenius coordinates [76], Fig. 10.40. This behavior is related to the Ti sublattice diffusion mechanism, which dominates Ti diffusion in the $D0_{19}$ structure of the α_2-Ti_3Al alloy [76].

Nb tracer diffusion reveals almost perfect Arrhenius-type temperature dependencies in binary α_2-Ti_3Al [132] and γ-TiAl [133] alloys. Moreover, Nb diffuses slower than Ti in binary aluminides by an order of magnitude. Due to the preferential solubility of Nb to the Ti sublattice [128,134], such a behavior is explained by the Ti sublattice diffusion mechanism of Nb atoms and a repulsive interaction of Nb atoms and Ti vacancies. According to the standard electrostatic model of vacancy–solute atom interaction [135], one would expect that addition of Nb to TiAl would produce a retardation effect on the diffusion rates. The activation enthalpy of Nb diffusion in both α_2-Ti_3Al and γ-TiAl is larger than that of Ti.

This is definitely not the case for the high niobium-containing alloys. Both Ti and Nb diffuse faster in the ternary alloy in comparison to the binary ones with similar activation enthalpies, see Fig. 10.40.

Thus, the experimental study yields the following, somewhat unexpected results:

1. The heavy alloying by Nb enhances both Nb and Ti diffusion in the γ-Ti–54Al–10Nb alloy with respect to the binary γ-Ti–56Al alloy.
2. Ti diffusion in the γ-Ti–54Al–10Nb alloy reveals an almost linear Arrhenius temperature dependence, whereas it is significantly curved in the binary γ-Ti–56Al.
3. Nb diffuses more slowly than Ti in all three titanium aluminides. However, the activation enthalpy of Nb diffusion, Q_{Nb}, is significantly larger than that of Ti, Q_{Ti}, in the binary aluminides. However, $Q_{Nb} \approx Q_{Ti}$ was measured in the ternary aluminide.

The experimentally observed behavior can be rationalized if one supposes that the heavy Nb alloying favors the vacancy jumps between the successive layers along the c axis and changes the relationship between the activation energies of the Ti sublattice diffusion mechanism, Q_{sb}, and that of the ASB mechanism, Q_{ASB}, in the $L1_0$ structure of the ternary alloy: $Q_{ASB} < Q_{sb}$. Indeed, it were the relation $Q_{ASB} > Q_{sb}$ and the large geometric factor related to the ASB mechanism, which resulted in the non-Arrhenius temperature dependence of Ti diffusion in the binary γ-TiAl alloy [20]. The relationship $Q_{ASB} < Q_{sb}$ will result in the dominance of the ASB mechanism in the whole temperature interval, resulting in an Arrhenius-type temperature dependence of Ti diffusion, which indeed is experimentally observed.

Since the addition of a large amount of Nb atoms decreases the c/a ratio in the ternary γ-phase with respect to the binary one [129,136], it is reasonable to suggest that the energy barriers for the corresponding ASB-type jumps are decreased, enhancing simultaneously the total diffusion rate.

Mechanical spectroscopy of high-Nb-bearing γ-TiAl-based alloy resulted in larger values of the activation enthalpy of the high-temperature damping background in comparison to the binary γ-phase [137]. This damping background was attributed to diffusion-assisted climb of dislocations [137]. However, the corresponding activation enthalpy, about 400 kJ/mol, is much larger than the activation enthalpy of Ti tracer diffusion Q_{Ti} in both the binary γ-TiAl, $Q_{Ti} = 270$ kJ/mol, and the ternary γ-TiAlNb, $Q_{Ti} = 304$ kJ/mol, alloys.

Heavy alloying by Nb was shown to enhance the creep resistance of TiAl alloys at service temperatures of about 700°C [138,139]. Extrapolating the Ti diffusivity in the ternary γ-Ti–54Al–10Nb and the binary γ-Ti–56Al alloys down to such temperatures, one can see that self-diffusion in the ternary γ-TiAlNb alloy occurs somewhat slower than in binary γ-TiAl, Fig. 10.40. Then the superior creep properties of Ti–Al–Nb alloys may probably be related to this bulk diffusion retardation. Moreover, at relatively low service temperatures grain boundary diffusion may play a decisive role and may

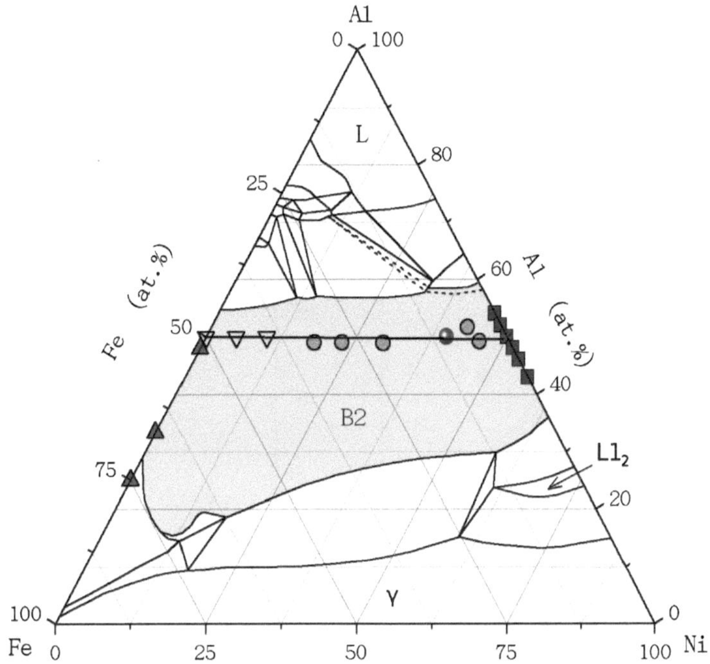

Figure 10.41 The NiFeAl ternary phase diagram at $T = 1000$ K, taken from [140]. The alloy compositions in which diffusion has been measured are marked by squares [57], circles [141], spheres [142], triangles up [110], and triangles down [143]. The B2 phase region at 1000 K is highlighted. The quasi-binary section $Fe_{50}Al_{50}$–$Ni_{50}Al_{50}$ is marked by a thick horizontal line.

profoundly affect the creep resistance. In this case the segregation of Nb atoms at grain boundaries can produce a retardation effect on grain boundary self-diffusion of Ti and thus on the creep rate.

10.6 DIFFUSION IN THE TERNARY SYSTEM NI–FE–AL

Fig. 10.41 presents the NiFeAl ternary phase diagram due to Petzow and Effenberg at $T = 1000$ K [140]. The quasi-binary section Al = 50% belongs completely to the B2 ordered region and diffusion of both Ni and Fe along this section was investigated in detail [141–143]. This section of the ternary phase diagram is interesting from a fundamental point of view, since a strong variation of the defect concentrations on different sublattices and even a change of the dominant point defects may be expected for the same structure. The study is naturally accomplished by the available data on Ni and Fe diffusion in binary end-members, NiAl and FeAl [57,110,118]. The compositions for which tracer diffusion was measured are indicated in Fig. 10.41 by different symbols. Here we will analyze diffusion of Ni in these alloys.

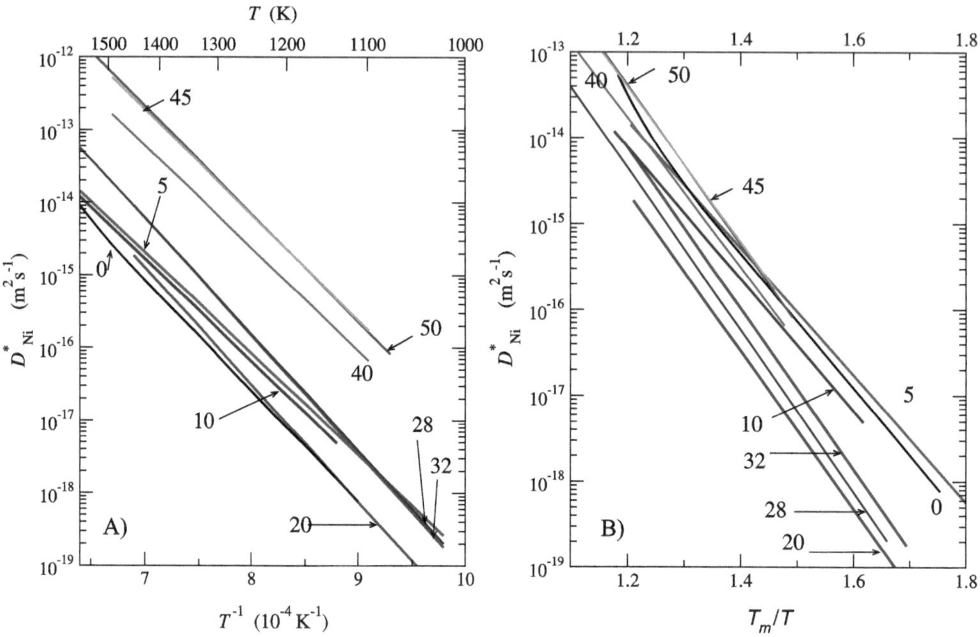

Figure 10.42 Arrhenius plots of Ni tracer diffusion coefficients, D^*_{Ni}, in the NiFeAl ternary alloys in comparison to the Ni diffusivity in binary $Ni_{50}Al_{50}$ [57] (black line) and $Fe_{50}Al_{50}$ [143] (red line) alloys as functions of the inverse temperature $1/T$ (A) and the inverse homologous temperature T_m/T (B). Here T_m is the melting point of the corresponding composition. The Fe content x in the $Ni_{50-x}Fe_xAl_{50}$ alloys is indicated. (For interpretation of the references to color in this figure legend, the reader is referred to the web version of this chapter.)

Within experimental uncertainties, nearby linear Arrhenius dependencies were measured for all compositions (at the moment we will not consider the upward deviations from the linear Arrhenius dependencies observed in binary B2 NiAl alloys [57]). In Fig. 10.42(A) the Ni diffusion coefficients in NiFeAl alloys [141–143] are compared with Ni diffusion rates in binary NiAl [57] and FeAl [118] alloys. At a given absolute temperature T, Ni tracer diffusion in FeAl is by two orders of magnitude faster than that in stoichiometric NiAl. The diffusion rates in $Ni_{50-x}Fe_xAl_{50}$ alloys are somewhere in-between. However, a non-monotonous composition dependence is explicitly seen – Ni diffusion in the $Ni_{50-x}Fe_xAl_{50}$ alloys with about $x = 20$ at.% is almost as large as in stoichiometric NiAl ($x = 0$), whereas it is remarkably faster in the material with $x = 5$ or 10 at.%.

In Fig. 10.42(B) the measured diffusion coefficients are replotted against the inverse homologous temperature, i.e. T_m/T, where T_m is the melting point of the corresponding alloy. The melting points were estimated from the projection of liquidus surfaces investigated in [144,145] and the corresponding values are given in Table 10.5. Such

Table 10.5 Arrhenius parameters of Ni diffusion in $Ni_{50-x}Fe_xAl_{50}$ alloys. T_m is the melting point estimated from the projection of liquidus surfaces investigated in [144,145]

Exact composition (in at.%)	T_m (K)	D_0 (m^2s^{-1})	Q (kJ/mol)	Reference
$Ni_{50}Al_{50}$	1949	3.50×10^{-5}	290	[57]
$Ni_{45.8}Fe_5Al_{49.2}$	1883	1.18×10^{-5}	267	[141]
$Ni_{40}Fe_{10}Al_{50}$	1838	1.29×10^{-5}	270	[142]
$Ni_{29.9}Fe_{21.2}Al_{48.9}$	1753	2.55×10^{-4}	309	[141]
$Ni_{23.1}Fe_{27.9}Al_{49.0}$	1729	1.39×10^{-3}	310	[141]
$Ni_{18.5}Fe_{32.5}Al_{49.0}$	1693	9.87×10^{-4}	306	[141]
$Ni_{10}Fe_{40}Al_{50}$	1623	5.0×10^{-4}	271	[143]
$Ni_5Fe_{45}Al_{50}$	1605	2.8×10^{-3}	278	[143]
$Fe_{50}Al_{50}$	1603	5.8×10^{-3}	284	[118]

scaling of the diffusion rates allows justified comparison of behaviors observed in different materials with the same crystalline lattice [52]. The Ni diffusion rates, when plotted using homologous temperature, changes systematically with Fe content – whereas similar diffusivities are seen for binary end-members (B2 NiAl and FeAl with 0 and 50 at.% of Fe, respectively), Ni diffusion first slows down with addition of Fe to stoichiometric NiAl and then increases again, Fig. 10.42(B).

Since the exact compositions of NiFeAl alloys deviated from the nominal values, see Table 10.5, we will use the composition parameter Fe/(Ni+Fe), where the corresponding amount of Fe and Ni (in at.%) are used. In Fig. 10.43 the Ni tracer diffusivity D^* (at two representative temperatures) and the corresponding activation enthalpy Q_{Ni} are plotted as functions of the composition parameter Fe/(Fe+Ni). Note that in the case of $Ni_{50-x}Fe_xAl_{50}$ alloys it is simply equal to $0.02x$. Rather non-monotonous dependencies are observed. The replacement of Ni by Fe atoms in the $Ni_{50-x}Fe_xAl_{50}$ alloys causes at first a relatively small effect on the Ni diffusivity in the temperature interval studied. Only if the Fe concentration exceeds that of Ni ($x > 30$), the Ni diffusivity increases considerably by orders of magnitude.

A minimum in the activation enthalpy Q_{Ni} is observed at about 5–10 at.%Fe and at 40 at.%Fe. The activation enthalpy significantly decreases with respect to the stoichiometric binary NiAl alloy, by about 30 kJ/mol if 10 at.% of Fe is added. Simultaneously, there exists a local maximum in D^* at about 5–10 at.%Fe (Fig. 10.43(A)). Thus, the enhanced creep resistance of the $Ni_{40}Fe_{10}Al_{50}$ [146] alloy is not directly related to the tracer self-diffusion properties of the constituent atoms.

The tracer diffusion measurements reveal a conspicuous maximum in the activation enthalpy Q_{Ni} of Ni diffusion close to the $Ni_{25}Fe_{25}Al_{50}$ composition, Fig. 10.43(B). This result likely indicates an additional ordering in the base B2 structure of the $NiFeAl_2$ alloy. In the perfectly ordered B2 structure the Al atoms occupy the centers of cubic

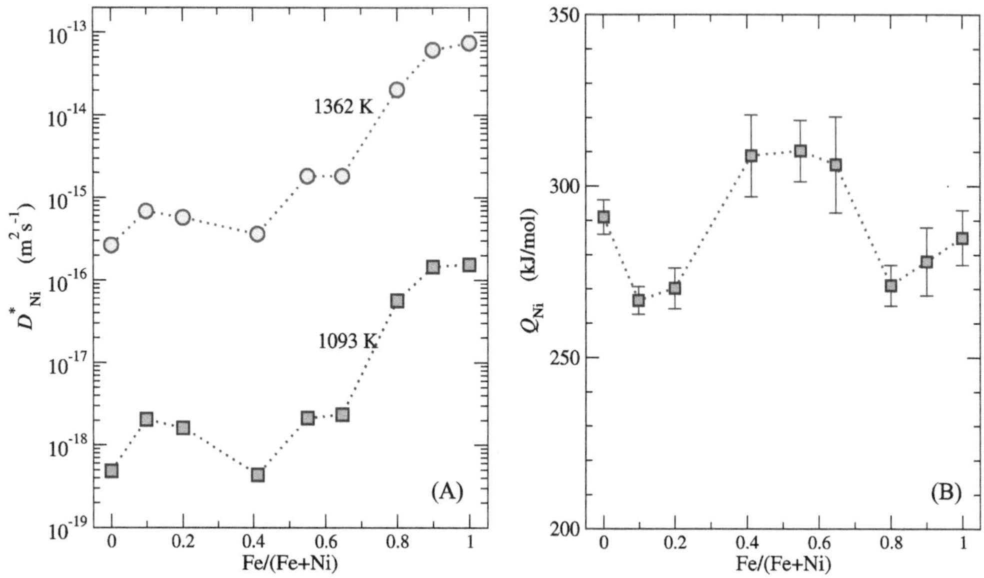

Figure 10.43 Ni tracer diffusivity D^*_{Ni} (A) and the effective activation enthalpy Q_{Ni} (B) in $Ni_{50-x}Fe_xAl_{50}$ alloys as functions of the Fe/(Fe+Ni) parameter. The experimental uncertainties of the measured values of D^*_{Ni} (if they are not shown explicitly) are less than the corresponding symbol size. The lines are drawn as a guidance.

unit cells, and Fe and Ni atoms are randomly distributed over the corner positions of these cells. The experimentally measured increase in the activation enthalpy of Ni diffusion suggests an additional order in the distribution of Fe and Ni atoms on their sublattice.

In order to analyze this behavior, the data in Fig. 10.43 are replotted using the homologous temperatures. In Fig. 10.44(A) Ni diffusion coefficients determined at two constant homologous temperatures, $0.77T_m$ and $0.59T_m$, along the quasi-binary section $Ni_{50}Al_{50}$–$Fe_{50}Al_{50}$ are presented. These homologous temperatures are arbitrary chosen for the values of $T_m/T = 1.3$ and 1.7. The above mentioned trend is clearly seen (sketched by dashed lines) – Ni diffuses with similar rates in binary end-members (when normalized on corresponding melting points), while there is a significant gap in the tracer diffusion rate in-between. In Fig. 10.44(B) the normalized activation enthalpies, Q_{Ni}/RT_m, are plotted as functions of the composition parameter Fe/(Ni+Fe). Obviously, Q_{Ni}/RT_m increases with replacement of Ni by Fe, keeping Al content constant, but there is a bump at intermediate concentrations which corresponds to elevated activation enthalpies of Ni diffusion in the alloys.

The ground states of NiFeAl alloys were studied by Fähnle and Lehermann by combining first-principle electron theory and statistical mechanics [147]. The zero-temperature ab initio calculations favor the so-called G1 and G2 structures for the

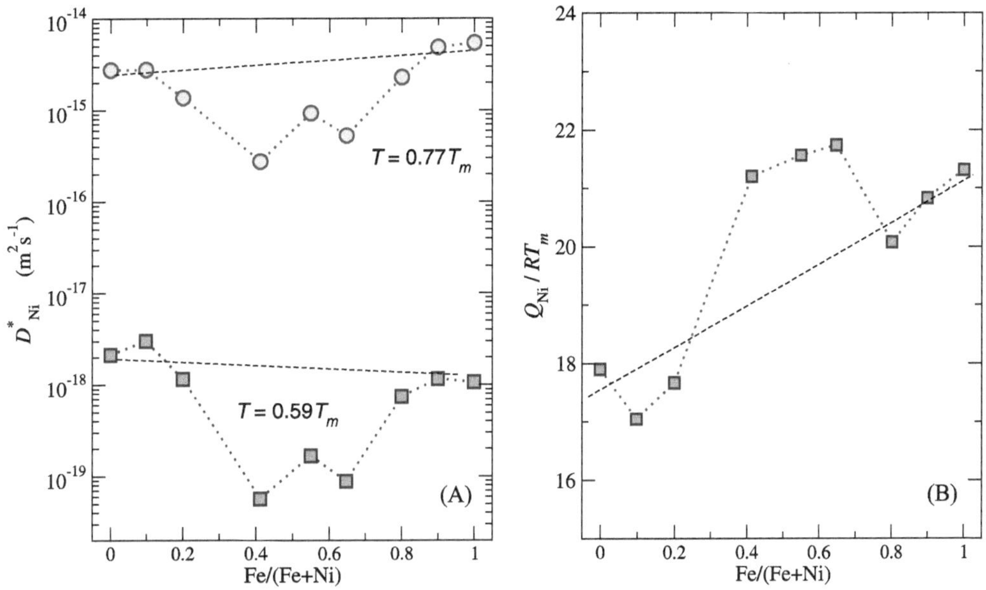

Figure 10.44 Ni tracer diffusivity D^*_{Ni} (A) at two selected reduced temperatures, $0.77T_m$ and $0.59T_m$, and the normalized activation enthalpy Q_{Ni}/RT_m (B) in $Ni_{50-x}Fe_xAl_{50}$ alloys as functions of the Fe/(Fe+Ni) parameter. T_m is the melting point of the corresponding alloy. The dotted lines are drawn as a guidance. The dashed lines represent anticipated trends of the diffusion coefficients or the Q_{Ni}/RT_m values vs. Fe content.

$NiFeAl_2$ alloy, which correspond to an ordered distribution of Ni and Fe atoms over their sublattice [147]. The calculated difference between the energies of $L2_1$ (random distribution on Ni/Fe sublattice in the B2 structure) and G1/G2 structures was relatively small, about 20 to 30 meV/at. Thus, the neglected contribution of atomic vibrations at finite temperatures may significantly change the relation between free energies of the different structures. However, the present experimental data on atomic mobilities in NiFeAl alloys seems to indirectly support the occurrence of additional ordering in the base B2 structure of these alloys around the $NiFeAl_2$ composition.

In Fig. 10.45 the tracer diffusion coefficients of Ni in $Ni_{50-x}Fe_xAl_{50}$ alloys are compared with the point defect concentrations, which were determined by the ALCHEMI method by Pike et al. [148]. Fe is known to substitute preferentially Ni sites [149]. However, a certain fraction of Fe atoms substitutes the Al sites and the replacement of Ni by Fe atoms in stoichiometric NiAl is known to result in the formation of structural vacancies in the transition metal sublattice [148]. The vacancy concentration approaches a value of several percents in $Ni_{50-x}Fe_xAl_{50}$ alloys with $x > 10$ at.%. There is no straightforward correlation between D^* and the vacancy concentration in these alloys. This fact indicates a complex diffusion mechanism of Ni atoms in the ternary NiFeAl alloys.

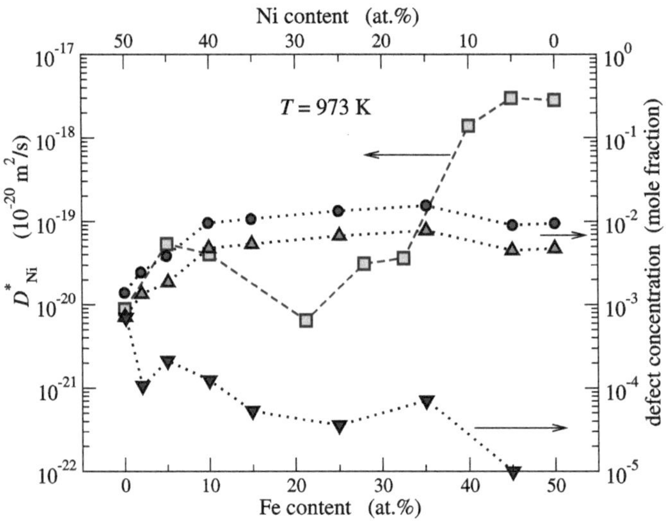

Figure 10.45 Ni tracer diffusivity D^* (full squares, left ordinate) as a function of the Fe content in $Ni_{50-x}Fe_xAl_{50}$ alloys at $T = 973$ K in comparison to the concentrations of point defects [148] (right ordinate): vacancies (circles), Fe antisites (triangles up), and Ni antisites (triangles down). The lines are drawn as a guidance.

It was shown in previous sections that Ni diffusion in Al-rich, nearby stoichiometric, and in Ni-rich binary NiAl alloys can satisfactorily be explained in terms of a dominant triple defect diffusion mechanism [57]. A theoretical study on the basis of the Bethe–Bragg–Williams approximation supported an important contribution of the triple-defect diffusion mechanism in B2 NiAl [150]. An increase in the Ni diffusivity in $Ni_{40}Fe_{10}Al_{50}$ alloy with respect to the stoichiometric $Ni_{50}Al_{50}$ compound was suggested to be related to a preferred formation of triple defects including Fe atoms in antistructure positions (i.e., on the Al sublattice) [142]. The similarity of activation enthalpies of Ni and Fe diffusion in the ternary $Ni_{40}Fe_{10}Al_{50}$ alloy [142] supports this conclusion.

10.7 GENERAL CONCLUSIONS

In this chapter tracer diffusion in ordered intermetallics is analyzed with a focus on aluminides of transition metals. It is shown that the point defects are not uniformly distributed on the sublattices of the given compound and the concentrations of defects are typically different on different sublattices. This feature largely determines diffusion rates and diffusion mechanisms in ordered compounds.

A number of mechanisms were suggested to describe diffusion in ordered intermetallics, such as six-jump cycles, sublattice mechanism, triple defect mechanism, just to mention the most important ones.

The following general conclusions can be drawn about diffusion in ordered aluminides of transition metals:

- Diffusion of the transition metal component (Ni or Ti, respectively) in Ni_3Al, Ti_3Al, and TiAl is dominated by nearest-neighbor jumps of vacancies in the transition metal sublattice (the sublattice diffusion mechanism). The triple defect diffusion mechanism is proposed to produce the main contribution in the nearly-stoichiometric NiAl alloys. Under specific conditions (high temperatures in TiAl, $T > 1450$ K, or large Ni content in NiAl, $c_{Ni} \geq 55$ at.%) the antistructure bridge mechanism contributes significantly.
- The transition metal sublattice diffusion mechanism superimposed by the ASB mechanism governs Al diffusion in Ni_3Al, Ti_3Al, and TiAl. Ga, which dissolves exclusively on the Al sublattice in Ni- and Ti-aluminides, is shown to represent a suitable solute to "mimic" Al diffusion in these compounds.
- Bulk self-diffusion in the antistructure-defect type Ni- and Ti-aluminides (Ni_3Al, Ti_3Al, and TiAl) reveals only a marginal compositional dependence, whereas a pronounced compositional dependence of D_{Ni} is observed in the triple-defect-type NiAl phase. This behavior is related to the fact that mainly the sublattice diffusion mechanism operates in Ni_3Al, Ti_3Al, and TiAl.
- Self-diffusion of the transition metal component in the closed-packed Ni- and Ti-aluminides, if measured at the same reduced temperature, decreases systematically when changing from pure metal to A_3B and to AB compound.

REFERENCES

[1] K. Girgis, in: R.W. Cahn, P. Haasen (Eds.), Physical Metallurgy, North-Holland, Amsterdam, 1983, p. 219.

[2] J.H. Westbrook, R.L. Fleischer, Intermetallic Compounds, Wiley, New York, 1995.

[3] G. Sauthoff, Intermetallics, VCH, Wertheim, 1995.

[4] G. Tammann, Über die Ermittelung der Zusammensetzung chemischer Verbindungen ohne Hilfe der Analyse, Z. Anorg. Chem. 37 (1903) 303–313.

[5] A.S. Kurnakov, Über die gegenseitigen Verbindungen der Metalle, Z. Anorg. Chem. 23 (1900) 439–462.

[6] G. Tammann, K. Dahl, Metallographische Mitteilungen aus dem Institut für physikalische Chemie der Universität Göttingen. CXI. Über die Sprödigkeit metallischer Verbindungen, Z. Anorg. Chem. 126 (1923) 104–112.

[7] J.H. Schneibel, G.F. Petersen, C.T.J. Liu, Creep-behavior of a nickel aluminide, Mater. Res. 3 (1986) 68.

[8] C.T. Liu, J.O. Stiegler, Ductile ordered intermetallic alloys, Science 226 (1984) 636–642.

[9] K. Aoki, O.J. Izumi, Improvement in room-temperature ductility of the $L1_2$ type intermetallic compound Ni_3A1 by boron addition, Jpn. Inst. Met. 43 (1979) 1190–1196.

[10] N.S. Stoloff, C.T. Liu, S.C. Deevi, Emerging applications of intermetallics, Intermetallics 8 (2000) 1313–1320.

[11] N. Cinca, C.R.C. Lima, J.M. Guilemany, An overview of intermetallics research and application: status of thermal spray coatings, J. Materials Research Technology 2 (2013) 75–86.

[12] J.M. Dubois, E. Belin-Ferré, Complex Metallic Alloys: Fundamentals and Applications, Wiley-VCH, Weinheim, 2010.
[13] J.W. Yeh, S.K. Chen, S.J. Lin, J.Y. Gan, T.S. Chin, T.T. Shun, C.H. Tsau, S.Y. Chang, Adv. Eng. Mater. 6 (2004) 299–303.
[14] M. Yamaguchi, H. Inui, K. Ito, High-temperature structural intermetallics, Acta Mater. 48 (2000) 307–322.
[15] L.N. Larikov, V.V. Geichenko, V.M. Falchenko, Diffusion Processes in Ordered Alloys, Amerind, Oxonian Press, New Delhi, 1981.
[16] H. Mehrer, Diffusion in intermetallics, Mater. Trans. Japan Inst. Metals 37 (1996) 1259.
[17] M. Koiwa, Diffusion in materials – history and recent developments, Mater. Trans. Japan Inst. Metals 39 (1998) 1169.
[18] Chr. Herzig, S.V. Divinski, Diffusion in Ordered Intermetallics, William Andrew, Norwich, 2005, pp. 173–238.
[19] H. Nakajima, K. Nonaka, W. Sprengel, M. Koiwa, Self-diffusion and interdiffusion in intermetallic compounds, Mater. Sci. Eng. A 239–240 (1997) 819.
[20] Chr. Herzig, T. Przeorski, Y. Mishin, Intermetallics 7 (1999) 389.
[21] Yu. Mishin, Chr. Herzig, Acta Mater. 48 (2000) 589.
[22] M. Eggersmann, H. Mehrer, Phil. Mag. A 80 (2000) 1219.
[23] E.W. Elcock, C.W. McCombie, Phys. Rev. B 109 (1958) 605.
[24] N.A. Stolwijk, M. van Gend, H. Bakker, Phil. Mag. A 42 (1980) 783.
[25] C.R. Kao, Y.A. Chang, Intermetallics 1 (1993) 237.
[26] S.V. Divinski, St. Frank, Chr. Herzig, U. Södervall, Solid State Phenomena 72 (2000) 203.
[27] S.V. Divinski, Chr. Herzig, Intermetallics 8 (2000) 1357.
[28] H. Okamoto, J. Phase Equilibria 14 (1993) 257.
[29] C. McCullough, J.J. Valencia, C.G. Levi, R. Mehrrabian, Acta Metall. 37 (1989) 1321.
[30] R. Kainuma, M. Palm, G. Inden, Intermetallics 2 (1994) 321.
[31] O. Kubaschewski, Iron Binary Phase Diagrams, Springer Verlag, Berlin, 1982.
[32] P. Scholz, PhD Thesis, University Göttingen, 2002.
[33] P.A. Korzhavy, A.V. Ruban, A.Y. Lozovoi, Y.K. Vekilov, I.A. Abrikosov, B. Johansson, Phys. Rev. B 61 (2000) 6003.
[34] A.F. Voter, S.P. Chen, Mat. Res. Soc. Symp. Proc. 82 (1987) 175.
[35] Y. Mishin, M.J. Mehl, D.A. Papaconstantopoulos, Embedded-atom potential for B2-NiAl, Physical Review B 65 (2002) 224114.
[36] J. Mayer, C. Elsässer, M. Fähnle, Phys. Stat. Sol. B 191 (1995) 283.
[37] J. Mayer, M. Fähnle, Defect Diffusion Forum 143–147 (1997) 285.
[38] A.A. Maradudin, E.W. Montroll, G.H. Weiss, I.P. Ipatova, Theory of Lattice Dynamics in the Harmonic Approximation, 2nd ed., Academic Press, New York, 1971.
[39] A.P. Sutton, Phil. Mag. A 60 (1989) 147.
[40] S.M. Foiles, Phys. Rev. B 49 (1994) 14930.
[41] R.D. Hatcher, R. Zeller, P.H. Dederichs, Phys. Rev. B 19 (1979) 5083.
[42] A. Glensk, B. Grabowski, T. Hickel, J. Neugebauer, Breakdown of the Arrhenius law in describing vacancy formation energies: the importance of local anharmonicity revealed by ab initio thermodynamics, Phys. Rev. X 4 (2014) 011018.
[43] Yu. Mishin, D. Farkas, Phil. Mag. A 75 (1997) 177;
Yu. Mishin, D. Farkas, Phil. Mag. A 75 (1997) 187.
[44] S.V. Divinski, L.N. Larikov, Self-diffusion in ordered intermetallic compounds with the L12 structure, Metal Physics and Advanced Technologies 19 (9) (1997) 31–37.
[45] S.V. Divinski, On mechanisms of self-diffusion in quasicrystals, Sov. Phys. Solid State 37 (1995) 2054–2058.

[46] S.V. Divinski, L.N. Larikov, On the diffusion mechanisms in quasicrystalline structures, Phys. Stat. Sol. A 154 (1996) 543–548.
[47] H.A. Domian, H.I. Aaronson, in: Diffusion in Body-Centered Cubic Metals, American Society for Metals, 1965, p. 209.
[48] M. Arita, M. Koiwa, S. Ishioka, Acta Metall. 37 (1989) 1363.
[49] D. Gupta, D.S. Lieberman, Phys. Rev. B 4 (1971) 1070.
[50] D. Gupta, D. Lazarus, D.S. Lieberman, Phys. Rev. 153 (1967) 863.
[51] I.V. Belova, G.E. Murch, Phil. Mag. A 82 (2002) 269.
[52] H. Mehrer, Diffusion in Solid Matter, Springer, Heidelberg, 2007.
[53] R. Hilgedieck, Chr. Herzig, Z. Metallkd. 74 (1983) 38.
[54] H. Mehrer, R. Galler, W. Frank, R. Blüher, A. Strohm, Diffusion in quasicrystals, in: H.-R. Trebin (Ed.), Quasicrystals – Structure and Physical Properties, Wiley-VCH, 2003.
[55] H. Wever, Defect Diffusion Forum 83 (1992) 55.
[56] H. Numakura, T. Ikeda, M. Koiwa, A. Almazouzi, Phil. Mag. A 77 (1998) 887.
[57] St. Frank, S.V. Divinski, U. Södervall, Chr. Herzig, Acta Mater. 49 (2001) 1399.
[58] S.V. Divinski, L.N. Larikov, J. Phys.: Condens. Matter. 35 (1997) 7377–7388.
[59] R.M. Ziff, S. Torquato, as reported by L. Kurzawski and K. Malarz, Rep. Math. Phys. 70 (2012) 163.
[60] M. Leitner, Thermodynamics of point defects and diffusion mechanisms in B2-ordered compounds, arXiv:1503.03798v1 [cond-mat.mtrl-sci].
[61] S.V. Divinski, St. Frank, U. Södervall, Chr. Herzig, Acta Mater. 46 (1998) 4369.
[62] Ch. Herzig, T. Przeorski, Y. Mishin, Self-diffusion in γ-TiAl: an experimental study and atomistic calculations, Intermetallics 7 (1998) 389.
[63] Chr. Herzig, T. Przeorski, M. Friesel, F. Hisker, S.V. Divinski, Intermetallics 9 (2001) 461.
[64] G.S. Collins, P. Sinha, M. Wie, Hyperfine Interactions (C) 1 (1996) 380.
[65] C. Wert, C. Zener, Phys. Rev. 76 (1949) 1169.
[66] G.H. Vineyard, J. Phys. Chem. Solids 3 (1957) 121.
[67] L.S. Darken, Trans. AIME 175 (1948) 184.
[68] J.R. Manning, Acta Metall. 15 (1967) 817.
[69] M. Salamon, S. Dorfman, D. Fuks, G. Inden, H. Mehrer, Defect Diffusion Forum 194–199 (2001) 553.
[70] T. Shimozaki, Y. Goda, Y. Wakamatzu, M. Onishi, Defect Diffusion Forum 95–98 (1993) 629.
[71] H. Numakura, T. Ikeda, H. Nakajima, M. Koiwa, Defect Diffusion Forum 194–199 (2001) 337.
[72] G.E. Murch, Phil. Mag. A 46 (1982) 575.
[73] I.V. Belova, G.E. Murch, Phil. Mag. A 78 (1998) 1085.
[74] G.E. Murch, I.V. Belova, Defect Diffusion Forum 194–199 (2001) 533.
[75] G.E. Murch, I.V. Belova, Defect Diffusion Forum 194–199 (2001) 541.
[76] J. Rüsing, Chr. Herzig, Intermetallics 4 (1996) 647.
[77] H.-E. Schaefer, K. Frenner, R. Würschum, Intermetallics 7 (1997) 277.
[78] W. Sprengel, H. Nakajima, N. Oikawa, Mater. Sci. Engng. 213 (1996) 45.
[79] R. Bormann, private communication, 1996.
[80] Chr. Herzig, M. Friesel, D. Derdau, S.V. Divinski, Intermetallics 7 (1999) 1141.
[81] Y.L. Hao, D.S. Xu, Y.Y. Cui, R. Yang, D. Li, Acta Mater. 47 (1999) 1129.
[82] J. Breuer, T. Wilger, M. Friesel, Chr. Herzig, Intermetallics 7 (1999) 381.
[83] H. Nakajima, M. Koiwa, S. Ono, Scripta Metall. 17 (1983) 1431.
[84] H. Nakajima, M. Koiwa, Titanium, science and technology, in: Proc. 5th Int. Conf. on Ti, 1985, p. 1759.
[85] M. Köppers, Chr. Herzig, M. Friesel, Y. Mishin, Acta Mater. 45 (1997) 4181.
[86] T. Ikeda, H. Kadowaki, H. Nakajima, Acta Mater. 49 (2001) 3475.

[87] C.-G. Lee, Y. Iijima, S.-E. Kim, Y.-T. Lee, H.-M. Kim, in: The 14th Japan Symposium Thermophysical Properties, 1993, pp. 395.
[88] C.-G. Lee, Y. Iijima, M.-H. Kim, S.-E. Kim, Y.-T. Lee, H.-M. Kim, in: The 15th Japan Symposium Thermophysical Properties, 1994, pp. 263.
[89] Y.L. Hao, R. Yang, Y. Song, Y.Y. Cui, D. Li, A. Niinomi, Intermetallics 12 (2004) 951.
[90] W. Sprengel, N. Oikawa, H. Nakajima, Intermetallics 4 (1996) 185.
[91] M.B. Bronfin, G.S. Bulatov, I.A. Drugova, Fiz. Metal. Metalloved. 40 (1975) 363.
[92] K. Hoshino, S.J. Rothman, R.S. Averback, Acta Metall. 36 (1988) 1271.
[93] Y. Shi, G. Froberg, H. Wever, Phys. Stat. Sol. A 191 (1995) 361.
[94] St. Frank, U. Södervall, Chr. Herzig, Phys. Stat. Sol. B 191 (1995) 45.
[95] T. Ikeda, A. Almazouzi, H. Numakura, M. Koiwa, M. Sprengel, H. Nakajima, Acta Mater. 46 (1998) 5369.
[96] M. Watanabe, Z. Horita, M. Nemoto, Defect Diffusion Forum 143–147 (1997) 345.
[97] K. Fujiwara, Z. Horita, Acta Mater. 50 (2002) 1571.
[98] G.F. Hancock, B.R. McDonnel, Phys. Stat. Sol. A 4 (1971) 143.
[99] S. Ochiai, Y. Oya, T. Suzuki, Acta Metall. 32 (1984) 289.
[100] Y. Minamino, H. Yoshida, S.B. Jung, K. Hirao, T. Yamane, Defect Diffusion Forum 143–147 (1997) 257.
[101] S. Frank, U. Södervall, Chr. Herzig, Defect Diffusion Forum 143–147 (1997) 245.
[102] S.B. Jung, Y. Minamino, H. Araki, T. Yamane, K. Hirao, S. Saji, Defect Diffusion Forum 95–98 (1993) 859.
[103] Y.A. Chang, J.P. Neumann, Progr. Solid State Chem. 14 (1982) 221.
[104] S. Shankar, Metal. Trans. A 9 (1978) 1467.
[105] S. Kim, Y.A. Chang, Metall. Mater. Trans. A 31 (2000) 1519.
[106] A. Steiner, K.L. Komarek, Trans. TMS-AIME 230 (1964) 786.
[107] S.V. Divinski, Chr. Herzig, Defect Diffusion Forum 203 (2002) 177.
[108] A. Lutze-Birk, H. Jacobi, Scripta Metall. 9 (1975) 761.
[109] Y. Minamino, Y. Koizumi, Y. Inui, Defect Diffusion Forum 194–199 (2001) 517.
[110] M. Eggersmann, H. Mehrer, Phil. Mag. A 80 (2000) 1219.
[111] L.A. Girifalco, J. Phys. Chem. Solids 24 (1964) 323.
[112] H. Bakker, N.A. Stolwijk, L. Van der Maij, T.J. Zuurendonk, Nucl. Mater. 20 (1976) 96.
[113] Zs. Tokei, J. Bernardini, P. Gas, D.L. Beke, Acta Mater. 45 (1997) 541.
[114] M. Eggersmann, PhD Thesis, University Münster, 1998.
[115] R. Weinkamer, P. Fratzl, B. Sepiol, G. Vogl, Phys. Rev. B 58 (1998) 3082.
[116] B. Sepiol, W. Löser, M. Kaisermayer, R. Weinkamer, P. Fratzl, H. Thiess, M. Sladecek, G. Vogl, Defect Diffusion Forum 194–199 (2001) 349.
[117] M. Stana, M.. Ross, B. Sepiol, Experimental evidence of effective <111> atomic exchanges in a B2 intermetallic alloy, arXiv:1606.06516v2 [cond-mat.mtrl-sci], 2016.
[118] S. Peteline, E.M. Tanguep Njiokep, S.V. Divinski, H. Mehrer, Diffusion of solute elements in Fe_3Al, Defect Diffusion Forum 216–218 (2003) 175–180.
[119] M. Weinhagen, B. Köhler, J. Wolff, Th. Hehenkamp, Defect Diffusion Forum 143–147 (1997) 449.
[120] Th. Hehenkamp, P. Scholz, B. Köhler, R. Kerl, Defect Diffusion Forum 194–199 (2001) 389.
[121] Zs. Tokei, J. Bernardini, G. Erdelyi, I. Gödeni, D.L. Beke, Acta Mater. 46 (1998) 4821.
[122] M. Lübbehusen, H. Mehrer, Acta Metall. Mater. 38 (1990) 283.
[123] Y. Iijima, K. Kimura, K. Hirano, Acta Metall. 36 (1988) 2811.
[124] I.S. Golovin, S.V. Divinski, J. Cicek, F. Stein, Study of atom diffusivity and related relaxation phenomena in Fe3Al-(Ti,Nb)-C alloys, Acta Mater. 53 (2005) 2581–2594.
[125] I.S. Golovin, A. Strahl, H. Neuhaeuser, Anelastic relaxation and structure of ternary Fe–Al–Me alloys with Me = Co, Cr, Ge, Mn, Nb, Si, Ta, Ti, Zr, Int. J. Mater. Research 97 (2006) 1078–1092.

[126] S.V. Divinski, F. Hisker, Chr. Klinkenberg, Chr. Herzig, Niobium and titanium diffusion in the high niobium-containing Ti–54Al–10Nb alloy, Intermetallics 14 (2006) 792.
[127] S.V. Divinski, C. Klinkenberg, C. Herzig, Tracer diffusion of niobium and titanium in binary and ternary titanium aluminides, J. Phase Equilibria Diffusion 26 (2005) 452–457.
[128] Y.L. Hao, D.S. Xu, Y.Y. Cui, R. Yang, D. Li, Acta Mater. 47 (1999) 1129.
[129] H. Erschbaumer, R. Podloucky, P. Rogl, G. Temnitschka, R. Wagner, Intermetallics 1 (1993) 99.
[130] R. Benedek, A. van de Walle, S.S.A. Gerstl, M. Asta, D.N. Seidman, Phys. Rev. B 71 (2005) 094201.
[131] T. Kawabata, T. Tamura, O. Izumi, Metal. Trans. A 24 (1993) 141.
[132] J. Breuer, T. Wilger, M. Friesel, Chr. Herzig, Intermetallics 7 (1999) 381.
[133] Chr. Herzig, T. Przeorski, M. Friesel, F. Hisker, S. Divinski, Intermetallics 9 (2001) 461.
[134] C. Woodward, S. Kajihara, L.H. Yang, Phys. Rev. B 57 (1998) 13459.
[135] A.D. Le Claire, J. Nucl. Mat. 69/70 (1978) 61.
[136] Z.C. Liu, J.P. Lin, S.J. Li, G.L. Chen, Intermetallics 10 (2002) 653.
[137] M. Weller, H. Clemens, G. Haneczok, G. Dehm, A. Bartels, S. Bystrzanowski, R. Gerling, E. Arzt, Phil. Mag. Lett. 84 (2004) 383.
[138] F. Appel, J.D.H. Paul, M. Oehring, H. Clemens, F.D. Fischer, Z. Metallk. 95 (2004) 585.
[139] S. Bystrzanowski, A. Bartels, H. Clemens, R. Gerling, F.P. Schimansky, G. Dehm, H. Kestler, Intermetallics 13 (2005) 515.
[140] G. Petzow, G. Effenberg, Ternary Alloys, MSI, 1993.
[141] S.V. Divinski, F. Hisker, W. Löser, U. Södervall, Chr. Herzig, Ni radiotracer diffusion in B2 ordered NiFeAl alloys, Intermetallics 14 (2006) 308–314.
[142] S.V. Divinski, Y.-S. Kang, W. Löser, U. Södervall, Chr. Herzig, Ni and Fe tracer diffusion in $Ni_{40}Fe_{10}Al_{50}$ ternary alloy, Intermetallics 12 (2004) 511.
[143] S. Peteline, PhD Thesis, Münster University, 2005.
[144] A.J. Bradley, Microscopic studies on the Iron–Nickel–Aluminium system. Part I. $\alpha + \beta$ alloys and isothermal section of the phase equilibrium diagram, J. Iron Steel Inst. (London) 163 (1) (1949) 19–30.
[145] A. Schrader, H. Hanemann, Aluminium Ecke der Aluminium–Isen–Nickel Phasendiagram, Aluminium 25 (1943) 339–342.
[146] M. Rudy, G. Sauthoff, Mater. Sci. Eng. 81 (1986) 525.
[147] F. Lehermann, M. Fähnle, J.M. Sanchez, Private communication, 2004.
[148] L.M. Pike, I.M. Anderson, C.T. Liu, Y.A. Chang, Acta Mater. 50 (2002) 3859.
[149] G.H. Bozzolo, J. Khalil, R.D. Noebe, Comp. Mater. Sci. 24 (2002) 457.
[150] R. Krachler, H. Ipser, Phys. Rev. B 70 (2004) 054113.

INDEX

E

J

K

L

M

N

O

P

T

FSC
www.fsc.org
MIX
Papier | Fördert
gute Waldnutzung
FSC® C083411

Kolibri 360 GmbH
Ferdinand-Jühlke-Straße 7
99095 Erfurt, Deutschland
produktsicherheit@kolibri360.de